# PETER DAEMPFLE

# ESSENTIAL
## BIOLOGY

## AN APPLIED APPROACH

# Kendall Hunt
publishing company

**Book Team**

Chairman and Chief Executive Officer Mark C. Falb
President and Chief Operating Officer Chad M. Chandlee
Vice President, Higher Education David L. Tart
Director of Publishing Partnerships Paul B. Carty
Product/Development Supervisor Lynne Rogers
Vice President, Operations Timothy J. Beitzel
Permissions Editor Melisa Seegmiller
Cover Designer Faith Walker

**Kendall Hunt**
publishing company

www.kendallhunt.com
*Send all inquiries to:*
4050 Westmark Drive
Dubuque, IA 52004-1840

## DEDICATION

*For my wife, Amy*

*For my children, Justina and Konrad*

*For my father, Tobias*

# BRIEF CONTENTS

# CONTENTS

## UNIT 3    We Are Not Alone!

## 7.    Evolution Gives our Biodiversity

# UNIT 4    The Dynamic Animal Body                              379

## 11. Animal Organization                                       381

# UNIT 5  A Small Hole Sinks a Big Ship – Our Fragile Ecosystem

## 17. Population Dynamics and Communities that Form

## UNIT 6    Biology and Society

## 20. The Evolution of Social Behavior: Sociobiology

# PREFACE

The purpose of *Essential Biology* is to improve biological literacy and advance the importance of scientific thinking. This textbook grew out of 20 years of science teaching. It applies learning strategies that work in the classroom by weaving biological themes alongside stories and social applications for understanding. My central goals for this textbook are to motivate students *to look at science in a passionate way*. In a sense to appreciate Aristotle's view that "in all things of nature there is something of the marvelous." This textbook seeks a new way of approaching biology by incorporating stories, social themes, and integrating non-science areas to augment student interest. The textbook is intended to enhance teaching methods by bringing social applications of biology to each unit and to each student. It retains the rigor of the traditional college biology curriculum. Its pedagogy uses many different techniques to motivate students:

- **How?** The textbook uses instructional methods that foster active student participation in the lecture. Each chapter ends with a set of discussion questions that stimulate classroom conversations and span all levels of Bloom's taxonomy. A "Biology and Society Corner" assessment section for each chapter leads students into an application of the content to their lives and to the societies around the world. It touches upon historical and philosophical ideas threaded alongside rigorous content. Simple and clear language is important, but an appeal to the sociological (e.g. bioethical, medical, and practical) underpinnings of biology drives student motivation. Through my teaching, I found that explaining the subject is only part of its passion. An equally important aspect of working with undergraduates is to make the content come alive through tapping into the societal parts of biology that integrate content into other disciplines and student interests. Students do not live in a science vacuum, where the world is one of internally motivated, empirical hypothesis testing. Instead, they live in a society and they have interests besides basic science fact-knowing. Through tapping into that real world of experiences, this textbook asks the reader to commit to the most important aspect of a textbook – *a student's desire to keep on learning*. I begin each chapter with a story to introduce the content and make a paradoxical and provocative application to familiar societal encounters. Then, the reader is shown how to think scientifically.

  Each chapter anchors the content briefly in its historical roots within the science community, showing how biology and scientists work together to create knowledge. This models scientific thinking. Knowledge presented is clear and organized, as well as embedded in practical applications, sociological–ethical–legal dilemmas, literary cases, and medically related and provocative reflection on issues.

  For example, molecular genetics is not merely described and defined, but the content is surrounded within a case of skin color discrimination in another culture different from our own. That theme is treated through the chapter until it concludes with readers applying their newly learned content knowledge with the thematic issues. Each chapter, as in each stage of a play, has a clear beginning, middle, and end to captivate the reader at different junctures. The text uses this approach by adding specific sections to touch the reader at strategic points in the writing to get them to want to read and reflect more.

Many within the incoming cohort of first-year college students are academically underprepared and/or unfocused for entry into the field of study. They require a book that cleverly motivates them. The starter course for such programs is introductory college biology, generally for non-majors. Textbooks that are thematically non-specific and not engaging to non-major students usually define these courses. Instead, this textbook particularly directs the reader, and thus the introductory biology courses, to highlight the importance of biology within other areas of knowing. The textbooks works through using an approach that taps the varied interests of the non-major – with historical and social issues, and literary and health-related applications infused throughout the chapters. Although the textbook is intended for all non-majors college biology courses, it will particularly appeal to pre-allied health students by tapping into the societally based content that medical professionals are concerned with.

- **New Strategies**: As a Ph.D.-level researcher who studies the tenuous transition between secondary and post-secondary biology programs and a professor for 20 years in the General Biology-to-Human Anatomy and Physiology sequence, I see a real need for such a textbook. The new product will better prepare and harness the energy of this non-major cohort to prepare them to become scientifically literate adults in society. Such students are often bored and unengaged with post-secondary general biology textbooks and courses that do not address their unique interests. The new textbook will:
  - Use language and applications, clever and clear to understand in a style and format particularly aligned in a societally applied approach.
  - Engage readers in case studies and critical thinking applications and assessment required for scientific literacy in a modern society.
  - Employ and guide students' study strategies for learning general biology content and applications.
  - Retain the breadth and rigor of general biology course content.
- **Enhance Classroom Conversation**: Instructors using this textbook, both online and in the traditional classroom, can expect their students to become more engaged in cooperative learning as they apply the social themes of each story starting the chapters. Students will follow the story throughout the chapters to reflect continuously on the content material. Each discussion will tap higher order reasoning skills in helping students understand biology and its place within society.
- **Why?** Because: *student motivation equals academic results*. To illustrate, in my classroom and in my books, instead of merely describing the endosymbiotic theory of eukaryotic development, I relate it to a story about how life could be if we judge people based on the health of their mitochondrial DNA; and I discuss its importance in mitochondrial division and muscle building.

**Four Ways to Organize Information**

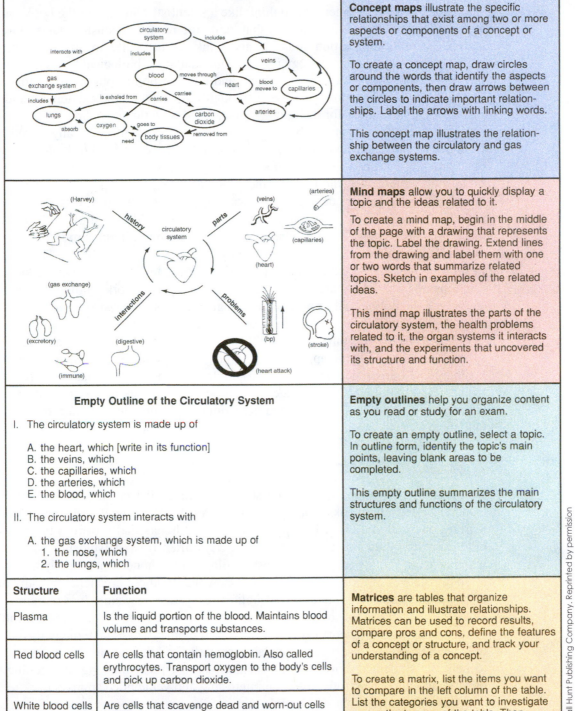

**Concept maps** illustrate the specific relationships that exist among two or more aspects or components of a concept or system.

To create a concept map, draw circles around the words that identify the aspects or components, then draw arrows between the circles to indicate important relationships. Label the arrows with linking words.

This concept map illustrates the relationship between the circulatory and gas exchange systems.

**Mind maps** allow you to quickly display a topic and the ideas related to it.

To create a mind map, begin in the middle of the page with a drawing that represents the topic. Label the drawing. Extend lines from the drawing and label them with one or two words that summarize related topics. Sketch in examples of the related ideas.

This mind map illustrates the parts of the circulatory system, the health problems related to it, the organ systems it interacts with, and the experiments that uncovered its structure and function.

**Empty outlines** help you organize content as you read or study for an exam.

To create an empty outline, select a topic. In outline form, identify the topic's main points, leaving blank areas to be completed.

This empty outline summarizes the main structures and functions of the circulatory system.

**Empty Outline of the Circulatory System**

I. The circulatory system is made up of

   A. the heart, which [write in its function]
   B. the veins, which
   C. the capillaries, which
   D. the arteries, which
   E. the blood, which

II. The circulatory system interacts with

   A. the gas exchange system, which is made up of
     1. the nose, which
     2. the lungs, which

| Structure | Function |
|---|---|
| Plasma | Is the liquid portion of the blood. Maintains blood volume and transports substances. |
| Red blood cells | Are cells that contain hemoglobin. Also called erythrocytes. Transport oxygen to the body's cells and pick up carbon dioxide. |
| White blood cells | Are cells that scavenge dead and worn-out cells and respond to tissue damage or invasion by microbes. Also called leukocytes. |
| Platelets | Are cell fragments. Play a role in blood clotting. |

**Matrices** are tables that organize information and illustrate relationships. Matrices can be used to record results, compare pros and cons, define the features of a concept or structure, and track your understanding of a concept.

To create a matrix, list the items you want to compare in the left column of the table. List the categories you want to investigate across the top row of the table. Then complete the table.

This matrix illustrates the components of the blood and their function.

**Figure I.5** Four ways to organize information. Use any of these methods at the end of each chapter to structure your thoughts and put them onto paper. A blank box is provided at the end of each chapter titled "concept" maps for your use for this purpose. This text is meant to be used with an active style of learning employed for each chapter. Research shows that if you can put all of the information from the chapter on one page, you have mastery of the material. From *Biological Perspectives*, 3rd ed by BSCS.

## Thinking Like a Scientist

The book shows the reader how to think like a scientist. Throughout the book, provocative biology examples are provided that guide the reader to consider facts more critically. The tools to question authority and think scientifically are given by exposing the reader to cutting-edge biology research, mathematics, biological history, integrated biology content, bioethical case studies, and science philosophy as roots to science literacy. In the many textboxes woven through the chapters, the excitement and optimism biology promises for the future is juxtaposed with threats to integrity.

*Essential Biology* should be used as a core text in courses for any introductory field of biology. It presents an applied approach to how each field of biology is developed and fits within larger areas of study. Its format is intended to treat the nature of biology in general and the major principles within each sub-discipline. The book is designed to help readers develop their own thinking about biology. It is able to do this because *Essential Biology* is one of the few texts that address all of the contributing disciplines that create biology as a rich subject. Its underpinnings present the reader with a view of biology as multifaceted and yet simply interdisciplinary. The influence of biology on societal development is a theme of the text and presents the sympatric relationships of biology areas within a sociological framework. Biology is portrayed as its own discipline, with rewards and integrity issues, hindrances, and a competing array of pseudosciences.

Biology is shown, throughout the book, as a dynamic part of a changing society. In short, the text's unique approach to learning biology leads the reader to real truths behind many life science phenomena treated. It develops an appreciation for the way in which we gain biological knowledge. The purpose of the book is thus to excite the reader's innate interests about biology and to recruit good people into science, on the whole. Our society needs this book.

## A Special Opportunity

Instructors' main challenges include student apathy and lack of student preparation. General biology textbooks that try to "cover" too much without focusing on actually learning the material exacerbate this. Very general books that flip from topic to topic but do not apply the content in meaningful ways often frustrate students. To illustrate, when general biology textbooks discuss cytology, mitochondria are merely memorized as "powerhouses of the cell." Applications showing that these organelles actually divide during muscle building, increase metabolic rates useful for weight loss in humans, are involved in many mitochondrial diseases, and are in fact inherited from our mothers (The Endosymbiotic Theory) are not treated well by the traditional biology textbook. In the above example, I use a case study and societally and thematically focused examples to help the student discover the information using critical thinking questioning. Few other textbooks consistently utilize an interesting and yet applied approach.

The above mitochondria example exposes the many missing pieces of interesting information that a non-majors student would want added to general biology textbooks. As an instructor of the course, I see what applications and explanations interest students and I know what motivates them to keep on learning. This motivation overcomes the lack of preparation because students have a desire to keep reading. This societally focused, relevant textbook will guide the reader to gain this energy.

# Audience

*Essential Biology* is a core text for non-major, Introductory/General Biology courses serving first- and second-year undergraduates at a two-year or four-year college.

There is also a growing and sustained influx of students into post-secondary allied health programs. A 63% increase in total number of degrees awarded in the allied health professions was documented in the 5-year span between 2004 and 2009, according to the National Center for Education Statistics (2010). This is warranted because the U.S. Department of Labor forecasts 3.2 million new positions in health fields between 2008 and 2018 in healthcare. Many within this incoming cohort of students are academically underprepared and/or unfocused for entry into the field of study. The starter course for such programs is introductory college biology, generally for non-majors. Textbooks that are thematically non-specific and not engaging to pre-professional students often define these courses.

Thus, as a secondary audience, this book is also useful to providing preparation in introductory biology for entering nursing degree and allied health students studying for programs in physical therapy, occupational therapy, physician's assistant, nutritionist, dental hygiene, medical lab technology, paramedic/emergency medical technician, and pharmacy. Its applied approach focuses on applications important in society such as medical and bioethical themes. The book also appeals to pre-professional students by using styles and content that mirror health science themes. For example, case-study utilization and medical science applications make it a perfect text to prepare students entering anatomy and physiology.

There is also a potential use for this book as a starter course in industrial and commercial training programs for employee advancement within sectors such as pharmaceutical, hospital, and government health organizations. This text is an excellent review or preparatory book for adult learners starting a career in the sciences.

# Unique Characteristics

Many introductory college biology textbooks rely on a breadth of content that is non-focused and very general to present the reader with as many factoids about life sciences as is possible. Many are cafeteria-style in that they attempt to draw students into too many areas of biology without a thematic alliance to the interest of a particular population. This book has that thematic focus on society and thus brings to the reader the many interesting applications as a result.

Unique aspects of the textbook will include but will not be limited to:

- Integration of biology content with everyday life, other science and non-science academic areas, and health-related applications.
- Ask divergent, open-ended questions (e.g. What is uric acid? vs. Why is uric acid is produced?) to get critical thinking practice in understanding life science phenomena and medical process. "What if" and "why" questions applied to interesting cases help infuse discussion into the classroom using this textbook.
- Use of "Paradoxes in Biology" (e.g. Dollar Bill Drop: A student is asked to catch a dollar bill in a certain way – this is impossible because their reaction time is too slow – too many synapses – to allow a successful catch; The "Case of out-of-place Color" is a twist on racial discrimination based on albinism in Tanzania). Instructional strategies such as these examples are provided to enhance teaching and on-line and in-class discussion.
- Mnemonics and Learning Strategies are included in each chapter. Study skills are incorporated into each chapter showing mnemonics, concept mapping, analogies,

and strategies for recall to organize new and old information. To illustrate, concept mapping as a technique to aid student learning is incorporated at the end of each chapter. This anchors old information from each past chapter to the new information. I found these methods to be very successful within my teaching acumen.

- A biology case study prefaces each chapter entitled "The Case of...". This is presented as a two-part series, with the first part as a case story from a fictitious situation and a second part includes case-based questions called a "Check Up" section, using content information about the case study. Each engaging story about a real-life experience integrates different branches of biology with societal themes and paradoxes. The case studies are unique, and titles can be viewed in each of the chapters in the Table of Contents. I have classroom tested these stories and they are quite appealing. Case studies range from science fiction and suspense stories to historical biology accounts, cataclysms, comedies and dramas.

- "Essentials" will start each chapter as a flow of cyclical pictures surrounding the stories and biology content to trace the main thematic ideas, acting as an advanced organizer for the chapter content.

- "Check In" sections list standards-based key points and are a prelude to the content presented in the chapter. The case stories and Check Ins address student knowledge at various levels of Bloom's taxonomy and mirror the biology settings as well as introduce key points for each chapter. This makes it thematically more appealing to focus the reader.

- "Key Terms" sections and "Summaries" emphasize collateral learning by linking terms and concepts within chapters, placed at the ending of each chapter. "Check In" points also link to end-of-chapter concept maps. This connects new and old information to develop new schemas for students, guiding students to develop a concept map at the end.

- "Check Out" end-of-chapter review questions and "key terms" connect based on the reading. These questions match the key points of the "Check In" at the beginning of each chapter.

- Special textboxes are boxes giving the gossip and lesser known type information about the stories behind commonly taught biology. For example, our Molecular Genetics chapter exposes the fact that Rosalind Franklin's X-ray image of DNA actually gave James Watson the idea for the double helix model.

- "Biology and Society Corner" sections at the end of each chapter give guide student application on key areas connecting topics that are challenging– e.g. endosymbiotic theory, muscle building, genetics of skin color, neurophysiology, sliding filament theory, environmental pollution, and nutrition – these guide discussion in online courses and in the physical classroom.

- Biology content is organized around themes and applications but is not compromised. The curriculum for Essential Biology is traditional in breadth and organization for the introductory college biology course.

## Environment

The textbook is useful in establishing the foundation of the course content for both traditional, online-only and hybrid introductory biology courses. Graphic images, detailed and beautiful biology artwork arranged with purpose and power point slides support any of these instructional formats. The traditional test bank, customized test generator, and instructor power point slides will be provided by Kendall/Hunt Publishers.

Web access to a tutoring site using Blackboard, Moodle, or Web CT platforms accompanies the textbook. These platforms include a comprehensive glossary, index, discussion questions, exam questions, and lecture slides to create a rich learning environment in all three methods of instruction. An eBook version of the text, with multimedia resources and study skill features is available.

# Learning Design

## *Overall experience*

As described in the sections above, a variety of methods are employed beyond the written word to engage students in biology. My primary goal is to facilitate learning by appealing to each student's abilities and thoughts through multiple methods within an integrated and applied approach. To illustrate, some students are visual learners who have had difficulty learning from the written word. The textbook taps this aspect of their learning strengths by having them develop concept maps and by presenting visuals alongside the content in just the right way to pique reader interest. The material is presented so that the student is not passive but actively uses the information in stories, textboxes, and critical thinking review questions. The student is able to draw relationships among ideas and organize their knowledge in a way *that makes sense to them*. Conversely, some students need hands-on interactions to supplement their learning while others are quantitative and need to see mathematical trends that underlie phenomena. The textbook uses graphs and data alongside hands-on activities and mneumonics to engage the reader. Within each chapter, the textbook uses multiple strategies to help students discover biology so that it may reach as many students as is possible. The Vision & Change core competencies are mirrored in this textbook by: 1) fostering communication on social applications of biology; 2) applying the scientific process early in the text; 3) integrating key areas of biology with one another and with societal applications; 4) actively involving students with group activities and discussion (particularly on social issues and in assessment strategies within the chapters); 5) applying math competencies alongside developing science literacy within biology (particularly in our first chapter on science literacy, unique only to this textbook).

## *Learning elements*

I have chosen "Check In," "Check Up," and "Check Out" sections at the start and end of each chapter to organize and parallel themes within each chapter and the book. The readings are organized around the essentials and key points and end with assessment strategies that link new and previously learned material (connections and concept mapping strategies). Thus, case studies, applied examples, clear writing style, and links to critical thinking/divergent questioning revolve around the key points. This kind of organization is vital in keeping the reader focused on what is important and maintains attention to the themes of this textbook. The ancillary materials also remain ordered around the central themes of biology.

## *Customization*

As seen by the "Table of Contents," the textbook may be easily customized by faculty because it is arranged around some integral units capable of existing individually or in combination. This is important because introductory biology courses for non-majors students range from half a semester to a full year. This textbook would be able to fit any of these permutations. The following units comprise the textbook:

- Welcome to Biology! is a brief overview of the biology highlights and scientific method.
- Unit 1 *That's Life...* is a basic reviewer of biology concepts.
- Unit 2 *Is It All in the Genes?* is a treatment of genetics with human applications.
- Unit 3 *We Are Not Alone!* is a primer of microbiology, plants, and animals and their applications.
- Unit 4 *The Dynamic Animal Body* is a study in the basic anatomy and physiology of the human body.
- Unit 5 *A Small Hole Sinks a Big Ship – Humans in the Ecosystem* acts as an introduction to ecology and human interactions within the biosphere.
- Unit 6 *Biology and Society* is a sociological treatise on evolution, animal behavior, and then a return from macrobiology to reflecting on the beauty of the organization of living systems.

Some portions of the book may be used to develop an alternate focus. For example, the microbial section could be emphasized with the ecology unit to form a general biology course thematically based on microbial ecology. There may be combinations from other units or even instructor-developed add-on materials to create a more "customized" approach.

It is my hope that the student and instructor will enjoy this textbook. The study of biology is a journey of scientific discovery and of understanding the complexity that underlies our existence.

Respectfully,

*Peter Daempfle*

# ACKNOWLEDGEMENTS

I thank Amy Daempfle for her writing of and research for (The Biosphere, Life Links to the Earth chapter 19), adding invaluable geochemistry writing and expertise. Amy is my life and my inspiration. I thank my father, Tobias Daempfle, who has had great impact on my life and whose conversations have contributed significantly to this book. To those who have died, and to the generations who have come before: Justine Daempfle, my mother and Josefa Nick, my grandmother. May there be a better place to meet again. To Betty Winter, Lina Klueber, and Gretl Volkart, for supporting my ideas and writing.

I also thank my teachers of the past, who helped form the foundations of thought to create this book. Their ideas were interdisciplinary, creative, and forward thinking. Their teaching was profound. They gave a great deal, more than they can know: Carole Demian, Robert F. Pospisil, Fr. John P. Rosson, Wendell W. Frye, William M. Elliott, Audrey B. Champagne, Julita Lambating, and Bill Vining. Let their teaching touch eternity. To my students, whom I have had the honor of teaching and sharing ideas. Let our paths cross again through this work.

I would like now to thank my friends through the years. Our good times and their support framed many stories and themes for this book: Thomas Hopcroft, esq., Chris (Tait) Miller, Laura Muller, Ralph, Jan and the Kroenkes, John Reeher and Marlene Hurley.

I appreciate the many reviewers who helped to revise and enhance this work, transforming it into a much better textbook for students:

Adele Register - Rogeres State University
Becky Brown, College of Marin
Caitlin Gille – Pasco-Hernando State College
Dan Fogell – Southeast Community College
Emily Lehman, Black Hawk College
Jennifer Wiatrowski – Pasco-Hernando State College
John Griffis - Joliet Junior College
John Reeher – SUNY Delhi
Karen Kendall-Fite, Columbia State Community College
Kate Lepore, Monroe Community College
Kiran Misra - Edinboro University of Pennsylvania
Liza Marie Mohanty, Olive Harvey College (City College Chicago)
Marirose Ethington – Genesee Community College
Mark Paulissen, Northeastern State University
Michael Rutledge - Middle Tennessee State University
Peter Kourtev, CMU (Michigan State University) -
Richard Gardner – Southern Virginia University
Scott Byington, Central Carolina Community College

Sue Hum-Musser, Western Illinois University
Tom Smith, Brigham Young University
Wendi Wolfram, Hardin Simmons University
William Mackay – University of Texas at El Paso

I appreciate the hard work and dedication of the Kendall Hunt Publishing group who were so valuable in making this book happen. My thanks to Paul Carty, for giving me the opportunity and to Lynne Rogers and the editorial staff for their organization, dedication, and energy. Thanks also to the supportive structure of the State University of New York (SUNY) system and SUNY Delhi, which supported this work through a sabbatical in the Spring of 2013. To the School of Nursing at SUNY Delhi, in appreciation for our many years of productive collaboration in sharing the teaching-learning experience with our allied health students, from which many of my ideas developed.

# ABOUT THE AUTHOR

Peter Daempfle was raised in Queens, NY in the 1970s and was a child of immigrants. He started early in life asking a key science question: "Why are things the way they are?" From the media's role in selling junk science to big tobacco's misuse of math to get people hooked on smoking, Dr. Daempfle explores good and bad science in his books.

He is a nature lover and enjoys the outdoors – hiking, swimming, and writing in the forest – the roots of his love for science. His singular goal is to help others see the excitement in studying the world around. His books help readers to evaluate science reports and empower them to be better consumers of science – he seeks to help them to think like a scientist.

As a child in the city, one of the few natural settings of the buildings was his eight-foot backyard. Here he learned about nature – he remembers thinking about the ants in the yard working hard to serve their colony. They seemed selfless and he wished for a world with their cooperation. His books show us why those ants are not so good as they appear and why science compels us to question. A theme of his books is to bring forth our scientific skepticism so the media and big business do not fool us. This text gives the foundation in biology to help students become their own advocates, from food and wellness to medical care.

He moved to the Catskills as a teen and obtained an education in science and biology to the doctoral level. His goal has been to improve biological literacy, contributing to the betterment of science education. Dr. Daempfle has taught in the sciences for over 20 years and is also the author of several books including *Good Science, Bad Science, Pseudoscience and Just Plain Bunk: How to Tell the Difference* and *Science and Society*. He is a science advisor in the standards-based reform effort, and has also authored several refereed journal articles, various science reviews, a laboratory manual, *Introduction to Anatomy and Physiology: A Guide to the Human Body*, and lectures to scientific and general audiences.

Dr. Daempfle has held faculty positions at Hobart and William Smith Colleges, Western New England University and is currently an associate professor of biology in the State University of New York, College of Technology at Delhi. He earned his Ph.D. in Science Education and his M.S. in Biology at the University at Albany, State University of New York; M.S. in Education from the College of Saint Rose; and B.A. in Biology from Hartwick College. He was class valedictorian of both Forest Hills High School, Queens, NY in 1988 and Hartwick College, Oneonta, NY in 1992 and graduated summa cum laude with departmental distinction in biology and German. Dr. Daempfle is on the Editorial Board of the Journal of Science and Popular Culture, Brisbane, Australia.

From 2001 to 2009, he was a science advisor/consultant to the Bush Administration's No Child Left Behind Act (NCLB) through Measured Progress, Inc. He focused on science content applications and test design in relation to standards development. He is known in the science literature for the publications focusing on the development of scientific reasoning, retention of students in STEM majors, studying the transition between high school and college biology programs, and human biology and microbiological applications. This new text contributes to the effort begun by the standards-based reform movement and the Vision and Change core competencies to improve national science literacy and advance the importance of scientific thinking.

Dr. Daempfle is dedicated to his family, his wife Amy, his children, Justina and Konrad, and his father, Tobias. They share an interest in exercise and fitness, as well as philosophy and natural studies. He is also a fan of old movies (Turner Classic Movies) and is an avid history buff.

See "Science Impact TV Show" for an interview with Peter Daempfle:
http://www.youtube.com/watch?v=XNDAKUrKxuE

# Welcome to Biology!

## ESSENTIALS

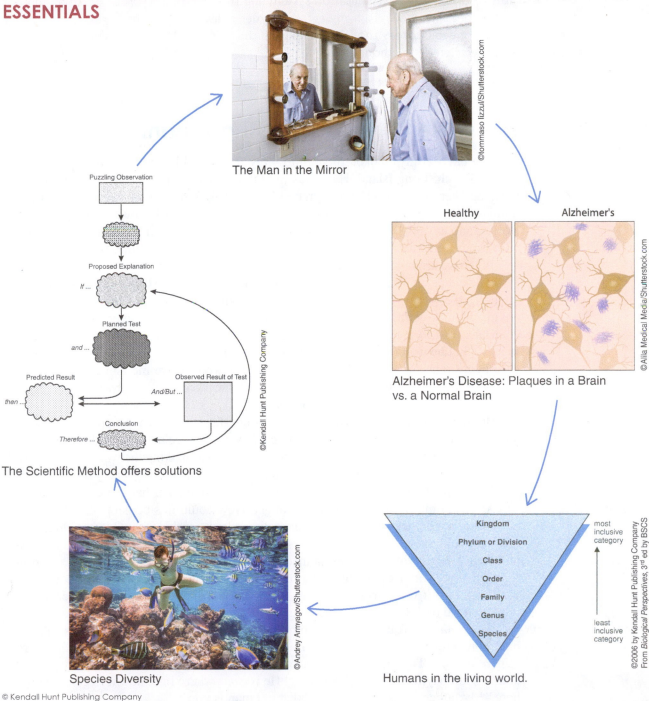

©tommaso lizzul/Shutterstock.com

The Man in the Mirror

Puzzling Observation

Proposed Explanation

If ...

Planned Test

and ...

Predicted Result

then ...

Observed Result of Test

And/But ...

Conclusion

Therefore ...

©Kendall Hunt Publishing Company

The Scientific Method offers solutions

Healthy    Alzheimer's

©Ailila Medical Media/Shutterstock.com

Alzheimer's Disease: Plaques in a Brain vs. a Normal Brain

Kingdom
Phylum or Division
Class
Order
Family
Genus
Species

most inclusive category

least inclusive category

©2006 by Kendall Hunt Publishing Company
From *Biological Perspectives*, 3rd ed by BSCS

Humans in the living world.

©Andrey Armyagov/Shutterstock.com

Species Diversity

## The Case of the Nonpaying Tenant

It was a day I always looked forward to – my father and I went to visit Uncle Hans in Mastic Beach, Long Island. Hans was my father's uncle, and we all enjoyed spending time together at the beach. We regretted that we saw him only once a year during the holidays. Uncle Hans had lived through World War II, built his own house, told many stories, and always liked seeing us on our visits. When I was a child, he was generous with candy and treats whenever we visited, which made it a fun time.

When we arrived at his house this time, we knocked on his door, but no one answered. We knocked again and finally went to the back door. There was Uncle Hans, sitting at the kitchen table, staring at the wall. We gestured to him, and he opened the back door, without much of a greeting. How strange that he was not smiling and hugging us as we entered.

Hans instead greeted us with a complaint: "That man is using my stuff. I told him to leave, but he won't go." "Who?" asked my father. "Why, the tenant in my house – he's not paying rent, he does what he pleases, and he won't leave me alone," explained an exasperated Hans. My father and I were very concerned . . .Who was this nonpaying tenant and why had we not heard about him?

But how odd; Hans had lived alone in his one-family house for many years. Had he taken in a boarder? We did not think that he really needed the money. Uncle Hans took us around the house, showing us all the mess the tenant was making: he was sloppy, used all of Uncle Hans' dishes, ate his food, and didn't pay his share of the grocery bills. We could see that the place had deteriorated since our last visit, and Hans had always been so neat.

At this point, my father and I were angry. How could this be happening? Why had our family not helped Uncle to take legal action to remove this tenant? It was elder abuse and we would not tolerate disrespect to seniors. "Uncle, where is this man right now? We will have a word with him!" I exclaimed.

"I'll show you. He's living in this room." Uncle Hans took us to what I thought was the bathroom. It was curious, and my father and I quixotically looked at each other. I think that we both knew that something was not right. No one *lives* in a bathroom.

Uncle Hans took us into the small room, but there was no one there. It was an empty bathroom with a towel on the mirror. Hans pulled off the towel and yelled into the mirror, "There he is! – That old man is using my toothpaste, my toothbrush, and my cologne, and he's not paying any rent or anything. He is everywhere I go; now let's get him out of here." My father said sadly to Hans, "that old man is you."

## CHECK UP SECTION

There are many examples of diseases caused by the aging process affecting our society. Lifestyle changes can help people cope with these challenges. Use our case in the story to research Alzheimer's disease. Explain two benefits and two drawbacks to the solutions you propose for coping with Alzheimer's disease. How do you see changes due to age-related diseases affecting our society in the near future?

# Getting to Know Biology

Hans was his own nonpaying tenant due to dementia brought on by Alzheimer's disease and aging. Changes in his brain chemistry, namely the accumulation of certain chemicals, caused Hans to become confused and prevented him from recognizing his own image in the mirror. Alzheimer's disease is associated with tangles or masses of proteins that form plaques along nerves in the brains of its victims. These plaques prevent proper transmission of nerve signals (Figure 1.1). Protein buildups cause fragmented thoughts and brain processes. There are many possible causes of Alzheimer's disease, ranging from traumatic head injuries and cardiovascular disease to genetics (family history). In fact, mutations of a certain gene are closely linked with higher risks of developing Alzheimer's disease.

Age-related dementia is a common symptom in elderly populations, but its prevalence does not make it less tragic. One in eight elders is afflicted with some degree of dementia due to aging, and Alzheimer's disease is the sixth leading cause of death among seniors. Nearly half of all seniors who are 85 years old and older experience Alzheimer's disease symptoms. The disease impacts their quality of life, their families, and their ability to contribute successfully to our society.

Everyone has an idea about the definition of biology, the study of life, but to really feel its effects in our everyday lives is another matter. Hans lives with his biology each day, struggling and trying to cope with its effects. Scientists work to study diseases to help people like Hans by finding cures or treatments for symptoms. Physicians treat conditions such as dementia but rely on research findings that scientists develop to help them fight the effects of diseases. Biology has many facets, affecting each of us uniquely.

## Alzheimer's disease

Progressive mental deterioration brought on by aging. The disease is associated with protein masses or tangles that form plaques along the nerves in the brains of the victims.

## Biology

Study of living creatures.

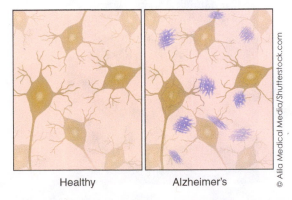

Healthy        Alzheimer's

© Alila Medical Media/Shutterstock.com

**Figure 1.1** Amyloid beta protein plaques on a nerve cell. Plaques block the transmission of nerve messages in the brain of Alzheimer's patients.

**Cell**

The structural and functional unit of an organism.

**Biophilia**

The affinity human beings share with other living creatures.

Biology encompasses many levels of study – from diseases to the basic unit of life, the cell, to how living beings interact with each other and their environment. It is a complex and exciting field, changing frequently as new species are discovered, new medical devices are developed, and old ways of looking at environmental problems are updated with new understandings and shifts in focus. Although many of you reading this chapter are not actively pursuing a career in one of the sciences, almost everyone has an interest in the natural world and the organisms we find in it. In fact, the affinity we share for other living creatures is termed biophilia, a term first coined by E. O. Wilson, a famous American biologist. Biophilia drives a questioning of how life works and how it relates to the world around us:

"How do birds fly south for the winter?"

"Why are we thirsty in the morning, after we wake up?"

"How does the *Dionaea muscipula* (Venus flytrap) trap flies?"

"Why does the *Amorphophallus titanum* (corpse flower) emit a horrible odor (Figure 1.2)?"

These questions and many others like them are important in understanding our relation to other living creatures and to the planet. They are based on principles in science that draw from fields outside of biology. A bird flies south in the fall because of visual and smell cues, but also because certain of the chemicals in its brain give it an ability to detect the Earth's magnetism (physics) to guide it. We are thirsty because we require water, with the right amount of salt, to bathe our cells (chemistry). The carnivorous diet of the Venus flytrap allows it to live in soils that are nutrient poor (geology). The odor of the corpse flower attracts sweat bees and beetles, which help spread its pollen to other flowers (ecology).

Asking questions such as those above helps to develop biological literacy, which comprises the many aspects of knowing about life: the facts, skills, ideas, and ways of thinking that enable a citizen to make decisions about and use biology and its technology. Everyone should be biologically literate. One does not need to perform specified tasks or understand complex instrumentation to be biologically literate. Biological literacy just means thinking like a scientist; that is the best descriptor of a biologically literate person.

**Biological literacy**

Is the ability to interpret, negotiate, and make meaning from the many aspects of knowing about life to make decisions and use biology and its technology.

© Paul Marcus/Shutterstock.com

**Figure 1.2** *Amorphophallus titanum* (corpse flower). The corpse flower has a smell similar to that of a decomposing animal.

Because biology is interrelated with concepts and terms branching from other areas, both science and non-science, we say that biology is interdisciplinary. This means that it draws from many fields of knowing. Science is often seen as a group of separate subjects, among them biology, geology, physics, and chemistry, but these subjects often overlap. The distinction between biology and other fields was discussed by the geographer Halford Mackinder in 1887, who stated, "The truth of the matter is that the bounds of all the sciences must naturally be compromises, knowledge . . . is one. Its division into subjects is a concession to human weakness." Biology should be understood in relation to the many fields that comprise it.

This chapter explores the scientific process used to understand big ideas in biology. We begin the textbook with a look at the characteristics of life, at how the Earth and its inhabitants came to be in the present, and at how scientists in all disciplines arrive at understandings about their fields of study; then in the next chapter, we will explore some of the chemistry and physics principles that drive life processes.

There are many branches of biology, each studying different aspects of the living world. A few include *microbiology*, the study of organisms not seen with the naked eye (99% of all living things); *pathology*, the study of diseases; *anatomy*, the study of structure, or how an organism appears; *physiology*, the study of function, or how an organism works; *genetics*, the study of inheritance and how characteristics are passed on between the generations; and *ecology*, the study of organisms and their interactions with the environment. Organization above the organism level is studied by *macrobiology*, which investigates how organisms interact with each other and within the environment. Macrobiology looks at animal behavior and nonliving environmental factors such as water in living systems, for example. Its emphasis is on human impacts within the greater ecosystem (or environment). Issues such as global climate change, acid rain deposition, overpopulation, and endangered species are macrobiology issues. Each of these studies is based on the same seven shared characteristics of living systems described in the next section.

## What Is Life?

All living things are composed of chemicals. Individually, the chemical substances are not living; however, in combination, they are the building blocks of life. What constitutes life? How does life differ from nonlife? What is it mean to be alive?

The difference between life and nonlife can be found in the organization of the substances that make up life. For instance, as substances become more complex, new properties emerge that are distinct from those of nonliving objects. There is no single difference between life and nonlife. Instead, there are a host of properties that biologists use to distinguish living from nonliving systems. These properties constitute the characteristics of life:

1) **Order**: Living organisms have a high degree of order, with complexity that is still being discovered and understood through increased testing and observation techniques. Consider the diatoms, a group of algae that contain intricate walls of silicon dioxide (the same material that sand is made of), arranged in a mosaic structure (Figure 1.3). Their ornate organization is a visual testament to how complex can be the arrangement of a simple living organism. Life is ordered in such a way that is can be divided into smaller and smaller categories. It has a hierarchical order, arranged from the smallest unit of life, the cell to a whole living organism.

**Interdisciplinary**

Involves two or more areas of knowledge.

**Macrobiology**

The study of how organisms interact with each other and within the environment

**Ecosystem**

A system that involves interaction of a biological community with its physical environment.

**Characteristics of life**

The seven features (adaptation, order, response to stimuli, growth, development, and use of energy, homeostasis, reproduction, metabolism, diversity) that differentiate between life and nonlife.

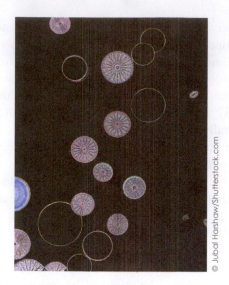

**Figure 1.3** Diatom cell wall structure. Note the complexity and almost artistic quality of the walls of the diatom.

© Jubal Harshaw/Shutterstock.com

## Homeostasis

The steady state maintained by living systems.

2) **Homeostasis**: Living systems maintain a steady state, termed homeostasis. They respond to and exchange materials with their outside environment to keep internal conditions stable. Temperature and salt and sugar levels in the blood are all maintained to promote stable workings within a living organism. Look to the simple *Stentor roeseli* in Figure 1.4, a single-celled creature that looks like a nonliving trumpet used in ancient battles. The *Stentor* carefully controls its internal environment by sensing chemical levels in its watery surroundings and then pumping out excess water through specialized internal pumps. Its nonliving appearance masks its many living characteristics.

3) **Growth, Development, and Energy Use**: All organisms acquire and use energy from their surroundings to grow and develop. A tree, an elephant, a mouse, and a single-celled bacterium each take in substances to obtain energy. They then conduct a series of chemical reactions, which are together termed metabolism. These reactions form new materials to grow and make changes in their structures. As

## Metabolism

The sum total of chemical reactions taking place in cells.

© Lebendkulturen.de/Shutterstock.com

**Figure 1.4** *Stentor roeseli*, a protozoan, resembles a trumpet.

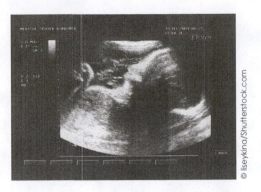

**Figure 1.5** Growth of a human fetus. The human embryo grows rapidly to a seven-month fetus in this ultrasound image.

shown in Figure 1.5, a human fetus grows larger and more complex as it develops in its mother's womb. This ability to acquire energy, grow, and develop makes living systems able to change.

4) Response to Stimuli: Living organisms are able to react to the world around them. Consider the mating behavior of *Rana pipiens*, the prevalent North American frog species. When a male *Rana* mates with a female, he jumps onto her back, and this initiates a croak reflex; the bottom frog croaks, indicating that she is actually a male (Figure 1.6)! The croak response tells the other male frog to get off, "I am a Male frog!" This signal allows more selective behavior by male frogs, helping them to determine male from female and thus prevents a possible altercation between two males.

Organisms respond to internal stimuli in addition to those in the environment. In order to maintain homeostasis, body systems monitor internal chemical balance, temperature, and even pressure and position. Maintaining balance, as when we are walking, requires cues from our eyes, inner ear structures, known as the semicircular canals, and specialized receptors found throughout our bodies that sense position, known as proprioceptors.

5) Adaptation: Populations of living things adapt to their surroundings and evolve or change as a group, with some organisms surviving and reproducing more

**Response to stimuli**

Ability to react to the various changes of the environment.

**Adaptation**

Populations of living things adapt to their surroundings and evolve or change as a group.

**Figure 1.6** Two glass frogs (teratohylamidas) mating. A male grabs hold of his female mate, stimulating her to produce eggs and his own sperm to be released.

**Figure 1.7**    English peppered moth: light variety hiding on an oak tree branch.

successfully than others. Consider the case of the English peppered moths in England, which contain a light, *Bistonbetularia f. typical*, and a dark variety, *Bistonbetularia f. carbonaria*. In the preindustrial era, more light moths were found across England. Light-colored moths blended with the light-colored trees of the countryside. However, as soot and pollution from factories created a darker environment during the Industrial Revolution of the 1800s, light-colored moths stood out, attracting predators. Thus, their numbers declined, and the dark variety became dominant. The dark moths blended in better with the changed backgrounds, helping them to survive more successfully. English peppered moths adapted to the Industrial Revolution by having two varieties that fluctuated with changing conditions (Figure 1.7). In Chapter 7, modern trends in English moth populations will be discussed further to elaborate upon adaptation of species in current times.

**Diversity**

The adaptation and evolving of organisms showing a great deal of variety.

6)    Diversity: The adaptation and evolving of organisms described in the above section resulted in a great variety of living creatures. Scientists have classified roughly 8.7 million nonbacterial species now living on the Earth. Living systems, considered a group, are diverse. A great deal of biodiversity – the variety of life forms in a given area – has yet to be uncovered because there are so many areas of the Earth that have not been sampled – deep sea vents, volcano interiors, and many polar regions, to name a few. The latest reports in *Nature* magazine state that almost 90% of marine and land species remain undiscovered. Unfortunately, the rate of extinction of species has increased a thousand fold in the past century, with about 20 species becoming extinct every minute in tropical rainforests. Some species may never be discovered before they become extinct.

**Biodiversity**

Variety of life forms in a particular habitat.

Coral reefs are one of the most diverse ecosystems of the world, containing almost 25% of all marine species and yet occupying only about 0.1% of the surface of the ocean. Like tropical rainforests, coral reefs are being seriously threatened by environmental and other changes that disrupt and sometimes destroy their fragile ecosystems.

**Reproduction**

The process of making new offspring.

7)    Reproduction: Living systems are able to reproduce themselves. While there is a great deal of variety of organisms, they all transmit the same set of hereditary

**Figure 1.8** The genetic trail. Baby ducks have hereditary information passed onto them from their parents.

material from generation to generation. Hereditary material is composed of genes, made of the chemical DNA (deoxyribonucleic acid). DNA is the genetic material of a cell that contains the instructions to life. Organisms use DNA to guide their growth and development. Genetic material is passed on to new offspring each generation.

While the same material is used among all of the species, how does one species differ from another? Why is a maple tree different from an oak tree? How is a baby duck similar to its mother (see Figure 1.8)? The answer lays in the subtle differences in the details of DNA's structure. Just minor differences in portions of DNA make organisms of different species, very different from each other. In fact, the genetic material of humans and chimpanzees is 99% the same, but small differences in hereditary codes cause big differences in the physical features of humans and chimps. Genetically, each human is 99.9% the same as all other humans, but small differences in the 0.1% of remaining DNA make each of us unique. Guidance by our genetic material helps us to carry out our life functions in a world with forces that often fight against us. Uncle Hans had bits of DNA that coded for plaques in his brain, which led to his problems with dementia. Often, simple changes inherited in our cells can lead to the many characteristics that make us unique.

All living things share each of the seven characteristics discussed in this section of the text. While there is incredible diversity in life across the Earth, all organisms need to be able to carry out these life functions to survive. This is one of the central themes of biology.

## Order in a Universe of Chaos

### Organizing Biodiversity: Hierarchy of Life

One characteristic of life is that it is ordered: it is composed of building materials that are very similar across organisms. In organisms, atoms are the smallest units of matter that maintain the properties of the larger sample. Atoms organize to form molecules,

**Atoms**

Are the smallest units of matter that can exist and maintain the properties of the larger sample.

**Molecules**

Atoms bonded together.

## Macromolecules

Molecules containing large number of atoms, which are the building blocks of living things.

## Organelles

Structures that carry out specific functions within cells.

## Membrane

A sheet-like structure that acts as a boundary in an cell.

## Tissues

Groups of cells having similar structure and performing similar functions.

## Organs

Specialized body parts that carry out specific functions for organisms.

## Organ System

A group of organs working together performing a united function.

## Organism

Living creature formed as a whole by organ systems.

## Population

A group of organisms of the same species living in a given area.

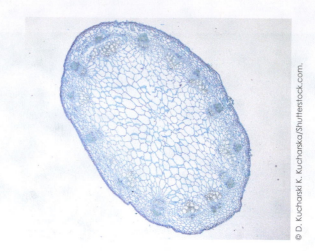

© D. Kucharski K. Kucharska/Shutterstock.com.

**Figure 1.9**   Robert Hooke's drawings of cork cells looked much like the image reproduced here from his Micrographia, 1665.

which are two or more atoms chemically combined. Water is an example of a molecule; it is composed of two hydrogen atoms attached to one oxygen atom. Some molecules join together to form macromolecules, which are the building blocks of living things: proteins, lipids (fats), carbohydrates, and nucleic acids. Proteins make up the structures and perform many functions in cells. Lipids store energy in the longer term for use by cells, and carbohydrates provide instant energy. Nucleic acids are the hereditary materials passed on from parents to offspring. Figure 1.10 shows the organization of chemical substances into a living cell. These macromolecules organize into living systems in such a manner as to produce life. Based on their chemistry, macromolecules orient themselves to form organelles, which are structures that carry out specific functions in living systems. Organelles organize together inside a membrane or a cell wall to form a cell, the fundamental unit of life.

There are over 200 types of different cells in the human body, each with unique structures and functions. Groups of cells that have similar structure and perform similar functions are called tissues. In humans, muscle, nerve, epithelial (covering), and connective tissues make up our structures. Tissues unite to form organs, which are specialized body parts. Organs carry out specific functions for an organism. For example, the kidneys filter blood, and the small intestines absorb food. Organs working together, such as the bladder, kidneys, and ureter tubes, make up an organ system. The digestive system, which processes food entering the body, is made up of several organs: the liver, gallbladder, intestines, stomach, and pancreas. Together the organ systems form a whole, living creature known as an organism.

As indicated in Figure 1.10, a group of organisms of the same species living in a given area is known as a population. A population of fungi in a forest is studied and counted as a discreet unit. Two or more populations in an area are termed a community. To extend our example, the community (or biocenoses) would include the pine trees, fungi, insects, birds, and chipmunks found in a forest. A community's interactions with the nonliving environment (air and water) comprise an ecosystem. The many effects of nonliving factors on a forest community are dynamic and often quite complex. All of the different ecosystems of the Earth interacting with their environment make up the biosphere.

Interactions of organisms with each other are graphically described as food chains and food webs. These show the way energy flows within an environment, with food chains showing energy flows from one organism to another, and food webs showing the

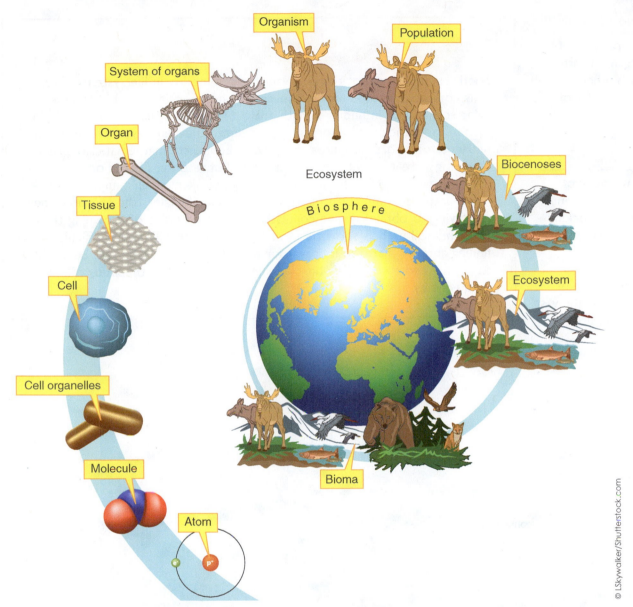

Organism

Population

System of organs

Organ

Biocenoses

Ecosystem

Tissue

Biosphere

Cell

Ecosystem

Cell organelles

Molecule

Bioma

Atom

© LSkywalker/Shutterstock.com

**Figure 1.10** Organizational levels of living systems. Life is ordered in a hierarchy, with increasing complexity from atoms and molecules to organ systems and whole organisms.

interaction of many organisms' energy flows with each other. Food chains are actually threads in a larger food web. An example of each is given in Figure 1.11.

## Taxonomy: The Science of Classification

The science of classifying the vast biodiversity described earlier is called taxonomy. Attempts to classify organisms into logical groups began thousands of years ago, as far back as the ancient Greek philosopher Aristotle, who organized life according to a scale of complexity, called *scala naturae*, meaning stairway of Nature. Aristotle's scheme is shown in Figure 1.12.

Later, in the 1800s, Carolus Linnaeus, a naturalist, standardized a naming system for living creatures that is still used today. Linnaeus' system of naming is known as

### Community

A group of living organisms living in the same area or having a particular characteristic in common.

### Biosphere

All of the different ecosystems of the Earth interacting with their environment make up the biosphere.

## Food chain

Interactions of organism with each other through the transfer of nutrients and energy.

## Food web

A network of interdependent and interlocking food chains.

## Taxonomy

The science of classifying the vast biodiversity.

Food resources are not the only valuable commodity in life systems. Organisms compete for the limited resource of the opposite sex. Either in a foraging scenario or in finding a mate, many organisms search far and wide for resources. In the process, organisms maximize their energy intake, using the nutrients of a particular area or "patch." Ecologists mathematically predict how long a creature will remain in a "patch" of resources based on a few factors. In searching for a female, for example, weaker, smaller, and, of course, more desperate amphibian males will hold onto females (the resource they are trying to obtain) during copulation (sex) to a point of drowning them! Mathematical models show that smaller amphibians will have a difficult time finding another mate, so it "pays" for them to stay atop a female. Being able to predict and even modify behaviors in animals is called the study of behavioral ecology.

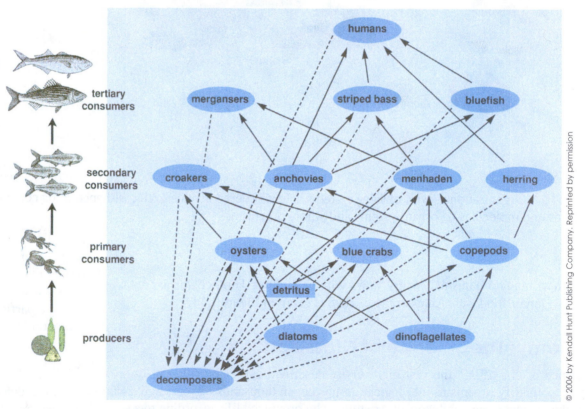

**Figure 1.11**    Different ways to show feeding relationships. A simple food chain in Chesapeake Bay on the left. Moreover, a simplified food web in the open water on the right. Both show various pathways of energy flow through living organisms in the environment. Decomposers eventually consume all living organisms. From *Biological Perspectives*, 3rd ed by BSCS.

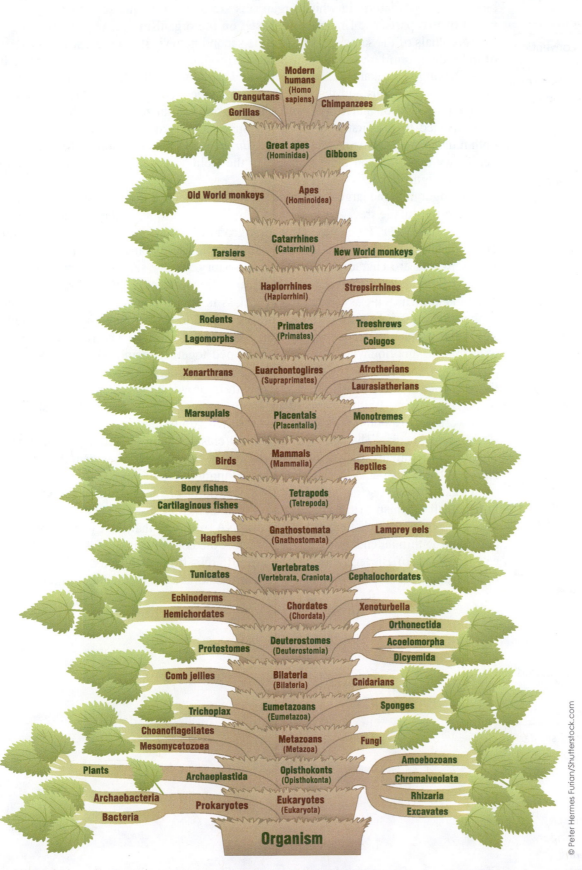

**Figure 1.12** *Scala naturae.* Aristotle's "Ladder of Life," with humans shown as dominant and more important than all other life.

## Binomial nomenclature

Naming convention for living creatures, in which organisms are given unique scientific name, composed of two parts. The first indicates the genus and the second the species.

## Genus

A group of individuals of the same species.

## Species

A group of individuals similar enough to be able to reproduce with one another to produce live, fertile young.

## Family

A group of genera consisting of organisms related to each other.

## Order

Used in the classification as a group of families.

## Class

A group of related orders.

## Phylum

Number of similar classes grouped together.

## Kingdom

The highest grouping under which living organisms are classified.

binomial nomenclature, in which organisms are given a unique scientific name, composed of two parts: the first name is based on the organism's genus, which is a group of individuals of the same species, and the second name is its species, which is a group of individuals similar enough to be able to reproduce with one another to produce live, fertile young. For example, *Drosophila melanogaster* is the fruit fly that belongs to the *Drosophila* genus and the *melanogaster* species. Note that the scientific name is always italicized or underlined, and the genus is capitalized. A scientific name may be shortened to the first letter of the genus, capitalized, and the full species name, both italicized: *D. melanogaster*. *Drosophila* are well known in biology for their simple hereditary make-up and their easy use in the field of genetics. Their mating ritual is shown in Figure 1.13.

A kingdom is the largest grouping used in Linnaeus' binomial nomenclature. Consider the red maple tree, *Acer rubrum*: Its species, *rubrum*, is the smallest grouping, and its genus, *Acer*, is the next largest. A group of genera is known as a family and a group of families is known as an order. Red maples are the members of the *Aceraceae* family, which are characterized by having watery, sugary sap. The *Aceraceae* family order is Sapindales, which are soapberry, usually wooden, plants. A group of related orders is known as a class. A red maple's class is Dicotyledonae, which means that all of these organisms have an embryo with two seed leaves (cotyledons). When a number of similar classes are grouped together, they form a phylum. The red maple phylum is termed Magnoliophyta. Many phyla grouped together form a kingdom. Magnoliophyta is grouped with other phyla into the plant kingdom. An easy way to remember the sequence of this classification scheme, from broadest to most specific, is by using the saying: "**K**ing **P**hillip **C**ame **O**ver **F**rom **G**erman **S**hores," with the first letter of each word corresponding to the first letter of Kingdom, Phylum, Class, Order, Family, Genus, and Species. The complete biological classification of red maples is shown in Figure 1.14. Humans, such as Uncle Hans, are *Homo sapiens*. How would you classify Hans? What is his genus? species? phylum? kingdom?

Organisms are also classified based on their domain. There are three known domains in which all organisms fit: Bacteria, Archaea, and Eukarya. Bacteria and Archaea are single-celled organisms that contain "naked" DNA, meaning that their genetic material is not found within an enclosed nucleus, which is the control center of the cell. DNA and cell structures of Bacteria and Archae are different from each other, classifying them as separate domains. Traditionally, organisms lacking a distinct nucleus and organelles

© Patricia Chumillas/Shutterstock.com

**Figure 1.13** *Drosophila's* fruit flies mating. 'Dancing Mate Ritual' *Drosophila* fruit flies engage in sexual foreplay through dancing to attract a mate. This leads to copulation (sex).

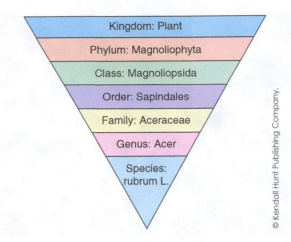

Kingdom: Plant

Phylum: Magnoliophyta

Class: Magnoliopsida

Order: Sapindales

Family: Aceraceae

Genus: Acer

Species: rubrum L.

© Kendall Hunt Publishing Company.

**Figure 1.14**   Classification schemes of red maples.

**Domain**

A division of organisms ranking above a kingdom in the systems of classification based on similarities in DNA and not based on structural similarities.

**Bacteria**

Single-celled organisms that have cell walls but lack an enclosed nucleus and organelles.

**Archaea**

Microorganisms that are similar to bacteria in size and structure but different in molecular organization.

**Eukarya**

One of the three domains of the biological classification system.

were called prokaryotes, but this classification is now informal. Both Bacteria and Archae are prokaryotes.

Over 99% of all organisms are classified as either Bacteria or Archaea. They inhabit all areas of the Earth, from boiling sulfur lakes to frozen arctic ice to the insides of our large intestines. Bacteria are heavier by mass (this is called *biomass*) than all other living organisms combined. Oddly, this is a world unseen by humans and yet it is vast, vibrant, and changing. Bacteria and Archaea live in such varied conditions that they are said to be omnipresent: on average, there are 100,000,000 bacteria (including Archaea) per square centimeter of surface at any place on the Earth.

Eukarya, or eukaryotes, are organisms containing organelles and a distinct, true nucleus with genetic material contained therein. Eukarya are composed of four different groups. The simplest Eukarya are the **protists or** Protista, a diverse group composed of both single-celled and multi-celled organisms, ranging from *Amoeba* to *Paramecium* to *Euglena*, as shown in Figure 1.15. A drop of pond water usually contains all of these creatures. Some protists are producers – they are able to produce their own food. Others, known as heterotrophs or consumers, acquire energy by eating other organisms. Protista are the oldest, evolutionarily, of all the groupings of eukaryotes.

The fungi (singular, fungus) are eukaryotes that secrete chemicals to break down other living or once-living materials. This allows fungi to consume these substances.

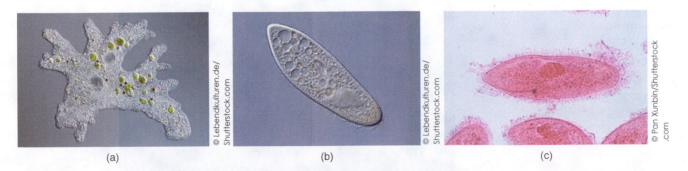

(a)    (b)    (c)

© Lebendkulturen.de/Shutterstock.com    © Lebendkulturen.de/Shutterstock.com    © Pan Xunbin/Shutterstock.com

**Figure 1.15**   a. *Amoeba*, b. *Paramecium*, and c. *Euglena*. Microorganisms depicted belong to the Protista kingdom. Each has characteristics that make them unique. An *Amoeba* appears blob-like, often engulfing prey as it moves; The *Paramecium* captures food as it beats its hair-like projections, creating waves to bring prey in toward it; A *Euglena* is able to make its own food from sunlight as well as capture prey.

## Nucleus

Control center of the cell.

## Prokaryote

Organisms that lack a distinct nucleus and organelles.

## Eukaryote

Organisms that contain organelles and a distinct, true nucleus with genetic material contained therein.

## Protista

A diverse group composed of both single-celled and multi-celled organisms.

## Producer

Organisms with the ability to make their own food.

## Heterotrophs

Also called consumers, these organisms acquire energy by eating other organisms.

## Fungi

Eukaryotic organisms that secrete chemicals to break down other living or once-living materials.

**Figure 1.16**   The Death Cap Mushroom. This is a deadly poisonous fungus. It likely killed Roman Emperor Claudius and Holy Roman Emperor Charles VI. It contains toxins that damage the liver and kidneys leading to fatalities.

Mushrooms are a common example of fungi, which are able to decompose dead organisms, but unable to make their own food (Figure 1.16).

Alternatively, all plants are able to obtain food by converting sunlight's energy to chemical energy through the process of photosynthesis. Carbon dioxide and water are rearranged by plants to produce the simple sugar, glucose, and oxygen, using sunlight as a source of energy. Plants are multicellular eukaryotes, with an ability to live on land, or in freshwater or saltwater. Their ability to survive without the need of energy from other organisms makes them **producers**, able to produce their own food (Figure 1.17).

Animals require the energy of other living creatures to survive. Animals are multicellular eukaryotes, which consume other organisms. Examples include herbivores, which eat plants, scavengers, which consume dead remains, parasites, which draw energy from a host organism while it is alive, and carnivores, which eat meat to survive. Animals are motile (able to move) in carrying out their life functions such as in obtaining food sources; plants are immotile and require other means to acquire food sources. For

**Figure 1.17**   Plants: A Giant Redwood in Yosemite National Park. Giant Redwoods are the largest trees on the Earth, with stems reaching over 75 meters (250 feet) tall.

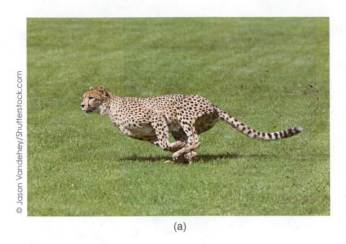

(a)            (b)

**Figure 1.18**   a. Cheetah running after prey; b. Maple tree in a field. While plants use sunlight to obtain energy, animals such as the cheetah pursue their prey to obtain energy. The cheetah runs after prey and a maple tree in a forest obtains light. Each performs life functions.

example, cheetahs, to obtain their prey are able to run up to 70 miles per hour, while plants use sunlight to obtain needed sugars (see Figure 1.18).

The three-domain classification scheme described earlier divides life into five different general **kingdoms**: Bacteria-Archaea, Protists, Fungi, Plants, and Animals. Figure 1.19 shows the five kingdoms of life. Although there is general consensus about how individual organisms should be classified, it is good to remember that classification schemes are human constructs, able to be reconsidered and changed. Molecular techniques are able to show new relatedness of organisms; the information from molecular data results in changing classifications almost every day. If anything, the new information increases the debate within the scientific community, but debate is an important part of the development of scientific findings.

## Asking Hard Questions

How did slime molds and muscle cells develop so that they can efficiently fulfill their intended function? Why do living systems sometimes fail, as in the case of Uncle Hans' dementia? Why do cells die and why do all living things die? These are difficult questions with answers that have been thought about over the millennia. Today, because of the contributions of many earlier scientists and philosophers and an impressive and ever-growing body of evidence, we know that the theory of evolution, the process of changes in species over time, explains how life developed. Theodosius Dobzhansky (1900–1975), the modern evolutionary biologist, explained that "Nothing makes sense in biology except in the light of evolution." We will begin our exploration of the answers to the questions that opened this section, along with hundreds of others, by tracing a little of the history of thought about how life began and progressed.

## The Development of Evolutionary Thinking

### Buffon and the Founding of Descent with Modification

While Aristotle developed a classification system of living creatures, it took over 2,000 years to accept that organisms could change over time. French scientist Georges-Louis de Buffon (1707–1788) was among the first to challenge commonly held views of the time by arguing that species were not the same as they when they first developed ages ago.

### Plants

Living organisms that are able to obtain food by converting sunlight's energy to chemical energy through the process of photosynthesis.

### Animals

Living organisms that feed on organic matter (other living creatures) for survival. Animals are multicellular eukaryotes and motile in nature.

### Evolution

The process of changes in a species over a period of time.

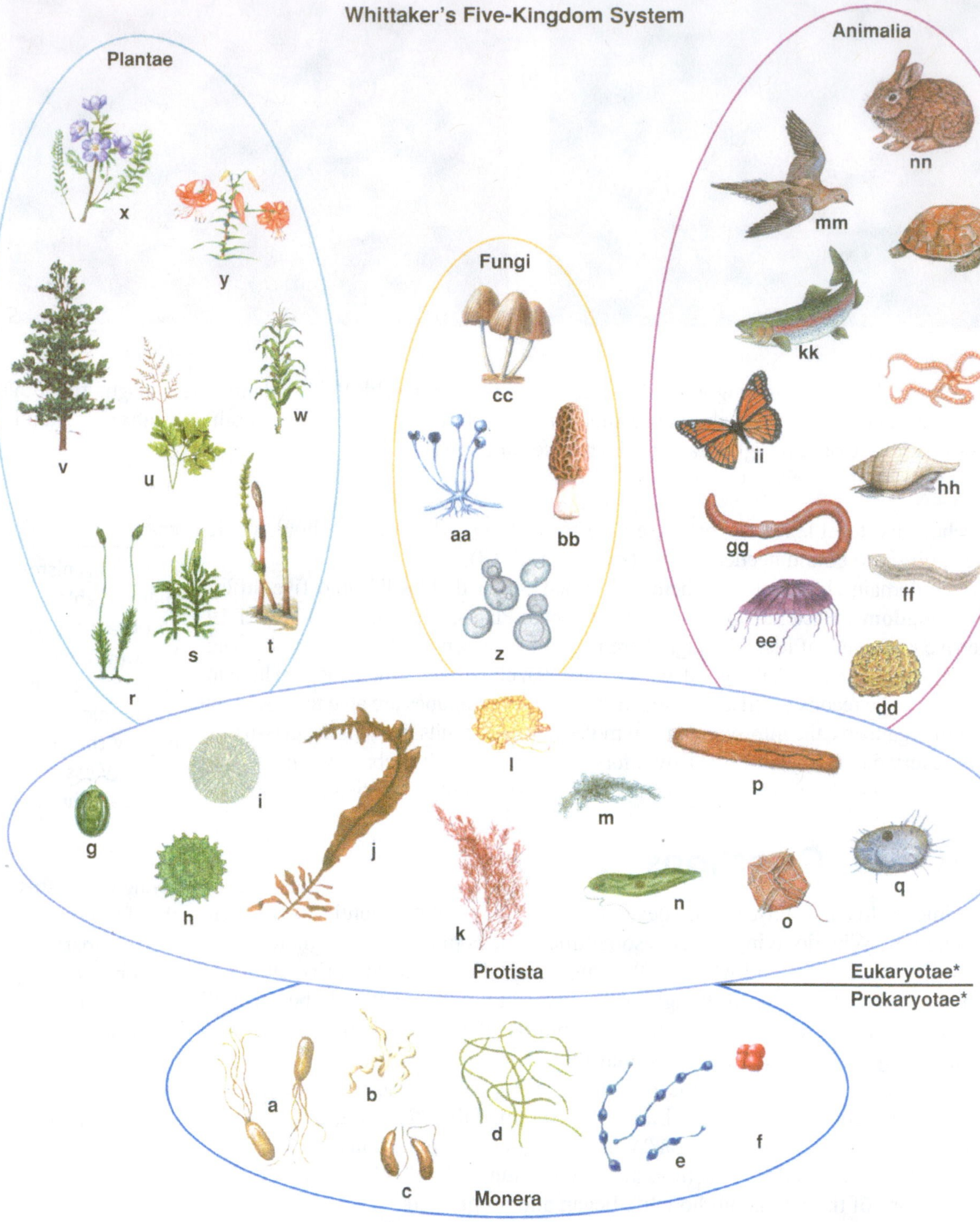

**Whittaker's Five-Kingdom System**

*Plantae*

*Fungi*

*Animalia*

*Protista*

**Eukaryotae\***

**Prokaryotae\***

*Monera*

**\*Prokaryotae** have prokaryotic cells that lack a membrane-enclosed nucleus, whereas **Eukaryotae** have eukaryotic cells that *do* contain a membrane-enclosed nucleus. Note that all of the bacteria (Kingdom Monera) are prokaryotic; all other organisms are eukaryotic.

**Figure 1.19**  Three domain systems. All three domains have a common ancestor. Bacteria and Archaea are more closely related and evolved earlier than Eukarya. Note Whittaker's Five Kingdom System of Classification. From *Biological Perspectives*, 3ʳᵈed by BSCS.

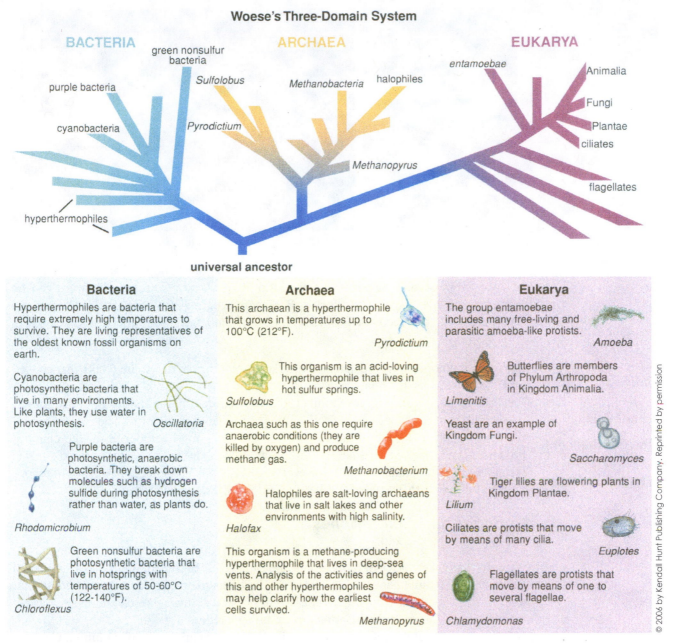

**Figure 1.19**    *(continued)*

He posed the degeneration hypothesis, stating that "There are lesser families conceived by Nature and produced by Time . . . improvement and degeneration are the same thing, for both imply an alteration of the original constitution." De Buffon, despite an unclear mechanism, was the first to state that changes in species occurred over time, explaining the vast diversity of life.

## Fossil Record

The evidence for evolution came from geologists, however. The gradual processes of wind, rain, and water flow lead to the deposition of sand and rocks in layers upon layers over the ages. Within these layers were remnants of life forms, discovered by early geologists. Geologist James Hutton (1726–1797) proposed that the Earth had been developed over a period of time through these geologic processes. This proposal contradicted

© 2003 by Corel Reprinted by permissions.

(a)                                                    (b)

**Figure 1.20**    Rock layers of the Earth. As layers of gravel and sand accumulate, fossils of once living organisms become embedded in the layers. These fossil layers can be dated to show the age of the fossils. Radioactive dating shows that older layers are found deeper in the Earth. A. From *BSCS Biology: A Human Approach*, 2nd Edition by BSCS.

Christian theologists, who calculated the Earth's maximum age to be 6,000 years based on Biblical records and analysis.

English surveyor, William Smith (1769–1839) noted, in his studies of caves, mines, and canals, that strata or layers of soil contained fossils of former life. The fossil record – the distribution of fossils in the Earth's layers – gave ample evidence for the changes organisms underwent over time. He stated that the deeper the rock layer is, the older it is (see Figure 1.20 for a depiction). However, it was not until the pieces of this fossil record were put together that a theory of the evolution of life was developed.

**Fossil record**

The distribution of fossils on the layers of Earth.

## Changes and Catastrophes

Charles Darwin's grandfather, Erasmus Darwin (1731–1802), questioned whether organisms were similar to their original forms during the Biblical creation period. Charles Darwin had never met his grandfather nor even held him in high regard, but Erasmus' ideas were similar to those developed much later by Charles. Both saw that animals may change in response to their environments and that offspring inherit those changes.

However, a number of theorists contributed to Darwin's ideas of evolution. Georges Cuvier (1769–1832), studied the fossil record, noting that 99% of all species that seemed to have lived were now extinct due to a variety of catastrophes. Catastrophism explained that new species formed after each destructive event, leading to a blossoming of new organisms. This idea shook the foundations of creationism because it held that forms of life were different from those found in the Garden of Eden of the Bible.

**Catastrophism**

The theory that explained that new species formed after sudden and violent catastrophes.

## Inheriting Acquired Traits

Like Cuvier, Jean Baptist Lamarck (1744–1829) noted that older rocks contained the fossils of simpler forms of life than newer rocks. Lamarck was the first to propose that individuals inherit traits from their parents. His hypothesis claimed that there was a natural progression or "evolution" dependent on the inheritance of acquired traits. Organisms acquired the traits of their parents through the use and disuse of those structures. For example, if trees became taller, their fruit would be higher, so giraffes adapted to the increased height by stretching their necks to reach their food. Those longer necks were passed down to their offspring (see Figure 1.21). Lamarck's view of evolution is

**Figure 1.21** Giraffes' necks are inherited; however, long necks did not develop because giraffes stretched them. Instead, they were a mutation that benefitted giraffes, enabling them to reach food at greater heights.

termed the **inheritance of acquired traits** through use or disuse of those traits. Lamarck was wrong because organisms cannot inherit characteristics of their parents developed during a parent's lifetime. For example, if a father loses his arm, his child is not more likely to be born without an arm. While Lamarck is now often disregarded because of his errors, he laid the foundation for Darwin's evolution because he noted that organisms *do adapt* based upon the environment of their ancestors.

## Darwin's Voyage: Natural Selection

Charles Darwin (1809–1882), author of the 1859 book *On the Origin of Species,* was the first to develop a full view of evolutionary theory. Through his voyages as a youth to the Galapagos Islands off the coast of Ecuador, he noted a variety of beak shapes and sizes in 13 different genera of finches. Each beak seemed "adapted" for the type of food on its particular island. One species of finch, for example, used a stick to pull insects out of bark but another pecked more easily through the softer wood. While all of the finches were of the same genus, they had minor differences based on their unique environments. These observations led Darwin to conclude that the environment had an effect on finch beak anatomy, seen in Figure 1.22.

It is safe to say that Darwin's voyage and the conclusions he made were influenced by the ideas of other, previous scientists, including perhaps his grandfather. His contribution is that he developed a synthesis of his own ideas and data with those of his fellow scientists, to comprise a working theory of how life developed over time. Darwin's theory of evolution can be summed into five steps:

1) All organisms overpopulate in any given area
2) Organisms then compete for the limited resources available to them due to that overpopulation
3) Individuals of a population have variations or differences that are inherited from generation to generation
4) Some organisms have an advantage in their variation over other organisms
5) An intense struggle for survival follows, leading to a "survival of the fittest" members of the population.

This process causes a change in the characteristics of organisms over time. It naturally selects out those best adapted to a particular situation and removes those individuals that are less well adapted. This process is termed natural selection and leads to the changes

**Variation**

Differences that are inherited from generation to generation.

**Natural selection**

The process whereby organisms better adapt to their environment survive and produce more off spring.

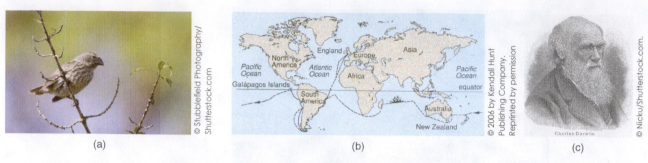

(a)                                    (b)                                    (c)

**Figure 1.22**    A. Anatomy of finch beaks. Darwin observed 13 different genera of finches on the Galapagos Islands. The shapes and sizes of each finch enable different feeding styles, with some able to eat insects and others leaves or fruit. This image shows a female medium ground finch from the Galapagos Islands. B. The map traces Darwin's trip to the Galapagos Islands. From *Biological Perspectives*, 3rded by BSCS. C. Darwin's origin of species was published in 1872. A portrait of Darwin is seen here.

or evolution of species over time. There were changes in dragonflies over their history in the fossil record (Figure 1.23).

### Evolution and Economic Systems

The ideas of scientists, like those of everyone, are influenced by their social and historical contexts. Darwin's description of the process of natural selection appears similar to a process seen in capitalist economies. To illustrate, a Wal-Mart Supercenter opens up on a street corner near a small deli. The deli cannot compete with Wal-Mart's lower prices, so it goes out of business. Similarly, the process of natural selection was well described in economic terms during Darwin's time by Adam Smith's (the 1750s), Thomas Malthus' (the 1790s), and David Ricardo's (the 1820s) theories of money and capitalism. It is surmised that growing up in a capitalist economic system in England of the 1800s influenced the development of Darwin's ideas, laying the foundation for his theory of biological natural selection and evolution.

These economists argued that an economic system of free markets and capitalism leads to a survival of the fittest businesses based on competition. They state that some businesses survive that are better able to outcompete others, doomed to fail. Their views claimed that businesses freely compete with each other for limited resources, with some being better suited than others, and thus a "struggle for the survival of the fittest" business. England in the 1800s was more purely capitalistic than any nation in the Western World today and probably best emulated nature in its natural selection of organisms. Darwin's ideas developed in a society that functioned in such a way that it influenced his own ideas and theory building.

> **Wallace Also Came Up with Evolution but Supports Darwin's Efforts**
> Alfred Wallace (1823–1913), a British geographer, worked independently of Darwin in developing his ideas on evolution from his knowledge of animal species in the tropics. In 1858, he became convinced that a process of evolution of organisms led to the diversity of animals he studied in the Amazon and in Southeast Asia. He was poor and a social activist, staunchly critical of the injustices he saw in capitalism of 19th-century England. Wallace actively wrote and spoke on the reality of evolution and published a review in 1867 "Creation by Law,'" which defended Darwin's thinking on evolution.

**Figure 1.23** Dragonfly size changed over time. The era of large insects ended as birds outcompeted larger insect species. Insects such as the dragonfly adapted with smaller sized forms of dragonflies increasing in dominance as time progressed.

One might argue that society was "ready" for the theory of evolution. Perhaps another person besides Darwin would have developed this theory in such an economy – to link capitalism to organismal change over time. Society had influenced Darwin, and Darwin has obviously influenced science and society. If he had not grown up within a capitalist economy, his world outlook might not have enabled the development of the theory of evolution.

Darwin's ideas were met with much opposition, especially from religious leaders, but his work revolutionized scientific thinking. "The theory of evolution," according to Ernst Mayr, "is quite rightly called the greatest unifying theory on biology." It explains all of the biodiversity seen in our world and the worlds that came before us; it shows how characteristics develop over time that are suited for one era but not necessarily for another; and it explains how Nature can lead to great life developments but also to mistakes. Perhaps there is a need for Alzheimer's disease to decrease the surplus population; perhaps Alzheimer's genes are an error that is recent because humans in the past did not live as long as they now do; perhaps Uncle Hans is a victim of evolutionary developments that occurred long before his own life. Regardless, evolutionary thinking explains many of our questions, while many remain unanswered.

# Scientific Thinking

## Scientific Literacy

All of the scientists mentioned in the previous section used scientific thinking to contribute to the community of ideas. Each of us is able to apply scientific thinking in our everyday lives to make decisions. Time after time, we encounter biology in statements made by the media or discussed with friends: "Echinacea prevents colds," "Smoked meats cause cancer," "Alzheimer's disease is caused by aluminum pots," and "Gay is genetic." All of these assertions require scientific analysis.

How do we practice science? It starts with science literacy, which is the comprehension of scientific concepts, processes, values, and ethics, and their relation to technology and society. In order to be able to really use biology, students should practice the science and its process. Science is not easily defined, but it is comprised of three parts: First, it is a **body of knowledge**, a set of facts that is extensive but continually changing as

**Science literacy**

The comprehension of scientific concepts, processes, values, and ethics, and their relation to technology and society.

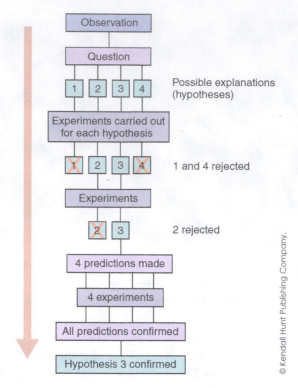

**Figure 1.24**  Scientific method steps.

### Scientific method

A procedure that has characterized natural science for centuries.

### Critical thinking

The analysis and evaluation of an issue to form a judgment

### Inquiry

Critical thinking used behind science to arrive at the truth.

new information becomes available. When Linnaeus classified organisms in his system of taxonomy, he constantly had to rearrange their order to add new species because so many were being discovered at the same time. Today, science facts are still changing rapidly with new technology and ideas.

Second, scientific thinking is based on a method. The scientific method has certain, specified steps, as shown in Figure 1.24. It is important to note that the scientific method is more complex than is shown in Figure 1.24. Many scientific discoveries occur by accident and many investigations require retracing of steps. Third, science is a way of thinking critically, being able to judge a claim and change one's reasoning about it if deemed necessary. The term "critical" comes from the Greek *kriticos*, to discern, or use judgment. Critical thinking takes practice and requires us to use all aspects of science from knowledge and method to reasoning about an issue at hand. The questions at the start of this section can be best answered using critical thinking.

Science, on the whole, is a way of thinking about the universe – a way of finding out the truth about phenomena. Inquiry is defined as the critical thinking used behind science. Inquiry follows a logical sequence of steps to arrive at truth but is also haphazard, backtracking in ideas and reformulating strategies. Science based on inquiry can be compared with the game of chess. Thomas Henry Huxley (1825–1895), an English biologist, described inquiry in the following excerpt from 1868:

> The chessboard is the world, the pieces are the phenomena of the universe, and the rules of the game are what we call the laws of Nature. The player on the other side is hidden from us. We know that his play is always fair, just and patient. But we also know, to our cost, that he never overlooks a mistake; or makes the smallest allowance for ignorance. To the man who plays well, the highest stakes are paid, with that sort of overflowing generosity with which the strong shows delight in strength. And one who plays ill is checkmated – without haste, but without remorse.

Science inquiry's unique process is well captured in this analogy to a chess game. Sometimes, you win and sometimes you lose, based on whether you play the game well enough to find out the truth behind your questions. Science is complex and multifaceted and requires creativity and practice. The purpose of the next section is to describe the process of scientific thinking and give you the skills to more critically evaluate scientific claims.

## Induction/Deduction

Evaluating a scientific question requires a piecing together of the facts. Science inquiry, like chess, requires both induction and deduction. Induction is a gathering of pieces of data to form a general conclusion, much as the fictional detective Sherlock Holmes investigates a crime scene. Deduction is the process of using a general premise to test and gather data, and eventually draw a conclusion. Both induction and deduction are used in the scientific method.

## Hypothesis Testing

Scientists begin with a "hunch," probably based on observations, about natural processes. Using inductive reasoning (stringing together the set of observations), scientific thinking may be ready to form a more solid guess about the phenomenon. A surfer may notice that changes in the ocean are happening, perhaps wondering if acid rain can lead to destruction of marine life. She or he might then look into what others know about the topic. A **critical review** of the existing literature should follow the selection of the problem to be studied. A critical review should be just that – looking at strengths and flaws in other studies, the methods and mathematics determining the conclusions, and new ways to investigate one's questions.

A surfer should know enough about his or her research problem to form a hypothesis, or possible explanation for the natural phenomenon. Any hypothesis should make sense and be based on a critical review of research. In the example, perhaps acid rain was shown to hurt diatom or algae populations in the ocean. Perhaps the surfer read that diatoms produce almost 80% of the world's oxygen and realized the importance of diatoms in our ecosystem.

A hypothesis must be empirically testable, meaning that the results must be measurable and logical and should address a question about a natural phenomenon. It should seek to explain and further science. A hypothesis is an unchecked idea and is really only a starting point in the scientific method. Forming tests of the idea is the true measure of the value of any hypothesis. A hypothesis is only an educated guess and as such is subject to change.

## Experimentation

Next, a test of the hypothesis is devised. There are many kinds of tests used in a scientific investigation, but the most powerful is the experiment. An experiment is a planned intervention, which analyzes the effects of a particular variable. The surfer's hypothesis, for example, requires that the effects of acid rain on diatom oxygen production be measured. A control group, which is the group given normal conditions, should be developed. In this case, diatoms could be placed in a test tube to develop under simulated normal marine conditions.

Next, at least one experimental group of diatoms should grow in a more acidic environment. An experimental group has one changed factor. That factor is termed the independent variable, which is the condition that the experimenter alters. In the diatom

---

**Hypothesis**

A possible or proposed explanation based on limited evidence for a natural phenomenon.

**Experiment**

A planned intervention that analyzes the effects of a particular variable.

**Control group**

A group in a study or experiment not receiving treatment by researchers and used as a benchmark to measure how other tested subjects do.

**Experimental group**

A group in a study or experiment that receives the test variable.

**Independent variable**

A variable that is altered by the experimenter.

**Dependent variable**

The results of the experiment.

experiment, what did we change? Yes, the acid levels. The dependent variable is the factor that is modified as a result of the independent variable having been changed. In other words, it varies or depends on the independent variable – it is what is measured through the course of an experiment. What was the dependent variable for the diatom experiment? Yes, oxygen production by diatoms. Simply put, the dependent variable always reveals the results of the experiment because it isolates the effects of one particular condition. A well-designed experiment seeks to keep all of the variables the same except for the independent variable. These are termed control variables, which are those factors that remain the same for all of the groups under study. The better controlled an experiment is, the stronger will be the study results. Careful preparation before an experiment is conducted may control the conditions.

## Data Analysis

**Data analysis (Qualitative and Quantitative)**

The process of evaluating information that is obtained by investigation. The reporting and use of non-numerical data is qualitative data analysis while reporting and use of numerical data is quantitative data analysis.

Information collected from an experiment is analyzed to form conclusions. Data analysis is the process of evaluating information obtained by an investigation. This may be either a qualitative or a quantitative process. Qualitative analysis is the reporting and use of data that are non-numerical in scope. It usually studies very few subjects or data pieces but looks at those in great depth. Observing changes in an ecosystem but taking notes on a chipmunk's behavior or tracing the movement of chemicals within a corn field may be qualitative studies.

Quantitative analysis is defined as the reporting and use of numerical data. This is a traditional scientific analysis and shows patterns from which to draw conclusions. Quantitative studies allow generalization of the results to a larger population. It requires a large number of individuals or units to sample. To illustrate, many trials of diatom testing need to be performed to make a conclusion that acid rain affects diatom oxygen production in marine environments. Small numbers of trials could lead to results that are just flukes. Enough numbers of individuals must be tested in quantitative studies for adequate statistical analyses. Quantitative analysis is what separates science from the many forms of pseudoscience.

## Math Gives Biology Power: Statistics

**Statistics**

The study of the collection, organization, analysis and interpretation of data.

Statistics is the study of the collection, organization, analysis, and interpretation of data. Quantitative analyses use statistics to analyze data. Statistics drives biological research by giving credibility to the claims a study makes from its data. For example, if diatom oxygen release is cut due to higher acid levels, "By how much?" and "Is it really significant?" should be questions asked and answered by the scientific community. Without math, scientists have little power to make recommendations or generalizations, as expressed by Galileo, the Italian natural philosopher of the 1600s. He understood that mathematics is the language of science.

**Null hypothesis**

The hypothesis that asserts that there is no effect or change due to a potential treatment.

"Philosophy is written in this grand book, the universe . . . It is written in the language of mathematics, and its characters are triangles, circles, and other geometric figures without which it is humanly impossible to understand a single word of it."

– Galileo Galilei

Experiments are set up using a null hypothesis, which is the opposite (or absence of relationship) of the experiment's hypothesis. A null hypothesis, represented as $H_o$, asserts that there is no effect or change due to a potential treatment. If the hypothesis states that variable #1 affects variable #2, the null hypothesis would state that variable #1 does not affect variable #2. When the null hypothesis is not supported by an experiment, then the real hypothesis can be accepted.

Statistically, the chance of error for supporting or failing to support a hypothesis is also calculated for an experiment's conclusions. The significance level is defined as a level of error that is likely, given the statistical analysis. It is written in the form of a decimal number and gives the percentage chance that the results are in error. For example, a significance level of .05 is equal to a 5% chance that the results are in error.

Many scientific investigations use the correlation, which is defined as a simple relationship between two variables. Variable "a" is somehow related to variable "b" as represented by the letter $r$, a correlation coefficient. Correlation coefficients range in value between $-1.0$ and $0$ and $+1.0$, with the negative values representing negative correlations and the positive values representing positive correlations. The two variables increase or decrease in tandem with one another in positive correlations and vary in opposing directions in negative correlations. Note that the closer the correlation to positive or negative 1.0 is, the stronger will be the linear relationship between two variables.

Correlations may show relationships but this does not tell anything about how the two factors are related or even if the relationship is important. In fact, variables other than the two given in a correlation may influence the relationship. Perhaps oxygen production in diatoms, for example, is not affected by acidity, but that acidity correlates with another chemical in the water. That chemical might be the real cause of the relationship between acid rain and oxygen production.

A more powerful statistical test, known as the ANOVA (Analysis of Variance), was developed, which identifies and isolates the independent variable to avoid such problems described in correlations. The ANOVA compares the mean (average) results of three or more groups in an experimental design. Very briefly, it is a process of isolating variables, showing the effects of the independent variable without effects of experimental error. ANOVA methods are beyond the scope of this text, but determining the effects of the independent variable is the point of any experiment or data analysis.

**Significance level**

The percentage chance that the results of a study are wrong.

**Correlation**

Relationship between two variables.

**ANOVA (Analysis of Variance)**

Is a powerful statistical method that compares the means of three or more groups.

## Results and Discussion

Publishing results involves objectively reporting the data and statistics of an experiment. A discussion of the results returns the investigation back to the subjective realm. The discussion interprets the data, explaining statistics based on the accepted literature, and makes recommendations for future research. It uses the intuitions of the scientist. It is the part of an investigation that is most creative and sometimes even speculative. However, it must be embedded in valid information – the information derived from the mathematics of the results. Conclusions are drawn from the analysis of the results in the discussion section.

Research findings first undergo a peer-review process, in which fellow scientists evaluate the research to determine if it is worthy of publication. A critical analysis of the results of research enables scientists to determine whether the results should be published for review and use to the scientific community as a whole. Published research enables other scientists to repeat the experiment and to rerun tests to determine the validity of the claims. Hypotheses are reformulated when results are unexpected or when confronting errors in the design of the study. In this way, research is continually changing and investigations are constantly being reformulated.

## BOX 1.2: ALZHEIMER'S DISEASE AND SMELL: IS THERE A LINK?

Is Alzheimer's disease able to be detected early in its development through a simple smell test? A question like this was posed by a variety of media outlets in 2004. Even Dr. Oz claimed that scientists could predict if a person will get Alzheimer's disease based on a smell test. Below is an excerpt of a report in *Senior Journal* describing the smell test for Alzheimer's disease. Use your critical thinking skills to judge the claims made, based on the accepted methods for scientific research described in this chapter.

December 13, 2004 – The inability to identify the smell of lemons, lilac, leather, and seven other odors predicts which patients with minimal to mild cognitive impairment (MMCI) will develop Alzheimer's disease, according to a study presented today at the American College of Neuropsychopharmacology (ACNP) annual meeting. For patients with MMCI, the odor identification test was found to be a strong predictor of Alzheimer's disease during follow-up, and compared favorably with reduction in brain volumes on MRI scan and memory test performance as potential predictors.

"Early diagnosis of Alzheimer's disease is critical for patients and their families to receive the most beneficial treatment and medications," says lead researcher D.P. Devanand, MD, Professor of Clinical Psychiatry and Neurology at Columbia University and Co-Director of the Memory Disorders Center at the New York State Psychiatric Institute. "While currently there is no cure for the disease, early diagnosis and treatment can help patients and their families to better plan their lives."

Smell identification test results from Alzheimer's disease patients, MMCI patients and healthy elderly subjects were analyzed to select an optimal subset of fragrances that distinguished Alzheimer's and MMCI patients who developed the disease from healthy subjects and MMCI patients who did not develop Alzheimer's. Results of the 10-smell test, which can be administered in five to eight minutes, were analyzed in Dr. Devanand's study, which evaluated 150 patients with MMCI every six months and 63 healthy elderly subjects annually, with average follow-up duration of five years. Inability to identify 10 specific odors (derived from the broader study) proved to be the best predictors for Alzheimer's disease: strawberry, smoke, soap, menthol, clove, pineapple, natural gas, lilac, lemon, and leather. from http://seniorjournal.com/NEWS/Alzheimers/4-12-13LemonSmell.htm

### Reflection Questions

1) What was the control in the experiment? What was the dependent variable? What was the independent variable?
2) Are there other factors that may contribute to a person's inability to smell these 10 scents?
3) For question #2, how does senescence affect olfactory (smell) abilities? What other age-related factors may lead to a decreased sense of smell?
4) Does the media report give statistical evidence for the ability of smell tests to predict Alzheimer's disease? Is it strong or weak? Why?
5) What recommendations for future studies do you recommend to validate or debunk the study shown in Box 1.2?

# Summary

The chapter began with dementia problems faced by an aging gentleman, Uncle Hans, who faces the challenge of his life due to the natural processes of aging. Through understanding life's characteristics and organization, we can better understand Hans' plight. While evolution and natural selection have made living systems better adapted, diseases such as Alzheimer's persist. New discoveries of ways to combat illnesses, made by using the techniques of scientific inquiry, hold promise for the future. Development of greater scientific literacy in our populace should help people to better understand the challenges we all face. Our biophilic relationship with other living things should help us to appreciate our link to them. Do dogs as well as humans get Alzheimer's disease as they age? Did Alzheimer's exist before modern times, when people did not live long enough to develop age-related problems? These kinds of questions can be answered through studying biology.

## CHECK OUT

**Summary: Key Points**

- Biology affects our lives in many ways, from diseases such as Alzheimer's to cures and solutions.
- All living systems have shared characteristics that set them apart from nonliving systems.
- Life's organization increases in complexity and connects living systems with each other and with their nonliving environment.
- Taxonomic systems classify organisms, many of which have not yet been discovered.
- Many scientists contributed to the development of evolutionary thinking.
- The process of natural selection and evolution has led to today's biodiversity and continues to change species.
- Research findings need to be based on properly controlled experiments and mathematical analysis for their results to be validated.

## KEY TERMS

Adaptation
Alzheimer's disease
Animals
ANOVA, Analysis of Variance
Atoms
Archae
Bacteria
Biodiversity

Biological literacy
Biology
Binomial nomenclature
Biophilia
Biosphere
Catastrophism
Cell
Characteristics of life

Class
Community
Control group
Correlation
Critical thinking
Data analysis, Qualitative and
Quantitative
Dependent variable
Diversity
Domain
Ecosystem
Eukarya
Eukaryote
Evolution
Experiment
Experimental group
Family
Food chain
Food web
Fossil record
Fungi
Genus
Heterotroph
Homeostasis
Hypothesis
Independent variable
Interdisciplinary
Inquiry
Kingdom
Macrobiology
Macromolecules

Membrane
Metabolism
Molecules
Natural selection
Nucleus
Null hypothesis
Order
Organelles
Organism
Organs
Organ System
Phylum
Plants
Population
Producer
Prokaryote
Protista
Reproduction
Response to stimuli
Science literacy
Scientific method
Significance level
Species
Statistics
Species
Taxonomy
Tissues
Type I error
Type II error
Variation

# Multiple Choice Questions

1. Many people feel a deep bond with their beloved dog. Which term best describes this relationship?
   a. biology
   b. biological literacy
   c. biophilia
   d. biodiversity

2. The maintaining of a steady carbon dioxide level in the blood is accomplished by:
   a. adaptation
   b. diversity
   c. homeostasis
   d. complement

3. Which is NOT a characteristic of living systems?
   a. adaptation
   b. order
   c. reproduction
   d. size

4. Humans and chimpanzees are _____% related genetically.
   a. 1
   b. 10
   c. 50
   d. 99

5. The statement, "The basic unit of life is the cell" is most likely a statement from this scientist.
   a. Lamarck
   b. Darwin
   c. Hutton
   d. Hooke

6. Which term includes all of the others?
   a. Class
   b. Order
   c. Hutton
   d. Hooke

7. The Galapagos tortoise, *Geochelone elephantopus*, is classified in the:
   a. class *Geochelone*
   b. phylum *Geochelone*
   c. genus *elephantopus*
   d. species *elephantopus*

8. A scientist discovers a multicellular, eukaryotic creature in the arctic, which is able to break down mosses in its root system but is not able to make food from sunlight. In which kingdom should the scientist place this organism?
   a. Animalia
   b. Fungi
   c. Plant
   d. Archaea

9. Which organism is able to obtain energy directly from dead moss in its surroundings?
   a. other moss
   b. fungi
   c. plants
   d. all of the above

10. An experiment studies human evolution, showing that sunlight leads to changes in skin color as a result of different levels of sunlight. Which is the dependent variable in the experiment?
   a. human evolution
   b. sunlight
   c. skin color
   d. both a and b are dependent variables

## Short Answer

1. Describe how Alzheimer's disease creates an imbalance in society and in the bodies of those afflicted.

2. List three ways a person can best become biologically literate.

3. A rock is not considered life, based upon the cell theory. Choose one postulate of the cell theory to defend why a rock is not life.

4. Name two kingdoms described in the six-kingdom system of taxonomy. Describe two differences between the two kingdoms. How are the two kingdoms the same?

5. Explain the drawbacks of a long, rectangular cell within living systems, in terms of the forces placed upon these systems.

6. Describe the development of evolutionary thinking through recent history. Be sure to include contributors: de Buffon, Adam Smith, Darwin, Lamarck, and William Smith.

7. Devise an experiment that tests the effects of changing salt concentrations on regeneration of brain cells in Alzheimer's patients. Define the independent variable, dependent variable, and controls you put in place for your investigation.

8. Which smells predict whether a person will develop Alzheimer's disease? Name four smells and explain why correlations such as between smell and Alzheimer's disease are "weak research results."

9. If a medical test's results claim that there is a significance level of .15, explain how this may affect a patient's interpretation of those results.

10. Compare and contrast the goals of the **results** and **discussion** sections of a scientific investigation. Be sure to include one way the sections have goals in common and one way the sections are different.

## Biology and Society Corner: Discussion Questions

1. Alzheimer's disease will affect a growing aging population in the United States over the next 25 years. Predict how this will impact healthcare, the economy, and family life patterns.

2. How does better scientific literacy improve a society's overall health and wellness?

3. In a democratically elected government, people vote for policy changes through electing their officials. If some people are not scientifically literate, should they still be allowed to vote? Why or why not? What measures would you consider ethical, if any, to ensure that the voting public is educated?

4. The rate of species loss is occurring at the greatest pace in human history. What are the dangers to increasing species loss? For human society? For natural ecosystems?

5. Write a plan to help those families afflicted with Alzheimer's disease. What are two ways the government can improve the quality of their lives? Name two ways families can best cope with this illness.

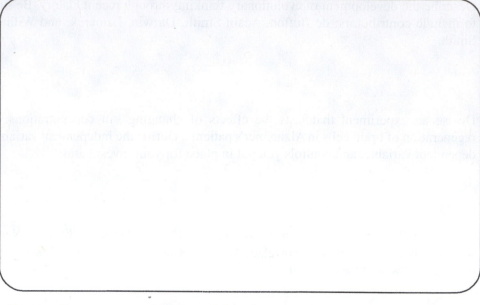

Figure – Concept map of Chapter 1 big ideas

# UNIT 1
## That's Life

# Chemistry Comes Alive

## ESSENTIALS

Village in China

There are many shapes of chemicals

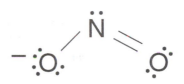

$NO_2$ ions travel throughout organisms

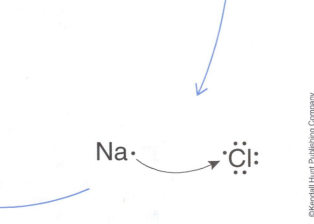

Chemicals react as they travel through living organisms

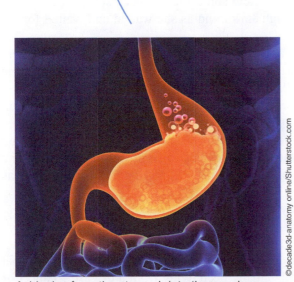

Acids rise from the stomach into the esophagus

# The Case of the Mysterious Killer:
## A 用硝酸处理; 硝化 Nightmare

"It is frightening – a force killing us off almost every day and we can't stop it," thought Jin, a villager in Lin Xian, China. People of the village lived in fear for their lives in this sleepy town, about 250 miles south of Beijing, China's capital city. Jin feared that the deadly force would strike down her little boy, only 8 years old. Everyone awaited the report from the detectives, who came from Beijing to help, and rumors were spreading that the killer was found!

"Why had Lin Xian been targeted? Why was it happening to us?" thought Jin. She remembers her brother getting killed by the force. First, it attacked his throat, and he could not eat. Soon he wasted away, unable to move and in pain every moment. Jin remembered how she hated the force and could not bear to see her brother hurt.

The force was something villagers could not see, but it could sneak up on them at any time. Jin had seen so many succumb to this killer and she knew how it started. When it hurt people, she watched, but she never spoke of it. Maybe by talking about it, she thought, it would find her. "Why was she able to avoid it?""Did she never meet it in the forest?" "Was she just lucky, or was there some she had been spared, but not her brother?" So many thoughts raced through Jin's mind as she waited and waited for the detectives to come to town.

Legend had it that over 2000 years ago, a curse had been placed on Lin Xian. There were many explanations as to why townspeople were attacked by the force, but no one really knew. Almost a quarter of all villagers died of the force eventually, and everyone blamed the cursed past of their ancestors. When the detectives finally reached town, Jin watched everything they were doing.

Detectives looked through the fields and forests, asked villagers about their food and how they lived. They had instruments and devices to fight the force, but no one really knew what was going to happen. "Perhaps it was too late and the force was growing too strong," thought Jin.

It took a long time; however, one day, Jin heard from one of the gossips that the detectives had found something. Some of the detectives came into her hut with a verdict. They looked very serious – the look on their faces meant they had information: They told her about the force – "It was 用硝酸处理; 硝化 (in English: nitrates)." They explained that the villagers, including her brother, had actually been dying from a disease – cancer of the esophagus (the muscular tube moving food from the throat to the stomach), probably caused by a chemical called nitrates, found in the food.

Detectives explained to Jin that food being grown by villagers was low in a substance called molybdenum, a soil nutrient for plants needed in only small amounts. Crops in the field pulled up more nitrates from soil to make up for low molybdenum levels. Nitrates in plants were being converted into nitrites and then into nitrosamines in the stomachs of residents; and nitrosamines are linked to various cancers, including esophageal and stomach.

They explained to Jin that low molybdenum levels also reduced vitamin C produced by plants. Low levels of vitamin C in the villagers' diets encouraged the conversion of nitrates into nitrites in our bodies, further increasing the risk of cancer. The detectives had a plan: (1) Villagers were going to be given vitamin C tablets to decrease their production of nitrites and (2) Villagers would coat their corn and wheat seeds with molybdenum to drive down plant nitrate levels. Jin, in an emotional reaction, said a short prayer for the victims of the deadly force…

---

### CHECK UP SECTION

Chinese detectives (scientists) in the story above studied the link between cancer and the high levels of nitrates in the food supply of Lin Xian. As a result of the recommendations by scientists, nitrite levels in vegetables have dropped 40% and vitamin C levels have risen 25% over the past two decades. Long-term results on esophageal cancer rates remain to be seen.

Choose a particular chemical that you find interesting that is found in *our* food supply. Research and explain how that chemical acts to cause benefits and/or harm to our environment and to the organisms living in our environment.

---

## Atoms and Elements that Make Up Life

The Chinese scientists in the story found that nitrates in the crops of local food growers caused the high rate of esophageal cancer in Lin Xian. What are nitrates? How do they form? How do their nitrosamine products cause diseases? We can answer all these questions and more by studying some key principles of chemistry, which is the study of matter. The nature of matter, which is defined as anything that has mass and occupies space, was studied throughout human history. Chemistry studies the composition and properties of matter, and the reactions by which matter is changed from one form to another. In order to understand the composition of both living and nonliving things, we need to begin with the smallest components, and then build a hierarchy. Chemicals drive the life functions described in the previous chapter. They form relationships with each other, build substances within living systems, and guide and direct all of an organism's activities. We begin with the simplest substance – an element – and its most basic unit – an atom.

This chapter moves from the composition of atoms to the ways in which atoms combine to form molecules. It looks at different types of chemical reactions, the bonds they form, and particularly how some chemicals change the environment of living systems as they donate hydrogen atoms. The larger chemicals of life will then be studied in the section called organic chemistry. A look to the foods we eat and their chemical make-up reveals that proteins, sugars, and fats play key roles in our health, alongside other chemicals such as nitrates depicted in our story.

**Chemistry**

Study of matter.

**Matter**

Anything that has mass and occupies space.

## Elements

Matter is composed of pure substances: nitrogen and oxygen, components of nitrates, are examples of pure substances, as are silver, lead, and iron (see Figure 2.1). Pure substances, known as elements, are those that cannot be broken down by ordinary chemical means. There are 92 naturally occurring elements. The **Periodic Table of Elements**, shown in Figure 2.2, displays all the elements, both those that are naturally occurring and those artificially made in laboratories. Elements are ordered on the table by increasing weight, shown with special abbreviations for each element (e.g. O = oxygen; N = nitrogen; Au = Gold; Fe = Iron; and Pb = lead). Note that some chemical symbols derive from Latin, with Au emanating from the Latin word *aurum* for the precious metal, gold; and Fe arising from the Latin word *Ferrum*, meaning iron.

## Atoms and Subatomic Particles

The smallest unit of any element that retains the unique properties of that element is the atom. The term *atom* comes from a word in Greek that means "indivisible." The characteristics of an element include: (1) how it acts with other elements and (2) how it appears at certain temperatures. These chemical and physical properties make each atom and element unique.

(a)    (b)

(c)    (d)

**Figure 2.1**  Examples of Elements. Elements are found in all matter, ranging from silver tea sets to iron beams, lead air gun pellets, and ammonia.

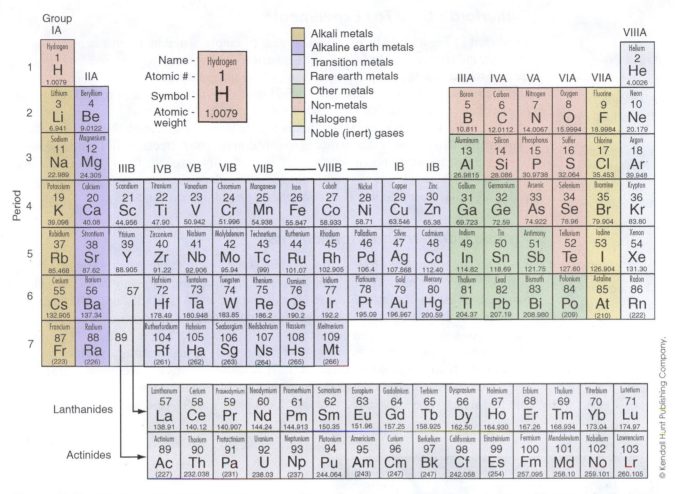

**Figure 2.2** The Periodic Table of Elements. The table shows the atomic mass and number of all of the elements known to humans.

Atoms are made of three subatomic particles: protons, which are positively (+) charged units (*charge* refers to the amount of electricity a chemical possesses), found in the central region of an atom known as the nucleus; neutrons, or neutral particles (0) with a zero charge that are also found in the nucleus; and negatively (−) charged particles called electrons, which move in orbits around the nucleus. An atom is considered negatively charged when it possesses a greater number of electrons than protons; and positively charged when it possesses a greater number of protons than electrons. The chart in Figure 2.3 shows the make-up of a variety of atoms: protons, electron, and neutrons (use the acronym PEN to help you remember the parts of an atom).

Electrons move around a nucleus in energy shells, which are layers of electron orbits circling the nucleus of an atom. The shell nearest the nucleus contains up to two electrons, and any additional shells contain a maximum of eight electrons. Electrons in the outermost shell form bonds (or chemical relationships) with other atoms.

The more energy an electron has, the farther from the nucleus its orbital shell. For example, an electron in shell #4 has more energy than electrons in shell #1. It takes energy to move an electron to higher shells because there is a force of attraction between an electron and its nucleus. Negatively charged electrons are attracted to the positively charged nucleus, and this attraction, in part, keeps subatomic particles together. While atoms are indivisible, their electron components can exchange with electrons of another atom, allowing atoms to react with each other. Electrons hold energy, which is exchanged during chemical reactions.

### Proton

A subatomic particle found in the nucleus, which is positively charged.

### Neutron

Particles with zero charge found in the nucleus.

### Electron

A negatively charged subatomic particle found in the orbit.

### Rutherford's Gold Foil Experiment

**Gold Foil experiment**

Also called Rutherford's gold foil experiment, is a series of experiments that showed an atom's structure.

As shown in Figure 2.3, there is a great deal of empty space between the electrons in an orbit and its nucleus. In fact, Ernst Rutherford, in his famous Gold Foil experiment, demonstrated that the atom is more than 90% empty space between orbiting electrons and the nucleus. In 1911, Rutherford published the results of his experiment, in which he shot helium atom particles through a solid but very thin sheet of gold foil (see Figure 2.3). He found that over 90% of helium particles passed through the gold foil. This experiment demonstrated that matter is mostly empty space. Because all living and non-living things are made up of atoms, it is theoretically possible for a person to walk through a door, with his or her atoms aligned with a door's empty space. Will you try to walk through a closed door and take the chance of passing through?

### Atomic Number and Atomic Mass

**Atomic number**

The number of protons in the nucleus of an atom.

Elements are numbered on the periodic table according to their atomic number, which corresponds to the number of protons in an atom. For example, nitrogen (N), which is a main component of nitrates, has an atomic number equal to 7, meaning that there are 7 protons in nitrogen. Another example is found in table salt, which contains the element Na or sodium. Na has an atomic number 11, indicating that it contains 11 protons. The atomic number also gives the number of electrons (when protons and electrons are equal, there is no charge on the atom, overall), because protons and electrons balance out to give an overall neutral charge to an atom. Thus, nitrogen's atomic number of 7 tells you that it has 7 electrons and 7 protons. Sodium thus has 11 electrons and 11 protons.

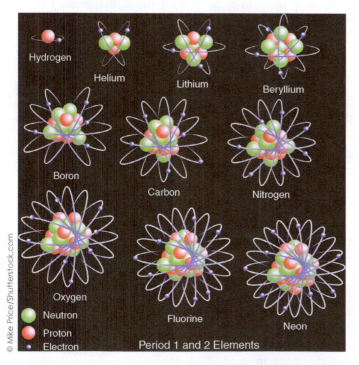

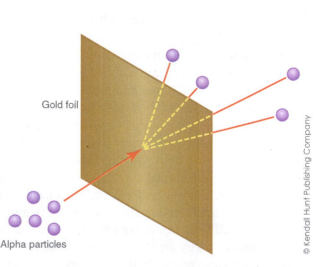

**Figure 2.3**    a. Protons, Electrons, and Neutrons in a variety of atoms. B. Rutherford's Gold Foil Experiment: Rutherford shot alpha particles through a very thin sheet of gold foil. He measured the number of particles that made it through the gold foil, finding that roughly 90% passed through the foil. This indicated that the atom is mostly empty space. In our story, nitrogen and oxygen forming nitrates (our killer) are mostly empty space.

Elements are identified by their atomic number, and no two different elements have the same atomic number. Chemists have also defined the atomic mass of an atom. Mass is a physics term that indicates the amount of matter in a substance. Atomic mass is the total matter within an atom; in other words, the mass of an atom is the combined weights of the subatomic parts that have weight. Protons and neutrons each have an atomic mass of 1 amu (atomic mass unit), but electrons have a negligible mass (1/1836 of an amu) and are not considered when calculating an atomic mass. For example, nitrogen has an atomic mass of 14, meaning that together, protons and neutrons in a nitrogen atom make up 14 amu units. Because nitrogen has 7 protons, based on its atomic number, it must have 7 neutrons to make up the total of 14 units. A rule of thumb is that atomic mass minus atomic number equals the number of neutrons in an atom. Nitrogen has a mass of 14 and a number of 7 ($14 - 7 = 7$ neutrons). What is the number of protons, electrons, and neutrons in a sodium atom? Use the Periodic Table to figure out its subatomic particle composition. Sodium contains 23 protons and neutrons together comprising its atomic mass, but only 11 protons, from its atomic number. Thus, it contains 12 neutrons ($23 - 11 = 12$ neutrons). Note that the number of protons and neutrons in an atom are often not the same.

**Atomic mass**

The mass of an atom is the combined weights of the subatomic parts that have weight.

## Ions

Frequently, atoms in living systems occur in the form of **ions**, which are particles with a charge. A charge, either positive or negative, occurs because of the addition or subtraction of electrons from an atom. If, for example, sodium loses one electron, as occurs when it is immersed in water in an organism's cells, it loses a negative charge because an electron is negative. Sodium therefore becomes slightly more positive by one unit and is called the $Na^{+1}$ ion. The "+1" is written as $Na^1$, for short. What happens when an atom gains an electron? It adds negative charges, one for each electron added. When chloride is immersed in water, it gains one electron and becomes the $Cl^{-1}$ ion. These ions are called "charged" because they have a number associated with their atom. The number equals the amount of charge an ion has. For example, $Na^{+1}$ has a charge of positive one.

You can compare the formation of ions to a party scene. In a party, when a negative person enters the room, the party feels a little more negative. When the negative person leaves the party, the party becomes a little happier – a little more positive. While ions are not humans, this is the same principle behind how ions form. Ions are very important in life processes because most substances in living systems are immersed in water and form ions. In the form of ions, atoms interact with one another. The sodium and chlorine ions described in the examples are important for proper nerve and muscle functions in humans. Nitrates, which caused so many problems for Jin in our story, are ionic substances. Their extra electrons make them reactive with substances within the body, often causing harm.

## Isotopes

Sometimes, atoms of the same element (those containing the same atomic number) have different atomic masses. These atoms are known as isotopes, which have the same number of protons but differing numbers of neutrons. For example, atoms of oxygen all contain sets of protons, as shown in Figure 2.4, but they can contain different numbers of neutrons. Isotopes, due to increased numbers of neutrons, are often less stable than their original atoms. Often, isotopes break down spontaneously, giving off energy known as radiation. Radioactive isotopes decompose spontaneously, losing particles and energy in the process. They can also cause cancer because of the destructive effects of radiation on cells. Isotopes of each other have the same physical characteristics as normal atoms and therefore act the same as one another in living systems.

**Isotope**

Are atoms of the same element having different atomic masses.

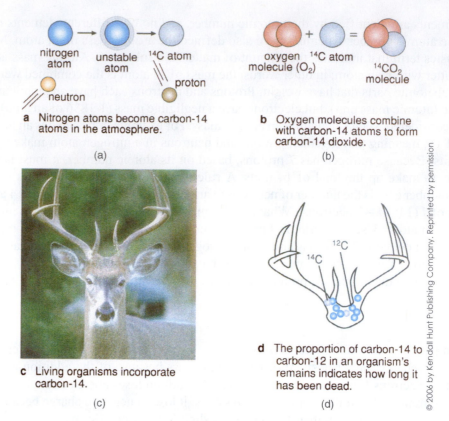

a  Nitrogen atoms become carbon-14 atoms in the atmosphere.

(a)

b  Oxygen molecules combine with carbon-14 atoms to form carbon-14 dioxide.

(b)

c  Living organisms incorporate carbon-14.

(c)

d  The proportion of carbon-14 to carbon-12 in an organism's remains indicates how long it has been dead.

(d)

**Figure 2.4**    Isotopes of Carbon travel through living systems. Atoms with the same atomic number (the number of protons in the nucleus) but different numbers of neutrons are isotopes. From *Biological Perspectives*, 3rd ed by BSCS.

While some isotopes can cause disease, others are used in the study of natural phenomena. For example, isotopes of oxygen are analyzed in ice cores to obtain historical climate data, giving an idea about the temperature of the Earth long ago. There exist three stable, naturally occurring isotopes of oxygen: oxygen-16, oxygen-17, and oxygen-18. The atomic number for oxygen is 8, so there are 8 protons and 8 electrons in oxygen-16. Using our formula for determining the number of neutrons, how many neutrons does oxygen-16 contain? Yes, 8 neutrons are found in oxygen-16. Oxygen-17 contains one extra neutron, bringing its total to 9 neutrons and raising the atomic mass to 17. Oxygen-18 has two extra neutrons, bringing its total to 10 neutrons and its atomic mass to 18. These three isotopes of oxygen occur in the Earth's atmosphere in the following proportions: Oxygen-16, 99.759%; Oxygen-17, 0.037%; and Oxygen-18, 0.204%. When combined with hydrogen, they form water ($H_2O$). Water containing the lighter isotope evaporates more readily than those containing the heavier isotopes; and heavier isotopes condense more readily as rainfall. Heavier isotopes of oxygen in water fall more easily as rainfall. Isotope proportions are measured in ice cores to determine the age of ice layers and the historic climate conditions of the Earth.

Other isotope examples include deuterium and tritium, which are less stable forms of the hydrogen atom. They are used in research to trace substances as they move in living systems. Radioactive isotopes are also needed for medical research, with Iodine-131 and Radium-226 used for cancer treatments. Isotopes of carbon are also studied to obtain the ages of once living material such as cloth and paper. Radioactive isotopes deteriorate at a certain rate, called a *half-life*, with half of the material changed into another substance in that period. For example, radioactive Carbon-14 has a half-life of roughly 5730 years. The proportion of C-14 left in a substance indicates the age of the material.

---

**BIOETHICS BOX 2.1**

The study of radiation and radioactive isotopes began by Marie Curie (1867–1934), who discovered its existence while studying geology. She died from the radiation found in the substances with which she worked. Several scientists continued Curie's work. Albert Einstein (1874–1955), a theoretical physicist, while never conducting experiments to split the atom, expressed mathematically that such a process would produce large amounts of energy. In his famous equation, $E = mc^2$, in which $E$ stands for energy released, $m$ for mass of a substances, and $c$ the speed of light (which is large: $3.0 \times 10^8$ meters per second), the relationship between matter and energy is shown. When applied by scientists, the discovery of nuclear fission and the atomic bomb was possible. Huge amounts of energy are released using very small masses of nuclear material. Is the use of nuclear weapons ever justified? Was it right to use the atomic bomb on Japan? Is Curie's discovery of radiation and Einstein's work leading to nuclear weaponry good or bad for society? Name an example of a benefit from their research.

Einstein was disturbed by the use of force to control people. He wrote a famous poem describing the horrors of groupthink and mind control by authorities to express his angst:

> By sweat and toil unparalleled
> At last a grain of the truth to see?
> Oh fool! To work yourself to death.
> Our party make truth by decree.
> Does some brave spirit dare to doubt?
> A bashed-in skull's his quick reward.
> Thus teach we him, as ne'er before,
> To live with us in sweet accord.

## Exposure to Radiation

Many substances that are radioactive occur naturally. To illustrate, radioactive carbon is found in trace amount throughout the Earth's atmosphere, and radioactive potassium is found in human bones. These isotopes give off small, harmless amounts of energy. However, some isotopes, such as radium-226 and uranium-238, decay and give off large amounts of dangerous energy. In nature, they occur in such small quantities that their effects are limited; however, in nuclear weapons, they are very destructive.

About 80% of the radiation to which living systems are exposed comes from the environment, primarily emanating from outer space cosmic rays, and is natural and harmless. Sources for the other 20% are cell phones, TVs, old fallout from past nuclear testing still in the soil, nuclear power plant leaks, and medical testing such as X-rays and CT scans. In fact, the radiation in one chest X-ray equals the natural exposure an airplane passenger receives on one round trip flight from New York to California. In the air, your body is exposed to more galactic cosmic rays at higher altitudes than on the Earth's surface. Airline workers have higher rates of cancer than the general public, indicating the negative effects of such exposure. While medical technicians are protected by law from excess exposure in their jobs, no such legal protection yet exists for airline workers.

### The Elements of Living Systems

Although living systems are complex, 96% of all living matter is made up of combinations of carbon, oxygen, hydrogen, and nitrogen (use the acronym COHN to help you remember this combination). Almost 99.9% of all living things are composed of only 10 different atoms! The chart in Figure 2.5 gives the relative proportions of all the elements found in the human body. Many of the elements found in lesser proportions in living systems, however play vital roles. As described earlier, sodium and chlorine have specific functions. Calcium is also an element found in small quantities that regulates enzyme activity, maintains our body temperature, and coordinates movement. Small disruptions in these minor elements may initiate serious imbalances. Heart failure may result from small changes in sodium, potassium, or calcium levels (see Figure 2.6).

## Substances Combine to Form Complex Systems

### From Atoms to Molecules

In the last section, we looked at the atom, the smallest discrete unit of a pure substance. However, both nonliving and living systems are combinations of larger chemicals, so we will look now at how atoms from pure substances combine with those from other pure substances to form new materials. Substances combine with one another through **chemical reactions**, and when atoms combine, a molecule is formed. Molecules may be the combinations of the same atom or different atoms. Molecules made up of different atoms are known as **compounds**. We can describe the formation of a molecule in a chemical equation. For example, the chemical equation for forming nitrates in our story is:

**Molecules**

Atoms bonded together.

$$N + 3O \rightarrow NO_3^-$$

On the left-hand side of a chemical reaction, the substances are termed **reactants** because they react with one another. N and O are reactants that form the nitrates in the soil in

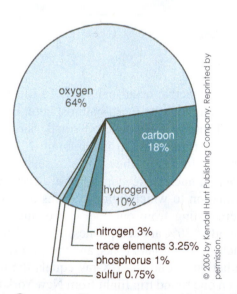

oxygen 64%

carbon 18%

hydrogen 10%

nitrogen 3%
trace elements 3.25%
phosphorus 1%
sulfur 0.75%

© 2006 by Kendall Hunt Publishing Company. Reprinted by permission.

**Figure 2.5**   The Most Common Elements in the Human Body: Carbon, Oxygen, Hydrogen, and Nitrogen comprise about 95–96% of all living organisms. From *Biological Perspectives*, 3rd ed by BSCS.

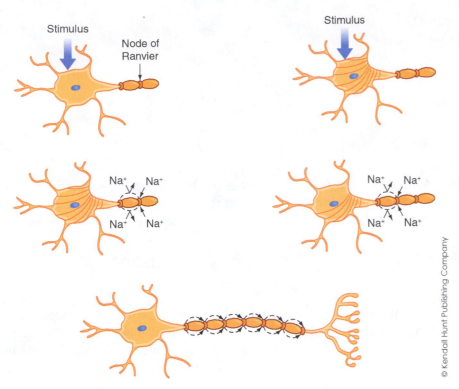

© Kendall Hunt Publishing Company

**Figure 2.6** Ions Flowing along a Nerve. Ions conduct nerve transmissions that sustain biological processes ranging from the beating of one's heart to thinking and breathing. The arrows show the direction in which chemicals travel to produce a nerve transmission. Disruptions in ion flow cause many problems, including heart failure.

Lin Xian. The substance or substances that appear on the right-hand side of a chemical equation are termed **products** because they are produced or formed from the reactants. $NO_3^-$ is the chemical formula for nitrate, and it may have led to disease in our story.

Nitrates ($NO_3$) are also found naturally occurring in the Earth's crust. They are needed for plant growth and development and are found in fertilizers, foods, and even explosives. Nitrogen and oxygen alone are harmless and in fact comprise over 90% of the atmosphere we breathe. However, when they react with each other to form nitrates, they may be toxic molecules linked to cancer in high doses in our foods. How do atoms that make up a molecule combine to form new substances with unique properties?

## Valence Electrons: How Matter Is Combined?

Chemicals combine using their outermost electrons. The chemical behavior of any atom is determined by distribution of electrons around it. Electrons in the outermost shell of an atom are called valence electrons. Valence electrons dictate the chemical activity of an atom. Chemical reactions occur when atoms share or exchange their valence electrons, forming bonds. Figure 2.7 gives an example of the exchange of valence electrons between three atoms of oxygen and one atom of nitrogen. The nitrogen atom obtains three electrons by forming bonds with three separate oxygen atoms, forming a molecule of nitrate. The molecule of nitrate in Figure 2.7 shows that nitrogen shares one electron with each of the oxygen atoms to complete their valence shells. Why does this process occur in such a regular and predictable manner?

**Valence electrons**

Electrons present in the outermost shell of an atom.

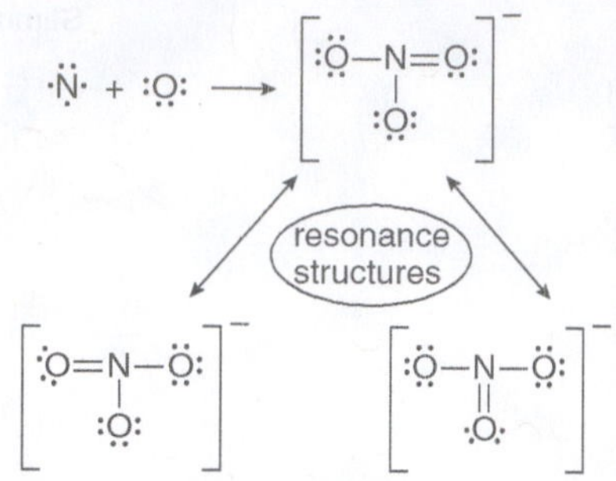

**Figure 2.7**  Bonds Form to Make Nitrates ($NO_3$). Nitrogen and oxygen satisfy the octet rule when they combine; forming bonds that shift in a variety of structures. All of these structures have the same chemical formula but feature shifting electrons.

**Octet rule**

A chemical rule that reflects how atoms react to attain eight electrons in their valence shell.

**Electronegativity**

The ability of an atom to attract electrons to itself.

**Polyatomic ion**

A special kind of ion, composed of more than one atom, forming a charge.

## Factors Influencing Chemical Reactions

Three factors govern chemical reactions:

1) The octet rule: atoms react to obtain eight electrons in their valence shell;
2) Electronegativity: the ability of an atom to attract electrons to itself varies; and
3) Electrons occur in **pairs**, which are represented as lines or bonds when molecules are drawn.

Let's apply these three rules of chemical reactions. As shown in Figure 2.7, a nitrogen atom has five electrons in its valence shell. In order to satisfy the octet rule, the atom needs three more electrons to complete its outer shell. Oxygen atoms have six electrons in their valence shells. Each oxygen atom requires two electrons to complete its valence. Thus, nitrogen shares a pair of electrons with three separate oxygen atoms. A free pair of electrons also rotates around the nitrogen atom, thus completing a full set of eight electrons around each of the atoms in the molecule. Nitrogen has a greater electronegativity than oxygen atoms, which pulls electrons into nitrogen's orbit more readily. In fact, nitrates form a special kind of molecule known as a polyatomic ion. Polyatomic ions are molecules that contain a number of atoms that together form a charge, with positive or negative, on their overall structure. Some common polyatomic ions are listed in Figure 2.8.

Another simpler chemical example, HF, hydrogen fluoride, is found in toothpaste. This molecule arrangement enables hydrogen to share its electrons with fluoride to allow

| Common polyatomic ions | | | |
|---|---|---|---|
| Ion | Name | Ion | Name |
| $NH_4^+$ | ammonium | $CO_3^{2-}$ | carbonate |
| $NO_2^-$ | nitrite | $HCO_3^-$ | hydrogen carbonate (biscarbonate is a widely used common name) |
| $NO_3^-$ | nitrate | | |
| $SO_3^{2-}$ | sulfite | | |
| $SO_4^{2-}$ | sulfate | $ClO^-$ | hypochlorite |
| $HSO_4^-$ | hydrogen sulfate (bisulfate is a widely used common name) | $ClO_2^-$ | chlorite |
| | | $ClO_3^-$ | chlorate |
| | | $ClO_4^-$ | perchlorate |
| $OH^-$ | hydroxide | $C_2H_3O_2^-$ | acetate |
| $CN^-$ | cyanide | $MnO_4^-$ | permanganate |
| $PO_4^{3-}$ | phosphate | $Cr_2O_7^{2-}$ | dichromate |
| $HPO_4^{2-}$ | hydrogen phosphate | $CrO_4^{2-}$ | chromate |
| $H_2PO_4^-$ | dihydrogen phosphate | $O_2^{2-}$ | peroxide |

**Figure 2.8**  The Most Common Polyatomic Ions.

© Kendall Hunt Publishing Company

both atoms to satisfy the octet rule. For example, fluoride, the most electronegative atom on the Periodic Table, pulls electrons so strongly that it unevenly shares an electron with hydrogen. Because fluoride has seven electrons in its valence, the extra electron from hydrogen completes its outer shell. Hydrogen is satisfied with the sharing of one electron because it takes only one pair of electrons to complete its shell. Hydrogen and helium are exceptions to the octet rule, because a single pair of electrons in their outer shell completes their valence. The first shell of any atom holds two electrons while all higher shells hold eight. Fluoride's strong electronegativity is what makes toothpaste able to kill oral bacteria, which cause dental caries (cavities). Fluoride literally pulls off electrons from key chemical reactions occurring within bacteria, preventing them from multiplying and causing damage to enamel. There is also some evidence that fluoride remineralizes the enamel on teeth, strengthening it and preventing cavities.

On the other side, some atoms are nonreactive, meaning that they have little ability to exchange electrons. These are known as noble gases – helium, neon, argon, krypton, xenon, and radon; they have full valence shells and little importance in living organisms. Radon, the last in our list, has importance as a chemical causing human health hazards. Radon is found in rock and soil particles from radioactive decay of the element, radium. After prolonged exposure to radon, which often seeps into basements, lung cancer may develop.

## Type of Chemical Bonds

Atoms gain, lose, or share electrons with one another to form chemical bonds, which are defined as electron relationships between atoms. Each bond contains energy in its arrangement of electrons between two atoms. Relationships between atoms are based on how electrons are exchanged. The more "pull," or **electronegativity**, an atom has for electrons, the more time electrons will spend around that particular atom. Thus, in the example of HF described earlier, which atom should hold the greatest time with the exchanged electrons? Yes, Fluoride, because it is most electronegative.

Let's explore the four major forms of bonding: **covalent**, **polar covalent**, **ionic**, and **hydrogen bonds**.

**Chemical bond**

Relationship between atoms.

### Covalent Bonds

Covalent bonds result from the equal sharing of electrons between atoms. We say that bonds are covalent when the bonding atoms have the same electronegativity, or the same pull, on the shared electrons. In cases such as carbon dioxide, in which there is an even pull on the electrons due to shape, there is also an equal sharing of electrons around the atoms of the molecule. The relationship could be compared to one in which both partners share all of the expenses and there is an even give-and-take between the two. Covalent bonding shares electrons completely evenly around the nuclei of the atoms comprising the molecule.

**Covalent bond**

Bonds that result from the equal sharing of electrons between atoms.

### Polar Covalent Bonds

Unequal sharing of electrons between atoms is known as polar covalent bonding. While electrons move around both atoms, they are not shared equally between atoms in a polar covalent bond. One atom has a greater attraction for electrons, or greater electronegativity, than another. This results in a slight positive charge on the atom that has less time spent with its shared electrons and a slight negative charge on the atom that has more time spent with its shared electrons. $NO_3^-$, or the nitrate ion, is an example of polar covalent bonding. Nitrogen has less of a connection with its electrons and has a relatively positive charge, while oxygen has more of a connection with electrons in its outer shell and therefore has a relatively negative charge.

**Polar covalent bond**

The unequal sharing of electrons between atoms.

## Ionic Bonds

**Ionic bond**

Bonds that result from complete transfer of electrons from one atom to another.

**Anion**

Negatively charged ion.

**Cation**

Positively charged ion.

Ionic bonds result from the complete transfer of electrons from one atom to another. A sharing of electrons does not occur in ionic bonding. An example is table salt (NaCl). Sodium gives one electron to chlorine and a bond forms to satisfy the octet rule for both. Atoms that compose a molecule in ionic bonding can have very different properties than the molecules they form. Sodium is a pliable metal and chlorine is a poisonous gas, but together they form table salt.

In an ionic relationship, one atom is a "taker" and one atom is a "giver." The taker is known as an anion, and the giver is known as a cation. In the case of table salt, sodium is a cation because it gives away one electron, and chlorine is an anion because it receives the electron. Anions become relatively negative in their charge because they have additional electrons, and cations become positive because they lose electrons. (You may recall the difference using the idea that cats (cations) are very positive to have as house pets).

When a molecule loses electrons, we say that the substance is **oxidized**. In the example of NaCl, Na loses an electron and is therefore oxidized in the ionic formation of a bond. **Reduction** is the opposite of oxidation and is defined as any reaction that causes the gaining of electrons. When a molecule gains electrons, we say that it is **reduced**. In the forming of NaCl, Cl is reduced because it receives an electron from Na. In our story, the dreadful disease was linked to the oxidation of villagers' cells by nitrosamine free radicals, the most likely force causing cancer in Lin Xian.

Ionic relationships are unstable because one atom is gaining and the other is losing electrons. In relationships that you know about: do you know a cation-type person? Do you know an anion? Are you one or the other type? This kind of bonding often breaks apart. Charges therefore occur on each atom in ionic bonds. This causes an attraction between other substances with opposite charge. For example, sodium ions having a positive charge are likely to attract a water molecule, which has a negative charge.

## Hydrogen Bonds

**Hydrogen bond**

Are fleeting bonds that form between hydrogen atoms and atoms of different structures. These bonds are based on attraction between positive and negative charges.

**Cohesion (or cohesive forces)**

The force that is formed when water molecules stick together due to hydrogen bonding.

Hydrogen bonds are fleeting bonds that form between atoms of different structures. Hydrogen bonds are defined as attractions between a hydrogen atom and another atom with higher electronegativity; in other words, they are based on attraction between positive and negative charges. Hydrogen in one water molecule may form bonds with atoms in other substances. For example, water forms hydrogen bonds with $Na^+$, when the positive charge of sodium attracts to the relatively negative charge of the oxygen in a water molecule. Atoms within water have polar covalent bonding, with one oxygen atom holding more tightly to electrons than hydrogen atoms. Hydrogen has a more positive charge within a water molecule and oxygen a more negative charge. The relative positive charge of hydrogen atoms attracts them to negatively charged ions. Hydrogen atoms will also find other bonds with negative oxygen atoms within water and therefore link water molecules to each other. When water molecules stick together due to hydrogen bonding, it is known as cohesion or cohesive forces. Cohesion is shown in Figure 2.9.

## The Importance of Water

We have already discussed some of the characteristics that make water so important biologically. It enables living processes to occur, all of which require a watery environment. Water allows many life functions to occur: (1) dissolving of ions; (2) hydration of plants; and (3) moderating of weather. Thus, hydrogen bonds formed by water described in the previous section are biologically very important. Let us elaborate on each of these.

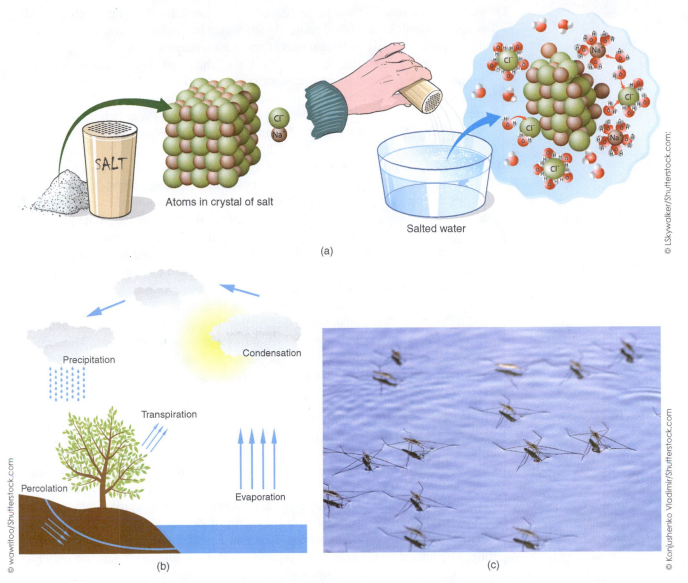

**Figure 2.9**    Water Has Many Features. a. Salt dissolving in water. Water surrounds salt's ions, making them "disappear" to the human eye. b. Transpiration in plants, an example of cohesion. Bonds form between water molecules, causing them to be "sticky", which allows them to be pulled up a plant. c. Cohesion (stickiness) between water molecules is caused by hydrogen bonding. These water striders (*Gerris*) walk on water.

**1)** Water is a **solvent**. Living cells are made of 60–80% water and dissolved substances. All movement within cells occurs in a watery environment. Dissolved materials are found within that watery world. When salt or sugar disappears in water, we say it has **dissolved**. The salt or sugar dissolves as a result of hydrogen bonds forming a wall of water around the dissolved ions. The resulting salty or sugary liquid is known as a mixture. The substance that is dissolved is called a solute and the substance doing the dissolving is termed a solvent. At the molecular level, when salt is added to a cup of water, as shown in Figure 2.9, sodium loses its valence electron to chlorine, and sodium becomes a positively charged ion. Then, a wall of water forms around the sodium ion, with relatively negative oxygen atoms surrounding the positive sodium ion. The ability of some molecules to dissolve is important for living systems to function, as we will see in later sections.

**Solute**

The component in a solution that is dissolved in the solvent.

**Solvent**

Substance that does the dissolving.

2) Water is cohesive. We saw in the preceding section that hydrogen bonds can link water molecules. In plants, roots absorb water by pulling it from the soil and into their cells. The cohesive forces of water molecules form a "sticky" line from the roots and soil all the way up to the top of the plant. Figure 2.9 shows transpiration, the continuous replacement of water molecules that evaporate from the leaves by water molecules continually moving upward from the soil. This characteristic of water molecules allows them to reach great heights in plants, upward of 300 feet in the case of giant sequoia trees. Water transport in plants will be discussed in more detail in Chapter 9.

3) Moderating Temperature. Water stores a great deal of energy because of its many hydrogen bonds. Energy is stored within each of its hydrogen bonds. Because of its stored energy, water has a high specific heat, the amount of energy required to raise the temperature of 1 gram of water 1 degree Celsius. This property, along with its ability to cohere, makes it critical to living systems.

   Without water living systems would not have developed and could not survive. Water moderates temperatures at coasts. When the weather is hot, hydrogen bonds in water absorb the heat and make the surrounding area less warm. When it is cold out, water's hydrogen bonds release heat, adding warmth to regions near it. Wild changes in the Earth's temperature do not occur because of the effects of water's specific heat on moderating temperature. This was a factor allowing life to develop on our planet. In fact, because of water's unique properties, life could develop within its oceans and lakes.

Usually, as temperatures cool, substances freeze and become denser or heavier. At 4 degrees Celsius (the freezing point of water in degrees Celsius is 0), water is at its heaviest. However, after this point, as it gets colder, it gets lighter. Ice floats because it forms a crystalline lattice structure that has more space between its molecules at zero degrees Celsius than at any other temperature. Ice serves as an insulator to the water underneath, from the colder air above. In this way, ice floating prevented the freezing of oceans and lakes from the bottom up, as would occur if water were denser at its coldest temperature. Life could thus develop within watery environments at deeper levels and be protected from freezing that was occurring at the top. Water also moderates body temperature due to sweating. The hottest molecules on an organism's surface evaporate, leaving only the cooler molecules of water. For life to exist as we know it, another planet would need to have water or another chemical with its unique properties to support and sustain life.

## Acids and Bases

The cohesive forces of water make it a biologically important molecule, but that is not the end of the story. Water also moderates the internal environment of living systems. Within a watery environment, ions of hydrogen from water form in different amounts. For example, when hydrogen is immersed in water, it changes into a positively charged ion, $H^+$. The amount of hydrogen ions in water changes the properties of water. Water has the ability to give up its hydrogen ions or take on more hydrogen ions depending on the conditions around it. In other words, at any time, a water molecule may surrender a hydrogen atom to its surroundings, or it may absorb one. Different parts of our bodies and even different parts of our cells require the right amount of hydrogen atoms to enable life functions.

When water yields more hydrogen into its surroundings, the resulting liquid is termed an acid, and when water absorbs more hydrogen from its surroundings, the resulting liquid is called a base. The amount of hydrogen in a solution is measured on a pH (power of hydrogen) scale. The scale ranges from 0 to 14, with pure water set at a pH of 7, shown in Figure 2.10. The pH scale shows the amount of acidity or base in a substance. This is so important because conditions for cells require pH homeostasis to survive.

**Acid**

The resulting liquid when water yields more hydrogen into its surroundings.

**Base**

The resulting liquid when water absorbs more hydrogen from its surroundings.

**pH scale**

A numeric scale that specifies the acidity or alkalinity of an aqueous solution.

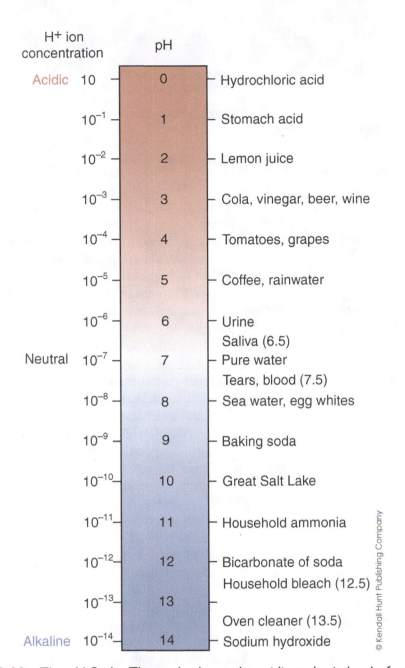

**Figure 2.10** The pH Scale. This scale shows the acidic or basic level of some common substances. The pH of a substance is based on the number of hydrogen ions it forms in a solution.

Recall that the term homeostasis is defined as maintaining stable internal conditions, from Chapter 1, including the proper pH.

The pH scale is based on the logarithmic amounts of hydrogen ions (H⁺) dissolved in water, meaning that each pH value represents a ten-fold increase or decrease in the amount of hydrogen ions. Water has the ability to break down into H⁺ and H⁻ ions. Thus, the lower the number on the pH scale is, the greater will be the acidity and the greater the amount of hydrogen ions found in solution. The higher the number on the pH scale, the more basic (or less acidic) will be the solution and the fewer hydrogen ions. The numbers of OH⁻ ions increase as a substance becomes more basic. Pure water, at a pH of 7, has a concentration of hydrogen ions that is $10^{-7}$ or .0000001 out of all the water molecules in solution. Pure water has an equal amount of H⁺ and OH⁻ ions. A concentration of 0.001

$$4H_2O \xrightarrow[\text{energy}]{\text{electric}} 2H_2 + 4OH^-$$

2 molecules of water    2 molecules of hydrogen    1 molecule of oxygen

**Figure 2.11** Dissociation of Water. Water breaks apart into hydrogen and hydroxide ions. From *Biological Perspectives*, 3rd ed by BSCS.

($10^{-3}$) would equal a pH of 3. At this pH level, it is considered very acidic, and there are many more hydrogen ions in such water than at a neutral pH.

As shown in Figure 2.11, while water is able to dissociate into both hydrogen ($H^+$) and hydroxide ($OH^-$) ions, certain chemicals added to water influence the number of $H^+$ and $OH^-$ ions in the solution. When a base is added to water, hydroxide ions ($OH^-$) increase in proportion and hydrogen ions decrease. An example of a strong base is NaOH (lye), which dissociates into $OH^-$ and $Na^+$. NaOH breaks into $Na^+$ and $OH^-$ (hydroxide). Hydroxide is able to absorb hydrogen ions and form water, lessening acidity: $OH^- + H^+$ yields $H_2O$. Hydroxide ions absorb hydrogen to and from water, lessening the proportion of $H^+$ in solution. Alternatively, HCl (hydrochloric acid) is a strong acid, which breaks apart and donates hydrogen to surroundings, making the solution acidic. Our stomach environment contains a low pH of between 2 and 3. A strongly acidic environment is needed to destroy the many bacteria and other pathogens that enter our bodies from the food we eat. Without such acidity, invaders would easily attack us from the inside out. Ulcers or open wounds in the intestines and esophagus sometimes form due to the extreme acidity leaking out of the stomach into these neighboring areas.

Acidity can become a chronic problem for some people – for example, gastroesophageal reflux disease (GERD) occurs when acid from the stomach travels up into the esophagus and causes damage. People suffering from GERD may take an antacid, in the form of the bicarbonate ion, $HCO_3^-$, to lessen the acidity. There are also surgical procedures to reduce the negative effects of acidity in the esophagus. In fact, many cases of untreated GERD lead to a condition called Barrett's esophagus, which is a change in the cells' structure in the esophagus due to stomach acid. Approximately 10% of people with Barrett's esophagus eventually develop esophageal cancer. In our story, nitrates caused this dreaded disease, but many factors may lead to cancer.

The bicarbonate ion in antacids used to treat GERD both gives off and absorbs hydrogen ions to stabilize pH of a solution. Bicarbonate uses special reactions, which occur in both directions, either toward products or reactants, to stabilize pH. Many reactions involving acid and base formation act in this way. Chemists use double arrows when showing chemical reactions that move in both directions. These reactions are called **reversible reactions**. One example of a reversible reaction is found in the blood, which maintains very stringent acid and base levels, called the carbonic acid-bicarbonate buffer system (Figure 2.12). A buffer maintains a certain pH. The blood's buffer system is an example of homeostasis, maintaining internal balance within a fairly narrow range.

$$CO_2 + H_2O \rightleftharpoons H_2CO_3 \rightleftharpoons H^+ + CO_3^-$$

Carbon dioxide  Water    Carbonic acid    Bicarbonate ion

**Figure 2.12** Acid–Base Buffering System in Blood. Carbonic acid acts as a buffer to maintain the pH of blood. At times it releases hydrogen ions and at other times it absorbs hydrogen ions. This buffering action regulates the hydrogen ion levels in blood and therefore pH.

Water, acids, and bases play an important role stabilizing the internal conditions of living systems. Chemical reactions take place within a stable environment but a guided by rules of chemistry that dictate their behavior. In the next section, we will explore how these chemicals come together to form larger molecules of life, the macromolecules. They are built with the same basic atoms described earlier, but use carbon as their backbone building material. Macromolecules are the most active chemicals or "living molecules" in cells.

# Why Carbon?

Life forms found on the Earth range from microscopic, unicellular organisms to those as large and complex as the blue whale. While water is an important component of all organisms, accounting for over two-thirds of their mass, carbon is the backbone. Complex carbon-based molecules are the molecules of life. A building is made of many materials, but its building blocks are usually uniform and have similar units. Our building block is the carbon atom.

Why carbon? Unlike many other atoms, it is very stable because it is generally neutral, sharing electrons equally. Carbon has four valence electrons and therefore forms four bonds with neighboring atoms (see Figure 2.13). This is a lot of bond potential. More bonds mean more connections with other atoms, more complexity in their arrangements, greater strength, and more possible structures. Scientists have isolated 13 million organic chemicals compared to only 300,000 inorganic (noncarbon) chemicals. Often, organic molecules form even larger numbers of compounds because they form isomers of one another. Isomers are substances with the same number and types of atoms as each other, but with different arrangements in their structure.

With such a large number of compounds, it might seem an impossible task to understand them. Fortunately, while they can undergo many different reactions, only one portion of the molecule, known as the functional group, is actually involved in reactions. Each functional group has a unique arrangement of atoms that acts a specific way in chemical reactions. Therefore, all of the compounds forming a particular functional group will react in the same way under a given set of conditions. A list of functional groups from our story is provided in Figure 2.14. Nitrates, seen in our story earlier in the

**Functional group**

A group of atoms that are involved in reactions.

**Carbon ATOM**

© Designua/Shutterstock.com

**Figure 2.13**   Why Carbon? Carbon forms four bonds with its neighbors to satisfy the octet rule. It is a stable and generally neutral atom, which makes it an ideal building block.

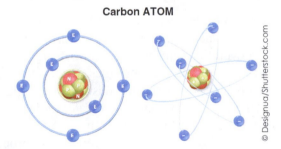

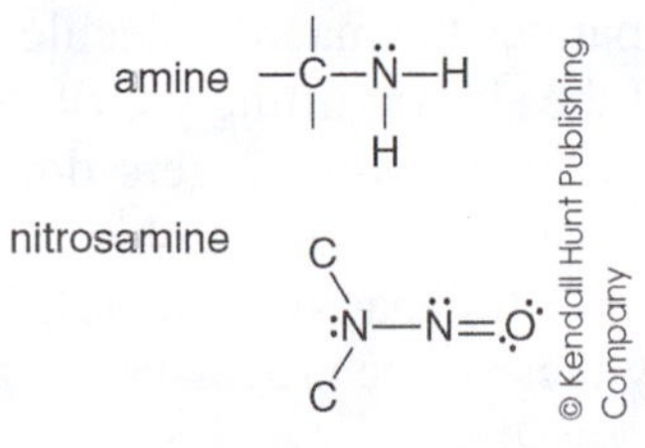

© Kendall Hunt Publishing Company

**Figure 2.14**   Functional Groups of Organic Molecules Including $-NH_2$, and $N-N=O$ (nitrosamine).

**Macromolecules**

Molecules forming the building blocks of living things.

**Carbohydrate**

Organic compounds providing "instant energy" for living tissues.

**Lipid**

Neutral fats, phospholipids, and steroids found in food and in living systems.

**Protein**

The most common macromolecule in living systems.

**Nucleic acid**

The genetic material of a cell.

**Dehydration synthesis**

A process in which hydroxyl and hydrogen atoms are removed from two organic compounds that merges them into one (covalent) bond.

**Hydrolysis**

The breakdown of a compound due to its reaction with water.

chapter, and their related nitrites and nitrosamines, are important and even destructive functional groups in living systems.

Functional groups are often very reactive, exchanging "free" or extra electrons with other substances. Nitrates are functional groups that react easily with other compounds to form nitrosamines. Some studies indicate that bacteria in human guts cause the dangerous nitrosamines to form from nitrates we consume from our foods. Substances containing free electrons, such as nitrates, are termed **free radicals**. Note the extra sets of electrons in the chemicals that scientists believed were plaguing Lin Xian in Figure 2.14.

In order to obtain extra electrons, free radicals oxidize (take away electrons) other substances and cause damage to parts of the body, including DNA in the nucleus and blood vessel walls. Antioxidants, as is obvious by the name, prevent oxidation of foods. Damage to DNA leads to many problems, including genetic defects, heart disease, and cancer. Damage to blood vessel walls can lead to clotting, stroke, heart attack, and other ailments. Antioxidants, which are found in many fruits and vegetables, prevent or slow damage from free radicals, so diet may play a role in disease prevention.

Recall that the scientists in our story recommended vitamin C to combat the effects of nitrosamines in Lin Xian. Foods high in antioxidants include garlic, blueberries, raspberries, onions, broccoli, carrots, and leafy greens. All of these high-antioxidant foods are fruits and vegetables. Vitamins A, C, and E (the acronym ACE will help you remember), all have antioxidant properties, are found in these foods to help fight disease.

## Macromolecules

Organic molecules form an incredible array of substances, from penicillin to petroleum products. In living systems, larger carbon-based substances, called macromolecules, carry out life's functions. There are four types of macromolecules in living organisms: carbohydrates, lipids, proteins, and nucleic acids (see Figure 2.17). Each of these macromolecules has specific roles in our life functions. They are able to form large strings of molecules that both support and carry out life's functions. They are assembled and disassembled for use by cells.

## Building Up and Breaking Down Macromolecules

How do these large macromolecules become so large and how do they break down once again? Through a process that is common to all of these chemicals: Dehydration synthesis and hydrolysis. Dehydration synthesis occurs when organic molecules link subunits together. This forms larger and larger molecules. During this process, a molecule of water is lost, allowing open bonds to become unstable and link up between nearby organic molecules. In Figure 2.15, the removal of a water molecule from two separate macromolecules leads to a bond formed between them.

Alternatively, macromolecules break down through the process of hydrolysis. Hydrolysis literally means the splitting of a chemical (lysis), using water (hydro). When water is added to the joined macromolecules (polymer) in Figure 2.15, two bonds form out of the one bond that originally joined them. Water breaks the two apart and hydroxide from the water molecule separates the macromolecules.

Take your hands and repeat the slogan, lifting your fingers up when saying: "Dehydration synthesis builds up," and point your fingers down when saying: "Hydrolysis breaks down." This slogan will help you to remember the difference between the two processes. We now explore each of the macromolecules, keeping in mind that they are all built up and broken down by these same processes.

### Carbohydrates

Carbohydrates are the "instant energy" macromolecule. They contain loads of readily available energy in the many covalent bonds linking their subunits. Carbohydrates are

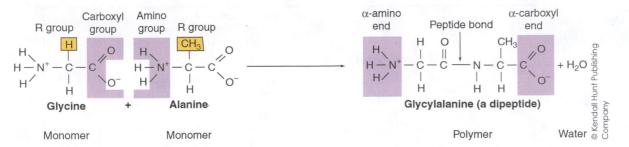

**Figure 2.15**    Dehydration Synthesis and Hydrolysis of Two Generic Monomers.

made up of a consistent ratio of atoms: one carbon to two hydrogen to one oxygen atom. This 1:2:1 ratio forms ring-shaped structures that are the building blocks of carbohydrates, called monosaccharides, or simple sugars. Common monosaccharides are glucose, galactose, and fructose (see Figure 2.16).

When two monosaccharides combine, they form disaccharides. Some examples include sucrose, lactose, and maltose. Sucrose is table sugar, lactose is milk sugar, and maltose is beer sugar. Glucose and fructose join together to form sucrose. Sucrose is shown in Figure 2.16.

The combination of three or more monosaccharides is known as a polysaccharide. Polysaccharides are long chains of simple sugars, with tremendous ability to store energy in the many bonds between the rings. The animal storage form of carbohydrate energy occurs mainly in glycogen, which is abundantly found in the liver and muscles. When energy is needed, the liver breaks off a piece of the polysaccharide for the body to use. In plants, the primary storage form of carbohydrate energy is starch. Starch is found throughout the plant's structure. Through photosynthesis, energy from sunlight is converted and stored as starch in roots, stems, and leaves. Seeds contain particularly high amounts of starch because they provide energy for the next generation of growing plants. The most abundant polysaccharide in plants is cellulose, which comprises much of the structure in stems and bark. Some polysaccharides are shown in Figure 2.16.

**Monosaccharide**

Ring-shaped structures that are the building blocks of carbohydrates.

**Disaccharide**

A class of sugars formed when two monosaccharaides combine.

**Polysaccharide**

The combination of three or more monosaccharides.

### 2.1 LACTOSE INTOLERANCE

Lactose, or milk sugar, is a way mammals give energy to their young in the form of milk. Between 30 and 50 million people in the U.S. are unable to digest lactose and therefore milk and milk products. Lactose-intolerant people experience bloating, indigestion, cramping, and diarrhea when ingesting lactose-containing foods. The evolution of lactose intolerance is evident in that people who come from the regions of the world that did not have domestication of animals (and thus did not use cow's, goat's, or the milk of another mammal as a food source) tend to have greater rates of lactose intolerance. Perhaps in these societies, the benefits of being able to drink milk were not present, thus not putting pressure on such populations to be able to digest lactose. Humans are, indeed, the only adult organisms that drink milk. However, recent studies report that lactose-intolerance problems may be more than 50% misdiagnosed. Other causes of the symptoms listed are often the culprit. It may be that lactose intolerance is an easy answer to more complex health issues.

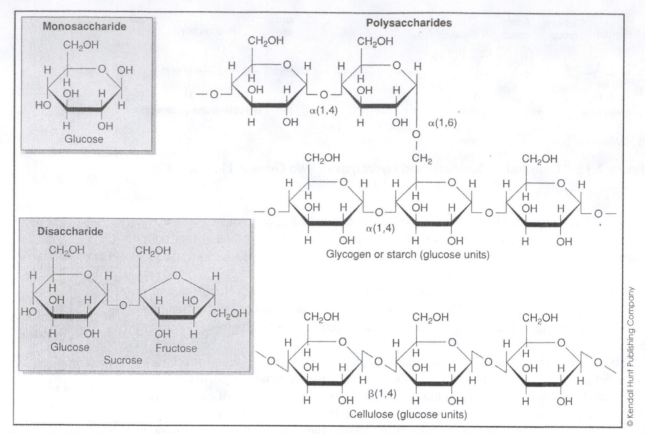

**Figure 2.16**  Carbohydrates Have Varied Types. Monosaccharides (glucose), Disaccharides, Polysaccharides (starch or glycogen). Several glucose molecules join together to form cellulose through dehydration synthesis.

## Lipids

**Lipids** are neutral fats (fats, waxes, and oils), phospholipids, and steroids found in our food and in our bodies. Lipids are stored in fat cells for use as long-term energy storage. They are found in all of our cell membranes – up to 50% of those membranes are made up of a type of lipid. They serve as hormones to help cells communicate, as components of cell parts, and as stabilizers or cushions for organs and tissues.

Much like carbohydrates, lipids contain carbon, hydrogen, and oxygen. However, they have a much higher amount of carbon and hydrogen than carbohydrates. Indeed, lipids are long chains of carbon skeletons with many bonds of hydrogen attached. When considering the large number of bonds, it is clear that energy storage, in the long term, is a main function of lipids. They are often used to cushion organs and are found side-by-side with the other macromolecules, as shown in Figure 2.17.

Oil and water do not mix – a chemical rule that shows how lipids work. Lipids are electrically neutral, meaning that they do not contain a charge. Water, on the other hand, is charged, as it contains polar covalent bonds. Charged substances mix with each, while neutral or uncharged substances mix only with other uncharged substances. Uncharged chemicals are known as hydrophobic, which translates into "water fearing" because of this rule. Water's charge drives hydrophobic substances away. Instead, substances that are charged dissolve in water and thus mix with water. These substances are known as hydrophilic, which means "water loving." Water is hydrophilic, which means that it sticks together in a cohesive way.

The hydrophobic nature of lipids drives their behavior within living systems. Lipids avoid water and other charged particles because of this aversion. In cells, lipids will arrange themselves away from water environments to form a cell's shape. Lipids

**Hydrophobic**

Compounds that do not dissolve in water (also called, water fearing).

**Hydrophilic**

Compounds that have the tendency to dissolve in or mix with water.

© Kendall Hunt Publishing Company

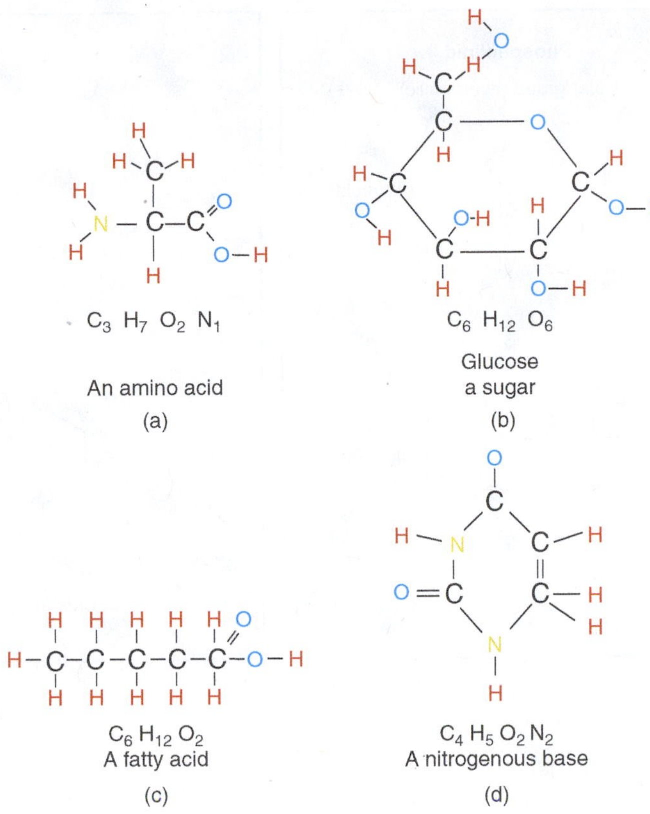

**Figure 2.17** The Four Types of Macromolecules. These molecules join together to form larger substances.

The amino acid structure is labeled $C_3 H_7 O_2 N_1$ — An amino acid (a)

Glucose is labeled $C_6 H_{12} O_6$ — Glucose a sugar (b)

The fatty acid structure is labeled $C_6 H_{12} O_2$ — A fatty acid (c)

The nitrogenous base is labeled $C_4 H_5 O_2 N_2$ — A nitrogenous base (d)

© Kendall Hunt Publishing Company

self-assemble, or arrange themselves, in accordance with the watery world surrounding them. This characteristic is important because lipids compose cell membranes, which self-assemble to surround cells. The cell membrane, for example, has a lipid layer on the inside of its membrane, which arranges itself away from the watery regions.

While lipids have similar behaviors overall, they are classified into three categories: neutral fats (**triglycerides**), phospholipids, and steroids.

## Triglycerides

**Neutral fats**, or **triglycerides**, are composed of three large fatty acids joined together by a short-chained glycerol molecule, as shown in Figure 2.18. The fats found beneath our skin, called subcutaneous fat, around our organs, and in our cells are neutral fats. There are many kinds of neutral fats; however, based on their bonding, they may be classified a few ways, as shown in Figure 2.18. Saturated fats are neutral fats that are literally saturated with as many hydrogen atoms as is possible in the carbon skeleton. Saturated fats come primarily from consuming animal products. Saturated fats are linked to heart disease and hardening of the arteries, called atherosclerosis. When a neutral fat contains only one double bond, it eliminates two hydrogen atoms from the carbon skeleton, forming a **monounsaturated fat**. This kind of neutral fat is associated with heart health because they are thought to eliminate fats from the walls of blood vessels and improving blood flow. Neutral fats are shown in Figure 2.18. **Polyunsaturated fats** are neutral fats that contain more than one double bond, as indicated by the name "-poly." Double bonds reduce the overall number of hydrogen atoms on the carbon skeleton. These fats are also associated with heart health, but monounsaturated fats are best. Unsaturated fats are associated with plants and plant products.

**Neutral fat**

A fat that is composed of three large fatty acids joined together by a short-chained glycerol molecule.

**Phospholipid**

A lipid composed of both a charged phosphate group and fatty acid chains.

**Steroid**

A type of fat that stabilizes the structure of cell membranes.

**Saturated fat**

Neutral fats that are literally saturated with as many hydrogen atoms as is possible in the carbon skeleton.

**Atherosclerosis**

Condition in which saturated fats are linked to heart disease and hardening of the arteries occurs.

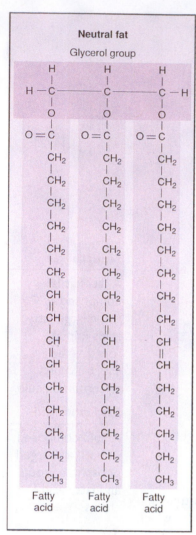
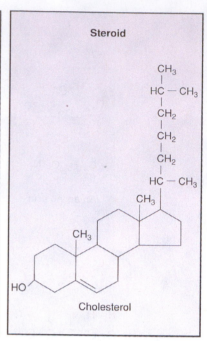

**Figure 2.18** Types of Lipids: Triglyceride is a neutral fat. Phospholipids in cell membranes, and cholesterol, a type of steroid.

### Phospholipids

**Phospholipids** are another category of the lipid group. They are composed of both a charged phosphate group and fatty acid chains. Figure 2.18 also shows the structure of phospholipids, which looks like a lollipop: it contains a circular phosphate head, with three negative charges attached to two sticks, or chains of fatty acids. In cell membranes, phospholipids arrange to form the bulk of its structure. The role of phospholipids in cells and in cell transport will be discussed in the next chapter.

### Steroids

Cell membranes also contain **steroids**, a type of fat that stabilizes their structure. As shown in Figure 2.18, these fats are very different in shape from the other types: they contain four flat hydrocarbon rings, which are made naturally by animals. Cholesterol is one example of a steroid. While some forms of cholesterol aid in disease-causing buildups of plaques on the walls of blood vessels (described earlier), it also serves an important role as a component of cell membranes. **Steroids** are also needed in the body as hormones, such as male and female sex hormones, testosterone, and estrogen. They

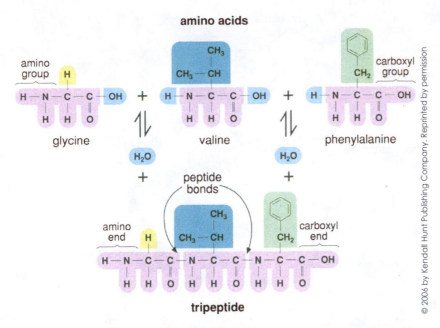

amino acids

glycine    valine    phenylalanine

peptide bonds

tripeptide

**Figure 2.19**    Amino Acids Are the Building Blocks of Proteins. This figure shows the basic amino acid structure, three types of amino acids, and several amino acids combining to form larger proteins (tripeptide). There are only 20 amino acid types used to make up all proteins in living organisms. From *Biological Perspectives*, 3rd ed by BSCS.

are also beneficial in use in different therapies to bring down inflammation, such as treatments for asthma and in back pain relief.

## Proteins

The most common macromolecule in living systems is protein. Proteins play a diverse role in our bodies, making up everything from hair, nails, and skin to having functions in chemical reactions within cells. The basic building block of proteins is the amino acid. Amino acids are composed of a central carbon bonded together with a hydrogen atom, a carboxyl group (–COOH), an amino group (–NH$_2$), and a variable group. Almost all of the amino acids in proteins come from a list of 20, similar in structure to those found in Figure 2.19. The variable group (those groups found atop of amino acids in Figure 2.19) determines the type of the amino acid because all of the other parts are the same. The amino group in an amino acid becomes nitrates in the body, the chemical discussed in our opening story.

The amino group, as a functional group, also serves in several chemical reactions as a way to form bonds between amino acids. When two amino acids combine to form larger proteins, a molecule of water is lost and a peptide bond forms. Peptide bonds form as a result of the H and OH leaving adjacent amino acids (see Figure 2.19). Through hydrolysis, water is added to break the two apart once again.

As a molecule of protein adds amino acids, it grows into a longer string of amino acids called a polypeptide. How proteins are organized is shown in Figure 2.20. The simple string structure is called its **primary** structure. Much like a power cord attaching your computer, this string has a long shape. However, the variable groups along the string of amino acids contain charges and chemical characteristics that allow them to bond with one another along the string. When bonding occurs, it results in two shapes at the **secondary** level of organization: an alpha helix and a beta-pleated sheet. An alpha helix is like a slinky, coiled together with bonds holding the structure. Take the power cord on your computer and wrap it around your finger. This is what the alpha helix

**Amino acid**

The building blocks of proteins.

**Polypeptide**

A long string of amino acids formed as molecules of protein adds amino acids.

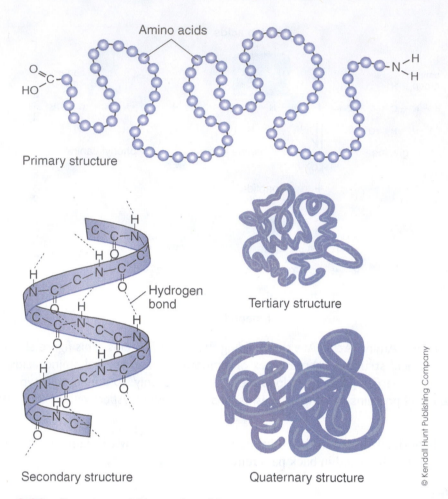

Amino acids

Primary structure

Hydrogen bond

Secondary structure

Tertiary structure

Quaternary structure

© Kendall Hunt Publishing Company

**Figure 2.20** Proteins and Hierarchy of Structures of Proteins. Proteins are organized from simple strings of amino acids at the primary level to more complex structures at higher levels within the hierarchy.

**Fibrous protein**

Structural compound that does not dissolve in water but remains solid support in parts of organisms.

**Globular protein**

A type of protein that is water soluble.

looks like. Beta-pleated sheets resemble a blanket and are found as keratin fibers in hair, clotting proteins in blood, and even as spider web silk. When secondary level polypeptides fold, forming more complex shapes as compared with lower levels, it is known as a **tertiary** protein. Proteins fold to give a distinct shape and function to them. Tertiary proteins combine with one another to form a unique shape called the **quaternary** structure. An example of a quaternary protein is the hemoglobin molecule which carries oxygen throughout animal systems. Its unique quaternary structure allows oxygen to be carried in large amounts.

Quaternary proteins may occur as either fibrous or globular. Fibrous proteins are structural, meaning that they do not dissolve in water but remain solid support in parts of organisms. Fibrous proteins appear as strands, such as collagen, which maintain cell structure, and keratin, which protects skin and nails in humans. Chemically active proteins that carry out life functions are termed globular proteins. Globular proteins are water soluble, meaning that they dissolve in water. They are "functional" proteins because they have specialized shapes that attach to other chemicals to perform reactions. Consider auxins, which are plant hormones that cause cells to grow, developing the root and vessel systems. Other globular proteins include antibodies to fight infection and clotting factors to prevent bleeding.

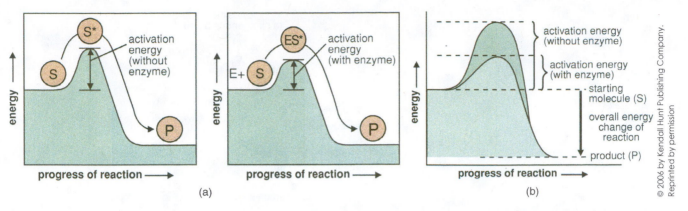

**Figure 2.21** Action of an Enzyme on Substrates. A. Enzymes lower the activation energy required to get chemicals to react. In the graph, substrates start at a higher energy level and after the reaction have less energy. B. An enzyme joins together substrates to form a new product. Enzymes also break down substrates. Lactase is an enzyme that breaks apart the lactose carbohydrate in milk. From *Biological Perspectives*, 3rd ed by BSCS.

## Enzymes

Another example of a globular protein is an enzyme, which is a specialized protein that speeds up chemical reactions. Enzymes are also called catalysts, which are substances that help chemical reactions to occur. Special shapes on each enzyme, called active sites, allow for binding to other chemicals, called substrates, to either bring substrates together or break them apart. Enzymes can be compared to a match-maker, or a meddling in-law, who breaks up or sets up their family members. Figure 2.21 shows the action of an enzyme in both cases, bringing forth a chemical reaction. When enzymes bind to substrates, they facilitate a reaction – they make it happen more quickly. Much like a match-maker, enzymes cannot make something happen that otherwise chemically would be impossible. Enzymes lower the activation energy required for the reaction to take place by bringing substrates together. Otherwise, the process might take much longer to occur. However, enzymes cannot force the substrates together. For example, lactose in our intestines does not get digested without the presence of the lactase enzyme. Lactase breaks down the lactose milk sugar into glucose, able to be absorbed by cells. Other enzymes include amylase, which breaks down carbohydrates in our mouths, and telomerase, which adds to our DNA to prevent damaged ends. Telomerase is associated with slowing the aging process, discussed later in the text. In each of these cases, enzymes are reusable, unchanged after reacting with substrates, and readily available once they facilitate a reaction.

Enzyme names usually end in –*ase*, with a prefix that indicates the type of substrate the enzyme acts upon. For example, lactose was mentioned as a form of milk protein. When lactase, the enzyme for lactose, acts upon milk, it causes milk to become digested into a form able to be utilized by animals. Enzymes often require specific environmental conditions to work: temperature, pH, and salt concentrations affect enzymatic activity.

Decomposing bacteria require the right pH to perform their role in breaking down living organisms. The optimal pH for decomposing bacterial enzymes is much higher than in peat bogs, allowing preservation of dead organisms in those environments. Bacteria could not act to break down the dead organisms because bacterial enzymes often do not work in an acidic bog.

**Enzyme**

Specialized protein that speeds up chemical reactions.

**Active site**

Special shapes on enzymes that allow for binding to other chemicals.

**Substrate**

A compound on which an enzyme acts.

**Activation energy**

The minimum amount of energy that the must be possessed by the reacting species to undergo a specific reaction.

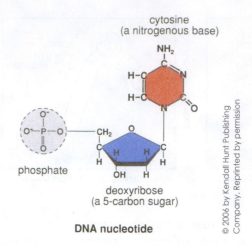

© 2006 by Kendall Hunt Publishing
Company. Reprinted by permission

**DNA nucleotide**

**Figure 2.22**   Nucleotide Structure. A phosphate, sugar, and base comprise the basic unit of nucleic acids, the nucleotide. From *Biological Perspectives*, 3rd ed by BSCS.

## Nucleic acids

**Deoxyribo-nucleic acid (DNA)**

A long macromolecule containing the information code that directs cellular activities in living organisms.

Nucleic acids are the genetic material of a cell. Genetic material stores information that (1) controls the cell and (2) passes that information on to new generations of cells. The main types of nucleic acids are DNA (deoxyribonucleic acid) and RNA (ribonucleic acid). DNA contains the information code that directs cellular activities in living organisms. RNA comprises a set of messenger molecules that carry out the orders given by DNA. In some viral types, RNA is the primary hereditary code, but this is rare. DNA and RNA each play a major role in the formation of proteins. Their code gives rise to proteins, which perform many vital functions for cells.

**Ribonucleic acid (RNA)**

A nucleic acid present in living cells.

Nucleic acids are long macromolecules, composed of repeating units of nucleotides, as shown in Figure 2.22. Each nucleotide is made up of three parts: a five-carbon sugar, a phosphate group ($PO_4^{-3}$), and a nitrogenous base. There are four nitrogenous bases, **adenine**, **guanine**, **cytosine**, and **thymine**. The arrangement of these bases in strings of sequences makes up the genetic information code. That code gives unique directives to cells and organisms.

- The details of the genetic code will be discussed in Chapter 5.

**Nitrogenous base**

A nitrogen containing molecule having the same chemical properties as a base.

ATP, or adenosine triphosphate, is a special nucleotide, holding large amounts of energy available for cell functions. It is composed of one **adenine** base combined with a **ribose** sugar forming an **adenosine** group. Adenosine has three **phosphate groups** attached to it, forming the ATP molecule, with energy contained within its phosphate bonds. The phosphate's energy within ATP drives cellular reactions, building up and breaking down the macromolecules, as shown in Figure 2.23.

The phosphate groups are held together by high-energy bonds, denoted by squiggly lines.

**Adenosine triphosphate (ATP)**

A special nucleotide that holds readily available energy for cell functions.

Phosphate bonds are unstable because of their high energy, causing these bonds to readily break and make energy available for cellular needs. When water is added to an ATP molecule, for example, hydrolysis occurs, releasing a free inorganic phosphate ($P_i$) along with energy and **ADP**, or **adenosine diphosphate** (ATP with one less phosphate group). ADP is recycled back and forth with ATP, as energy is formed and released in accordance with cell needs. This reaction is shown in the following chemical equation.

$$ATP + H_2O \rightarrow ADP + Pi + energy$$

ATP provides immediate and accessible energy for all cell functions. Its role is vitally important in almost every biological process.

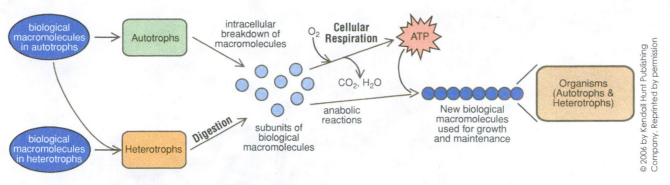

© 2006 by Kendall Hunt Publishing Company. Reprinted by permission

**Figure 2.23** Macromolecules Are Broken Down and Built Up during Metabolic Reactions. ATP (adenosine triphosphate) is used to make and break bonds. The energy from ATP, in high-energy phosphate bonds, drives cellular reactions. Autotrophs make their own food, and heterotrophs eat other life for food exchange matter and energy to perform their life functions. From *Biological Perspectives*, 3rd ed by BSCS.

### ARE NITRATES SO BAD FOR US? WHAT – BOLOGNA TOO!

Just as in the opening story, is there a killer lurking in our own diets? Should we be cautious about eating processed meats, for example, which contain nitrates like the food in Jin's village? There is a link between processed meats and cancer. Sodium nitrate has been added to processed meats such as bologna, hot dogs, ham, and bacon for years (Figure 2.24). It acts as a preservative, prevents botulism and provides longer refrigerator life.

Because of nitrates' link to various digestive cancers, as described in our opening story, in the 1970s, the US government set a limit of 120 part per million (ppm) for the amount of sodium nitrate allowable in processed meat. As also shown in Lin Xian, scientists also learned that adding 550 ppm of vitamin C or erythorbic acid (a relative of vitamin C) can prevent the formation of nitrosamines (known to cause cancer in lab animals) by bacteria in animal guts. Therefore, more recently, meat-packaging companies have been adding vitamin C to meats to protect from nitrosamine formation.

Sodium nitrate itself is not bad – we get it from vegetables too – like celery, lettuce, beets, radishes, and spinach, which absorb it from the soil. A person eating about 2.5 cups of vegetables might acquire as much sodium nitrate as if eating 10 hot dogs! However, vegetables contain other compounds including vitamin C that prevent nitrosamine formation.

More research is needed to establish the link between processed meats and cancer. The problem is that with processed meats, there are many variables involved. The link is complex because the smoking process, salt, fat, and chemicals in red meat also have a link to cancer. Additional research is necessary to sort out the variables. Further chemical models, just like the one depicted in our story, need to be developed to tease out these variables.

© Blue Pig/Shutterstock.com

**Figure 2.24** Processed Meats Contain Nitrates as a Preservative. These foods are a part of many of our diets but are linked to a variety of digestive cancers.

## Summary

Chemistry impacts our lives in many ways. Our story showed that the most likely mystery killer, a simple chemical, unseen and undetected for over 2,000 years, harmed the health of those villagers in Lin Xian. Chemicals make up our surroundings, including each of our cells. Reviewing the organization of matter: atoms join together to organize and form larger substances, including macromolecules, which comprise living systems. Atoms are mostly empty space, and their electron arrangement determines how they react. Exchanges of energy, through several types of reactions, lead to many new products. Complex conditions within organisms are maintained to allow chemicals to work together to perform life's functions. When macromolecules organize into living systems, each performs a vital role in an organism's survival.

---

### CHECK OUT

**Summary: Key Points**

- Chemistry affects our lives in many ways, from diseases such as esophageal cancer to solutions in medicine and in every day needs.
- Chemicals may be classified into types of matter: atoms, elements, ions, isotopes, molecules, and compounds.
- The atom is mostly empty space, with valence electrons orbiting a central nucleus containing protons and neutrons.
- Atomic mass and atomic number on the Periodic Table of Elements allow the calculation of the subatomic particles within atoms and ions.
- There are four types of bonds: covalent, polar covalent, ionic, and hydrogen bonds.
- Living systems carry out their processes through chemical reactions that keep pH, water balance, and the right set of conditions.
- Organic molecules make up the backbone of living structures.
- Organic molecules build up through dehydration synthesis and break down through hydrolysis.
- Research findings make many claims about the role of chemicals in human health and disease, all of which must be supported by the scientific method.

## KEY TERMS

| | |
|---|---|
| acid | ionic bond |
| activation energy | isotope |
| active site | kinetic energy |
| adenosine triphosphate (ATP) | lipid |
| amino acid | matter |
| anion | macromolecule |
| atherosclerosis | molecule |
| atom | monosaccharide |
| atomic mass | neutron |
| atomic number | nitrogenous base |
| base | nucleic acid |
| carbohydrate | neutral fat |
| cation | octet rule |
| chemical bond | organic chemistry |
| chemistry | phospholipid |
| cohesion | pH scale |
| covalent bond | polar covalent bond |
| decomposition reaction | polyatomic ion |
| dehydration synthesis | polypeptide |
| deoxyribonucleic acid (DNA) | polysaccharide |
| disaccharide | potential energy |
| dissolve | protein |
| electron | proton |
| electronegativity | reversible reaction |
| element | ribonucleic acid (RNA) |
| enzyme | saturated fat |
| fibrous protein | solute |
| functional group | solution |
| globular protein | solvent |
| Gold Foil experiment | substrate |
| hydrogen bond | steroid |
| hydrolysis | synthesis reaction |
| hydrophilic | triglyceride |
| hydrophobic | valence electrons |

# Multiple Choice Questions

1. What role did scientists suspect nitrates of playing in affecting the health of villagers in Lin Xian?

   a. Nitrates caused higher rates of antioxidants.
   b. Nitrates caused higher rates of heart disease.
   c. Nitrates caused higher rates of vitamins A and C.
   d. Nitrates caused higher rates of esophageal cancer.

2. Helium, a gas in the atmosphere, is:

   a. unreactive and a compound.
   b. unreactive and an element.
   c. reactive and a compound.
   d. reactive and an element.

3. Which BEST describes protons?

   a. They are positively charged but without mass.
   b. They are positively charged and have a mass of 1 amu.
   c. They are negatively charged but without mass.
   d. They are negatively charged and have a mass of 1 amu.

4. Rutherford's Gold Foil experiment shows that all matter is:

   a. mostly empty space.
   b. mostly dark matter.
   c. 50% protons and 50% electrons by mass.
   d. 99% dark energy.

5. Isotopes of carbon contain differing numbers of:

   a. protons
   b. neutrons
   c. electrons
   d. both a and b are true

6. Which term best describes the forming of covalent bonds?

   a. polarity
   b. sharing
   c. weak
   d. heavy

7. Using the Periodic Table of Elements, calculate the number of neutrons in an atom of phosphorous (P), found in large amounts within our cell membranes:

   a. 15
   b. 16
   c. 31
   d. 46

8. Water loses hydrogen ions in milk, giving it a pH of 6.2. You would classify milk as:
   a. slightly acidic
   b. slightly basic
   c. very acidic
   d. very basic

9. Which reaction forms large starch granules within potatoes?
   a. dehydration
   b. dehydration synthesis
   c. hydrolysis
   d. hydrolysis synthesis

10. An experiment shows that coconut oil increases levels of bad cholesterol in humans. Which component of coconut oil is most likely the cause of the increase?
    a. saturated fats
    b. monounsaturated fats
    c. polyunsaturated fats
    d. both b and c are true

## Short Answer

1. Describe how chemicals in a natural ecosystem, such as a lake or pond, could give diseases to humans.

2. List the following terms, from larger to smaller in size, of the following substances: compound, atom, neutron, molecule, electron and matter.

3. What is the number of neutrons found within an ion of potassium ($K^+$)? Show your work.

4. Compare how ionic, covalent, and polar covalent bonds differ from each other. Be sure to include the following terms in your comparison: electronegativity, polarity, and stability.

5. For question #4 above, list and draw an example of compound formed by each of the bonds described. Include in the picture the electron arrangement around the atoms.

6. The pH of the stomach of humans is roughly 2–3. What processes enable this low pH to form in stomach? Be sure to include an equation with water as the reactant.

7. ATP plays a major role in life's processes. Describe how the activities of the single-celled` *Amoeba* are dependent upon ATP.

8. Why is carbon the backbone of life? Describe the characteristics of carbon that make it our unique building block.

9. Explain the differences between fibrous and globular proteins. Which are important in blood clotting? Why?

10. Maltase is a protein that acts within plants, bacteria, and yeast. Explain the role of maltase in these organisms by:
    a. classifying maltase as a special type of protein.
    b. describing how maltase changes the activation energy of a reaction?
    c. identifying its substrate.
    d. describing its role in changing its substrate.

## Biology and Society Corner: Discussion Questions

1. Research and then predict how the role of high-fructose corn syrup within our food supply will affect human health. Explain how this will impact healthcare, the economy, and overall human impacts on the environment.

2. Isotopes are important chemicals found in nature. Trace the history of the discovery of radioactivity. What role do you think radioactivity will play in forming public policies and in influencing our role in world politics? Use either Iodine-131 or Radium-226 to frame your thesis.

3. In a democratically elected government, people vote for policy changes through electing their officials. Should governments ban foods using certain chemicals? Nitrates?

4. The European Union (EU) has banned over 1,000 chemicals in cosmetics while we ban only 13 of them. What do you think of the EU's policy as compared with our more limited ban? Are we justified in allowing more chemicals in cosmetics?

5. PCBs (polychlorinated biphenyls) are an industrial waste product contaminating many areas of the world. Research the effects of PCBs on environmental health of animals within North America. Write a plan to help mitigate the effects of PCBs on water systems within North America.

Figure  – Concept Map of Chapter 2 Big Ideas

# The Cell As a City

**3**

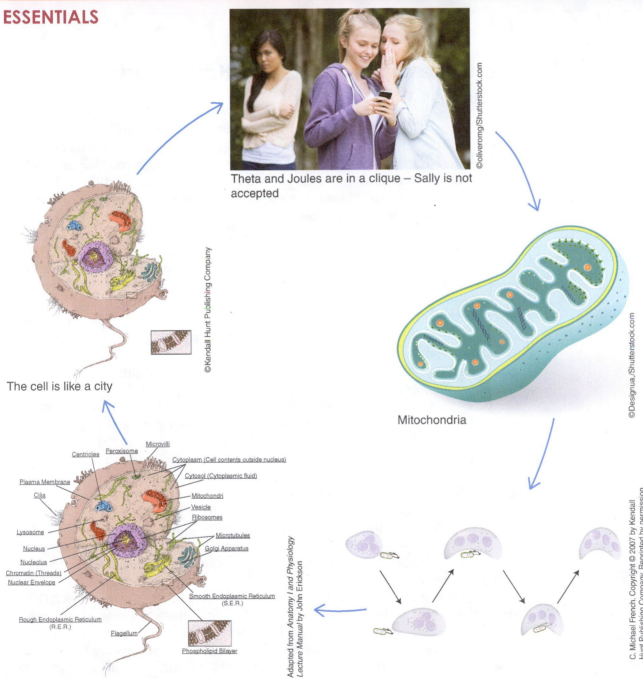

©oliveromg/Shutterstock.com

Theta and Joules are in a clique – Sally is not accepted

©Kendall Hunt Publishing Company

The cell is like a city

Mitochondria

©Designua./Shutterstock.com

A cell with its organelles

Centrioles — Peroxisome — Microvilli
Cytoplasm (Cell contents outside nucleus)
Plasma Membrane
Cytosol (Cytoplasmic fluid)
Cilia
Mitochondri
Vesicle
Ribosomes
Lysosome
Nucleus
Microtubules
Nucleolus
Golgi Apparatus
Chromatin (Threads)
Nuclear Envelope
Smooth Endoplasmic Reticulum (S.E.R.)
Rough Endoplasmic Reticulum (R.E.R.)
Flagellum
Phospholipid Bilayer

*Adapted from Anatomy I and Physiology Lecture Manual by John Erickson*

Primitive cells absorb mitochondria-like organisms

C. Michael French. Copyright © 2007 by Kendall Hunt Publishing Company. Reprinted by permission

© Kendall Hunt Publishing Company

## CHECK IN

**From reading this chapter, you will be able to:**

• Explain how differences caused by inherited organelles could have societal implications.
• Describe how the characteristics that are valued change from culture to culture and over time.
• Outline the cell theory, list and describe types of cells, and explain endosymbiosis.
• List and describe the organelles found in a cell, and explain their main functions.
• Explain the processes of diffusion, osmosis, facilitated diffusion, active transport, and bulk transport.

# The Case of the Meddling Houseguest: A Friendship Divided

Theta and Joules liked their friend Sally, but when they entered college, they learned that Sally was different. When they were all young, they played together on the block, went to each other's birthday parties, and had some great sleepovers. "We had a lot of fun with Sally in sixth grade . . . I wish she could join our sorority," said Theta. Aghast at the thought, Joules replied, "Don't even say it – you know what that would mean for us. We shouldn't even admit that we know her."

"Why can I not hang out with people I like? . . . Am I not allowed to be Sally's friend because of some test?" thought Theta. "There is no law against me being friends with Sally!" exclaimed Theta, after a long pause. Joules dismissed Theta smugly, "You know you can't do it. It will never happen." They were expecting Sally to come into the dorm any minute. Sally was expecting to hang out with them as usual. But on this day, their friendship had to end. On this day, Joules and Theta were going to pledge their new sorority . . . and Sally did not have the mark.

It was an advanced society, in 2113 with all of the comforts – space travel beyond the solar system, teleporting, and no more diseases that the ancients had; instead there were life spans approaching two centuries for the marked people. Humans had it better than ever, and teens had the world in their hands. Everyone with parents that had any sense had a mark on their children to denote their superior genetic lineage. People in the line of descent from genetically modified mitochondria had an "M" on the inside of their ears. Their life expectancy was much higher and their health much better than those without the mark. Finding out about one's mitochondrial DNA was easy, with tests dating back over 100 years to trace the origin of one's genes.

**Mitochondria**

Is the organelle that makes energy for a cell.

Mitochondria are organelles that make energy for a cell; they are inherited from mother to children because they have their own genetic material and divide on their own. Mitochondria are, in fact, separate structures existing within our cells. They were absorbed some 2.5 billion years ago, with their own set of DNA, making them houseguests in our bodies.

**Organelle (subcellular structure)**

Structures that function within cells in a discrete manner

The genes in the mitochondria stay intact from generation to generation. "This is why the mark was so important – the health benefits," thought Theta. Mitochondrial DNA with modified genes of a particular line of mitochondria made people much healthier, free of many diseases in the society of this story. Mitochondria are the meddling houseguests in the title because defects in them cause a range of diseases. For example,

## CHECK UP SECTION

The exclusion of people in our futuristic science fiction story reflects a theme in human society and history. As a result of cell differences between Theta and Sally, their friendship ended — each possessed a different type of mitochondrion.

Choose a particular situation in which a social stratification (layering) system is set up in a society, in which one group thinks it is better than another. You may choose a present system or one of the past. Is the stratification system reasonable? Is the system based on cell biology? What are the system's benefits? What are its drawbacks?

mitochondrial defects in the 21st century were responsible for many ailments, ranging from heart disease and diabetes to chronic sweating, optic nerve disorders, and epilepsy.

Joules told Theta, "People without the mark are jealous of us because they die earlier and have a worse life with more diseases. You know Sally would never understand us. Sally's genes are still from the 21st century." But something still bothered Theta: *She liked Sally*. Sally came into the dorm and Joules explained that they were leaving for the sorority. Sally knew what that meant and said good-bye. Theta looked deeply at Sally, realizing that their past was gone and that they would not see each other again as friends. Sally and Theta both had a single tear in their eyes and they knew they were part of each other's youth . . . and that meant something.

But Theta looked back one last time and said thoughtfully to herself, "She's not one of us."

## Culture, Biology, and Social Stratification

Culture plays an important role in defining what is desirable and valued in society. Often decisions on what it means to be "better" are based on cell biology. Our genetic material makes each of us unique and guides the workings of our cells. We all have the same set of cell structures or organelles, but, as in our story, genetic variations give each person unique characteristics. While the opening story is science fiction, its possibilities are real. Gene technology is improving human health and has the potential to "design" human genes and organelles, possibly leading to social issues like those described in the conflict faced by Sally, Joules, and Theta.

Biological differences may lead to social changes based on what a society values at any one time. For example, research shows that certain biological features are used to decide social value of people: symmetry of one's face, body fat distribution in both genders, and musculature in males; smooth skin, good teeth, and a uniform gait. These are all biologically determined, based on how our cell structures work together. Much as mitochondrial inheritance, described in the story dictates health and organismal functioning, all cell structures give living systems their characteristics.

Historically, all cultures have used biology to classify people. Humans are susceptible to group messages, such as the one that influenced Theta's and Joules' final decision to abandon their friendship with Sally. The average American is exposed to about 3,000 marketing messages per day. This sets up a value system that requires us to reflect on how biology and society can affect our thinking.

## BODY ART AND SKIN BIOLOGY IN SOCIETY

Body alterations in the quest for physical beauty are as old as history. Egyptians used cosmetics in their First Dynasty (3100–2907 BC). Hairstyles, corsets, body-weight goals, and body piercing and tattooing trends have changed through human history. Scars have been viewed as masculine and a mark of courage, and tattoos were drawn and carved in ancient European, Egyptian, and Japanese worlds.

Body art was popular in modern western society among the upper classes in the early 19th century. It lost favor due to stories of disease spreading because of unsanitary tattoo practices. Only the lower classes adopted body art to show group affiliation. Tattooing has recently gained popularity; but body art has been used as a symbol of self-expression and as a social-stratification mechanism in many cultures: Indian tattoos mark caste; Polynesians used marks for showing marital status; the Nazis marked groups from their elite SS to concentration camp prisoners; and U.S. gangs use it to show group membership. Tattooing has been firmly established in societies and continues to grow in popularity in the United States.

The canvas for tattoos is skin, which is part of the integumentary system and has a variety of functions in humans (Figure 3.1). It

- maintains temperature;
- stores blood and fat; and
- provides a protective layer.

We will discuss this important system in a later chapter.

© FXQuadro/Shutterstock.com

**Figure 3.1**    Tattoos and body art. Dyes penetrate into the skin cells of a tattoo.

In this chapter, we will look at the structure and function of the eukaryotic cell. We will see that, while there are marked differences between plant and animal cells, the basic processes carried out at the cellular level are remarkably the same, as are those of simple, unicellular organisms. We will compare the organelles (structures) of the cell to functions of a city to emphasize that all parts are needed. Each organelle has its own duties, and the parts work together to make an efficient machine. We begin by looking at the development of the microscope, without which our understanding of cells and how they function would be incomplete.

# Exploring the Cell

## The Microscope

The human body is composed of over 10 trillion cells, and there are over 200 different types of cells in a typical animal body, with an amazing variety in sizes (see Figure 3.2). Despite the variety in size, all of these cells and the structures within them are too small

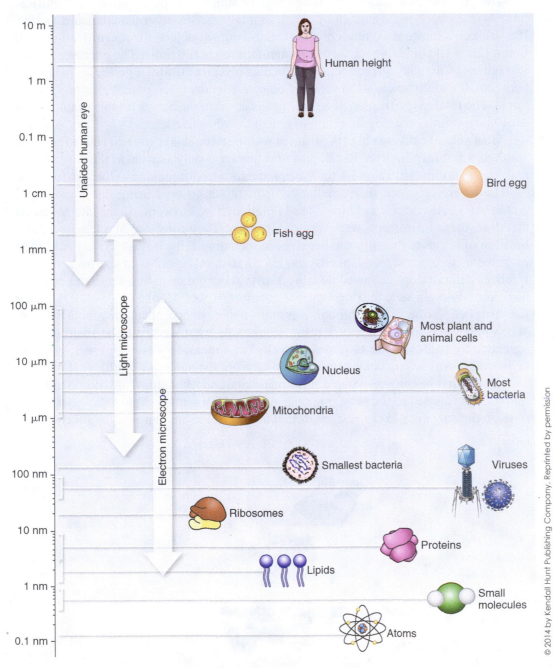

**Figure 3.2**   Biological size and cell diversity. When comparing the relatives' sizes of cells, we use multiples of 10 to show differences. The largest human cell, the female egg, is 100 µm, while the smallest bacterial cell is 1000 times smaller at 100 nm. Most cells are able to be seen with the light microscope. The smallest object a human eye can see is about 1 mm, the size of a human egg cell (or a grain of sand). From *Introductory Plant Science*, by Cynthia McKenney et al.

to be visible to our naked eyes and can only be identified by using microscopes to magnify them.

There are several types of microscopes; perhaps the one with which you are already familiar is the compound light microscope. The compound light microscope uses two lenses: an ocular and an objective lens. Each of these is a convex lens, meaning that its center is thicker than its ends. Convex lenses bring light to a central, converging point to magnify the specimen. A microscope's parts are seen in Figure 3.3.

The purpose of a microscope is to magnify subcellular parts. What is magnification? Magnification is the amount by which an image size is larger than the object's size. If a hair cell's image is 10 times bigger than its original object, the magnification is 10 times. If it is 100 times bigger, then the magnification is 100 times. The microscope uses two lenses to magnify the specimen: an ocular (eyepiece), which generally magnifies between 10 and 20 times, and a series of objective lenses (each with higher magnifications). The total magnification of a specimen is equal to the ocular (in this example let's use 10 times) times the magnification of one of the objective lenses.

Most animal cells are only 10–30 μm in width. It would take over 20 cells to span the width of a single millimeter. Recall that a millimeter is only as wide as the wire used to make a paper clip. See Table 3.1 for measurements used for looking at living structures.

How were cells and their smaller components discovered using the microscope? Anton van Leeuwenhoek and Marcello Malpighi built microscopes in the late 1600s. At this time, those instruments were very rudimentary. They consisted of a lens or a combination of lenses to magnify smaller objects, including cells. Both scientists used their instruments to observe blood, plants, single-celled animals, and even sperm. Van Leeuwenhoek's microscope is shown in Figure 3.4. At about the same time that van Leeuwenhoek and Malpighi were making their observations, Robert Hooke (1635–1703) coined the term *cell*, as he peered through a primitive microscope of his own construction. When he viewed tissues of a cork plant, Hooke saw what seemed to be small cavities separated by walls, similar to rooms or "cells" in a monastery (see Figure 3.4). These cells are defined as functioning units separated from the nonliving world.

Although it has progressed in design, materials, and technology, the compound light microscope is based on the same principle as in the 17th century: light bends as it passes through the specimen to create a magnified image. Some amount of light always bends

**Compound light microscope**

Microscope that uses two sets of lenses (an ocular and an objective lens).

**Magnification**

Is the amount by which an image size is larger than the object's size.

© RTimages/Shutterstock.com

**Figure 3.3**   Compound light microscope – its parts and internal lens system.

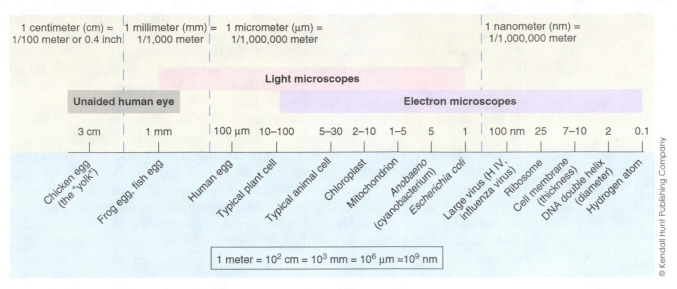

| 1 centimeter (cm) = 1/100 meter or 0.4 inch | 1 millimeter (mm) = 1/1,000 meter | 1 micrometer (µm) = 1/1,000,000 meter | | | | | | | 1 nanometer (nm) = 1/1,000,000 meter | | | | |
|---|---|---|---|---|---|---|---|---|---|---|---|---|---|

| | | | **Light microscopes** | | | | | | | | | | |
|---|---|---|---|---|---|---|---|---|---|---|---|---|---|
| **Unaided human eye** | | | | | | **Electron microscopes** | | | | | | | |
| 3 cm | 1 mm | 100 µm | 10–100 | 5–30 | 2–10 | 1–5 | 5 | 1 | 100 nm | 25 | 7–10 | 2 | 0.1 |
| Chicken egg (the "yolk") | Frog egg, fish egg | Human egg | Typical plant cell | Typical animal cell | Chloroplast | Mitochondrion | Anobaeno (cyanobacterium) | Escherichia coli | Large virus (H IV, influenza virus) | Ribosome | Cell membrane (thickness) | DNA double helix (diameter) | Hydrogen atom |

$$1 \text{ meter} = 10^2 \text{ cm} = 10^3 \text{ mm} = 10^6 \text{ µm} = 10^9 \text{ nm}$$

© Kendall Hunt Publishing Company

**Table 3.1** Measurements Used for Microscopy. The units of measurement used in the study of molecules and cells correspond with methods by which we are able to detect their presences.

when hitting the edges of the lens, causing scattering in a random way. The random scattering of light, called diffraction is bad for getting a clear focus on the image. Diffraction also limits the resolution of the image. Resolution is defined as the ability to see two close objects as separate. (Think about looking at two lines on a chalkboard that is very far away; chances are they blur together and look like one messy line.) In fact, the human eye has a resolving power of about 100 µm or 1/10th of a millimeter for close-up images. In other words, two lines on a paper closer than 1/10th of a millimeter apart look blurry to us. The light microscope is limited in the same way by diffraction because the diffracted rays create blurry images.

**Diffraction**

The random scattering of light.

**Resolution**

Is the ability to see two close objects as separate.

**Figure 3.4** Hooke's microscope from the 1600s and van Leeuwenhoek with his microscope. These simple microscopes led to the first descriptions of cells. Van Leeuwenhoek's microscope consisted of a small sphere of glass in a holder.

Higher magnification under the microscope leads to greater diffraction. This is the reason a compound light microscope can magnify only up to 1000–1500 times (under oil immersion), after which there is too much diffraction for a clear image to be formed. To overcome the effect of diffraction and achieve clarity at higher magnifications, oil is placed on the slide. However, even with oil immersion, only the large nucleus of a cell can be seen; other organelles appear as small dots or not at all.

So how did the more complex world of even smaller structures within cells get discovered? The 1930s saw the development of the electron microscope that allowed for magnifications of over 200,000 times greater than that of the human eye. There are two types of electron microscopes: transmission electron microscope (TEM) and scanning electron microscope (SEM). Transmission electron microscopy allows a resolving power of roughly 0.5 nm (see Table 3.1) that visualizes structures as small as five times the diameter of a hydrogen atom. Electron microscopes use electrons instead of light, which limits diffraction and increases resolution. Magnets instead of lenses focus electrons to create the image. The electrons pass through very thin slices of the specimen and form an image.

A SEM looks at the surfaces of objects in detail, while a TEM magnifies structures within a cell. The SEM has a resolving power slightly less than the TEM, at 10 nm. (A depiction of an electron microscope is shown in Figure 3.5.) Electron microscopy has led to many scientific developments, uncovering subcellular structures to help us understand cell biology. Seeing a mitochondrion enables us to better understand diseases and perhaps, if our opening story becomes reality, improve societal health through its use.

**Transmission electron microscope (TEM)**

A type of electron microscope that magnifies structures within a cell.

**Scanning electron microscope (SEM)**

An electron microscope that looks at the surfaces of objects in detail by focusing a beam of electrons on the surface of the object.

## Cell Theory

Fairly recent advances in microscopy have allowed scientists to learn about the structure and function of even the tiniest components of cells, but the cell theory, which states key ideas about cells, developed a long time ago. We have seen that scientists began studying cells in the early 1700s. About a century later, in 1838, a German botanist named Matthias Schleiden (1804–1881) concluded that all plants he observed were composed of cells. In the next year, Theodor Schwann (1810–1882) extended Schleiden's ideas,

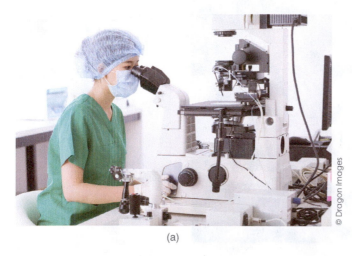
(a)

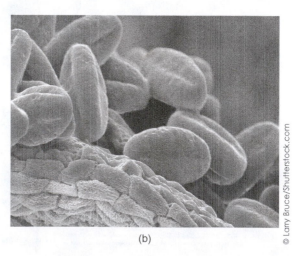
(b)

**Figure 3.5**   a. A researcher sits at a modern electron microscope. b. Apple tree pollen grains on cells, an electron micrograph.

observing that all animals are also made of cells. But how did these cells come to survive generation after generation? The celebrated pathologist Rudolf Virchow (1821–1902) concluded in 1858 that all cells come from preexisting cells (He wrote this in Latin: "*Cellula e cellula*"). These scientists contributed, together, to the postulates of the **cell theory**. The cell theory is a unifying theory in biology that places the cell as the center of life and unifies the many branches of biology under its umbrella. The cell theory states that:

**1)** All living organisms are composed of cells.
**2)** The chemical reactions that occur within cells are separate from their environment.
**3)** All cells arise from other cells.
**4)** Cells contain within them hereditary information that is passed down from parent cell to offspring cell.

The cell theory showed not only that cells are the basic unit of life, but that there is continuity from generation to generation. Genetic material is inherited in what we refer today as the cell.

## Types of Cells

Microscopes allowed researchers to examine differences between organisms that had previously been impossible to determine. A current classification of organisms defines five kingdoms, with organisms in those kingdoms having similar types of cells (There is some debate arguing inclusion of Archaea bacteria as a separate kingdom, and a six-system classification scheme is thus also accepted). Cells of organisms in the five kingdoms each have many internal differences, as summarized in Table 3.2. Images of some organisms of each kingdom are given in Figure 3.17 as examples.

**Prokaryotes** (bacteria) are composed of cells containing no membrane-bound nucleus and no compartments or membranous organelles. They are much smaller than eukaryotes, by almost 10 times. Prokaryotic genetic material is "naked," without the protection of a membrane and nucleus. They are composed of very few cell parts: a membrane, cytoplasm, and only protein-producing units called ribosomes. Even without most structures found in other organisms, prokaryotes contain genetic material to reproduce and direct the functions of the chemical reactions occurring within its cytoplasm.

**Table 3.2**    Differences in Cell Structure within the Five Kingdoms: Plants, Animals and Prokaryotes.

| Group | Domain | Cell Type | Cell Number | Cell Wall Component | Energy Acquisition |
|-------|--------|-----------|-------------|---------------------|--------------------|
| Bacteria | Bacteria | Prokaryotic | Unicellular | Peptidoglycan | Mostly heterotrophic, some are autotrophic |
| Protists | Eukarya | Eukaryotic | Mostly unicellular, some are simple multicellular | Cellulose, silica; some have no cell wall | Autotrophic, heterotrophic |
| Plants | Eukarya | Eukaryotic | Multicellular | Cellulose | Autotrophic |
| Animals | Eukarya | Eukaryotic | Multicellular | No cell wall | Heterotrophic |
| Fungi | Eukarya | Eukaryotic | Mostly multicellular | Chitin | Heterotrophic |

From Introductory Plant Science by Cynthia McKenney et al. Copyright © 2014 by Kendall Hunt Publishing Company. Reprinted by permission.

Prokaryotes have a simple set-up, but all of the needed equipment to carry out life functions. Bacteria have a rapid rate of cell division and a faster metabolism than eukaryotes. Most organisms on Earth, in terms of sheer number, are prokaryotes.

- As indicated in Chapter 1, prokaryotes include organisms in the Bacteria and Archae domains. These organisms will be discussed further in Chapter 8.

All other organisms (plants, animals, fungi, and protists) are **eukaryotes**. Cells of eukaryotes are complex, containing a membrane-bound nucleus that houses genetic material. Eukaryotic cells comprise compartments that form a variety of smaller internal structures, or organelles. Eukaryotic cells are the focus of this chapter, which will give an overview of the primary organelles and their functions (Figure 3.6).

Eukaryotes may be examined by dividing into its four groups: plants, animals, fungi, and protists. Plants contain cells that are surrounded by a cell wall, a rigid structure giving its organisms support. Plant cells contain chloroplasts, which enable plants to carry out photosynthesis, using energy from sunlight to make food.

**Photosynthesis**

The process by which green plants use sunlight to synthesize nutrients from water and carbon dioxide.

- Plant cell walls contain cellulose, which gives structure to plants as discussed in Chapter 2. The process of photosynthesis, producing food for plants, will be further discussed in Chapter 4.

Plants also have large vacuoles or storage compartments to hold water and minerals for a plant's functions. While both plants and animals have a cell membrane, animal cells are

(a)

© 2006 by Kendall Hunt Publishing Company. Reprinted by permission

**Figure 3.6**    a. Differences between prokaryotes and eukaryotes. Prokaryotes have a generally simple structure (see top cell in figure above), while eukaryotes (the lower cell in figure above) have multiple organelles and membranes forming complex compartmentalization. From *Biological Perspectives*, 3rd ed by BSCS. b. Differences between plants and animals. Plant and animal cells perform different functions, and their subcellular structures are also different. Plant cells have chloroplasts to produce sugar and a cell wall to give added strength. The animal cell shown has no cell wall or chloroplasts but possesses centrioles. From *Biological Perspectives*, 3rd ed by BSCS.

**cell membrane:** semifluid cell boundary that controls passage of materials into and out of cell; made of two lipid layers with proteins on the surface and embedded within the layers

**nucleus:** large organelle that is the control center of cell; enclosed by double membrane with pores; contains most of cell's genetic information coded in DNA; contains one or more **nucleoli**, where some RNA is made

**cytoskeleton:** network of proteins throughout cytoplasm that provides internal organization and shape and facilitates movement

**lysosomes:** small organelles containing enzymes that carry out breakdown of ingested material and worn out organelles

**centrioles:** pair of tubular structures that play a role in cell division

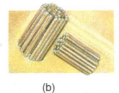

**generalized animal cell**

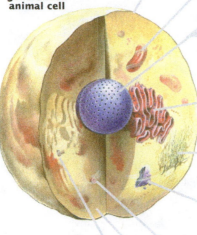

(b)

**Figure 3.6**    (*Continued*)

**cell wall:** outer boundary of plant cells that provides rigidity; formed of cellulose fibers

**mitochondria:** small organelles that are the site of energy-releasing reactions in most cells; enclosed by double membrane with inner membrane much folded

**generalized plant cell**

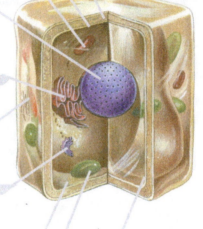

**endoplasmic reticulum:** tubular membrane system that compartmentalizes the cytoplasm and plays a role in the synthesis of various macromolecules

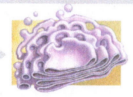

**Golgi apparatus:** system of flattened sacs that modifies, sorts, and packages macromolecules for secretion or for delivery to other organelles

**cytoplasm:** semifluid material surrounding organelles; site of many cellular reactions

**chloroplasts:** small organelles in plant cells that are the site of all reactions of photosynthesis; enclosed by double membrane; third, inner membrane forms layered structures

**vacuole:** large organelle in plant cells that stores nutrients and waste products; may occupy more than 50 percent of volume in plant cells

(b)

**Figure 3.6** (*Continued*)

### Centriole

Minute cylindrical organelles found in animal cells, which serve in cell division (not given in bold in text).

less rigid, surrounded only by a cell membrane and lacking a cell wall for support. Both plants and animals contain membrane-bound organelles, but animals also contain a set of small structures called centrioles, which serve in cell division. Animal cells are also quite complex, as we will see. While lacking certain organelles, such as cell walls and chloroplasts, they have flexible strategies to perform many functions.

Fungi have cell walls but no chloroplasts. They are not able to make their own food and, instead live off of dead and decomposing matter as well as other living organisms, to obtain energy. Mushrooms and yeasts are familiar types of fungi, which will be discussed in Chapter 7.

Some species of protists are a bit animal-like in that they are able to move; other species are a bit plant-like in that they have chloroplasts. Protists such as *Amoeba* in Figure 3.7 have varied environments. *Amoeba* live in freshwater and, in a rare infectious disease, grow and destroy human brain cells. We will discuss protists in more detail in a later chapter.

The kingdom **Plantae** contains the plants—multicellular organisms composed of eukaryotic cells that live by manufacturing high-energy organic molecules through the process of photosynthesis. Included are **r** moss (*Polytrichium*); **s** club moss (*Lycopodium*); **t** horsetail (*Equisetum*); **u** grape fern (*Botrychium*); **v** pine (*Pinus*); **w** corn (*Zea*); **x** Jacob's ladder (*Polemonium*); **y** tiger lily (*Lilium*).

The kingdom **Fungi** includes eukaryotic, typically multicellular, organisms such as mushrooms and molds that live by absorbing organic molecules from the environment. Examples shown are **z** yeast (*Saccharomyces*); **aa** bread mold (*Rhizopus*); **bb** morel (*Morchella*); **cc** mushroom (*Coprinus*).

The kingdom **Animalia** consists of animals—multicellular organisms composed of eukaryotic cells that live by consuming other organisms as their source of organic molecules. We show **dd** sponge; **ee** purple jelly; **ff** planarian (*Dugesia*); **gg** earthworm (*Lumbricus*); **hh** snail; **ii** viceroy (*Limenitis*); **jj** brittle star; **kk** trout (*Salmo*); **ll** turtle (*Gopherus*); **mm** mourning dove (*Zenaida*); **nn** rabbit (*Silvilagus*).

The kingdom **Protista** (sometimes called **Protoctista**) includes all eukaryotic organisms that are not classified as plants, animals, or fungi. Protists range in size from microscopic, one-celled organisms such as *Paramecium* to large, multicellular organisms such as the seaweed kelp. Multicellular algae like the kelp are usually classified as protists because they do not develop from embryos surrounded by parental tissues. Many biologists, however, feel they should be classified in the kingdom Plantae. Shown here are two green algae **g** (*Chlamydomonas*) and **h** (*Pediastrum*); **i** diatom (*Stephanodiscus*); **j** brown alga (*Alaria*); **k** red alga (*Polysiphonia*); **l** slime mold (*Physarum*); **m** entamoeba (*Amoeba*); **n** flagellate (*Euglena*); **o** dinoflagellate (*Gonyaulax*); two ciliates, **p** (*Spirostomum*) and **q** (*Euplotes*).

The kingdom **Monera** (now usually called **Prokaryotae**) contains the bacteria. Bacteria are one-celled microorganisms whose cell structure is prokaryotic. The examples shown are two gram-negative bacteria **a** (*Pseudomonas*) and **b** (*Spirullum*); **c** sulfur bacterium (*Desulfovibrio*); **d** cyanobacterium (*Oscillatoria*); **e** purple bacterium (*Rhodomicrobium*); **f** gram-positive bacterium (*Micrococcus*).

**Figure 3.7** Cells of the five kingdoms. While the cells of organisms in all of the kingdoms perform similar life functions, their individual structures enable differing functions unique to each kingdom. From *Biological Perspectives*, 3rd ed by BSCS.

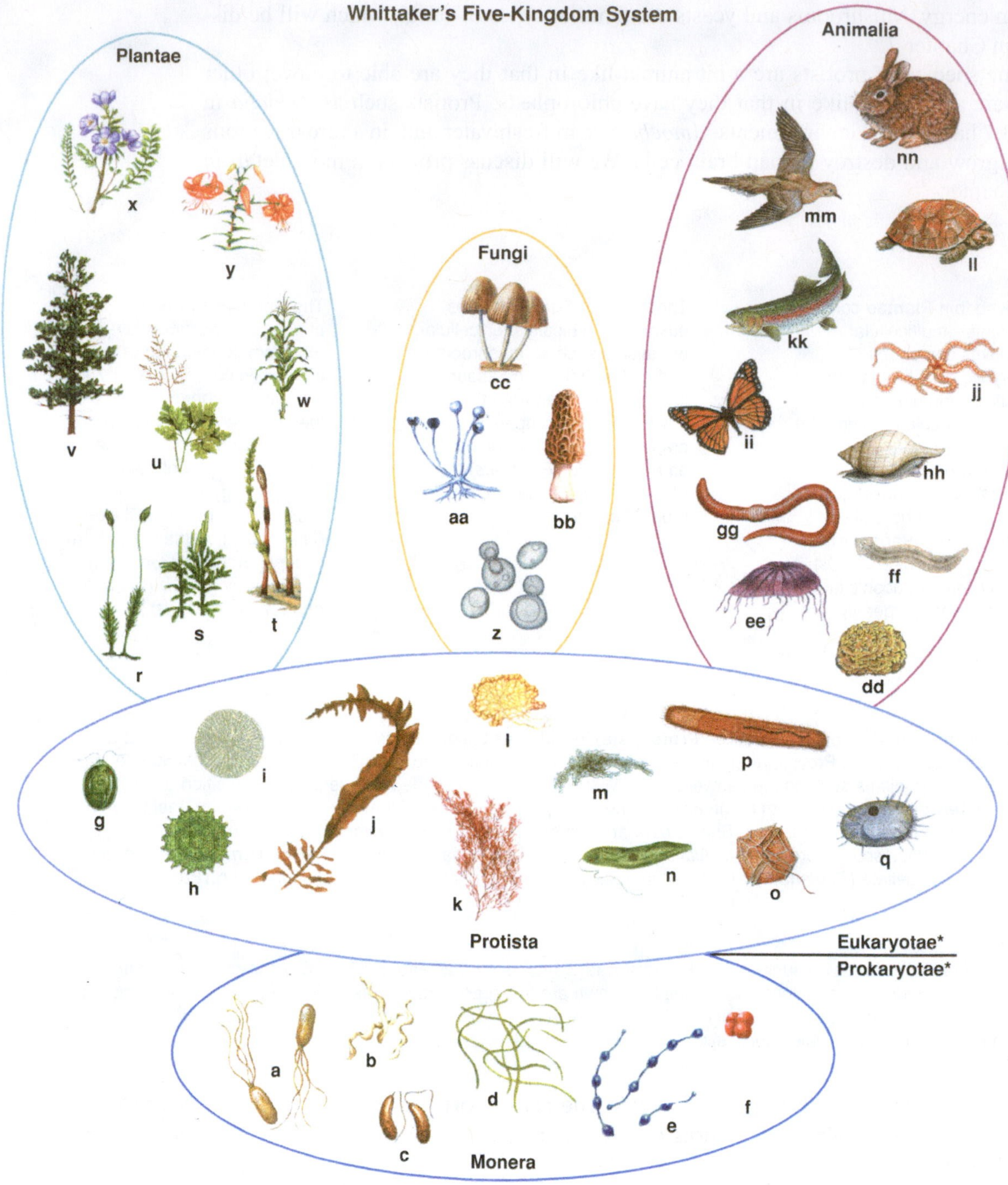

**Whittaker's Five-Kingdom System**

*****Prokaryotae** have prokaryotic cells that lack a membrane-enclosed nucleus, whereas **Eukaryotae** have eukaryotic cells that *do* contain a membrane-enclosed nucleus. Note that all of the bacteria (Kingdom Monera) are prokaryotic; all other organisms are eukaryotic.

**Figure 3.7** *(Continued)*

# The Role of Inheritance

The stratification system depicted in our opening story is based on the inheritance of cellular components. We know that organelles are structures that carry out functions within a cell. In fact, organelles work in concert with one another, coming together to

## THE OXYGEN REVOLUTION

Have you ever tried to imagine the Earth in its early stages? After millions of years during which the Earth was a mass of molten gases, those gases began to cool into layers that became landmasses, while water vapor formed seas. There were as yet no animals and only a few prokaryotic forms living in the waters. One form of bacteria, known as cyanobacteria, is believed to have been among the first organisms to photosynthesize (convert light energy into chemical energy to drive cellular activities). Since a by-product of many forms of photosynthesis is oxygen, over several million years, more and more oxygen was released into the atmosphere – the oxygen revolution.

The oxygen revolution occurred, according to geologists, about 2.5 billion years ago. It led to an availability of oxygen that could be used by cells that evolved to use it. The advantage of using oxygen to yield larger amounts of energy led to an increase in the number of aerobic cells. Aerobic cells, or those cells that use oxygen as the fuel for obtaining energy, contain mitochondria to provide large amounts of energy production. Over the course of many millions of years, single-celled eukaryotes, with more complex cellular processes than prokaryotes, evolved, then multicellular eukaryotes, and eventually organisms became larger and more complex. It is important to realize that oxygen is a key player in the development of organisms since it can be used to produce fast, plentiful energy.

**Oxygen revolution**

The biologically induced appearance of dioxygen in Earth's atmosphere 2.5 billions of years ago.

**Aerobic**

Occurring in the presence of oxygen or require oxygen to live.

form a complex, dynamic cell. Mitochondria, so important in the society in our story, are the powerhouses of the cell, providing energy for a cell's functions.

All organelles are built and controlled by inherited genetic material. Thus, the way a cell works is based upon its genetics. But some organelles are inherited separately from the others. There are three ways to inherit organelles, including mitochondria: 1) maternal inheritance, in which organelles are inherited from mothers; 2) paternal inheritance, in which organelles are inherited from fathers; and 3) bi-parental inheritance, in which organelles are inherited from both mothers and fathers. The inheritance type varies among species and for different organelles. For example, chloroplasts are paternally inherited in the giant redwood *Sequoia* but maternally inherited in the sunflower *Helianthus*. Most animal species have maternal transmission of mitochondria, as seen in the story, because female eggs hold most of the mitochondria in their large cytoplasmic cells. Sperm contributes very little cytoplasm or organelles in human species, although there are exceptions among other organisms; for example, green algae *Chlamydomonas* has paternal transmission of mitochondria.

## Endosymbiosis

Eukaryotes appeared in Earth's history about 1.5 billion years after prokaryotes. In their 2.0 billion years on Earth, eukaryotes have evolved into living systems that range from butterflies to beavers, crocodiles to humans. As you progress through this text and the course, you will learn about how this amazing diversity evolved.

Eukaryotes have two types of organelles – those that evolved as membranes and those from other, simpler organisms as precursors, called endosymbionts. Endosymbionts

**Endosymbionts**

Any organism living in the body or cells of another organism.

have their own genetic material and are semi-independent within the cells of eukaryotes. Endosymbionts include two types of organelles – mitochondria and chloroplasts. Mitochondria are the energy producers of animal cells, and chloroplasts are solar power transformers of plant cells. Mitochondria divide independently and have an internal environment that is different from that in the rest of the cell.

Evidence tells us that endosymbionts were once independent prokaryotes that were somehow incorporated into eukaryotic cells. Lynn Margulis, a well-known evolutionary biologist, formulated the endosymbiotic theory, which states that some organelles in eukaryotes were descendants of ancient bacteria that were absorbed by larger cells. (*Symbiosis* refers to a mutually beneficial relationship of organisms living together; *endo* means within.) The larger cell gave ancient bacteria absorbed by larger cells a "home." The bacteria received protection and in return gave their unique set of chemical reactions to the larger host cell.

Analysis of endosymbionts have particular uses in today's society: Crime scene investigations can test for mitochondrial DNA; ancestry can be traced using mitochondrial DNA; and research on diseases inherited due to faulty mitochondrial genetic material may yield medical treatments. In our story, Joules and Theta had a different form of mitochondria than Sally. Joules, Theta, and Sally all inherited their mother's mitochondria, but Sally's inheritance included a predisposition for a number of diseases.

We have said that mitochondria can be thought of as the power plants of the cell. We say this because mitochondria carry out the series of energy-producing reactions that convert food energy into ATP (defined in Chapter 2; the form of energy used to drive cell functions). This set of reactions is called cell respiration.

Chloroplasts may have originated from cyanobacteria, which were able to transform light energy into usable sugar for energy in the process called photosynthesis – the making of food (glucose) from sunlight, carbon dioxide, and water. Thus, precursor mitochondria provided instant ATP energy for animal host cells, and ancient chloroplasts made stored glucose available for longer-term use in plants. Modern chloroplasts use sunlight to rearrange carbon to form food in the form of sugars, for cell usage. They are the solar power plants of cells because they trap sunlight and generate ATP energy.

- These energy-obtaining processes, cell respiration, and photosynthesis will be discussed in greater detail in Chapter 4.

Evidence for mitochondria and chloroplasts as endosymbionts is considerable:

1) Mitochondria and chloroplasts are similar in size and shape to bacteria, roughly 7 μm in length.
2) Both contain their own genetic materials and divide in the same way as prokaryotes, through binary fission (splitting in half).
3) Both contain the same type of 70S ribosomes (described below; small organelles that make protein) as bacteria, whereas eukaryotes contain 80S ribosomes.
4) Mitochondrial and chloroplast DNA are more related to bacterial than to eukaryotic DNA. See the proposed process on endosymbiosis in Figure 3.8.

We will now look at each of the structures of the cell, using the analogy of a city, with the organelles being structures critical to the smooth functioning of the "city." Organelles work in concert with each other to carry out life functions. Chloroplasts and mitochondria carry out key energy-producing and releasing functions to drive cell activities. However, there is an intricacy to cell biology akin to the workings of a large and dynamic city.

**Endosymbiotic theory**

The theory that states that some organelles in eukaryotes were descendants of ancient bacteria that were absorbed by larger cells.

**Cell respiration**

A series of energy-producing reactions that convert food energy into ATP.

**Chloroplast**

A part of plant that contains chlorophyll and conducts photosynthesis.

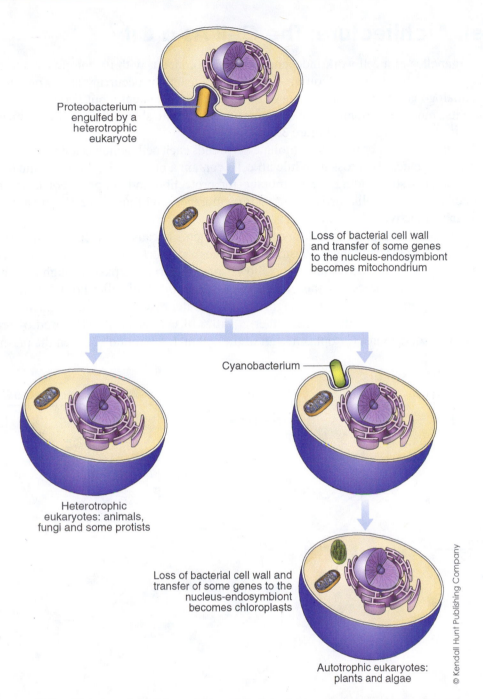

Proteobacterium engulfed by a heterotrophic eukaryote

Loss of bacterial cell wall and transfer of some genes to the nucleus-endosymbiont becomes mitochondrium

Cyanobacterium

Heterotrophic eukaryotes: animals, fungi and some protists

Loss of bacterial cell wall and transfer of some genes to the nucleus-endosymbiont becomes chloroplasts

Autotrophic eukaryotes: plants and algae

© Kendall Hunt Publishing Company

**Figure 3.8**   This model shows how mitochondria and chloroplasts wound up in eukaryotic cells. Evidence for endosymbiosis. Chloroplasts and mitochondria are similar in size and shape to bacteria. The ribosomes in bacteria and mitochondria and chloroplasts are "70S." Most bacteria have a size of roughly 7–10 μm and 70S ribosomes, both characteristics are similar to mitochondria and chloroplasts.

# Cell Architecture: The Cell As a City

Plasma (cell)
membrane

A biological membrane
that separates the
cell's interior from the
outside environment.

Selectively
permeable

A condition in which
the membrane allows
some materials to pass
through cells but not
others.

Cytoplasm

A semisolid liquid
that holds organelles
suspended within it.

The organelles of a cell work independently but in concert, with its components inter-acting actively, having many thousands of chemical reactions occurring at any one time. In an analogous way, the components of a city work independently but cooperatively to ensure its smooth functioning. The structures in the cell are shown in the cartoon image depicting the cell as a city in Figure 3.9.

A cell membrane or plasma membrane surrounds each cell, which is a wrapping that allows some materials across it. While all cells contain a plasma membrane, some cells have structures surrounding the membrane for protection and support. For example, plant cells have cell walls surrounding their membranes, and fungi have chitin barriers, which are protective polysaccharides.

Plasma membranes are important to living systems because they control the mate-rials entering and leaving them. Plasma membranes are selectively permeable. Selective permeability means that the membrane allows some materials to pass through cells but not others. Within the cell is the cytoplasm, a semisolid liquid that holds organelles suspended within it. Cytoplasm is roughly 60–80% water by volume, and chemical reac-tions occur within its medium. The other 20–40% of cytoplasm is composed of pro-teins and dissolved ions. Cytoplasm is all of the cell material found between the plasma

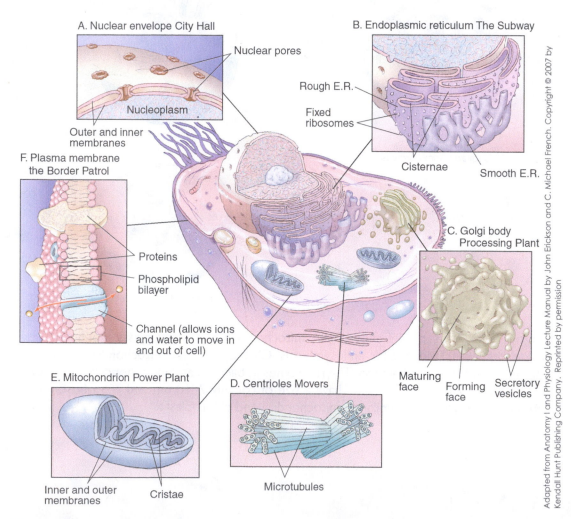

A. Nuclear envelope City Hall
Nuclear pores
Nucleoplasm
Outer and inner membranes

B. Endoplasmic reticulum The Subway
Rough E.R.
Fixed ribosomes
Cisternae
Smooth E.R.

F. Plasma membrane the Border Patrol
Proteins
Phospholipid bilayer
Channel (allows ions and water to move in and out of cell)

C. Golgi body Processing Plant
Maturing face
Forming face
Secretory vesicles

E. Mitochondrion Power Plant
Inner and outer membranes
Cristae

D. Centrioles Movers
Microtubules

Adapted from Anatomy I and Physiology Lecture Manual by John Erickson and C. Michael French. Copyright © 2007 by Kendall Hunt Publishing Company. Reprinted by permission

**Figure 3.9**   A cell  is like a city. Its parts work together to perform a cell's function.

membrane and the nucleus. A nucleus, or control organelle, serves as the city hall of the cell, usually centered within cytoplasm. It contains the genetic material that produces and controls the cell's parts.

A cell's architecture is very complex, with continual activity within many compartments of each cell. Compartmentalization of a cell is accomplished by a series of membranes throughout its cytoplasm. These membranes allow for a surface to perform chemical reactions such as those carried out by enzymes. The layers of membranes also separate different chambers of the cell so that conditions may be different within each chamber. An acidic pH in one compartment, for example, may be suitable for cell functioning in one section, while a basic pH might be needed in another section. Throughout the cell, ropes (such as collagen and elastin) and membranes maintain a cell's organization. Figure 3.9 gives you an idea of the complex architecture of the cell.

## Plasma Membrane: The "Flexible" Border Patrol

A border surrounds all cells: in eukaryotic animal cells, the border is the plasma or cell membrane. This dynamic covering is very complex, with parts that continually move to allow certain materials into the cell and keep other substances out. The plasma membrane is composed of a **phospholipid bilayer** in and around which are **membrane proteins** (see Figure 3.10).

- Recall from Chapter 2 that phospholipids are molecules with a phosphate (hydrophilic) head and two tails made of carbon and hydrogen atoms.

The phospholipids are arranged as a bilayer, with the heads facing to the outside and tails to the interior of the cell. The membrane proteins are part of the plasma membrane border, moving in between cholesterol molecules and phospholipids. Although it may seem that they are arranged randomly, in fact, they form a pattern. The membrane is often referred to as a fluid mosaic (see Figure 3.11) because it is made of different pieces that form a pattern and seem to float and move in the watery environment. Note the phospholipid bilayer in Figure 3.11.

There are two types of membrane proteins suspended within the phospholipid bilayer: integral proteins, which span the entire lipid bilayer; and peripheral proteins, which station either inside or outside of the membrane. Integral proteins are also known as **transmembrane proteins**. These membrane proteins serve to anchor cells to each

**Nucleus**

The central and the most important part of a cell and contains the genetic material.

**Compartmentalization**

The formation of cellular compartments.

**Fluid mosaic model**

A model that describes the structure of cell membranes.

**Integral protein (transmembrane or carrier protein)**

A type of membrane protein permanently embedded within the biological membrane (not given in bold in text).

**Peripheral protein**

Is a protein that adheres only temporarily to the biological membrane with which it is associated.

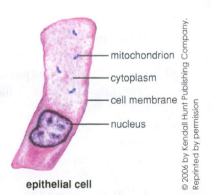

mitochondrion

cytoplasm

cell membrane

nucleus

© 2006 by Kendall Hunt Publishing Company.
Reprinted by permission

**epithelial cell**

**Figure 3.10** Every living cell is surrounded by a cell membrane. From *Biological Perspectives*, 3rd ed by BSCS.

other, move materials, much like a ferry, across the membrane, receive chemicals such as hormones, and transport ions through pores within them. These proteins orient themselves according to the bonds they make with surrounding chemicals.

Integral proteins in the fluid mosaic model serve four functions:

**Receptor**

A structure of the cell's surface that selectively receives and binds a specific substance.

**Recognition protein**

A protein type functioning as binding site for hormones.

1) As receptors, allowing chemicals to bind to cell surfaces. Insulin, a hormone-regulating blood sugar, binds to cells to increase the absorption of sugar from blood into cells. Insulin's special shape matches to the shape on integral proteins. The docking of the two elicits chemical reactions within the cell to maintain sugar balance;

2) As recognition proteins, giving the immune system a code that informs it that a cell is its own and not a foreign substance. Cells with the wrong recognition proteins are rejected, as occurs in cases of organ transplants whose codes do not match up with those of the recipient;

3) As **enzyme surfaces**, facilitating various chemical reactions occurring on the surface of a cell. Enzymes assist in the digestion of certain nutrients, the first step in allowing nutrients to be absorbed; and

4) As **transport proteins**, moving material across a membrane.

Transport proteins act as a sort of flexible border-patrol system, which allows some materials to pass through the membrane while keeping others out. This border-patrol system is an important part of the cell as a city. The size, shape, and chemistry of substances attempting to move into and out of cells determine which materials pass through the plasma membrane. Fatty materials pass through the membrane easily. For example, ethanol in alcoholic drinks easily passes through membranes because it also dissolves in fats. It is absorbed through the phospholipid bilayer and quickly giving a "high" to a person after drinking. The interior of the lipid bilayer is hydrophobic, so it avoids water and dissolves other substances that avoid water. Thus, only other hydrophobic, usually fatty or fat-soluble, materials may pass easily through the bilayer.

Integral proteins of the membrane allow certain nonfatty materials, including smaller charged or polar particles to move through the lipid bilayer directly. These include chemicals such as oxygen ($O_2$), carbon dioxide ($CO_2$), and ammonia ($NH_3$). The quick movement of these materials is essential for life functions. For example, $NH_3$, a nitrogenous waste that builds up within all cells, needs to be rapidly removed.

Thus, integral membrane proteins serve to facilitate movement of materials that cannot easily pass across the membrane. Integral proteins move nonfatty materials across the membrane, such as sodium ions $Na^+$ and potassium ions $K^+$. Larger polar molecules, such as amino acids and glucose, also move through polar (or charged) channels within integral proteins. Channels that are polar are thus hydrophilic, allowing materials with a charge to pass through. Some substances use carrier proteins as pumps for transport, as is the case for the very important ions sodium ($Na^+$), potassium ($K^+$), and calcium ($Ca^{+2}$).In the case of the movement of water ($H_2O$), the most abundant chemical in living systems, it was recently determined from studies on **aquaporins**, or integral proteins with channels within them, that water presses its way through the bilayer using integral proteins. Sometimes water moves directly across the lipid bilayer by "wiggling" type motion to press itself through. Physical forces drive this movement.

As shown in Figure 3.11, short carbohydrate chains jut out from the plasma membrane. These serve as a recognition code for the immune system of animals. Embedded within the phospholipid bilayer of animal cells, cholesterol binds together molecules, helping to maintain the flexibility and motion of the membrane.

Consider for a moment, how many different pieces make up the plasma membrane, and you can see why it is called a mosaic. It is called a fluid mosaic because it

Plasma Membrane Structure

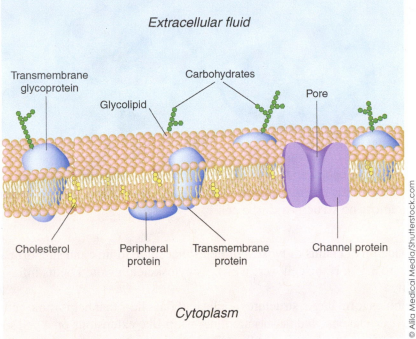

**Figure 3.11** Cell membrane structure. The cell membrane is composed of a phospholipid bilayer with proteins embedded. Phospholipids are composed of two parts – a phosphate head which attracts to water (hydrophilic) and fatty acid tails which repel water (hydrophobic). Proteins are embedded within the lipid bilayer and each type performs a specific function.

is constantly moving, docking proteins, transporting materials, and changing its shape with the movements of a cell. Membranes are also found throughout the cell's structure, with transport occurring continually across different areas. The types of transport will be discussed later in the chapter. First, let's explore the structure surrounding plasma membranes in some organisms.

## Walls of the City

What other borders exist around cells? Cell walls are found outside of plasma membranes in plants, fungi, prokaryotes, and many protists, particularly in algae. For example, single-celled algae known as dinoflagellates acquire their shape from their cell walls. Cell walls give protection to the material within cells and add to their structure. Cellulose, composed of large combinations of glucose units, polysaccharides, and lignin (a type of chemical cement), makes up the stiffening agent of plant cell walls. While plants contain cellulose, fungi cell walls contain chitin, also found in insect bodies and shells of some marine organisms. Cell walls are not found in animal cells, which have no need for a stiff structure that would prevent movement and flexibility. The cell walls of plants are shown in Figure 3.12.

## Cytoskeleton: The City Scaffolds

The shape of a cell is also determined by the skeleton within a cell's cytoplasm, called its cytoskeleton. You may be surprised to learn that the cell itself has a skeleton made up of three types of fibers: microtubules, microfilaments, and intermediate fibers. All three

**Microtubule**

Is a larger filament structure that helps whole cells move.

**Microfilament**

A cytoskeletal fiber used for muscle movement.

**Intermediate fiber**

The smallest fibers of the cytoskeleton, which circulate materials within a cell.

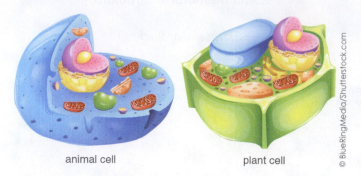

animal cell                    plant cell

© BlueRingMedia/Shutterstock.com

**Figure 3.12**    Plant cell wall. A rigid cell-wall composed of lignin and other stiffening materials maintains structure in plants. Animal cells have only a thin cell membrane, which gives almost no structure to the cell.

**Cilia**

Are short extensions that help cells move.

**Flagella**

Are long, whip-like extensions on a cell's surface that help in the movement of cells.

**Intracellular transport**

A process in which microfilaments circulate materials within cells.

fibers are made up of chains of twisted proteins that serve to link and support parts of a cell (see Figure 3.13). If the cell were a city, its cytoskeleton would be its infrastructure or scaffolding.

Microtubules form larger structures – cilia and flagella – that help whole cells move. Cilia are short extensions, and flagella are long, whip-like extensions on a cell's surface. Many types of organisms use cilia or flagella. A particular bacterium, *Proteus mirabilis*, has numerous and extravagant flagella arrangements. Flagella are found in humans only on sperm cells, to propel sperm toward their goal: the female egg.

Microfilaments, another cytoskeletal fiber, are used for muscle movement. Muscles contract with the help of microfilaments moving in a sliding motion. Microfilaments also circulate materials within cells, in a process called intracellular transport. Intracellular transport moves materials around the cell in a wavelike motion, adding

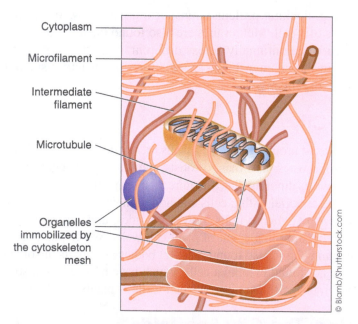

Cytoplasm

Microfilament

Intermediate filament

Microtubule

Organelles immobilized by the cytoskeleton mesh

© Blamb/Shutterstock.com

**Figure 3.13**    The internal skeleton of the cell. The cytoskeleton is composed of three types of fibers: microtubules, microfilaments, and intermediate fibers. Cell parts are held in place by the cytoskeleton and chemical reactions that occur on its scaffolding. The cartoon image in the figure shows the cytoskeleton of an animal cell.

efficiency to internal transport. Microfilaments greatly aid how materials are moved inside cells.

Intermediate fibers, the third type of cytoskeletal fiber, join cells together, especially for muscles such as the heart, where there is a great deal of pressure. Intermediate fibers are made of fibrous proteins, whereas, the other cytoskeleton fibers are composed of globular proteins.

- Fibrous proteins are not easily disassembled as described in the protein section of Chapter 2.

All three fibers anchor organelles in place, such as the mitochondria discussed in our opening story, within the cytoplasm. The cytoskeleton gives a solid framework on which chemical reactions take place. Some chemicals move along their fibers from one location in a cell to another. But overall, the cytoskeleton is not a permanent structure, as the name "skeleton" implies. Much like the plasma membrane, the cytoskeleton changes and reforms with the needs of a cell.

## Nucleus: A City's City Hall

The most important membranous organelle is the nucleus, the control center of the cell. In our analogy of the cell as city, the nucleus is City Hall. The source of its leadership is DNA (deoxyribonucleic acid), the master planner for the organism. The nucleus is especially protected more than any of the other organelles, guarding its vital components. Thus, it is composed of a double membrane, called the nuclear envelop, with special pores or holes that have an octagonal shape and are highly selective. Only certain materials may pass through the nuclear membrane because the pores are very narrow.

In order to communicate with the cytoplasm and the outside world, DNA in the nucleus sends its messenger chemical RNA (ribonucleic acid) out through nuclear pores. RNA directs the synthesis of proteins, which in turn controls cell functions and the building of cell structures. A nucleus is shown in Figure 3.14.

In most organisms, DNA is inherited from both father and mother through genetic material stored and transmitted from nucleus to nucleus. Unlike in our story, which deals with the maternal lineage of mitochondrial DNA, nuclear DNA is obtained from the egg of one's mother and the sperm of one's father in sexually reproducing organisms. When cells are not dividing, DNA is coiled around **histone proteins**, which are a group of five small round proteins bound together with DNA in eukaryotes, together forming chromatin (Figure 3.15).When a cell divides, its DNA coils more tightly to histones, forming a shorter and thicker mass called chromosomes.

### Nuclear envelop

The double membrane that protects the nucleus.

### Chromatin

Is a complex of macromolecules found in cells and consist of protein, RNA, and DNA.

### Chromosome

A thread-like structure formed when a cell divides and its DNA coils more tightly to histones.

## Ribosomes, the City's Factory

The nucleus also contains large bodies called nucleoli (see Figure 3.16), which are special regions of DNA that clump together and produce small, spherical organelles known as the ribosomes. Ribosomes are made up of **RNA**. Ribosomal RNA directs the production of proteins, guided by the RNA message sent out of the nucleus from DNA. Ribosomes are the protein factories of the cell and thus you might think of them as the manufacturing plants of the cell city. RNA carries a message that is translated on the ribosome. This message gives the type of protein ordered by DNA in the nucleus. These processes of protein manufacturing will be discussed further in a later chapter.

All organisms require proteins and thus all living systems contain ribosomes. Bacterial ribosomes are lighter than eukaryotic ribosomes, but all ribosomes make proteins in the form of strings of amino acids. Proteins comprise so many cell activities and structures that ribosomes are essential organelles found in all living organisms.

### Nucleoli

A small, dense round structure found in the nucleus of a cell.

### Ribosome

Small, spherical organelle that is the site protein synthesis.

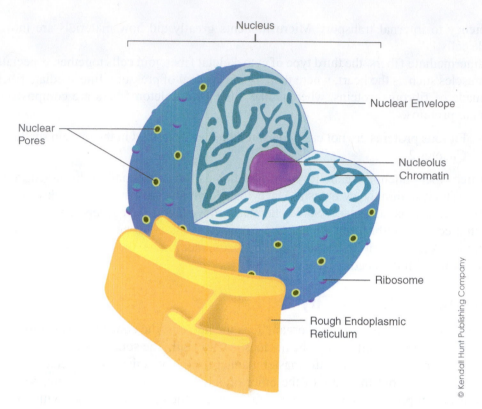

Nucleus

Nuclear Envelope

Nuclear Pores

Nucleolus
Chromatin

Ribosome

Rough Endoplasmic
Reticulum

© Kendall Hunt Publishing Company

**Figure 3.14** The cell nucleus and its components. The nucleus is the control center of a cell. It communicates with the rest of the cell by directing the production of proteins. Proteins serve many roles, particularly in chemical reactions. Vesicles transport materials from the nucleus to other parts of the cell and to the exterior through its endoplasmic reticulum.

© BlueRingMedia/Shutterstock.com

**Endoplasmic reticulum (ER)**
───────
Is a system of interconnected membranes that form canals or channels throughout the cytoplasm of a cell.

**Figure 3.15** DNA coils around histone proteins, with large amounts of genetic materials becoming packaged into very small units.

## Endoplasmic Reticulum: A City's Subway

Ribosomes may be found freely within the cytoplasm or attached to membranes of other organelles. One organelle that often has ribosomes on its surface is the endoplasmic reticulum or **ER** for short. You can think of the ER as the subway system of the cell city.

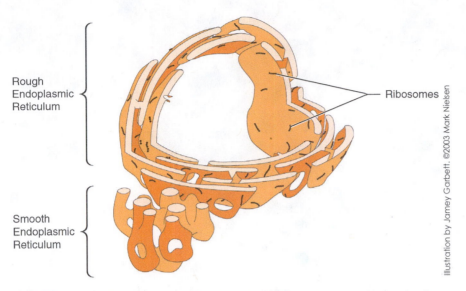

Rough
Endoplasmic
Reticulum

Ribosomes

Smooth
Endoplasmic
Reticulum

Illustration by Jamey Garbett. ©2003 Mark Nielsen

**Figure 3.16** Smooth and rough ER. Rough ER has ribosomes attached to it, and smooth ER does not have ribosomes attached. Both act as the subway of the cell, transporting materials.

The ER is a system of interconnected membranes that form canals or channels throughout the cytoplasm of a cell. Materials are transported rapidly through the ER. Ribosomes found on ER surfaces dump newly produced proteins into the channels, allowing them to be transported to areas of a cell that need them. ER with ribosomes attached is called **rough ER** because it is rough in appearance. ER without ribosomes attached is termed **smooth ER** due to its undotted appearance. Figure 3.16 shows images of both smooth and rough ER

Rough ER is found in cells that export large amounts of protein, such as pancreatic cells that produce the protein-based hormone insulin. Smooth ER is a main component of organs that produce lipids, because it has enzymes that help in their manufacture. Organs (both testes and ovaries) that produce lipid-based sex hormones have smooth ER. Cells that manufacture and export lots of fatty material have elongated smooth ER.

## Golgi Apparatus: A City's Processing Plant

When materials are transported out of the ER, they are often not fully processed and need the addition or removal of parts. The Golgi apparatus is the processing plant of the cell city that refines the materials passing through it. The Golgi apparatus consists of a series of four to six flattened sacs, resembling a stack of pancakes. The Golgi apparatus is shown in Figure 3.17.

As processing occurs, the Golgi apparatus rearranges bonds, adds carbohydrates, or places a lipid on materials moving through. For example, the Golgi apparatus adds phosphate (glycerol) heads to lipid chains to form the phospholipids that make up plasma membranes. At the end of Golgi processing, vesicles or sacs are pinched off the export side. Vesicles are now ready for transport and contain needed materials for either within or outside of the cell. Certain vesicles are also made into another organelle, called the lysosome.

**Golgi apparatus**

Is the processing plant of the cell city that refines the materials passing through it.

Illustration by Jamey Garbett. ©2003 Mark Nielsen

**Figure 3.17**    The Golgi apparatus processes materials such as a phospholipid as they pass through its sacs. A phosphate is added to a lipid molecule to produce a phospholipid for the plasma membrane. The golgi apparatus is the purifier of substances within a cell.

## Lysosomes: A City's Police Officer

**Lysosome**

A small sac filled with digestive, hydrolytic enzymes enclosed in a membrane.

The lysosome, a small sac filled with digestive, hydrolytic enzymes, may be thought of as the cell's police officer.

- The term *hydrolytic* refers to hydrolysis, the breakdown of substances (described in Chapter 2).

Lysosomes are found throughout the cytoplasm. They act as defenders of cells, just like police officers do. When invading microbes enter cells, lysosomes fuse with the "enemy" to digest it. Their contents break down materials within cells, and thus they are responsible for intracellular digestion. Consider the *Amoeba* (Figure 3.18) that is taking in food. Lysosomes act as its entire digestive system, fusing with a food particle and breaking it down.

**Intracellular digestion**

The process of breakdown of substances within the cytoplasm of a cell.

Thus lysosomes are the police officers of the cell city, continually breaking down those substances that need to be removed and cooperating with the immune system to defend against invaders.

Lysosomes also play a critical role in reorganizing material. For example, did you know that our fingers and toes were fused together, or webbed, during our fetal development? Lysosomes digest the tissue between our digits, allowing fingers and toes to form. In roughly 1% of the population, lysosomes fail to fully digest this tissue, and fused toes or fingers are the result.

Tadpoles lose their tails during metamorphosis to adult frogs due to lysosome action that digests tails. Normal cell aging, or senescence, is a result of deterioration of lysosomes, according to one hypothesis. The deterioration of lysosomes releases their enzymatic contents, thereby digesting our cells from the inside out. Lysosomes normally have a double membrane, similar to the nucleus, giving double protection that limits such a disaster. The double membrane, however, cannot stand the effects of time, resulting in membrane failure and aging.

**Figure 3.18** Amoeba digesting food. The amoeba surrounds its prey, and lysosomes digest it.

There are 30 known inherited human diseases associated with the abnormal functioning of lysosomes. These disorders are called lysosome-storage diseases because in each case lysosomes lack a particular enzyme that breaks down substances that would normally be removed. These waste products build up in lysosomes, accumulating in cells. Tay–Sachs disease, commonly found in Ashkenazi Jewish populations, is a progressive and fatal disease in which fatty substances build up in tissues and nerve cells of the brain. There is blindness, mental deficiency, and death at an early age due to Tay–Sachs disease. Like most lysosome storage diseases, Tay–Sachs disease has few effective treatments.

**Lysosome storage disease**

A group of 30 known inherited human diseases associated with the abnormal functioning of lysosomes.

## Vacuoles: A City's Warehouse

Storage in cells is accomplished by **vacuoles**, which are single membrane structures that hold materials in a cell; they are the warehouses of the cell city. There are three types of vacuoles: food, water, and waste vacuoles, each holding what their name indicates. **Food vacuoles** bring materials to and from the plasma membrane, depending upon a cell's need for nutrients. Adipose or fat cells in humans contain food (fat) vacuoles that comprise up to 80% of the cell, with enormous capabilities for energy storage. Food vacuoles shrink or expand based on the amount of fat stored in adipose cells.

**Waste vacuoles** are temporary storage compartments that send materials directly to the plasma membrane for export. Nitrogen-containing compounds are regularly stored in waste vacuoles. Their accumulation is dangerous, which is a major reason untreated kidney failure is often fatal after only a few days. One nitrogenous waste, ammonia ($NH^3$), is particularly poisonous, as you can attest if you have handled household ammonia. Wastes are therefore exported from a cell as quickly as they accumulate.

**Water vacuoles,** which are storage vesicles for water, can also grow very large, especially in plants. They serve to give shape to a plant cell, exerting pressure from within the vacuole onto the plant cell wall. A special type of water vacuole, found in simple-celled organisms such as the *Paramecium* is called the **contractile vacuole**. The contractile vacuole has the unique ability to actively pump water out water of cells, acting as a mini-kidney. Its shape is shown in Figure 3.19, with a star-like appearance denoting its structure.

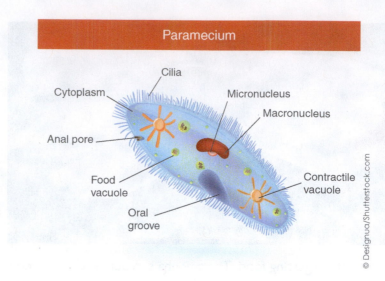

**Figure 3.19**  The *Paramecium* has many of the structures found in protists. The contractile vacuole pumps excess water out of the paramecium.

## Plastids: The Cell City's Paint Shops

**Plastid**

Organelle, found only in plants and algae, which store special substances.

Plastids are organelles that store special substances. They are found only in plants and algae. Colored substances called pigments are stored in some plastids. **Chloroplasts,** which contain chlorophyll, are the most well-known plastid.

- Chlorophyll is a green-colored pigment that will be discussed in Chapter 4.

Chromoplasts store yellow and orange pigments that give colors to trees, flowers, fruits and vegetables, such as carrots. **Leukoplasts** store starch and other food storage materials. Leukoplasts are usually found in roots, as in turnips and potatoes. We eat the starch-filled roots of these and several other foods.

**Chromoplast**

An organelle that contains any plant pigment other than chlorophyll.

## Cell Junctions: The City's Bridges

**Desmosome**

One of the three types of connections between cells, is a cell structure specialized for cell-to-cell adhesion.

In the same way that cities have bridges to connect and support their parts, cells contain cell junctions. Cell junctions serve a cell by providing areas of extra support, enabling communication between cells and preventing leaking through their borders.

There are three different types of connections or bridges between cells, each playing a needed role in particular cases. The first type of cell junction  called desmosomes are anchoring junctions or spot welds between cells. You can think of desmosomes/anchoring junctions as the reinforcing rods embedded in concrete slab bridges of a city. Within desmosomes, intermediate fibers entangle with linker proteins and attach to a thickened region of the plasma membrane to give extra support in holding cells together. Heart or cardiac cells contain many desmosomes to withstand the pressure of millions of heartbeats.

**Tight junction**

A specialized cell junction that fuses areas together to prevent leaking and act as sealants.

The second type of cell junction, tight junctions, fuses areas together to prevent leaking and act as sealants. Intestinal cells have tight junctions to prevent their contents from leaking out of the bodies of animals.

**Gap (communicating) junction**

Are channels that run from one cell into another to allow rapid transport helping cells communicate with other cells.

The third type known as gap or communicating junctions are channels that run from one cell into another to allow rapid transport. These bridges help cells communicate with

© Vadim Ratnikov/Shutterstock.com

**Figure 3.20**    Cell junctions are connections between two different cities.

other cells. Embryonic cells have gap junctions because, in the absence of a developed vessel system for transport, an embryo compensates with gap junctions. The structures of cell junctions are shown in Figure 3.20.

## Cell Shape and Size

Most cells in the body are round or spherical, except when a cell wall shapes a cell to conform to the forces around it. There are several reasons for the spherical shape, but resisting torque is a main one. Round surfaces resist the damaging effects of torque forces (circular forces). For example, consider a square cell, with edges that are likely to break off or become damaged when torque forces are applied to such cells. Any shape with a corner is more likely to be damaged than something that is spherical.

In senescence studies, the **wear-and-tear hypothesis** contends that damage due to forces placed upon living systems is a main reason for the aging of cells and the loss of cellular function. As a cell ages, it is more and more likely to succumb to the dangers and damages of the outside environment. Senescence is the loss of cell functions and in our story of Uncle Hans in Chapter 1, his dementia-related symptoms were a result of the loss of proper transmission of nerve signals within his brain. Senescence studies are continually looking at the damages to cells in attempting to reverse or slow the aging process.

There are many types of cells within multicellular organisms. As stated earlier, humans are constructed of about 200 different kinds of **somatic** or body cells. While they are each of very distinct types, with different characteristics, what is so surprising is their similarity. As discussed earlier, every cell has a control center to maintain the operations of the cell on a daily basis: a **membrane** that separates itself from the outside environment; and a semi-solid plasma, called **cytoplasm**, in which cellular activities occur. Most animal cells are between 10 and 30 µm in width, but can be as large as

© BlueRingMedia/Shutterstock.com

**Figure 3.21** Surface area of large cube vs. several smaller cubes of equal size. The surface area of the many smaller cubes is much greater than that of a single cube with a comparable volume.

100 µm (the size of a single grain of sand), as seen in the human egg or up to 1,500 µm as in frog's eggs. Eggs tend to be very large because they support a growing organism.

The small size of cells is explained by the **surface-to-volume hypothesis**, which states that the surface area (surface that is in contact with the outside environment) decreases rapidly in proportion to the volume of the cell. Thus, if the volume of a cell is too great, it cannot exchange enough of the materials (for example, oxygen) to support its size. The surface area of a variety of cubes that the little boy is hammering is shown in Figure 3.21. Note that the surface area of the single large cube is so much smaller than that of the aggregate of the many smaller cubes.

Another reason for limits on a cell's size and shape is its ability to control a large cell from a nucleus. Just as a long distance relationship is difficult due to distance, a cell too large has a more difficult time regulating activities of its structures from farther away. There are some organisms that have developed ways to adapt to these problems. A slime mold, shown in Figure 3.22 is, in fact, one giant, thin cell with thousands of nuclei strewn throughout its cytoplasm. In this way, it avoids the constant-volume problem by remaining thin for exchanges to occur and has control centers in enough areas to keep control efficient. Human muscle cells also are multinucleated and very large to perform their motility functions.

## The Moving Crew: Rules and Procedures are City Law

The description of the cell's organelles and their functions provides a glimpse at how complex their structure is and how many functions they carry out. Physical laws govern how they interact to transport materials. Let's look now at a few of the basic laws of the city. The different transport types will be discussed according to their requirement for energy: 1) No energy needed: simple diffusion, facilitated diffusion, osmosis; 2) Energy needed: active transport, bulk transport (phagocytosis, pinocytosis, receptor-mediated endocytosis, and exocytosis).

© Kucharski K. Kucharska/Shutterstock.com

**Figure 3.22** Slime molds. These organisms have multiple nuclei to control their large sizes. During some phases of their life cycles, they appear as gelatinous slime, giving them their infamous name.

## Passive Transport

Passive transport involves the movement of materials without the use of cellular energy. Recall that the cell is bathed in water. Most atoms and molecules therefore exist as ions. Ions move constantly, jiggling, hitting one another, and forcing themselves farther and farther apart to fill the available space. The motion of each ion is driven by kinetic energy, or energy of motion. This kind of movement is called passive transport because it occurs naturally and spontaneously and again, requires no cellular energy.

Substances move apart during passive transport until they spread evenly through an area. This even level of dispersion is known as equilibrium. You have experienced this action of molecules if you have ever been in a bus or subway car with someone whose deodorant has failed on a hot day. The underarm odor is first detected by nearby passengers, but the smell can travel until it reaches even those people at the other end of the vehicle. Sweat odor molecules may travel in any direction at the same time. However, there is a net movement of smell molecules from an area of higher concentration to an area of lower concentration. The movement from a higher concentration to a lower concentration is called diffusion. In living systems, substances spontaneously diffuse from higher to lower areas of pressure, electrical charge, or concentration without the use of cellular energy. Figure 3.23 shows the process of diffusion in this subway situation.

The difference between higher and lower concentration areas is known as a gradient. Many examples of diffusion are seen throughout this text. Oxygen, carbon dioxide, alcohol, and vitamins move through a cell via diffusion. The kidneys work by regulating body levels of ions using diffusion principles. Simple cutting of an onion often makes us cry – not because we care about an onion's rights to live – but because sulfuric acid, an irritant, diffuses from the cut onion toward our eye membranes. Passive transport processes within cells can describe movement for many of the chemicals discussed in the previous chapter.

The mitochondria in our opening story use gradients to generate energy for a cell to use. Joules and Theta were fortunate to have more efficiently developed mitochondria to help their physiological functions. Movement of substances within the mitochondria

**Passive transport**

The movement of substances across cell membranes without the need of energy expenditure by the cell.

**Equilibrium**

The even level of dispersion of substances.

**Concentration, higher and lower**

The presence of a large amount of a specific substance in a solution or mixture is higher concentration. Any solution containing fewer dissolved particles is lower concentration.

**Diffusion**

The net movement of molecules from higher concentration to a lower concentration.

**Gradient**

The difference between higher and lower concentration areas.

High concentration

© a katz/Shutterstock.com

**Figure 3.23**    Diffusion in a subway. Odor moves from a higher to a lower concentration in a subway car. Molecules bounce off of each other, moving farther and farther apart in diffusion.

will be discussed in greater detail in the next chapter, energetics. Next, we will look at a special case of passive diffusion of water across membranes, osmosis.

## Osmosis: A Special Case of Diffusion

Because water is critical for life, it must be able to move freely into and out of cells – in other words, across membranes. Diffusion of water across a **selectively permeable membrane** – one that allows the transfer of only some substances – is termed osmosis. Osmosis occurs spontaneously, so it requires no cellular energy and is thus a special type of passive transport. It occurs only when water moves across a membrane. Most molecules within a cell cannot simply diffuse across a plasma membrane. If they could, their contents would readily spill out and a separation from the outside world would not be possible.

While the integrity of the cell is critical, it is also critical that materials other than water move across the membrane. A unique property of water – its ability to dissolve other substances – makes this possible. The cell's cytoplasm is roughly 1% dissolved materials, and the right amount of dissolved material in the cell is required for proper cell functioning.

If a cell has a higher concentration of solute as compared with its surrounding environment it is hypertonic; if it has less solute, it is hypotonic. Osmosis is the mechanism by which the internal and external environments equalize. An example of what can happen when conditions are abnormal occurs when an athlete drinks a large amount of water in a short period of time following exercise and suffers from hypernatremia, or ion imbalance, which can be fatal (Figure 3.24).

The term **hyper**tonic can be remembered because "hyper" refers to a large amount of something, as in a hyperactive child who has too much energy. For example, if a cell with a concentration of 0.9% NaCl were placed in a beaker with a concentration of .01% NaCl, the cell would have the greater amount of solute. The cell, with more solute, has less water, and water would flow from the fluid outside into the cell. When that happens, the cell swells and could even lyse (break open) if too much water enters it.

Hypotonic solutions are those that have less solute dissolved than their external environment. The term **hypo**tonic can be remembered because "hypo" rhymes with $H_2O$, which implies that there is more plentiful water in a hypotonic situation. To illustrate,

**Osmosis**

The process of diffusion of water through a semipermeable membrane that allows the transfer of only some substances.

**Hypertonic**

A cell having a higher concentration of solute as compared with its surrounding environment.

**Hypotonic**

A cell having a lower concentration of solute as compared with its surrounding environment.

**Lyse**

Breakdown of cell membrane.

**Isotonic solutions**

Cells retain their normal size and shape in isotonic solutions (same solute/water concentration as inside cells; water moves in and out).

**Hypertonic solutions**

Cells lose water by osmosis and shrink in a hypertonic solution (contains a higher concentration of solutes than are present inside the cells).

**Hypotonic solutions**

Cells take on water by osmosis until they become bloated and burst (lyse) in a hypotonic solution (contains a lower concentration of solutes than are present in cells).

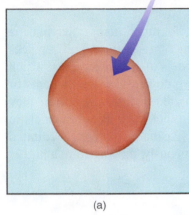

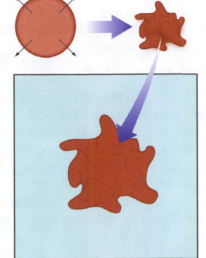

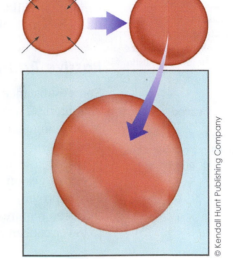

© Kendall Hunt Publishing Company

(a)     (b)     (c)

**Figure 3.24**   Red blood cells in hypotonic, hypertonic, and isotonic solutions. Red blood cells shrivel up in a hypertonic environment and blow up in a hypotonic environment. They are normal in an isotonic situation. Note the changes in a red blood cell's shape in each of the conditions.

when a cell with a concentration of 0.9% NaCl is placed in a beaker with a concentration of 10% NaCl, the cell has a lower amount of solute than the solution outside. The cell in this case is hypotonic, but the outside is hypertonic. Thus, water will flow out of the hypotonic cell along its concentration gradient. A good way to remember where water flows in cells is by the golden rule: "water follows solute."

Of course, no net movement of substances occurs if a cell has the same concentration of solute as its environment. It is then said to be isotonic (iso = the same as) to its outside. There is an even concentration of solute and water on either side of the plasma membrane. A cell might have a 0.9% concentration of NaCl both inside and outside of the cell. In this case, solute and water are both traveling in and out of the cell, but their net or overall movement is the same toward either side.

**Isotonic**

Even concentration of solute and water on either side of the plasma membrane.

## Special Cases in Osmosis

Gradients drive the movement of solutes and water in each of the examples above. Single-celled organisms, such as the *Paramecium* described earlier in this chapter, live in hypotonic environments. Their surroundings contain a high concentration of water. Water is therefore always attempting to enter the *Paramecium* because it contains more dissolved solutes. This requires a contractile vacuole to act as a constant pump, moving the excess water out of the cell. This would make sense, because *Paramecium* is a freshwater protist, which needs dissolved solutes within its cell to survive.

Plants also use gradients to maintain their structure and stand upright. My grandmother often served older lettuce, as she was quite frugal. It wilted in the refrigerator, but after she placed the lettuce in a bowl of water, it perked up and looked as good as new. Lettuce cells were hypertonic to the outside, which was hypotonic to normal tap water. Water thus rushed into the cells of the lettuce leaves along their concentration gradient. Water entered water vacuoles within the plant cell, creating what is called turgor pressure, pressure exerted against the walls of a plant cell. Turgor is any pressure on a cell wall that gives the wall rigidity. My grandmother used the principles of osmosis to save money. When a plant cell is placed in a hypertonic environment, what do you predict will happen? It will lose water to its outside, causing it to shrink. Water's movement across a membrane not allowing solutes, as in our lettuce example, is shown in Figure 3.25.

## Passive Transport with a Helper

Some substances require the help of carrier proteins to move across the plasma membrane. This final type of passive transport is called facilitated diffusion. Substances move along their concentration gradient through the channels of carrier proteins, which facilitate their movement. Sugars such as glucose, and amino acids, as components of proteins, both move into and out of cells through facilitated diffusion. Because they have charges on them they cannot cross the membrane on their own. Polar channels in carrier proteins enable these ions to be carried across the plasma membrane. Facilitated diffusion requires no cellular energy and it occurs spontaneously, moving from a higher to lower concentration.

**Turgor pressure**

The pressure exerted against the walls of a plant cell when water enters the water vacuoles of plant cell. Vacuole: Single membrane structures that hold materials in a cell.

**Facilitated diffusion**

A type of passive transport requiring a carrier protein.

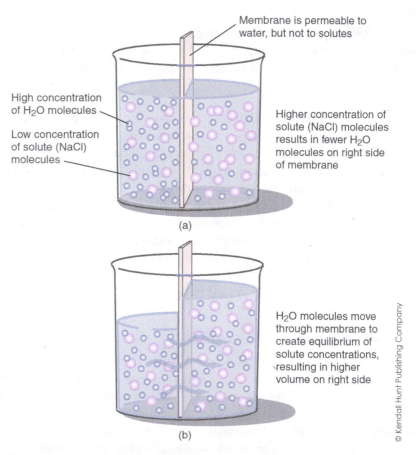

Membrane is permeable to water, but not to solutes

High concentration of $H_2O$ molecules

Low concentration of solute (NaCl) molecules

Higher concentration of solute (NaCl) molecules results in fewer $H_2O$ molecules on right side of membrane

(a)

$H_2O$ molecules move through membrane to create equilibrium of solute concentrations, resulting in higher volume on right side

(b)

© Kendall Hunt Publishing Company

**Figure 3.25**   Water moves across the membrane between two sides of the beaker but does not permit the movement of dissolved substances. This leads to a column of water rising as water follows solute into the right side of the beaker. Lettuce, in our example, has increased turgor pressure as water rushes into its cells.

Our story discussed mitochondria, which have much movement of ions within its inner membranes to make ATP energy. Next we will look at how ATP energy generated by mitochondria drives another form of movement: active transport.

## Active Transport

Some forms of movement occur with the addition of energy to drive its actions. Active transport is the movement of substances against a concentration gradient, from a lower concentration to a higher concentration, which requires cellular energy. Active transport does not occur spontaneously; it requires ATP energy to overcome the nonspontaneous movement of materials against a concentration gradient. Consider our subway example: Would it not be strange if all of the smell molecules aggregated toward the armpit of an unsuspecting person on the subway, making him or her smell terribly? Could smell molecules throughout the subway car attack this person? It cannot happen unless there is an active transport mechanism at work.

In order for cells to carry out some processes, they must use active transport. For example, the sodium–potassium pump is an integral protein that uses ATP energy to move sodium ions ($Na^+$) out of the cell and bring potassium ions ($K^+$) into the cell. Keeping the right concentrations of $Na^+$ and $K^+$ is vital in nerve and muscle activity.

A concentration gradient forms as a result of three $Na^+$ pumped out of the cell for every two $K^+$ pumped into the cell. This specific movement results in an electrical and concentration gradient across the membrane. More sodium winds up outside of the cell and more potassium inside. The environment outside the cell becomes more positive than inside it because there are more positive ions being pumped out. Potential or stored energy develops and drives many cellular activities much as a dam on a river stores energy that is converted to electrical power for our use. Figure 3.26 shows the workings of the sodium–potassium pump.

Secondary active transport is the movement of substances using stored energy. The sodium–potassium pump creates a store of electrical and potential energy for later use by cell processes. Instant energy for moving one's arm or riding a bicycle is obtained through secondary active transport by using stored ATP. Let's give a look in the next section to the active movement of bulk materials in and out of the cell.

## Bulk Transport: A Bigger Moving Van

At times, cells as a city require transport of larger materials – sometimes, whole organisms – to carry out their life functions. Bulk transport is the movement of large amounts of material across the plasma membrane. It is a form of active transport obtaining energy from ATP. Movement of single ions or molecules is often not enough for the many cell reactions occurring at any one time. Cell activities often require removal or inputs of large resources.

The moving van of the cell is the vesicle, containing the bulk materials. Bulk transport can occur by moving materials into the cell (called endocytosis) or by moving material out of a cell (exocytosis). (*Endo-* sounds like "in" and *exo-* sounds like "exit," which should help you recall the terms.) There are three types of endocytosis: phagocytosis, pinocytosis, and receptor-mediated endocytosis.

Phagocytosis is defined as the movement of solid particles into a cell. It is accomplished by a cell's cytoplasm projecting around the material to be dissolved. This action appears much like a blob, engulfing its prey. The substance may be a whole organism or a particle of organic matter. Immune cells, particularly macrophages, a special kind of white blood cell, attack foreign invaders by phagocytosis. Once the material is

**Active transport**

Is the movement of substances against a concentration gradient, from a lower concentration to a higher concentration, which requires cellular energy.

**Sodium–potassium pump**

An integral protein that uses ATP energy to move sodium ions out of the cell and bring potassium ions into the cell.

**Secondary active transport**

The movement of substances using stored energy.

**Bulk transport**

Is the movement of large amounts of material across the plasma membrane.

**Endocytosis**

The process of moving materials into the cell.

**Exocytosis**

The process of moving materials outside the cell.

**Phagocytosis**

The movement of solid particles into a cell.

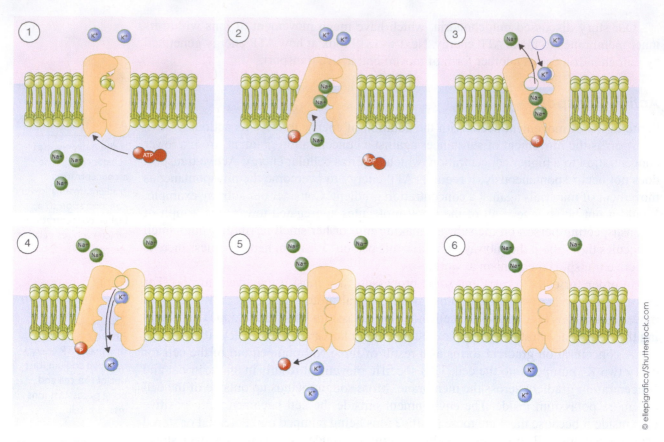

© ellepigrafica/Shutterstock.com

**Figure 3.26** The workings of the sodium–potassium pump. With each turn, the pump moves three sodium ions out of its cell and two potassium ions into its cell. A gradient is set up with more sodium ions outside than inside a cell. The inside of the cell becomes relatively more negative than the outside because more positive charges are pumped to the outside of the cell.

fully engulfed through this process, lysosomes fuse with the newly formed vesicle containing the material. Lysosome's hydrolytic enzymes digest the particle for use by a cell or release it as a waste product.

Cells also move large amounts of liquid, along with their dissolved solutes, into their cytoplasm through pinocytosis. Cell drinking or pinocytosis requires a furrow, or invagination, in the plasma membrane of a cell to input liquids. Many animal cells obtain their nutrients through pinocytosis, with large human egg cells developing multiple furrows to obtain nutrients from the surrounding cells.

A form of endocytosis that requires a specific binding of a receptor protein to cell membrane is known as receptor-mediated endocytosis. In this case, bulk materials move into a cell by docking with a specifically shaped receptor. The receptor is an integral protein jutting out of a plasma membrane. It has a specific shape onto which a substance matching its shape may bind. A signal stimulated by the joining of receptor to substance starts to bring the plasma membrane inward. The membrane forms a vesicle filled with both the substance bound to the receptor and the materials dissolved, as the membrane moves inward. Hormones, nutrients, and neurotransmitters, such as acetylcholine (described in Chapter 1), move in and out of a cell through this process.

Because receptors are proteins, they are genetically determined. For example, in cases of hyperlipidemia, high amounts of cholesterol build up in the blood. Receptors that dock with vesicles carrying cholesterol are either malformed or available in

**Pinocytosis**

The mechanism by which cells ingest extracellular fluid and its contents.

**Invagination**

The process of being folded back on itself to form a pouch (not given in bold in text).

**Receptor-mediated endocytosis**

A form of endocytosis that requires a specific binding of a receptor protein to cell membrane.

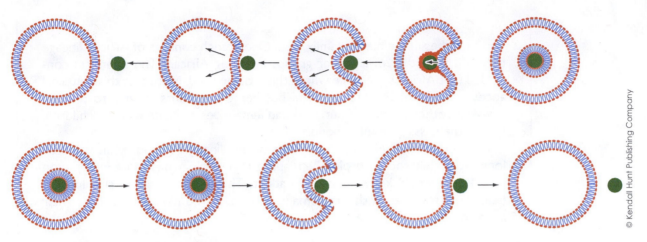

**Figure 3.27**  Endocytosis and exocytosis.

insufficient amounts in hyperlipidemia. Lipids cannot be removed from the blood as easily, reducing rates of receptor-mediated endocytosis and leading to buildup of fats in the blood. It is a dangerous disease due to its close link with heart attacks and stroke. Many other diseases are associated with improperly formed receptors, including diabetes, depression, and susceptibility to viruses.

All bulk movement out of a cell is accomplished by exocytosis. Any material to be exported – wastes, hormones, mucous or enzymes – is removed from a cell by exocytosis. Exocytosis removes either waste products or materials for export from the cell. Vesicles containing these materials fuse with the plasma membrane and are expelled. Figure 3.27 shows the types of bulk transport.

All of the forms of bulk transport are used by mitochondria described in our opening story. For a mitochondrion to function, much like any other organelle, it communicates with the nucleus, cell membrane, and other cell parts to perform its role in helping the cell to function. Joules and Theta probably did not realize the complexity of cell movement within their mitochondria that contribute to its survival. Perhaps if they had, they would have accepted their friend Sally. Knowing that mitochondrial DNA is not the only contributor to an individual cell's survival would have opened their eyes to the many unknowns in making each of us unique. Cell biology might have made a difference in the outcome of our story and in the friendship of Joules, Theta, and Sally.

### IT'S ALL ABOUT EVE: EVERYONE HAS THE SAME MOTHER IN OUR HISTORY.

Our story opens with the importance of ancestral mitochondrial DNA in social organization. However, evidence shows that all humans have the same maternal mitochondria, derived from the same ancestral mother.

A group of genetic researchers published a study in the journal Nature revealing a maternal ancestor to all living humans called "mitochondrial Eve." Researchers concluded that every human on Earth now could trace lineage back to this single common female ancestor who lived around 200,000 years ago.

Mitochondrial DNA was taken from 147 people across continents and ethnic groups. It was determined that people alive today have lineage on one of two branches in the human family tree. One branch consists of African lineage only and the other of all other groups including African. The scientists concluded that Africa is the place where this woman lived. Note that this maternal ancestor was not the first woman, but her descendants survive to present day while other women of her time had female descendants with no children, halting the passing on mitochondrial DNA.

Thus, Joules, Theta, and Sally were of one lineage, bound by an ancestor long ago. While society broke them apart, their biology had more in common than Joules would know or care to admit. Their commonalities should have held them together, with the humanness of a shared heritage. Perhaps their ancestor is one of us now . . .

## Summary

Chemistry is the language of structure and movement within living systems. Physical and chemical principles determine the behavior of cell structures. Developments in microscopy led to the discovery of the detailed workings of cells. While most organelles evolved through invaginations of membranes, mitochondria and eukaryotes developed by becoming absorbed into larger cells. Together, organelle interactions with one another and the outside environment comprise a living organism. The fluid mosaic model shows how cells transport materials across their many membranes. Transport systems accomplish varied goals from moving small particles directly through cells to moving materials in bulk. All of the needs of cells are accomplished by organelles communicating through use of the different transport systems. Our story showed that cell parts not only dictate only life processes, but influence human health and the social structure of our society.

## CHECK OUT

**Summary: Key Points**

- Cell biology affects our lives in many ways, from diseases such as lysosome storage disorders to the way society is structured based on human differences.
- Organelles have specific roles within cells, communicating with each other and the outside environment through membranes.
- The endosymbiotic theory explains the origin of mitochondria and chloroplasts in eukaryotic cells based on chemical and biological evidence.
- The development of microscopy led to the discovery of cells and cell parts.
- The plasma membrane has a unique structure that enables transport across a cell.
- Cells of the five kingdoms diverge based on just a few key differences.
- Passive and active transport mechanisms are used to move materials within a cell.
- Different chemicals use transport mechanisms differently to move across a cell.

## KEY TERMS

active transport
aerobic
bulk transport
cell respiration
centriole
chloroplast
chromatin
chromosome
chromoplast
cilia
compartmentalization
compound light microscope
concentration, higher and lower
cristae
cytoplasm
desmosome
diffraction
diffusion
endocytosis
endoplasmic reticulum (ER)
endosymbionts
endosymbiotic theory
equilibrium
exocytosis
facilitated diffusion
flagella
fluid mosaic model
gap (communicating) junction
golgi apparatus
gradient
hypertonic
hypotonic
integral protein (transmembrane or carrier protein)
intermediate fiber
intracellular digestion
intracellular transport
invagination
isotonic

leucoplast
lyse
lysosome
lysosome storage disease
magnification
microfilament
micrometer
microtubule
mitochondria
nanometer
nuclear envelop
nucleoli
nucleus
organelle (subcellular structure)
osmosis
oxygen revolution
passive transport
peripheral protein
phagocytosis
photosynthesis
pinocytosis
plasma (cell) membrane
plasmolysis
plastid
receptor
receptor-mediated endocytosis
recognition protein
resolution
ribosome
scanning electron microscope (SEM)
secondary active transport
selectively permeable
sodium–potassium pump
stoma
thylakoid membrane
tight junction
transmission electron microscope (TEM)
turgor pressure
vacuole

# Multiple Choice Questions

1. How might a chloroplast disease, preventing its key processes, affect human society?
   a. There would be no more food.
   b. There would be no more light.
   c. There would be no more water.
   d. There would be no more carbon dioxide.

2. If a plant cell absorbs water and becomes stiff, it becomes _____ through the process of _____.
   a. crenated; passive transport
   b. turgid; osmosis
   c. rigid; active transport
   d. reactive; facilitated diffusion

3. Which sentence BEST describes a ribosome?
   a. It is a large organelle that makes protein.
   b. It is a small organelle that makes protein.
   c. It is a large organelle that breaks down protein.
   d. It is a small organelle that breaks down protein.

4. Which type of microscope can best view the cristae of a mitochondrion?
   a. compound light microscope
   b. phase contrast microscope
   c. scanning electron microscope
   d. transmission electron microscope

5. Which term best describes the plasma membrane?
   a. static
   b. dynamic
   c. organized
   d. rigid

6. Which term best differentiates between active transport and passive transport?
   a. polarity
   b. facilitation
   c. concentration
   d. energy

7. Which is NOT a passive transport mechanism across a plant cell?
   a. turgor pressure
   b. diffusion
   c. osmosis
   d. receptor-mediated endocytosis

8. When a cell builds a new Golgi apparatus, new materials need to be brought into the cell, sometimes against a concentration gradient. Which process best accomplishes the task of building a new Golgi apparatus?

   **a.** diffusion
   **b.** osmosis
   **c.** exocytosis
   **d.** active transport

9. In question #8 above, which chemical drives the process for the correct answer?

   **a.** sodium
   **b.** water
   **c.** ATP
   **d.** adenine

10. An experiment places a skin cell within a petri dish containing pure, distilled water. Predict what will happen to the skin cell within minutes:

   **a.** The skin cell will shrink as water leaves the cell.
   **b.** The skin cell will lyse as water leaves the cell.
   **c.** The skin cell will shrink as water enters the cell.
   **d.** The skin cell will lyse as water enters the cell.

## Short Answers

1. Describe how senescence is linked to lysosomes. Explain how lysosomes are both linked to human diseases and to our proper functioning. Give an example of each.

2. Define the following terms: passive transport, active transport, and receptor-mediated endocytosis. List one way each of the terms differs from the others.

3. Compare the two types of electron microscopes, explaining how their images are similar and how they are different. Use a drawing to make the distinctions clear. Show your art work. Which was more important in discovering subcellular structure?

4. There are several ways chemicals move within cells. At times they use passive transport and at times they use active transport. List the types of passive transport and explain how each works. Be sure to include the following terms in your explanation: gradient, concentration, and ATP energy?

5. For question #4 above, list and draw an example of an active transport mechanism within a cell. Include in the picture the electron arrangement around the atoms.

6. Name two differences and two similarities between cells found in the matched kingdoms:
   a. fungi–protist
   b. animal–plant
   c. plant–protist
   d. bacteria–animal

7. ATP plays a major role in life's processes. How does ATP energy function in secondary active transport processes? Explain how the sodium–potassium pump plays a role in secondary active transport.

8. Draw the fluid mosaic model of the cell membrane. Describe how and why a charged molecule such as sugar that needed by every cell enters through the fluid mosaic.

9. Explain the differences between passive and active transport. Explain the difference using the following terms: ATP, spontaneous, gradient, concentration, ions.

10. A battle between an amoeba and a single-celled protist, *chilomonas* ensues. The amoeba wins, but the large piece of *chilomonas* is too big to pass through the plasma membrane of the *amoeba* to eat. Describe the process by which the amoeba absorbs and digests the *chilomonas*.

## Biology and Society Corner: Discussion Questions

1. Many scientists argue that our most pressing resource crisis is not food shortage or climate change, but water scarcity. How is water important for a cell? Predict how the role of a limited water supply will affect society. Explain how the water crisis is already impacting policy changes, the economy and overall human impacts on the environment.

2. How would van Leeuwenhoek be surprised by discoveries since his viewing of cells? What role do you think microscopy will play in future scientific developments? How did microscopy affect society in the 20th century? Use either the scanning electron microscope or transmission electron microscope to frame your thesis.

3. If a person is stranded on a desert island, without freshwater, what are the dangers of drinking sea water. Explain the transport mechanism at work, which leads to problems for a human cell in contact with seawater.

4. Mitochondrial diseases are attributed to human health problems, from sweating and epilepsy to diabetes and heart disease. Do you support biotechnology research to help people with these diseases? Do you support its use in potentially making "better" mitochondria to help people live better lives, as seen in our story. What restrictions would you place, if any, on such developments?

5. Plasma membranes are delicate and can weaken in some cases causing human health disorders. Research the effects of plasma membrane problems on the health of humans and animals. Choose one of the organelles discussed in this chapter. Describe what happens to the organelle and human or animal health as a result of the delicate nature of plasma membranes.

Figure – Concept Map of Chapter 3 Big Ideas

# Energy Drives Life

## 4

## ESSENTIALS

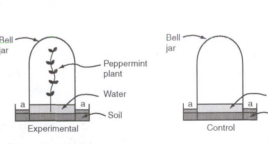

Mrs. Green in her garden

Mrs. Green's White Pine Tree

Bell jar

Peppermint plant

Water

Soil

Experimental

Bell jar

Water

Soil

Control

©Kendall Hunt Publishing Company

Plant experiments

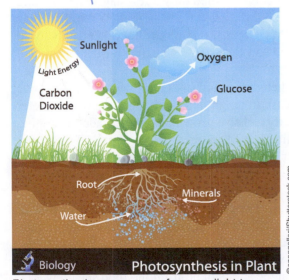

Sunlight

Oxygen

Light Energy

Glucose

Carbon Dioxide

Root

Minerals

Water

Biology

Photosynthesis in Plant

Photosynthesis uses energy from sunlight to produce carbohydrates

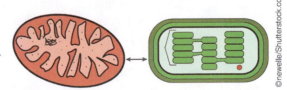

Chloroplast and Mitochondria share a close relationship

# The Case of a White Pine Memory

"It was a time to remember," thought Ms. Green about the days when she and her father worked on their land. She could remember when it was just a corn field that her father had plowed. But that was almost 80 years ago and how time flies, she thought. The birds in the sky floated with the wind. She spotted them and thought ". . . time flies away like the birds."

There it was – so wide and so impressive – she had never forgotten the day her father planted the tree. It was a white pine tree she and her daddy planted so many years ago. The image of the pine traveled with Ms. Green through her life. She was just eight years old on the day her father brought the tree home from the store. He said that he wanted shade when he worked in the field. Daddy planted the white pine, *Pinus strobus* he called it, right in the center so it would tower over the other trees. And at 80 feet tall, it really did tower over all the other trees in the area.

But he would not live to see its shade; her daddy died only a few days after planting the pine. He was the love of her life. He believed in her and he believed in life. "He planted the pine for more than just shade," Ms. Green thought. She knew her daddy loved to nurture nature and other people; and she had loved how he cared for his family and his field.

Ms. Green was known in the town for her garden and its central white pine. The pine had grown rapidly and continued to increase in height and width, adding over a meter and thousands of kilograms per year. The city had also grown over the decades, changing from a farm town to a thriving municipality. But Ms. Green's field remained the same; except that the other crop fields around her land had become buildings and tarred streets. Ms. Green, everyone knew, would never sell her land, but builders kept building around her just the same.

Each day, Ms. Green worked in her garden, always looking up at the pine with fondness. Everyone she knew through her life had to join her in her garden. Her friends quickly realized, if they wanted to stay her friend, they needed to work alongside Ms. Green in the field. She built a nice stone wall around her garden, with stones from the land. She had any vegetable one could imagine and cooked from the food she grew. Ms. Green loved nature and loved her field.

It was only two acres, but tending the garden became harder and harder as the years passed. She was, after all, over 80 years old now. Then one day, as she worked in the garden pulling out weeds, she knew she could go on no more. "It was her time," she accepted, "to end." She was very sad because the life she knew was slipping away. She looked up at the pine and knew they would soon part.

The white pine would live for many more years, but her good-bye she knew would come sooner. "It wasn't fair . . . time was cruel," protested Ms. Green to the inflexible passage of time. Separation from all she loved was too hard to take. But as she cried, she spied the birds flying overhead. Was it true, or had her eyes deceived her? A nest high in its branches sat atop the majestic white pine. The eagles soared toward the treetop nest. Suddenly, she felt a sense of peace, and a smile grew across her face. She was letting go, but it would be all right: A family had taken over for her.

---

### CHECK UP SECTION

The processes occurring in the white pine described in our story not only help plants to grow but are vital for human existence. Research the following questions: 1) How are plant processes necessary for human society? 2) Are there any environmental threats to plant energy processes? Choose a particular example in which a plant's processes are threatened in nature. Discuss how such a threat may impact human health.

---

## Discovering Energy Exchange

In this chapter, we will explore the ways organisms harness energy from the sun and liberate that energy from foods. Organisms use resources from their environment to survive. Some organisms, such as the white pine in our story, use sunlight to manufacture food. Other organisms, such as Ms. Green, cannot make their own food, and obtain energy by eating plants and other animals. In both plants and animals, energy is transferred in a series of chemical reactions. The different stages that take place to make food from sunlight and into available energy for cells will be our focus.

What processes make some trees, like the white pine in the story grow so large and live so long? Do plants absorb food from the soil, just as animals eat food from their surroundings? Until about 350 years ago, scientists believed that plants obtained all of their energy from the ground. Jan Baptista van Helmont (1577–1644) contradicted this widely held view through an experiment. In it, van Helmont grew a baby willow tree in a pot for 5 years, noting the initial weight of the tree and the soil. He added only water and at the end of this period was surprised to find that the soil increased in weight by 57 grams, but the willow increased in weight by 74,000 grams! Where did all of this matter come from? Van Helmont concluded that the mass must have come from the added water. However, water could not be an agent of organic matter (recall from Chapter 2); water is composed of hydrogen and oxygen atoms. Where is the carbon that is needed for sugar production? While van Helmont's experiment didn't answer this question, it is important because it was one of the first carefully designed experiments in biology.

Adding to the mystery of plant growth, Joseph Priestly (1733–1827), an English clergyman and early chemist, conducted an experiment to determine the effects of plants on air quality. He placed a sprig of mint in a glass jar with a candle. The candle burned out, as was expected but after the 27th day, Priestly discovered that another candle could once again burn in the same air in the jar – somehow the presence of the plant caused the air to regenerate. Priestly concluded that vegetables ". . . do not grow in vain." He proposed that plants cleanse and purify the air. In actuality, we now know that plants give off oxygen and remove carbon dioxide gases. While Priestly's experiment could not be replicated at the time by others scientists (or by his own laboratory), it laid the foundation for the discovery of the other secret ingredients to photosynthesis. Priestly's experiment is shown in Figure 4.1.

It was not until a Dutch physician, Jan Ingenhousz (1730–1799), later replicated Priestly's work that the importance of sunlight for plants was recognized. Ingenhousz added that restoration of air by plants only took place in sunlight. He concluded that "the sun by itself has no power to mend air without the concurrence of plants." At the same time that Ingenhousz performed his work, Antoine Lavoisier (1743–1794), an extraordinary chemist of his time, studied how gases are exchanged in animals. He confined a guinea pig in a jar containing oxygen for 10 hours and measured the amount of carbon dioxide it released. Lavoisier also tested gases exchanged in humans as they exercised. He concluded that oxygen is used to produce energy for animals and that "respiration is merely a slow combustion of carbon and hydrogen." Unfortunately, Lavoisier's life ended early; his intellect threatened the government during the French revolution, and he died by guillotine on May 8, 1794. But he was able to show the overall equation for cellular respiration:

**Cellular respiration**

The process through which most organisms break down food sources into useable energy.

$$C_6H_{12}O_6 + 6O_2 \rightarrow 6CO_2 + 6H_2O + energy$$

Cellular respiration is the process through which most organisms break down food sources into usable energy. As shown in the equation, simple sugar (glucose) is broken down or oxidized to give energy,with carbon dioxide and water as byproducts.

Ingenhousz quickly used Lavoisier's deductions, realizing that plants absorb the carbon dioxide that is later burned for energy, "throwing out at that time the oxygen alone, keeping the carbon to itself as nourishment." Building upon this, Nicholas Theodore de Saussure (1767–1845) revealed the final secrets of photosynthesis – that equal volumes of carbon dioxide and oxygen were exchanged during photosynthesis. Thus, a plant gains weight by absorbing both carbon dioxide and water and releasing oxygen. All of the elements of the equation for photosynthesis were now identified – carbon dioxide, water, sugar, oxygen, and light to give:

**Photosynthesis**

The process by which green plants (plus some algae and bacteria) use sunlight to synthesize nutrients from water and carbon dioxide.

$$6CO_2 + 6H_2O + energy \rightarrow C_6H_{12}O_6 + 6O_2$$

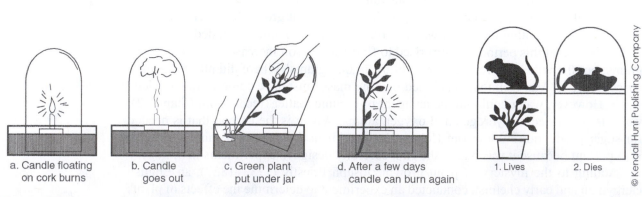

a. Candle floating on cork burns    b. Candle goes out    c. Green plant put under jar    d. After a few days candle can burn again    1. Lives    2. Dies

© Kendall Hunt Publishing Company

**Figure 4.1** Priestly's experiment. Priestly showed that plants regenerate the air surrounding them.

Photosynthesis is the process by which some organisms trap the sun's energy, using carbon dioxide and water, to make simple sugars (glucose). As shown in the equation on the previous page, oxygen is a byproduct of photosynthesis.

Both plants and animals carry out cellular respiration to obtain energy from food sources. But only those organisms carrying out photosynthesis produce their own food sources. These processes comprise the key reactions in **cell energetics**, which is the study of the energy exchanges within a cell. In order for the white pine to grow so large in the opening story, exchanges of energy between chemical players in cell energetic processes took place over many years. Its growth is a characteristic of life that shows how tiny chemical reactions may lead to large changes in organisms.

The two processes of photosynthesis and cellular respiration, in their overall equations, are indeed the reverse of one another: photosynthesis is the taking in of energy to yield food, and cellular respiration is the taking in of food to yield energy. The specifics of the processes, however, differ in this comparison. Also, while plants, most algae, and some bacteria produce their own food, all other life must obtain energy by consuming products of photosynthesis. We will examine these processes in greater detail after looking at the physical laws that describe the flow of energy.

## Rules for Energy Exchange: Energy Laws

The opening story demonstrated the flow of energy from sunlight to plants and finally to Ms. Green as she ate her vegetables (see Figure 4.2). While large amounts of energy enter Earth through sunlight, about one-third of sunlight is reflected back into space. The remaining two-thirds is absorbed by Earth and converted into heat. Only 1% of this energy is used by plants, an impressive fact because that fraction drives most life functions. With just a few exceptions, everything that is alive in some way uses the sun's energy, and humans owe their existence to plants' use of this small sliver of harnessed energy.

The flow of energy through our environment and in our cells is explained by thermodynamics, the science of energy transformations. As the sun's energy moves from object to object and organism to organism, it follows the same rules. The first rule, called the first law of thermodynamics, states that energy can be changed from one form to another

**Thermodynamics**

The science of energy transformations that explains the flow of energy through environment and in cells.

**First law of thermodynamics**

A law that states that energy can be changed from one form to another but cannot be created or destroyed.

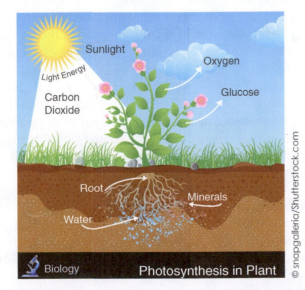

**Figure 4.2**  Ms. Green's garden. Energy is first brought into the garden by plants using sunlight to form sugars.

but cannot be created or destroyed. The total energy of a system remains constant. While 99% of sunlight entering the Earth is lost to organisms, it is actually reflected toward space or changed to heat; it is still conserved. The first law of thermodynamics is also called the law of conservation of energy. While newly formed sugar molecules from photosynthesis contain potential energy, which is energy of stored position, it is not newly created. Organisms, to drive life functions use potential energy, stored in the bonds of sugar molecules. In accordance with the first law of thermodynamics, sugar's energy was transferred from the sun to the plant.

**Second law of thermodynamics**

A law that states that all reactions within a closed system lose potential energy and tend toward entropy.

**Entropy**

Randomness or any increase in disorder.

The second law of thermodynamics states that all reactions within a closed system lose potential energy and tend toward entropy, which is randomness or any increase in disorder. A good example of entropy is your room or house: if you do not regularly tidy it (expend energy), it gets messier and messier. Natural processes tend toward randomness and energy release. In living systems, cellular respiration ($C_6H_{12}O_6 + 6O_2 \rightarrow 6CO_2 + 6H_2O$ + energy) releases 3.75 kcal of energy per gram of glucose. Cells, to drive cellular processes, use this energy.

Energy is exchanged in cells through the action of the ATP or adenosine triphosphate molecule, which contains two high energy bonds.

- As discussed in Chapter 2, ATP transfers its high-energy phosphates by breaking or making bonds between its three phosphates.

When ATP loses a high-energy phosphate, two phosphates remain, and the molecule is called ADP, or adenosine diphosphate. If an ADP molecule gains a high-energy phosphate, it again contains three phosphates, forming ATP. When a high-energy phosphate is transferred to another molecule, it brings with it the potential energy of its bond. Higher energy states change the molecule onto which an ATP's phosphates attach. These changes drive many cell reactions, such as cellular respiration.

Cellular respiration is very efficient at obtaining energy from food sources. Over 40% of the energy in glucose bonds is converted into useful ATP for a cell, with between 30 and 32 ATP per glucose molecule. In comparison, over 75% of energy from bonds in gasoline is lost as heat through the combustible energy of an automobile, and only 25% is converted into useful forms for a car's driving.

Photosynthesis started the flow of energy through the system in our opening story. Plants in Ms. Green's garden manufactured food, using sunlight. Plants were able to efficiently use these nutrients through cellular respiration. Then, Ms. Green was able to obtain energy from plants by consuming them and breaking their stored energy through cellular respiration. The flow of energy begun by photosynthesis and traced in a simple system resembles the flow in our environment.

Photosynthesis uses 3.75 kcal of energy to produce 1 gram of glucose. In this special case, its product (glucose) has a higher potential energy than reactants (carbon dioxide and water). Glucose is more organized and has less entropy than its gaseous reactants, with a ring of chemicals. Does photosynthesis violate the second law of thermodynamics? It does not, because the system in photosynthesis includes both the Earth and the sun. The sun is slowly losing its power; its reactions cause it to have less potential energy and more entropy as time passes. Thus, the glucose gains the energy that is lost by the sun. Eventually, the sun will lose enough energy that it will die out, ending life as we know it. There is no cause for immediate alarm, however; the sun is not expected to die for about 20 billion years.

Thus, life processes are driven by a sun that is running down. Its loss of energy is our gain, and photosynthesis is the gateway reaction to tap this resource for the benefit of living things. As plants capture solar energy and transform it into glucose, the sugar is used by mitochondria to produce usable energy. Some energy is transferred to heat in the process but reactants are reused readily.

**Figure 4.3** A hummingbird in Ms. Green's garden The humming bird derives its energy from products made by a tree's capture of sunlight. Sugars in nectar are a nutritious source of food.

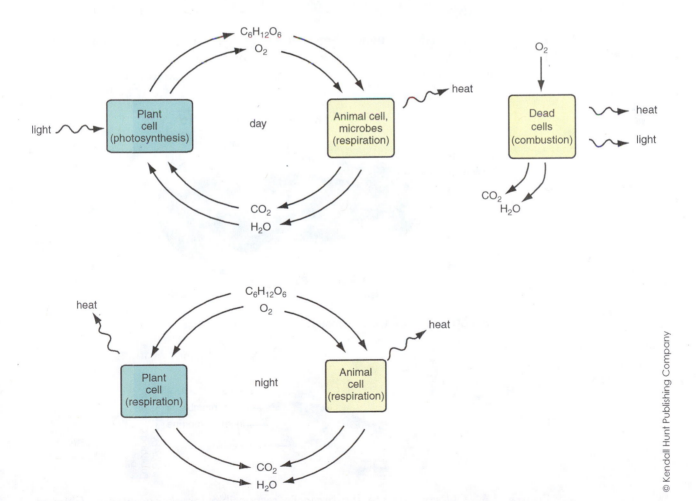

**Figure 4.4** Biological energy moves along: plants and animals have interdependent reactions.

# Photosynthesis: Building Up Molecules of Life

**Light reactions**

A reaction that traps energy from sunlight using special pigments.

**Pigment**

A naturally occurring special chemicals that absorb and reflect light.

**Calvin cycle**

A set of chemical reaction absorbing carbon dioxide and making glucose, taking place in chloroplasts during photosynthesis.

The process of making sugar from sunlight via photosynthesis uses carbon dioxide and water and liberates oxygen. Photosynthesis occurs in two stages: Light reactions, which trap energy from sunlight within special pigments, and the Calvin cycle (once called dark reactions), which uses carbon dioxide to make the glucose structure (see Figures 4.10 and 4.11). The two parts of the word *photosynthesis* describe these two stages: "photo" refers to light energy that is converted to chemical energy during light reactions; "synthesis" refers to the making of glucose during dark reactions.

## Chloroplasts: Where the Action Takes Place

The processes of photosynthesis occur in **chloroplasts,** which are specialized organelles found only in organisms that carry out photosynthesis. Each chloroplast contains a series of special membranes called **thylakoid membranes,** within which are molecules of the pigment chlorophyll (see Figures 4.5 and 4.6). Chlorophyll contains electrons that become excited by light energy from the sun and transfer that electron energy into a series of photosynthesis processes. Sunlight has special wave properties that stimulate photosynthesis in chloroplasts. These characteristics of light waves enable plant and algae cells to transform light wave energy into usable sugars and other products.

### What Is Light?

**Electromagnetic energy**

A type of energy released by into space by stars (sun).

**Radiant energy**

A type of energy travelling by waves or particles.

Photosynthesis transforms light energy into complex macromolecules. Sunlight is a form of energy known as electromagnetic energy or radiant energy. Electromagnetic energy travels in waves, carrying with it bundles of energy in the form of photons. The

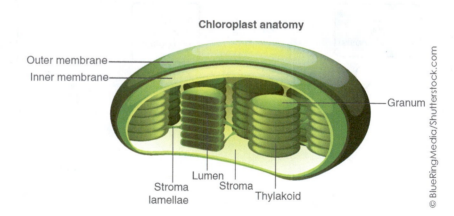

**Chloroplast anatomy**

Outer membrane
Inner membrane
Granum
Stroma lamellae
Lumen
Stroma
Thylakoid

© BlueRingMedia/Shutterstock.com

**Figure 4.5**   Structure of a Chloroplast.

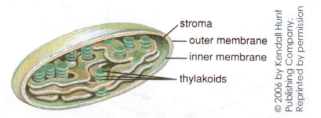

stroma
outer membrane
inner membrane
thylakoids

© 2006 by Kendall Hunt Publishing Company. Reprinted by permission

**Figure 4.6**   Chloroplasts are the organelle responsible for photosynthesis. Chloroplasts have interdependent reactions. From *Biological Perspectives,* 3rd ed by BSCS.

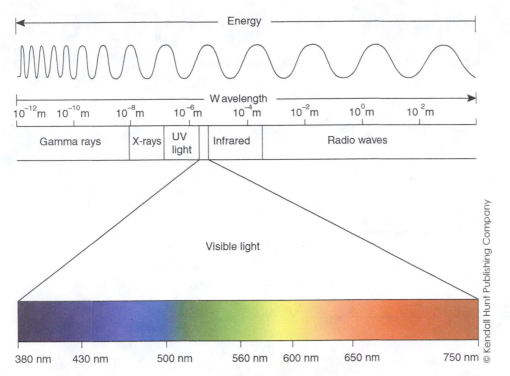

**Figure 4.7** Wavelengths of the electromagnetic spectrum. Only a narrow range of wavelengths are visible light, used for photosynthesis.

wavelength of light, which is the distance between the wave crests, is related to the amount of energy a wave carries (see Figure 4.7).

Each wavelength range appears as a certain color on the rainbow, corresponding to the amount of energy it carries. Visible light (see Figure 4.7) has a wavelength range of 380–750 nm. Note that the frequency of each wave in Figure 4.7 is the number of wave crests per second. The more frequent the wave crests, the higher the amount of energy in a light ray. When light hits an object, it is either absorbed or reflected. When it is absorbed it disappears from our sight, and when it is reflected, we see it. Thus, in a green leaf, very little green light is absorbed or used by a plant because it is reflected.

## THE AUTUMN LEAVES OF COLOR

Light that is reflected gives color to an object. Chlorophyll appears green because it uses very little green light for photosynthesis. When autumn begins and temperatures cool in many areas, the leaves of some plants change colors. This color change occurs because the plant is shutting down for the winter, ceasing chlorophyll production in its leaves. Only the yellow-orange colors of carotenoid pigments and the red color of anthocyanin pigments remain, giving trees their beautiful foliage. It is, however, a concession that plants make to living in colder climates, as will be discussed in a later chapter. Leaf drop is a big waste of energy but is necessary. In our story, Ms. Green's white pine did not shed needles during the winter because pines are adapted to withstand harsh conditions.

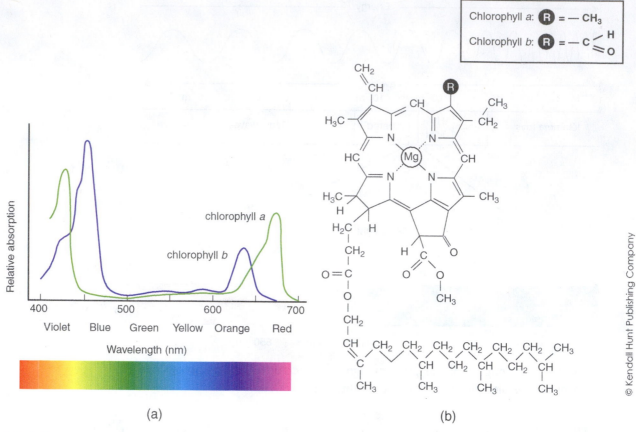

(a)                                                           (b)

**Figure 4.8** The absorption spectra for chlorophylls a and b. Green and yellow wavelengths are used least in photosynthesis and red and purple wavelengths are used most effectively.

## Pigments

**Photon**

Discrete unit of light energy that when hits a pigment in chlorophyll transfers its energy to electrons in the pigment.

Plants and algae both contain **pigments**, special chemicals in chloroplasts that absorb and reflect certain visible wavelengths of light. Pigments include green-colored chlorophyll *a* and *b* as well as other pigments. The structure of the pigment chlorophyll is shown in Figure 4.8. Violet-blue and red wavelengths are most effectively absorbed by chlorophyll pigments. The absorption spectra for chlorophylls *a* and *b*, two types of chlorophyll, are given in Figure 4.8. From Figure 4.8, which colors besides green are least used by chlorophyll?

## The Light Reactions

**Ground state**

The lowest state of energy of a particle.

**Excited state**

A state of a physical system that is higher in energy than in its normal state.

When photons, or discrete units of light energy hit the pigment in chlorophyll, photon energy is transferred to electrons in the pigment, and those electrons begin moving more rapidly; in technical terms, they become excited to a higher energy state. In other words their electrons move from a ground state to a higher excited state.

The excited state of electrons in chlorophyll makes them unstable and loosely held within the pigment. An excited electron can either return to its ground state or be tossed to a nearby molecule. Some electrons fall back to their ground state, producing energy as they move to the lower energy state, as shown in Figure 4.9a. Some electrons shoot out like pinballs to get accepted by another molecule, which then has more energy than it had before. Both of these paths of electron excitement are the "photo" part of photosynthesis, also called the **light reactions**, in which energy is captured and passed along (Figure 4.9b). The capturing of light energy is step one in the process.

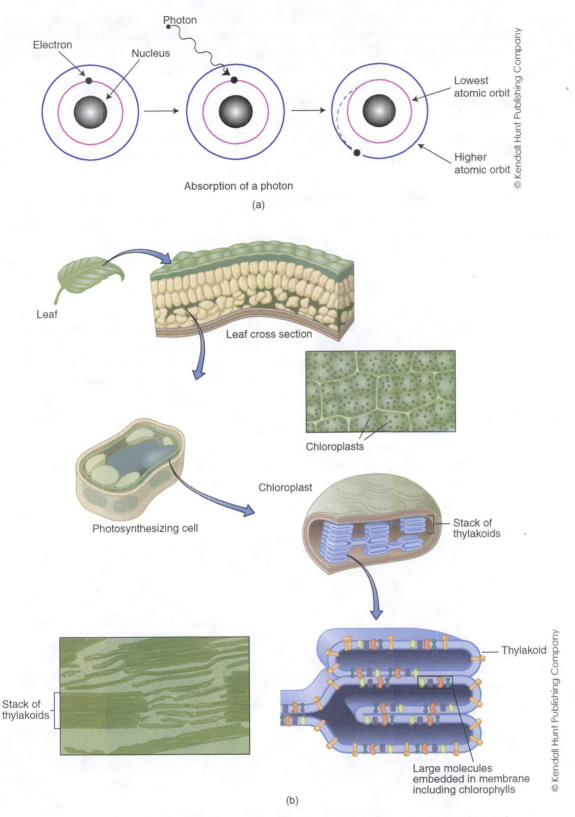

**Figure 4.9**    a. Electrons fall to lower energy levels after they become excited by light energy. b. Light reactions take place along the inner membrane of chloroplasts.

If you inspected needles from Ms. Green's pine tree with an electron microscope, you would see within the chloroplasts many thylakoid membranes, which look somewhat like stacks of coins (see Figure 4.9). Each thylakoid membrane contains bundles of chlorophyll and other pigments. These light-capturing bundles are called photosystems. There are two photosystems, **Photosystem II**, which we will call the water-splitting photosystem, and **Photosystem I**, the nicotinamide adenine dinucleotide phosphate (NADPH)-producing system. Photosystem II works first in the process of photosynthesis, and then photosystem I takes over. (Although photosystem I occurs after photosystem II, it bears its "I" name because it was discovered first.)

## The water-splitting photosystem

The process starts when light is captured in the water-splitting photosystem (II). Water molecules from fluid within chloroplasts donate electrons to the photosystem, releasing oxygen and hydrogen ions ($H^+$). Light energy causes the released electrons to move to the excited state. Excited electrons return their ground state, but give off energy they gained to neighboring pigment molecules.

As energy spreads through the collection of pigment molecules, it reaches the center of a photosystem. There, energy is captured by chlorophyll *a*, a special molecule in a photosystem that does not move its electrons back to the ground state. Instead, excited electrons in chlorophyll *a* are transferred to a neighboring primary electron acceptor.

Now begins a game of a pinball, in which excited electrons are moved from chlorophyll *a* to the primary electron acceptor, losing energy just a bit with each transfer. Much like a pinball bouncing around a pinball machine, electrons move from place to place, losing energy with each hit. This energy is eventually captured in ATP.

To understand the many steps of photosynthesis, follow the pinball of energy (look again at Figure 4.10) as it moves from place to place in the chloroplast. The pinballs or electrons are too energized to remain in one place for very long. They are transferred to

**Photosystems**

A light capturing bundle of pigments which absorbs light for photosynthesis.

**Chlorophyll *a***

A special molecule in a photosystem that does not move its electrons back to the ground state.

**Primary electron acceptor**

An electron acceptor in a particle that can be reduced by gaining an electron from some other particle

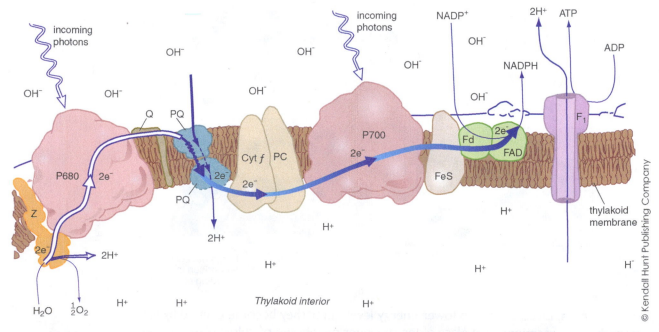

**Figure 4.10**    A detailed look at the photosystems. Photosystems obtain electrons from water to produce energy molecules. ATP and NADPH pigments hand off electrons to their primary electron acceptors developing an electrochemical gradient across the membranes. This gradient drives the production of energy.

cytochromes, which are special electron-holding carrier proteins. When excited electrons are moved from neighboring cytochrome to neighboring cytochrome, held only for a short while by each, electrons pass down what is termed an electron transport chain (ETC). ATP and NADPH are high-energy molecules produced as electrons fall to lower energy levels in the ETC. Figure 4.10 shows how this process proceeds, with electrons moving in an orderly and continual progression toward lower energy states.

In order to replace electrons lost from a photosystem, water is split to yield free electrons and hydrogen ions. This is called the photolysis **of water** and is required to maintain a constant supply of electrons for a photosystem. Ms. Green's pine tree needs water each day to replenish its lost electrons. Electron replacement is a reason all photosynthetic organisms require water to grow and survive.

With each handoff along the electron-transport chain, electrons give up a little bit of energy. This energy is used to pump protons (the $H^+$ ions mentioned above) from the stroma into the thylakoid stack. The stroma is the liquid region surrounding the thylakoid sac in a chromosome. $H^+$ ions are found throughout the stroma that are able to be used by the photosystem. Eventually, as Figure 4.10 shows more hydrogen ions accumulate inside the thylakoid membrane, creating an electrochemical gradient. That is, more positive charges on hydrogen ions and more hydrogen are on one side of the membrane than on the stroma side. As a result, potential energy is stored in the hydrogen ion difference across the thylakoid, much as a dam stores water for later use – with more hydrogen ions on one side of the membrane as compared with the other side. As hydrogen ions pass back into the stroma and down the electrochemical concentration gradient, energy is released to form ATP from ADP. The stored potential energy resulting from the concentration difference is transferred into the energy of the phosphate bond in ATP.

## The NADPH-producing photosystem

While photosystem II, the water-splitting photosystem, starts the light reactions of photosynthesis, the ETC links it with photosystem I, the NADPH-producing photosystem. Chlorophyll *a* molecules in the water-splitting photosystem absorb light best at a wavelength of 700 nm. Light energy entering the NADPH-producing photosystem is absorbed at 680 nm, beginning the photooxidation of chlorophyll once again.

Electrons from the water-splitting photosystem move along the ETC to supply vacancies or empty places within a cytochrome, created in the NADPH-producing photosystem. Electrons are at a low enough energy state to enter into photosystem I. Cytochromes only allow electrons with certain energy states to become attached to them. As in a game of pinball, when the ball has lost its energy, it passes through the flippers into the drain of the game. This occurs when electrons are at their lowest energy state. A pinball or an electron may be shot out again in another game of pinball or photosystem energizing. This second game is the NADPH-producing photosystem. The lower energy electrons are re-excited in the NADPH-producing photosystem by entering light.

The NADPH-producing photosystem has the same steps as the water-splitting photosystem: It also has electrons that become excited, are accepted by a primary electron acceptor, and fall down to lower energy levels within an ETC. However, electrons in Photosystem I are eventually passed to a molecule of **NADP⁺**, or nicotinamide adenine dinucleotide phosphate and form NADPH. $NADP^+$ is a high-energy electron carrier that transfers the energy of a high-energy electron from one part of a chloroplast into another part. Electrons travel with an assistant in this form, the hydrogen ion. When $NADP^+$ finally accepts electrons at the last step of Photosystem I, $NADP^+$ adds two H atoms (with their electrons) to become reduced NADPH. NADPH is a high-energy electron carrier molecule that carries electrons to be used in the next set of reactions in the stroma to build sugar.

---

**Cytochrome**

Hemeproteins that contain heme groups and are responsible for ATP generation through electron transport (not given in bold in text)

**Electron transport chain (ETC)**

A chemical reaction in which reactions are transferred from a high-energy molecule to lower-energy molecule.

**Photolysis**

The process in which water is split to yield free electrons and hydrogen ions to replace electrons lost from a photosystem.

**NADPH**

Nicotinamide adenine dinucleotide phosphate is used as reducing agent in reactions.

# How is Sugar Made?

**Light-independent reactions**

Chemical reactions that convert carbon dioxide into glucose (not given in bold in text)

**RUBISCO**

An enzyme present in chloroplast of plants.

Carbon dioxide in the atmosphere provides the building materials for sugar construction in the next step of photosynthesis. Ms. Green's garden required elements from the atmosphere to survive; its plants could not produce sugar with sunlight and water alone. The "synthesis" portion of photosynthesis produces a six-carbon glucose molecule by using carbon from $CO_2$. Through a set of light-independent reactions known as the **Calvin cycle**, named after Melvin Calvin, an American chemist who discovered its steps, energy from ATP and electrons from NADPH drive a cycle of reactions that lead to sugar. The specific steps are given in Figure 4.11.

The Calvin cycle takes place within the stroma of chloroplasts. It is initiated by an enzyme of the Calvin cycle called RUBISCO, which unites carbon dioxide from the atmosphere with chemicals in the cycle. In fact, enzymes in each step of the Calvin cycle make each reaction happen. Enzymes bring molecules of the cycle together in such a way that the entering carbon dioxide is eventually reorganized into a glucose molecule.

In the Calvin cycle, getting pulled into cells using RUBISCO to facilitate the process incorporates gaseous $CO_2$ molecules. RUBISCO acts much like a sponge, with a

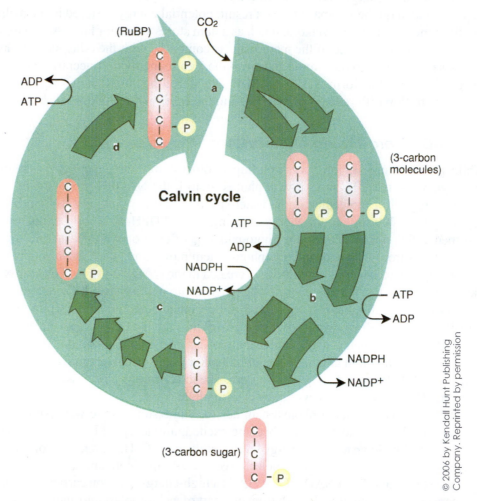

**Figure 4.11**   The Calvin cycle. Making sugar enables a plant to function. The light reactions are linked to the Calvin cycle. Energy from ATP and NADPH are used to drive the Calvin cycle, producing sugars from carbon dioxide and water. These were once called the "dark reactions" of photosynthesis because they do not require direct sunlight to function. They may occur in light or dark conditions. From *Biological Perspectives*, 3rd ed by BSCS.

great absorptive power to suck up $CO_2$ from air surrounding a plant. It "fixes" carbon onto another molecule, RuBP, or or ribulose 1,5biphosphate. RuBP is a five-carbon molecule. Thus, the Calvin cycle is also known as carbon fixation because carbon is literally fixed into position to grow a molecule of glucose. As indicated in Figure 4.11, the first part of the Calvin cycle is the fixation portion.

In the second stage of the Calvin cycle, chemical reorganization happens. Reorganization requires energy in the form of ATP and NADPH to rearrange bonds. The end result of this chemical reshuffling is G3P, or glyceraldehyde 3-phosphate (the 3-carbon sugar in Figure 4.11). A molecule of G3P is combined with another G3P to form glucose. Glucose is later transformed into any of the macromolecules through cell energetics. It takes three turns of the Calvin cycle and three molecules of $CO_2$ to form one, three-carbon G3P. It takes six turns to generate enough material to make one glucose molecule.

In the final stage of the Calvin cycle, as shown in Figure 4.11, some remaining G3P is used to regenerate the original five-carbon molecule of RuBP. This regeneration process requires ATP energy to reorganize G3P back into a five-carbon chain. All of the molecular players in this game of sugar production are reused. Thus, the Calvin cycle acts like a water wheel, continually turning to crank out sugar, using energy from ATP and NADPH. The synthesis portion of photosynthesis requires nine molecules of ATP and six molecules of NADPH from light reactions to make one molecule of G3P. Carbon dioxide is used to build a sugar molecule by the Calvin cycle, forming other macromolecules to allow Ms. Green's white pine to grow into such a large tree. Its great width and its height of 80 feet were possible because of the molecular players reused in photosynthesis.

## Some Like it Hot

Ms. Green's white pine tree functioned successfully in her garden, with ample water and optimal conditions. Her pine carried out the most common form of photosynthesis called the C3 pathway. This pathway is called C3 because it uses a three-carbon molecule in the Calvin cycle. Plants using the C3 pathway keep their stomata, small holes

**RuBP**

The first chemical in the Calvin Cycle, which combines with carbon dioxide.

**Carbon fixation**

The conversion process of carbon dioxide to organic compounds by living organisms.

**G3P**

Also known as glyceraldehyde 3-phosphate, is a chemical substance occurring as a product of the Calvin Cycle.

**C3 pathway**

The most common form of photosynthesis that uses a 3-carbon molecule in the Calvin cycle.

**Stomata**

A minute pore found in the epidermis of a plant's leaf or stem through which gas and water pass.

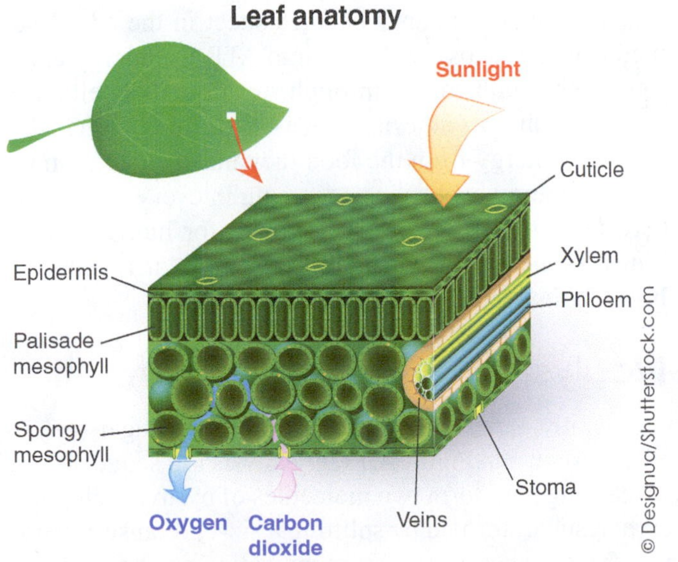

**Leaf anatomy**

Sunlight

Cuticle

Xylem

Phloem

Epidermis

Palisade mesophyll

Spongy mesophyll

Stoma

Oxygen   Carbon dioxide   Veins

© Designua/Shutterstock.com

**Figure 4.12** Structure of a leaf. The cross section of a plant leaf shows that its upper and lower layers are a protective waxy surface while its internal, mesophyll cells carry out photosynthesis. The vascular bundle transports water and food throughout the plant. Stomata, openings on the underside of a leaf, allow gas exchange between a plant and its environment.

on the underside of their leaves, open to obtain needed carbon dioxide gas. Stomata in C3 plants close in the night to conserve water, but remain open in the daytime to obtain needed chemicals for photosynthesis.

A drawback to open stomata is that some water evaporates from the plant, although in climates with sufficient rainfall this evaporation has little effect on the plant. In Ms. Green's garden, which she watered regularly, the C3 pathway of the pine tree functioned well, adding carbon mass every day. Over 95% of plants use the C3 pathway for photosynthesis.

However, some environments are harsher; they are hot and dry, with little rainfall. Some plants are able to survive in these areas through adapting two alternate forms of photosynthesis: the C4 pathway and the CAM pathway. The C4 pathway of photosynthesis uses a very absorbent sponge, an enzyme called phosphoenolpyruvate (PEP) carboxylase, to suck up carbon dioxide instead of RUBISCO. As a result, stomata may be only partially open and still obtain the required gas. Less water is lost by evaporation through stomata in C4 plants. However, because the C4 pathway uses a series of reactions to fix carbon into the Calvin cycle, it takes extra energy – this is a disadvantage. Overall, though, the C4 strategy is better suited for hot and dry conditions. C4 plants include corn, sugar cane, sorghum, and Bermuda grass.

Some desert plants such as orchids, pineapples cactuses, and even the Jade plant, a common houseplant, use the CAM pathway. The CAM pathway works at night, keeping their stomata closed in the day but open only at night. This method incorporates carbon dioxide into organic acids located in vacuoles during night time hours. When stomata are closed all day to prevent water loss, they may still obtain needed carbon dioxide at night with cooler temperatures and less evaporation. Carbon fixation occurs all day, with stomata closed, to produce glucose.

**C4 pathway**

A method used by plants to pull carbon dioxide into the Calvin Cycle more easily.

**CAM pathway**

A type of photosynthesis working at night and exhibited by plants that inhabit warm and dry areas.

# Cellular Respiration: Breaking It All Down

We've been looking at photosynthesis in plants – the process by which plants turn sunlight into energy. Now we turn to cellular respiration – how organisms turn food into energy that drives cellular processes. Cellular respiration occurs in a series of stages that may be compared with an accountant's balance sheet in the end. Energy is accounted for as it is changed from a glucose molecule into ATP, the energy currency of the cell.

Most living systems obtain energy through some form of cellular respiration. Even organisms that carry out photosynthesis, such as Ms. Green's pine, also carry out cellular respiration to obtain energy from the food they make. Energy in the form of ATP is used most easily, with energy exchanges occurring in every step of the many reactions in a living cell. To obtain energy from fuel, which for humans includes glucose and other carbohydrates, as well as proteins, and fats, cellular respiration occurs in three steps: 1) glycolysis, 2) the Krebs cycle, and 3) the ETC.

## Step 1: Glycolysis, the Upfront Investment

**Glycolysis**

Is a sequence of chemical steps in which glucose is rearranged to form two molecules of pyruvic acid, or pyruvate.

Glycolysis, which literally means the "splitting (-*lysis*) of sugar (-*glyco*)," occurs in the cytoplasm of cells. As shown in Figure 4.13, glycolysis is a sequence of chemical steps in which glucose is rearranged to form two molecules of pyruvic acid, or pyruvate. Pyruvic acid is a three-carbon sugar, formed by splitting a six-carbon sugar molecule. Much like a match lighting a fire, it takes a little bit of activation energy to get glycolysis started. Energy is used after eating a large meal because ATP is required to get glycolysis going.

Cellular respiration is a game of accounting; that is, counting numbers of energy molecules gained or lost in the processing of sugar through a cell. You can keep track of ATP gains and losses to see how much energy is obtained through cellular respiration.

As can be seen in Figure 4.13, the first steps of glycolysis require an input of one molecule of ATP energy to disrupt the sugar molecule enough to make it split into two. The first part of glycolysis requires an energy investment, and this part of the process is called the **energy- investment phase**. The 2 ATP investment is small compared to the payoff of energy in the long run of about 30–32 ATP per glucose molecule. **Aerobic respiration** results in the large ATP payoff, using oxygen as a final step to obtain this energy. Like using a match to light a fire, the energy gained by the end of the process is worth the small investment (the cost of the match).

The second part of glycolysis is the **energy-yielding phase**. Two molecules of pyruvic acid are produced by splitting glucose, and both go through the next series of reactions. As seen in Figure 4.13, a gain of 4 ATP energy molecules results from the processing of these two molecules. Because 2 ATP were used in the energy investment phase, a net gain of 2 ATP molecules results from glycolysis. This is enough energy for some organisms, which use glycolysis as their only energy source. When energy processing stops at this point, it is called anaerobic respiration, which does not use oxygen to complete glucose breakdown. Instead, organisms using this system obtain only a net of 2 ATP molecules per glucose. Bacteria on the roots of Ms. Green's tree, for example, use anaerobic respiration, only a modified form of glycolysis, for energy.

Rearrangements of glucose also give electrons to NAD⁺, or nicotinamide dinucleotide, to produce 2 NADH molecules. Electrons travel in pairs associated with hydrogen atoms and reduce NAD⁺. A molecule of NADH is a high-energy electron carrier, which later converts its potential energy to ATP energy. However, the remaining carbon skeleton of pyruvic acid needs to be reformed to allow it to move into mitochondria.

**NADH**

Nicotinamide adenine dinucleotide is a naturally occurring biological compound, which is converted to energy (not given in bold in text).

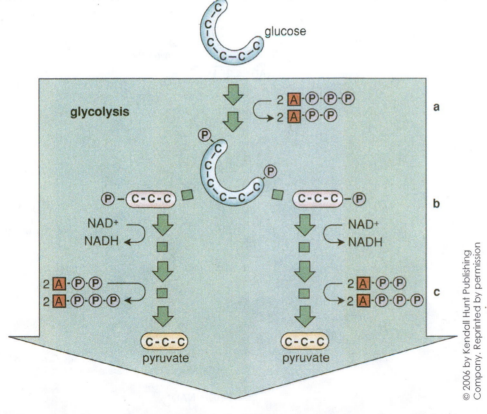

**Figure 4.13** Glycolysis: investment phase (a to b) and yield phase (b to c). In the investment phase, glycolysis uses 2 ATP molecules to destabilize a molecule of glucose. The yield phase produces a total of 4 ATP and 2 NADH energy molecules. From *Biological Perspectives*, 3rd ed by BSCS.

## Step 2: Moving Money

### The Energy Shuttle

While the steps of glycolysis take place in the cytoplasm, pyruvic acid must be transported into mitochondria to be further processed. Mitochondria are like a bank, exchanging energy much as a bank exchanges money. Pyruvic acid needs to be changed into acetyl-coenzyme A (*acetyl-CoA* for short), a form that is acceptable to the mitochondria bank. This process is called the acetyl-CoA shuttle system.

For the conversion to acetyl-CoA, pyruvic acid transfers its high-energy electrons to NAD⁺, producing NADH (as you can see at the top of Figure 4.14). This is the only energy produced by the shuttle system. Carbon dioxide is also released from the carbon skeleton in the process, which we exhale in our breath and a tree such as Ms. Green's releases into the atmosphere. Coenzyme A, a very large molecule sitting within the cytoplasm, acts as a shuttle for the remaining carbon chain. Carbon dioxide attaches to pyruvic acid, losing a high-energy electron pair (along with hydrogen) to form NADH and acetyl-CoA and enters into the mitochondrion. It costs the cell about 2 ATP to shuttle acetyl-CoA and its high-energy electrons in NADH into the mitochondrion.

**Krebs cycle**

A series of enzyme-catalyzed reactions forming an important part of aerobic respiration in cells.

## Step 3: Breaking Bonds and Giving Credit

### The Krebs Cycle

Acetyl-CoA is a two-carbon sugar that enters a series of eight steps known as the Krebs cycle, (also called the citric acid cycle). Bonds in acetyl-CoA store energy that needs to be transformed into something more usable. To do this, acetyl-CoA enters the Krebs

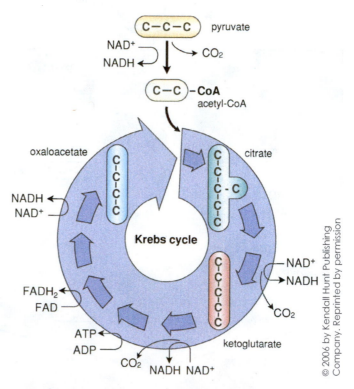

**Figure 4.14**   The Krebs Cycle (Citric Acid Cycle). NADH energy molecules and carbon dioxide gas are main products of the Krebs cycle. From *Biological Perspectives*, 3rd ed by BSCS.

cycle, which breaks its bonds, producing NADH molecules. NADH is later exchanged for ATP energy, much as foreign money is exchanged for U.S. currency in banks. The ATP (money) is later used for cell activities (to buy goods and services).

The two-carbon acetyl-CoA enters the Krebs cycle by attaching to a four-carbon molecule, oxaloacetic acid. Together, they form a six-carbon citrate. Citrate undergoes a series of bond changes that produce a large amount of high-energy electron carriers: six NADH and two $FADH_2$, or flavonoid dinucleotide molecules. Carbon dioxide and two ATP molecules are also generated by this cycle. The original oxaloacetic acid is also regenerated to continue the process over again, as shown in Figure 4.14. With each turn of the Krebs cycle, two carbons enter as acetyl-CoA, and two carbons leave as carbon dioxide. The carbon chain from the original glucose molecule is no more, but its bond energy is exchanged for credit (rather than direct ATP) in the form of NADH and $FADH_2$.

A large amount of carbon dioxide is formed by the Krebs cycle. In plants, carbon dioxide is used again in photosynthesis to reform new molecules of sugar. Ms. Green's pine tree has a convenient set up, with its products of cellular respiration readily reusable for carbon fixation in photosynthesis. Most energy from the Krebs cycle is still in the form of NADH and $FADH_2$. These credits are not usable by a cell until they are transformed into ATP, the energy currency of the cell. Bonds from entering acetyl-CoA have been transformed into high-energy molecules. How is this credit exchanged for ATP cash? You will see in the next section that the ETC exchanges NADH and $FADH_2$ for ATP energy.

## Step 4: Cash is King – Getting Money Exchanged

### Electron Transport Chain

The real energy payoff for an organism happens in the ETC, located on the inner membranes of the mitochondria. The ETC is a collection of molecules embedded in the cristae, the inner membrane of the mitochondria. The mitochondria contain two regions: the inside space within the cristae is called the matrix; and the material outside of cristae is called the intermembrane space (see Figure 4.15). The process is similar to that occurring in chloroplasts. In both systems, energy is produced as electrons fall to lower and lower energy levels. The energy currency of cells, ATP, is able to pass its energy as

**Cristae**

A fold in the inner membrane of the mitochondria.

**Matrix**

The inside space within the cristae.

**Intermembrane space**

The material found outside of cristae.

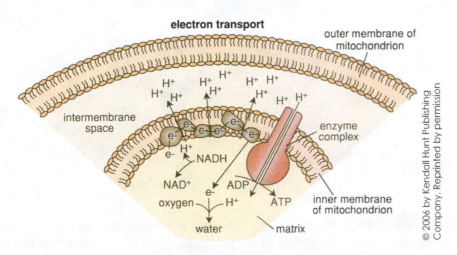

© 2006 by Kendall Hunt Publishing Company. Reprinted by permission

**Figure 4.15** Mitochondrial membranes. The electron transport chain occurs on the inner membranes of mitochondria. From *Biological Perspectives*, 3rd ed by BSCS.

phosphate bonds between molecules. So far, very little net gain of ATP has happened: 2 ATP from glycolysis and 2 ATP from the Krebs cycle.

Most of the molecules of the ETC in cristae are cytochromes, electron-holding carrier proteins. As shown in Figure 4.16, as electrons enter and move along cytochromes in the mitochondria's membrane, $H^+$ ions are pumped out, forming a gradient. Each cytochrome holds electrons at different energy states. Higher energy electrons enter the ETC at higher levels and lower energy electrons enter at lower levels. NADH carries electrons with the most energy, entering at a level higher than $FADH_2$. Upon entering the ETC, both NADH and $FADH_2$ pass an electron pair to a carrier protein. Recall that electrons travel in pairs with two hydrogen atoms associated. NADH and $FADH_2$ are then recycled back into the Krebs cycle.

Electron pairs in the ETC fall from carrier to carrier, each time releasing a bit of energy. At the lowest energy step, electrons (along with their $H^+$ companions) are passed onto a molecule of oxygen, $O_2$, which combines with hydrogen to form water. We exhale some water vapor as a byproduct of cellular respiration. Again, plants such as Ms. Green's pine release water, sometimes from the oxygen they themselves produce. The energy released from the ETC is a result of what happens as electrons fall. As each electron pair (traveling with hydrogen) drops down the chain, carriers pump hydrogen ions out of the matrix and into the intermembrane space. These pumps are shown in Figure 4.16.

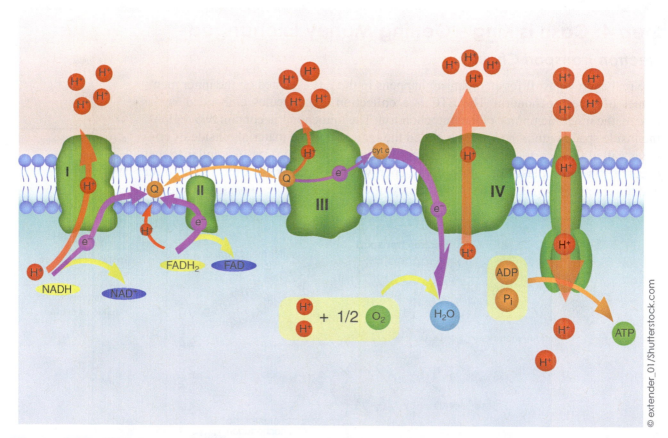

© extender_01/Shutterstock.com

**Figure 4.16** The electron transport chain. Hydrogen atoms from macromolecules (sugars, fats, proteins) travel along the membrane (with their electrons), leading to the production of ATP. Oxygen keeps the process flowing continually to give cells energy.

Because there are more hydrogen ions in the intermembrane space than in the matrix, a concentration gradient forms. Hydrogen ions, with more now outside than inside in Figure 4.16, flow down this concentration gradient into the matrix. This gradient leads to a force driving hydrogen ions to move across the membrane. Much potential energy is stored across the cristae (think of the force built up by a dam across a river). The force in this case is the strength of the H⁺ ions flowing through the membrane, back into the matrix. Movement of protons through the membrane transfers the stored energy into ATP molecules to be used by the cell.

Hydrogen ions, H⁺, are essentially protons without their electrons. As they accumulate in the intermembrane space, the matrix becomes relatively negative and the inner-membrane space, relatively positive. The stored energy across the cristae drives a proton motive force to push electrons across the cristae. The cristae function as a dam, allowing a flow of H⁺ whenever there is an opportunity. An enzyme embedded in the cristae, **ATP synthase** is the only place through which H⁺ may flow, containing special channels for H⁺. As H⁺ flows through ATP synthase, ADP is transformed into ATP by adding high-energy phosphates. Figure 4.16 illustrates the production of ATP from this proton-motive force.

Energy stored in NADH translates into 3 ATP molecules, and FADH₂ is worth about 2 ATPs. In an accounting of ATP produced by the ETC, with 10 NADH and 2 FADH₂ molecules funneled into the ETC, a total of 30 ATP are made from NADH and 4 ATP are made for each glucose molecule processed by cellular respiration. ETC itself garners a total of 30–32 ATP for a cell. Thus, over 90% of a cell's usable energy comes from the ETC. The maximum amount of energy derived from a glucose molecule is 36 ATP. Figures 4.17 and 4.18 track the energy and chemical exchanges during cellular respiration.

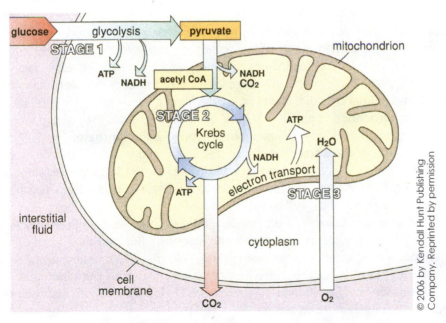

**Figure 4.17**  Overview of cellular respiration. To obtain energy from food, glucose moves through three stages: from blycolysis, to the Krebs cycle, and finally to the electron transport chain. From *Biological Perspectives*, 3ʳᵈ ed by BSCS.

Maximum Energy Produced for one Molecule of Glucose

| | | | |
|---|---|---|---|
| In the cytoplasm | | 2 ATP → 2 ATP = 2 ATP | |
| In mitochondria | | | |
| From glycolysis | 2 NADH → | 6 ATP → 6 ATP = 6 ATP | |
| Pyruvic acid → acetyl CoA | 1 NADH → | 3 ATP (×2) → 6 ATP = 6 ATP | |
| Krebs | | 1 ATP (×2) → 2 ATP | |

3 NADH → 9 ATP (×2) →               18 ATP = 24 ATP

1 FADH$_2$ → 2 ATP (×2)→                4 ATP

TOTAL = 44 ATP – (8 ATP lost as waste) = 36 ATP net gain

**Figure 4.18**   An accountant's balance sheet for cellular respiration: counting ATPs produced through the process of cell respiration. Courtesy Peter Daempfle.

Challenge Question: Trace the steps of cellular respiration by placing the following numbers in their correct order: 1) pyruvic acid, 2) ATP made in large amounts, 3) CO2 released in large amounts, 4) glucose, 5) Acetyl CoA, 6) entrance into mitochondria, 7) CO2 first released, and 8) ATP first used.

## CYANIDE AS A KILLER

Some chemicals interfere with the flow of electrons traveling down the ETC. Cyanide, a poison found in crime scenes of yesteryear, has greater pull on electrons than ETC cytochromes. Cyanide is thus able to pull electrons from the ETC preventing its flow to oxygen. This stops energy production from the ETC, and animal cells die. Plants rarely die from cyanide poisoning. While they contain mitochondria and an ETC just as animal cells do, they also contain an enzyme that breaks down cyanide, beta-cyanoalanine synthase.

Fluoride is also a toxic substance that is harmful in large doses to humans. When fluoride was first added to toothpastes in 1914, its use was not supported by the American Dental Association (ADA). Fluoride in toothpaste was widely rejected as well by many consumers.

Proctor and Gamble, a pharmaceutical company, worked feverishly in the 1950s to show both the uses and the safety of fluoride as a part of daily hygiene. Then, after intense testing, in 1960 the ADA issued a statement approving fluoride toothpaste. Their research supported the claim that fluoride was beneficial to humans in small doses. Evidence shows it also works by remineralizing enamel on teeth.

Fluoride is, in fact, poison to all living systems including humans. However, the fluoride in toothpaste is in such small doses that it is harmless. Fluoride as a toothpaste additive helps to inhibit bacterial growth by literally "sucking up" electrons from bacteria's biochemical pathways. This works in the same way as cyanide described earlier. As you recall from Chapter 2, fluoride is the most electronegative element. In toothpaste, it is used to pull electrons from the ETC of bacteria in our mouths, killing the bacteria and preventing the acid production that causes tooth decay.

In 2006, the biotech company, BioRepair, began testing the first toothpaste with the additive hydroxyapatite to prevent dental caries. Hydroxyapatite works differently from fluoride to prevent caries. Hydroxyapatite adds an extra layer onto the enamel of a tooth. The extra enamel protects a tooth from bacterial acid. Hydroxyapatite adds strength to bone material. This breakthrough may supersede fluoride's effects to change dental health.

# Bioprocessing: Where does the Cash Get Used?

Once there is available ATP energy, cells are able to build whatever resources they require from raw materials. Some materials are needed for growth, some for reproduction, and some materials are used to restructure or reorganize parts of cells. Evolution has developed pathways for living systems to change glucose and intermediates of cellular respiration into any macromolecule.

The sum total of all the reactions in a living system is known as its metabolism. You may have heard the term before referred to in diets – perhaps to describe a person as having a "fast" or "slow" metabolism – but metabolism is a very complex series of energy exchanges. There are two forms of metabolism: anabolism and catabolism. Anabolism is the series of reactions that builds up complex molecules using stored energy. Photosynthesis is an example of anabolism because it uses energy and raw materials to produce a larger glucose molecule. The process does not happen spontaneously; it requires an input of energy. Catabolism is the series of reactions that break down complex molecules to yield energy. Cellular respiration is an example of catabolism because it breaks a molecule of glucose down, releasing its stored energy. It occurs spontaneously, without a net energy input. There are trillions of metabolic reactions occurring at any one time in humans, classified as either anabolic or catabolic. Both anabolism and catabolism work together to perform life functions.

The building up (anabolism) and breaking down (catabolism) of macromolecules are together collectively known as bioprocessing (see Figure 4.19). When macromolecules such as lipids, carbohydrates, and proteins are needed for energy, they undergo catabolism. Alternately, when macromolecules are in short supply, cells will produce more of them through anabolism. Both are vital for cell functioning.

Carbohydrates, as you recall from chapter 2, are long chains of simple sugars. In order to obtain energy from carbohydrates, the chains must be broken apart and processed in the steps of cellular respiration. The same sequence of steps occurs, with carbohydrate products added at different points in cellular respiration. When carbohydrates are needed, cells will form longer chains from shorter chains of simple sugars through anabolism.

When proteins are broken down, their toxic nitrogen groups are eliminated by cells. Their carbon skeleton is reused, either forming new amino acids or being shuttled into cellular respiration for breakdown and energy. Figure 4.20 shows the process of nitrogen removal from amino acids, called **deamination**. In humans, deamination occurs in the liver, where urea forms, then is expelled as urine.

**Metabolism**

The sum total of all the reactions taking place in a living system.

**Anabolism**

A series of reactions that builds up complex molecules using stored energy.

**Catabolism**

A series of reactions that break down complex molecules to yield energy.

**Bioprocessing**

The process of building up (anabolism) and breaking down (catabolism) of macromolecules.

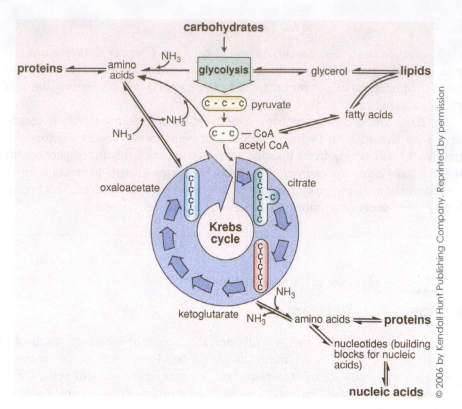

**Figure 4.19** Bioprocessing. Fats, carbohydrates, and proteins move through the same set of chemical reactions to release energy. From *Biological Perspectives*, 3rded by BSCS.

When lipids are broken down by a process called **lipolysis**, they form fatty acids and glycerol. Fatty acids are inserted into the Krebs cycle for breakdown, and glycerol is input into glycolysis, as shown in Figure 4.19. Fat catabolism releases much energy from its bonds. Building up of fats, called **lipogenesis**, is also a needed process. When sufficient ATP and glucose are available, the required fats are made into triglycerides. These are later converted into different forms of fat.

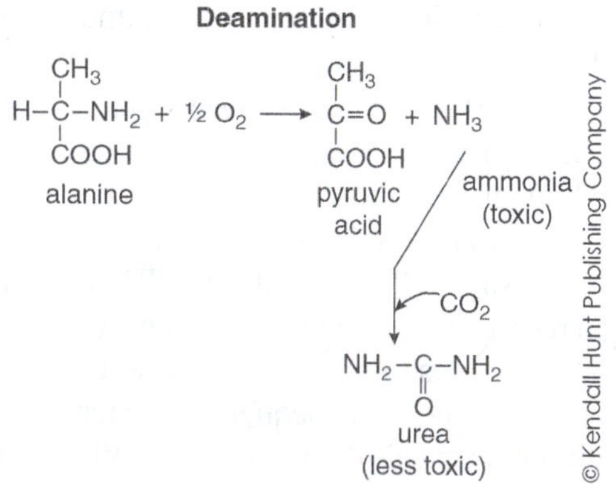

**Deamination**

$$CH_3 \qquad\qquad CH_3$$
$$H-C-NH_2 + \tfrac{1}{2}O_2 \longrightarrow C=O + NH_3$$
$$COOH \qquad\qquad COOH$$

alanine           pyruvic acid      ammonia (toxic)

$$CO_2$$

$$NH_2-C-NH_2$$
$$\parallel$$
$$O$$
urea (less toxic)

**Figure 4.20** Deamination. Deamination is a process in which an amine group is removed from protein, causing toxic nitrogen-containing materials such as ammonia to be formed within living systems. When ammonia is combined with carbon dioxide in the liver, a less toxic nitrogen-containing compound is produced, called urea, which can be excreted safely from the body.

# Beer, Wine, and Muscle Pain

Glycolysis releases up to 25% of the stored energy in glucose. Much of this energy is not immediately available, as it must first pass to the mitochondrion for processing. But for some organisms, glycolysis is their only energy-yielding process. These organisms usually use glycolysis only when there is no oxygen present.

Glycolysis requires no oxygen and is often a part of anaerobic respiration. Anaerobic respiration or fermentation, mentioned briefly earlier in the chapter, is the series of reactions that form alcohol from sugar. Its steps give off a little bit of energy in the process, enough to sustain cells. However, most of the energy obtained by anaerobic organisms is lost as the alcohol waste product. This is why aerobic respiration, which goes through all three phases of cellular respiration (glycolysis, Krebs, and the ETC) yields so much more energy using oxygen. The process of aerobic respiration described above is more involved but is also much more efficient.

## Anaerobic respiration

Have you ever had pain in your muscles during intense exercise? Try to do a wall sit for about five minutes, and a burning sensation will spread through your upper leg muscles (quadriceps). This sensation is due to anaerobic respiration. A lack of available oxygen forces cells to do the next best thing – obtain energy through glycolysis. Because lactic acid is its by-product, the pH of muscles decreases as lactic acid accumulates. Lactic acid reduces the ability of muscle fibers to contract and causes muscle fatigue.

After completing an intense exercise, however, sensations of burning stop after a short period. Aerobic respiration proceeds to allow enough oxygen to get to all cells. Lactic acid breaks back down, by the liver and into energy. Lactic acid also attracts mosquitos, which is why sweating during exercise outdoors can make us appeal to our insect friends.

Glycolysis is also able to sustain life functions in many single-celled organisms such as yeast and bacteria. Anaerobic respiration, at least, yields 2 ATP molecules to keep its cells going. There is a cost: the waste product discards much unused energy. Some other organisms that carry out anaerobic respiration to produce lactic acid include *Streptococcus mutans*, a bacterium that dissolves tooth enamel to cause dental caries (cavities); and *Lactobacillus acidophilus*, a bacterium that curdles milk and makes cheese and yogurt, both use lactic acid fermentation as their source of energy (Figure 4.21).

## Fermentation

Consider alcohol, in our beverages and foods. Yeast cells carry out fermentation, a special kind of anaerobic respiration yielding low amounts of energy from sugars, when oxygen is not present (see Figure 4.21b). These cells are capable of more efficient aerobic processes, but will carry out fermentation in the absence of oxygen. Yeast converts pyruvic acid, made by glycolysis, into acetaldehyde, and in the process releases bubbles of carbon dioxide. Acetaldehyde rearranges, recycling $NAD^+$ while producing ethanol. Ethanol is used as a fuel source, in spirits to give a kick, and in cleaning products, such as rubbing alcohol. Depending upon the type of food that is fermented, different forms of alcohol are produced. Grape fermentation produces wine, fermentation of a germinating barley plant produces beer, and potato fermentation makes vodka. The same process of fermentation occurs regardless of the food source and alcohol product.

**Anaerobic respiration**

A series of reactions that form alcohol from sugar.

**Fermentation**

A special kind of anaerobic respiration yielding low amounts of energy from sugars, when oxygen is not present.

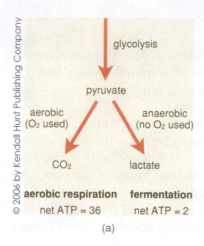

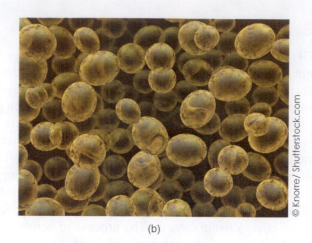

**Figure 4.21** a. Anaerobic respiration: human lactic acid system. While it provides very little energy for a cell, a small amount of energy from anaerobic respiration is better than no energy. Yeast's alcohol fermentation. Alcoholic fermentation in beer is accomplished by Saccharomyces, a type of yeast carrying out anaerobic respiration to produce ethanol. From *Biological Perspectives*, 3rded by BSCS. Reprinted by permission. b. This photo shows Baker's yeast. It carries out anaerobic respiration to produce ethanol.

## Alcohol and Cellular Respiration: Is it OK for Me to Drink Heavily Just in College?

In college, a social life is important, and alcohol remains the drug of choice at parties as well as school-sanctioned social functions. Understanding the effects of alcohol is important to maintain health. Alcohol affects several processes involved in cellular respiration and causes organelle changes and organ damage.

Alcohol's effects on the liver are the main problems of heavy drinking. The liver breaks down toxic substances, including alcohol. Alcohol, in the form of ethanol ($CH_3CH_2OH$), is catabolized by the liver to form acetaldehyde ($CH_3CHO$). Acetaldehyde (the good guy) stimulates the release of brain chemicals that give us pleasure. The next time you are at a party, suggest this, and say ". . . you actually want a glass of acetaldehyde." This is sure to win you friends! Acetaldehyde breaks down into carbon dioxide and water vapor, which are exhaled.

Let's review the steps of cellular respiration: Recall that the first set of reactions in cellular respiration, glycolysis, makes sugar into pyruvic acid and reduces $NAD^+$ to NADH. Second, pyruvic acid is shuttled into the Krebs cycle to make more $NAD^+$ into NADH. The third step in getting energy from food, the ETC, converts the NADH into usable energy. Alcohol slows down the first two steps (glycolysis and Krebs cycle) but greatly increases the third step (electron transport).

What is the problem? Extra hydrogen from the ethanol is removed to form acetaldehyde. Extra hydrogen (with the associated electrons) attaches to $NAD^+$, preventing free $NAD^+$ from being used in cellular respiration. This prevents food stuffs from being broken down.

The extra hydrogen atoms are the bad guys; they are the culprits in liver disease. Hydrogen from ethanol occupies the $NAD^+$ that would otherwise be used for glycolysis and the Krebs cycle. Instead, with $NAD^+$ no longer available, macromolecules (proteins, carbohydrates, and fat) in the liver sit idle and turn into fat. Foods do not go through the three steps (glycolysis, Krebs cycle, and electron transport).

Fats accumulate in the liver cells (also called a fatty liver), and cells die due to malfunctioning in this strange situation. Dead liver cells trigger an inflammation called alcoholic hepatitis. More and more liver cells die in this inflammation, causing scarring known as cirrhosis. Cirrhosis of the liver is the ninth leading cause of death in the United States.

Evidence for this mechanism is in the abnormal structure of liver tissue. Livers of heavy drinkers have enlarged mitochondria because of the exaggerated processes of electron transport occurring with extra NADH. Liver endoplasmic reticulum, which processes the excess fat onto proteins, also enlarges in such livers, illustrating the effects of increased fat deposits in cirrhotic livers.

The effects of alcohol on the liver are dangerous, but alcohol is also related to numerous other health problems. Long-term usage effects are high-blood pressure; heart and kidney disease; a weakened immune system; cancers of the esophagus, stomach, mouth, and liver; obesity; and muscle loss. Short-term effects include, of course, the hangover. Alcohol's effects on cell energetics are worthy of supporting the argument against excess alcohol usage.

You may be thinking, "For all this to happen it must take a long time. Thank goodness I have time to tone it down." But the research shows otherwise . . . yes, bad news. In a study conducted by Lieber and colleagues at the Bronx Veterans Administration Hospital and the Mount Sinai School of Medicine in New York City, in a very short time (18 days) of heavy drinking (six 10-ounce drinks of eight to six proof/per day) an eightfold increase in fat deposits in the liver was seen. These subjects were human volunteers fed a high-protein, low-fat diet to see if a good diet mattered. The myth of eating a good diet to protect from alcohol's effects was not supported by this study.

## BUDDHA'S TREE: FICUS RELIGIOSA GIVES AN ENLIGHTENMENT – BODHI

Buddhism, a religion with 300 million believers, seeks to find peace through a life of good actions. One tenet of Buddhism is an appreciation for other life – to respect it and care for other organisms – which results in good karma, or fortune, and a release after death to a better life. The spiritual leader of Buddhism, known as Buddha, is said to have achieved enlightenment or "Bodhi," under a large and old sacred fig tree, *Ficus religiosa* in Bodh, India over 2,000 years ago.

This same tree still grows today at the Mahabodhi Temple in Bodh Gaya, India. It is a sacred fig tree believed to be a sapling cut from the historical tree under which Buddha became enlightened. This tree, planted in 288 B.C. is the oldest living human-planted tree on Earth. It has a known date of planting making the tree, Jaya Sri Mahabohdi, over 2,300 years old. This tree is a frequent destination for Buddhist pilgrims and uses cell energetics processes in our chapter to grow and survive for so long.

The enlightenment experienced by Ms. Green under her white pine at the end of our opening story parallels the kind of connection to life Buddha felt in his experience at the *Ficus religiosa*. Ms. Green expresses an acceptance of life's ending but finds peace in her continuity with other life on Earth, namely the new family of birds atop her white pine.

The peaceful end of Ms. Green in the story is a goal of Buddhism, to enable one to transition to the next life, perhaps in the form of other animals or of other humans. Buddhism teaches that life may change to other forms after death but does not end. Her good karma from the garden and the pine prepared her for the life that was yet to come. Through giving to other life, she is, at the end of the story, free to "fly with the birds."

# Summary

Cell energetics comprise a complex interaction of steps occurring within organisms. Photosynthesis and cellular respiration, the two key processes in cell energetics, manufacture energy, store and release that energy when needed. The discovery of the ingredients and mechanism of these two processes required a confluence of many scientists' work. Physical and chemical principles determine the way cell energetics take place. Sunlight is the ultimate source of life, provides base nutrition for life on our planet. As energy flows through the environment, chemical interchanges form and reform molecular players. Energy and atoms are recycled to perpetuate life. Some organisms use only portions of cell energetics for their energy, such as anaerobic bacteria. Some organisms use both photosynthesis and cellular respiration in their processes, such as plants.

---

## CHECK OUT

### Summary: Key Points

- Cell energetics affect our environment and human health in many ways, from regenerating our air to processing the food we eat.
- The discovery of its processes of cell energy exchanges took scientists from van Helmont and Priestly to Lavoisier.
- The first and second laws of thermodynamics determine how energy is exchanged within cells and through the universe.
- Chloroplasts have unique properties that enable it to fix carbon from sunlight, carbon dioxide gas, and water.
- Mitochondria have unique properties that enable it to extract energy from glucose molecules using oxygen.
- Evolution of photosynthesis to CAM and C4 systems has resulted in advantages for some plants.
- Evolution of cellular respiration from anaerobic to aerobic systems has resulted in an advantage in energy extraction for eukaryotes.
- Bioprocessing changes materials taken in by organisms into many forms.

---

## KEY TERMS

| | |
|---|---|
| anabolism | cellular respiration |
| anaerobic respiration | chlorophyll a |
| autotroph | cristae |
| bioprocessing | cytochrome |
| C3 pathway | electromagnetic energy |
| C4 pathway | electron transport chain (ETC) |
| CAM pathway | entropy |
| Calvin cycle | excited state |
| carbon fixation | fermentation |
| carnivore | first law of thermodynamics |
| catabolism | G3P |

| | |
|---|---|
| glycolysis | photooxidation |
| ground state | photosynthesis |
| herbivore | photosystems |
| intermembrane space | pigment |
| Krebs cycle | primary electron acceptor |
| light reactions | primary consumer |
| light-independent reactions | producer |
| matrix | proton motive force |
| metabolism | radiant energy |
| NADH | RUBISCO |
| NADPH | RuBP |
| omnivore | second law of thermodynamics |
| photolysis | stomata |
| photon | thermodynamics |

# Multiple Choice Questions

1. How do the products of photosynthesis improve conditions on Earth for humans?
   a. There is more oxygen for cellular respiration.
   b. There is more carbon dioxide for photosynthesis.
   c. There is more water vapor for bioprocessing.
   d. There is more CAM and C3 forms of photosynthesis.

2. Which scientist measured the growth of plant matter to conclude that water was the source of it mass?
   a. Lavoisier
   b. Priestly
   c. de Saussure
   d. van Helmont

3. Which term BEST describes the breakdown of glucose?
   a. anabolism
   b. catabolism
   c. photosynthesis
   d. metabolism

4. A cheetah, which eats deer as its prey is classified as:
   a. a carnivore
   b. a herbivore
   c. a producer
   d. an autotroph

5. If a chemical reaction spontaneously gathers raw materials to produce an organized cluster of chemicals, it would violate:
   a. diffusion
   b. light-dependent reactions
   c. first law of thermodynamics
   d. second law of thermodynamics

6. Which represents a logical flow of higher energy electrons to lower energy electrons in photosynthesis?
   a. photosystem II → photosystem I → chlorophyll a → water
   b. photosystem I → photosystem II → chlorophyll a → water
   c. water → photosystem I → chlorophyll a à → photosystem II
   d. chlorophyll a → photosystem II → Photosystem I → water

7. Which is the source of energy, driving the Calvin cycle?
   a. NADH
   b. $NAD^+$
   c. chlorophyll a
   d. RUBISCO

8. When a plant keeps stomata closed all day long, it is a sign that the system of photosynthesis it is carrying out is:

   a. C3 photosynthesis.
   b. C4 photosynthesis.
   c. CAM photosynthesis.
   d. Light-dependent photosynthesis.

9. In question #8 above, which chemical reactions are occurring at night?

   a. Calvin Cycle
   b. Photolysis of water
   c. Photosystem I
   d. Photosystem II

10. In question #8, which process for a plant directly obtains the MOST ATP energy from a molecule of glucose?

    a. Calvin cycle
    b. Photosystems II
    c. Glycolysis
    d. Electron transport chain

## Short Answers

1. Describe how cell metabolism affects the processing of a pear as it moves through the process of cellular respiration. Be sure to list each step of cellular respiration and account for the energy released from the pear at each step.

2. Define the following terms: anabolism and catabolism. List one way to explain how each of the terms differs from each other in relation to cellular respiration and photosynthesis.

3. Describe the experiments of two scientists: Joseph Priestly and Jan Baptista van Helmont. Use a drawing to make the descriptions clear. Show your art work. How did each discover an aspect of photosynthesis? How did their knowledge build upon one another's?

4. Trace the flow of carbon within the process of photosynthesis. Be sure to include the following terms in your description: NADPH, ATP, Calvin cycle, RUBISCO, G3P.

5. For question #4 above, how are the light reactions of photosynthesis connected to the Calvin cycle?

6. If a green plant is exposed to only green light in a laboratory, predict what will happen to the green plant. Why?

7. Explain the advantages and disadvantages of the C4 pathway for photosynthesis. Under which conditions would a C4 plant have an advantage or a disadvantage?

8. Trace the flow of a carbon atom from glycolysis to the Krebs cycle. Be sure to include the following terms: glucose, acetyl CoA, NADH, pyruvic acid, mitochondrion, and cytoplasm. Why is there no need for carbon in the electron transport chain?

9. Explain how 40 ATP are produced from the processes of cellular respiration and yet only about 36 ATP are actually extracted.

10. A yeast cell produces beer for a beer enthusiast. He works in his basement to concoct the beverage. What processes occur to make his beer? Under what conditions do you recommend he place his yeast to make beer?

## Biology and Society Corner: Discussion Questions

1. A slice of pizza contains drizzled cheese and oils. There are 298 calories per slice, with 37% fat, 47% carbohydrates, and only 14% protein. Compare this with a serving of deer meat, which contains only 32 calories per ounce and has 18% fat, 0% carbohydrates, and 82% protein. Which types of processing result more from a diet high in cheese pizza as compared with deer meat?

2. How would van Helmont have used information from this chapter to help his hypothesis about plant growth? Why?

3. If a person would be able to live as long as a tree, Ms. Green in our story would not have died before her white pine. Senescence is the study of aging. Research the characteristics of pine trees that scientists believe allow its longevity. Based on your research, what part of a plant cell should future research look into to discover how humans might live as long. Why?

**4.** Acid rain is a danger to photosynthetic plants as well as other organisms within the environment. How is acid rain affecting photosynthesis within phytoplankton? Based on Priestley's early results, how might its effects harm humans and other organisms?

**5.** A newspaper claims: "Who cares about trees? . . . They have less impact on our environment's air quality than other organisms." Defend this statement . . . then also refute this statement. Use your knowledge of photosynthesis and cellular respiration to answer.

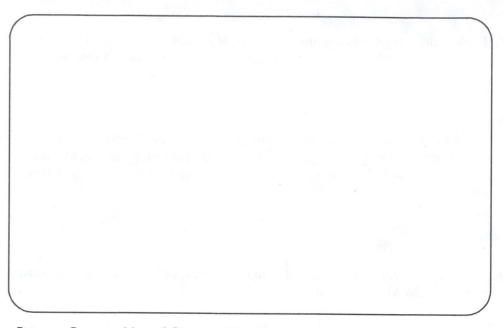

Figure – Concept Map of Chapter 4 Big Ideas

# UNIT 2
## Is it all in the Genes?

# Molecular Genetics

## ESSENTIALS

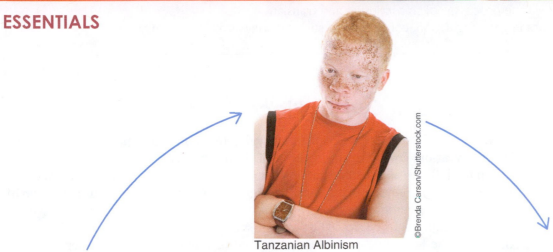

Tanzanian Albinism

©Brenda Carson/Shutterstock.com

Melanin Pigment in skin provides a dark color. It darkens skin as sunlight penetrates its layers

©dean bertoncelj/Shutterstock.com

DNA provides the instructions for life, including skin color

©Dabarti CGI/Shutterstock.com

**DNA replication**

DNA polymerase
Original DNA
Topoisomerase
Lagging stand
Okazaki fragment
RNA primer
Primase
Helicase
Parent DNA
Leading stand

©Designua/Shutterstock.com

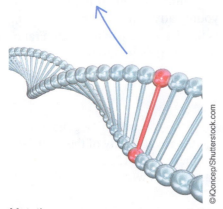

Mutation on a gene

©iQoncep/Shutterstock.com

DNA Replication makes more of itself passing itself onto new generations

## CHECK IN

**From reading this chapter, students will be able to:**

- Examine how genetics affects society and our everyday lives.
- Explain the scientific development of big ideas in genetics and life's origins.
- Describe and draw DNA structure and compare it with RNA.
- Explain the process of DNA replication.
- Use base sequences to view DNA as the universal language of genetics, and connect DNA to protein production via the processes of transcription and translation.
- Analyze errors in gene regulation, connecting to such diseases as albinism and cancer.

## The Case of Out-of-Place Color

"It was a terrible sight," recalled the emergency room attending physician, Dr. Franc. Fourteen-year-old Joyce Carl had been rushed into the ER last week with bruises and cuts on her left arm and shoulder, after a man tried to abduct her while she was walking home from school. Joyce resisted, and the aggressor tried to cut off her arm. A group of onlookers helped her to escape.

**Albinism**

Is a noncontagious disease that is genetically inherited and results from a lack of pigmentation.

Even uninjured, Joyce would be a strange spectacle here in Dar es Salaam, Tanzania. An albino, she had white skin, blonde hair, and pale blue eyes, making her stand out dramatically among her black peers. Albinism is a noncontagious disease that is rare in most parts of the world but fairly common in Tanzania, affecting one in 2,000 people. Genetically inherited from both mother and father, it results from a lack of pigmentation. Eyes, skin, and hair are without color, and individuals with the disorder are highly susceptible to skin cancers and burns.

Dr. Franc also knew that in Tanzania, as well as other parts of sub-Saharan Africa, albinos are believed to have magical powers. It is not a compliment to their difference. Witch doctors sell albinos' hair, skin, bones, and internal organs on the open market as ingredients for potions that are supposed to make people rich. With an arm going for $2,000, about 20 Tanzanian albinos are killed each year for their body parts. This is a great deal of money in developing economies, equivalent to over USD $200,000. Joyce was almost a victim, and an estimated 170,000 albinos live in fear.

Dr. Franc told Joyce she would leave tomorrow morning, given that her condition was improving. She was a friendly and well-adjusted young lady and he wished her well. But as he was leaving the room she asked him, "Why is this happening to me?" Dr. Franc knew that human genetics is a powerful force in society as well as in our bodies.

## Early Ideas about Genetics

At times, a very obvious family trait is handed down from generation to generation. Consider the distinctive facial features of the Hapsburgs, the royal family of the Austrian Empire in Europe, which dominated the political scene there from 1282 to1918. Many of its members had a protruding lower lip that became associated with the wealthy upper class of old Austria (Figure 5.1).

While it was easy to observe certain physical characteristics, like the protruding lower lip, which have passed from one generation to the next, understanding how

## CHECK UP SECTION

There are examples of past and present discrimination in the U.S. society based on genetic differences. Choose a particular case to research. Explain parallels you see between it and the discrimination seen for Tanzanian albinos.

characteristics were passed on came only recently. Life arises by using information passed down from parents. The information chemical was not discovered until much later in our history. This chapter will explore the structure and function of this inherited information.

Underlying the question of inherited characteristics is a more basic one: how did life arise? Some thought we spontaneously developed without a need for parents: that life simply arose on its own, nurtured within a womb. The theory that life could arise from nonlife is termed *spontaneous generation*. This question has been pondered and answers posed at least since the Greek philosopher Aristotle. The first recorded scientific consideration for the question of how life began came from Aristotle. He believed that a male's semen was an imperfect mixture of materials that, when united with "female semen," would combine to make a more perfect human offspring. Nothing more than this was known about how our species was formed.

© Anton_Ivanov/Shutterstock.com

**Figure 5.1** Kaiserin (Empress) Maria Theresia of the Austrian Empire—The Hapsburg Royal Family. The protruding lips of the Hapsburg family members clearly identify them as related to one another.

**Animalcules**

The dated term for a microscopic animal, we now know of as microorganisms.

Ideas on spontaneous generation went generally unchallenged for over 3,000 years. Then, in 1677, Anton van Leeuwenhoek, the Dutch lens maker (discussed in an chapter 3), observed living material under his newly developed microscope. He described the "little animals or animalcules" in the many water samples he took from lakes. These are what we now know to be microorganisms or microbes. He saw bacteria and fungi on cheese, bacteria in his saliva, and sperm from his own samples (his own sperm sample he did not publicly disclose due to religious rules at the time). Van Leeuwenhoek described the objects he saw as "little eels or worms, lying all huddled up together and wriggling. …This was for me, among all the marvels that I have discovered in nature, the most marvelous of all." But it was Robert Hooke who first coined the term "cell" to describe structures composing the plants he observed, which looked to him like a monk's cell or quarters in a monastery (as discussed in earlier chapters).

The sperm van Leeuwenhoek and several contemporaries observed, were described as little humans encased in a special cell with a tail. Thus, they reasoned, any resemblance of a child to its mother was due to her internal chemicals influencing the fetus' development - that a fully formed human came from fathers. Figure 5.2 depicts the image van Leeuwenhoek and his contemporaries claimed to have seen under the microscope. We now know that these images were sperm cells with a nucleus and not a fully formed organism. The sperm's composition is quite a bit different from what van Leeuwenhoek supposed. Ideas about the start of life changed over the centuries and remain a source of public debate, as discussed in Bioethics Box 5.1.

Further experiments on plants and animals yielded a change in thinking. From work on the pollination of flowers and trees, it became clear that both male and female parents contributed to the next generation of plants. Knowledge about animal reproduction advanced to show that it took a fusion of egg and sperm to create a new organism. This understanding raised the question: What exactly *is* being inherited by the offspring? Spontaneous generation was disproved in an experiment by Louis Pasteur in the mid-1800s. In his experiment, he constructed special flasks with elongated necks to keep out microbes. He showed that life would appear only from other life. However, a clear mechanism explaining how organisms inherited information from parents had not yet been suggested.

Gregor Mendel, an Austrian monk working in his garden in the late 1800s, studied pea plants and their changing characteristics through successive generations. He noted that certain patterns of inheritance emerged, and that predictions about offspring could

© Kendall Hunt Publishing Company

**Figure 5.2**   Homunculus, (little man) a future human being, preformed in a human sperm.

be made from observations about the mating parents. His studies are considered to mark the birth of genetics and give Gregor Mendel the title of "founder of genetics" as a modern discipline of study. Mendel's conclusions will be discussed in detail in the next chapter. The laws of inheritance described by Mendel can explain Joyce Carl's albinism.

At about the same time that Mendel was making observations about patterns of inheritance, other scientists had observed structures within cells that moved apart when a cell divided. These structures were called chromosomes, which are compact bodies that are inherited from one cell generation to the next (as discussed in chapter 3). Experimentation showed that chromosomes were made up of two substances: proteins and deoxyribonucleic acid (DNA).

The discovery of DNA was made by Friedrich Miescher, a German chemist who in 1869 extracted a substance from the nuclei of cells he was working with. This substance was white, slightly acidic, and contained phosphorous; Miescher called it an organic acid. However, little work was done on DNA for decades, because it was not known to be the agent of heredity. Only in the mid-20th century did biochemists find out that DNA was hereditary material; this discovery unleashed a flurry of research to determine its molecular structure.

To do this, early in 1928, Frederick Griffith, a British microbiologist, conducted experiments on mice to determine which material is inherited: proteins or DNA? He studied the effects of a pathogenic (disease-causing) strain of *Streptococcus pneumonia* bacteria. He unexpectedly noticed that when he injected mice with a set of different strains of the same bacteria, sometimes they would not get sick. There were two forms of *S. pneumonia*: the R- strain, which lacks a polysaccharide coat making it appear rough and an S-strain, which had a coat, making it appear smooth in shape. He determined that those bacteria having a surrounding polysaccharide coat (the S-strain) were pathogenic, or disease causing. The coat must have conferred some sort of protection from the immune system of the mouse that allowed coated bacteria to cause the disease. When Griffith heated the coated S-strains to kill them and injected the dead bacteria into the mouse, the mouse lived. Bacteria were not able to hurt the mouse because they were dead. When Griffiths mixed heat-killed virulent S-strain bacteria and live nonpathogenic R-strain bacteria, however, the mouse died. How could a dead, nonpathogenic bacteria kill the mouse? Griffith concluded that a chemical possessed by the heat-killed virulent bacteria must be a transforming agent. It changed the living R-strain bacteria from one type to another, into a S-strain type (see Figure 5.3). In 1944, Oswald Avery, a professor at the Rockefeller University, isolated and analyzed this chemical, determining it to be deoxyribonucleic acid. The story of Joyce Carl is based on this mystery material and how it leads to skin color. We will explore the role of DNA in determining our unique characteristics as this chapter unfolds.

Thus, it became clear, through a series of experiments in the early part of the 20th century, that 1) DNA was inherited; and 2) DNA was the chemical that directed new cell production as well as all cellular activities. However, it was unclear what this new molecule looked like and how it actually worked.

Many scientists contributed their ideas and expertise to discovering the shape of DNA. U.S. chemist Linus Pauling showed that protein chains of amino acids were helical in shape, like a slinky, and were held together by hydrogen bonds between successive turns. He suggested that the DNA molecule could resemble such a structure, and he was right. X-ray diffraction showed that there were turns in the DNA molecule and that certain chemicals, known as nitrogenous bases, occurred in regular, repeating patterns.

- Nitrogenous bases, as described in Chapter 2, are an important component of nucleotides, the building blocks of DNA.

**Chromosomes**

Structures within cells that move apart when a cell divides.

Genetic Transformation

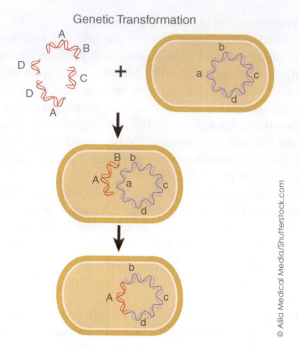

© Alila Medical Media/Shutterstock.com

**Figure 5.3**   Griffiths Experiment. Bacterial transformation, as show in the figure, was the key to Griffith's experiement. Bacteria ingest genetic material from the environment and begin to exhibit those chracteristics of that genetic material. As he studied vaccines for pneumonia, Griffith discovered that bacteria could mutate quickly. One dead bacterial cell (heat-killed S-cell) could be consumed by another bacteria (R-cell), transforming it into another type of S-cell. It is a bit like a horror story, eating another creature and becoming just like it.

Bases adenine and thymine occurred in the same proportions, and guanine and cytosine also occurred together in the same proportions. This finding led to further discovery about DNA structure. The repeating patterns of the bases indicated that they must be paired together in a certain way.

When English scientist Francis Crick and American James Watson were working in Cambridge University in 1953, they used information from several sources to develop a new model for DNA. The 23-year-old James Watson, a newly minted PhD traveled to London to visit the lab of Maurice Wilkins at King's College. There he discovered an X-ray image of DNA taken by Rosalind Franklin, Wilkins's colleague (Figure 5.4). Watson studied the image to deduce the shape of DNA to be of a certain size and shape, and from that developed a clue for the model. Franklin's work indeed gave Watson the idea for his model, but she did not receive credit in the publication describing the arrangement of DNA, as described in the Bioethics Box 5.2.

Watson and Crick put all the research from the varied sources together, figuring out just how DNA is inherited from generation to generation. This work represented the birth of molecular genetics, a new field that united biology, chemistry, and genetics, to study inheritance at the chemical level. Inheritance could now be explored at its most elemental level.

Watson and Crick's model is used to explain many aspects of chemical inheritance, such as: the way in which DNA reproduced itself; the way it is transformed into protein for a cell's use; and how DNA directs the many activities within the cell. Watson and Crick's discovery is more than a simple description of a chemical; it is the basis for explaining how information is passed on from generation to generation and within cells. Their model will be used throughout this chapter to describe that information flow.

**Molecular genetics**

A new field that united biology, chemistry and genetics, to study inheritance at the chemical level.

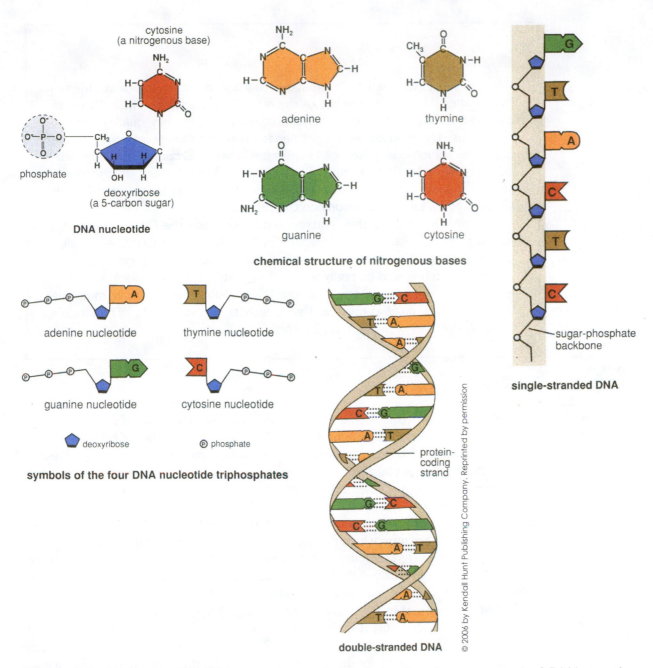

**Figure 5.4** DNA uncoiled. DNA contains weak bonds holding the two strand of DNA together. These break easily allowing DNA to unzip and expose the base pairs. The sugar-phosphate backbone of the DNA molecule provides support to the structure. Note its base pairs, which, when exposed, gives DNA its unique informational message. From *Biological Perspectives*, 3rd ed by BSCS.

# DNA As an Inherited Substance

## The Structure of DNA

Watson and Crick showed that DNA resembles a twisted ladder, with sugars and phosphates on the vertical parts of the ladder and bases, making up the rungs of the ladder. The sugars and phosphates comprise the backbone of the DNA molecule. This type of structure is known as a "double helix." As discussed in Chapter 2, DNA

## CRICK'S PERSPECTIVE ON DNA

Watson and Crick used their inductive abilities to gather information from many other scientists and publicize the model we use today. Their discovery captured the interest of the scientific community and they quickly became celebrities. This interest helped get needed research money into their developing field of molecular biology. However, it takes a certain humble appreciation for the molecule as a thing of beauty, to really understand its meaning in society and not the scientist's fame. Consider the excerpt below from Francis Crick in 1974, for his reflections on this process:

Rather than believe that Watson and Crick made the DNA structure, I would rather stress that the structure made Watson and Crick. After all, I was almost totally unknown at the time and Watson was regarded, in most circles, as too bright to be really sound. But what I think is overlooked in such arguments is the intrinsic beauty of the DNA double helix. It is the molecule which has style, quite as much as the scientists. - Francis Crick, "The Double Helix: A Personal View," *Nature*, 26 April 1974.

**Nucleotide**

The basic functional unit of a DNA molecule.

**Ribose**

The sugar backbone found in RNA.

**Deoxyribose**

The sugar backbone found in DNA.

**DNA**

A long macromolecule containing the information code that directs cellular activities in living organisms.

**Adenine**

A purine base that is a component of RNA and DNA.

**Thymine**

A pyrimidine base that is found in DNA but not RNA.

**Guanine**

A purine base that functions as a fundamental constituent of RNA and DNA.

**Cytosine**

A type of base found in DNA.

**Uracil**

A pyrimidine base that is one of the fundamental components of RNA.

is a **nucleic acid**, a macromolecule that stores information – the "code of life" – in strings of base sequences. Each of these bases constitutes a code to guide the cell's activities. The exact structure of DNA, as sketched out by Watson and Crick, is shown in Figures 4.5 and 5.5. Sugars and phosphates hold the up-down portions of the molecule together while bases pair with each other in the horizontal levels of the DNA in the figure.

The basic functional unit of a DNA molecule is the nucleotide, which is made of three parts: a **sugar** backbone, either ribose, found in **RNA**, or deoxyribose, found in DNA; a **phosphate group**, which contains four oxygen atoms bound to a central phosphate and is negatively charged and very acidic; and a nitrogen-containing molecule that is the **base**. Bases make up the genetic code. There are four types of bases: adenine, thymine, guanine and cytosine, commonly written as **A, T, G**, and **C, but with U** (uracil) **instead of T (thymine) in RNA**. Bases are held together by weak hydrogen bonds that are able to be taken apart relatively easily when they are used to direct the cell's activities or make new DNA. (Figure 5.4 and Figure 5.7 shows all of these components.) Bases comprise a set of directions for the cell much like a set of directions used to find a particular location or address.

Just as you may arrive at the wrong place if there is a mistake in the directions, a cell may have problems when its DNA has errors. In the case of cells, an error in a base in the DNA sequence is called a **mutation** and can lead to problems for living cells. Mutations are responsible for a number of diseases that will be treated in the next chapters. Mutations are sometimes caused by environmental factors. They are an example of how the environment has an influence on how DNA becomes expressed. For example, factors in the environment, such as radiation or harmful substances, increase the risk of mutations and therefore change the genetic structure (Figure 5.6). Changes in DNA's structure also lead to changes in organisms' characteristics. Mutations are what led to Joyce Carl's albinism because a simple change in nucleotide sequence is the cause of her lighter skin.

Bases couple together specifically: adenine always pairs with thymine (A-T) and guanine always pairs with cytosine (G-C), for example. This special coupling is called

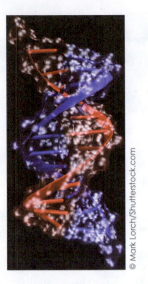

© Mark Lorch/Shutterstock.com

**Figure 5.5**  DNA's structure.

complementarity. You can remember this particular base pairing by recalling the phrase: "**AT** the **G**rand **C**anyon," where the initials AT connect adenine and thymine, and the Grand Canyon's initials G and C stand for linked guanine and cytosine.

Just what material do we inherit from one generation to the next? We inherit genes. We have all heard someone saying, "It's in your genes!" A gene is a discrete bit of data on the DNA molecule, usually a series of nucleotides that code for information to be used by the cell. A gene is the functional unit of heredity and the main player in information transfer within cells. It carries the codes for all of our characteristics. Humans have between 20,000 and 25,000 genes in their cells' nuclei. A single gene has over 100,000 nucleotide pairs, and a DNA molecule contains over 200 million base pairs. A full set of human DNA is estimated to contain over 3 billion base pairs, or enough information to fit into 600,000 printed pages of 500 words each. In essence, that is equivalent to 1000 library books! How can all this information fit into a single cell when it is thousands of

**Complementarity**

The specific coupling of bases.

**Gene**

A portion DNA sequence serving as the basic unit of heredity.

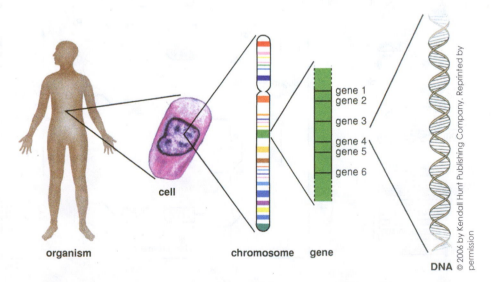

organism                    cell                    chromosome    gene                    DNA

gene 1
gene 2
gene 3
gene 4
gene 5
gene 6

© 2006 by Kendall Hunt Publishing Company. Reprinted by permission

**Figure 5.6**  Melanin mutation on the genetic code. A mutated melanin (gene #3 in this hypothetical example) gene does not code for a proper functioning melanin pigment molecule. From *Biological Perspectives*, 3rd ed by BSCS.

### BIOETHICS BOX 5.1: WHY WAS DR. ROSALIND FRANKLIN, A BRILLIANT SCIENTIST, IGNORED BY HER COLLEAGUES?

The truth behind how DNA was discovered starts with King's College in London, 1953. Dr. Franklin was born in 1920 and earned a doctorate in physical chemistry at the age of 26, against her father's wishes. She worked at Kings College, refining X-ray diffraction to produce the "Photograph 51" that helped Watson and Crick develop their model of DNA.

When Dr. Franklin was a colleague of Maurice Wilkins, she was treated as a mere helper and suffered gender discrimination in a male-dominated field. Although her work was published in the same issue of the journal *Nature*, as the Watson and Crick paper, in April, 1953, she was not given credit for her contributions to the model.

Tragically, she died of ovarian cancer in 1958, at the age of 37, four years before Watson and Crick received the Nobel Prize for their work. It is likely that she died for the model . . . She worked hundreds of hours to perfect her photographs of crystallized DNA, exposing herself to large doses of radiation. The photo below shows scientist Rosalind Franklin's X-ray image of DNA. Most scientists are aware of her contributions today, and Dr. Franklin is given posthumous credit in this textbook (Figure 5.7).

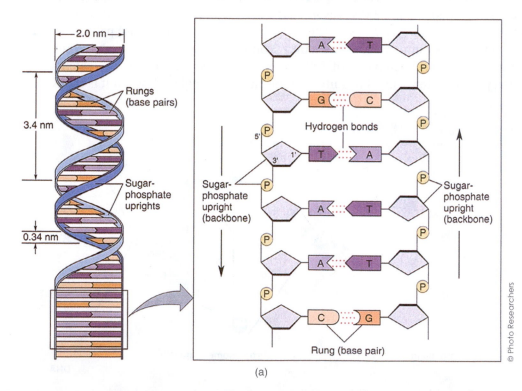

(a)

**Figure 5.7**    a. and b. Watson and Crick Model of DNA. c. X-ray image of DNA by Franklin. Franklin helped Watson and Crick to develop their model of DNA structure in the 1950s.

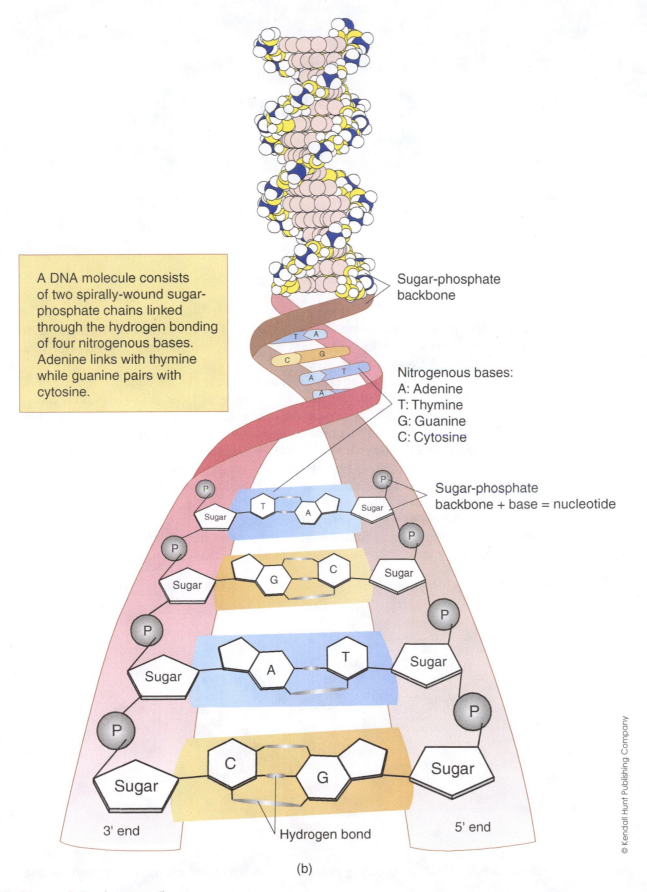

A DNA molecule consists of two spirally-wound sugar-phosphate chains linked through the hydrogen bonding of four nitrogenous bases. Adenine links with thymine while guanine pairs with cytosine.

Sugar-phosphate backbone

Nitrogenous bases:
A: Adenine
T: Thymine
G: Guanine
C: Cytosine

Sugar-phosphate backbone + base = nucleotide

3' end

Hydrogen bond

5' end

(b)

**Figure 5.7** (*continued*)

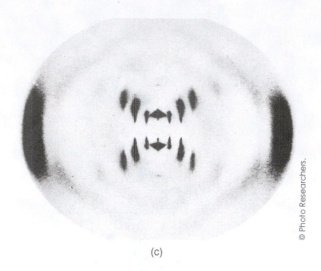

(c)

**Figure 5.7** *(continued)*

**Histone**

Group of basic
proteins in chromatin.

times longer than the cell itself? The answer is that DNA is **supercoiled**, (Figure 5.8) which means that it is packaged together very tightly around histone **proteins**, like a shoelace wrapped around your fingers.

DNA is common to all living creatures: humans, bacteria, mushrooms, and oak trees all contain the same type of macromolecules. A full set of DNA in an organism is called its *genome*. In bacteria, as in all prokaryotes, the genome is naked and circular; it is not surrounded by a nucleus and occurs as a continuous series of nucleotide bases. Eukaryotes contain genomes packaged into discrete units as chromosomes. Each species has a unique, set number of chromosomes. Fruit flies have only 8, corn has 20, and dogs have 78. All eukaryotes contain chromosomes in pairs and sometimes organisms of different species have the same number of chormosomes.

Humans have a total of 46 chromosomes, with 23 pairs of them. In humans, chromosomes are inherited, one from a mother and another from a father. Before a cell can divide, chromosome pairs must double in number so that each new cell will contain a

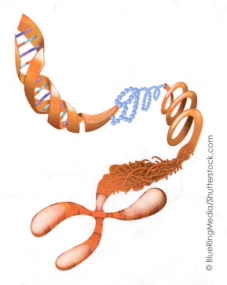

**Figure 5.8** DNA supercoiling genes on DNA are coiled extensively around histone proteins.

full set. The chromosomes then move to two separate new cells during cell division. How the DNA makes itself into two full sets was finally answered by Watson and Crick's model.

# How Does Eukaryotic DNA Reproduce Itself?

## Mitosis

During roughly 90% of an average cell's life cycle, it is actively conducting the many normal cell functions described in Chapter 3. This period of time is known as the interphase for a cell. In the remaining 10% of the time, the cell divides via mitosis.

The cell cycle, or the life span phases a cell goes through, includes mitosis (see Figure 5.9). The cell cycle involves the division of the cytoplasm and nucleus to produce two new identical daughter cells from one original cell. During interphase, a cell gets ready for mitosis by. doubling its genetic material, increasing its number of organelles and its cytoplasm size.

There are three phases of interphase: $G_1$ phase, S Phase, and $G_2$ phase (see Figure 5.9). During the $G_1$ or growth-1 phase, a cell grows rapidly in size, forming new organelles and proteins for future daughter cells. Centrioles, a special microtubule units used for mitosis, is made during $G_1$ of interphase. During the S, or synthesis, phase, chromosomes are synthesized to duplicate the genetic material. It is just before the start of the S phase that a cell "decides" to divide or not. Why at this point? Because once a cell enters the S phase, the large investment in doubling the DNA is too great to turn back; a cell must continue to divide into two daughter cells. The decision to either become nondividing or begin DNA synthesis depends on two major factors: 1) a cells' cell-to-volume ratio. A cell needs a certain amount of cytoplasm to be able to function normally. As discussed in Chapter 3, that ratio is set to allow cells to transport materials to every region; and 2) the presence

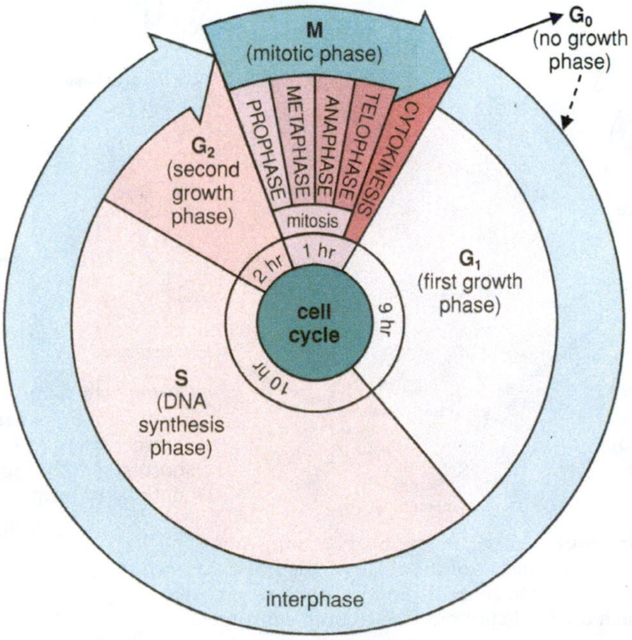

**Interphase**

The stage in cell development following two successive mitotic or meiotic divisions

**Cell cycle**

The life span phases a cell goes through.

**S phase**

A period in the cell cycle in which DNA is replicated.

**$G_1$ phase**

A period in the cell cycle in which a cell grows rapidly in size, forming new organelles and proteins for future daughter cells.

**$G_2$ phase**

A period in the cell cycle in which growth of the cell's cytoplasm and organelles is completed and final preparations for mitosis takes place.

**Figure 5.9**   The cell cycle. Cells undergo a series of phases throughout their life cycle; growing and dividing, with checkpoints regulating the process. From *Biological Perspectives*, 3rd ed by BSCS.

of MPF, mitosis-phase promoting factor. This chemical triggers mitosis by activating proteins that help in the cell-division process. Finally, during the $G_2$ phase, growth of the cell's cytoplasm and organelles is completed and final preparations for mitosis takes place.

Mitosis then occurs, with cells dividing into identical new cells. **Mitosis** is defined as the process by which the nucleus and nuclear components divide, resulting in two new identical cells. Mitosis produces new cells for healing cut skin, making new organs in an embryo, or giving rise to a newly created single-celled organism such as an *Amoeba*. The stages of mitosis are shown in Figure 5.10.

Cells are constantly being reproduced in the human body. As body cells wear out, they need to be replaced or repaired. Consider human bones: Did you know that all of our bones are replaced every five to ten years? Bones remodel continually according to a variety of factors: whether there is enough calcium for their development or whether forces are placed upon the bones through exercise to build more mass, to name a couple. Thus, cells divide to accomplish the life functions of an organism.

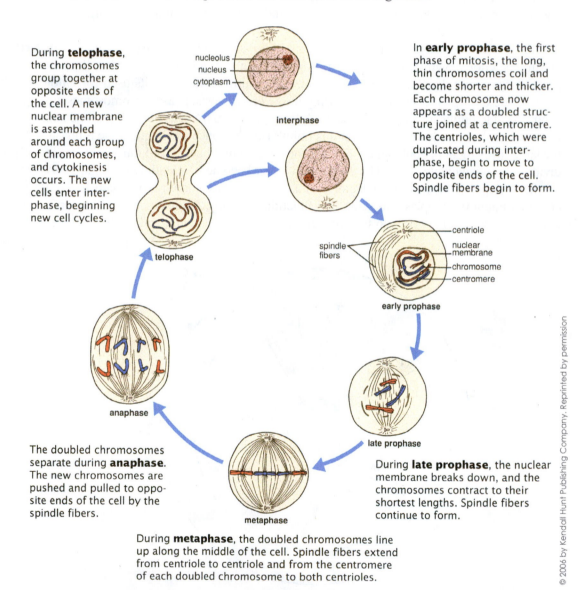

During **telophase**, the chromosomes group together at opposite ends of the cell. A new nuclear membrane is assembled around each group of chromosomes, and cytokinesis occurs. The new cells enter interphase, beginning new cell cycles.

nucleolus
nucleus
cytoplasm

interphase

In **early prophase**, the first phase of mitosis, the long, thin chromosomes coil and become shorter and thicker. Each chromosome now appears as a doubled structure joined at a centromere. The centrioles, which were duplicated during interphase, begin to move to opposite ends of the cell. Spindle fibers begin to form.

telophase

spindle
fibers

centriole
nuclear
membrane
chromosome
centromere

early prophase

anaphase

The doubled chromosomes separate during **anaphase**. The new chromosomes are pushed and pulled to opposite ends of the cell by the spindle fibers.

metaphase

late prophase

During **late prophase**, the nuclear membrane breaks down, and the chromosomes contract to their shortest lengths. Spindle fibers continue to form.

During **metaphase**, the doubled chromosomes line up along the middle of the cell. Spindle fibers extend from centriole to centriole and from the centromere of each doubled chromosome to both centrioles.

**Figure 5.10**   Stages of mitosis. A cell with four pairs of chromosomes is shown here, dividing. The stages of mitosis and their respective characteristics are given for each phase. From *Biological Perspectives*, 3rd ed by BSCS.

Mitosis occurs in an orderly manner, with set steps for eukaryotic cells. The first phase of mitosis is prophase (see Figure 5.10 to follow the steps of mitosis). Prophase is characterized by chromatin being packaged into chromosomes in a cell's nucleus. Chromatin is the thin, strewn about form of DNA in the nucleus. Chromatin condenses to form chromosomes. Chromosomes, as dense bodies, can then be transported as discrete packages into new cells as the cell divides. During prophase, the nucleus disintegrates and centrioles move to opposite ends of the cell. The function of centrioles is not completely understood, but they are thought to organize a network of spindle fibers. Spindle fibers are made up of microtubules, later used for pulling chromosomes to opposite ends of the cell.

The next phase is metaphase, during which chromosomes line up at the middle of the nucleus, the equator, attaching to the spindle fibers. Sister chromatids (identical strand of the duplicated chromosomes) attach to spindle fibers via a kinetochore, which is like a protein handle, securing chromosomes by way of a centromere, to the microtubules of the spindle, as shown in Figure 5.10. During anaphase, identical chromosomes move apart while attached to the spindle fibers. This physically divides genetic material to opposite sides of a cell, called poles. The mechanism believed to occur is by microtubules shortening, which pulls their attached chromosomes to the poles. At this point, genetic material at the poles is identical to that of its parents.

The last phase is called the telophase, during which time there is a reversal of the events occurring during prophase. Two new nuclei start to reform at the poles, chromosomes elongate, forming chromatin once again, and the cell's cytoplasm pinches, forming an indentation along the equator of the cell. This final process, whereby the division of the cytoplasm takes place, is called cytokinesis. In animal cells, the first sign of cytokinesis is the formation of a shallow groove along the equator of the cell. This pinching of cytoplasm is actually a cleavage furrow formed by a pulling of microfilaments. Quickly, much like purse strings, the cleavage furrow deepens to divide cytoplasm, forming two new cells. In plant, algae, fungal, and some bacterial cells, a cell plate forms at the equator, with vesicles from the Golgi apparatus coalescing to form two new plasma membranes and later, two new cell walls.

**Prophase**

A stage that is characterized by chromatin being packaged into chromosomes in a cell's nucleus.

**Metaphase**

A phase in which chromosomes line up at the middle of the nucleus, the equator, attaching to the spindle fibers.

**Anaphase**

A cell division stage in which chromosomes split into two identical groups and move toward the opposite poles of cells.

**Telophase**

A phase during which time there is a reversal of the events occurring during prophase.

**Cytokinesis**

The division of cell cytoplasm.

---

### A WAY TO REMEMBER THE PHASES OF MITOSIS

To help you remember these phases, mitosis can be analogous to the dating process. When your date arrives during the *prophase*, do you clean up your room or keep it sloppy? Of course, you tidy up your clothes and try to make a good impression. Much like in dating, a cell organizes its chromatin by making it into chromosomes. Then in the middle of the date, or *metaphase*, you have a good time; then you talk about too much biology, and Ana, in the *anaphase*, starts running away from you. The same thing occurs in the anaphase of cell division, in which chromosomes are running away to opposite poles of the cell. Have you ever experienced that heinous call or even text message which states: "It's over between us . . ." and dead silences or no responses show it is really the end? This series of events is analogous to the telophase. Ana is breaking up with you on the telephone just like the cell breaks up during telophase. One would hope that a breakup would be via telephone and not a mere text! This analogy may help you remember the salient aspects of the mitotic phases.

# Molecular Processes during Mitosis

**Initiation sequence**

A sequence of bases that starts the unwinding of DNA during replication.

**Helicase**

The enzyme that untwists the double helix so that replication can occur.

**DNA polymerase**

Special enzymes that add new bases onto the exposed DNA strands.

**Semi-conservative model**

A mode by which DNA replicates as half-new and half-old DNA.

**Nucleoside triphosphate**

A molecule that contains a nucleoside bound to three phosphates.

**Replication fork**

Molecules with both its sides exposed for adding bases.

In order to carry out mitosis, just before the start of cell division, DNA doubles itself during the **S (synthesis) phase**. This doubling process is termed **replication** and follows a series of specified steps. First, during the **unwinding phase**, the vertical strands unwind beginning at a certain sequence of bases called the Initiation sequence. Helicase is the enzyme that untwists the double helix so replication can occur. This untwisting is much like a zipper unzipping. This step exposes bases within the DNA so that new bases may be added onto the exposed strands by special enzymes called DNA polymerases. In the next **rebuilding phase**, the exposed base strand allows a new layer of nucleotides to form along the existing base sequence. Each of the single strands becomes a double strand. To accomplish this, another set of enzymes, DNA polymerases, carry complementary bases to the exposed regions, so that, for example, whenever an A is exposed a T binds to it and whenever a G is exposed a C binds to it. In this way, each new strand contains an old set of nucleotides and a new set of nucleotides. The end result is two double-stranded DNA molecules, both identical to its parent strand: half original material and half newly placed by the rebuilding phase. Because of this half-new and half-old DNA structure, this process of replicating DNA is termed semi-conservative. See Figures 5.11 and 5.12, which illustrates this doubling of DNA.

Energy needed to add bases during replication is obtained by DNA polymerases through hydrolyzing nucleoside triphosphate (a relative of ATP). DNA polymerases move along the DNA molecule in a certain way. DNA polymerases add bases in a certain direction on the DNA molecule, as shown in Figure 5.11. Notice that the unzipped DNA looks like a fork, – in fact it is called a replication fork – with two sides of the molecule exposed for adding bases. DNA polymerases can add nucleotides only where there are already existing nucleotides in place. A **primer** is laid down to start the process. This is a segment of RNA molecules 10 nucleotides in length. DNA polymerase in Figure 5.12 recognizes this sequence and begins adding nucleotides. In this part of the replication fork, the DNA polymerase moves

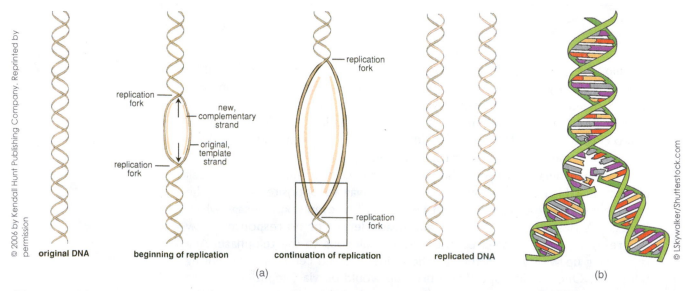

original DNA        beginning of replication        continuation of replication        replicated DNA

(a)        (b)

© LSkywalker/Shutterstock.com

**Figure 5.11**    Semi-conservative model of DNA showing replication fork. a. During replication DNA polymerase lays down a new set of nucleotides in a replication fork. From *Biological Perspectives*, 3rd ed by BSCS. b. The new neucleotide strand is complementary to an old strand.

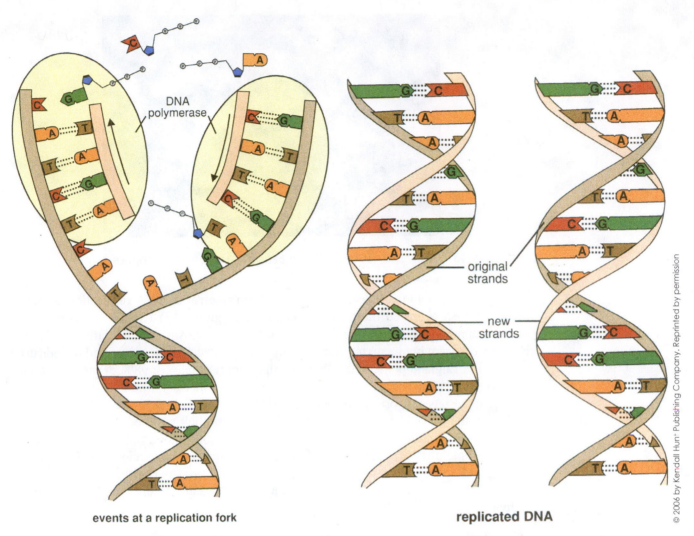

events at a replication fork      **replicated DNA**

- original strands
- new strands

**Figure 5.12**    DNA replication showing a chemistry clip of replication fork adding complementary bases. From *Biological Perspectives*, 3rd ed by BSCS.

smoothly to produce newly added segments. As the fragments are laid down, primers are detached and DNA ligases combine the segments of DNA to create one smooth DNA molecule.

There are approximately six billion base pairs in our 46 chromosomes. In order for replication to occur, multiple areas are untwisting at any one time, with nucleotides being added constantly and rapidly. While there are so many possibilities for mutation or error, only 1 in 100,000 errors actually occur. While this may appear like a small number, when considering how much DNA a cell has, three billion base pairs, errors would result in more than 120,000 mistakes every time a cell simply divides. To remedy these problems, there are about 50 different types of DNA repair enzymes, which remove mutated nucleotides and replace them with correct complementary ones. DNA polymerases and DNA ligases both work together with DNA repair enzymes to create a new strand of DNA, moving along the DNA molecule at a speed of about 20 base pairs per second in humans. Of course, errors in replication still happen, leading to mutation and gene changes. Joyce Carl experienced the effects of these mutations on select genes controlling albinism.

**DNA ligase**

A type of enzyme that joins DNA strands together.

**Figure 5.13**    A light variation of *Biston betularia*, the English peppered moth.

Another example of a disorder resulting from an error in DNA replication is found in the English peppered moth, *Biston betularia* (Figure 5.13). There are two varieties of this peppered moth: light- and dark-colored. It was shown that the dark color arose because of a rare but recurring mutation in some moths. During the period of industrialization in England in the 1800s, dark moths comprised almost 98% of the population in the city of Manchester, which was known for its sooty conditions. Why might this change in variety proportions have occurred? Studies by H.B.D Kettlewell showed that darker moths thrived because they blended better with their polluted surroundings than lighter colored moths. He hypothesized that lighter moths stood out to predators. This is an example of how mutations can help some members of a species to survive. Having a genetic variation in populations because of mutations is healthy for any species' survival. Changing environmental conditions sometimes allow survival of at least some individuals.

## Why Go through It All? Prokaryotic Cell Division Is More Simple

**Binary fission**

The process by which a cell divides directly in half.

**Circular genome**

Genetic material in a circular form found in prokaryotes.

Bacteria do not go through the many steps of mitosis to reproduce. Instead, they divide using a process called binary fission. Genetic material is in a circular form in prokaryotes, within what is termed a circular genome. Prokaryotes do not have a nuclear envelop to protect their genetic material. Thus, DNA in prokaryotes is called "naked DNA," without a nuclear covering. Circular, naked DNA divides to form two new circular DNAs, each attaching to two different areas of the cell membrane. As a prokaryotic cell grows, it pulls the genetic material to opposite ends of the cell. Cytokinesis then occurs, forming the physical separation between the cells. Separate, smaller circles of genetic material, called plasmids, carry information for specific activities within a cell. Plasmids replicate independently of the circular genome, also moving into new daughter cells.

How does genetic diversity get maintained in prokaryotes when parents are identical genetically to offspring? Mutations during replication are a primary source of difference between prokaryotes. Diversity is also promoted through bacterial genetic exchange or sometimes called "bacterial sex" because DNA is getting transferred between organisms. In this process, genetic material is exchanged through pili, or hair-like structures, that connect two bacterial cells through which DNA is exchanged.

Prokaryotes have been on Earth almost since its origin – for over 3.9 billion years. Prokaryote genetic resilience is astounding. Bacterial diversity is maintained through mutation because there are many cell divisions and many chances for errors and mutations. Consider that, on average, it takes a generation, or 25 years, for humans to reproduce. It takes bacteria only minutes to form new organisms. This rapidity in reproduction allows for more chances for mutation; and thus more chances for genetic variation. In our story, Joyce suffers for her genetic variation, but without it, any species would die off quickly, unable to withstand changing environmental conditions. Mutations can range from being unnoticed to creating monstrous results, depending on how the gene sequence is transmitted to the rest of the cell.

## DNA Is the Universal Language

While Watson and Crick's model of DNA established it as the hereditary material and explained the mechanism for replication, it did not explain how the instructions are carried out. The way the genetic code is read and expressed in living systems will be discussed in this next sections. In short, we need to answer the question: "How are genes expressed?"

Genes control the functions of any cell by giving directions or orders to carry out. The orders are found in its nucleotide sequence that is read to allow the genes to be expressed. How does DNA accomplish this? The four types of nucleotide bases in DNA form a nearly infinite number of possible combinations to create a language of information to transfer from DNA to cell structures. DNA has complex instructions for building cells and organelles for every type of organism. All of life's diversity depends on this molecule and the many possible arrangements of nucleotides that create its language.

After Watson and Crick's model described its structure, DNA was studied to determine what it actually does. Linus Pauling reasoned that disease was due to differences in chemicals between normal and afflicted people. In particular, Pauling studied hemoglobin protein differences between people with sickle-cell anemia, carriers for the disease, and normal individuals. Sickle-cell anemia is a disease that leads to abnormally shaped red blood cells, poor oxygen carrying capacity, and a host of complications such as blood clots and organ damage. Carriers for sickle-cell anemia only have one copy of the sickle cell gene and sometimes show symptoms, but usually only under extreme circumstances (e.g., severe dehydration, physical exhaustion, or high altitude). Those afflicted with the disease carry both copies of the sickle cell genes.

Pauling used **electrophoresis**, a process of separating organic materials based on their electric charges, and found a difference in hemoglobin's proteins: the hemoglobin of normal people carries a stronger negative charge than that of sickle cell sufferers (Figure 5.14). The hemoglobin of sickle-cell carriers has a charge somewhere in between. This was the basis for determining that genes affect protein structure. Thus, Pauling deduced that, because proteins are found everywhere in the body doing so many varied things, DNA must somehow affect protein structure. Pauling's ideas eventually led to the discovery that our genes give directions to make up to 2 million different proteins. In the example of sickle-cell anemia, some years after Pauling, Vernon Ingram showed that there was only one difference in proteins between sickle cell hemoglobin and normal hemoglobin. One in 300 amino acids was changed, which was enough to reorient the entire structure leading to a sickle-shaped red blood cell.

The carrier's (and an afflicted person's) hemoglobin shape was later determined to confer some degree of immunity to a tropical disease called malaria. Malaria is an infectious disease spread by mosquitos carrying a parasite that invades red blood cells and reproduces in them. It causes flu-like symptoms, ranging from fever and chills to

**Sickle cell anemia**

A disease that leads to abnormally shaped red blood cells, poor oxygen carrying capacity, and a host of complications such as blood clots and organ damage.

**Malaria**

An infectious disease spread by mosquitos carrying a parasite that invades red blood cells and reproduces in them.

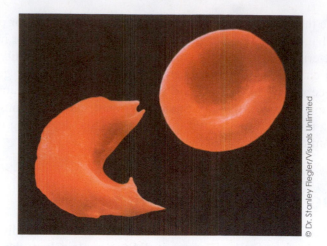

© Dr. Stanley Flegler/Visuals Unlimited

**Figure 5.14**    Genetic differences between normal and sickle-cell gene sequences lead to changes in red blood cells shown above, with the abnormal red blood cell shaped like a sickle. Only one base pair change (from GAG to GUG) between normal and sickle cell DNA causes one amino acid change and, thus, the disease. From *Biological Perspectives*, 3rd ed by BSCS.

respiratory problems, coma, and death, if left untreated. How does an immunity to malaria manifest? Basically, the alternate hemoglobin shape of sickle cell gene holders (carriers and afflicted individuals) causes enough 3-D changes in their red blood cells to prevent the malarial parasite from getting into red blood cells and dividing. *Plasmodium* is the protist that infects human red blood cells, causing this malady. Possessing gene copies for sickle-cell anemia prevents *Plasmodium* from entering the human red blood cell.

There are more than 300 million cases of malaria each year, and it kills about 1 million people in Africa and Asia annually, according to the World Health Organization 2010 World Malaria Report. Sickle-cell carriers have 1/10th the likelihood of contracting the most dangerous form of cerebral malaria. Thus, being a carrier for sickle-cell anemia in tropical areas, where the disease is most likely to spread, is beneficial. The carriers' benefits come at a price though, because some individuals in the population are going to have sickle-cell anemia. This is an example of how a harmful mutation, such as sickle-cell anemia, can persist in a population: when there is a benefit to survival, such as in this case (in warm areas affected by the disease), the mutation will continue to be expressed for thousands of years.

## What Do Proteins Do?

As discussed in Chapter 2, proteins perform almost every aspect of an organism's means for maintaining an existence. Feel your skin – it is **keratin** protein that protects you. Insects use chitin protein for their protection. An analysis of our hormones, such as insulin, shows a variety of very specific proteins perform very specific functions. Proteins are also enzymes, carriers of oxygen such as hemoglobin, and make up half of cell membranes; they compose hair, hold cells together, receive hormones and chemicals for cells, and move muscles, to name a few functions. Joyce Carl had albino skin coloration because she could not produce the protein melanin, a skin-color molecule which makes cells a darker shade. Proteins therefore express the essence of being alive because they carry out both structural and functional aspects of life functions. Figure 5.15 shows the varied roles that proteins play in living systems.

**Melanin**

The pigment that gives color to human eyes, hair, and skin

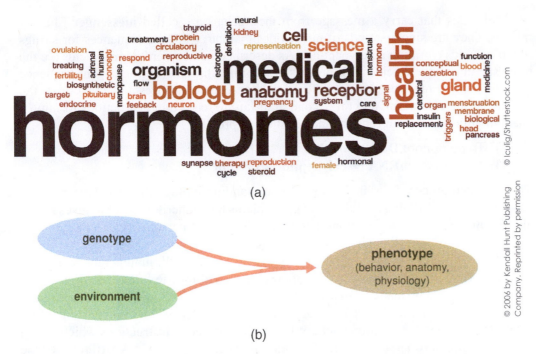

© 2006 by Kendall Hunt Publishing Company. Reprinted by permission

© Iculig/Shutterstock.com

**Figure 5.15** Varied roles of proteins in living systems a. Roles of proteins in living systems. b. How do proteins affect our appearance? From *Biological Perspectives*, 3rd ed by BSCS.

## Gene Expression: How Proteins Are Made

Gene expression is defined as the ability of a gene to carry its information to the rest of a cell and perform its directives. The way our genes accomplish gene expression is to make the variety of proteins described in the previous section. Gene sequences produce proteins to carry out orders found in the genetic code.

However, not all genes are made into protein. Only about 5% of our genes give rise to proteins. In fact, more than half of our genes contain information that simply repeats thousands to hundreds of thousands of times, with no information to be passed on. These sequences are noncoding, or do not produce proteins. These nucleotide sequences do not become expressed as proteins because they are spliced out. In other words, they do not have a chance to code for proteins and therefore influence an organism's traits. When genes do get expressed, however, it results in the production of protein and therefore has the potential to affect organisms' characteristics.

Thus, while the totality of our genes comprises our genotype, or genetic make-up, only those genes leading to or coding for a protein will result in our observable characteristics. Our protein make-up results in our observable traits, or the "way we look," which is termed our phenotype. Joyce Carl's phenotype was albinism, but her genotype led to the condition. How are there changes between the genotype and the expressed phenotype? Simply put, in order to be expressed, a gene must be able to code for a protein. If Joyce's gene mutation for albinism had occurred on a portion of genes that do not code for proteins, she would not have developed albinism.

The production of proteins from DNA begins in the nucleus of eukaryotes. Eukaryotic organisms have cells with chromosomes protected by a nuclear membrane. DNA remains protected in the cell, which prevents potential damage to those chromosomes. Because we know from Chapter 3 that ribosomes make proteins, where do you predict the message will be sent from the nucleus? Yes, messages are sent from the nucleus to the ribosomes to express a message into a protein form.

**Gene expression**

The ability of a gene to carry its information to the rest of a cell and perform its directives.

**Genotype**

The genetic makeup of a cell.

**Phenotype**

The observable traits of an organism.

Molecules that carry a message from the nucleus are called **messenger** RNA or mRNA. They are single strands of nucleotides that carry coding sequences for strings of amino acids (forming a polypeptide). As described in Chapter 2, amino acids are the basic subunit of all proteins. This process of information transfer from a gene sequence to a polypeptide requires two steps. The first, transcription, moves a message from the nucleus to the cytoplasm, and the second, translation, reads that instructions forming a chain of amino acids that reorient and constitute a protein:

1) Transcription: **DNA → mRNA**
2) Translation: **mRNA → Polypeptide à Protein**

This two-step process is so important to understanding biology that it is known as the Central Dogma **of Biology** (Figure 5.16). It explains how inherited material gives rise to all our unique structures, functions, assets (like a pleasant personality or a high intelligence), and liabilities (like disease).

With 20 different types of amino acids within a living cell, innumerable combinations are possible for making any type of protein needed for life functions. How can so many amino acids be made from a set of only four types of bases? The answer is that it takes a sequence of three bases in the DNA and mRNA to "code" for a single amino acid. When DNA codes for an amino acid, it has a specific set of instructions which match to the production of an amino acid. Consider the DNA sequence, AAA (three adenine bases) that codes for UUU (three uracil bases) on the mRNA molecule. UUU delivers a phenylalanine amino acid to a growing polypeptide. The flow of information is given as a simple equation below:

$$AAA \rightarrow UUU \rightarrow Phenylalanine$$

The set of instructions, in the case above AAA is a sequence of three bases (carried by its respective nucleotides), called a **triplet** sequence on DNA. Its corresponding

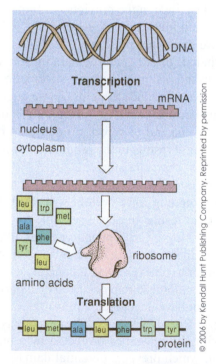

**Figure 5.16**   Central Dogma of Biology. Gene expression is a two-step process in which DNA is transcribed into RNA which is then translated into proteins. From *Biological Perspectives*, 3rd ed by BSCS.

| First Base | Second Base | | | | Third Base |
|---|---|---|---|---|---|
| | **U** | **C** | **A** | **G** | |
| **U** | phenylalanine | serine | tyrosine | cysteine | **U** |
| | phenylalanine | serine | tyrosine | cysteine | **C** |
| | leucine | serine | stop | stop | **A** |
| | leucine | serine | stop | tryptophan | **G** |
| **C** | leucine | proline | histidine | arginine | **U** |
| | leucine | proline | histidine | arginine | **C** |
| | leucine | proline | glutamine | arginine | **A** |
| | leucine | proline | glutamine | arginine | **G** |
| **A** | isoleucine | threonine | asparagine | serine | **U** |
| | isoleucine | threonine | asparagine | serine | **C** |
| | isoleucine | threonine | lysine | arginine | **A** |
| | (start) methionine | threonine | lysine | arginine | **G** |
| **G** | valine | alanine | aspartate | glycine | **U** |
| | valine | alanine | aspartate | glycine | **C** |
| | valine | alanine | glutamate | glycine | **A** |
| | valine | alanine | glutamate | glycine | **G** |

**Figure 5.17** Genetic code (for amino acids) table. This table shows the genetic code in the mRNA codes for specific amino acids, during translation. Start codons and stop codons do not specify a particular amino acid, instead they signal the cell to start or stop translation. From *Biological Perspectives*, 3rd ed by BSCS.

sequence on mRNA, in our case given, UUU, is termed a codon. Figure 5.17 shows the combinations of codons that code for their respective amino acids. A specific codon leads to a particular amino acid placed onto a protein that forms in the process of a gene's expression. Thus, the order of nucleotides in the DNA and resulting mRNA strands determine the combination of amino acids in a protein for which the genetic material codes. There are 64 possible codons, and 61 are known to code for the 20 amino acids in existence. The other 3 codon triplets serve as "start" and "stop" signals for the process of protein production. Neither of these code for amino acids except for the start codon, AUG, which also serves to code for the amino acid methionine. The sequence of amino acids within a protein is important because it determines the three-dimensional structure, orientation, and function of respective proteins. Use the chart in Figure 5.17 to trace the nucleotide sequences that produce their specified amino acids. Which amino acid develops from an original triplet DNA sequence, CCT?

If you answered glycine, you correctly traced the origins of the amino acid to its DNA code. Most of the time, amino acids are correctly coded for by their triplet and codon sequences.

At times, however, if one specific amino acid is misplaced, an incorrectly formed protein results. Such a scenario may lead to disease. To illustrate, consider our earlier

**Codon (triplet)**

Normal genetic code in which a sequence of three nucleotides codes for a specific amino acid.

exampler sickle-cell anemia is due to a single amino acid error. A change in a single base (thymine to adenine) in the DNA strand leads to amino acid number 6 switching from glutamic acid to valine. This single amino acid difference leads to the structural changes described earlier in the hemoglobin.

The goal of transcription is to copy a sequence of nucleotide bases correctly into an mRNA molecule. While mRNA is somewhat different than DNA (see Chapter 2 to review the differences), it is able to match up to the DNA molecule to produce a complementary set of nucleotides. There are three phases of transcription: **Initiation**, Elongation, and Termination (Figure 5.18b). The first step in initiation is to again, as in replication, unwind the DNA molecule. Instead of DNA polymerase and helicase accomplishing this, RNA polymerases bind to a specific sequence to unwind the DNA at certain sites and start transcription. These special sites are called **promoter sites**, which are composed of a series of base sequences. This region occurs at the beginning of each gene and ensures that the mRNA is made from that point forward. RNA polymerase moves from the promoter and along the DNA molecule, adding complementary bases in the elongation phase. Bases are added in the same manner as during replication, with one exception: **Uracil**, a complementary base is found in mRNA (which replaces thymine), is matched with adenine bases on the DNA molecule to produce mRNA. For example, if a sequence of DNA being copied is

### Elongation

One of the three phases of transcription in which nucleotides are added to the growing RNA chain.

### Termination

The phase in which RNA polymerase will reach a sequence of DNA that tells it to stop.

ATTGCCACC

The mRNA sequence will have a complementary strand of

UAACGGUGG

Again, note that it is the same type of complementary base pairing as in replication, but uracil is found in RNA in the place of thymine. The other base pairings remain the same. Eventually, RNA polymerase will reach a sequence of DNA that tells it to stop, called a **termination sequence**. This is the end of the gene, and RNA polymerase detaches from the DNA strand. During the termination phase, mRNA is released from the DNA molecule and the separated DNA strands reform into a double helix. A cap and tail is then added to the mRNA molecule before it leaves the nucleus for protection. This is called RNA processing and protects the mRNA information, much as a cover protects a book. The mRNA makes its journey out of the nucleus through nuclear pores because ribosomes are found within the cytoplasm. It was at this point in Joyce Carl's transcription process that her albinism first became expressed. Joyce probably had Oculocutaneous albinism type I, which results in the transcription of a mutated gene on chromosome 15. The code was brought out into the cytoplasm to be made into protein.

If her gene had not been transcribed, Joyce would not have developed albinism. Some genes are not expressed into proteins causing organisms' traits. Eukaryotic cells process these noncoding sequences, called introns, by splicing them out and leaving only coding sequences, exons. The message then gets sent through the pores of the nuclear envelope into the cytoplasm to be "read" and made into proteins.

### RNA processing

The process in which cap and tail is added to mRNA before it leaves the nucleus (for protection).

### Intron

A nucleotide sequence removed by RNA splicing.

### Exon

A segment of RNA or DNA that contains information coding for a protein.

## Reading the Message: Translation

When mRNA leaves the nucleus through nuclear pores, it attaches to ribosomes. Like workers in a mini-factory, ribosomes work to read the message on mRNA coming out of the nucleus. There are many "workers," each with a specialized task in the transfer of genetic information on mRNA into the amino-acid sequence found in all proteins. Translation is also composed of three phases, termed **Initiation, Elongation, and Termination**, the same names used for the phases of transcription (Figure 5.18b). We've seen that codons on mRNA either give start or stop directions or code for amino acids.

Amino acids are brought into the ribosome because they match complementary codons on the mRNA molecule. Many amino acids have more than one codon on the mRNA that draws it into the ribosome. For example, using the genetic code table in Figure 5.17, the amino acid alanine is shown to have four different codons that code for it: GCA, GCC, GCG, and GCU. The codons draw amino acids to the mRNA to form proteins.

However, mRNA does not directly produce amino acids. It requires a host of other "workers" to make proteins. **Transfer RNA** (tRNA) molecules are shaped like a clover, carrying specific amino acids on one side of their shape and on the other, binding with specific sequences on the mRNA molecule. In this way, amino acids are brought in to match the sequence found on the mRNA molecule. The process occurs with the assistance of the **ribosomal RNA** (**rRNA**). Ribosomes are composed of two subunits consisting of rRNA. Ribosomal subunits have specific shapes to hold the mRNA strand while amino acids are added together. See Figure 5.18 for the structure of tRNA, rRNA, and mRNA.

**tRNA**

Small RNA molecules that carry amino acids to ribosomes for protein synthesis.

**rRNA**

RNA component of ribosome.

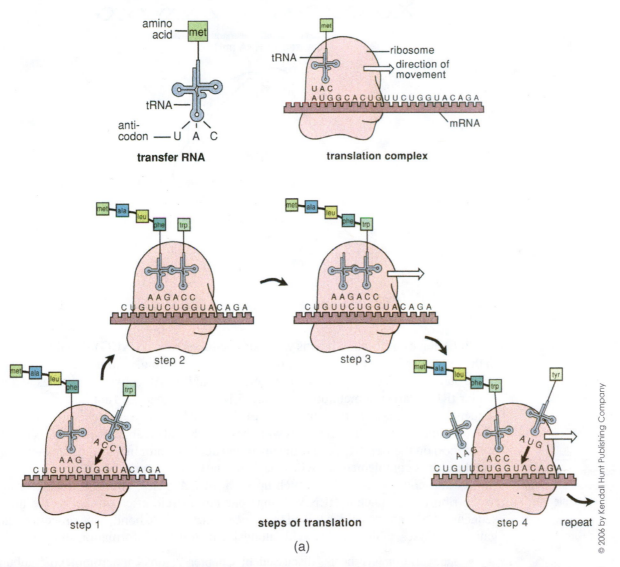

© 2006 by Kendall Hunt Publishing Company

**Figure 5.18** a. Methionine tRNA starts the process of translation, shown as the first to arrive at a ribosome in the figure. Then, other tRNAs bring amino acids to ribosomes during translation. During translation, a string of amino acids (forming a protein) is made along the surface of a ribosome. From *Biological Perspectives*, 3rd ed by BSCS. Reprinted by permission. b. The steps of transcription: the movement of RNA polymerase along a DNA molecule, making mRNA. From *Biological Perspectives*, 3rd ed by BSCS.

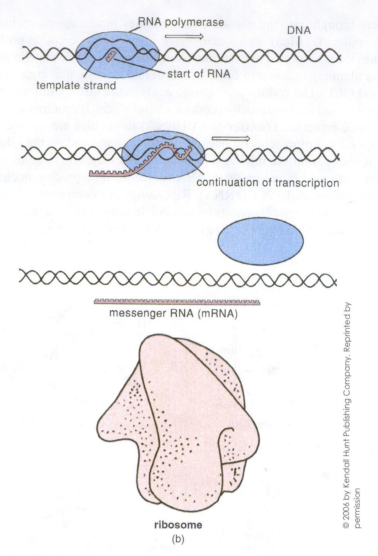

© 2006 by Kendall Hunt Publishing Company. Reprinted by permission

**Figure 5.18**  (*continued*)

Initiation of translation begins when **start codons** (always **AUG**) signal the location in the mRNA to begin translation. A tRNA carries a matching, complementary sequence to the start codon. A start codon on the mRNA is always AUG, which matches to a particular tRNA carrying methionine. Figure 5.18a shows that the first amino acid in the sequence being made is met (methionine) Every tRNA has a sequence of bases on it called an anti-codon, which matches with the bases found on an mRNA. **UAC** is the start anti-codon on the first tRNA to bring an amino acid, methionine, to the ribosome. UAC matches (or is complementary with) the AUG start codon on mRNA.

As anti-codons on tRNAs match up with mRNA bases, amino acids are brought to the ribosome. Matching tRNAs brings one amino acid after another to the mRNA sequence. Each amino acid is linked to the next via a peptide bond. This process of elongation continues, adding amino acids alongside the mRNA information strand.

- Dehydration synthesis, discussed in Chapter 2, links macromolecule subunits together to form larger amino acid chains.

Enzymes on the ribosome catalyze the process, removing water to form peptide bonds between amino acids. The growing polypeptide chain continues to elongate until tRNA

**Anti-codon**

A sequence of three nucleotides in transfer RNA molecule.

reaches a **stop codon** of several types: **UAA, UAG, UGA** that signals the end of translation. Stop codons do not code for any amino acids. At this point along the mRNA, called termination, tRNA, the ribosome, and mRNA detach from each other; the amino acid polypeptide chain is released simultaneously. It is a multipart process in which many specialized "workers" come together for a moment, guided by a message far away in the DNA, to make a whole protein. Figure 5.19 depicts the many molecular players involved in transcription and translation, forming proteins.

Once released from the ribosome, the polypeptide chain reconfigures and adopts its unique three-dimensional conformations based on the chemistry of the amino acid subunits. Large molecules may form, such as the hemoglobin shown in Figure 5.20.

We began the chapter describing Joyce Carl's struggle with albinism. Essentially, we now know that albinism is a problem with the making of enzyme proteins that produce the pigment, melanin. Melanin gives skin, eyes, and hair their color and protects the skin from damage by ultraviolet (UV) light. It shields the nucleus of cells to prevent damage to chromosomes and the vitally important DNA sequence within the nucleus. This is why sunlight is so dangerous for people like Joyce, because the sun's rays can mutate skin cells to make them cancerous. Albinism is originally caused by a mutation on either chromosome number 5, 9, 11, or 15, each leading to its own type of albinism. This mutation is transcribed and then translated into precursor enzymes that are unable to lead to normal melanin pigment production. In Joyce Carl's form of albinism, the most common in Sub-Saharan African and African-Americans, hair is yellow or ginger in color and eyes are often gray-blue. With limited melanin, skin often freckles with moles over time due to sun exposure.

In fact, skin color on the whole is due to only 10 different genes out of our total of up to 25,000 genes. Skin color evolved because of environmental benefits to individuals in the past. To illustrate, sunlight can be devastating for our skin's health, as in the case of the Joyce Carl or as you may have learned if you experienced a serious sunburn. UV light has been shown to deplete folic acid from the skin. Folic acid is a very important nutrient in a fetus's brain and spinal cord development. Thus, in the distant past in a world lacking nutrition, melanin provided the needed protection so that folic acid could be spared for proper fetal development. Sunlight has been shown to deplete folic acid. In sunnier climates, it was beneficial to have darker skin not merely to protect against skin cancer, but to protect folic acid from sunlight and thus retain needed folic acid for the growth and development of the next generation. Thus, evolution favored darker skin tones in climates with more sunlight and thus, the evolution of darker skin colors.

On the other hand, sunlight facilitates the body's production of vitamin D. Because melanin blocks UV light, less vitamin D synthesis takes place in people with darker skin tones. Fifty thousand years in the past, therefore, when food choices were limited, it would have been better to have lighter skin to absorb sunlight and produce vitamin D. However, because skin color is a result of only about 10 different genes, there are many combinations and color types. A random shuffling of genes can result in a set of very light color genes for one fraternal twin of a pair and a set of darker color genes for the other. What determines "race?" and Does race even exist? are questions to ask when classifying people based solely on the 10 genes of skin color.

**Folic acid**

A water-soluble vitamin and a very important nutrient in a fetus's brain and spinal cord development.

**Vitamin D**

A fat-soluble vitamin that promotes that is essential for the absorption of calcium.

# Gene Regulation

Not all our genes are expressed in our phenotype. Cells in the kidney do not express hair color, for instance – there is no need. Cells produce only what is necessary, as we've seen elsewhere in the text. In fact, genes are active only 5–10% of the time in a normal living cell. The ability to shut certain genes off and turn some genes on, like a light switch in a room, is termed gene regulation. Overproduction of materials is unnecessary.

**Gene regulation**

The ability to shut certain genes off and turn some genes on.

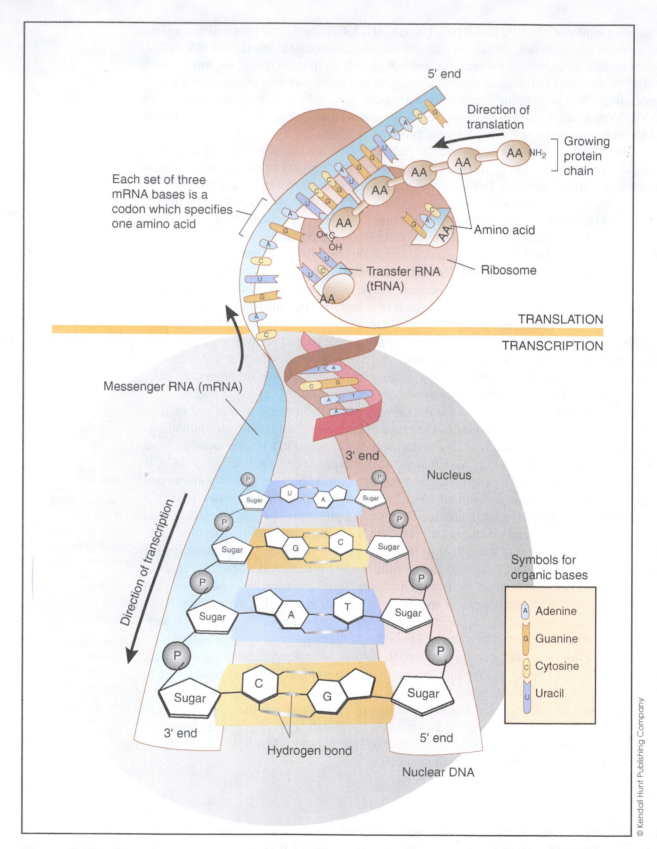

**Figure 5.19** Steps in the process of translation: initiation, elongation, termination (a top) and transcription (at the bottom).

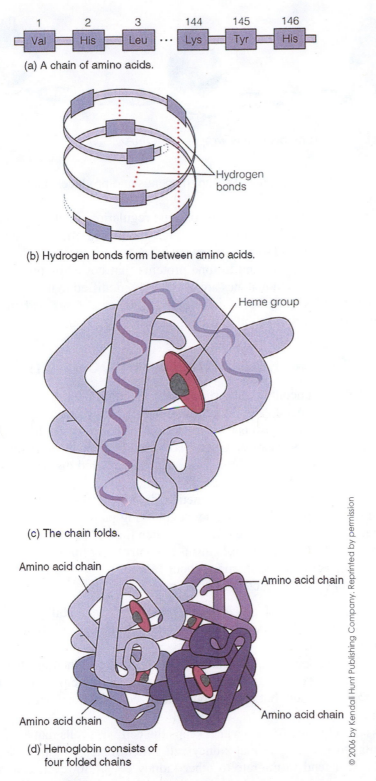

(a) A chain of amino acids.

Hydrogen bonds

(b) Hydrogen bonds form between amino acids.

Heme group

(c) The chain folds.

Amino acid chain

Amino acid chain

Amino acid chain

Amino acid chain

(d) Hemoglobin consists of four folded chains

© 2006 by Kendall Hunt Publishing Company. Reprinted by permission

**Figure 5.20** Hemoglobin molecule. Three-dimensional shape gives hemoglobin its function. There are four polypeptide chains together forming a quarternary structure. The specific orientation of amino acids allow for oxygen carrying "heme" groups to sit within the molecule and hold enormous amounts of oxygen. The unique shape of hemoglobin is shown in the image. In the next chapter, this shape is the impetus for the suspense story. From *Biological Perspectives*, 3rd Edition by BSCS.

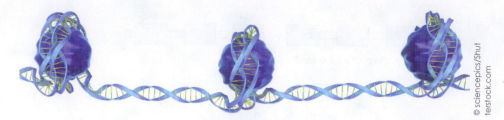

© sciencepics/Shut
terstock.com

**Figure 5.21**     A histone protein is wrapped in DNA.

Remember, the cell is efficient and cheap. It spends and produces only when it is necessary, a theme of this text.

There are two primary mechanisms for gene regulation. First, the promoter region, discussed earlier in our transcription section, may be turned off. For example, the addition of chemical groups to DNA may prevent RNA polymerase from binding to DNA. Second, DNA is wrapped around histone proteins that cover the promoter region, as shown in Figure 5.21. When histones are chemically modified to unwind DNA strands, the gene can be expressed. However, under tight coiling conditions, without access to the promoter site, no transcription and thus no translation can occur.

# Errors in Gene Regulation: A Focus on Cancer

At times, cells divide uncontrollably when gene regulation fails. Unchecked and unregulated by their own genetic "stop" mechanisms, cells grow out of control. Abnormal, uncontrollable cell division results in a tumor, or an abnormal growth of cells. The most common cause of tumors is cancer. Cancer is caused by gene changes that prevent normal rates of mitosis. Cancer is a complex of over 200 related diseases and is a leading cause of death in the world.

Cancer was first thought to be of genetic origin by Theodore Boveri, who studied pedigrees of families with cancer and noted emerging patterns of inheritance. He thus proposed that normal cells become cancerous when their chromosomes become altered in some way to prevent the usual mechanisms of control over mitosis. Today, his research is shown to be correct in its assumptions about cancer.

There are four major characteristics of cancer cells:

1)   A loss of contact inhibition, which is the cell's normal ability to come into contact with its neighbors while dividing and inhibit its growth based on the limited spacing around it;

2)   Dedifferentiation, which is a loss of the specialized functions that normal cells perform. For example, a normal kidney cell will participate in filtering materials from the blood, but a cancerous kidney cell will not filter or function like a kidney cell;

3)   Loss of **cellular affinity**, which keeps the cell with cells that are histologically similar to itself. A normal kidney cell, when mixed in a petri dish with liver cells, will tend to migrate to other kidney cells. However, cancerous kidney cells lose this affinity and attach to liver cells instead of other kidney cells. This is a most dangerous characteristic because it allows cancer cells to metastasize, or spread to other parts of the body. When a cell is determined to be capable of spreading it is termed malignant. Malignant cells enter either the lymphatic system or the blood stream to migrate to other parts of the body and grow. It is this growth that gets in the way of other organ functions and leads to serious health consequences and sometime death; and

**Cancer**

A tumor caused by an uncontrolled division of cells.

**Contact inhibition**

Cell's normal ability to come into contact with its neighbors while dividing and inhibit its growth based on the limited spacing around it.

**Dedifferentiation**

Is the loss of the specialized functions that normal cells perform.

**Metastasize**

The process in which cancer cells spread to other parts of the body.

**Malignant**

The ability of a cell to spread.

**4) Immortality** – Cancer cells may live for an eternity if given the right conditions such as nutrients and water. They do not age as other cells do; instead an enzyme called telomerase rebuilds their DNA ends, or telomeres, so that the cells may replicate forever. In normal cells, telomeres wear out after about 100 DNA replications, at which point the cell can no longer divide, and it dies.

**Telomerase**

An enzyme that rebuilds the DNA ends of cancer cells.

**Telomere**

A compound structure found at the end of a chromosome.

Is simply injecting telomerase the fountain of youth? Preliminary studies show injections of telomerase into experimental animals, instead of leading to longevity, caused tumorogenesis, or increased tumor formation, and thus premature death. In other words, injecting telomerase caused increased cancer rates in animals. This dampened the enthusiasm for telomerase as a fountain of youth. The inheritance of cancer is under some degree of genetic control. Cancer may strike at any age and any socioeconomic class. For example, Hollywood star Christina Applegate (see Figure 5.22) fought breast cancer in her thirties and is now cancer-free.

# Summary

The chapter began with the plight of a young person, Joyce Carl, who endured social discrimination and physical harm as a result of the natural processes described: transcription, translation, mutation, and gene expression. The processes of replication and mitosis serve to produce new, identical cells. In order for genetic material to direct the activities of a cell, it is transcribed and translated into proteins. These proteins then act within a cell to carry out its activities. With a change in melanin production, for example, our chapter story shows that lives are changed within our society. The processes described in this chapter yield our many characteristics that enable life functions. Those traits will be described in more detail in the next chapter. It is the hope that greater knowledge about genetics and its promise for improving human health will lead to a more understanding world.

## BIOETHICS BOX 5.2: IMMORTALITY OF HELA CELLS – SINCE 1951 AND GOING STRONG!

A best-selling book by Rebecca Skloot, *The Immortal Life of Henrietta Lacks*, chronicles a unique 1951 case in which a woman's cervical cancer cells were harvested and grown, although without her knowledge. The patient, Henrietta Lacks, died of cancer the same year, but given nutrients her cancer cells are still kept alive in labs around the world, where they are used in experimentation and observation. Recently, these HeLa cells (named after the patient from whom they were drawn) were studied to determine a relationship between human papilloma virus and cervical cancer. This resulted in a vaccine to prevent transmission of cervical cancer.

Of course, cancer cells do die when the host organism dies, because they lack nutrients from the body. Cancer cells are subject to the same kinds of needs as any other cell, but they look different. Their structure is different, mitosis occurs more frequently, and dedifferentiation is obvious. The heirs of Henrietta Lacks are in court to determine the legal ownership of HeLa cells and, of course, who benefits financially from HeLa's scientific results. The legal results remain to be seen.

**Figure 5.22** Actress Christina Applegate was diagnosed with breast cancer in 2008.

---

### MOBY DICK WAS AN ALBINO

Our chapter starts with the story of human albinism in Joyce Carl. However, albinism can occur in animals, including whales. Years before classic author Herman Melville wrote his fictional work, Moby Dick, about a great white whale, whalers were captivated by another great white ... "Mocha dick." This large albino sperm whale was named after the Chilean island of Mocha in the Pacific. From 1810 through the 1830s Mocha Dick had numerous encounters with whalers – attacking and damaging numerous ships, leaving some men dead. Mocha Dick was likely not the only albino whale in the sea, but he was certainly a notable inspiration to a classic tale. Our next chapter begins with a story about vampirism, also an inherited characteristic, with a history long ago emanating from parts of Eastern Europe.

---

### CHECK OUT

**Summary Key Points**

- Differences in traits may lead to serious social discrimination issues.
- DNA is the hereditary agent of transmission.
- Genes are sections of DNA that contain instructions for making proteins.
- Replication is the way DNA divides itself.
- Transcription is the way DNA is made into messenger RNA. Translation is the way messenger RNA is made into proteins.
- Skin color evolved due to the benefits and drawbacks of the environments of the times.
- Cancer is an inability to regulate the genetics of mitosis.

## KEY TERMS

Albinism

Adenine

Anaphase

Animalcules

Anti-codon

Binary fission

Cancer

Cell cycle

Central Dogma

Chromosomes

Circular genome

Codon (triplet)

Complementarity

Contact inhibition

Cytokinesis

Cytosine

DNA

DNA Polymerase

DNA ligase

Dedifferentiation

Deoxyribose

Elongation

Exon

Folic acid

$G_1$ phase

$G_2$ phase

Gene

Gene expression

Gene regulation

Genotype

Guanine

Helicase

Histone

Initiation

Interphase

Intron

Malignant

Malaria

Metaphase

Metastasize

Melanin

Molecular genetics

Nucleotide

Nucleoside triphosphate

Phenotype

Prophase

Replication fork

Ribose

RNA

RNA processing

mRNA

tRNA

rRNA

S phase

Semi-conservative model

Sickle-cell anemia

Telomerase

Telomere

Telophase

Termination

Thymine

Transcription

Translation

Uracil

Vitamin D

# Multiple Choice Questions

1. How many genes determine skin color in humans?

   a. 10
   b. 100
   c. 1,000
   d. 1,000,000

2. A person is often judged by his or her appearance. Which most affects how a person is perceived in society?

   a. exons
   b. genotype
   c. phenotype
   d. complement

3. Miescher, Griffith, and Avery each sought to explain heredity based on Mendel's laws. Which did they each focus upon?

   a. pea plants
   b. dominance
   c. recessiveness
   d. chemicals

4. Which of the following are the same in every DNA molecule?

   a. ribose
   b. ligase
   c. polymerase
   d. phosphate

5. Which portion of the nucleotide is most important in transmitting information?

   a. deoxyribose
   b. ribose
   c. phosphate
   d. adenine

6. Which occurs when a mismatched nucleotide is expressed in a gene sequence?

   a. a changed protein
   b. a changed mRNA
   c. a mutation
   d. all of the above

7. How many amino acids are produced from a gene sequence containing the following bases: TTAACGCCCCTA. Assume that all of the genes are expressed as amino acids and no noncoding or start/stop sequences are included.

   a. 1
   b. 4
   c. 12
   d. 24

**8.** DNA polymerase serves to:
   **a.** lay new RNA nucleotides
   **b.** lay new DNA nucleotides
   **c.** fuse DNA fragments
   **d.** fuse RNA fragments

**9.** Which best describes the genetic cause associated with sickle-cell anemia?
   **a.** a missing piece of chromosome 14
   **b.** an elongated piece of chromosome 14
   **c.** a mutation from thymine to adenine
   **d.** a mutation from valine to glutamic acid

**10.** Which most likely linked to evolutionary changes in melanin production over the past 50,000 years of human evolution?
   **a.** folic acid
   **b.** Plasmodium frequency
   **c.** complementarity of bases
   **d.** cell affinity

## Short Answers

**1.** Draw a molecule of mRNA derived from the DNA sequence: TTAGGCCACCTC.

**2.** List three differences between a strand of DNA and a strand of RNA.

**3.** Draw a diagram of the Watson and Crick double helix and label its parts, including deoxyribose, phosphate, adenine, guanine, cytosine, and thymine. Show hydrogen bonds using dotted lines.

**4.** Name two enzymes that are needed for replication. Explain the role of each enzyme in doubling the DNA.

**5.** What is meant by the term "semi-conservative?" Use a drawing with colors to explain how DNA is made using this term.

**6.** Explain the process of DNA transcription to a friend. What are the main results of the process? Do the same for translation.

**7.** Place the following terms in the correct order, starting from the beginning to the end, tracing the flow of materials through the central dogma:

> Promotor
> mRNA
> DNA
> DNA polymerase
> DNA ligase
> rRNA
> tRNA
> Methionine
> RNA polymerase

**8.** Transcribe the following sequence of DNA: TTAACGCC

**9.** Which of the following sequences cannot exist for an mRNA?
  **a.** ATTGCC
  **b.** UTTCCT
  **c.** AAAATT
  **d.** CCCCCC
  Explain your answer above.

**10.** Antibiotics are used to kill bacteria by stopping the ribosome from functioning. Based on the central dogma of biology, why is this so deadly for bacteria?

# Biology and Society Corner: Discussion Questions

**1.** What was Watson and Crick's main purpose for making a model of DNA? How does it lead to the information given in this chapter? Would we have still been able to develop this chapter without their model of DNA? Why or why not?

**2.** Based on the readings in this chapter, is there such a thing as "race" in humans? Explain your answer.

**3.** Who should own the rights to cells harvested from people during medical procedures? Why?

**4.** What could be done by the Tanzanian government to prevent discrimination of African albinos in a culture which holds beliefs that endanger their lives? How can African albinos best improve their social integration into society?

**5.** Write a plan to help African albinos, who are in fear for their lives, cope with the existing social discrimination. What are four ways albinos can improve the quality of their lives?

Figure – Concept Map of Chapter 5 Big Ideas (Below is a sample of a concept map for this chapter - You may draw your own in the box provided above to help you make your personal connections)

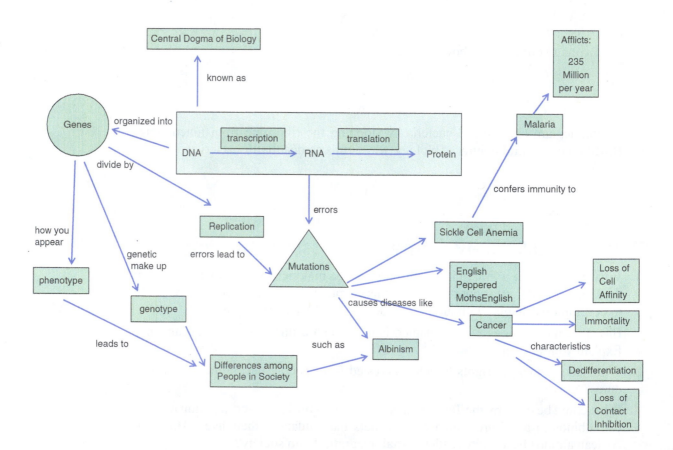

# Inheriting Genes

## 6

## ESSENTIALS

Vincent Van Gogh, the famous artist, was believed to have been afflicted with porphyria

©irisphoto1/Shutterstock.com

The porphyria gene is on a chromosome

©udaix/Shutterstock.com

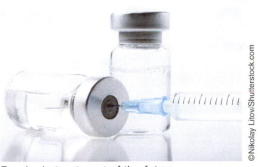

Porphyria treatment of the future

©Nikolay Litov/Shutterstock.com

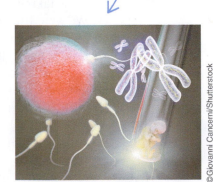

Porphyria gene on chromosome separates into sperm and egg

©Giovanni Cancemi/Shutterstock

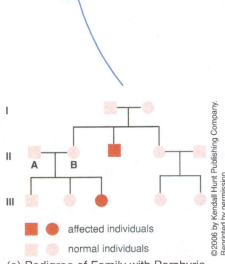

affected individuals

normal individuals

(a) Pedigree of Family with Porphyria

©2006 by Kendall Hunt Publishing Company. Reprinted by permission

(b) A Town in Transylvania. From *Biological Perspectives*, 3rd ed by BSCS

# The Case of the Vampire Diary

Diary entry: February 13, 2013.

Last month, during our college semester abroad, we experienced something I will never tell another person. I can't be sure, and maybe I am crazy, but I know it happened… and it changed my life forever.

It all began when we spent a month in Europe. My mother's family came from what was once the Austro-Hungarian Empire. They immigrated to America many years ago and did not speak much about their lives in the old country. The town they came from, Sibiu, was now in Romania. Before the wars, the area was German and was called Hermannstadt; before 1918, it was in the province of Transylvania.

My friend and I rented a small car, a Yugo, and made our way to Sibiu from Vienna. The day we left was hectic, and the sun was very bright. I did not like the bright sun; it always made my skin ache. It was just the two of us taking a weekend away from the rest of our class, which stayed back in Vienna. We were friends, in a way more like acquaintances. He was bored of the party scene in Vienna and wanted to immerse in the local culture. So he decided to accompany me to my ancestor's home town. He sure got what he wanted.

It was a cold night when we got to Romania, with clouds quickly moving overhead, making the moon appear ominous. Keep in mind, I wasn't scared at all – I had no idea of what was yet to come. All of a sudden, the Yugo started to sputter. The car shut down and my friend yelled, "You dummy, you forget to add gas to this thing!" I was embarrassed and really felt bad about letting him down. I knew it was the sunlight that confused me when I picked up the rented Yugo.

We were tired, and there were no houses along the road. "At least it isn't snowing," I said meekly to try and break the cold mood between my friend and me. There was no response as we walked through the fields. There was also no road – it had ended at an open field with no sign of civilization. In the darkness, over on a hill in the distance, we spotted an old house. As we came closer, it was more like a hut, with clapboard walls and a rundown porch. I told my friend, "Let's keep on going . . . Sibiu couldn't be too far off." I knew this was a lie but I had a bad feeling about the place. There was no response, and I knew my friend was bent on going to the house for gas.

We knocked on the door, with enough force to make it heard. After a long time with no answer, we started away. As we were leaving, an old lady opened the door. "Come in out of the cold. You must be Americans." I told the lady that my family had come from this area a long time ago. The lady was the last of the Germans left in Transylvania. "You are one of us then!" she exclaimed. She came very close to my face, looking deep into my eyes. She remarked inappropriately, "You look like my father did when he was young."

As we sat in her parlor we explained that we needed just a bit of gas to get us to the next town. The room was creepy, but the lady was very agreeable. "I'll get my brother, Herbie, to fetch some gas." After she left the room, we waited and waited, but no brother. Then, my friend felt something behind the couch – it was a man lying on the ground! "I see you have met Herbie," said the old lady. "He's been drinking and needs his bed. Would you help him up?"

This was getting to be too much, but each of us grabbed a limb to carry him. At that moment in time, we froze, looked at each other, and knew something we dared not say – *this man was dead. His flesh was cold, and his skin bloated.* Herbie's skin was scarred, teeth were fangs, and his face appeared almost wolf-like. He looked just like a vampire. My friend and I looked at each other but said not a word.

We brought Herbie up to his bed and laid him down for one last rest. It was then that he sat up, looked at us and thanked us. He looked at me and said, "You look like my father!" I ran out of the house as fast as I could, maybe 15 miles to the town of Sibiu.

I now know that my family was from vampires; maybe their father was my grandfather or maybe I inherited their vampire ways, somehow. But I knew one thing – I am a vampire too.

## CHECK UP SECTION

In the story, Herbie has a blood disorder called **porphyria**. It is an inherited disease, occurring in about 1 in every 25,000 people. Enzymes that produce parts of his red blood cells, called hemes (which carry oxygen) are not formed properly. More specifically, heme groups, or substances that store oxygen in blood cells, are not formed correctly in porphyria. Without these enzymes, porphyrins (parts of hemes in red blood cells) build up, causing lesions in the body.

Symptoms include sensitivity to light (photosensitivity); craving for blood (due to a lack of heme groups); receding and bloody gums making teeth look like fangs; scabs and lesions from sun; organ damage; and rampant growth of hair in body parts to appear wolf-like. Porphyria sufferers need blood transfusions to replace their deficient hemes. We cannot be sure if the college student who narrates the story has inherited porphyria. Its symptoms usually appear during late adolescence. However, it is possible to manifest later in life.

Study porphyria to determine its genetic and/or environmental causes in more detail. How might porphyria have contributed to the myth of vampires in our society? Do you think the narrator in the story had porphyria, based on your research of the disease?

## Unraveling the Mystery of Inheritance

Chapter 5 described the molecular players in gene transfer; in this chapter, we look at the processes underlying inheritance. We begin in a small garden monastery in the 1800s. Gregor Mendel (1822–1884), an Austrian monk who failed out of a science teaching major in college, discovered how we pass traits onto the next generation.

By the mid-1800s, it was generally accepted that ova and sperm both contribute genetic information to new offspring. Most biologists believed, at the time, that inheritance from parents occurred as a blending of characteristics. In this view, traits from both parents averaged together to produce new, unique offspring.

Seeking to discover if there were specific patterns in the inheritance process, Mendel devised a set of experiments using pea plants as his subject. Using pea plants to study inheritance was not original, but his approach to understanding how we inherit our traits was unique. The passing of characteristics from parent to offspring is known as heredity. Through his experiments, Mendel was able to successfully develop the basic principles of heredity.

Mendel's experiment was successful for a few reasons:

**Heredity**

The passing of characteristics from parent to offspring.

1) The garden pea plant Mendel chose was commercially grown at the time, reproduced quickly, and possessed traits easily measured by simple observation. The garden pea plant self-pollinated, meaning that egg and sperm from the same plant would unite. The pea plant's sexual structures were enclosed by a petal capsule, preventing cross-pollination from other plants. Therefore, Mendel could control cross-breeding with select plants and not worry about accidental cross-pollination.

2) Mendel chose measurable variables to study; those that were clearly discernible. He selected seven pea plant traits to study because they were clearly one of two alternatives. These seven traits included shape of seeds, color of seeds, shape of pods, color of pods, height of plant, color of flower, and position of flower. For example, plants had either round or wrinkled peas; and either yellow or green peas.

3) He used mathematics to measure and expose patterns in his results. Figure 6.1 shows the results of Mendel's experiments. Note that the frequency of plant characteristics shows distinct proportions resulting from the crosses in each generation.

4) Mendel's experiment was careful, logical, and sequential. His steps were meticulous and well thought out. Mendel credited any successful science experiment to certain attributes, stating in his original paper, "The value and utility of any experiment are determined by the fitness of the material to the purpose for which it is used."

**Figure 6.1**   Gregor Mendel is the father of genetics and was an Austrian monk who discovered the laws governing patterns of inheritance.

While Mendel's findings were groundbreaking, they were unrecognized for 35 years. In 1865, Mendel reported his experiments and results in a paper presented to the Natural Historical Society in Bruenn (now Brno, Czech Republic), the Austro-Hungarian Empire. Scientists in the audience dismissed his findings. Afterward, Mendel returned to his monastery, tending to his priestly duties; he was ignored and unappreciated by the scientific community. It was not until after his death, in 1900, that biologists began to build upon Mendel's paper. Mendel's work eventually was recognized and discredited the blending of traits perspective. Instead, his findings showed that traits were inherited as discrete units from each parent. Mendel thus began a scientific revolution in the field of genetics. Gregor Mendel is now recognized as the father of genetics for his work on pea plants. Let's take a closer look at the laws of heredity that Mendel formulated so long ago.

# Mendel's Laws

## Law of Dominance

Let's revisit Mendel's work: First, he chose to crossbreed certain plants. For example, Mendel noted that one variety of plant always produced yellow peas, while another produced green. He took the male anther portion of a yellow pea plant and dusted the female stigma of a green. He called these original parents the $F_0$ generation. When he crossed the two, all of the offspring were still yellow and none of them were green. This first cross Mendel called the $F_1$ generation. Any trait that appeared in the $F_1$ generation he called dominant for that characteristic. He surmised that any dominant trait covers up the alternative characteristic of an organism in the $F_1$ generation. Next, Mendel crossed the organisms in the $F_1$ generation in the same manner and their offspring were analyzed, the $F_2$ generation. Mendel conducted what is now termed a monohybrid cross. This is a mating between two organisms, each having both characteristics for a particular trait – in this case both yellow and green. It is termed "mono-" because it looks at the inheritance of only one trait. Mendel surmised that although all of the plants of the $F_1$ generation were yellow, they harbored a hidden green characteristic able to be given to offspring.

Mendel formed a hypothesis: the covered-up trait would reappear in the $F_2$ generation. Indeed, he predicted correctly that some offspring of all-yellow plants would be green. He was correct; the covered-up trait always reappeared in the $F_2$ generation, bred from parents that did not exhibit the trait. The idea that a dominant trait covers up another is known as the law of dominance. He called the characteristic that is covered up the recessive trait. In his experiment, the yellow trait was dominant, and the green trait was recessive. The original parents, the $F_0$ generation, he deduced, were each pure – the yellow parent had only dominant yellow characteristics and the green parent had only green characteristics – but that these characteristics would pass along in a predictable manner through each generation.

## Law of Segregation

When Mendel analyzed the $F_2$ generation, he found that a certain proportion always appeared in his data. Note in Figure 6.2 that the $F_2$ generation for all seven characteristics he chose had a roughly 3:1 ratio of dominant to recessive characteristics.

Because the appearance and disappearance of traits occurred in constant proportions, Mendel inferred that traits must be inherited as two separate, discrete units. We now call these units alleles. Alleles are alternate forms of the same trait. For example, if a pea has a yellow or green color possible, then either a yellow or green allele is

**Dominant**

The trait that covers up other forms of the characteristic.

**Monohybrid cross**

The mating between two organisms, each having both characteristics for a particular trait.

**Law of dominance**

The idea that a dominant trait covers up another.

**Recessive**

The trait that is covered up by a dominant trait.

**Alleles**

An alternative form of the same trait.

| P generation | F1 generation | F2 generation | Ratio |
|---|---|---|---|
| Long × Short | All Long | 787 Long, 277 Short | 2.84:1 |
| Purple × White Flowers | All Purple | 705 Purple, 224 White | 3.15:1 |
| Axial × Terminal flowers | All Axial | 651 Axial, 207 Terminal | 3.14:1 |
| Green × yellow pods | All Green | 428 Green, 152 Yellow | 2.82:1 |
| Smooth × constricted pods | All Smooth | 882 Smooth, 299 Constricted | 2.95:1 |
| Yellow × Green seeds | All Yellow | 6,022 Yellow, 2,001 Green | 3.01:1 |
| Round × Wrinkled seeds | All Round | 5,474 Round, 1,850 Wrinkled | 2.96:1 |
| Total | All Dominant | 14,949 Dominant, 5,010 Recessive | 2.98:1 |

© Kendall Hunt Publishing Company

**Figure 6.2**    Results of Mendel's experiments with pea plants F$_1$ and F$_2$ generations.

**Law of segregation**

The hypothesis that states that there are two separate, discrete alleles that could be inherited separately.

responsible for it. The hypothesis that there are two separate, discrete alleles that could be inherited separately is known as Mendel's law of segregation.

We are now able to trace the movement of alleles from parent to offspring using a Punnett square. While this square was not actually used by Mendel, it derives from Mendel's law of segregation. A Punnett square is a diagram based on the law of segregation that is used to predict the probability of inheritance of alleles between parent and offspring. Figure 6.3 uses a Punnett square to show how alleles are discretely passed on to a new generation. The mother's alleles appear on one side of the box and the father's on the other side. Each parent in the figure has two possible alleles based on their genetic make-up. Capitalized alleles are dominant and lower case alleles are recessive in the Punnett Square. Alleles from each parent have a 50:50 chance of segregating into an egg or sperm, eventually forming a new organism with a new genetic make-up.

- To recall from Chapter 5, an organism's genetic make-up is known as its genotype. The expression of that genotype is an organism's phenotype. In other words, how an organism appears is its phenotype; and what comprises inside an organism's genes is its genotype.

The Punnett square gives the probability of producing an organism with a particular genotype within each box. Each box of the Punnett square represents a 25% chance that an organism's genotype will appear in the offspring generation. Figure 6.4 shows the process of allele transfer between parents to offspring in porphyria. In our story, acute intermittent porphyria (AIP) is a dominant trait, meaning that if a person has one allele for it then he or she will have the disease.

**Porphyria**

An inherited disease which is characterized by abnormal metabolism of the blood hemoglobin.

## Law of Independent Assortment

In another set of experiments, called a dihybrid cross, Mendel mated plants tracing two different traits – pea shape and color. In a **dihybrid cross** both parents possess dominant and recessive characteristics for a particular trait. It traces the inheritance of two separate traits at the same time. The term "di-" is used because it looks at the

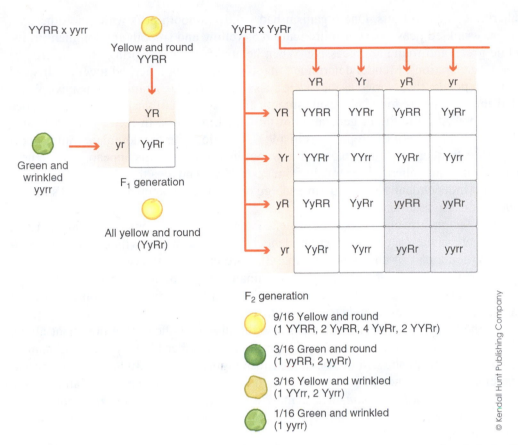

Figure 6.3 Principle of segregation. Alleles separate into opposite ends of the cell. Alleles move independently of each other, according to Mendel's laws, with equal chances of being transmitted to offspring. Note that any one of the four gametes produced by parents in the figure could be transmitted to the offspring generation. The Punnett square shows the relative probability for each gamete to give rise to its genotype.

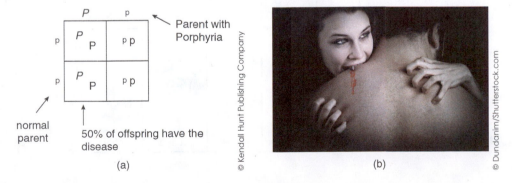

Figure 6.4 a. A Punnett square for porphyria, a cross is shown between a parent with a dominant gene for porphyria and a normal parent for the F₁ generation. The Punnett square shows that 50% of offspring will exhibit the disease. b. The disease porphyria is the likely basis for the legend of vampires.

inheritance of two traits. One organism had yellow, smooth peas while the other had green, wrinkled peas. As shown in Figure 6.5, yellow and smooth are dominant traits, while green and round are recessive. When he analyzed the offspring of these crosses (the $F_1$ generation), Mendel determined that traits were not inherited together. Instead they independently assorted as they were passed from one generation to the next. Wrinkled and round were found alongside smooth and green. All of the possible types of pea plants showed up in the $F_1$ generation of this dihybrid cross. In fact, these organisms also showed a pattern of proportions in a 9:3:3:1 ratio, with the dominant traits occurring most frequently (Figure 6.5). All new combinations of traits appeared in the next generation, each inherited separately from each other. The idea that each pair of alleles is sorted independently when sperm and egg are formed is known as Mendel's law of independent assortment.

Two factors are inherited separately, one from a mother and one from a father. Thus, once together, they occur as either an identical pair or as a pair with different components. When a pair of alleles is the same, they are called homozygous. When both are dominant forms, they are homozygous dominant. When both are recessive, they are homozygous recessive. When alleles in a pair are different from each other, they are called heterozygous or hybrids for that trait.

In the porphyria case seen in our story, the disease is held on a dominant allele. Thus, if a person possesses an allele for porphyria, whether homozygous dominant or heterozygous, he or she will get the disease (see Figure 6.4). Only a homozygous recessive individual does not exhibit the disease. If *P = the allele for porphyria* and *p = the allele for the normal condition*, then an individual with PP, homozygous dominant or Pp, heterozygous will have the porphyria trait. Only a homozygous recessive, pp will have a normal blood condition.

**Law of independent assortment**

The idea which tells that each pair of alleles is sorted independently when sperm and egg are formed.

**Homozygous**

The condition in which a pair of alleles is the same.

**Heterozygous**

The condition in which alleles a pair are different from each other.

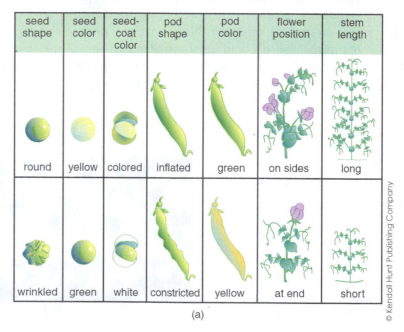

| seed shape | seed color | seed-coat color | pod shape | pod color | flower position | stem length |
|---|---|---|---|---|---|---|
| round | yellow | colored | inflated | green | on sides | long |
| wrinkled | green | white | constricted | yellow | at end | short |

(a)

© Kendall Hunt Publishing Company

**Figure 6.5** a. traits of pea plants, with the top row dominant and the bottom row recessive; b. Punnett square for a dihybrid cross for pea color and shape in pea plants. A cross is shown between two parents with both traits. The Punnett square below shows a 9:3:3:1 ratio in offspring characteristics.

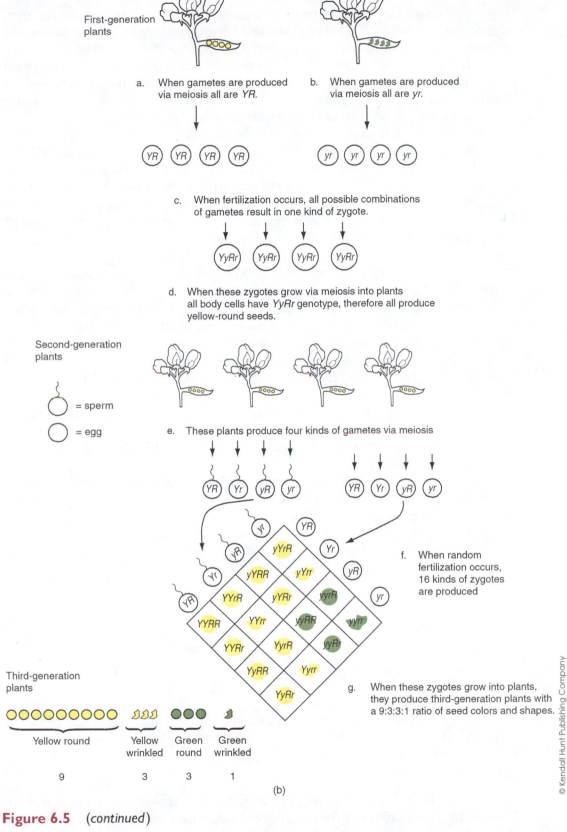

First-generation plants

a. When gametes are produced via meiosis all are *YR*.

b. When gametes are produced via meiosis all are *yr*.

YR  YR  YR  YR        yr  yr  yr  yr

c. When fertilization occurs, all possible combinations of gametes result in one kind of zygote.

YyRr  YyRr  YyRr  YyRr

d. When these zygotes grow via meiosis into plants all body cells have *YyRr* genotype, therefore all produce yellow-round seeds.

Second-generation plants

= sperm

= egg

e. These plants produce four kinds of gametes via meiosis

YR  Yr  yR  yr        YR  Yr  yR  yr

f. When random fertilization occurs, 16 kinds of zygotes are produced

Third-generation plants

Yellow round    Yellow wrinkled    Green round    Green wrinkled

9          3          3          1

g. When these zygotes grow into plants, they produce third-generation plants with a 9:3:3:1 ratio of seed colors and shapes.

(b)

**Figure 6.5**    (*continued*)

© Kendall Hunt Publishing Company

# Testcross

How do we determine the genotype of an organism?—, it is not always obvious from its appearance. Consider a green pea plant that inherits two green recessive alleles, one from each parent. It is green in its phenotype, indicating that it inherited two green alleles. If the plant had inherited one green and one yellow allele, it would have been yellow. When a yellow plant appears, it is more difficult to know its genotype without knowing its history. A yellow pea plant has a dominant allele, but does it have a recessive that is covered up, or is the other allele dominant as well?

**Testcross**

A known homozygous recessive organism is mated with a dominant organism.

Through using a testcross, the genotype of the yellow pea plant is explored. In a testcross, a known homozygous recessive organism, for example, a green pea plant is crossed with a yellow phenotype. The green pea plant, we know, has two green (recessive) alleles. But what is the yellow plant's genotype? In this case, we do not whether the plant with yellow peas is homozygous dominant or heterozygous. In this testcross, a hidden recessive is most likely to be revealed. The homozygous recessive individual (the green plant) has the best chance of passing all its recessives to the next generation. Figure 6.6 shows a testcross between a yellow pea plant and a green pea plant to determine whether or not green peas will result in their offspring. The testcross helps to determine the true genotype of the yellow pea plant.

The appearance of a recessive in the testcross's progeny is the only definite proof that the unknown genotype was indeed a heterozygote. In other words, if one of its offspring is green, then alleles coming from both parents must have been green. For the new offspring to have become a homozygous recessive, one recessive allele had to come from the yellow parent plant. However, if there is no individual with a recessive trait for pea color in the $F_1$ generation, it may mean simply that the recessive allele may get expressed in another generation. Perhaps that recessive allele simply did not get passed along this time around. It is impossible to know for sure, but a testcross gives the best chance of being able to reveal the true genotype of an individual with a dominant phenotype, such as the yellow plant.

This is also the reason recessives are so difficult to study and/or remove from a group. They are hidden, and only chance dictates whether or not an allele will become expressed. Many times recessive traits are deleterious, or cause harm to an organism having them. Many diseases are recessives and it may take several generations for a recessive disease to appear. It is hard to track recessives for this reason. For example, a family may be surprised that sickle-cell anemia is in their genetic history. Family members

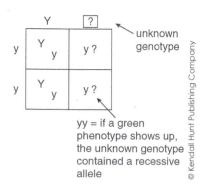

**Figure 6.6**  Testcross for color on pea plants. A testcross always uses a homozygous recessive to attempt to reveal the recessive of its dominant mating partner. If even one of the offspring shows recessive characteristics, then the dominant partner harbors a recessive allele. In the example shown Y = yellow and y = green coloration in pea plants.

may think it is not a risk because no one has had sickle-cell anemia, for as long as they can remember. However, it may have been hidden in the heterozygote condition for a period of time and was unexpressed. Deleterious recessives are a difficult but common thread in most groups. Some forms of porphyria are recessive, showing up many generations after they are thought to be gone. Did our narrator in the story experience the reappearance of a long-silent recessive allele?

## Meiosis: How Sex Cells Are Formed

We have seen how traits are passed on from one generation to the next; now we will examine how organisms reproduce sexually. During Mendel's time, it was accepted that parents transfer their hereditary information through a process called reproduction, to form a new organism. The central step in reproduction is fertilization, when a male and a female sex cell, both called gametes, unite. The female sex cell is the egg; the male, the sperm. Each contributes half the total genetic material that unites and recombines in the zygote. If the offspring receives genetic material from both parents, how is it that the offspring contains the same number of chromosomes as the parents? The answer is meiosis, which is a special form of cell division in which the newly produced daughter cells contain only half the number of chromosomes of the parent. This half-quantity is called the haploid or N condition, while the full complement of genes in all of our other cells, called somatic cells, is known as diploid or 2N. If a sex cell were not haploid, then the genes in the sex cells would double with each successive generation.

Every species has a set number of chromosomes. A mosquito has six chromosomes per cell; a sunflower, 34; a human, 46; a dog, 78; and a little goldfish, an impressive 94 chromosomes. In contrast, gametes of each of these species contain only half of these numbers: a mosquito gamete has 3 chromosomes, a sunflower, 17; a human, 23; a dog, 39; and a goldfish, 47 chromosomes.

In a diploid cell, each chromosome has a partner, much like a pair of shoes (see Figure 6.7. The chromosome partners are known as a homologous pair. Homologs have one maternal and one paternal copy of a chromosome. Alleles on each homologous chromosome code for the same trait. An allele for eye color, for example, on one chromosome codes for eye color alongside the allele on its homologous pair. Figure 6.7 shows the

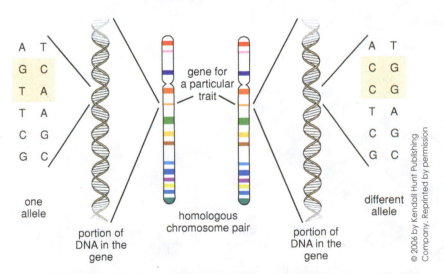

**Figure 6.7**    Homologous genes, knows as alleles, occur at the same location and code for the same traits. From *Biological Perspectives*, 3rd ed by BSCS.

**Fertilization**

Is the process in which male and female sex cells unite.

**Gametes**

Reproductive cells (not given in bold in text).

**Zygote**

A fertilized egg cell.

**Meiosis**

A special form of cell division in which the newly produced daughter cells contain only half the number of chromosomes of the parent.

**Haploid (N)**

The half number of chromosomes of the parent.

**Somatic cells**

The full complement of genes in all of other cells.

**Diploid (2N)**

The full complement of chromosomes in all body cells (except sex cells).

**Homologous**

The chromosome partners in a diploid cell.

alleles on a chromosome (in varied colors). Before meiosis and mitosis take place, homologous chromosomes are duplicated. Thus, each replicated pair is composed of two sister chromosomes, identical to each other. Each set of duplicated homologous chromosomes contains four strands altogether: two original homologs and two duplicated strands.

Each homolog of the pair contributes one allele for a trait to its offspring. As shown in Figure 6.8, homologous chromosomes separate into four gametes during production of sex cells. Whether the individual homologue gets into a sperm or egg depends upon

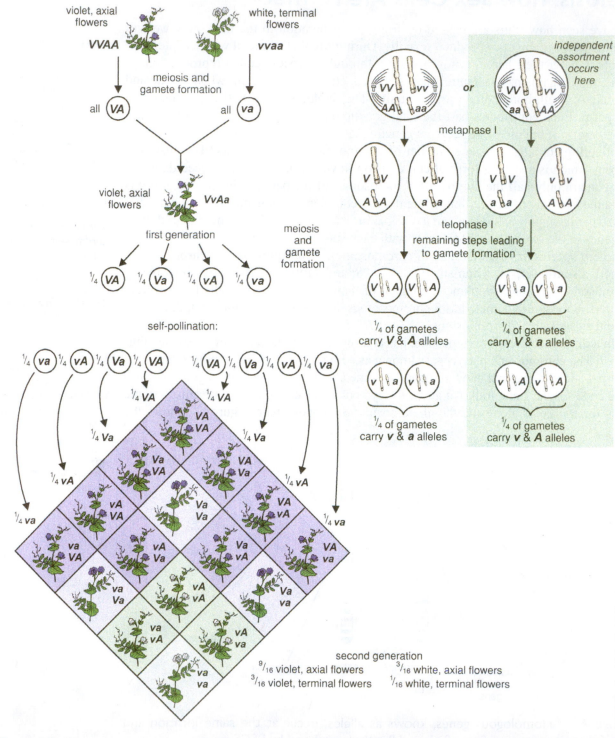

**Figure 6.8** Meiosis. Homologous chromosomes separate eventually into four sex cells (gametes). The doubling of genetic material takes place before the parent cell is able to divide. From *Biological Perspectives*, 3rd ed by BSCS.

chance, as described by Mendel. During meiosis, homologous chromosomes separate and move into one or the other of the gametes produced. They have an equal chance of entering a newly formed gamete because chance determines their entrance. Homologous chromosomes are inherited separately, as shown by Mendel's law of segregation. Trace the movement of replicated chromosomes in Figure 6.8 to find the gametes' destination.

The diploid number (2N) of chromosomes in a parent cell is divided equally into the sex cells during meiosis. Thus, the halving effect on the chromosome number occurs during gamete formation. The result is a set of haploid (N) sex cells. This halving effect counteracts fertilization, which unites genetic material from two sex cells into one somatic cell, the **zygote** or fertilized egg cell. The result is a unification of N + N = 2N.

As demonstrated in Figure 6.9, meiosis follows a series of stages similar to those seen in mitosis. Indeed, the names are also the same for the phases in both mitosis and meiosis. There are a few differences:

1) In meiosis, there are two sets of the same series of stages, meiosis I and meiosis II; but only one series in mitosis. This results in two cell divisions in meiosis and only one cell division in mitosis.

2) In meiosis, four new daughter cells are produced as a result of the two divisions, while only two are produced by mitosis.

3) Each daughter cell contains only the haploid number of chromosomes in meiosis, but daughters in mitosis contain the diploid.

4) Gametes contain a variety of genetic possibilities, in part because homologous chromosomes separate into one or another of the sex cells, forming innumerable combinations.

**diploid parent cell**

**a**  This cell is diploid—that is, it has two pairs of chromosomes.

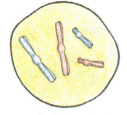

**replication of chromosomes**

**b**  Just before this diploid cell begins meiosis, DNA synthesis occurs and each chromosome is duplicated.

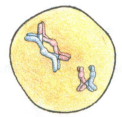

**prophase I**

**c**  The homologous pairs of duplicated chromosomes become closely aligned and join in several places. At these junctions, equivalent pieces of the paired chromosomes may exchange places. This exchange process is called **crossing over.**

**metaphase I**

**d**  The paired chromosomes line up along the middle of the cell. Spindle fibers attach to each duplicated chromosome.

(a1)

© Allia Medical Media/Shutterstock.com

**Figure 6.9**  a. Phases of meiosis. There are two stages of meiotic cell division, I and II. The end result of meiosis is the production of four haploid gametes (sex cells). Meiosis occurs in eight stages with descriptions of each stage given in the figure. b. Mitosis occurs in one division and results in two identical cells.

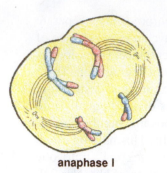
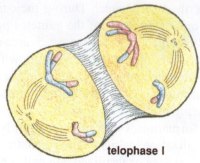

**anaphase I**

**e**   During the first cell division, spindle fibers pull apart each pair of duplicated chromosomes.

**telophase I**

**f**   Each new cell resulting from this division contains two duplicated chromosomes, one from each homologous pair.

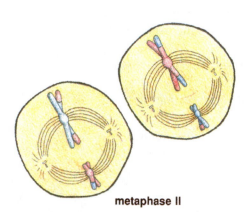

**metaphase II**

**g**   A second cell division now takes place with no further DNA synthesis. During this cell division, there are no chromosome pairs. Instead, the duplicated chromosomes line up in single file.

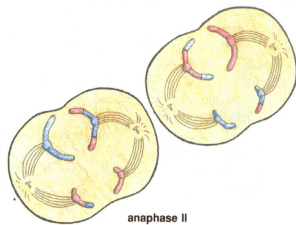

**anaphase II**

**h**   During the second cell division, spindle fibers pull apart each duplicated chromosome.

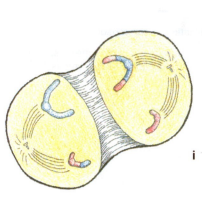
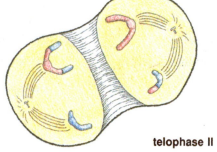

**telophase II**

**i**   Each of the four resulting offspring cells is haploid—that is, it has only one chromosome from each pair. Mature human gametes form from the products of meiosis.

(a2)

**Figure 6.9**   (*continued*)

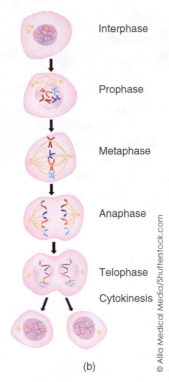

Interphase

Prophase

Metaphase

Anaphase

Telophase

Cytokinesis

© Alila Medical Media/Shutterstock.com

(b)

**Figure 6.9** *(continued)*

**Table 6.1**    Comparison of meiosis and mitosis

|  | **Meiosis** | **Mitosis** |
|---|---|---|
| Number of cells | Four new cells produced | Two new cells produced |
| Number of divisions | Two cell divisions | One cell division |
| Genetics of cells | Haploid cells made | Diploid cells made |
| Compared to parents and each other | Different (variability) | Identical |

Courtesy Peter Daempfle.

## The Phases of Meiosis

In a period before meiosis, the interphase carries out functions similar to those during the interphase before mitosis: cells grow in size, organelles duplicate and grow, and genetic material doubles in the nucleus. When genetic material doubles during interphase, two pairs of homologous chromosomes are formed. The purpose of meiosis is to produce daughter cells capable of fertilization. To do this, fertilization requires two haploid cells with haploid genetic material to unite.

During the first series of stages of meiosis, called meiosis I, homologous chromosomes separate (refer Figure 6.9 to see each stage). Just like the first stage of mitosis, when a cell begins meiosis, nuclear material condenses, transitioning from chromatin to chromosomes, its nuclear envelope disappears, and chromosomes attach to a spindle fiber. Unlike mitosis, the first stage of meiosis I, called prophase I, homologous chromosomes in proximity to each other exchange genetic material through a process called crossing over. In crossing over, segments of one chromosome swap with segments of another pair. Crossing over enhances the genetic combinations possible in gametes, as shown in Figures 6.8 and 6.10. Areas that are crossed over randomly swap genetic material, leaving each homolog with a unique set of DNA.

In metaphase I, the homologous chromosomes line up as pairs, which later separate and move to opposite poles during anaphase I. Spindle fibers pull the pair of duplicated homologs into the center. In the next phase, anaphase I, homologous chromosomes

**Meiosis I**

The process of cell division by which homologous chromosomes separate and new cells are haploid.

**Prophase I**

Also called the first stage of meiosis I, in which homologous chromosomes in proximity to each other exchange genetic material through a process called crossing over.

**Crossing over**

The exchange of genes between chromosomes.

**Metaphase I**

The stage of mitosis and meiosis that follows the prophase stage and precedes the anaphase stage (not given in bold in text).

**Anaphase I**

The stage of cell division in meiosis in which homologous chromosomes separate.

separate to opposite ends of the cell. They are pulled apart in a random manner. A paternal homolog may be pulled onto one side while a maternal homolog may be pulled onto another side. At this point, the developing cells are haploid – with half the number of a complete set of chromosomes. With 23 sister chromosomes pairs, there are $2^n$ possible new combinations. Thus, with 23 pairs of chromosomes in humans, there are $2^{23}$ new possible genetic combinations in each newly formed gamete: $2 \times 2 \times 2 \times 2 \ldots 23$ times! The genetic variation produced by random assortment is enormous.

Mendel hypothesized this random segregation of chromosomes, long before an understanding of the phases of meiosis. Thus, three sources of genetic variation among organisms are seen: 1) meiotic segregation of chromosomes; 2) random mutations in genes as discussed in Chapter 5; and 3) crossing over, as discussed earlier in this section. The processes of obtaining genetic variation are shown in Figure 6.10.

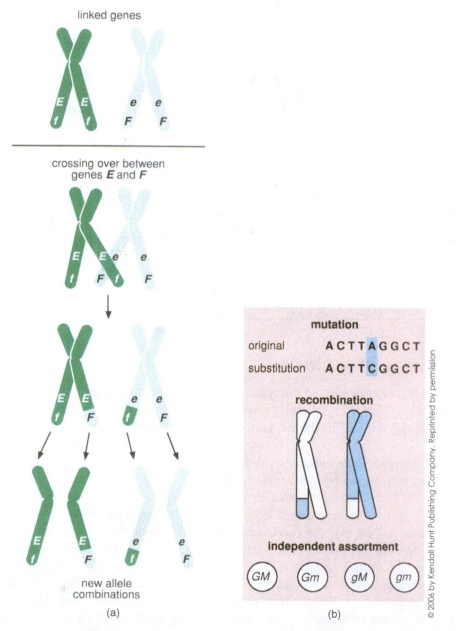

**Figure 6.10**  Genetic variation is introduced in species, especially during meiosis. a. Crossing over. b. Mutation. Recombination and independent assortment during segregation of alleles. All of these mechanisms add genetic diversity to cells and organisms. From *Biological Perspectives*, 3rd ed by BSCS.

After this, telophase I and cytokinesis reform the nuclear envelop, with two new daughter cells containing their own nucleus. These new cells are haploid or N, containing only half the original number of chromosomes. When homologs are pulled apart during meiosis I, sister chromosomes are placed in daughter cells. The genetic composition of sister chromosomes is identical for the two. Thus, for these daughter cells, it is like getting two left shoes instead of a right and left. The two daughter cells of meiosis I are haploid, but contain a double set of half of the chromosomes.

Separating of sister chromosomes occurs during the next series of stages of meiosis, called meiosis II. A short period separates meiosis I and II in a brief interphase. In this time, there is no new duplication of genetic material and quickly cell division resumes into prophase II. Chromatids reorganize, coiling tightly once again as chromosomes, in preparation for the pulling apart process. During metaphase II, chromosomes line up singly and then the two sister chromatids (which are identical **In anaphase II**, identical chromosomes separate, pulled apart by spindle fibers to opposite poles. In the last phase, **telophase I**, nuclei reform and chromosomes become tightly coiled once again. The physical separation of cytoplasm takes place during **cytokinesis**, as it pinches off to become two new cells. The end result of telophase II and cytokinesis, in which two new nuclei and cells form, is a total of four new haploid or N daughter cells.

As a result of meiosis, each human gamete contains only a haploid, 23 single strands of chromosomes, much like having 23 "left" shoes. It is fertilization by another gamete, containing 23 "right" shoes, that gives new life with a full diploid set of 23 pairs of chromosomes. Figure 6.11 shows chromosomes during meiosis represented as shoes.

## Male and Female Gametes

In animals, male meiosis produces four new sperm; and in females, one egg and three polar bodies form. All four gametes are haploid as products of male and female meiosis. Their nuclear material is evenly divided in both sexes. However, cytoplasm is unevenly divided in females. During a female's telophase I, most of the cytoplasm is retained in one daughter cell, leaving the other three with very little cytoplasm.

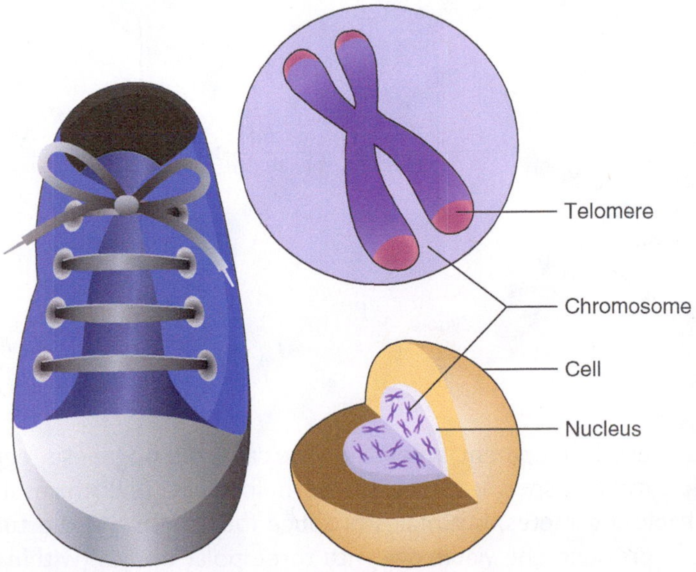

**Telophase I**

The stage resulting in the forming of a set of new cells.

**Cytokinesis**

The division of cell cytoplasm following mitosis or meiosis.

**Meiosis II**

The stage in which sister chromosomes are separated.

**Metaphase II**

The stage in which chromosomes line up singly and then the two sister chromatids separate and move to opposite poles of the cell.

**Prophase II**

The first stage of meiosis II.

**Telophase II**

The last stage in the second meiotic division of meiosis.

© Kendall Hunt Publishing Company

- Telomere
- Chromosome
- Cell
- Nucleus

**Figure 6.11** Chromosome separations of "shoes."

- Although note that not all animals reproduce sexually. For example, in ants, bees and wasps, a virgin birth (parthenogenesis) takes place to produce males, which will be discussed in Chapter 20.

During the next meiotic division, another unequal partition of cytoplasm happens. In most female animals, the result is a set of three small sex cells and one large sex cell. The daughter obtaining most of the cytoplasm becomes the female egg, while the others become polar bodies. Polar bodies generally disintegrate quickly and are not viable for fertilization. In human males, four gametes are made per meiotic division (Figure 6.12). However, many divisions occur simultaneously, continuously producing large numbers of gametes. The average ejaculation contains about 225 million sperm. In females, there is generally only one egg in a cycle. A great deal of energy is placed into egg production, but sperm are made en masse.

In most plant and animal species, the female gamete contains most of the cytoplasm. Can you deduce why? The egg will provide most of the resources, both nutrients and organelles, for a developing zygote. Once fertilized, an egg has the full complement of genetic material from unification with a sperm. Its cytoplasm provides an excellent

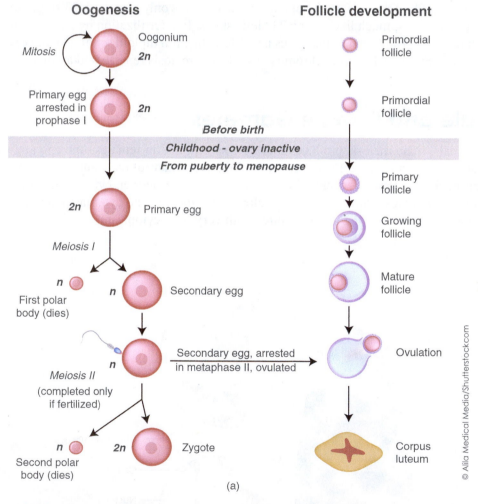

© Alila Medical Media/Shutterstock.com

**Figure 6.12**   Gamete development. a. Females who carry out oogenesis (egg formation) and b. males, who carry out spermatogenesis (sperm formation). Both result in the production of four haploid gametes, but males produce four sperm (in the tubules of the testes) and females produce one viable egg with three polar bodies (within ovaries).

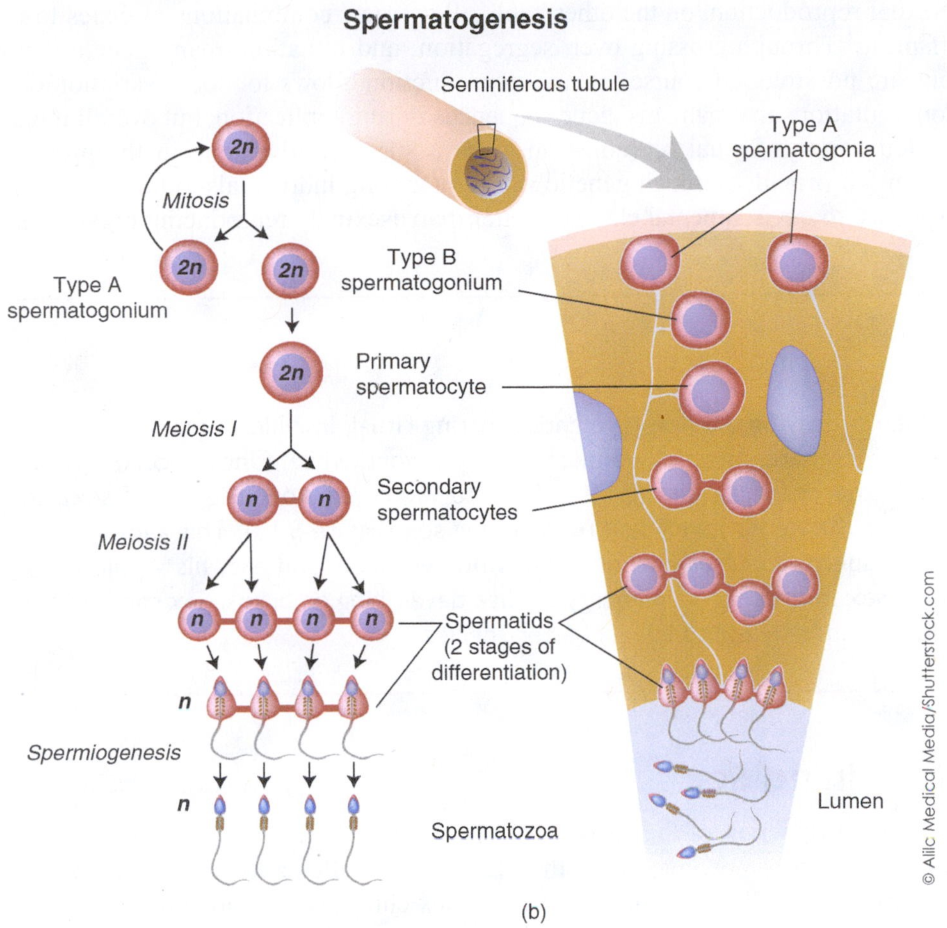

**Spermatogenesis**

(b)

**Figure 6.12** (*continued*)

nutrient resource for the new organism's survival. Also, by having instant organelles on hand for growth and development, the new embryo has an advantage. This is why mitochondria and chloroplasts are so beneficial to simply inherit, in accordance with the endosymbiotic theory – instant energy and food production for a new organism.

## Sex: A Cost–Benefit Analysis

Why sex? It has its advantages and its disadvantages for a species. Asexual reproduction is more efficient and requires less cell machinery. Prokaryotes reproduce by binary fission, simply splitting in half to form two new organisms. The main disadvantage of asexual reproduction is limited genetic variation. Asexual reproduction perpetuates the genotypes of its parents, changing very little from generation to generation. Sexual reproduction instead, leads to many varieties of offspring, enabling some organisms to survive during changing conditions.

If, for example, a change in the environment should occur, as in the potato famine in Ireland in the 1800s, all asexually produced offspring will respond in the same manner. In Ireland, all of the potato plants at the time were grown asexually from the same original plant. The organisms produced, with the same genetic variety, were susceptible to the same fungal-like protist, causing them to decay and leading to famine. Genetic variation allows for differences in a group so that at least some will survive. Variety in potatoes in Ireland at the time would have saved over two million lives.

Sexual reproduction, on the other hand, allows new combinations of genes to form in offspring. Through crossing over, segregation, and mutation, many genetic combinations are possible. Of course, asexual reproduction allows for some variation due to random mutations in organisms' gene sequences during replication, but overall it results in limited variety. Sexual reproduction gives a survival advantage in the process of evolution – it provides enough genetic variation among individuals to help them adapt, as a species, to environmental changes better than asexually reproducing organisms.

---

### THE BENEFITS OF SEX IS DEBATABLE

When observing the praying mantis's mating ritual, in which a male has innate fear of its mate, one still wonders if it is all worthwhile. One hypothesis contends that a female bites his head off during copulation (the act of sex), in order to "ease his mind" and relax during sex (Figure 6.13). This allows more of his sperm to enter into her. She is not wasteful, and eats his whole body after sex in order to gain energy for her developing embryos. Sex can be very efficient in its quest to build a better species.

---

## Determining Sex

**Sex chromosome**

The final smallest pair of the 23 pairs of chromosomes in humans.

**Y chromosome**

A sex chromosome that is found only in males (not given in bold in text).

Upon closer inspection of the 23 pairs of chromosomes in humans, the final smallest pair are the sex chromosomes. The other 22 pairs are called **autosomal chromosomes**, which carry out a cells' life functions. Human sex chromosomes are either X or Y chromosomes, and these determine the sex of an organism. If a human has both an X and Y chromosome, or is XY, it is male; and if it has two X chromosomes, or is XX, it is female. The sex chromosomes differ from each other in a number of ways: a Y chromosome is much smaller than an X; a Y chromosome carries very little genetic information; and a person can survive without a Y chromosome. After all, human females carry only two X chromosomes. A karyotype, which shows a visual map of a set of chromosomes for an organism, is given in Figure 6.14. In some disorders, chromosomes fail to separate

© 2happy/Shutterstock.com

**Figure 6.13**    Praying mantis sex. Soon after copulation, she will bite off his head and consume his body for the energy to raise their young.

Human karyotype

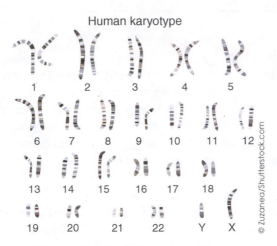

**Figure 6.14**   Human karyotype.

and an abnormal number of chromosomes are seen. For example, in disorders such as Down syndrome, an extra chromosome #21 is found after the failure of that chromosome to separate during meiosis.

In ants, bees, and wasps, in the order hymenoptera, queens produce haploid males and diploid females, making females more related genetically to each other than to males. In fact, they share 75% similarity in DNA, because females have all of the same genes in common from their fathers. The father is haploid and has only one set of the same chromosomes to give to all of his daughters. This phenomenon is known as haplodiploidy, in which some offspring are haploid and some are diploid. This is a basis for close relationships in ants, bees, and wasp societies: they share duties very closely within colonies. In fact, most females within a colony give up sex altogether, remaining sterile castes whose main purpose is to serve the queen master (Figure 6.15). Many plants and all earthworms have both male and female parts; they produce both male and female gametes. Sometimes simple temperature determines sex, as in turtles, lizards, and reptiles. In turtles, cooler temperature eggs become males, while warmer temperatures elicit females.

**Figure 6.15**   Worker ants helping their queen. Loyalty is strong for a queen who controls all aspects of ant society. In this image, worker ants move their queen's eggs, serving both their queen and their future sisters who will hatch from those eggs.

### SEX IS NOT SEXUAL PREFERENCE

Most research supports a strong genetic basis of sexuality. Behavioral genetics is the research specialty that studies the genetic basis of behavior, including sexual preferences. Sexual drive and desire vary across a continuum in most animal societies from asexual (no sex) to hypersexual (excessive sex). It is not a simple like or dislike of certain attributes in the opposite sex. Studies of monozygotic (identical) twins show high contributions of genetic influences for sexual preferences.

Biological bases for sexuality lie in two factors in animals: 1) activity of the medial preoptic area of the brain (MPOA) and 2) DRD4 dopamine receptor gene. Dopamine is a neurotransmitter found in the brain. Neurotransmitters are chemicals that affect different parts of the brain. In humans and rats, for example, the greater the activity in the MPOA area and the greater the number of DRD4 receptors, the higher the sexuality rates in humans and rats.

A range in sexual drives and behaviors makes sense evolutionarily. Hypersexuality, or having many sex partners, may appear favorable for enhancing one's reproductive success (more offspring with more partners), but this is not so—quality also counts. Consider that after fertilization, in many animals a seminal plug forms after a male ejaculates. If another partner enters, the plug is dislodged and this next partner is also able to produce a viable offspring. In promiscuity, the final partner is equally likely to father the child as compared with the first partner. Usually the last partner in is weaker, older, and has poorer quality genes than the first. In animal systems, hypersexuality is therefore selected against, with many partners leading to weaker offspring. Experimental evidence shows that hypersexual behavior in rats leads to decreased reproductive success for the female.

At the other extreme, asexuality, which is a lack of sex drive, is observed in about 1% of humans. Why do such genes persist? One obvious answer is that a lack of sexual attraction does not mean lack of sexual behavior.

On the other side, one would also expect homosexuality to be selected against as it does not lead to new offspring. Another hypothesis as to why "gay" genes remain in our gene pool is based on kin selection. Kin selection is the theory that evolution favors helping between family members or kin to augment the transmission of their related genes. People who do not have their own children are more likely to help their nephews and nieces (kin), who are 25% identical to them. This behavior perpetuates their own genes more than not having any children. Thus, there is strong evidence for a genetic basis of sexual preference and helping behaviors.

**Neurotransmitter**

Are chemicals that affect different parts of the brain.

**Hypersexuality**

The condition in which one has many sex partners.

**Asexuality**

The lack of sex drive.

**Kin selection**

The theory that evolution favors helping between family members or kin to augment the transmission of their related genes.

## Mendelian Traits: Single Gene Characteristics

**Mendelian characteristic (single-gene trait)**

Traits that are determined by instructions on a single gene.

Mendel did not yet know about molecular structures and the chemical idea of the gene discussed in chapter 5, but his explanation for their transmission was remarkably accurate for many traits: that there is pattern to their heredity and that they are inherited as discrete units. Traits that are determined by instructions on a single gene are called Mendelian characteristics, or single-gene traits. There are more than 9,000 single-gene human traits that follow the principles of Mendelian genetics. These are either-or characteristics: an organism has either one type or the other.

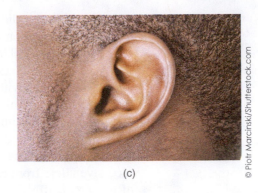

(a)    (b)    (c)

**Figure 6.16**   Examples of single-gene traits. A variety of characteristics are controlled by a single gene pair. Tongue rolling, for example, is dominant over not being able to roll one's tongue and attached ear lobes are recessive. What Mendelian characteristics do you have? a. Colin Farrel, shown here with his sister, has a widow's peak. b. This father and son both exhibit the tongue rolling ability. c. This man's ear lobes are attached.

Consider being able to roll your tongue or not roll your tongue; having a widow's peak or not having a widow's peak; and having albinism or not having albinism. Each is determined by whether one has dominant or recessive sets of alleles. A person who has a widow's peak has a dominant allele dictating that the characteristic will show up. Figure 6.16 illustrates a few single-gene traits.

There are three possible patterns of inheritance of single-gene traits leading to an organism's outward appearance: 1) **autosomal dominant**, in which the dominant allele gets expressed, 2) **autosomal recessive**, in which both recessive alleles are present for a person to get the recessive trait, and 3) sex-linked, in which the X chromosome determines the characteristic (Figure 6.17). Each pattern follows Mendel's rules, expressing the dominant allele in the phenotype. Examples of traits for each pattern are given in Figure 6.17.

Autosomal Dominant. Diseases that are autosomal dominants are expressed when even one allele is contained within a genotype. For example, in Huntington's disease, a degenerative and progressive muscular illness, the trait is inherited as a dominant allele. If a person receives the autosomal dominant Huntington gene, she or he will develop its related disease. Symptoms usually develop after an age of 30 years, well after she or he could pass it onto children.

Singer Woody Guthrie, who sang "This land is Your Land," died from the disease at an age of 55 years, 13 years after symptoms appeared. He was the father of singer

**Sex-linked**

One of the three possible patterns of inheritance of single-gene traits in which the X chromosome determines the characteristic.

**X chromosome**

A sex chromosome that is found twice in females and singly in males (not given in bold in text).

**Autosomal dominant**

The patterns of inheritance of single-gene traits in which the dominant allele gets expressed.

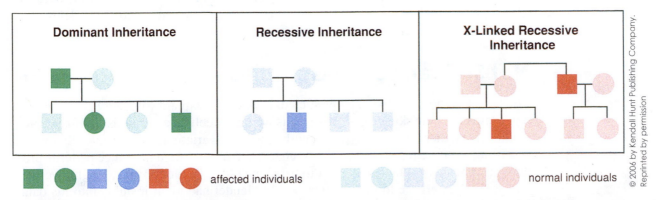

affected individuals    normal individuals

**Figure 6.17**   Examples of traits in three patterns of inheritance: autosomal dominant, autosomal recessive, and sex-linked traits. Each method of inheritance depends upon the expression of genes. Pedigrees for each pattern of inheritance give affected and normal individuals in each generation. From *Biological Perspectives*, 3rd ed by BSCS.

**Autosomal recessive**

The patterns of inheritance of single-gene traits in which both recessive alleles are present for a person to get the recessive trait.

Arlo Guthrie, who did not inherit the disease from his father. Arlo had a 50:50 chance of getting Huntington's disease. Its origin is thought to have arisen from a small town in Venezuela. About 30,000 Americans suffer from the disorder today.

Autosomal Recessive. Most diseases are carried on recessive alleles. Recessive alleles stay hidden within a genotype without being expressed for longer periods of time than autosomal dominants. As discussed earlier in this chapter, a person may harbor a recessive allele without knowing it is present; the dominant allele covers its effects within the genotype. Thus, deleterious recessives persist in groups.

For an autosomal recessive trait to be expressed, an individual must inherit one recessive allele from each parent. Thus, two unaffected individuals have a 25% chance of having an affected child. In Xeroderma pigmentosum, lack of DNA repair enzymes due to recessive alleles leads to skin lesions and skin cancers

While certain forms of porphyria as described in our story are inherited in an autosomal dominant pattern (AIP), other forms occur through an autosomal recessive pattern (congenital porphyria). In both forms, those affected lack enzymes to produce heme groups in red blood cells. Because oxygen is carried throughout the body by heme groups lack of heme causes damage to body systems. (Human systems will be discussed in later chapters.) Both dominant and recessive porphyria are difficult to treat because insufficient blood causes irreversible damage to vital organs. In January 2013, it was reported that the remains of the mad King George III of England were discovered. His mental health as a leader was in question throughout his reign. King George III likely suffered from porphyria. His mental deterioration and decline are chronicled in the 1994 film, *The Madness of King George*. Many of the royal families married kin; increasing chances for inheriting harmful genes, such as porphyria.

Sex-Linked. Sometimes males have a greater chance of inheriting a trait than females. This occurs in sex-linked traits, in which a trait is determined by a gene located on a sex chromosome, making inheritance patterns different between males and females. In sex-linked traits, such as in color-blindness, often the disease-causing allele is recessive. Most genes are found only on the X chromosome, so it determines the expression of a trait.

If a female has one gene for color blindness, for example, she will not become color blind if she has another dominant, normal gene on her other X chromosome. The dominant allele masks the recessive allele causing color blindness. Alternatively, the same situation in a male would result in color blindness. A male does not have two X chromosomes to hide the one troublesome, recessive gene. Because a male has a Y chromosome, which has very little genetic information, it does not hide the effects of the normal dominant allele. Sex-linked traits are more common in males than in females because of this pattern (see Figure 6.18). Females have greater opportunity to hide alleles with genes from their other X chromosome.

## Not So Mendelian Genetics

Most traits do not act as Mendel predicted. How do we explain why there is not simply one or two possible skin colors? If all traits were Mendelian, all organisms of a species would have either one phenotype or another, with no variations in between. Obviously, this is not the case for most organisms' characteristics. Other inheritance patterns produce the phenotypes most common to us: skin color, IQ, blood types, height, weight, and sexual preference to name a few. While Mendel had great insights into his data, most of our genetic expression is more complex than the seven pea plant traits he chose to study.

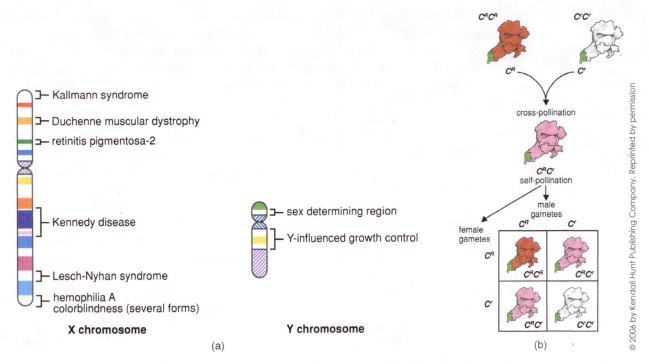

**X chromosome**
- Kallmann syndrome
- Duchenne muscular dystrophy
- retinitis pigmentosa-2
- Kennedy disease
- Lesch-Nyhan syndrome
- hemophilia A colorblindness (several forms)

**Y chromosome**
- sex determining region
- Y-influenced growth control

(a)

(b)

**Figure 6.18** a. Sex-linked traits. Inherited on the X chromosomes, they are more likely to appear phenotypically in males. b. Snapdragons. Red and White Cross of F1 generation results in pink plants. 50% of the offspring exhibit incomplete dominance, showing a pink coloration. This phenotype was not seen in its parents. From *Biological Perspectives*, 3rd ed by BSCS.

## Incomplete Dominance

**Incomplete dominance**

A genetic situation in which one allele does not completely dominate another allele.

Incomplete dominance results from two different alleles contributing to gene expression. Snapdragon plants, for example, occur in red and white varieties, but may produce pink flowers when mated together. A cross between a white and red Snapdragon plant is shown in the Punnett square in Figure 6.18. The red and white alleles are equally expressed in snapdragons, resulting in a pink color.

## Multiple Alleles

**Multiple alleles**

A series of three or more alternative forms of a gene, out of which only two can exist in a normal, diploid individual.

Some traits are controlled by several genes, each expressing a particular phenotype. These traits are examples of **multiple allelism**. Individuals still carry only two of the multiple alleles at any one time, one from a father and one from a mother. However, the traits are all expressed within a population. In human blood groups, there are three alleles controlling blood types: allele A, allele B, and allele O. Alleles A and B are codominant, or share dominance with each other, and allele O is recessive. When allele A or B is present with O, as in AO or BO, the result is a blood type of A or B, respectively. When A and B are inherited together, a blood type AB results, and when allele O is homozygous with OO as the genotype, the result is blood type O.

Alleles code for antigens, or special proteins on plasma membranes of red blood cells: allele A codes for an A antigen, allele B codes for a B antigen, and allele O codes for no antigen. Antibodies are chemicals made by the immune system that initiate an attack on foreign bodies. When blood types with foreign antigens mix, antibodies are made against antigens found on red blood cells.

To apply this, blood type O may be donated to any other blood group because it contains no antigens on its red blood cells for which to attack. Blood type O is therefore called the universal donor. Blood type AB contains both A and B antigens on the red blood cells. Therefore, a person with blood type AB is able to receive all other blood types because they appear non-foreign to an AB immune system – all of the antigens are already on its red blood cells. Blood type AB⁺ is therefore called the universal recipient. Blood type A cannot donate to blood type B and vice versa. Blood type A has A antigens and makes antibodies for B (because B appears foreign to it). Blood type B has antigens and makes antibodies for A (because A appears foreign to it).

Note that "+" and "−" have been used to describe blood types. Blood is classified as either positive "+" or negative "−" because of a surface protein marker on red blood cells, called the Rh factor. If blood contains an allele coding for the Rh marker, then its blood is considered positive. Type A+ blood contains at least one allele for the A antigen and one allele for the Rh factor. The Rh marker is another substance for immune cells to recognize and attack. Those with Rh positive blood types are able to receive Rh negative blood. Those with Rh negative blood, however, are not able to receive Rh positive blood. Rh positive blood contains the Rh marker, which would be recognized and rejected by immune cells of an Rh negative person. Figure 6.19 shows the four blood types along with their antigens and the red blood cells associated with each blood type.

## Polygenic Inheritance

Most of an organism's characteristics are polygenic traits, which are traits with patterns of inheritance determined by more than one gene and influenced by the environment. These include height, skin color, eye color, weight, hypertension, cancer, and heart disease. Polygenic traits are said to be continuous, with many levels expressed along a bell-shaped curve. Figure 6.20 shows the curve for height in athletes as they have changed in the past century. Both exhibit a polygenic bell shape, but the average has increased considerably. What factors in society have changed to increase average height in our society?

Dominance or recessive expression is not so clear cut for polygenic traits. We are not either short or tall, strong or weak, a smart or a bad student or even a brown or blue eye color. There are many variations in between these extremes. Most individuals cluster around an average with very few found at the extremes.

**Universal donor**

A person of blood type O who may donate blood to any other blood group because the blood group contains no antigens on its red blood cells.

**Universal recipient**

A person of blood type AB who may receive blood from any other blood group because the blood group contains all antigens on its red blood cells.

**Polygenic traits**

Are traits with patterns of inheritance determined by more than one gene and influenced by the environment.

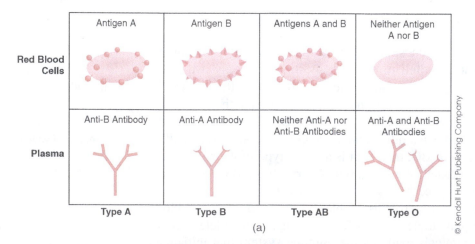

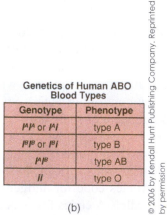

**Genetics of Human ABO Blood Types**

| Genotype | Phenotype |
|---|---|
| $I^A I^A$ or $I^A i$ | type A |
| $I^B I^B$ or $I^B i$ | type B |
| $I^A I^B$ | type AB |
| $ii$ | type O |

© Kendall Hunt Publishing Company

© 2006 by Kendall Hunt Publishing Company. Reprinted by permission

**Figure 6.19** Codominance and multiple alleles. a. There are four discrete blood types in humans: A, B, AB, and O. Three different alleles determine blood type. Blood type is expressed as codominance with alleles sharing a phenotypic expression. b. Genetics of the human ABO blood groups. From *Biological Perspectives*, 3rd ed by BSCS.

**Figure 6.20** People are often categorized by their height. The mean height of men today is 5'10", whereas in 1913 it was 5'8". The photo from 1913 shows a group of college students categorized by height. Note that the categories follow a bell-shaped curve, a characteristic of polygenic traits. What factors do you think contributed to the change in average height over the past century?

Polygenic traits are influenced by the environment because genes alone do not explain the variation in phenotypes. They are called multifactorial traits because they have many factors that affect their expression. Environment interacts with genes to form a phenotype. Obesity, a polygenic trait, was studied to determine the effects of genes and the environment on its expression in humans and mice. Identical twins, which have the exact same genotypes because they arise from the same fertilized egg, were studied. Twin studies often measure how much a polygenic trait is due to genetics. Obesity had a concordance rate of 70%, meaning that 70% of the time obesity is found in both twins, regardless of what they ate.

The mouse Ob gene encodes for a weight-controlling hormone, leptin, produced in fat cells. Figure 6.21 shows two mice, one overweight, with a mutated Ob gene, and one normal weight, with a normal Ob gene. The human gene for leptin is on chromosome #7 and its mutation increases the risk for developing obesity. However, a mutation of the leptin gene is not the only contributor to obesity. Obesity is a complex disorder, involving the interaction of several genes with the environment. Indeed, scientists have detected genes for obesity in humans on chromosomes: #2, #3, #5, #6, #7, #10, #11, #17, and #20. Research on this multifactorial condition continues.

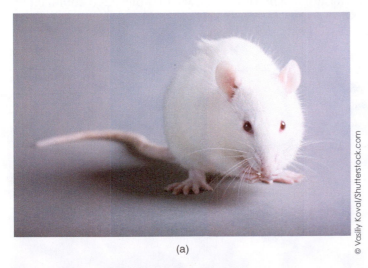

(a)

(b)

**Figure 6.21** (a) Normal vs. (b) chubby rat, the ob gene has its effects on weight in rats (normal rat on the left and obese rat on the right.)

In our story, porphyria symptoms emerge from genetic and environmental factors. While there is a genetic component, stress, smoking, alcohol, and sun exposure trigger symptoms of porphyria. It is also shown that garlic aggravates porphyria symptoms, possibly the root of the assertion that garlic keeps vampires away.

There are eight enzymes involved in heme biosynthesis. Each enzyme has genes that code for it. If any one of these genes is mutated, abnormal heme production results. Thus, the disease has genetic roots as well as environmental triggers. It is multifactorial because many (or multiple) factors affect its expression.

Some polygenic traits are due to gene–gene interactions with very little environmental input. For example, eye color is influenced by about 16 different genes, with less than 1% of its phenotype due to the environment (Figure 6.22). You may have assumed that eye color is an either/or scenario, but in fact it is a polygenic trait, with a continuum of colors possible. Have you ever wondered how hazel or green eyes develop? It is a matter of pigments. The more genes inherited for pigmentation in the eye's iris, the darker the coloration. If there are no alleles for pigment production in one's genotype, eyes will be blue; if there is one or two genes, eye color will be green; if there are three or four alleles for pigment, coloration will be hazel, and more alleles for pigment give varying shades of brown.

## Pleiotropy

**Pleiotropy**

The condition in which one gene affects more than one trait.

When one gene affects more than one trait, this effect is called pleiotropy. Several species of farm birds – chickens, turkeys. – exhibit a "frizzle" mutation on one of their genes. The frizzle allele causes bird feathers to be stringy and weak, providing poor insulation. More seriously, the mutated frizzle allele affects the bird's heart, kidneys, and thyroid and impairs its overall health. Pleiotropy is seen in many characteristics from phenlyketonuria (PKU) in humans, with effects on brain and skin functions, to multiple congenital deformities in rats. All of the associated features of the disorders are due to a single-gene effect on multiple traits.

## Tracing Gene Flow in Families: Pedigree Analysis

Pedigrees are diagrams of genetic relationships among family members through different generations; they are used to trace gene flow through a family (see Figure 6.23 as an example). They show patterns and help figure out whether one has a dominant or

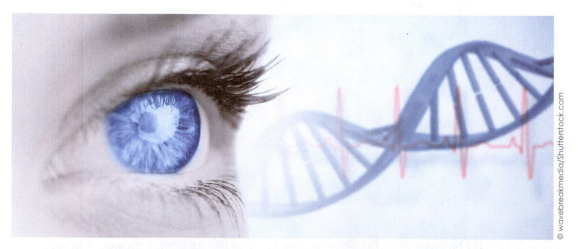

**Figure 6.22**  Eye color genotypes and phenotypes. Eye color is mostly written in our genes.

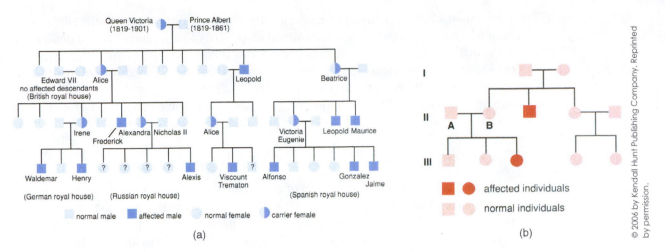

**Figure 6.23**    A pedigree shows the genetics of a family tree. a. Symbols are used to create the family tree. This pedigree shows the inheritance of hemophilia by the royal families of Europe. Hemophilia is a sex-linked trait. The Bettmann Archive. b. A pedigree of a family with congenital porphyria, a recessively linked trait. From *Biological Perspectives*, 3rd ed by BSCS

recessive allele, based on one's parents. The pedigree diagram uses circles to indicate females and squares for males. No shading indicates unaffected individuals, shaded are affected and half shaded are known carriers or heterozygotes. Horizontal lines between circles and squares show mating and vertical lines show descent. Several traits shown in Figure 6.23 indicate how genes are expressed through generations: pedigrees for hemophilia and a family with porphyria are given.

## Tracing Gene Flow in Groups: Population Genetics

How do genes move between villages, cities, and continents? If you compare groups that are separated geographically do you find different characteristics? Is there such a concept as "race," genetically separating different groups of humans? These questions have answers in the branch of genetics called population genetics. A population is defined as a group of individuals able to breed with each other in a given area, producing fertile offspring. The study of patterns of gene flow from one group to another and within groups is known as **population genetics**.

Among other things, population geneticists investigate how diseases are carried in a population of organisms. Mathematical calculations determine the frequency of alleles in a group. These numbers help determine trends in gene flow over time. Porphyria was found to be in high proportions in populations in the old Austro-Hungarian Empire's province of Transylvania, where the myth of vampirism originated in our opening story. Further studies are being done to determine the exact origins. In the example of Huntington's disease, however, population geneticists determined that the gene arose from one woman in a small town in Venezuela, according to records dating back to the 1700s. Scientists collected information from 90,000 people and developed pedigrees to chart gene flow. They tested blood samples to detect the disease and plotted its movement through the years. Though Huntington's disease is inherited, a 2001 study indicated that roughly 10% of cases result from new, random mutations.

Understanding how genes move within a population can help explain why certain genes persist in that population, and this in turn enables us to better understand diseases

**Pedigree**

Are diagrams of genetic relationships among family members through different generations; they are used to trace gene flow through a family (not given in bold in text).

**Population genetics**

The study of patterns of gene flow from one group to another and within groups.

and organism characteristics. For example, by mapping out where cystic fibrosis is located geographically, scientists determined its benefits to immunity against cholera.

It is difficult to determine the exact number of carriers in a population because carriers exhibit a normal phenotype. However, scientists may use a mathematical formula to estimate the probability of occurrence of a recessive allele in a population. The Hardy–Weinberg quadratic equation for equilibrium shows the relative proportion of alleles in a population through counting the number of recessive individuals:

$$p^2 + 2pq + q^2 = 1$$

In the equation, $p$ equals the proportion of dominant alleles in a population and $q$ equals the proportion of recessive alleles within a population. Homozygous dominant organisms are given as $p^2$ and homozygous recessive are given as $q^2$. Heterozygotes or carriers are given as $pq$. Through counting the number of dominant individuals, which one is able to detect through observation, $p$ is calculated. Then, $q$ is solved for, and the rest of the equation's letters are calculated using the quadratic equation. This is a quadratic equation set equal to one. It assumes that a population is not evolving or changing in its allele frequency. It assumes no immigration, emigration, natural selection, or mutations that alter normal gene frequency.

Obtaining data through use of the Hardy–Weinberg equation helps determine the risk of having a particular gene within one's population, helps understand if a population, such as a stand of red maple trees, is undergoing a change in gene flow, and examines how populations compare with each other based on genetic factors. For example, with respect to the alleles for sickle-cell anemia: African American populations with West African ancestry have a 12.5% prevalence of sickle-cell anemia, but West African populations have a 20–40% prevalence. This indicates that the populations have diverged in their overall genetic compositions.

Genealogy is the study of family history. It is related to population genetics, using pedigrees to investigate one's family history. New tests are available that allow one to send in a blood or saliva sample and have it analyzed to trace genetic origins. For instance, tests identify over 400 different ethnics groups in Africa from which our genes may be compared to determine origins. Is this useful or does this further divide people based on the social construct of race?

## INBREEDING: TOO CLOSE FOR COMFORT OR A GOOD STRATEGY?

Consanguinity, or sharing blood through mating with close relatives, such as brothers and sisters, has been shunned by most societies throughout history. The cultural taboo has a practical origin: *inbreeding depression,* or the loss of heterozygotes and at the same time, the acceleration in the number of recessive alleles in a population that are often harmful. The Hardy–Weinberg equation shows the increase of both recessives and their related diseases in studies of inbreeding groups.

Individuals in the same family share many genes in common. The recessive genes that would otherwise be covered up by the dominant allele are more likely to become expressed when recessives occur more frequently. It is likely that pockets of porphyria existed in Medieval Europe, where intermarriage was somewhat common. Porphyria would have been more pronounced in such areas, where dominant normal alleles for heme formation were less prevalent. When close relatives mate, both are part of a lineage that has the potential of sharing more of the same harmful genes in common. Examples may include sickle-cell anemia, cystic fibrosis, or even cancer.

On the other hand, recent research shows that a certain amount of inbreeding can produce healthier children. In a study of Iceland's family history lineage, marriage between third and fourth cousins produced the most numerous and healthiest children over the past 1,000 years. It is hypothesized that outbreeding, or mating with someone too different from one's own genotype, may also lead to health problems in children. In fact, about 20% of marriages worldwide occur between first cousins. This practice is illegal in many of the United States.

Outbreeding too far also has negative consequences, though. One such example occurs for the Rh factor, cited earlier in the chapter. Rh is a set of protein markers on red blood cells that need to match between mother and child for a healthy baby to be born. If the mother is Rh negative, and the father is Rh positive, then the blood of the second fetus who is Rh positive (from the father) will be recognized as an invader by the maternal immune system. Presently, Rhogam is a treatment given to pregnant mothers to prevent mismatched blood from causing a problem (Figure 6.24). Without modern technology, however, such a match would be disfavored. Thus, there is an optimal level of inbreeding for reproductive success. However, third and fourth cousins have only about 1/256 to 1/512 genes in common with one another, so the chances of revealing recessive alleles is quite low.

© Lisa S./Shutterstock.com

**Figure 6.24**  Rhogam is used to treat Rh incompatibility between mother and fetal blood types.

# Gene Technology: Solving Problems Using Genetics

**Biotechnology**

The branch of science that uses biological knowledge and procedures to produce goods and services for human use and financial profit.

**Gene technology**

The technology that modifies plants, bacteria and animals to create products for society.

**Genetically modified organism**

Are organisms in which DNA is genetically altered via genetic engineering techniques.

**Genetic engineering**

The process in which an organism's genes are manipulated in a way other than is natural.

**Recombinant DNA technology**

The process by which DNA is extracted from nuclei of organisms and treated with restriction enzymes.

Biotechnology is the branch of science that uses biological knowledge and processes to produce goods and services for human use and financial profit. Its techniques manipulate genetic sequences in organisms to produce medical drugs and develop weather- and pest-resistant crops, to name a few examples. One significant sub-branch is gene technology, which modifies plants, bacteria, and animals to create products for society. First, the genome of a specific organism is modified by inserting a gene from another organism into the subject organism's already existing DNA. The resulting organism is called a genetically modified organism (GMO) and it is classified as transgenic because it contains genes from another species. Inserted genes produce proteins, for which the inserted gene codes. Human proteins such as insulin, to help diabetics, human growth hormone or HGH, to help in dwarfism, and factor VIII to help hemophiliacs are produced by these GMO organisms. Transgenic tobacco plants produce HGH, as shown in Figure 6.25.

Before gene technology, the available means of collecting these proteins had many drawbacks. HGH was collected from dead bodies and could cause disease when injected into patients, for example. Hemophiliacs, who suffer from life-threatening blood loss due to the lack of a blood clotting factor, were dependent on blood transfusions, which carry a risk of containing infected blood. Before AIDS was discovered in the early 1980s, many hemophiliacs were infected with HIV from blood transfusions. Gene technology changed their treatment options, leading to less risk. Hemophilia is now treated with genetically produced clotting factor VIII. Lessened risk from disease-causing agents is a great step forward for society due to gene technology.

GMOs are produced through genetic engineering, which is the manipulation of an organism's genes in a way other than is natural. This manipulation is accomplished through using a technique called recombinant DNA technology (Figure 6.26). Recombinant DNA technology is the process by which DNA is extracted from nuclei of organisms and treated with restriction enzymes. Restriction enzymes cut DNA at specific sequences. A bacterial plasmid, which is a circular strand of DNA, is also cut with

© Vasiliy Koval/Shutterstock.com

**Figure 6.25** Transgenic tobacco plants. These plants are being used to produce human growth hormone (HGH) to treat human growth disorders. A gene has been inserted into these plants to produce HGH.

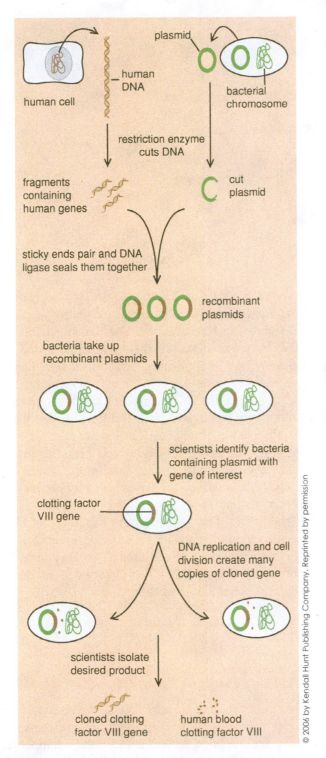

**Figure 6.26** Genetic recombination techniques. They are steps used in producing a genetically modified, transgenic organism. Note that the restriction enzymes cut DNA at specific locations, allowing plasmid DNA to attach and become a "part" of the DNA of the newly created transgenic organism. In this figure, the clotting factor VIII gene is inserted into bacteria in order to produce factor VIII *en masse* for human use. The bacteria made by genetic recombination are genetically engineered "transgenic" organisms. From *Biological Perspectives*, 3rd ed by BSCS.

© Nikolay Litov/Shutterstock.com

**Figure 6.27**   Panhemin Vial.

the same restriction enzyme. Bacterial and human DNA fragments are mixed together, causing them to link with each other. The bacterial plasmid now contains the human gene that will be used for coding new proteins. The plasmid is then transferred into a new bacterial cell. This bacterial cell expresses the newly inserted gene to make the desired protein. It divides over and over, forming new cells that make the a product. The bacterium with its newly inserted gene is said to have been recombined.

In our story, Herbie might benefit if biotechnology treatment options were available for porphyria. To date, porphyria is treated with limited success, with symptoms and long-term problems plaguing its sufferers. An area of study that holds promise for more successful treatment of porphyria and other diseases is gene therapy. Gene therapy is the insertion of genes into an organism to treat its disease. In the past two decades, gene therapy has had mixed success. Future research may find a way to insert a gene into porphyria patients such as Herbie, that blocks the mutated gene, which is unable to produce normal heme groups. Another advance for porphyria sufferers would be in the area of blood production. Presently, blood transfusions restore deficient heme in the blood of porphyria sufferers. Panhemin is also a drug used today to treat porphyria by limiting the liver's production of porphyrins (Figure 6.27). Both treatments are derived from human blood and have risks of carrying infectious agents.

**Gene therapy**

The process in which genes are inserted into an organism to treat its disease.

## ARE PRODUCTS OF BIOTECHNOLOGY HELPFUL OR HARMFUL TO SOCIETY?

Many products are made available through the use of biotechnology. Transgenic crops, for example have greater resistance to herbicides, and viral and fungal diseases. They are modified to withstand cooler temperatures longer and grow faster with larger fruits and vegetables. Soybeans, corn, cottonseed, and canola crops have seen large increases in transgenic numbers in the past decade, as shown in Figure 6.28. Over 93% of all soybeans and cotton crops are genetically modified in the United States. Eighty six percent of all corn, a major staple for cattle and humans, is produced by GMO organisms. If these organisms were not permitted to contribute to our food supply, would we be able to sustain our need to produce food, as a world population?

There have been big increases in farm production since the development of GMO foods. Crops are hardier and more productive, but it is a hotly debated area of study. The greater abundance of food means that fewer people go hungry. However, some GMO foods may also be linked to disease. A 2012 study in Europe shows that a corn variety, NK603 containing genes making it more resistant to the weed killer Roundup, was shown linked to cancer-causing effects when fed to a group of mice. Owing to this "cancer corn," some European nations are placing restrictions on transgenic products. Is this fear of NK603 corn justified?

The public has been consuming GMO products for over 15 years. No known ill effects have been confirmed by the scientific community in this time. What effects will be shown in 10, 20, or 30 years from now, is yet to be determined? Long-term results are not available because GMOs have not been around long enough.

## The Things We've Handed Down: Should We Tamper With Our Genes?

HGH produced by gene manipulation for the past 25 years helps extreme cases of growth disorders. Before recombinant DNA techniques were available, HGH was extracted from the pituitary glands of cadavers and carried the risk of contaminating patients. HGH is now fast and easy to produce, without contamination risks, making it more commercially available.

This brings ethical and practical medical questions into play: Should a preteen male, predicted to grow to a height of about 5' 4", take the drug? What if it is against the doctor's advice, which is based on the American Medical Associations guidelines to restrict the drug only for extreme cases? What is an extreme case? What are the side effects? These are difficult questions to answer.

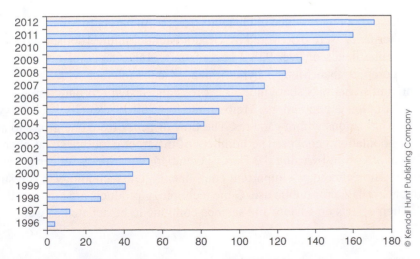

© Kendall Hunt Publishing Company

**Figure 6.28** Graph showing relative increase in transgenic crops over the past 20 years.

HGH has known side effects – from abnormal growth of joints to chronic pain, disfigurement and even death. HGH is used more and more in society to help teens reach a desired height. Some people are very happy with the results and other are devastated. There is uncertainty in medical procedures and treatments and their risks and benefits should be weighed.

What are the social issues involved in being short? How many female readers would date a person taller than they are? I presume many would. How many would date a person shorter than they are? I am not sure. How many male readers would date a person shorter than they are? I presume most. How many male readers would date a person taller than they are? I am not sure.

The reader should weigh the pros and cons of using technologies that scientists have made available to them. It is difficult to judge one another's decision without understanding the social implications of the medical treatment.

Ethically, will another doctor help the patient if one doctor denies treatment? What is ethical for doctors to do if a patient desperately wants treatment to grow taller? These are just a few of the provocative questions about HGH that some young Americans face every day.

---

### MYTH, FABLE, AND SUPERSTITION HAVE EXPLAINED MALADIES INCORRECTLY

In the story at the beginning of the chapter, the character ponders his identity as a vampire. Throughout history, unexplainable illnesses have often been linked with folklore of the occult, witchcraft, and creepy creatures. Epidemics of the plague, consumption (tuberculosis), and the like often led to exhuming bodies and labeling one or more victims to some sort of myth or fable based upon fear. All of these illnesses had biological origins, as shown in our opening story, which shows the power of myth in shaping a person's mental health and outlook—Was the narrator in the story a victim of his own superstitions?

---

## Summary

Heredity is the study of inheritance of characteristics from parent to offspring. Predicted patterns of inheritance were discovered by Gregor Mendel in the 1800s. Mendel's three laws describe inheritance of over 9,000 human traits. Inheritance of genetic information is more complex than Mendel hypothesized. Genes interact with each other, the environment, and sometimes share in their expression. Sexual reproduction results in a great deal of variation in populations. Meiosis, the forming of sex cells, produces unlimited genetic combinations within gametes. The flow of genes from one group to another is studied by population genetics. The numbers of different genotypes and phenotypes in a population are given by using the Hardy–Weinberg equation, with certain assumptions accepted. Biotechnology's important component, gene technology, has resulted in many products available for public use. Gene technology products are continually being developed. Their effects on society and science continue to be debated as well.

# CHECK OUT

## Summary: Key Points

- Heredity affects our physical characteristics, our environment and our future generations.
- The discovery of inheritance by Gregor Mendel explains many of life's characteristics.
- Inheritance can be explained in future generations by probability using Punnett squares and in populations using the Hardy–Weinberg equation.
- The stages and products of meiosis explain how sexual reproduction leads to great genetic variation.
- Many traits in organisms are non-Mendelian, explained by codominance, polygenic inheritance, multiple alleles, and pleiotropy.
- Pedigrees clarify gene flow within families.
- Population genetics studies gene flow between and within populations.
- Biotechnology has advances to provide products for human use, with debatable effects.

## KEY TERMS

| | |
|---|---|
| alleles | law of segregation |
| anaphase I, II | meiosis, meiosis I, meiosis II |
| asexuality | Mendelian characteristic, single-gene |
| autosomal dominant | trait |
| autosomal recessive | metaphase I, II |
| biotechnology | monohybrid cross |
| crossing over | multiple alleles |
| cytokinesis | neurotransmitter |
| diploid (2N) | pedigree |
| dominant | pleiotropy |
| fertilization | polygenic traits |
| gametes | population genetics |
| gene technology | porphyria |
| gene therapy | prophase I, II |
| genetic engineering | recessive |
| genetically modified organism (GMO) | recombinant DNA technology |
| haploid (N) | sex chromosome |
| heredity | sex-linked |
| heterozygous | somatic cells |
| homologous | telophase I, II |
| homozygous | testcross |
| hypersexuality | universal donor |
| incomplete dominance | universal recipient |
| kin selection | X chromosome |
| law of dominance | Y chromosome |
| law of independent assortment | zygote |

# Multiple Choice Questions

1. Which is an inherited disorder?

   a. porphyria
   b. obesity
   c. Huntington's disease
   d. all of the above

2. Which of Mendel's laws was derived from the presence of 100% yellow phenotypes in the $F_1$ generation?

   a. law of dominance
   b. law of independent assortment
   c. law of continuity
   d. law of segregation

3. The way in which an organism appears is its

   a. genotype
   b. phenotype
   c. pleiotropy
   d. codominance

4. If two heterozygous parents mate both carriers for a recessively inherited form of porphyria) what is the chance that their offspring will have porphyria?

   a. 0
   b. 25
   c. 50
   d. 100

5. Which stage of meiosis involves the separation of homologous chromosomes?

   a. anaphase I
   b. anaphase II
   c. prophase I
   d. prophase II

6. Which represents the correct flow of stages in meiosis?

   a. prophase II→metaphase I→anaphase I→telophase I
   b. prophase I→metaphase I→anaphase I→telophase I
   c. anaphase II→prophase II→telophase II→metaphase II
   d. telophase I→anaphase I→metaphase I→prophase I

**7.** Which is the source of hemophilia for Prince Frederick using the pedigree in the figure below?

   **a.** grandmother
   **b.** grandfather
   **c.** mother
   **d.** uncle

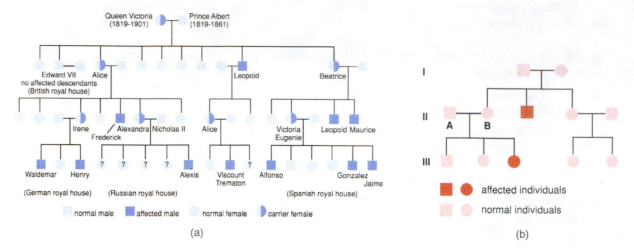

(a)                     (b)

**8.** In the Hardy–Weinberg equation, if the frequency of recessive alleles is 5% of the population, what is the number of recessive individuals in that population?

   **a.** 25 out of 10
   **b.** 25 out of 100
   **c.** 25 out of 1,000
   **d.** 25 out of 10,000

**9.** In question #8 above, what is the frequency of dominant genes in the population?

   **a.** 5%
   **b.** 25%
   **c.** 75%
   **d.** 95%

**10.** Which statement best describes the benefits of GMOs to society?

   **a.** Photosynthesis decreases greenhouse gas effects.
   **b.** The food supply can support the population.
   **c.** Nonnative species are kept in check.
   **d.** GMOs kill many species of insects.

## Short Answers

**1.** Describe how porphyria affects the health of those inheriting it. Describe the mechanism by which porphyria causes damage. How does porphyria get portrayed as vampirism in history? Is it justified? Why or why not?

2. Define the following terms: phenotype, genotype, and pleiotropy. List one way each of the terms differ from each other in relation to heredity. Give an example found within fowl to make this clarification.

3. Describe the experiments of Gregor Mendel leading to the law of independent assortment. How does this law relate to genetic diversity within offspring?

4. In question #3, describe two other mechanisms by which genetic diversity is increased in populations through sexual reproduction.

5. Draw a Punnett square for a cross between two heterozygous tongue rollers? What percentage of their offspring are heterozygotes? Homozygous dominant?

6. List the stages of meiosis I and II, indicating the point at which a cell becomes haploid. Why does it become haploid at this point?

7. Describe a testcross used to determine genotype in a pedigree. What is an advantage of a testcross to determine genotype in a pedigree? Are its results certain? Why or why not?

8. One in 22 people in the United States are carriers for cystic fibrosis. What is the percentage of individuals who actually have this disease, using the Hardy–Weinberg equation? Show your work.

9. Describe the process of recombinant DNA technology. Use the following terms to write its description: restriction enzyme, bacterial plasmid, vector, human DNA, protein.

10. Define the process of inbreeding. What are disadvantages of inbreeding? Are there any advantages? Explain your answers genetically.

## Biology and Society Corner: Discussion Questions

1.  Diseases such a porphyria manifest in ways that give them a bad reputation. What other inherited traits have a bad reputation in society? Choose one trait and discuss how it is treated by the dominant culture. How are people with the trait treated differently? Suggest ways to improve the lives of people with this trait within our society.

2.  The genetic basis of sexual preference was advocated for in this chapter. With which side do you agree, genetic or environmental in cause? What factors do you think limit or enhance the acceptance of alternative sexual preferences in society? Does the idea of a genetic basis have an impact in this acceptance?

3.  If a society decided to remove all of the harmful recessive genes, such as cystic fibrosis, within its population, what would be its ethical difficulty? What would be its practical difficulty, based on the Hardy–Weinberg equation? Explain you answer fully.

4.  Race is used in decision-making regularly in the U.S. organizations. Why is race such an important factor in society? Do you think it should be so? Is there a genetic basis to human race classifications?

5.  A health food guru claims that GMOs are making people fat. Explain why this statement is false. How have GMO foods helped society? How have GMO foods harmed society? Is your answer certain? Why or why not?

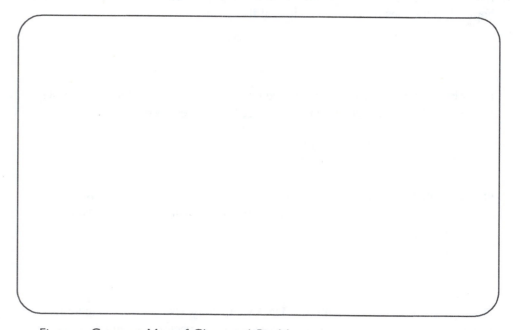

Figure – Concept Map of Chapter 6 Big Ideas

# UNIT 3
## We Are Not Alone!

# Evolution Gives our Biodiversity

**7**

Father and Son Sailing

The island once had many moths

Bats evolved echolocation to prey on insects such as moths

Island without moths; Ecosystem low in diversity…today

# The Case of the Quiet Island

I was a young student studying biology in Wales when I talked with my father into sailing with me. I had seen my friend sail a small boat earlier in the season, and it looked fairly easy. So my father and I set off to visit an island that our family once vacationed on when I was a child. The wind was at our backs, so we zipped along toward the island with ease. I thought, "I should buy a sailboat and make this my new hobby; I am really great at sailing."

We docked the boat within a few minutes and began walking the island. It was just a few miles around On an earlier trip I had noticed that the island had many different creatures – plants, birds, bats, and lots of moths, one of my least favorite insects. My father noticed that it was very quiet – very peaceful and a great place to read – I think he was telling me I should get away to this island and quit bothering him with my studies.

We hiked up its small hills and took leaves and plants for study. We noticed that there were very few insects buzzing around us. "How great," I thought, "no bugs around to bother us." I had always disliked the arthropods, all of the insect classes, in fact. I recalled times when deer flies bit through a shirt into my neck when I was gardening. "They are good for nothing," I reminded myself, happy to be relieved of them for at least this walk. There were only small farms on the island, which consisted of quiet countryside cottages. Oddly though, on our walk there were neither birds nor insects to make noise–well, peace at last.

The island had developed a great deal of farming, and the crops looked very healthy. My father commented, "This is what England needs, productivity. Big farms like this one will make Britain strong again!" I had read in a journal article that two-thirds of Britain's 337 large moth species are in significant decline. It was evening, and I again appreciated that there were no moths buzzing around our heads by the lamplights on the road.

"A nice quiet night but where were the moths that I once watched in the lamps along the road?," I envisioned, recalling their bulbous bodies. As a biology student, I knew that moths were an insect class, Lepidoptera, with 150,000 known species. Moths were not beautiful like butterflies and were pretty gangly, throwing themselves at lights. I would never be an entomologist, who studies arthropods for a living.

My teacher made us read an article reporting that there had been a 99% decline in common garden moths, *Marcaria wauaria*, in the past few decades. The total number of large moths was down by almost 50% in southern England. Three species of moths had disappeared from southern England in the past decade: Orange upperwing, *Jodiacroceago*; Bordered gothic, *Heliophobus reticulate*; and Brighton wainscot, *Oria musculosa*. "Good riddance, life goes on without them; but I wonder why so many were gone?" I pondered.

When we started back to the coast, I realized the wind was against us. I didn't really know what to do when the wind did not have our sail. My father and I struggled to keep the sail straight and steered helplessly through the waves that had developed while we were on the island. It had become a nightmare. I frantically tried to steer and pull as the boat went out of control. The boom hit my father's head in the confusion. He yelled, "You idiot, you'll kill us yet! Why didn't you tell me you had never sailed?" He was bleeding and I felt terrible. How was I supposed to know that sailing could get so out of control? It seemed easy when the conditions were just right. It dawned on me that the slight shift in wind direction, much like one fluctuation in moth populations, could usher in significant change leading to disastrous effects.

Moths are an indicator species, meaning that the state of the environment is first indicated by moth population health. Fewer moths mean less food for birds and bats, which eat moths. Those organisms eating birds and bats also are affected. One change in the environment can have profound impacts on the whole ecosystem.

The island was quiet like the sea when we arrived. There were few insects and few birds to make sounds, but the quiet island had spoken – and it was quiet no more.

---

### CHECK UP SECTION

In the story, our character is at first happy about the loss of biodiversity on the island. By the end of the story, it dawns on him or her that there may be more to the quiet island. Changing environmental factors, much like sailing conditions, can be unpredictable and get out of control. Moths in England declined in numbers in part due to habitat loss: large-scale farming destroyed hedges lining smaller farms, an area where moths thrive; pesticides also were shown to kill off many moths.

Study the life history of the *Marcaria wauaria*, noting its prevalence, habitat use, and the purported reasons for its decline in southern England. Some species of moths saw population increases in southern England. The least carpet moth increased by 75,000%. Research why this occurred: How might changes in moth prevalence impact on our society? Do you think the narrator in the story had a change of heart about moths, about the environment?

---

## What Are the Origins of Life?

Life originated about 3.5–4.1 billion years ago, giving rise to the great diversity of organisms we see today. The origins of our biodiversity emanate from a small set of species of prokaryotes. Stacks of sediment made by colonies of bacteria, called stromatolites, are evidence of our primitive ancestors. Found in Africa, Australia, and the Bahamas, stromatolite layers contain carbon from bacteria dating back to early Earth. Since Earth was formed 4.6 billion years ago, life began relatively early in Earth's history.

How did life originate from our molten ball of Earth chemicals? Early scientific thinkers believed that life originated from nonliving matter. The idea that life appeared from nowhere, called spontaneous generation, was held firmly by scientists for many centuries.

**Spontaneous generation**

The idea that states that life appeared from nowhere.

The 17th-century scientists hypothesized that organic matter in food automatically generated maggots and all associated life when coming into contact with air. (You can make the same observation if you leave food at room temperature for a long-enough period of time; you will likely see mold and flies at the least.) Then, Francesco Redi (1636–1697), an Italian naturalist, became the first to disprove spontaneous generation.

Redi devised an experiment that involved placing a piece of meat into a glass jar. The jar was covered with gauze, which allowed air flow to the meat but no other agents larger than the holes in the gauze. A second control jar was left uncovered to allow contact with any external agent. Redi's experiment is shown in Figure 7.1.

Redi's results showed that the gauze-covered jar did not have maggots, but that the uncovered jar did. Realizing that some other agent had caused maggots and not the meat itself, Redi's experiment was the first to disprove the idea of spontaneous generation. We now understand that flies were the cause of new life on decaying meat, with maggots growing from eggs laid on the organic material.

There were no microscopes in the 1600s to view the developing fly eggs on meat. The mechanism for new species growth on food was therefore unknown. However, Redi was criticized because new growth spoiled foods in both his control and his experimental jars – we now know the bacteria of decay cause the food to spoil. Thus, debate continued on whether life could arise spontaneously. Scientists also sparred over what caused milk and beer to sour. French biologist, Felix Pouchet (1800–1873) believed that microorganisms spontaneously arose in some foods, such as milk and beer, which had the right combinations to create life.

Pouchet heated flasks of hay brews to 100° C. He sealed the flasks, but even though they were sterilized, bacteria formed. Pouchet thus concluded that organisms could arise from a good mixture of materials, as found in beer and other fermentation environments. He tried many times to sterilize the flasks, but in every case within a short period of time, he observed a sea of bacteria. Pouchet thus reopened the defense of spontaneous generation.

Louis Pasteur (1833–1895) proposed an alternate hypothesis, arguing that bacteria were everywhere. He claimed that Pouchet's experiment was contaminated when he

JAR 1          JAR 2

© Kendall Hunt Publishing Company

**Figure 7.1** Redi's experiment. After covering jars, Redi determined that a covered jar does not lead to maggot formation on meat. It was a first attempt to disprove the spontaneous generation of life.

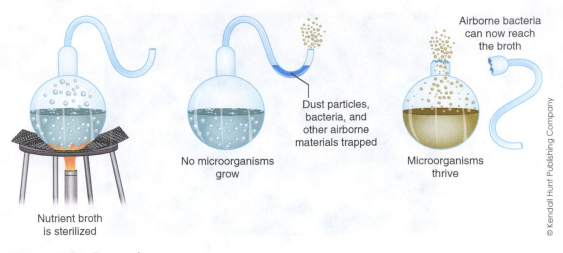

Airborne bacteria
can now reach
the broth

Dust particles,
bacteria, and
other airborne
materials trapped

No microorganisms
grow

Microorganisms
thrive

Nutrient broth
is sterilized

© Kendall Hunt Publishing Company

**Figure 7.2**   Pasteur's experiment.

placed lids onto the flasks. Pasteur designed special long-necked flasks to keep broth placed inside free from bacterial growth as he sterilized its contents. Air was still able to get through to the broth; this eliminated the criticism that life might not have air to breathe, as in Redi's sealed jars. The lower part of the neck of the flask trapped the heavier dust particles and microbes. His flask design is shown in Figure 7.2.

With no external agent, Pasteur reasoned correctly that the "trap" in the neck kept out microbes, and no new life formed in his flasks. When Pasteur tipped a flask to allow broth to touch the trap, bacteria appeared in the broth in a few days. His rejection of Pouchet's results brought Pasteur membership in and an award from the French Academy of Science. Pasteur's experiment showed how his critical thinking led to a solid disproof of spontaneous generation – and led to the birth of microbiology as a discipline.

Personally, Pasteur was a devout Roman Catholic. He performed the experiment to emphasize the sanctity of life. He reasoned that if spontaneous generation were true, then there would be no need for a creator God to exist. His disproof of spontaneous generation was actually a movement against atheism. It worked toward a resurgence of religious faith in the 1800s. While Pasteur promoted his work as an example of pure science, it may show how one's personal beliefs, even as a scientist, influences thinking about scientific research.

The Pasteur–Pouchet debate illustrates how scientific arguments continue through the centuries. The germ theory of biology, which places a focus on sterile techniques to prevent microbial disease spread, led to important improvements in medicine. The widespread use of sterile techniques decreased deaths, especially during childbirth.

The origin of life is of continual interest to scientists. In 1953, physical chemists Stanley Miller (1930–2007) and Harold Urey (1893–1981) devised an experiment demonstrating that precursors to life could have formed from the right mixture of chemicals. Conditions on early Earth were simulated in a glass tube containing methane, hydrogen sulfide, hydrogen gas, and water vapor. The experiment in Figure 7.3 shows the design of Miller and Urey's experiment.

An electrode was placed in the glass tube which simulated X-rays, ultraviolet light, and lightning of the early Earth. The environment in Miller and Urey's glass tube was an oxygen-free system, just like on early Earth. When an electric charge was applied to this primordial mixture of chemicals, *organic* molecules formed. Fats, sugars, proteins, and genetic material were produced from the simulation.

Organic molecules make up life, and as discussed in Chapter 2, are able to self-assemble based on their chemistry. It is hypothesized that droplets of organic material

**Germ theory**

The theory that places a focus on sterile techniques to prevent microbial disease spread, led to important improvements in medicine.

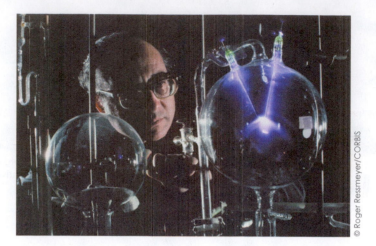

**Figure 7.3**  Miller and Urey apparatus. Gases in the glass tubes reacted to form organic molecules, precursors to life.

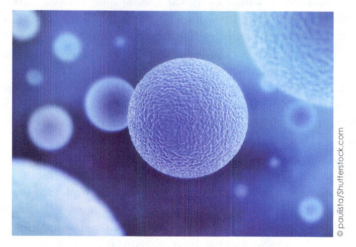

**Figure 7.4**  Prebionts. These tiny spheres share some characteristics of life, including a separate membrane to keep it distinct from its environment.

**Prebiont**

A sphere of organic material that led to first living cells.

formed from the newly made organic products. As shown in Figure 7.4, the droplets formed a sphere that was separate from its environment. This sphere of organic material is called a prebiont and was the first form of new life. It was capable of replicating, absorbing genetic material, and forming new prebionts. These new "cells" further developed into prokaryotes found as fossils within stromatolites. But how did so much life originate from such a simple prebiont?

# Natural Selection and Biodiversity

**Reproductive success**

Is the number of viable offspring an individual produces.

Take a moment to ponder the fact that living organisms, in all their magnificent diversity, emerged from a simple assortment of chemicals on early Earth. Chapter 1 discussed Darwin's principles of evolution; we will expand on some of those principles here.

When populations have more individuals than an environment can support there is inevitably a struggle for survival. Individuals have varied characteristics, with some better able to survive than others. These more successful organisms reproduce more and thus have better reproductive success (RS), defined as the number of viable offspring an individual produces. The frequency with which genes appear in a population change based upon these different RS rates. Organisms with a successful RS increase their relative proportions of genes in a population.

The change in gene frequencies in a population, over time, is defined as evolution. However, evolution acts only upon phenotypes, or the physical characteristics of one's genes. Those organisms with traits better adapted for a particular environment will increase in numbers. The driving force behind evolution is thus natural selection, or nature selecting for or against certain attributes. It results from an interaction between organisms and their environments. Consider the polar bear, *Ursus maritimus*, which has white fur. Over time, those bears with a light color as camouflage were better able to blend in with their snowy surroundings. Contrast the darker colors of the brown bear, *Ursus americanus*, which blends better within darker forests of North America (Figure 7.5).

Their respective environments influence the phenotypic traits that are selected for and against. To illustrate, the dark color of a brown bear roaming the polar ice caps would stand out like a sore thumb, making it easy prey for its enemies and obvious predators to its prey. Thus, different environmental conditions favor different phenotypes at different times. If the ice caps were to melt, becoming forests, polar bears would no longer have an advantage with respect to fur coloration. In our story, the characters witness changes in moth populations on the island. Moth species thus experience a change in gene frequencies within their populations, with some decreasing and some thriving. With declines in the V-moth, *Marcaria wauaria* and extinctions of three moth species in England, Orange upperwing, *Jodia croceago*; Bordered gothic, *Heliophobus reticulate*; and Brighton wainscot, *Oria musculosa*, natural selection is at work. Changed environmental conditions, such as loss of habitats and harmful pesticides in farming, contributed to moth species changes (Figure 7.6).

Why the changes in moth populations in England? Consider that currants and gooseberries, once very popular and found in many gardens, lost favor across households. Currants and gooseberries are a big part of V-moth diets. Without easy access to these foods, V-moth populations declined substantially. On the other hand, some conditions favored certain species of moths. In fact, one-third of moth species experienced increased numbers in the region discussed in our story. The reason was the improvements in air pollution and acid rain led to a rise in lichens, which are fungi–algae colonies. Moth species that increased in numbers had one common feature – they all fed on lichens. This is an example of natural selection occurring before our eyes – changes in moth species in our opening island story due to a response from environmental factors.

**Evolution**

The change in gene frequencies in a population, over time.

**Natural selection**

The natural selection for or against certain attributes.

(a)                              (b)

**Figure 7.5**   a. Brown bear. b. Polar bear.

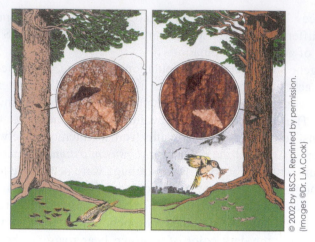

**Figure 7.6**    There are light and dark moths on both the light and dark trees. Which moths do you think are more likely to be spotted and eaten by predatory birds? Why? From *BSCS Biology: An Ecological Approach*, 9th Edition by BSCS.

In order for natural selection to occur on a particular trait, three conditions must exist: 1) there needs to be genetic variation in a population; 2) variation in traits must be heritable, that is, able to be inherited from one generation to the next; and 3) individuals with one trait must have better RS than individuals with another trait. In the case of the British moths in the story, their characteristics must be different from those of other species in some ways, and these differences must be inherited for natural selection to work.

In addition to losing important food sources, the declining species differ in their appearance and in their habitat requirements from other moths. Large farms recently emerged, reducing their shrubby habitats. Without places to deposit eggs and food for their offspring, their numbers dwindled. In contrast, it is widely believed that climate change and increased temperatures allowed many of their competitor moths to colonize the island, thus forcing them out. The three moth species experienced low RS, given the changed environmental conditions, resulting in their extinction from England. Changes in the environment may be temporary, but if the gene frequencies within a population also change, sometimes evolution has irreversible results.

## Types of Natural Selection

If a person has a harmful trait such as porphyria seen in the story in chapter 6, it affects that person's survival. If another person is a carrier for porphyria, then his or her phenotype is normal, and natural selection does not affect survival. Natural selection acts only on phenotypes because the environment only works on those traits expressed by an organism. While nature acts on one's phenotype, phenotype emerges from one's genes. The genes within an organism give rise to its physical appearance. Thus, gene frequencies change when a population is evolving due to natural selection.

There are three types of effects by natural selection: directional selection, stabilizing selection, and disruptive selection (Figure 7.7). Directional selection occurs when individuals at one extreme of the range of variation in a population have a higher degree of fitness. If a group of dogs is bred, allowing only those with an aggressive disposition to mate, the offspring are likely to exhibit more aggression. Vicious dog breeds are commonly used as attack dogs by owners. The idea that behavior can be modified by selecting for certain characteristics is a theme of behavioral genetics.

**Directional selection**

The process that occurs when individuals at one extreme of the range of variation in a population have a higher degree of fitness.

## EVOLUTION DOES NOT CAUSE THE BEST ORGANISMS TO SURVIVE

Evolution is not the survival of the best organisms, only those best adapted to their particular surroundings at any one time. Evolution is a product of the pressure by nature to select out the weak and keep those organisms best adapted for a particular environment. The strongest do not always survive. Consider dinosaurs, which were very strong, according to fossil prototypes, but were selected out; they were not the best adapted at some point in the past. Dinosaurs are believed to have died as a result of a giant meteor impact that added dust and debris to the atmosphere, causing a cool down.

The extinction of dinosaurs is a hotly examined topic. There is a debate between geologists and biologists as to the cause of their extinction. It has long been held that an asteroid hit the Earth about 65.5 million years ago, causing a major shift in climate. Dust from the impact led to less sunlight, fewer plants, and thus less food for dinosaurs and other species. Fossil evidence dating back to that period shows higher than normal amounts of certain materials, including iridium, indicating meteor-like hits. The layer of soil containing these particles is known as the K–T boundary (Cretaceous–Paleocene boundary) and correlates with a high extinction rate for many species types. The large crater on the Yucatan peninsula is thought to be evidence for this meteor impact.

There is an alternate hypothesis that microbial infections spread throughout dinosaur populations leading to their extinction. Emerging theories of disease or infection as the cause implicate a viral or other parasitic infection as the reason for the dinosaur extinction. The hyper-disease theory of dinosaur extinction states that a microbe evolved rapidly to kill off other living creatures during the time period. While weather-related or biological causes led to their destruction, dinosaurs disappeared from the Earth due to the forces of natural selection, despite their impressive physical strength.

Death from global infectious diseases has had major impacts on society more recently in human history. For example, over 70% of Native American Indians died due to diseases brought by Europeans and not through battles. They lacked a natural immunity to those infectious pathogens such as influenza, small pox, and bubonic plague. The power of epidemics to destroy human cultures and other species has historical grounding.

**Hyper-disease theory**

The theory that states that a microbe evolved rapidly to kill off other living creatures during the time period.

**Stabilizing selection**

Occurs when individuals at mean or average range of variation in a population have higher fitness.

**Disruptive selection**

The process in which individuals at extremes of the variation spectrum experience higher fitness than at the middle.

Stabilizing selection occurs when individuals at the mean or average range of variation in a population have a higher fitness. In human birth weight, for example, the average newborn is 7.1 pounds or 3.3 kg. This is also the weight associated with the lowest infant mortality and is thus selected for in nature. At other ends of the spectrum, infants with a low birth rate suffer more health complications without the required body fat; and at the higher end, birthing is difficult due to the large size of the baby. Modern medical treatments are allowing greater variation in birth-weight survival.

In disruptive selection, individuals at extremes of the variation spectrum experience higher fitness than at the middle. In fish, larger males are stronger and able to

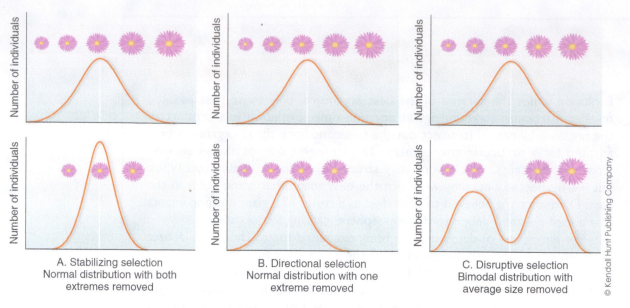

A. Stabilizing selection
Normal distribution with both
extremes removed

B. Directional selection
Normal distribution with one
extreme removed

C. Disruptive selection
Bimodal distribution with
average size removed

© Kendall Hunt Publishing Company

**Figure 7.7**  Graphs showing three types of natural selection.

obtain females more readily. Oddly, smaller males, known as sneaker males, also have better chances of survival than those of intermediate size. Sneaker males are able to "sneak" into a nest and impregnate a female without the larger male detecting him. It is a peculiar strategy for survival and works to help smaller sized fish persist in populations.

In our story, certain phenotypes result from a winning combination of genes. To illustrate, a set of genes determines the odd pattern on the eyed hawk-moth, *Smerinthus ocellatus*. Its coloration allows it to remain camouflaged along bark when it is at rest, and when disturbed, it displays a set of "eyes" that startle its predators, allowing it time to escape. As shown in Figure 7.8, a phenotype enabling greater survival chances for an organism such as the eyed hawk-moth increases those gene frequencies, causing the species to evolve that trait. Natural selection pushes changes in gene frequencies in certain directions based on their efficacy in an environment, allowing organisms to adapt.

© Henrik Larsson/Shutterstock.com

**Figure 7.8**  Eyed hawk-moth is hidden by its surroundings.

# Speciation Increases Biodiversity

In Chapter 1, a *species* was defined as a population of organisms that are able to interbreed and produce live, fertile offspring. Speciation is defined as the process by which natural selection drives one species to split into two or more species. It occurs when the new groups of species cannot interbreed with each other. Their inability to mate is called reproductive isolation. Several conditions may lead to reproductive isolation: mating songs may be so different that organisms don't mate; changes in the genetic composition of two groups of organisms may make their offspring unviable, as in the case of the sterile mule, which is the offspring of a female horse and a male donkey; and divergence of groups into new geographic areas may prevent members of the new group from mating.

It is likely that the eyed hawk-moth evolved from an ancestor that lacked "eyes." This beneficial moth phenotype, its "eyes," enabled greater survival. Those organisms with the trait lived on to become the eyed hawk-moth, while those without the trait were at a disadvantage. Other moths of the Lepidoptera family do not possess this unique phenotypic advantage. They survive, with other adaptations to help their success. The result is a number of different species, each with differing characteristics.

How do new species emerge from an ancestral species? There are two types of speciation: allopatric and sympatric speciation (Figure 7.9). Allopatric speciation refers to the development of new species when there is a physical barrier separating members of a group of organisms. You can remember this as "all apart" speciation, because groups of new species form when they are all apart in different environments.

When members of a population undergo different environmental pressures, new species may result because different traits are selected. Speciation arises due to mechanisms that isolate groups, allowing nature to act. Natural selection works differently when there are different conditions in two distinct environments. The different environments may lead to enough changes to develop separate species.

In the case of the changes in diversity of moths on the island in our story, if V-moths decline as a result of a loss of food, perhaps a set of survivors will persist. Perhaps these survivors inherited genes that enable them to eat lichens on the island or some other

**Speciation**

The process by which natural selection drives one species to split into two or more species.

**Reproductive isolation**

The inability to mate.

**Allopatric speciation**

The process of development of new species when there is a physical barrier separating members of a group of organisms.

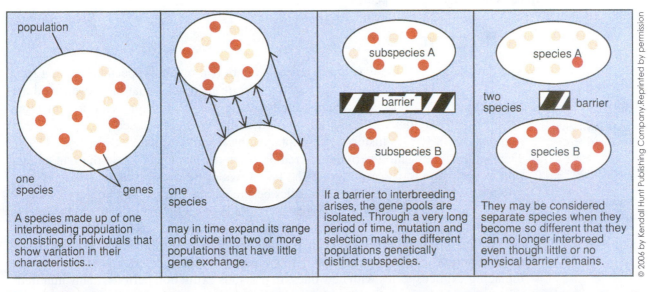

population

one species                                genes

A species made up of one interbreeding population consisting of individuals that show variation in their characteristics...

one species

may in time expand its range and divide into two or more populations that have little gene exchange.

subspecies A

barrier

subspecies B

If a barrier to interbreeding arises, the gene pools are isolated. Through a very long period of time, mutation and selection make the different populations genetically distinct subspecies.

species A

two species     barrier

species B

They may be considered separate species when they become so different that they can no longer interbreed even though little or no physical barrier remains.

© 2006 by Kendall Hunt Publishing Company. Reprinted by permission

**Figure 7.9** Allopatric speciation is shown above. When populations are geographically separated into different species it is termed allopatric speciation; when populations occupying different ecological zones develop new species, it is termed sympatric speciation. From *Biological Perspectives*, 3rd ed, by BSCS.

food not in decline. This group may survive and thrive, while the non-lichen eaters die off completely.

V-moths in other parts of Europe, away from the island, might undergo selection pressures that are different from those experienced by individuals remaining on the island. Consider two groups of V-moths, one on an island in southern England and one in mainland Europe. When enough changes occur in each group's genetic make-up, the two groups of the same species of moth might diverge, forming two new species. This phenomenon is called adaptive radiation, in which a population of a species changes as it is dispersed into a series of different habitats. Adaptive radiation often occurs after 1) the extinction of a competitor, which helps increase the size of a population of species; 2) finding a new habitat, also allowing a group of species to thrive; or 3) after new genes give new advantages to a group of organisms. In each case, adaptive radiation results in many species emerging from a common ancestor. Recall from Chapter 1 that adaptive radiation occurred when new species of finches developed on the Galapagos Islands.

Another form of speciation results in new species due to behavior patterns. When new species emerge while living within the same geographical areas, it is known as sympatric speciation. It may result from 1) changes in genetic material among organisms, as in polyploidy or extra sets of chromosomes in plants, 2) use of different aspects of the same habitat, preventing organisms from interacting with each other, or 3) inability to reproduce with each other. At times, reproductive barriers separate species by preventing them from mating with one another. In sympatric speciation, members of the same species no longer reproduce with one another, because of either a change in mating behaviors or use of different habitats for food, for example. Sympatric speciation occurred in the meadow grasshoppers, *Chorthippus biguttulus* and *Chorthippus brunneus*. The two species are similar in body size and shape, but they are reproductively isolated. The calling songs of each to attract a mate of the opposite sex have different vibrations, so potential mates may not recognize each other. As a result, natural gene flow between groups ceases, and each group changes with its environment as a separate entity. In the field, the two species of grasshopper do not mate, but in the laboratory, members of the two groups are capable of mating with each other. These premating barriers to gene exchange occur simply due to one minor difference in vibration in mating calls between the two species.

## Extinction

The development of biodiversity in our ecosystem is a result of speciation over a long period of time. However, as stated earlier, 99% of all species that have ever lived are now extinct. Extinction is defined as the loss of a species forever, with no remaining organisms to maintain its population reproductively. There have been five great extinction periods in Earth's history, each explained by environmental causes (Figure 7.10). The last great extinction period occurred 65 million years ago. The sixth great extinction is occurring today, at 80 times the rate at which species go extinct as in nonextinction periods. Only 1–2% of all species became extinct in the past century, but their extinction is permanent. Human impacts are believed to be the primary cause of this new great extinction era. Climate change and habitat loss, seen as responsible for moth declines on the island in our story, are primary drivers of environmental change and species extinction. The trouble with modern extinction is that speciation processes are not replacing extinct species with new ones. In previous extinction periods, many hundreds of thousands of years passed with species lost and gained, but in the current human-derived extinction period, occurring only in the past few centuries, new species do not have time to emerge. The danger to our fragile ecosystem is human unwillingness to

**Adaptive radiation**

The changes that occur in a group of organisms to fill different ecological niches.

**Sympatric speciation**

The emergence of new species while living within the same geographical areas.

**Extinction**

The state in which a species is lost forever, with no remaining organisms to maintain its population reproductively.

## EVOLUTION DOES NOT CAUSE INCREASING COMPLEXITY

A misconception that evolution leads to increasing complexity is not supported. The best adapted creatures are often the simplest, and many complex organisms – for example, dinosaurs – have become extinct. As discussed at the start of this chapter, prokaryotes have been around for most of Earth's history, 3.5 of the 4.1 billion years of the planet. Prokaryotes remain very competitive because of their simplicity in design. They contain very few organelles and very simple genetic information, with only a primitive nucleus, a cell membrane, and some cytoplasm. There are no fancy organelles as found in plants and animals. This limits the amount of things that can go wrong with these creatures. Evolution does not lead to perfect organisms, only those best able to adapt to our world.

Consider the old VW beetle, which had no air conditioning, no power locks, windows, or brakes. It ran and ran for decades without problems. The VW is much like a prokaryote. Alternatively, more expensive and complicated automobiles contain many features that can break down. Anyone who has had a check engine light turn on appreciates the aggravation in finding the small problem causing an emissions issue. The complexity of humans and other creatures can be their downfall. Over 99% of all organisms that have lived on this planet are now extinct! Complex life does not necessarily survive better, and certainly, I would predict that prokaryotes will outlive humans by billions of years.

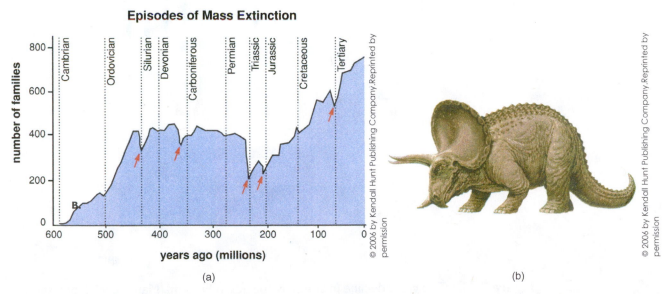

(a)                                    (b)

**Figure 7.10**    a. Record of mass extinction periods in geologic history; five mass extinction events in Earth's history occurred in relatively short periods of geologic time. The latest and most rapid extinction period is blamed on human society and its effects on the environment. From *Biological Perspectives*, 3ʳᵈ ed, by BSCS. b. Dinosaurs, such as this Triceratops became extinct at the end of the Cretaceous Period, roughly 65 million years ago. It is hypothesized that a large meteor hit Mexico, leading to climate changes that did not support the life of dinosaurs. From *Biological Perspectives*, 3ʳᵈ ed, by BSCS.

alter environmental conditions so that species can emerge. Perhaps creating and enforcing policies that limit climate change effects or curb habitat loss will slow the sixth great extinction.

In our story, it is easy to see how quickly environmental changes can have unintended impacts. Other species besides the moths - namely butterflies, have experienced declines in England as well. Obviously, with a 99% decrease in the number of Black Hairstreak butterflies, the species is on its way to extinction there (Figure 7.11). With over 2,500 moth species reported in England, of which 900 are described as larger moths, it is concerning that the total numbers of larger moths has declined 38%, overall. Many biologists have suggested that species declines could lead to losses in biodiversity and extinctions of other species.

In our story, the main character's final realization that there is a delicate balance in nature is heartening. The character finally shows an interest in learning about declining biodiversity and its effects on the environment. While difficult to detect upon simple observation, species losses affect human society as well other organisms in unintended ways. For example, while some species of moths increased in numbers, most declined. Moths are pollinators and are a food source for animals such as birds, bats, and small mammals. Along with moth species' decreases, butterflies, bees, and carabid beetles also are declining in number. Biologists point out that this might be a part of a greater contraction in biodiversity in the insect classes as a whole – that England is only a prototypic ecosystem suggesting larger changes.

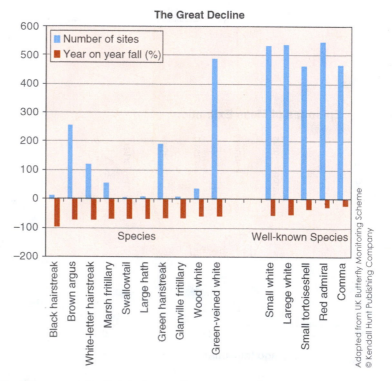

**Figure 7.11**   The great decline in butterfly species in England. Many species of insects declined in 2012, with 300, 000 fewer sightings of butterflies in one year. Changing environmental conditions, including weather patterns such as increased rainfall are thought to be contributing factors.

**Figure 7.12** Ecosystem of island in story – each of these organisms will be affected by a loss of moth species – which means that many other species are also affected.

Changes in biodiversity are difficult to predict; the moths in our story are food for many species of birds and bats, which are eaten by wolves, owls, and eagles, to name a few. A delicate balance is maintained in our ecosystem (Figure 7.12). While natural selection is natural, humans impacts are not.

### ARE DINOSAURS EXTINCT?

Yes and no . . . While dinosaurs as a set of distinct species have been removed from the community of life, their DNA is very similar to that of birds. In fact, many biologists argue that birds are direct descendants of dinosaurs. Feathers were found on dinosaur fossils as far back as 1860. This link between feathers and dinosaurs was found in over 20 species of dinosaurs. Studies of fossils from dinosaurs, namely the *Tyrannosaurus rex,* show that birds are more closely related to dinosaurs than to other organisms such as reptiles, amphibians, or humans. Through analysis of collagen fibers, ropes of proteins in the soft tissues of animals, genetic relatedness between bird and dinosaur species was shown. Obviously, dinosaurs are no longer roaming the planet (Figure 7.13). However, much of their DNA remains – in birds.

**Figure 7.13**    Dinosaur extinction: Did a meteor hit the Earth and cause their extinction?

## Extinction and Biodiversity

The majestic coral reefs of the tropical waters are the largest living creatures in the world.

Moreover, they provide large habitats for many ocean species. They are, however, in decline, and their decline threatens the species who live among them. They have been adversely affected by boating, pollution, acid rain, and change in climate. Much like the moths in our story, they are a marine indicator species showing oceanic health.

Corals are resilient to a degree; for example, even if a small portion (or polyp) of the reef is destroyed, the organism can survive. But given the variety of adverse conditions they face, they are slowly dying, and when they die, so will many of the other species that inhabit them.

Skeletons of coral reefs, made of calcium carbonate, are a major component of the soil of the many archipelagos and islands in the tropics. Because coral reefs are so immense, they are not likely to die off from single isolated attacks. Many features of living systems are adapted for their survival (Figure 7.14). Coral reefs are large, living edifices but their survival is not guaranteed.

Coral reefs are just one example of how people often react to changes when it is too late. Our story showed how declines in moth populations could lead to a rocky sea of change for the environment, as the characters experienced in their journey are back from the island. Will it take a seaside sailing expedition for us to see how fragile our environment is?

## Evidence for Evolution

The extinction and emergence of new species are not new in Earth's history. Organisms have changed throughout their time on Earth, as shown in the evidence for their evolution.

**Figure 7.14**    Colors of the coral reef: The vast and varied structure of coral reefs provide a habitat for a large number of species.

The evidence for evolution is given by four sources: 1) modern day examples of evolution – recent natural selection in organisms based on environmental pressures; 2) the fossil record, which shows organisms of the past in rock layers; 3) homology, or common ancestry; 4) biogeography, or the way species are distributed; and 5) molecular evidence.

## Modern Day Evolution

Organisms are exposed to environmental factors affecting their survival. We do not, however, evolve as individuals, changing with the times. Instead, evolution is a population concept. That is, it is change over time in the genetic composition of a population. An individual will not develop gills because the Earth becomes an ocean, as in the film *Water World*. Instead, one mutant human could be born with some sort of gills that help it survive. While far-fetched, such an adaptation would likely take generations to emerge within a population. Also, it would develop only by random chance. In fact, gills slits in humans are not likely to develop. Usually, species go extinct and do not have the chance to survive with such a random mutation.

Evolution usually occurs slowly over many generations in life's history. However, there are times when we are able to see it happen before our eyes. For example, bacteria reproduce very rapidly, within minutes in some species. The chance for mutation and adaptation among bacteria species is high within our lifetimes because one human generation (25 years) equals many thousands of generations of a bacterium.

The mutations of $H_1N_1$ influenza virus are examples of an organism changing in response to the environment. Each year, its viral strains cause great suffering to the humans: 225,000 people are hospitalized each year due to influenza, and between 5,000 and 50,000 die of flu. In influenza-causing bacteria, recognition proteins on the surface of the membrane surrounding the influenza virus are either H, or hemaglutinnin, or N, neuraminidase. When viruses attach to their host to attack, they use these recognition proteins to dock. As they mutate, some forms of N and H are better able to resist our immunizations. It becomes a continual struggle for modern medicine and pharmaceutical research to keep these mutating viruses at bay. During some years, influenza hits harder, and the vaccines are said to be less effective than in other years. This happens because the N and H recognition proteins have mutated enough to withstand the effects of the vaccines. Influenza thus evolves before our eyes, forming new strains with new shapes of recognition proteins. As vaccines knock out strains of influenza, the survivor viruses, with new recognition proteins, become resistant to current vaccines.

**Fossil record**

One of the four sources of evidence for evolution, which shows organisms of the past in rock layers.

**Homology**

Common ancestry.

**Biogeography**

The way species are distributed.

Many species of bacteria have become resistant to antibiotics in a manner similar to influenza. Indeed, when penicillin was introduced in 1944, over 90% of strains of staphylococcus, a skin bacterium, were susceptible to the drug; by 1950, only half were, and today only 30% of staphylococcus strains are susceptible to penicillin. The infamous MRSA, methicillin-resistant *Staphylococcus aureus* is increasing in frequency in hospitals across the United States and is the cause of a variety of difficult to treat infections. These "superbugs" evolved from earlier strains that mutated to resist the ability of penicillin to damage their cell walls.

Biologists have also observed the ongoing evolution of two genuses, the *Heliconius* genus of long-wing butterfly and the *Passiflora* genus of passion flower plant, each evolving defenses against the other in a process termed coevolution (Figure 7.15). In coevolution, two organisms evolve traits based on their relationship with one another. The *Heliconius* butterfly lays its yellow eggs on the plant, and when the eggs hatch, the caterpillars eat the plant. The *Passiflora* evolved yellow spots on its leaves to mimic eggs. This strategy protects the plants since butterflies will not lay eggs on a plant that already has eggs on it. Over time, the *Heliconius* developed more acute vision that enabled the sharp-eyed individuals to detect the fraudulent yellow spots, and once again use *Passiflora* as a nursery and food source for the larvae. In response, the *Passiflora* developed mutations that coded for extra-floral nectarines, which attracted ants onto the plant to defend it from developing butterflies. Ants eat the *Heliconius* eggs. This example of coevolution clearly illustrates a logical battle, using natural selection to develop new phenotypes between two species. However, changes are always based on random mutations that may or may not benefit each organism. In this case, plants with yellow spots were better able to survive and reproduce, just as butterflies with more acute vision were able to pass on their genes to offspring that had an ample food source. Because of these features, brought about through random mutations, some were more likely to survive than other individuals of the same species.

## The Fossil Record

As the changes on the island in our story illustrated, characteristics of populations change slowly over time or rapidly. Some changes are slow enough that they are measured in eons – periods of geologic time. Geologists can track these changes by

**Figure 7.15** Heliconium butterfly and Passiflora flower, a case in coevolution. This Heliconium caterpillar is feeding on a leaf of the Passiflora plant. Passiflora combats this by "faux" eggs. Heliconium eggs resemble those "faux" on the leaf of the Passiflora plant. These "faux" eggs dissuade the Heliconius from laying more eggs on the plant.

**Figure 7.16** Soil layers and fossils found within them. There are many organism classes (from prokaryotes to modern humans) found in layers of the Earth, connecting their existence to different time periods. Note that the rocks with embedded fossils in this layer are from Whitby, England.

inspecting the fossil layers of the Earth. The fossil evidence for evolution supports the predictions by Darwin that there were changes in species over time. The fossil record shows that prokaryotes did indeed precede all other life and that animal classes developed in predicted taxonomic ways.

The sequence of development shows that first fish fossils, then amphibians, then reptiles then mammals, then birds, and finally humans emerged on Earth's scene. This sequential order of appearance contrasted with the sudden creation hypothesis that predated evolutionary thinking. Because fossils appeared over many hundreds of millions of years, the fossil record supports species change over time.

Paleontologists study the fossil record by measuring isotopes in the soil, dating layers back to when they were first deposited (Figure 7.16). In the past, before radioactive dating of isotopes, fossils were dated based on how deep in the Earth they were. The deeper the fossil is, the older the layer of soil. However, earthquakes have disturbed layers, mixing fossils of different time periods. This was a primary criticism of the evidence for evolution, in which chronology in development of organisms was confusing.

## Homology

Darwin hypothesized that all species evolved from a common ancestral species. Thus, he surmised that they would have characteristics similar to those of their common ancestor. Indeed, scientists have since shown, through analysis of the fossil record, that Darwin was correct. Among several types of evidence, scientists have studied homologous structures – similar structures found in different species (Figure 7.17). Bones such as the femur or thigh bone have the same general shape and relative size in many different species: humans, whales, bats, and birds, for example. While the fin of a whale is used for swimming and the wing of a bat for flying, the bones supporting these structures appear similar to human arm structure. Very often, evolution does not change design entirely. As discussed in Chapter 1, life is efficient, and when a good design works, it persists in many species.

At times, structures that once had a purpose but no longer appear to be functional, called vestigial organs, are found across different species. Snakes, for example, retain their vestigial walking bones such as a pelvis and leg bone, but no longer are able to

**Homologous structures**

Similar structures found in different species.

**Vestigial organs**

Structures that once had a purpose but no longer appear to be functional.

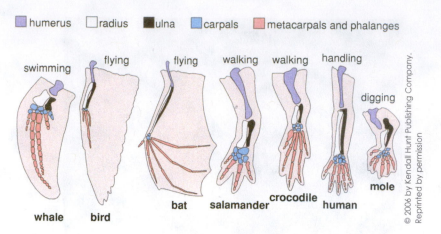

**Figure 7.17** Homologous structures in several species. These structures are similar because they are derived from the same ancestor. From *Biological Perspectives*, 3rd ed, by BSCS.

walk. Our tail bone or coccyx, the final bone along our back, is our vestigial tail. The presence of vestigial organs is one sign of common ancestry.

Embryos also show commonality among related species. Pharyngeal pouches in the throat regions of vertebrates exist in embryos but not in adults (Figure 7.18). These pouches all appear similar upon inspection of the embryos but develop into either gills in fish or Eustachian tubes of the middle ear in humans. These structures show that efficient systems were worked out during our common embryological homology, but that different purposes evolved at some point in the past to direct development in other ways. A saying, "ontology recapitulates phylogeny" means that development of embryos (ontology) reiterates phylogeny (classification based on evolutionary evidence), and

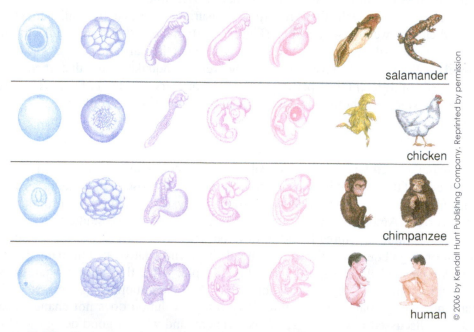

**Figure 7.18** Embryos of several vertebrates: salamander, chicken, chimpanzee, and human. The embryos are similar, passing through the same anataomical stages. Comparative embryology is evidence that they arise from a common ancestor. From *Biological Perspectives*, 3rd ed, by BSCS.

emphasizes that the similar embryologic structures show our common ancestry and thus is strong evidence for evolution.

## Molecular Evidence

Molecular evidence is perhaps the strongest evidence of all for evolution. Consider that all organisms have the same genetic structures – DNA and RNA – to carry out life functions. The central dogma, DNA ➔ RNA ➔ protein, defines the common characteristic found in all living species. Moreover, all organisms use the same general form of DNA because it is efficient.

Molecular DNA data confirm findings from the fossil record and from analysis of homologous structures. Molecular analysis of DNA across species shows that the more DNA organisms have in common, the more closely related the species. Molecular biologists and other scientists study amino-acid sequences in different proteins to demonstrate the degree of relatedness of species . . . Figure 7.19 shows the commonalty of amino-acid sequences in hemoglobin among several species. Common molecular homology is a strong confirming piece of evidence supporting evolution and relatedness of species.

## Biogeography

The geographical arrangement of organisms he observed first gave Darwin evidence for his ideas on evolution, as described in his Galapagos Island experiences (see Chapter 1.) Different beak characteristics in finches developed according to different environmental conditions on their respective islands.

When organisms in different environments develop the same characteristics, although unrelated, it shows that environment plays a major role in developing traits. Consider the sugar glider of Australia and the flying squirrel of North America (Figure 7.20a). Both organisms have a gliding lifestyle, with wing-like structures, but they are unrelated. They resemble each other but are only distantly related to one another. The sugar glider is more closely related to kangaroos, for example, while the flying squirrel is more closely related to other mammals. Sugar gliders and kangaroos are both marsupials, meaning that they complete their embryological development in a pouch. Flying squirrels are mammals, meaning that they complete their development internally in a uterus (Figure 7.20b). Sometimes similar physical appearance does not always imply closeness in two organisms' evolutionary histories.

In our story, the island had a different climate and different species than in other parts of Britain. Different environmental conditions affect species differently based on their genotypes and thus they display different traits. Island biogeography often gives indications for evolution, because each island may have its own set of environmental

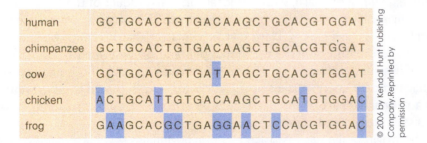

| human | GCTGCACTGTGACAAGCTGCACGTGGAT |
| chimpanzee | GCTGCACTGTGACAAGCTGCACGTGGAT |
| cow | GCTGCACTGTGATAAGCTGCACGTGGAT |
| chicken | ACTGCATTGTGACAAGCTGCATGTGGAC |
| frog | GAAGCACGCTGAGGAACTCCACGTGGAC |

© 2006 by Kendall Hunt Publishing Company. Reprinted by permission

**Figure 7.19** Similar molecular DNA comparisons across species. Genetic similarities in hemoglobin structure show very few base differences between humans and other species. Humans and chimpanzees show a very high degree of genetic relatedness. From *Biological Perspectives*, 3rd ed, by BSCS.

(a)                                                    (b)

**Figure 7.20**    a. Sugar gliders. b. Flying squirrels.

conditions and organisms. Islands provide a good setting to study how organisms change over time. Island biogeography in our story gives first indications for how the larger ecosystem may change in the future.

## Evolutionary Design: There is No One Right Answer

As stated earlier in this chapter, evolution does not favor the strongest or the smartest. Instead, it is based on a response by organisms to different environmental conditions at different times and in different places. A phenotype that works at any one

### DOES EVOLUTION EXPLAIN THE ORIGIN OF LIFE OR THE ORIGIN OF THE UNIVERSE?

Evolution does not explain either of these fascinating beginnings; in fact, evolution only explains life *after* it evolved. The Big Bang Theory attempts to explain how the universe formed, from an extremely hot ball of gases some 13.75 billion year ago (Figure 7.21). The universe continues to expand in what is termed an inflationary epoch. Evidence for the Big Bang Theory is based on observations by scientists of movement and particle attraction. Matter is thought to be exploding and expanding outward, with our local sun a by-product of the Big Bang.

Physicists have indirectly detected that over 95% of the universe is either dark matter or dark energy. Dark matter is matter that is cannot be seen but only detected based on its gravitational pull and other evidence for its existence. Physicists estimate that dark matter makes up 23% of the universe and its affiliated dark energy makes up 72%. Matter that is known to us composes only a very small part of the universe. It is strange to think that most of our universe is unseen and undetected by us.

The origin of life itself is traced to our beginnings – we are made of stardust from the big bang, but before that time, there is no explanation of the forming of matter as we know it.

(a)

(b)

**Figure 7.21**    a. The Big Bang is an explosion that is theorized to have begun matter and the universe. b. The TV show "Big Bang Theory".

time may be ineffective at another time. Evolution does not produce a perfect or even the best organism. Contrary to popular belief, evolution does not continually make species better. In fact, at times organisms can get much worse. Selection for large peacock feather to attract a mate actually hampers their lifestyle, preventing them from being able to outrun predators (Figure 7.22). On many farms, chubby chickens have been bred to be so large and tasty that they can no longer engage in sexual relations. They need to be artificially inseminated to produce live young, so selection in this case is artificial and not natural. These are extreme examples of selection that decreases an organism's quality of life. It is important to note that evolution acts to change organisms in response to their environmental conditions and not to make them better or worse creatures.

© Shawn Hempel/Shutterstock.com

**Figure 7.22** Peacock's large tails attract mates but often interfere with its ability to walk and carry out everyday functions.

### EVOLUTION OF DIABETES: BLESSING OR CURSE?

Diabetes is a sugar-sparing chemical system. It is not a new disease – it has been a genetic characteristic of humans since the origin of our species. Diabetes is, in fact, a useful and natural sugar-sparing characteristic that in prehistoric times conserved energy in times of starvation. Genes linked to diabetes are termed "thrifty" genes because they keep sugar levels high in the blood.

Diabetes is defined as a disease with higher than normal levels of glucose (sugar) in the blood (80–120 mg of glucose/100 mL blood). Under normal conditions, sugar in the form of carbohydrates is consumed by an animal, triggering the pancreas to respond by making more insulin hormone. Insulin stimulates the glucose to be ingested into body cells and causes the liver to store it. When blood sugar levels drop too low, the pancreas secretes the hormone glucagon, which stimulates the liver to release sugar. This negative feedback mechanism maintains homeostasis of sugar to around 90 mg/100 mL blood consistently through a lifetime (Figure 7.23).

Diets high in sugar and simple, refined carbohydrates are linked to the development of Type II diabetes because excess sugars "wear out" insulin receptors. Intakes of food with refined sugars, such as donuts and cakes, elicit a surge in insulin and a docking with cells in the body. This wearing-down process is known as *down regulation*. Each cell in the body has proteins on membranes to which insulin attaches. After docking, insulin causes the target cell to take up glucose. Diets high in sugars wear out insulin receptors and cause insulin resistance. Diabetes Type II (also known as adult-onset diabetes) works in this way; not to confuse it with Type I diabetes, which is an autoimmune attack on pancreatic cells that produce insulin. Both result in hyperglycemia (high sugar levels) but have very different mechanisms.

Damage to tissues and organs occurs whenever glucose levels are too high or too low. In Type II diabetes, hyperglycemia results in diabetes, with insulin unable to allow sugars to be used by cells. This keeps blood sugar levels high, and cells do not obtain needed energy; thus they are starving. Clearly, these diabetic bouts damage nerves, blood vessels, heart muscle, and other organs. Alternatively, at low levels, hypoglycemia takes place, with individuals suffering from weakness, disorientation, and even unconsciousness as the brain is deprived of needed sugars. This condition would be an evolutionary disadvantage for prehistoric individuals who regularly missed meals because of the difficulty of getting food in harsh environments.

A mutation to prevent the conversion of glucose to glycogen was beneficial at one point in the past. The diabetes mutation maintained sugar levels during starvation conditions. These "thrifty" genes spare sugars in the blood, keeping it available for use. Consider the Neolithic diet, in which calories may be hard to find at certain times of the year; for example, when there is little game to be found. The individual with "thrifty" genes would benefit because normal circulating levels of sugar would be maintained longer for hunting and gathering to find food.

James V. Neel, a geneticist at the University of Michigan, discovered the "thrifty gene" sequence in some human populations. When faced with starvation conditions, natural selection would have favored such gene sequences. The Pima Indians of Arizona, he found, were not only resistant to insulin's effects of taking up sugar from the bloodstream but also retained more fat in storage. This helped the Pima endure longer periods of reduced food availability and starvation conditions.

The advantage of the "thrifty" genes to store more fat and keep circulating available blood sugar was an important survival strategy for populations during pre-agriculture times. However, in modern society, with availability of food much increased and a change from hunter–gatherer lifestyles that required more energy, people with "thrifty" genes are more prone to obesity and diabetes. In fact, about one-half of the Pima Indians have diabetes, and about 95% of those diabetic individuals are obese. Are diabetes and obesity on the rise due to the dissonance between evolved metabolism and modern diets? Can lifestyle changes improve these maladies?

Studies indicate that a return to more traditional lifestyles and diets could improve the health of individuals. Pimas practicing traditional lifestyles in isolated parts of the Sierra Madre mountains of Mexico have significantly lower rates of diabetes (8%) and obesity (rare) compared to the modern U.S. Pima Indian population (Figure 7.24). This may be a case study to guide changes in our approach to combat obesity and diabetes. A diet rich in variety and whole grains, fresh vegetables and fruit, and low-fat protein sources has abundant support in science as a recommended diet.

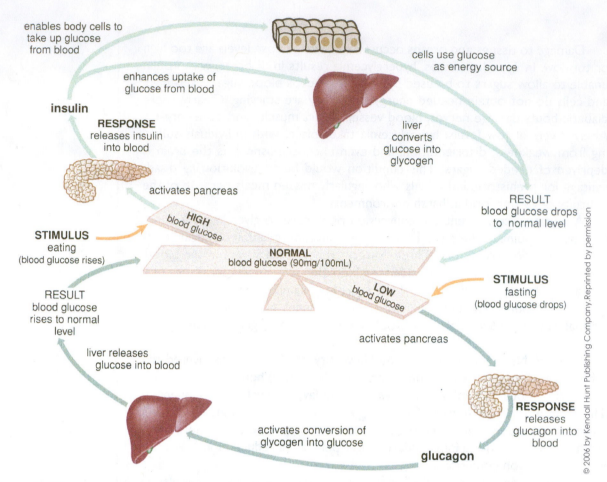

enables body cells to take up glucose from blood

cells use glucose as energy source

enhances uptake of glucose from blood

**insulin**

liver converts glucose into glycogen

**RESPONSE**
releases insulin into blood

activates pancreas

**RESULT**
blood glucose drops to normal level

**HIGH**
blood glucose

**STIMULUS**
eating
(blood glucose rises)

**NORMAL**
blood glucose (90mg/100mL)

**LOW**
blood glucose

**STIMULUS**
fasting
(blood glucose drops)

**RESULT**
blood glucose rises to normal level

activates pancreas

liver releases glucose into blood

**RESPONSE**
releases glucagon into blood

activates conversion of glycogen into glucose

**glucagon**

**Figure 7.23**   Negative feedback controls sugar levels. Insulin and glucagon control glucose levels in the blood to keep blood sugar at optimal levels. From *Biological Perspectives*, 3rd ed, by BSCS.

(a)

(b)

**Figure 7.24**   a. Pima Indians in Sierra Madre Mountains in the 1800s. b. Pimas in Gallup, New Mexico, U.S., 2013.

# Sexual Selection

Natural selection not based on a struggle for survival but instead based on a struggle for the opposite sex is called sexual selection. In most animal societies, females choose their mates. They choose a male based on one of two factors: his available resources or his appearance. The more resources in a particular male's territory, the more appealing the male. More resources indicate that the male will be better able to care for their young. A male's appearance also plays a role in a female's decision. If a male is more symmetrical, then the male generally has a better genetic quality. Consider the quest for physical beauty, discussed in Chapter 3, showing the importance of appearance in finding a mate in human society.

While female choice is a primary determinant in mating, male aggression is also important. There are two forms of male aggression in sexual selection: passive sexual selection and active sexual selection. Passive sexual selection involves the development of charms and appearance to attract a mate. It is passive because it is used to attract rather than actively obtain a mate. The mating songs of grasshoppers (discussed earlier) and the feathers of a male Peacock are examples of passive sexual selection. Active sexual selection involves aggression by males to obtain a female (Figure 7.25). Antlers on deer, physical strength to fight, and tusks on elephants are examples of weapons used in active sexual selection. Sexual selection and the development of sex were discussed in Chapter 6 to show how variation helps species' survival. Sexual selection drives the most fit organisms to be perpetuated in a population.

**Sexual selection**

Is the natural selection not based on a struggle for survival but instead based on a struggle for the opposite sex.

**Figure 7.25** Active sexual selection is an example of Bighorn Rams fighting for dominance over females in the herd.

# Summary

Organisms change over time in response to the environment. Change is a part of life's history whether it is expressed in the emergence of new species, an increase in numbers, a decrease in numbers, or complete extinction. Life's origins harken back to early Earth conditions, but changes in species have occurred throughout its history. The moth populations described in this chapter show how environmental factors affect species even in a short period of time. Natural selection leads to species changes with extinction as a final step in species loss and speciation as an outcome of adaptive radiation. There is ample evidence for evolution of species throughout Earth's history.

## CHECK OUT

### Summary: Key Points

- Humans play a role in the evolution of other species within the environment, and are intrinsically linked to them.
- The process of discovering life's origins point to conditions on Earth that were favorable for spontaneous development of organic molecules and later, primitive cells.
- Natural selection favors certain traits and increases them in a population and disfavors other traits, decreasing them in a population over time.
- Natural selection may lead to speciation, which increases the number of species and thus biodiversity.
- Extinction of species leads to permanent decreases in biodiversity.
- Evidence for evolution is based on the current examples of modern-day changes in species, the fossil record, homology, organisms' biogeography and molecular evidence.
- Sexual selection explains how organisms compete for the opposite sex to obtain the best adapted offspring.

## KEY TERMS

| | |
|---|---|
| adaptive radiation | hyper-disease theory |
| allopatric speciation | natural selection |
| biogeography | prebiont |
| directional selection | reproductive isolation |
| disruptive selection | reproductive success |
| evolution | sexual selection |
| extinction | speciation |
| fossil record | spontaneous generation |
| germ theory | stabilizing selection |
| homologous structures | sympatric speciation |
| homology | vestigial organs |

# Multiple Choice Questions

1. Which caused decreases in V-moth populations in southern England?

    a. Farming
    b. Acid rain
    c. Pollution
    d. Fishing

2. Which scientist used a long-necked flask to disprove spontaneous generation?

    a. Redi
    b. Pasteur
    c. Oparin
    d. Urey

3. Which best describes how moth populations changed in southern England in recent years?

    a. All moths declined in their numbers.
    b. All moths increased in their numbers.
    c. Some moths became new types of moths.
    d. Some moths increased in numbers and other moths decreased in numbers.

4. Disruptive natural selection may lead to:

    a. extinction
    b. speciation
    c. convergence
    d. directional selection

5. Which type of natural selection is most likely to cause adaptive radiation?

    a. Disruptive
    b. Stabilizing
    c. Unidirectional
    d. Bidirectional

6. Allopatric speciation:

    a. causes a new species to form.
    b. causes extinction.
    c. has the opposite effect of sympatric speciation.
    d. is less effective than sympatric speciation.

7. Which term is NOT associated with increases in biodiversity?

    a. Allopatric speciation
    b. Homology
    c. Adaptive radiation
    d. Sympatric speciation

8. Two organisms are very closely related, with 98% of genes in common. This evidence for evolution is based on

    **a.** fossil records.
    **b.** biogeography.
    **c.** homology.
    **d.** molecular data.

9. In question #8 above, which provides the best evidence for showing the organisms' relatedness?

    **a.** Molecular data
    **b.** Anatomical data
    **c.** Geographical data
    **d.** Speciation data

10. If two deer fight over a female using their antlers, this is an example of:

    **a.** passive sexual selection.
    **b.** active sexual selection.
    **c.** natural selection.
    **d.** adaptive radiation.

## Short Answers

1. Describe how human society affects species diversity in your own neighborhood. Use one example of species that raises your concerns.

2. Define the following terms: adaptive radiation and disruptive natural selection. List one way how each of the terms differ from the other in relation to biodiversity.

3. Describe the experiments of two scientists: Francisco Redi and Louis Pasteur. Use a drawing to make the descriptions clear. Show your art work. How did each discover an aspect of spontaneous generation? How did their knowledge build upon one another's?

4. Draw the Miller and Urey apparatus. What were the products of their simulation? How did Miller and Urey reignite debate on spontaneous generation?

5. For question #4 above, how might you argue that life could emerge from nonlife?

6. List three ways evolution can be verified. Which piece of evidence is the most convincing to make the case that species today are a result of evolution?

7. Explain the process by which moth species declined in England in the past decades? How did some species of moths increase at the same time?

8. Some organisms base their choice for a mate on physical appearance. In barn swallows, an experiment showed that after making a male swallow less symmetrical, fewer females chose him. Explain how evolution favors this result in female choice. How will offspring change over time, given the results?

9. Name the type of speciation that results when a species cannot mate due to a change in their use of a habitat. Explain how it results in speciation.

10. A bacterial cell becomes resistant to a type of antibiotic. Predict the outcome for the population of this species of bacteria.

## Biology and Society Corner: Discussion Questions

1. Moth populations in the chapter's story show rapid changes in frequencies. Describe two steps that the English government could take to help prevent unwanted biodiversity loss in England. Are there any drawbacks to your suggested approaches?

2. How did Pasteur's religious views affect his scientific outlook and methods? Are personal or religious views justified in propelling scientific thought? Why or why not?

3. If a person has a nonheritable form of intelligence that enables her to read people's minds; what is this trait's likelihood for changing the direction of evolution in society. Justify your answer.

4. Acid rain is a danger to some organisms and a benefit to others. In the moth species described in this chapter, explain how this is so.

5. An Internet site claims that people are getting smarter and smarter with each generation due to evolution. Is this claim valid? Why or why not?

Figure – Concept Map of Chapter 7 Big Ideas

## 8

# Before Plants and Animals: Viruses, Bacteria, Protists, and Fungi

## ESSENTIALS

Louis Pasteur

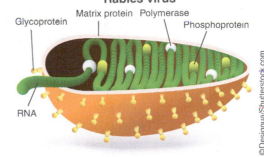

**Rabies virus**

Glycoprotein  Matrix protein  Polymerase  Phosphoprotein

RNA

Rhabdovirus (rabies virus)

The Death Cap Mushroom is the world's deadliest poisonous mushroom

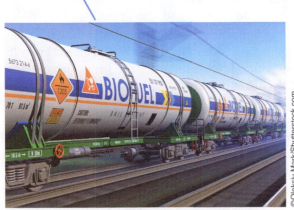

Many biofuels are made by bacteria

Yogurt is a product made by several types of bacteria through fermentation of sugars in milk

## CHECK IN

**From reading this chapter, you will be able to:**

- Explain how the diversity of organisms affects our health and society.
- Discuss the discovery of pathogens, how they cause disease, and how this knowledge affected the medical profession.
- Describe the characteristics of viruses, types of viruses, and their related diseases.
- Compare the lytic and lysogenic life cycle of viruses.
- Describe the characteristics of prokaryotes, types of prokaryotes, their biological roles and relationships with humans.
- Describe the characteristics of protists, types of protists, their biological roles and relationships with humans.
- Trace the evolution of protists and fungi from ancient prokaryotes.
- Describe the characteristics of fungi, types of fungi, their biological roles and relationships with humans.
- List and describe the diseases caused by viruses, prokaryotes, and protists and fungi, evaluating their impacts on human society.

## The Case of the First Rabies Survivor

Andre played with his dog almost every day, but this time it was different. His friendly dog was strange; it seemed almost unaware of what was going on. Andre went to his beloved dog to help him get through the door of the barn. Suddenly, the dog lunged forward and bit Andre's arm. This had never happened before.

In the course of the day, Andre's mother noticed the bite on his arm. She inquired to find out what had happened. Andre explained, "The dog bit me. He was very odd, not himself, and looked like he had foam coming from his mouth." "What's wrong, mother?" Andre asked. His mother was crying, knowing Andre would have a horrible death that would come soon.

"Doctor, we brought the boy in just minutes ago. He was bitten by a rabid dog!" exclaimed the nurse. Dr. Louis Pasteur, a different kind of doctor who tried out-of-the-box techniques, was Andre's last hope. Pasteur had a reputation, and many were afraid that his procedures were too far from mainstream medicine.

Dr. Pasteur was determined that no person would again die of rabies in the future. He had seen so many succumb to the infection. Pasteur worked in the 19th century, in a hospital that had none of the technology found in hospitals today. Still, Pasteur knew the symptoms of rabies that were awaiting Andre: fever, headache, tiredness, drooling and death. He was angry at organisms that he could not even see. "I cannot let another person die of rabies!" Pasteur protested.

Pasteur consulted with his colleagues, Dr. Vulpain and Dr. Grancher, to ask for their recommendation . . . should he attempt a drastic and experimental procedure, never done before? Both responded, "You should try it. Otherwise, this child will die; there is nothing to lose." "But is it going to work – Is it even ethical to do?" Pasteur wondered. His research on rabies-infected dogs showed Pasteur that living infected dogs survived when injected with a vaccine made of ground up spinal cords from dead rabid dogs. But humans were not dogs, and Pasteur was nervous – "How macabre," he thought, "to inject a human with ground up spinal cord from a dog."

The only other choice was death. Pasteur proceeded, injecting Andre with a spinal cord mixture. The night was long and emotional; as daybreak passed, Andre was still alive, perhaps a bit stronger than the day before. Each day for the next three months, Pasteur injected Andre with small bits of spinal cord. At the end of the treatment, Andre emerged as the first person in history to ever survive a rabies infection. Upon seeing Andre stand, Pasteur recalled the words of mathematician and scientist Blaise Pascal: "Man is but a reed, the most feeble thing in nature; but he is a thinking reed . . . All our dignity consists, then, in thought . . . by thought, I comprehend the world." Pasteur reflected, "The frailty of the human condition is overcome by the strength of human thought." He had succeeded in overcoming rabies with his planning and his thoughts. It was a glorious day for Andre and his family – and for biology.

Louis Pasteur's 1800s discovery of the rabies vaccine shows the human struggle to use the power of the mind to overcome nature's adversity. The seeds for medical progress and biological research are born of this quality, as shown by Louis Pasteur's scientific reasoning used to cure Andre.

## CHECK UP SECTION

This story embodies the essence of a way of thinking about the natural world, a movement from the power of physical strength to the power of the mind. This story highlights the struggle to overcome nature through innovation in the history of science.

The future of science lies in its past: the passion of the great scientists, their struggles against nature's challenges, and their creativity in discovery. All point to the characteristics needed to propel scientific thought into our future.

Study the rabies infection, caused by the **rhabdovirus**, to determine its causes, symptoms, and treatment by vaccine in more detail. How were rabies and other microbial infections discovered, while being unable to see the organisms that cause it? What are some examples of modern diseases that are being studied to better humanity?

# Discovering Pathogens and Ways to Treat Them

The world of living organisms previously unseen by human eyes was discovered by microscopy. Anton van Leeuwenhoek described **microbes**, organisms that cannot be seen with the naked eye, very clearly and accurately in the 1600s as discussed in other chapters. This chapter focuses on the many organisms revealed by microscopy: viruses, prokaryotes, protists, and fungi (see Figure 8.1). Each of these groups has great variety; some organisms are beneficial to humans and some are harmful. Organisms that are harmful to humans are called pathogens.

Pathogens were not generally recognized as agents of disease until the late 1800s. The work of Louis Pasteur, described in our story, as well as that of other scientists, such as Robert Koch who studied tuberculosis in humans, pointed to microbes as infectious agents in plants and animals. Scientists showed that pathogens have a few ways to attack their host:

**Pathogen**

Organisms that are harmful to humans.

1) Some directly attack the host, destroying cells, as is the case for viruses and many bacteria. Viruses, including the rabies virus, invade cells and destroy them from the inside. They grow using their host's resources as fuel and shelter.

**Diversity of Life on Earth**

| some major groups of organisms | insects | other animals | higher plants | fungi | protists | bacteria | viruses | total identified | total estimated to exist |
|---|---|---|---|---|---|---|---|---|---|
| number of species identified (approximate) | 950,000 | 281,000 | 270,000 | 72,000 | 40,000 | 4,000 | 4,000 | 1,750,000 | 4–112 million (working number = 14 million) |

(a)

(b)

From *Biological Perspectives*, 3rd ed. by BSCS. © 2006 by Kendall Hunt Publishing Company. Reprinted by permission

**Figure 8.1**    a. There is great diversity of life on Earth. Some major groups of organisms and their respective numbers of species are given in the figure. Insects comprise the largest proportion of species diversity. As seen in our story in Chapter 7, insects play an important role in our environment. b. Some less appreciated forms of life; a. arachnids; b. fungi.

**2)**  Some pathogens cause disease indirectly, by producing substances that harm the host. For example, diphtheria is caused by the bacterium *Corynebacterium diphtheria,* which is inhaled and lodges in the upper respiratory tract. There it makes a toxin, or poison, that prevents human cells from making needed proteins and thereby kills them.

**3)**  Some organisms cause damage by eliciting an immune response by the host. In some forms of pneumonia, the bacterium *Streptococcus pneumonia* causes such overwhelming release of fluids in response to its presence that the fluids fill the air sacs of human lungs, causing the host to have difficulty breathing. The inflammation caused by the pathogen causes damage instead of the pathogen itself.

**4)**  Many pathogens cause multiple symptoms, as in the familiar cases of strep-throat (see Figure 8.2). The disease-causing agent for strep-throat is often *Streptococcus pyogenes*, which causes throat pain, a fever, and in 0.5% of cases, damage to heart valves, joints, and other tissues of the body. For this reason, strep-throat is considered more dangerous than a normal sore throat. Jim Henson, creator of the Muppets, died of infected heart valves from a streptococcus bout.

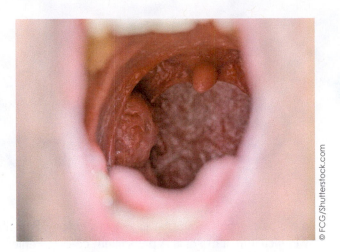

**Figure 8.2**   Strep-throat is a common infection. It is the inflammation of the tonsils and surrounding tissue at the back of the throat, resulting in redness and soreness.

By recognizing that some microbes caused disease, doctors and scientists began taking measures as far back as the 1800s to prevent their transmission. In medical procedures, sterile techniques began to be used to reduce their numbers. Improvements in hygiene, hospital cleanliness, removal of public waste, and water treatment contributed to cleaner surroundings as a result of recognizing microbial illnesses. In the past, for example, a major killer of women and children was childbed fever. It was spread by visiting doctors who carried streptococci from patient to patient as they examined them. The sterilization of instruments and introduction of hand-washing reduced patient deaths from microbes. Before these procedures became common practice, upward of 50% of children died within their first years of life. Since the advent of sterile techniques and vaccinations, the infant mortality rate has declined dramatically over the past century (see Figure 8.3).

The discovery of the first vaccination is described in our historical opening story. Bacterial and viral infections are treated and prevented through immunization, which

**Immunization**

The technique that uses dead pieces of disease-causing agents to strengthen immunity against a disease.

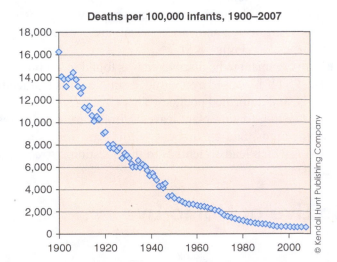

**Figure 8.3**   Infant mortality from 1900 to 2007. Dramatic decreases in infant mortality over the past century were a result of the discovery of antimicrobial techniques such as vaccines and penicillin.

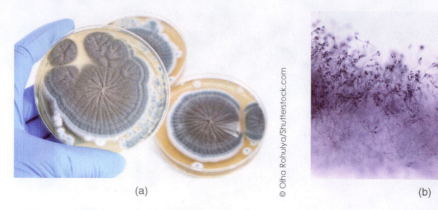

(a)

(b)

© Olha Rohulya/Shutterstock.com

© Christopher Meade/Shutterstock.com

**Figure 8.4**   Penicillin is a blue-green mold that grows in a colony. Penicillin is a type of antibiotic produced by Penicillium, a mold. a. A penicillium colony. Penicillium produces round spores at the tips of its reproductive structures. It provides a special chemical, penicillin, that has saved millions of human lives. b. Chains of round tips at the ends of its reproductive structures are shown in this figure.

uses dead pieces of disease-causing agents to strengthen immunity against a disease. (Immunizations will be discussed in Chapter 15.)

In 1935, sulfanilamide, a new "wonder drug" was discovered in Germany, one of the first to control bacterial infections. Penicillin, which attacks cell walls of bacteria, was discovered by Alexander Fleming in 1940. Penicillin is a type of antibiotic, which is defined as any chemical that stops the growth of microorganisms (see Figure 8.4). Many antibiotics are extracted from bacteria or fungi that produce a natural defense against competitor microbes, harvested in antibiotics. Today, antibiotics and vaccines are also produced in laboratories. Antibiotics changed people's lives and their life spans, as the most dramatic, large scale health improvement of the century.

Many microorganisms, such as the rabies virus, were undetectable under the microscope. They were discovered by indirect means, much as Pasteur detected and treated rabies. Because they cause so many plant diseases, viruses were first discovered in plants. Viruses are composed of combinations of nucleic acid, serving as their genetic material, and a surrounding protein coat. Their life cycle centers on their disease-causing ability, (discussed in the next section). Viruses cause many diseases ranging from the common cold to herpes, influenza, and cancer. There are a variety of types of viruses, each with differing shapes (see Figure 8.5).

The first virus-like organism to be discovered was found in studying PSTV, or potato spindle-tuber disease. PSTV is caused by a viroid, a simple virus that leads potato plants to produce long, gnarly, and deformed potatoes. This viroid also infects tomatoes, stunting their growth and twisting tomato plant leaves. Viroids are also causes of disease in citrus trees, chrysanthemums, and cucumbers. Viroids are very simple, composed of only a strand of RNA – which leads scientists to the question: "Are viruses and viroids actually living organisms?"

## Viruses: To Live or Not to Live . . .

### Features

Viruses are intracellular parasites, meaning that they invade host cells and live within them. They are not cells and are not included in the classification schemes of living organisms. They are obligatory parasites, unable to live outside of a host cell. Thus, they

**Penicillin**

An antibiotic obtained from the molds of Penicillium genus.

**Antibiotic**

Any chemical that stops the growth of microorganisms.

**Intracellular parasite**

Living organisms that invade host cells and live within them.

**Obligatory parasite**

Organisms that are unable to live outside of a host cell.

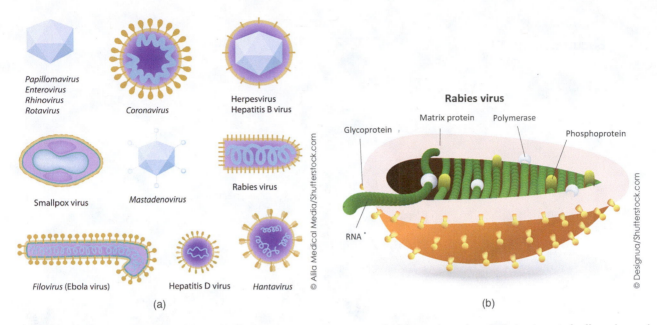

**Figure 8.5**   a. Morphology (shape) of some common viruses. b. Note that the rhabdovirus is bullet-shaped. It possesses receptor proteins on its coat (yellow buttons on the outside of the Rhabdovirus in image) which dock with host cells. Rhabdoviral DNA is spiral in shape and is protected within a protein coat.

are not considered to be living organisms. Viruses do not have an independent metabolism and cannot carry out life functions while outside of host cells.

Neither are viruses nonliving matter: they have genetic material, carry out some life functions, make proteins, mutate, and are able to reproduce. The structure of a typical virus includes a set of genetic material – either DNA or RNA – surrounded by a protein coat or capsid. The genetic material is simple, ranging from 1 to 100 genes along either a double or a single strand. A typical virus is shown in Figure 8.6.

Viruses are usually species-specific, with one type of virus affecting only one species of host. A human virus therefore cannot infect a cat and vice versa. The reason for this is that viruses use docking proteins to attach to surface receptors on host cells. As discussed in Chapter 3, cell membrane proteins attach to docking chemicals. Viruses also use this docking system to attach to host cells. There are obvious exceptions, such as rabies in our story, in which Andre is bitten by a member of another species that transmits the virus. The docking and transmission between species is able to occur because rabies proteins match many species.

Some viruses have tail ends, shown in Figure 8.7, which enable them to attach specifically to a host. The exception to the species-specificity rule occurs when mutated forms of viruses change enough to cross over to infect a new species. Scientists believe that this occurred in the spread of AIDS (acquired immunodeficiency syndrome), caused by HIV (human immunodeficiency virus), which mutated from nonhuman primates in sub-Saharan Africa in the late 19th or early 20th century.

**Capsid**

The protein coat that surrounds structure of a typical virus.

**Species-specific**

Limited to or found in one species.

## Size of viruses

To show perspective, a virus size is very small, about .2 μm, while a bacterium such as *Escherichia coli* is 3 μm, and a human liver cell is large, at about 20 μm. Viruses often exist outside of their hosts as crystals. They remain able to be activated, so catching a virus from a door handle or a toilet seat is its mode of transmission. It remains dormant in the crystalline state and activates upon the first opportunity to enter into a host.

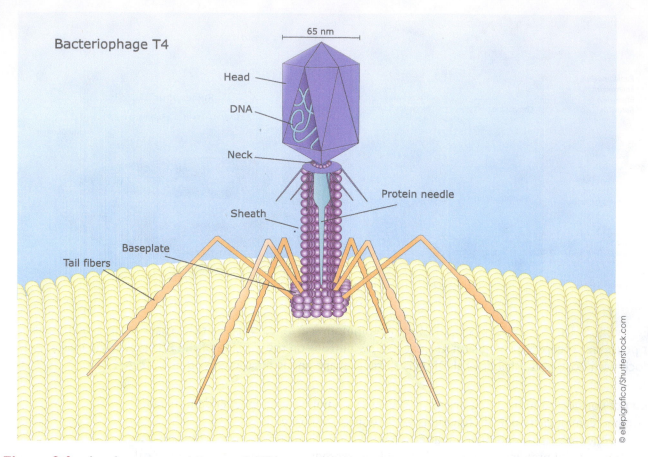

**Figure 8.6**    A color-enhanced image of a T4 virus. It has a protein capsid head surrounding genetic material and helical tail with fibers and needle to insert its DNA. The virus uses its tail fibers to attach to a host bacterial cell wall.

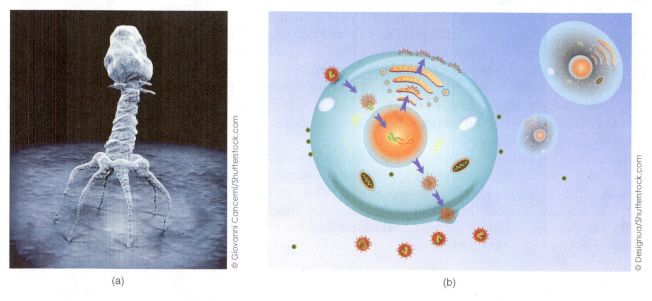

(a)

(b)

**Figure 8.7**    a. Microscopic view of a virus docking to host membrane protein. There is a specific-fit between a viral protein and receptors on the surface of host cells makes connections very exact. b. A virus then gains entry in a host cell, allowing it to conduct its strategies. In the figure given, viruses use the host cell's machinery to manufacture new viruses.

Whether airborne through the respiratory membranes of a human throat or via a cut in the skin, a virus returns to "life" when it enters another organism.

Some nonliving infectious creatures are prions, which are simple strands of proteins that "liven up." Prions cause brain infections such as mad cow disease and chronic wasting disease in deer. People who ingest meats infected with prions may exhibit no symptoms for decades, and suddenly prions reemerge, eating through tissues of the brain. No person has ever lived beyond one year after symptoms emerged from a prion illness. Prion diseases cannot be treated, and the best way to avoid their transmission is to eliminate brain or spinal cord meats from one's diets. At times, such meats are ground up in hamburgers and in other processed food, with customers unaware of the true contents.

## Viruses: The Internal Terrorist

There is no magic bullet to kill a virus because most drugs do not work on a virus-induced illness. There are natural immune defenses, which will be discussed in Chapter 15, but viruses have unique actions that make them a difficult enemy. In addition, while antibiotics directly attack bacteria, viruses lodge themselves within our cells. This prevents their targeted destruction because host immune systems must then also destroy their own body cells. There are ways to attack a virus, discussed in the next section, but defenses against viruses are problematic.

Viruses are internal terrorists because they invade host cells, undetected by the body's defenses, and incorporate themselves into its host's genetic material. Viruses are successful much in the way that a spy infiltrates its enemy – by remaining unseen and unnoticed. If a host's immune system detects a virus, it is destroyed and the virus is easily defeated. As long as a virus remains incognito, it can be successful at taking over its host cells.

There are two types of life cycles for viruses: the lytic life cycle, which results in a virus's immediate destruction of a host cell, and the lysogenic life cycle, during which a virus inserts its genes into a host and waits for a time in the future to destroy the host. During the lytic life cycle, viruses attach to their host cells by grabbing onto their membrane proteins (see Figure 8.8). In the example of a **bacteriophage**, which is a virus that invades a bacterium – for example, an *Escherichia coli* bacterium, tail fibers on the virus specifically match the shape of host membrane proteins. Often the base of the virus contains special enzymes that bore holes to the inside of invaded cells. Some viruses contain a coil that acts as an injection needle to thrust the virus's DNA or RNA into a host cell. Once inserted into a host, viral genes take over the nucleus of the cell, directing the production of hundreds and thousands of new viruses. Viruses are terrorists because they use the cell machinery of their hosts to destroy them. Viruses are then released, breaking their host cells – this is the "lytic," or breaking, part of the cycle. The rabies virus seen in our story uses a lytic life cycle to rapidly kill its host. Andre would have suffered great pain and death from this process because damage occurs instantly when a virus breaks apart cells in the lytic life cycle.

During the lysogenic life cycle, a virus injects its genes into a host cell (see Figure 8.9). Viral genes are incorporated into a host's DNA or RNA, and every time an infected host cell divides, viral genes divide along with it. It is an unstable relationship because at any one point in time, a set of viral DNA may activate and begin all of the steps of the lytic life cycle. The lytic life cycle is the same as the lysogenic life cycle, with one exception: the lytic process begins after a virus activates its genes within a host genome, while the lysogenic life cycle includes the period of dormancy of viral genes.

**Prions**

A small infectious particle believed to be the smallest disease-causing agent.

**Lytic life cycle**

A reproduction cycle which results in a virus's immediate destruction of a host cell.

**Lysogenic life cycle**

A reproduction cycle during which a virus inserts its genes into a host and waits for a time in the future to destroy the host.

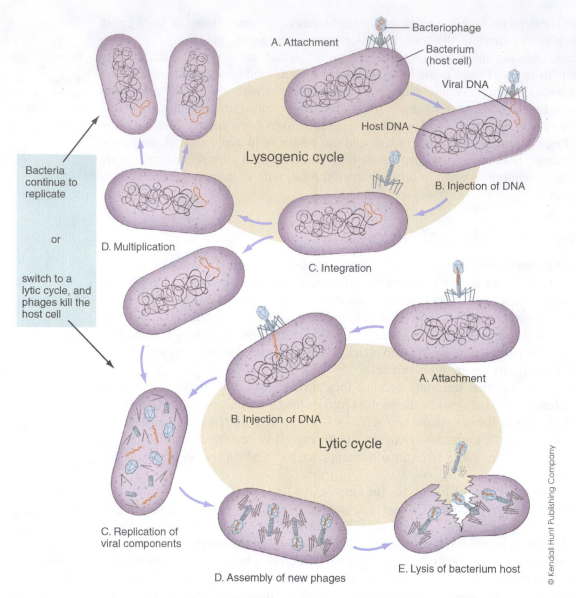

**Figure 8.8**    Lytic life cycle of a bacteriophage invading an *E. coli* cell. During this part of a virus's existence, it takes over the cell's nuclear machinery. The virus in the picture uses a host's DNA to produce more of its own genetic material and proteins coats, making many new viruses.

Many factors can bring on the start of a lytic life cycle: sunlight, chemicals, or stress are three such factors. In the case of Herpes Simplex I, which causes fever blisters around the lips, anxiety or sunlight stresses may bring a virus out of dormancy of the lysogenic life cycle. This results in cell destruction and, of course, fever blisters. The focus of our story in Chapter 15 is on our immune system's defenses against the herpes virus's biology. Because the virus is lodged around nerve cells, fever blisters on lips are very painful, affecting nerve sensations.

You may remember these different life cycles as the one with "lytic" means to lyse or break open, and the one with lysogenic, refers to "genic" or genes that are dormant but may lyse or break a host cell at a later point.

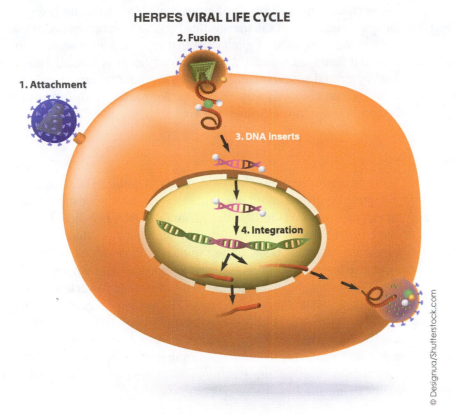

**HERPES VIRAL LIFE CYCLE**

© Designua/Shutterstock.com

**Figure 8.9**    Lysogenic life cycle of Herpes Simplex I.

## Some Interesting Viruses

### Herpes Virus

Herpes viruses come in two types: Herpes Simplex I and Herpes Simplex II. Discussed above, Herpes Simplex I is not sexually transmitted, is localized around the lips of humans. and remains dormant in cells in a lysogenic life cycle (see Figures 8.9 and 8.10). Its viruses recognize skin and nerve cells because they come from the same embryo layer – the ectoderm – which will be discussed in Chapter 16. Stress brings out herpes because the immune system usually keeps it in check.

**Herpes simplex I**

An inflammatory skin disease characterized by the formation sores around the lips.

**Herpes simplex II**

A sexually transmitted and is characterized by genital sores.

(a)    © Sergii Chepulskyi/Shutterstock.com

(b)    © jurgenfr/Shutterstock.com

**Figure 8.10**    a. Cold sores are symptoms of a Herpes Simples I infection. Eighty percent of adults are infected with the virus causing oral herpes. b. Genital herpes is sexually transmitted and is painful, which frequently recur in some individuals.

When a stress is placed upon the body, herpes reemerges from its lysogenic life cycle, entering the lytic phase. Herpes is transmitted by direct contact between two people. The sores it causes are composed of giant cells that are filled with white blood cells that ingest invaders and damaged materials. Herpes Simplex II is sexually transmitted and is characterized by genital sores (see Figure 8.10). They are recurrent and reemerge during stress on the body in the same way as Herpes Simplex I. Both forms of herpes viruses are delicate and cannot travel through the air.

## Rhabdovirus

**Rhabdovirus**

A bullet- or rod-shaped RNA virus found in plants and animals.

Rabies is 100% fatal and is an exception to host specificity; it is able to be transferred from one species to another, as described in our opening story. As mentioned earlier, the rhabdovirus causes rabies, its structure is bullet shaped, and its DNA is helical. When a bite occurs, saliva containing the rhabdovirus transmits it into a new organism. The rhabdovirus moves along the nerves of the body, up the spinal cord toward the brain.

In the brain, the rhabdovirus multiplies, causing symptoms and irreversible damage. Symptoms include fever, headache, hallucinations, intense muscular activity such as an arched back, and difficulty swallowing. Owing to the last symptom mentioned, a person with rabies shies away from water with a fear of swallowing. Foaming at the mouth, paralysis, and heart and lung failure leading to death are certain without treatment. In our story, before Pasteur's discovery, society lived in fear of a bite from a rabid animal. It was a certain and horrible death. Rabies vaccination directly acts on the rhabdovirus, with special proteins or **antibodies** that attack viruses, stopping their action. Rabies vaccine is a form of immunization that acts directly on pathogens and without the help of a host's immune system.

## Rhinovirus

**Rhinovirus**

The most common viral infectious agent that causes common cold in humans.

Less serious but nonetheless annoying is the common cold, caused by the rhinovirus. The rhinovirus kills ciliated cells of the upper respiratory tract in humans. The destroyed cilia are replaced with other cells along the throat. Once cilia are damaged, sufferers cannot adequately filter air. The mucous that forms in the common cold is a result of damage to these cells. The body attempts to cover up cell losses with mucous, leading to nasal congestion and respiratory upset. The common cold rarely kills, unlike the rabies virus in our story, but almost everyone is a victim of the rhinovirus because of its frequent occurrences and uncomfortable symptoms (see Figure 8.11).

## Myxovirus

**Myxovirus**

Any group of RNA-containing viruses.

**Neuraminidase**

A protein found in Myxovirus that digests through mucous membranes.

**Hemagglutinin**

A type of protein that enables Myxovirus to bind with its host.

One of the illnesses caused by the myxovirus is influenza, which affects three to five million people worldwide each year and causes upward of 500,000 deaths. Its symptoms include fever, muscle aches, pains, coughing, weakness, and fatigue. Sometimes a secondary infection with bacterial pneumonia occurs, the result of an immune system weakened by myxovirus. The myxovirus contains eight single strands of RNA, each able to mutate. This is a large amount of genetic material for a virus and explains in part how the influenza virus changes each year, with mutations in RNA sequences making new strains harder to vaccinate against. Myxovirus contains two types of proteins: neuraminidase (N), which digests through mucous membranes, and hemagglutinin (H), which enables the virus to bind with its host. Whenever genetic material of the myxovirus mutates, it changes its N and H proteins. Influenza infections have

**Figure 8.11** The common cold, caused by rhinovirus, is a nuisance but rarely kills its sufferers.

spurred pandemics, killing people worldwide. In the United States, more people died in the 20th century from influenza than from all of the major wars combined, as show in Table 8.1.

## Papillomavirus

Human warts are benign tumors of the skin caused by viruses. Papillomavirus, the cause of human warts, contains double-stranded DNA surrounded by an icosahedral or 20-sided protein coat. Human papillomavirus (HPV) is associated with cervical cancer and is the basis of the relatively new vaccine against cervical cancer. It is spread by human contact, and in the case of cervical cancer through sexual transmission. There is debate as to whether or not to vaccinate all girls at age 11–12 to prevent most forms of cervical cancer. The debate occurs because it is a sexually transmitted infection, and people view teen-age sex through a variety of lenses. There was no debate when Pasteur saved victims from rabies, as described in our story – only celebration.

**Papillomavirus**

A group of virus that cause papillomas or warts.

**Table 8.1** Influenza outbreaks occur due to mutated strains of the myxovirus. In the 20th century, there were more U.S. deaths from the 1918 flu outbreak than due to all of the major wars combined. Worldwide it is estimated that 30–50 million people died in the pandemic. Each year in the 20th century, the flu killed an average of about 36,000 people.

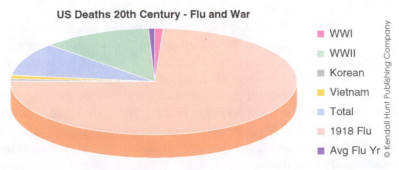

## THE WAR ON VACCINES

Long forgotten are the benefits of vaccination to human society: fatalities from small pox fell from 2 million per year in 1967 to 0% by 1980; the Salk and Sabin vaccines prevented more than 5 million cases of paralytic polio; and vaccinations against infectious childhood diseases prevent more than 3 million deaths in young people each year.

Yet if you do a web search on the harmful effects of vaccines or the autism-vaccine link, thousands of sites are listed, all convincing their readers that vaccines are bad news. Hollywood stars such as Jenny McCarthy advocate for ending vaccines in the face of rising autism cases.

The risks and side effects of vaccines are unproven or extremely low. Many people do not recognize the benefits of vaccinations because they did not live through the many fears of life without them — polio, influenza, mumps, measles, hepatitis, small pox, to name a few. Our life span increased enormously from 38 years of age in 1850 to roughly 75–85 years today, in part because of vaccines. In the time before vaccines, people died early or during childhood or child birth without the benefits of vaccines to keep harmful microbes at bay.

The movement against vaccines is resulting in the reemergence of some diseases. For example, when an outbreak of measles swept across Europe in 2010–2011, 48,000 people were hospitalized and 28 people died unnecessarily. When less than 90% of the population is unvaccinated, dangerous spreading of disease becomes more likely. Over 80% of those infected in this outbreak were not vaccinated. In fact, being unvaccinated endangers the most susceptible among us — children under 5 years of age.

The press reported two studies in the late 1990s that linked vaccines with autism: One was a false link between mercury-containing preservatives in vaccines and autism and a second was a discredited claim based on a study of only 12 children claiming that the measles-mumps-rubella (MMR) vaccine was linked to autism. The dangerous outcome of such reports is not in questioning the side effects of vaccines but the panicked decision by the public to avoid vaccination. It engenders a long-term danger that diseases may reemerge or harmful microbes may thrive and even mutate, and become stronger, making their prevention and treatment more difficult. There is little debate on the value of the rabies vaccine. Its direct link to surviving rabies, as seen in our story, is undisputable. We have come a long way, but perhaps we've come a bit too far for public health.

(source: *Wall Street Journal*, "Rolling Back the War on Vaccines," February, 2013)

## *Oncovirus*

**Oncovirus**

Any virus that carries a gene associated with cancer.

**Oncogene**

A normal gene that under certain circumstances can cause cancer.

Any virus that carries a gene associated with cancer is known as an oncovirus. It is able to insert its genes, called oncogenes, into a host genome, potentially causing cancer in the host and in the host's offspring. Its genes are inherited, and this is a basis for evidence for the genetic cause of cancer. Oncogenes from an oncovirus become activated due to an event. The event might include an environmental stimulus like smoking or drinking, which causes oncogenes to turn on, resulting in cancer. Oncogenes could take decades to emerge or not show up at all in one's phenotype, unlike the guaranteed death that rabies affords. Prevention of cancer at the genetic level of research holds the most promise in tackling the root cause of the disease.

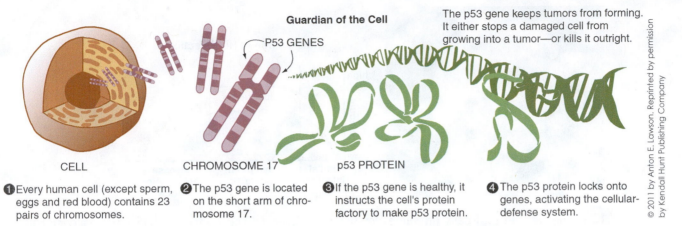

**Guardian of the Cell**

P53 GENES

The p53 gene keeps tumors from forming. It either stops a damaged cell from growing into a tumor—or kills it outright.

CELL          CHROMOSOME 17          p53 PROTEIN

❶ Every human cell (except sperm, eggs and red blood) contains 23 pairs of chromosomes.

❷ The p53 gene is located on the short arm of chromosome 17.

❸ If the p53 gene is healthy, it instructs the cell's protein factory to make p53 protein.

❹ The p53 protein locks onto genes, activating the cellular-defense system.

**Figure 8.12**    The oncogene theory. Oncogenes are mutations from normal genes which then cause cancer. For example, in human bladder cancer, small changes in the position of base pairs on a gene create oncogenes. As a result of a Chromosome #17 base pair change in the p53 shown in the figure, cancer develops. From *Biology: An Inquiry Approach*, 3rd ed by Anton E. Lawson.

The characteristics of cancer cells were discussed in previous chapters, but oncogenes point to their genetic origin. In a study by Weinberg, who isolated genes from bladder cancer cells, he found that only one base differed between normal cells and cancer cells: base #35 in normal bladder cells contained a guanine nucleotide and in an oncogene, base #35 contained a thymine nucleotide. Such a small difference genetically can lead to large changes in an organism's health and survival. See Figure 8.12 for a visual description of oncogene theory and the p53 oncogene.

## Retrovirus

AIDS is caused by HIV, which destroys a human immune system's fighting capabilities. HIV is a retrovirus, which is a virus containing RNA and the enzyme reverse transcriptase. Reverse transcriptase retroactively makes RNA into DNA, the opposite of the central dogma described in Chapter 5. The invasion of a host cell by HIV begins with its attachment to T-helper cells, which are key immune cells in humans. T-helper cells have a CD4 receptor on the cell membrane that matches with HIV-docking proteins. When the two attach, their membranes fuse, and RNA from HIV comes out of the virus and into the host's cytoplasm. HIV is unique in its ability to convert its single-stranded RNA into single-stranded DNA, able to carry out life processes. Once viral DNA migrates to the nucleus, it uses host enzymes to make double-stranded DNA. In this form, it is integrated into the host's genome, now "invisible" to the human immune system in a lysogenic life cycle. All progeny of T-helper cells are now infected with this dormant viral DNA. At a point in time, it reemerges and shunts into the lytic life cycle, causing the symptoms of AIDS. The course of the disease and its symptoms are shown in Figure 8.13 and Table 8.2. In the graph, you can see that at first the number of T-helper cells increases due to the host's immune system fighting the infection. However, the numbers decline substantially as T-helper cells become more and more infected. With the immune system compromised, AIDS patients become susceptible to numerous illnesses, including pneumonia and rare cancers such as Kaposi's sarcoma. Eventual death often results along with mental deterioration as the brain finally becomes infected. Its life cycle is slower than the rapid rabies advance on the brain that would have happened to Andre in our story.

Multiple drug treatments have result in marked improvement in the treatment of HIV and AIDS. A drug cocktail containing AZT, azidodeoxythymidine, inhibits reverse

**Retrovirus**

A virus containing RNA and the enzyme reverse transcriptase.

**Reverse transcriptase**

An enzyme that generates complementary DNA from a RNA template.

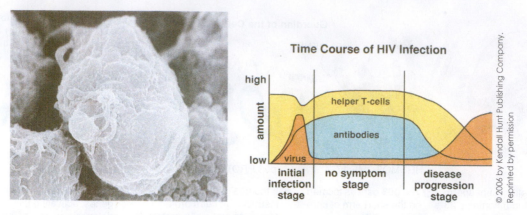

**Figure 8.13** The decline in T-cells (immune defense cells, seen to the right) in AIDS patients over the course of the disease. After infection with HIV, T-cells are invaded by the virus. At some point, HIV enters a lytic life cycle and destroys T-cells. T-cells usually defend humans against infections. When T-cell numbers decline, susceptibility in AIDS patients becomes dangerous, almost always leading to death if left untreated. Infections occur later on in the disease's course. From *Biological Perspectives*, 3rd ed by BSCS.

**Table 8.2** Symptoms of HIV infection.

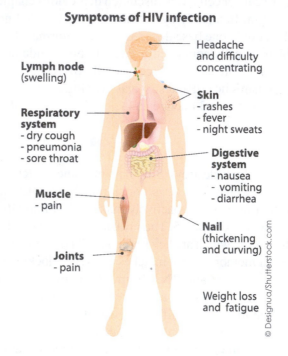

transcriptase in HIV and has been shown to be effective in halting the disease course and extending life for HIV patients for many years.

# Prokaryotes: The Little Things in Life

## Features

From an evolutionary perspective, the most successful kingdom is the **prokaryotes**, commonly known as **bacteria**. They have survived over 3.5 billion years, are adapted to almost every environment on Earth, and reproduce rapidly, allowing their colonies to change to adapt to new conditions in a short period of time. They are found in hot springs, in deep-sea vents beneath the ocean, under arctic ice, in our guts, and on our teeth. As discussed in Chapter 7, their simple, efficient ways of dividing and mutating help them to change with the times and inhabit just about any environment.

Classification of the 4,000 different species of bacteria is based on appearance or upon metabolic, chemical reactions rather than evolutionary relationships, as opposed to naming systems in other organisms. There are testing methods, such as the API testing strips, used to determine which organic molecules are metabolized by bacteria. Micro-biologists classify bacteria into two different groups: archaebacteria, or ancient forms of bacteria, with only a few surviving branches; and bacteria, which were once called eubacteria, and are the modern prokaryotes. Continual developments in molecular tech-niques make the field of classifying bacteria changing. At this point, the two domains of prokaryotes are most commonly accepted.

Bacteria have small sizes compared to eukaryotes, with diameters less than 5 μm. A typical bacterial cell is many million times smaller than a human being, but it is much larger than the nanometer size of viruses such as the rhabdovirus of our story.

While tiny in size, prokaryotes outnumber all other life, making up more than 99% of all living creatures by sheer mass. Their circular genes are simple and, unlike those of eukaryotes, are not surrounded by a nucleus. They contain no organelles except for ribo-somes. Eukaryotes, which comprise all other life, are more complex. Eukaryotes contain all of the organelles discussed in Chapter 3 (see the comparison of cells in Figure 8.14). Their cells have genetic material surrounded by a nuclear membrane, heavier ribosomes, more complex DNA, and larger diameters ranging from 10 to 100 μm.

**Archaebacteria**

Ancient forms of bacteria, with only a few surviving branches.

**Bacteria**

Single-celled microorganisms that are found everywhere.

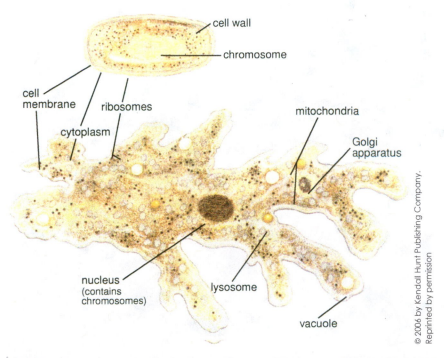

© 2006 by Kendall Hunt Publishing Company.
Reprinted by permission

**Figure 8.14** Comparison of prokaryotes and eukaryotes. The simple prokaryote has few parts but is able to outcompete eukaryotes because of its simplicity. The many organelles of eukaryotes (see Chapter 3) are shown in this figure with few in prokary-otes besides genetic material and ribosomes. From *Biological Perspectives*, 3rd ed by BSCS.

Decomposer
food chain

1. mustard-yellow
   polypore (shelf
   fungus)
2. oyster
   mushroom
3. amanita
4. yellow morel
5. snail
6. sow bug
7. centipede
8. wood roach
9. springtails
10. mite
11. ant
12. carrion beetle
13. soil fungi
14. soil
    protozoans
15. earthworm
16. inorganic
    compounds
17. soil bacteria

**Figure 8.15** Decomposition and recycling of nutrients by bacteria. Bacteria, along with other organisms, recycle organic matter, such as dead animals and plants, into reusable products. Several organisms described in this chapter work together to break down organic matter and return to useable materials in the environment. From *BSCS Biology: An Ecological Approach*, 9th Edition by BSCS.

Only a few prokaryotes cause diseases in humans; most are beneficial to us and to the environment. Life on Earth would not exist without the activities of bacteria. Bacteria return nutrients to the Earth through decomposition (see Figure 8.15). Prokaryotes are used in the production of many chemicals, such as acetone and butanol, in fingernail polish, and cleaning agents. They make vitamins and antibiotics, milk, and cheese. As described in Chapter 6, they are used in gene technology to manufacture insulin for diabetics and human growth hormone (HGH) to treat dwarfism. In Chapter 4, we saw how the bacteria *Lactobacillus acidophilus* is used in yogurts through the process known as fermentation In fact, *L. acidophilus* bacteria are used to treat gastrointestinal illness because they are normal inhabitants of the human digestive tract. As they grow, *L. acidophilus* bacteria outcompete potentially troublesome bacteria in our bowels and produce lactic acid, which keeps other bacteria from growing. The roles of bacteria in human society are diverse, and many things in our environment require the workings of prokaryotes. While viruses like rabies create mostly human harm, we require bacteria for our existence.

## Shapes, Sizes, and Types

The morphology (shape) of bacteria allows a general identification of its many strains. There are roughly 10,000 types of bacteria. They may be classified based on their metabolism and their shape and arrangements. They may be round cells called coccus, rod-shaped, called bacillus, or spiral, spirillum. Bacteria may be arranged either singly or in groups. When bacteria are found in chains, they are classified with the prefix strep-; when they are found in clusters, they are classified as staph-. Figure 8.16 shows the shapes and arrangements of bacteria.

**Morphology**

The form of an organism that allows a general identification of its many parts.

**Coccus**

Round-shaped bacteria.

**Bacillus**

Rod-shaped bacteria.

**Spirillum**

Spiral-shaped bacteria.

**Strep**

The prefix given to bacteria that are found in chains.

**Staph**

The prefix given to bacteria that are found in clusters.

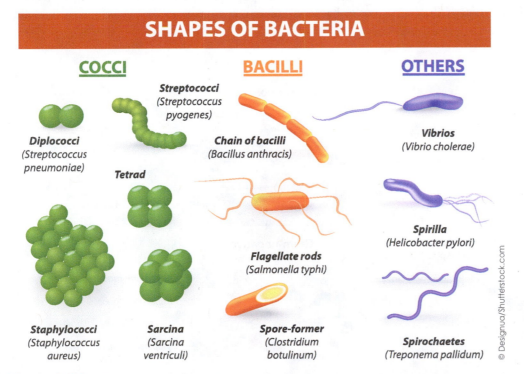

## SHAPES OF BACTERIA

**COCCI**

**Diplococci**
(*Streptococcus pneumoniae*)

**Tetrad**

**Staphylococci**
(*Staphylococcus aureus*)

**Sarcina**
(*Sarcina ventriculi*)

**Streptococci**
(*Streptococcus pyogenes*)

**BACILLI**

**Chain of bacilli**
(*Bacillus anthracis*)

**Flagellate rods**
(*Salmonella typhi*)

**Spore-former**
(*Clostridium botulinum*)

**OTHERS**

**Vibrios**
(*Vibrio cholerae*)

**Spirilla**
(*Helicobacter pylori*)

**Spirochaetes**
(*Treponema pallidum*)

© Designua/Shutterstock.com

**Figure 8.16** Common shapes of prokaryotes: Spiriulum (spiral), Baccilli (rod), and Cocci (round). The arrangement of chains (staph) and clusters (staph) of bacteria is also shown.

## MR. TOOTHDECAY AND MR. PIMPLES ARE HARMFUL TO OUR HEALTH

Strains of Streptococcus are leading causes of tooth decay, such as *Streptococcus mutans*, which deposit acids on the enamel of teeth. Staphylococcus strains cause pimples on skin surfaces, as well as boils and other serious skin infections. Some strains of *Staphylococcus aureus*, for example, cause necrotizing fasciitis or flesh-eating bacteria syndrome. While skin is not actually "eaten" by bacteria, it is destroyed rapidly by a spreading infection that, if left untreated by surgery and antibiotics, is fatal.

**Peptidoglycan**

Are a type of protein found in bacterial cell walls.

**Gram stain**

A dying technique that identifies bacteria as being one of two categories.

**Gram-positive bacteria**

A group of bacteria that retains the dye in Gram staining method of bacterial differentiation.

**Gram-negative bacteria**

A group bacteria that lose the crystal violet dye in Gram staining method of bacterial differentiation.

Almost all prokaryotes have cell walls, like plants, but the structure of the cell wall is different from that of plants. Bacterial cell walls contain peptidoglycans, which are a type of protein known as glycoproteins. Sugars cross-link with each other to hold peptidoglycans together. There are two types of bacterial cell walls that identify bacteria as being one of two categories, based on using a dying technique called the Gram stain: Gram-positive bacteria are colored purple by the staining technique, and Gram-negative bacteria are colored pink. Gram-positive bacteria have simpler cell walls, with a thick layer of glycoproteins. This layer retains the dye from a Gram stain and appears purple. Gram-negative bacteria have more complex cell walls, with less peptidoglycan, and gram stain washes away, making cells appear pink (see Figure 8.17). Gram-negative bacteria also have an outer membrane with lipopolysaccharides that protect their cells. Even though they are simple, bacteria are much more complex in structure than the rabies virus infecting Andre.

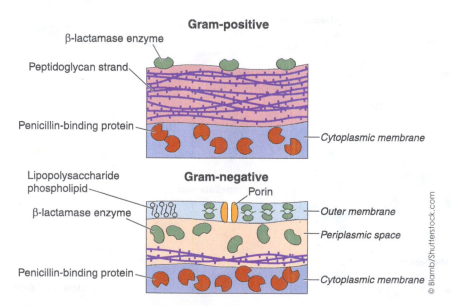

**Figure 8.17**   Gram-negative (pink) and Gram-positive (purple) bacteria cell walls. Each type of bacteria stains differently due to the thickness of their layers of peptidoglycans. Gram-positive bacteria have thick peptidoglycan layers while Gram-negative bacteria have thin peptidoglycan layers.

Penicillin works well in treating Gram-positive bacteria because it prevents cross-linking of peptidoglycans in the cell layers. Gram-positive bacteria walls thus fall apart due to penicillin. However, in Gram-negative bacteria, penicillin cannot move through the outer layer, preventing it from working to kill bacteria.

Many strains of bacteria have a **capsule** surrounding them, allowing them to prevent water loss and live in dry areas, such as in deserts and on our skin surface. Capsules are sticky and help bacteria to adhere to surfaces and to other bacteria.

Bacterial surfaces often have pili, surface hairs that allow bacteria to bind with each other. Through pili, bacteria exchange substances, including genetic material. This is bacterial sex; while not very elaborate, it serves to give bacteria greater genetic variation. This exchange of genetic material through pili is known as conjugation. Pili are also used to help bacteria bind to surfaces. *Neisseria gonorrhoeae*, for example, fastens itself onto mucosal genital regions in humans. It is the cause of the sexually transmitted disease, gonorrhea. In fact, bacteria and microbes cause a number of sexually transmitted diseases.

**Pili**

Surface hairs that allow bacteria to bind with each other.

**Conjugation**

The process of exchange of genetic material through pili in bacteria.

## Prokaryote Nutrition

More than half of all prokaryotes have motility. They move by means of flagella, filaments around their outer cell walls (see Figure 8.18). They also move by gliding, secreting chemicals on surfaces to move quickly. Flagella may be arranged as either scattered units, in tufts, or as a single length. *Salmonella typhimurium*, which causes the foodborne illness Salmonella, has scattered flagella leading to uncoordinated movement. Motile bacteria are characterized by **taxis**, which is movement toward or away from a stimulus. In **chemotaxis**, for example, bacteria move toward or away from food or oxygen sources. In **phototaxis**, bacteria move toward light.

Bacteria are also metabolically diverse. In phototaxis, some bacteria move toward light in order to obtain food via photosynthesis. These bacteria use a process of photosynthesis that is quite different from plants. MIT Technology Review recently announced the bioengineering of photosynthetic bacteria to produce up to 30 times more sugar per

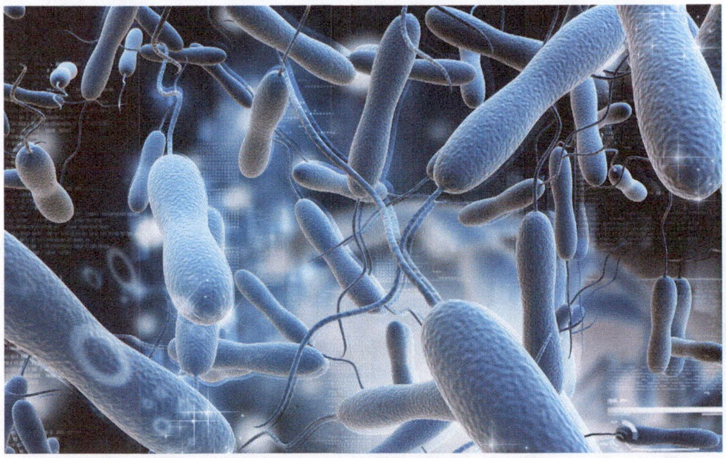

**Figure 8.18**   Flagella arrangements around bacteria. Flagella may be arranged as single tails as shown in the image of the bacterium that causes cholera in humans, *Vibrio cholera*. Bacteria cells also contain flagella grouped in tufts or as strewn throughout their surfaces.

acre than sugarcane plants for biofuel. When grown in transparent containers, these organisms use sunlight to produce ethanol from sugars to be used in cars as fuel (see Figure 8.19).

When bacteria use inorganic chemicals as energy, they are known as chemoautotrophs. These bacteria use inorganic molecules such as hydrogen sulfide and ammonia rather than food or sunlight to produce energy. They are independent, able to make their own food, and use carbon dioxide as their source for organic molecules. Chemoautotrophs do not require sunlight or oxygen. As shown in Chapter 7, the environment of early Earth provided an environment in which chemoautotrophs could have survived.

It is thus believed that these were the first organisms on Earth. Organisms in our intestines, which produce sulfur odors and methane gases, fall under this category. Nitrogen-fixing bacteria, living in root nodules of bean, pea, and clover plants, use gaseous atmospheric nitrogen to produce ammonia ($NH_3$) in their reactions to obtain energy. In the process, ammonia is made available as soil nitrogen, which is vital for plants. Most bacteria are heterotrophs, such as decomposers, which use dead organic matter for energy. They use dead matter to release carbon dioxide into the atmosphere as a product to be fixed later by plants in the Calvin cycle (described in Chapter 4).

**Chemoautotroph**

Bacteria that use inorganic chemicals as energy.

## Bacterial Reproduction

Prokaryotes reproduce by splitting in half through the process of binary fission. A circular bacterial chromosome divides, attaches to its cell wall, and as the bacterial cell grows, the replicated chromosomes separate into opposite ends of the cell. Plasmids, small circular strands of DNA, also replicate, moving into two new cells. Daughter cells form after parent cell cytoplasm pinches inward.

Most of a prokaryote's DNA codes for proteins, unlike in eukaryotes, in which 90% of the genes are not used in protein synthesis. Thus, binary fission is efficient and productive once a new cell forms. Fission can occur very quickly, within 20 minutes, resulting in over 20 billion cells in only 12 hours! It is able to maintain genetic variation through mutations, with so many chances for changes because of the many times a prokaryote divides. As discussed in the previous section, conjugation also affords unique combinations of genetic material to recombine in prokaryotes.

Two other processes contribute to their genetic diversity. Transduction occurs when a virus, known as a bacteriophage, invades a prokaryote, inserting its genes into

**Binary fission**

The process of reproduction by splitting in half.

**Transduction**

The process that occurs when a virus invades a prokaryote, inserting its genes into the host.

© www.sandatlas.org/Shutterstock.com

**Figure 8.19**    Biofuels made by crops of photosynthetic bacteria. A possible solution to the energy crisis? Shown, is a pond filled with photosynthetic bacteria in Hawaii.

### Genetic Transformation

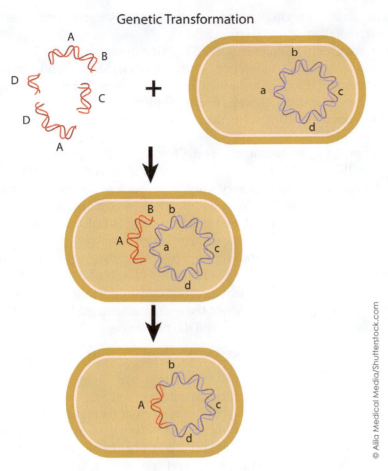

© Ailia Medical Media/Shutterstock.com

**Figure 8.20** Transformation in bacteria. Foreign DNA from a dead bacterium enters its host bacterium. Once DNA is exchanged, this results in a new cell with dead DNA incorporated. This adds to genetic diversity in prokaryotes. New DNA is a new combination for organisms to use and their offspring to inherit.

the host. Genetic recombination, as discussed in Chapter 6, is accomplished using a bacteriophage in the procedure. Insulin and HGH are manufactured in this manner. Transduction results in a transgenic prokaryote, with new genes added from the virus. Imagine a rhabdovirus transduced within a new bacteria. While fictitious, this would form a transgenic new organism, capable of causing rabies through bacteria transmission!

In addition, when prokaryotes absorb DNA from their environment, either through eating dead or dying bacteria or scattered matter, the genetic material is added to their own genome. This process is called transformation because the newly inserted DNA from the environment changes or transforms a bacterial cell into a new genotype (see Figure 8.20). Recall that Griffith's discovery of DNA in chapter 5 studied transformation in pneumonia-causing bacteria.

**Transformation**

The process in which a newly inserted DNA from the environment changes or transforms a bacterial cell into a new genotype.

## Prokaryote Diversity

### Archaebacteria

Archaebacteria are a type of prokaryote that is now classified as a separate domain from other bacteria are surrounded by cell walls that lack peptidoglycans, have unique cell membranes, contain RNA polymerase that resembles eukaryotes rather than other bacteria, and live in extreme environments, resembling early Earth conditions – in short, they appear different than other bacteria.

### Methanogens

Organisms that react to oxygen as a poisonous substance.

There are three groups of archaebacteria. The first group is the methanogens, which react to oxygen as a poisonous substance. Methanogens use hydrogen to reduce carbon dioxide into methane, $CH_4$. They must therefore live in areas that have no oxygen such as marshes and guts of animals. Sewage treatment plants and landfills use underground methanogens to convert garbage to methane. This causes the typical sulfur odor of a landfill.

### Halophiles

Are organisms that grow or live in very salty conditions.

The second group of archaebacteria is the halophiles. Halophiles are "salt lovers," which means that they are able to live in very salty conditions that few other organisms can withstand. Halophiles living in very salty lakes, such as Mono Lake in California (see Figure 8.21), use "pumps" to flush out the salt. Few other organisms can compete with halophiles in hypersaline areas such as Mono Lake in California, the Dead Sea in Israel, and the Aral Sea between Kazakhstan and Uzbekistan.

### Thermacidophiles

Organisms that thrive in strongly acidic environments at high temperatures.

The third group of archaebacteria is thermacidophiles, which "love it hot and acidy." They exist in temperatures from 60 to 80°C and in pH levels of 2–4. Sulfobus, found in sulfur springs in California, is one such organism that thrives in the extreme conditions of the hot springs. The extreme lifestyle of the archaebacteria leads scientists to believe that they were our most distant ancestors and the precursors to all life.

## Bacteria

Most prokaryotes are bacteria, with varied morphology and functions. Some interesting bacteria follow, which show the variety of forms of bacteria that play a role in human health and the environment. One type of bacteria is the actinomycetes which have filament strands and resemble fungi. Actinomycetes are decomposers recycling dead organic matter. As heterotrophs, actinomycetes rely on dead material to obtain food and therefore return organic materials back to the soil (see FIgure 8.15 for the decomposition process).

### Actinomycetes

A type of bacteria having filament strands and resemble fungi. They are decomposers recycling dead organic matter.

### Cyanobacteria

Are photosynthetic bacteria and contain bacteriochlorophyll.

Cyanobacteria, such as Anabena and Nostoc, are photosynthetic and contain bacteriochlorophyll (chlorophyll pigment found only in bacteria). These bacteria produce much oxygen in our atmosphere. Upon closer inspection, cyanobacteria contain large cells, called **heterocysts**, which contain nitrogen-fixing complexes that return nitrogen to the soil for plant use. Cyanobacteria reproduce by splitting at their heterocysts, breaking them open to produce new chains (see Figure 8.22).

© lilyling1982/Shutterstock.com

**Figure 8.21**    Mono Lake, California.

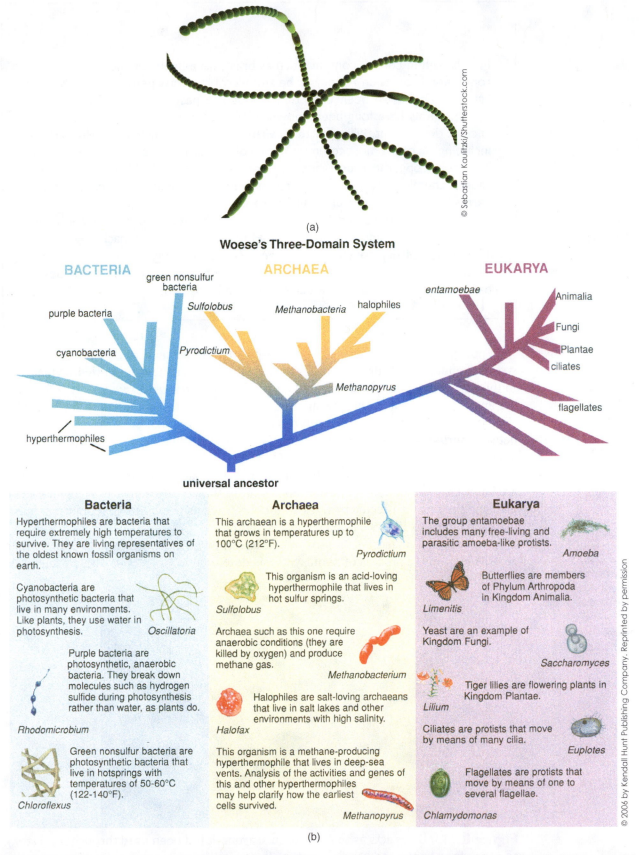

© Sebastian Kaulitzki/Shutterstock.com

(a)

## Woese's Three-Domain System

**BACTERIA**

green nonsulfur bacteria

purple bacteria

cyanobacteria

hyperthermophiles

**ARCHAEA**

*Sulfolobus*

*Pyrodictium*

*Methanobacteria*      halophiles

*Methanopyrus*

**EUKARYA**

*entamoebae*

Animalia

Fungi

Plantae

ciliates

flagellates

**universal ancestor**

### Bacteria

Hyperthermophiles are bacteria that require extremely high temperatures to survive. They are living representatives of the oldest known fossil organisms on earth.

Cyanobacteria are photosynthetic bacteria that live in many environments. Like plants, they use water in photosynthesis.

*Oscillatoria*

Purple bacteria are photosynthetic, anaerobic bacteria. They break down molecules such as hydrogen sulfide during photosynthesis rather than water, as plants do.

*Rhodomicrobium*

Green nonsulfur bacteria are photosynthetic bacteria that live in hotsprings with temperatures of 50-60°C (122-140°F).

*Chloroflexus*

### Archaea

This archaean is a hyperthermophile that grows in temperatures up to 100°C (212°F).

*Pyrodictium*

This organism is an acid-loving hyperthermophile that lives in hot sulfur springs.

*Sulfolobus*

Archaea such as this one require anaerobic conditions (they are killed by oxygen) and produce methane gas.

*Methanobacterium*

Halophiles are salt-loving archaeans that live in salt lakes and other environments with high salinity.

*Halofax*

This organism is a methane-producing hyperthermophile that lives in deep-sea vents. Analysis of the activities and genes of this and other hyperthermophiles may help clarify how the earliest cells survived.

*Methanopyrus*

### Eukarya

The group entamoebae includes many free-living and parasitic amoeba-like protists.

*Amoeba*

Butterflies are members of Phylum Arthropoda in Kingdom Animalia.

*Limenitis*

Yeast are an example of Kingdom Fungi.

*Saccharomyces*

Tiger lilies are flowering plants in Kingdom Plantae.

*Lilium*

Ciliates are protists that move by means of many cilia.

*Euplotes*

Flagellates are protists that move by means of one to several flagellae.

*Chlamydomonas*

© 2006 by Kendall Hunt Publishing Company. Reprinted by permission

(b)

**Figure 8.22**   a. Cyanobacteria microscope photo (Anabaena). b. Kingdom Monera: The evolutionary tree of Monera shows its vast diversity from a common ancestor. Cyanobacteria are only one example of the many species within the domains Archae and Bacteria. From *Biological Perspectives*, 3rd ed by BSCS.

## A GERMOPHOBE'S NEW PEN

Metals high in copper content, such as brass, have been certified by the Environmental Protection Agency to be antimicrobial. Brass pens, doorknobs, and light switches have been used throughout the past centuries (Figure 8.23). These metals have long been known to work against diseases. Copper is an ancient disinfectant, killing bacteria, viruses, and other disease-causing agents, including six of the most common strains of bacteria such as *S. aureus* and *E. coli*. The more copper in a substance, the more it is antimicrobial in nature. Copper causes chemical reactions in microbes leading to oxidative damage: copper harms bacterial cell membranes and proteins, preventing microbe functioning.

In studies conducted by the University of Southampton, copper alloys were shown to eradicate influenza A, $H_1N_1$, and various stomach bugs within 10 minutes of dry contact on copper surfaces. This finding has motivated a switch to copper and brass products in some hospitals, to keep microbes at bay among the sick. Should we all purchase a brass pen?

### Endospore-forming bacteria

Are Gram-positive, flagellated rods that form endospores to endure harsh, dry conditions.

### Mycoplasma

The smallest known bacterium.

### Phototrophic anaerobic bacteria

A group of bacteria that do not release oxygen in their photosynthetic-like processes because the photolysis of water does not occur.

Endospore-forming bacteria are Gram-positive, flagellated rods. They form endospores to endure harsh, dry conditions such as those found in deserts or dried-up marshes. In this form, spores as old as 250 million years, from *Bacillus permeans*, were found in the ancient salt sea underneath Carlsbad, New Mexico, and successfully revived. This ability to survive unfavorable conditions and revive in optimal ones marks a similarity between viruses and spore-bearing bacteria, but viruses like the rhabdovirus must find a host in order to reproduce.

Bacteria range in sizes, with the smallest known bacterium, the mycoplasma. They are between 100 and 250 nm in size, approaching some viruses in size. They lack a cell wall and are thus unaffected by most antibiotics, such as penicillin. Mycoplasmal pneumonia is a serious illness caused by this bacterium.

Phototrophic anaerobic bacteria reduce $NADP^+$ with electrons from $H_2S$ rather than $H_2O$, as seen in the process of photosynthesis described in Chapter 4. These bacteria do

**Figure 8.23**  Brass acts as an antimicrobial agent. It has been used throughout history as pens, door knobs and handles, and as plates on light switches. Studies show that the chemistry of brass keeps microbes at bay.

not release oxygen in their photosynthetic-like processes because the photolysis of water does not occur. Instead, they break-up $H_2S$, which releases sulfur gas. This gas gives areas with these bacteria, for example, southern U.S. marshes, the characteristic smell of sulfur.

**Enteric bacteria** are found in animal digestive tracts. They are Gram-negative and include E. coli, for example. They thrive in conditions free of oxygen, without competition from aerobic bacteria in other areas. Enteric bacteria are beneficial, providing vitamin K for humans, among other things.

Our interactions with bacteria should be studied in part to understand the human role in their life cycles. As seen in our story of Andre, our place in the ecosystem and possibilities for disease prevention are better implemented through researching life cycles of other organisms.

## The Misfit Kingdom: Protista

Protists are eukaryotes, are mostly unicellular, have asexual and sexual reproduction, and move via cilia or flagella. Other than these common features, protists share few similarities, and may be viewed as a group of misfits because of their dissimilarities. Protista is the most diverse kingdom: its members have very different structures, metabolisms, and ecological roles. They range in types of organisms, from the strange trumpet-like Stentor to the giant kelp, which grows 150 feet "long at a rate of 2" per day. With 60,000 species of brown algae alone (of which the giant kelp is one), protista comprise a very diverse kingdom (see Figure 8.24).

Molecular evidence shows that Protista were the first eukaryotes, emerging roughly 1.5 billion years ago. About 1 billion years prior to their appearance on Earth, the oxygen revolution resulted from the photosynthetic activity of cyanobacteria. Prokaryotes evolved as some of the life forms able to use oxygen in energy processing. As discussed in Chapter 3, the endosymbiotic model explains that eukaryotes then developed as prokaryotes absorbed oxygen-using creatures and evolved larger, more complex cells. Mitochondria and chloroplasts helped these new eukaryotes to obtain energy.

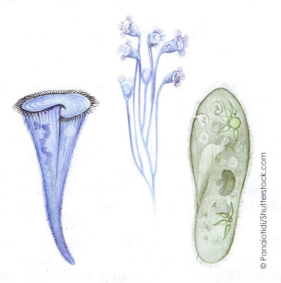

© Panaiotidi/Shutterstock.com

**Figure 8.24** Images of Protista. They are the most varied of all the kingdoms. Some are motile and some sessile; some are plant-like and others animal-like. Depicted in this image is the sessile. Stentor, far left and the very active *Paramecium*, far right.

According to the autogenous model of protist formation, primitive protists also evolved by invaginations of their membranes to form some of the membranous organelles such as the endoplasmic reticulum. The nucleus probably formed by invaginations around existing genetic material in primitive cells. This added benefit helped protists protect their DNA from environmental conditions. The complexity of protists helped them to survive and reproduce, competing with the simpler prokaryotes.

The first of these cells to develop were the Protista, which developed over 1 billion years after the oxygen revolution began. However, the protists of 1.5 billion years ago would not resemble those found today. Evolution and speciation have led, over these past 3 billion years, to speciation and the vast diversity in Protista. Biologists agree, however, that all other kingdoms – fungi, plants, and animals – originated from early protists.

## Classification

New techniques in electron microscopy, molecular analysis of DNA and new biochemical techniques make classification of Protista a changing field. The number of phyla, divisions of Protista and relationships within the kingdom are being debated. Biologists generally agree that there are three broad categories: 1) those that resemble plants – algae such as the giant kelp; 2) those that resemble animals – protozoans such as the stentor; and 3) those that resemble fungi – slime molds such as *Physarum cinereum* that creep across turf grass on lawns.

## Algae

Algae comprise a wide variety of what is commonly known as seaweeds. Algae are multicellular organisms that are photosynthetic, and they contain a variety of pigments such as chlorophyll a and b (green), carotenoids (yellow-orange), phycobillins (red and blue), and xanthophyll (brown). These pigments increase the photosynthetic output of algae, making them responsible for over 50% of all oxygen production by the process on Earth. Pigments also give algae their definitive colors, which are used in their classification.

The giant kelp, mentioned earlier, is a member of the **brown algae** or phaeophyta, which has roughly 2,000 species. They are vast organisms, growing rapidly to form large regions of seaweed in temperate areas of the ocean in North America, South Africa, and the South Pacific. Giant kelp resemble sea forests with their dense mats of plant-like leaves, which provide a home to many hundreds of marine species – from other protists, fish and snails, to larger marine mammals such as the sea otter and gray whales.

Diatoms, or bacillariophyta, are a major producer of oxygen via photosynthesis. Diatoms live in oceans, as well as freshwater rivers, lakes, and streams. Their unique arrangements, due to complex silicon dioxide shells, make them a beautiful, ornate organism (see Figure 8.25). Diatoms are part of the many protists that make up phytoplankton, which are all of the aquatic organisms that absorb carbon dioxide and release oxygen into the atmosphere. Phytoplankton form large and dense layers of organisms in water systems. Alone, they comprise about 25% of all oxygen production on Earth. Their ecological role in climate change and pollution is one of the most important dynamics of ecological study because of their contribution of oxygen in our atmosphere.

Other colors and forms of algae are found in bodies of water (see Figure 8.27): 1) **Red algae**, or rhodophyta, are red due to their phycoerythrin (red pigment), found along tropical coasts; 2) chlorophyta, or **green algae**, comprise about 7,000 species and are similar in cell structure and pigments to modern plants. Thus, they are believed to be ancestors to the first plants, discussed in Chapter 9. Chlorophyta are also a component of lichens, which are green algae or cyanobacteria living in association with fungi found

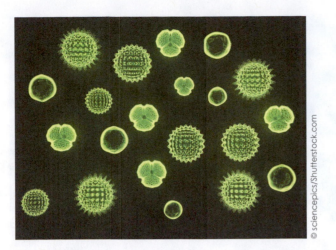

**Figure 8.25**   Phytoplankton, a type of diatom shown in this image. Diatoms serve an important role in aquatic habitats, from producing large amounts of oxygen to providing a food source for many organisms.

throughout most ecosystems (see Figure 8.26). Lichens are partnerships of organisms, with algae providing food from photosynthesis and fungi providing a physical place for algae to live. They are widespread – there are over 24,000 species of lichens, named by their type of fungal species, which is usually an acomycete fungus variety – and they can survive dry, harsh conditions, requiring only light, air, and a few minerals. 3) Chryso-phyta or **golden-brown algae** contain carotenoids and xanthophylls to give them their rich color. They form cysts that can survive very harsh conditions. They are colonial organisms and are biflagellated, with two flagella to propel them around; 4) Euglena or euglenophyta are unique in that they are both plant and animal-like. They contain an eye spot, which helps them to detect light and move toward it to carry out photosynthesis. Euglena, while able to make their own food, are also somewhat heterotrophic, requiring vitamin B-12, which they obtain through eating particles via phagocytosis.

**Euglena**

A green single-celled, motile freshwater organism.

**Figure 8.26**   There are over 24,000 species of lichens. Their algal layer produces food from sunlight and their fungal mycelium protects algae and anchors it to underlying substrates. Two species of lichens are shown on this branch.

**Figure 8.27**    A sample of Protista diversity: types of colored algae. A brown and red algae bed is shown beneath the ocean.

## Protozoans

Protists that have motility and are heterotrophic are characterized as protozoans. They include single-celled organisms such as the *Amoeba*, in the phylum Sarcodina. Amoebas are able to extend their cytoplasm, in the form of pseudopods (false feet), to move or obtain food. Sarcodina are the simplest protozoans, but their simple appearance does not mean that they are simple. They are able to behave in complex ways to obtain food, for example. When an amoeba senses its prey, through chemical detection, it extends its pseudopods toward and around its victim. It will also pursue its prey as long as it is close enough; amoebas judge the size of their prey, sending out just the right amount of pseudopod to ensure a meal (see Figure 8.28a).

*Paramecia,* in phylum Ciliophora, move using their fully ciliated bodies, by creating waves in the liquid of their surroundings. They obtain food by beating their cilia to make currents to bring organisms into their oral grooves along the side of their cells (see Figure 8.28b). *Paramecia* are capable of both sexual and asexual reproduction.

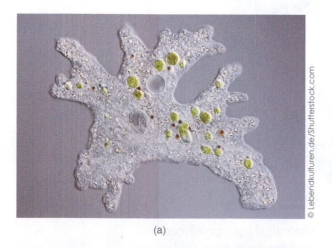

(a)

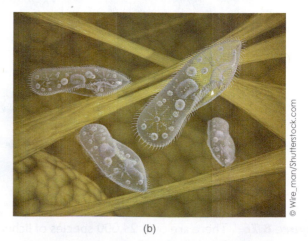

(b)

**Figure 8.28**    a. An amoeba uses its pseudopods to move and also to engulf prey. b. Paramecium consumes food, usually entering through their oral groove.

Those organisms in the phylum Mastigophora, such as *trypanosoma gambiense*, are parasites, causing sicknesses in the organisms they inhabit. Trypanosoma, which causes African sleeping sickness, is transmitted in the bite of a Tse tse fly. This illness may lead to kidney dysfunction, nerve problems, and eventual death. Trypanosoma is a reminder, as in our story, that microbes may cause serious disease among humans, with scientific research required to continually search for cures.

Another mastigophore is Trichonympha, which lives in the guts of termites, allowing cellulose, a plant substance, to be broken down for energy. This organism provides its host the ability to digest wood. However, most mastigophores are parasites, causing sickness in humans.

## Slime Molds

Slime molds are often referred to as lower fungi because they resemble fungi.

However, they are different from Fungi in terms of their structure and biological processes. They are slimy because they release an oozing substance that flows along surfaces to engulf prey, such as other protists, bacteria, and fungi (see Figure 8.29). They are similar to fungi in that they both live in cool, dark, and moist places, such as forest floors and shower drains. All slime molds are heterotrophic and use spores to reproduce by forming fruiting bodies that release those spores.

**Figure 8.29**   Slime molds form fruiting bodies used in reproduction. The red-colored spheres in the image contain spores that are used to spread new mold.

### THE AGGRESSIVE AMOEBA

In a terrible twist of events, a young boy, 12 years old, was admitted to the hospital in central Texas. He attended a summer camp but went home "not feeling well." He had been participating in water sports in a nearby lake to the camp, along with the other children. His mother took him to the hospital after days went by and he continued to have fever, loss of smell, and flu-like symptoms. After admittance, several diagnoses were tentatively made: meningitis, pneumonia, and bacteremia. Then the horrible discovery in his cerebrospinal fluid: amoebas. The boy died only 5 days after being admitted to the hospital.

The temperature of the lake in Texas in which the boy was swimming in the summer of 2007 was 84.4°F (29.1°C). This was warm enough to support the life cycle of amoeba parasites. The story is based on a real report by Texas health authorities chronicling the events surrounding the death of a Texas boy. The infection was a result of *Naegleria fowleri*, a species of freshwater amoeba, which enters the brain and aggressively destroys it. While it is a rare illness known as amoebic meningoencephalitis, there are regular cases in which pond water enters into a patient's nasal passages. Amoebas pass through the very small opening between the nasal passage and the frontal region of the brain, which houses the olfactory bulbs, used for smell. The parasite lodges itself there, right above the nasal passages, dividing and causing large lesions. Symptoms first include a loss of smell or a sense that something is burning or rotting. Amoebas rapidly advance through the brain, dividing and damaging all areas until the victim's death.

The course of the disease takes only between 3 and 12 days. Its survival rate, with aggressive medical treatment, is very low at only 3%. Other symptoms of amoebic meningoencephalitis include headache, fever, nausea, vomiting, and a stiff neck but later result a loss of balance, seizures, hallucinations, and death.

There have been only 160 cases since 1960, including two from using tap water through a neti pot irrigation system. The CDC (Centers for Disease Control and Prevention) report that warm pond water as well as some well water and municipal drinking water, may become infected with amoebas. While drinking infected water is harmless, when it travels through a person's nose, then the potential exists for amoebic meningioencephalitis.

The disease remains a protist mystery, with unanswered questions: Why are some susceptible while others go unharmed, swimming in the same waters? How does the amoeba work to move across nasal membranes and into the brain? With such a small number of cases, is the public outcry about the illness an overreaction? Should people stop swimming in warm pond water?

# A Favorite Fungus

What is your favorite fungus? . . . Mushrooms for making soup? Yeast for bread? Penicillin for your illness? or perhaps Athlete's foot, found in over 25% of the U.K. population? Each of these is an example of a fungus; they are found everywhere from feet to forest floors, usually in moist, dark places.

## Features and Types

Many people think of mushrooms when they imagine a fungus. However, fungi are diverse, as shown in our examples. Fungi were once classified as plants, but structurally and chemically, they are more closely related to animals than plants. They do not contain chlorophyll, and their cell walls contain substances found primarily in animals. Their classification is not fully agreed upon among biologists, but most of their species fall into two phyla: Ascomycota, or decomposers, and Basidiomycota, commonly seen as mushrooms and puffballs. While most species of fungi are multicellular eukaryotes, some, such as yeasts are unicellular. Molds, which are difficult to remove from homes and cause poor air quality due to their spores, can also be beneficial, for example, as a source of the antibiotic penicillin. There are approximately 100,000 species of fungi with many more expected to be discovered. Fungi emerged relatively recently in Earth's history, about 500,000 years ago. An aquatic protist was most likely its ancestor.

The body of a fungus is generally composed of a mass of filaments called myce-lium. Each individual fungal filament is called a hypha, and the hyphae together create a mycelium mat (see Figure 8.30). These mats anchor deep into material that it is decom-posing in order to absorb its matter. Hyphae are one cell-layer thick, making it easy for fungi to absorb substances from their surroundings. A septum divides hyphae filaments into compartments, each septum having openings for communication. The cell walls of hyphae are composed of chitin, a polysaccharide found primarily in insect exoskeletons, their hard outer covering. Chitin is a strong substance, giving support to fungal cell walls. Chitin is rarely found in plants, an example of their distant relationship with fungi.

Fungi are not motile, but instead move rapidly through growth of their hyphae. Some fungi produce more than a kilometer of new hyphae in a single day. They have high rates of growth. For example, the yellow honey mushroom fungus, in Oregon grows so quickly that it covers an area of over 4 square miles. Their growth allows them to exert a great deal of pressure (1,200 psi) against other organisms, enabling them to penetrate deeply into plants, for example. Tips of fungal hyphae have the ability to penetrate hard surfaces, such as plant cell walls, insect coats, and even human skin. When given the chance, studies show that fungal hypha grow directly through human skin.

Fungi play a critical role in nature. As heterotrophs, fungi obtain nutrients through absorption of dead matter. Fungi eventually consume all living things. Thus, fungi act as decomposers, along with bacteria, to recycle organic matter through our ecosystem (see Figure 8.15).

Fungi also occur in relationship with other organisms, sometimes living on other organisms as in Athlete's foot, ringworm, jock itch, and beard itch. Fungal diseases are very contagious because their spores survive for long periods on surfaces.

At times, fungi exist in a positive relationship with other organisms, as seen in our lichen example earlier in this chapter. In another **symbiotic** arrangement, root fungi or mycorrhizae live in root nodules of plants. Fungi benefit by obtaining sugar from the plant, and plants benefit by obtaining nitrogen and phosphorous from fungi, extracted from the soils around them. Many plant diseases are also fungal in origin, such as the

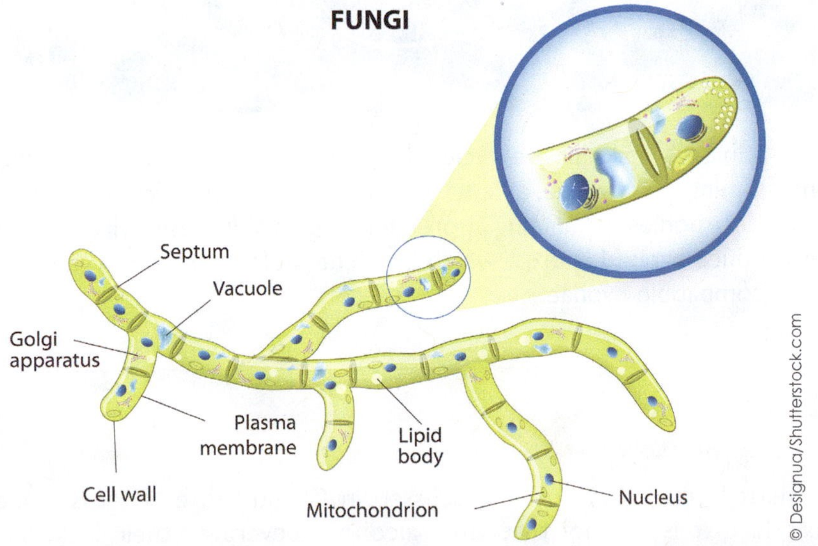

**Mycelium**

The mass of filaments that form the vegetative part of a fungus.

**Hypha**

Each of individual threads that make up the fungal mycelium.

**Septum**

A partition that separates two chambers of tissue in an organism.

**Figure 8.30** The entire strand of fungal cells is knows as its mycelium. Mycelium is composed of hyphae that are divided into separate cells. The fungal cell's components are shown in this figure.

fungus *Cryphonectria parasitica*, causing a blight that led to the destruction of the American chestnut tree.

Fungi reproduce both asexually and sexually. In asexual reproduction, a piece of mycelium breaks off, creating a new organism, in a process called fragmentation. In sexual reproduction, spores form in a tightly packed set of hyphae. A mushroom is actually a reproductive structure produced by a fungus to develop and release spores (see Figure 8.31). Mushrooms occur as dikaryotic structures, meaning that their cells have two haploid nuclei that do not fuse. Some parts of a mushroom become diploid, with their haploid nuclei fusing. These diploid portions produce haploid spores through meiosis, which are released into nature. These combine with other spores to result in a new diploid organism. Sexual reproduction gives more variation to fungal species, unlike mutations, the sole source of variation in the rabies virus discussed in our story.

**Fragmentation**

The stage in asexual reproduction in which a piece of mycelium breaks off, creating a new organism.

**Figure 8.31**   In the reproductive cycle of a fungus, spores form under the gills of mushrooms, forming new hyphae, able to mate and produce new fruiting bodies. Mushrooms are fruiting bodies, producing spores for a fungus. Meiosis within gill cells of a mushroom produce haploid spores, which combine to form a new organism through the mating of compatible hyphae.

## ROOT BEER

When yeast ferments sugars, as discussed in Chapter 4, ethanol is made as a by-product. It is ethanol that gives alcoholic beverages their kick. When fermentation is stopped before alcohol fermentation really takes off, large amounts of sugar are left and only small amounts of alcohol. The result is the popular drink root beer. It was originally produced from the root of a sassafras plant or bark. Roots are used as a source of many soft drinks. In the 19th century, farmers used yeast to ferment sugars in the sassafras root to make root beer that contained a small amount of alcohol. It was popular during their gatherings and a light-alcohol beer alternative.

The first commercially produced root beer was sold at the Philadelphia Centennial Exhibit in 1876, by pharmacist Charles Hires. It became quite popular during the Prohibition era of the 1920s and early 1930s, as a substitute for beer. In 1960, the main ingredient of the sassafras root, was found to be carcinogenic and was banned by the FDA. Since then, artificial sassafras acts as a key ingredient, giving root beer its unique taste. Using and understanding natural products, as Pasteur did to discover a rabies vaccination, is a key to advances in health science.

# Summary

The variety of organisms that are not plants or animals have many important functions in our ecosystem, but certain forms can also be harmful. Viruses have unique characteristics that classify them as an in-between living and nonliving state. They are intracellular parasites that seize a host cell's machinery. Prokaryotes, much larger cells, carry out all life functions and play key roles in human health and in the ecosystem. They are our earliest ancestors evolutionarily. Their simplicity contributes to their success. Protists, the most diverse of life's kingdoms, and quite a bit more complex than prokaryotes, emerged as our closest eukaryotic ancestors. They were the first eukaryotes, containing organelles and contributing to the evolution of higher plants and animals and fungi. Protista emerged about 1.5 billion years ago after the oxygen revolution made adaptations to use oxygen beneficial. Fungi emerged recently, only about 500,000 years ago, from protists. Fungi play a key role in the ecosystem, recycling dead matter. Fungi have many human uses ranging from medicines, wine, and breads to beer and delicate foods.

## CHECK OUT

**Summary: Key Points**

- Viruses, prokaryotes, protists, and fungi affect our environment and human health in many ways, from recycling chemicals and providing food and medicine to a variety of diseases.
- The discovery of the many non-animal/plant organisms showed their many characteristics, shared evolutionary history and role in the environment.
- Viruses are intracellular parasites, with two major types of life cycles: the lytic and the lysogenic.
- Prokaryotes are simple and evolutionarily successful ancestors to eukaryotes.
- Protista have three general groups: algae, protozoans, and slime molds, each of which contains a variety of organisms with varied characteristics.
- Fungi are decomposers; molds, yeasts, mushrooms, mycorrhizae, and lichens are types of fungi, which are heterotrophic, and absorb nutrients from their environment.

- Fungi and Protista evolved from prokaryotes after developing membranous organelles, such as mitochondria.
- Diseases caused by organisms in this chapter include viral: rabies, herpes, influenza, cancer, the common cold; bacterial: necrotizing fasciitis, food poisoning, pneumonia; protist: amoebic meningio-encephalitis, African sleeping sickness; fungal: Athlete's foot, ringworm, and jock itch.

## KEY TERMS

| | |
|---|---|
| actinomycetes | mycelium |
| algae, red, green, brown, golden-brown | mycoplasma |
| antibiotic | myxovirus |
| archaebacteria | neuraminidase |
| autogenous model | obligatory parasite |
| bacillus | oncogene |
| bacteria | oncovirus |
| binary fission | papillomavirus |
| capsid | pathogen |
| chemoautotroph | penicillin |
| coccus | peptidoglycan |
| conjugation | phototrophic anaerobic bacteria |
| cyanobacteria | phytoplankton |
| diatoms | pili |
| endospore-forming bacteria | prions |
| euglena | protozoan |
| fragmentation | retrovirus |
| Gram-negative bacteria | reverse transcriptase |
| Gram-positive bacteria | rhabdovirus |
| Gram stain | rhinovirus |
| halophiles | septum |
| hemagglutinin | slime molds |
| herpes simplex I and II | species-specific |
| hypha | spirillum |
| immunization | strep- |
| intracellular parasite | staph- |
| lichens | symbiosis |
| lysogenic life cycle | thermacidophiles |
| lytic life cycle | transduction |
| methanogens | transformation |
| morphology | |

# Multiple Choice Questions

**1.** Which organism is LEAST useful to human society and the environment?

   **a.** Virus

   **b.** Protista

   **c.** Bacteria

   **d.** Fungi

**2.** Which is a process by which pathogens work to harm host cells?

   **a.** They cause inflammation.

   **b.** They directly attack host cells.

   **c.** They produce toxins.

   **d.** All of the above.

**3.** Species specificity states that a virus has this many host species:

   **a.** 1

   **b.** 2

   **c.** 3

   **d.** 4

**4.** The lysogenic life cycle holds viral _____ dormant within host cells:

   **a.** proteins

   **b.** genes

   **c.** carbohydrates

   **d.** fats

**5.** A cluster of spiral bacterial cells would be classified as:

   **a.** staphylococcus

   **b.** staphylospirillum

   **c.** steptobacillus

   **d.** coccus

**6.** Which represents a logical order, from early to later, in the evolution of organisms?

   **a.** protista → fungi → archaebacteria → bacteria

   **b.** archaebacteria → bacteria → protista → fungi

   **c.** fungi → archaebacteria → bacteria → photosystem II

   **d.** protista → fungi → bacteria → archaebacteria

**7.** A motile, eukaryotic, and heterotrophic organism is discovered on a distant island, with cilia surrounding its unicellular structure. Which is its best classification?

   **a.** Slime mold

   **b.** Bacteria

   **c.** Algae

   **d.** Protozoa

**8.** Which term includes all of the others?

   **a.** Hypha

   **b.** Mycelium

   **c.** Chitin

   **d.** Septum

9. In question #8 above, with which kingdom are the terms most associated?
   a. Protista
   b. Fungi
   c. Prokaryote
   d. Virus

10. In question #8, which process helps these organisms to obtain needed ATP energy
    a. Photosynthesis
    b. Absorption
    c. Exocytosis
    d. Species specificity

## Short Answers

1. Describe three ways in which prokaryotes benefit humans. List three ways in which prokaryotes harm humans. Be sure to list and describe each.

2. Define the following terms: transduction and conjugation. List one way each of the terms differ from the other in relation to genetic variation and biodiversity.

3. Describe the rabies experiment of Louis Pasteur discussed in the story. Research how Pasteur's injections cured Andre. How do rabies immunizations work today?

4. Name three characteristics of viruses. Are viruses living or nonliving? Defend your answer.

5. For question number 4 above, list two types of viruses and describe their life cycles.

6. List the three groups of archaebacteria. How are they different from bacteria? Which would most likely be found in a hot liquid with a pH of 3? Why?

7. Explain the structure and function of a mushroom. Use the following terms in your answer: spores, haploid, dizygotic, diploid.

8. A disease destroying mycorrhizae in forests concerns a group of biologists. Why are they worried about the effects of the disease on the ecosystem? On humans?

9. Explain the role of phytoplankton in our environment and in human society.

10. Diseases due to viruses are plentiful. Name three diseases caused by viruses in humans. Which are not species specific? Why?

# Biology and Society Corner: Discussion Questions

1. In order to prevent amoebic meningioencephalitis, measures should be taken to reduce risks by government agencies. Research amoebic meningioencephalitis and list three recommendations you might make to improve public health with respect to the disease. Are your recommendations justified? Why or why not?

2. Louis Pasteur took a great risk with another person's life. Give an example of an experimental procedure that you have heard about which is controversial. Are such risks in medicine justified? Why or why not?

3. The overuse of penicillin and antibiotics is well documented in medical research articles. Based on your research of these articles, why is it bad practice to prescribe antibiotics such as penicillin to patients with a common respiratory illness? Should you or your loved ones request these drugs during a patient-doctor visit? Why or why not?

4. Bioengineering of photosynthetic bacteria increases ethanol production greatly. Describe one way that this procedure might have unintended negative effects on human society and the environment.

5. A lumber company claims that dead trees need to be removed from forest floors to keep the forest clean and healthy. Defend their statement, and then also refute their statement. Use your knowledge of bacteria and fungi to formulate your answer.

Figure – Concept Map of Chapter 8 Big Ideas

# Getting to Land: The Incredible Plants

**9**

## ESSENTIALS

A group of bryophytes live in a village

Mosses form little villages

A population of mosses live on a forest floor

A large maple towers over the conifer and mosses

A conifer towers over the mosses

# The Case of the Wet Village

We are all short and physically weak; our neighbors are all tall and strong. Perhaps we are a bit jealous of those who dominate among us. Before the neighbors came to our area, our shortness did not matter – I once heard this axiom from a friend: "In the absence of what you are not, what you are isn't." In other words, if there is no one tall around us (what we are not), then what we are (short) isn't. Our height never mattered until 'the others' came along. We were the most efficient group of settlers before them and our methods worked the best. Now things are different.

When our ancestors first settled the village, they knew that water was the most important resource that would help the settlement to survive. Water is still the resource that is most scarce – while some say it is oil or gold – we all need water to survive and many towns and nations die off without access to clean water. We are environmentalists, in that sense; we are also against the pollution of our air and water. Villagers are also energy conscious – each settler has their own solar power plants, producing energy from sunlight. When the others are not overshadowing us, we are efficient at obtaining energy from the sun.

There have been many floods, wiping out entire villages of our neighbors. But floods never harm our villages. We are a hardy group – we love the rains – it gives us new life. In fact, our lifestyle helps to prevent flooding for everyone. We use water so quickly that the neighbors benefit from our flood-control practices. Indeed, during rainstorms our community celebrates and procreates, in a local "fertility festival." The watery streets bring the village together, and there are always new births after the rainstorms.

Along "shady lane," a northern district of the neighborhood, there is a densely populated group of villagers. Our population is growing in that direction, perfectly content to live in wet and shady conditions, where fewer of our larger neighbors overshadow us. At times, a larger animal, feline in character, lays its reign over the town. We are all scared, especially when it digs around in our village, but we survive. It likes our village because we are so comfortable, providing a soft area of rest visitors.

We all hear about and see climate change occurring before us – it has led to some villages drying out and dying off. We cannot live, or produce the new generation, without lots of water. We are scared of what the climate changes could do to us; and without us, the larger community will suffer.

We comprise 12,000 species found throughout the world, from forests to deserts and peat bogs. Our villages grow best in shady, moist areas, or bogs but we reach from Antarctica and the arctic to the tropics. Villagers are never more than 15 centimeters in height and absorb minerals and water from soils via diffusion. Diffusion, as you recall, is a slow process that moves materials only so high, requiring that we have such small heights. We do not have structures to allow water and minerals to be transported far up our bodies.

Humans use our villages to enhance other soils, in the form of peat moss. We colonize areas, controlling floods and make the soil ready for larger plants. We do fear humans as well, because we are first indicators of their pollution, and we are so sensitive to changes in the environment.

This is the story of "our town." A bryophyte's perspective . . .

---

## CHECK UP SECTION

This story shows the biology of a group of plants called bryophytes, which comprises modern mosses, liverworts, and hornworts. Their place within our ecosystem is vital and their role should be protected.

Study the life cycle and contributions of bryophytes to our ecosystem and to human society. A number of species of bryophytes are considered endangered. Which are they? What factors are leading to this problem? What suggestions would you make to help protect their communities?

---

## The Village's Move to Land: A History

Imagine a primordial sea of one-celled algae crowded together, exhausting minerals found in the sea such as nitrogen, sulfur, and phosphorous. This is how algae existed before their evolutionary move to land. The sea was a place of intense competition, and the quest for limited resources is always expensive, taking energy from other life functions.

Thus began the transition of plants, including bryophytes, from the sea to the land. It occurred about 400 million years ago. All plants appear to have emerged from a single group of green algae, chlorophyta, (discussed in Chapter 8). The transition to land probably occurred at the coasts, where green algae were swept ashore. Tidal seaweeds were exposed to land at low tides and gradually became permanently land-based.

Along coastal land areas there were likely rich deposits of minerals that were washed up from ocean and river bottoms. Mutations enabling green algae to persist on coastal land were very valuable to exploit these mineral reserves (see Figure 9.1). Perhaps an extension helping a cell to anchor to the coastal land or a protective layer to help it from drying out enabled algal cells to make the transition to land areas possible. Biologists also surmise that a symbiotic fungus might have aided algal cells to colonize coastal lands. While the bryophyte village in our story had challenges, namely from their larger neighbors, the land afforded them many advantages.

**Figure 9.1**    Plants emerged from the sea. Primitive algae, as show in this image, aggregated along coasts to attach to dry land.

The land had a number of benefits for algae and the succeeding primitive plants:

1)  Most importantly, there were no competitors for the algal cells. No other organism had emerged on land to overshadow or out-produce them. The land was an open ecological space.
2)  There was ample carbon dioxide in the air to allow plenty of carbon fixation.
3)  There was continual light on land, unlike the filtered light obtained through multiple layers of water in aquatic environments.

In the story, the emergence of the "others" who were taller eclipsed some of these advantages. Today, taller trees and pollution by humans stunt bryophyte growth (see Figure 9.2), making conditions less favorable than they were 400 million years ago. But the move back to the sea would be ill advised – the pros of being land-based outweigh the cons, as we will see.

What were the major drawbacks to the move onto land? There was less water than in the sea, so drying out of plant cells on land limited survival and reproduction. Bryophytes in the story could grow only so high, limited by the pull of gravity in getting water to all of their cells. Bryophytes rely on diffusion for movement of materials, such as water and

**Figure 9.2**    Mosses in a forest live side-by-side with other plant life.

minerals, through their bodies. Later, larger plants evolved a **vascular system**, which consists of vessels that transport water and minerals up plants more efficiently than via simple diffusion.

Standing upright also presented challenges to land plants. Before, in water, they could remain horizontal, absorbing light. On land, plants needed to grow taller, in part to compete as more plants evolved to reach sunlight for photosynthesis.

A lack of water on land led to other adaptations. To cope with it, all plants exhibit an alternation of generations, in which there are haploid and diploid phases of their life cycle function in reproduction. This life cycle allows gametes to move through watery surroundings, mating with other organisms. (We will discuss alternation of generations in more detail later in this chapter.) Plants developed a number of strategies for preventing water loss. For example, a thick layer such as a cuticle surrounding a leaf prevents water loss (see Figure 9.3). Stomata that open to exchange gas then close to prevent evaporation also helped plants to transition to land (see chapter 4).

**Alternation of generations**

The life cycle of plants in which haploid and diploid phases of their exist for survival and reproduction.

## Evidence for Green-Algae Ancestry

Several pieces of evidence lead biologists to believe that green algae are the ancestors of plants. First, both green algae and plants contain chlorophyll a and b and beta-carotene, both of which are used for photosynthesis. Second, both plants and green algae contain cell walls made of cellulose, a strong structural carbohydrate that allows plants to grow to great heights. Third, both plants and green algae carry out cytokinesis using a cell plate instead of cytoplasmic pinching, which is seen in other organisms.

The first plants appeared in the fossil record about 475 million years ago. They were likely mere patches of low-growing green plants, without more complex roots, stems, or leaves seen in modern plants. These primitive plants were colonies of cells that could survive on land but had no upright structure.

After the transition to land, plants diverged into two lineages roughly 75 million years later: **bryophytes** and tracheophytes, or vascular plants. The difference between the two groups is that tracheophytes have a developed vessel system and bryophytes do not. Bryophytes first appeared in the fossil record during the start of the Devonian period, 350 million years ago, but they were probably present on Earth millions of years before. Bryophyte fossils appear very similar to modern-day bryophytes, indicating that there was little change over this long period. Vascular plant fossils were very different

**Tracheophyte**

Vascular plant; plants having a well developed vessel system.

**Figure 9.3**   Prickly pear cactus; note the thick cuticle on a leaf of this desert cactus. It prevents water loss and allows cacti to lie in very dry conditions.

from those seen today, and experienced a tremendous divergence. These changes led to many branches of tracheophytes that will be discussed later in this chapter.

## What are Plants?

All plants are eukaryotes and multicellular organisms that do not move once their seeds take hold. Most species of plants produce their own food through photosynthesis using chlorophyll and sunlight to convert carbon dioxide and water into sugar. There are over 280,000 species of plants ranging in size from 1 mm (0.04 inches) to 117 m (380 feet). Plants provide many resources for human society ranging from food, building materials, and medicines. They trap energy to build the carbon-based mass of which their bodies are composed. Through photosynthesis, they recycle the air, adding oxygen, and removing carbon dioxide. Plants structures also provide a home for other organisms, such as birds and insects. When plants die, their organic matter is recycled by decomposers to enrich their soils with nutrients and layers of new soil.

Most plants adhere to the described definition, but there are a few exceptions. The Dodder plant, for example, lacks chlorophyll and derives its nutrients by living parasitically on other plants. The Ghost Orchid (Epipogium aphyllum), is another plant without chlorophyll. Because it cannot make its own food from photosynthesis, it uses nutrients from a network of fungi (mycorrhizae) beneath its roots. It grows underground for most of its life cycle but emerges from the soil only to flower. Both are examples of how some plants adapt to their living conditions.

Plants are most closely related to algae evolutionarily, which are also photosynthetic. The difference is that plants have the ability to live on dry land and algae must remain in watery environments. First to evolve were the bryophytes. Plant evolutionary history traces how they emerged from watery environments onto land. After bryophytes, the first of the tracheophytes to evolve was *Rhynia major*, now extinct but found in the fossil records estimated to be 400 million years old. While adapted to land just like the bryophytes, *Rhynia major* differed in that it had a central tube of vascular tissue. This may be seen in the fossils recreation of Rhynia shown in Figure 9.4, containing a central vessel that allowed it to transport materials up its structure. While primitive, this structure provided a major advantage over bryophytes, which were limited in size due to their need to rely on diffusion for transport. *Rhynia major* could grow taller and be the biggest among the plants of that time, obtaining the most sunlight and thus food. Rhynia major plants were "in the absence of what it was not. . ." – other taller plants. They overshadowed the bryophytes of our story.

## Plant Structure Refinements to Help Them Live on Land

**Phloem**

A series of tubes that carry sugars and dissolved organic materials down a plant.

**Xylem**

A series of tubes conducting water and dissolved minerals up a plant.

**Root system**

The parts of plants below the surface.

The centralized vessel allowed plants to grow larger and larger. Over time, plants developed more and more elaborate conducting vessels to transport water and nutrients. The conducting system of modern tracheophytes consists of two types of tissues: the phloem, which is a series of tubes that carry sugars and dissolved organic materials down a plant, and xylem, which is a series of tubes conducting water and dissolved minerals up a plant (see Figure 9.5). These tissues carry needed materials to every cell in a plant. Their vessels are always within diffusion's distance from any plant cell. In aquatic worlds, there is no need for these structures because watery nutrients bathe every cell.

Second, root systems, the parts of plants below the surface, formed to absorb water and minerals from soils (see Figure 9.5). Sugars are not the only source of food for a plant. Many required nutrients are also absorbed through the soil. Recall in Chapter 2 that proteins require nitrogen to build amino acids, and ATP requires phosphorous for

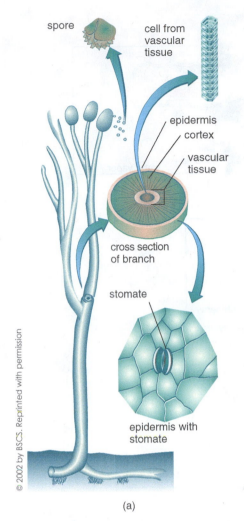

(a)

(b)

**Figure 9.4** a. *Rhynia major* From *BSCS Biology: An Ecological Approach*, 9th Edition by BSCS. vs. b. Mosses Rhynia major was one of the earliest known vascular plants with primitive stems containing vessels (vascular tissue). Its stems were photosynthetic. Rhynia major lacked leaves and roots but had sporangia that produced spores. It had an advantage over the mosses – it could grow taller because its vessel system transported water to higher heights.

high-energy bonding. At times, roots reach far beneath the surface, tapping needed water supplies. The wild fig tree at Echo Caves, Ohrigstad, Mpumalanga, South Africa, has a root system 400-feet deep.

Third, plants have a shoot system, stems that support the leaves that carry out photosynthesis (see Figure 9.5). The shoot also gives height to a plant, enabling it to maximize sunlight. Shoots are strengthened by lignin, a stiffening substance that supports plant cell walls as they grow recall from Chapter 3.

Fourth, plants have mechanisms to prevent water loss. A thick cuticle, for example, surrounds desert cactus leaves for protection from evaporation. Stomata on the underside of its leaves limit a plant's loss of water through evaporation.

Fourth, protection from predators is vital because plants cannot move to escape conflicts. Their methods of response to the environment will be elaborated upon later in the chapter. Prickly thorns found on a rose bush or a rash caused by poison ivy is no accident – these defenses function evolutionarily to prevent predators from killing plants.

**Shoot system**

The system that consists of stem, leaves, lateral buds, flowering stems, and flowering bud.

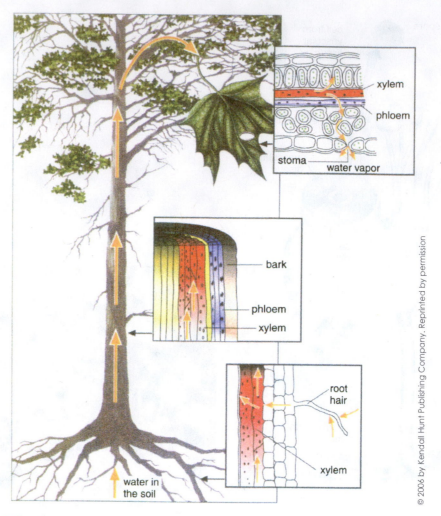

**Figure 9.5**    The plant body. Roots absorb water from the soil shoots have leaves for photosynthesis; and the vascular layers, xylem (on the inside) and phloem (on the outside) transport water and nutrients. From *Biological Perspectives*, 3rd ed by BSCS.

Finally, the fifth adaptation of plants to land is their life cycle with its alternation of generations. To elaborate, on dry land gametes (sex cells) cannot easily travel from one organism to another, as they can in an aquatic environment. So, plants evolved a system of travel adapted for land in their alternation of generations. Plants never meet to have sex by walking around and socializing, and instead rely upon a separate structure – a gametophyte generation, which we will discuss in the next section, that allows plants to bring their sex cells to one another. In some plants, they use animals to carry their gametes; in others, wind or water does the task.

# Divisions of Plants

## Bryophytes

As discussed earlier, when plants first arrived on land, they quickly separated into two divisions: bryophytes (nonvascular plants) and tracheophytes (vascular plants). In the opening story, vascular plants had a big advantage over nonvascular plants: they could carry water higher and thus grow taller to obtain sunlight. Bryophytes include mosses, liverworts, and hornworts, comprising about 12,000 species of plants. Bryophytes grow only in a moist environment, and they tend to be smaller in size. Their need for water to reproduce requires that there is enough moisture for sperm to swim from one organism to another through the environment.

The bryophyte life cycle shows an alternation of generations, which is specifically defined as a life cycle that contains a period of time in which there is a multicellular haploid phase and another multicellular diploid phase. The mossy village seen in the story or in nature is composed of adult mosses that are haploid organisms, also known as gametophytes because they produce the gametes (see Figure 9.6).

Gametophytes are adult moss plants that are either male or female, with respective parts. Gametes, sperm and egg, are produced by these reproductive structures. When water is sufficient in their surroundings, mosses release their male gametes, the sperm swimming through the community in the "fertility festival" described in our story. When a sperm reaches a female moss plant, it fertilizes an egg, dividing to form an embryo. The embryo develops from a female forming a new structure, called a sporophyte. This structure elongates from a female body and produces haploid spores. Spores land in moist areas, again leading to another gametophyte generation and the production of new adult mosses.

Adult mosses, gametophytes, contain only cells with half the number of chromosomes or N. The haploid condition, with its genetic components, was discussed in Chapters 5 and 6. They do not have a complete set of DNA. The sporophyte generation is diploid or 2N, and represents a complete genetic organism. It is strange to think that the adult mosses in the village in our story consisted of organisms with only half of their full set of DNA. In bryophytes, the gametophyte generation is thus said to be dominant because most of their life cycle is spent in this haploid period. This trend is common among plants, all exhibiting a haploid state at some point in their life cycle, to cope with their move to land and potentially dry conditions (see Figure 9.6). Spores are able to tolerate a dry or hostile environment.

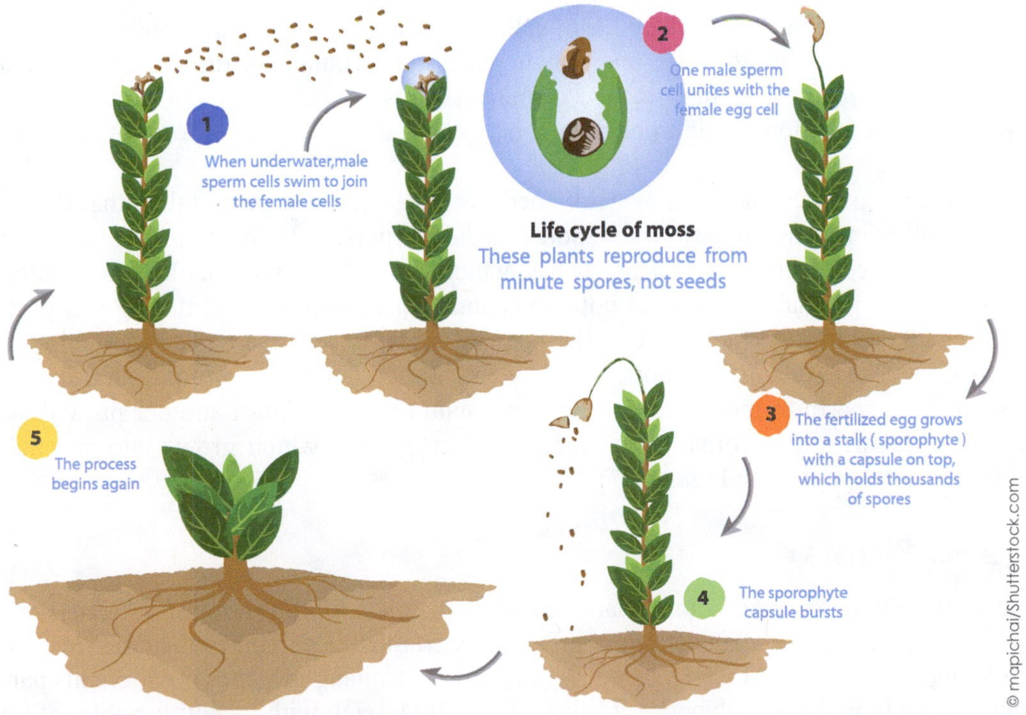

**Gametophytes**

Haploid organisms that produce the gametes.

**Sporophyte**

The diploid organism, producing spores.

**Figure 9.6**   Moss life cycle: haploid and diploid phases alternate in the life of a moss. Spores comprise the haploid phase while spore-forming structures (sporophytes) comprise the diploid phase.

## Tracheophytes

**Seed**

An embryonic plant with its own internal and protected supply of water and nutrients, also led to another division of plants: seedless and seeded.

Vascular plants, or **tracheophytes**, are those plants containing a vessel system for transport of materials. They differ from the more primitive nonvascular bryophytes in this way. However, the evolution of the seed, an embryonic plant with its own internal and protected supply of water and nutrients, also led to another division of plants: seedless and seeded. Seedless plants, of which bryophytes are one type, are described in our story. Bryophytes use spores for longer distance movement away from their parents. Spores are haploid and contain only DNA, RNA, and some proteins. Spores grow up to become a gametophyte. While bryophytes and some tracheophytes have spores that transmit their haploid genotypes, most tracheophytes have seeds. Seeds offer an advantage to longer distance travel because they contain all of the required materials to help an encapsulated diploid organism grow where the seed lands.

## Seedless Plants

Seedless tracheophytes evolved as an advantage over bryophytes. Seedless plants include Pteridophyta (ferns), Sphenophyta (horsetails), and Lycophyta (club mosses). While spores instead of seeds are used for reproduction in each of these phyla, they are better adapted to land than the mosses in our story. Each seedless plant also has a primitive vessel system, while bryophytes have none. However, the seedless plant vessel system is the most primitive of all the tracheophytes. Nonetheless, this enables seedless plants to grow to greater heights than bryophytes, less limited by water uptake. However, ferns also have a branching vessel system from their central canal that brings water and nutrients within diffusion distance of all its cells. This system is more elaborate than that of other seedless plants, enabling ferns to grow even taller and be more efficient at transport of water and minerals.

Seedless plants require water for reproduction, in a similar way to the bryophytes. Ferns have sporangia on the underside of their leaves in which spores are produced. Adult fern plants are composed of the sporophyte generation, containing diploid cells. When water is available, ferns transport their gametes via a watery environment. However, the sporophyte generation is said to be dominant because this phase is the form of the adult fern.

**Prothallus**

The gametophyte generation of ferns.

Because of their vascular system, seedless tracheophytes grow taller than bryophytes, allowing wind to carry their spores farther from parents. When a spore lands in a moist surface, it grows into a prothallus, which is the free living haploid generation of ferns. The prothallus produces both eggs and sperm; sperm travel during wet periods, much as in the moss village in the opening story, to fertilize eggs within another prothallus. The prothallus is short, heart-shaped, and is much like a moss plant in that it allows its sperm to move through water to obtain an egg within another prothallus. Fertilization within a prothallus produces another embryo which grows into an adult fern, the sporophyte (see Figure 9.7).

## Seed Plants

Seeds were the great adaptation to land—enabling plants to traverse the globe. Seeds have food and protection to make a long journey away from their parent. In the film, *Failure to Launch*, a 35-year old man, played by Matthew McConaughey, does not leave his parent's home to make an independent life for himself. A seed, with its comfortable set-up containing nutrients and protection, does not experience a "failure to launch."

Thus, the gametophyte generation is greatly reduced in seed plants. There is no need for spores because seeds contain a whole diploid organism within its protective covering.

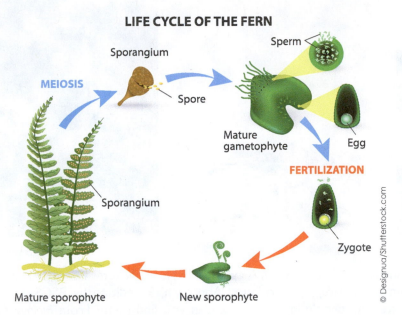

**LIFE CYCLE OF THE FERN**

© Designua/Shutterstock.com

**Figure 9.7** An alternation of generations is still evident in the fern life cycle but its gametophyte generation is smaller than in mosses. The gametophyte is a heart-shaped structure which produces egg and sperm cells.

This allows offspring of seeded plants to be transported as fully formed organisms to their next home in one swift move. There are several approaches plants use to transport seeds, as will be discussed later in this chapter. Seeds are diploid, containing genetic material from both parents, and grown in the right conditions of a new environment, develop into a new organism. Moving longer distances from parents lessens competition and inbreeding between parent and offspring.

There are two groups of seed-bearing plants: gymnosperms, plants with seeds that do not develop in an ovary, usually cone producing; and angiosperms, which are flowering plants with seeds developed in an ovary. Seed plants have a life cycle of alternating generations as well, but their sporophyte generation is dominant. Adult seed plants are sporophytes, but they possess gametophyte male and female reproductive structures that produce sperm and egg, respectively. Animals, as will be discussed in the next chapter, also have a reduced gametophyte generation consisting of sperm and egg cells.

## Gymnosperms

Pine trees like the one in Chapter 4's story, can inspire awe in their admirers. Gymnosperms are the tallest, oldest, and thickest of all plants. They comprise the imaginary forests of Snow White and of medieval eras in our minds – images evoked are a dark forest with soft pine needle ground. Gymnosperms consist of roughly 1000 species of four major groups: conifers, cycads, gnetophytes, and ginko plants (see Figure 9.8).

**Conifers**, cone-bearing plants such as pines and firs, were the first plants to completely evolve away from reliance on water. Their use of seeds about 160 million years ago helped them to populate land first, before the angiosperms invaded. Because they did not need water for reproduction, their sperm encapsulated in pollen, they could grow in many more areas than bryophytes. No longer relegated to low-lying and wet conditions, gymnosperms took advantage of new environments. The dinosaurs probably roamed in solely coniferous forests, with flowering plants evolving only much later, about 35 million years ago.

**Gymnosperm**

Plants with seeds that do not develop in an ovary, usually cone producing.

**Ovary**

A female reproductive organ containing ovules in which eggs develop.

**Angiosperm**

Are flowering plants with seeds developed in an ovary.

(a)    (b)    (c)

**Figure 9.8** What kinds of gymnosperms live today? a. A cycad called *Encephalartos woodii*, b. A gnetophyte called *Welswitschia mirabilis*, From *BSCS Biology: An Ecological* Approach, 9th edition by BSCS. c. The giant sequoia tree is a conifer. There are two people in this photo. Can you find them? From *Biology: An Inquiry Approach*, 3rd ed by Anton E. Lawson.

Gymnosperms have familiar appearances, as conifers or evergreens – pines, spruces, firs and redwoods, commonly found in temperate regions – all have needles forming their leaves. But gingko, also a gymnosperm plant, looks very different, with large leaves, very unlike a pine tree. There is only one species of gingko plant, but there are 9,000 conifers.

Needles on conifers, with their thick cuticles, are adapted to protect cells beneath from cold and water loss. This allows most conifers to retain their needles throughout the winter in temperate areas, enabling photosynthesis to occur all year. Tamaracks are one of the few species of conifers that lose their leaves in response to winter.

Sap in conifers contains chemicals that act as antifreeze and enables flow in freezing temperatures. This benefit enables gymnosperms to grow in cold as well as warm areas. Sap allows transport in vessels even at sub-zero temperatures. Conifers are found in every non-frozen region on Earth. They may grow for many decades. Conifers contain woody bark that is resistant to diseases caused by the many organisms described in the last chapter. Many woody plants are also resistant to herbivore attacks, growing too tall for other organisms to reach their leaves.

So valued by the lady in our story in Chapter 4, the largest and oldest trees are often gymnosperms. The tallest tree in the world is the coast redwood at 380 feet and the oldest tree, a Great Basin bristlecone pine, is the Methuselah tree, at 4,800 years (see Figure 9.9). Some root systems are said to be over 8,000 years old, regrown after shoot death occurs. The greatest predator of trees is human society, which harvests wood for many uses; wood is the third largest globally traded commodity, with only oil and gas ranking higher.

## Angiosperms

The vast botanical beauty of our surroundings is composed of angiosperms, or flowering plants. Most of the flowering plants on Earth resemble their early forms according to the fossil record, changing little since they first evolved. There are 250,000 plant species of angiosperms, each with flowers with unique shapes, sizes, and colors (see Figure 9.10). Flowering plants include trees, bushes, and grasses of many types and occupy over 90% of the Earth's vegetative surface. Almost all crops that provide us

(a)                                                                    (b)

**Figure 9.9** a. Methuselah tree (oldest known tree) and b. Coast Redwood (tallest tree); Conifers have reached the oldest ages and the greatest heights of all plants.

food are angiosperms, as well as large trees and all flowering plants. It is the angiosperms that make mosses in our story most jealous – they have so many characteristics that outshine bryophytes.

## Flowers, Fruit, and Plant Reproduction

What makes angiosperms so special? Angiosperms have flowers to attract insects and birds. All flowers have certain common features. First, flower structures are generally the same. All flowers are supported by a stem with modified leaves, called **petals**. Petals may be colorful or flashy to attract other organisms. Angiosperms have male and/or female structures, most often on the same plant. The male reproductive structure is called the stamen, which is composed of an anther, or capsule which hold pollen grains supported by a stalk or filament. The female reproductive structure is called the carpel, which is composed of a stigma, or sticky flat surface on which pollen grains land, a style, which extends downward to an **ovary**, which contains ovules in which eggs develop. Mosses seen in our story do not have flowers, with only a feathery end that produces gametes. Angiosperms possess male and female reproductive structures, much more complex and often very beautiful.

Some flowers have both male and female parts, called monoecious; and some have only a male or a female part, called dioecious flowers, with stamen and carpals on separate plants. Corn plants are monoecious, with "ears" as clusters of carpals and tassels as stamen. Date palms are dioecious, with a few males able to provide hundreds of females with pollen.

Movement of pollen from one plant to another is called pollination. In pollination, pollen is placed on the stigma of a carpal. Often, many pollen grains are released during pollination, with very few obtaining a place on the stigma. While many pollen are wasted in the effort, pollination has persisted among plant species through their evolutionary history. As indicated in Figure 9.11, the stamens in a flower are shorter than the carpel of a female. This limits the chances of pollen landing on the stigma and self-pollinating.

For both gymnosperms and angiosperms, plant gametophytes consist of pollen grains, the male gametophyte, and ovules, the female gametophyte. When pollen grains land on a female reproductive structure, the pollen grows a tube down into the ovule.

**Stamen**

Male reproductive structure.

**Anther**

A reproductive structure that holds pollen grains.

**Filament**

Thread-like structure supports the anther.

**Carpel**

An organ found at the center of a flower and bears one or more ovules.

**Stigma**

A sticky flat surface on which pollen grains land.

**Style**

The part of carpel that extends to ovules in which eggs develop.

**Monoecious**

Flowers that have both male and female parts.

**Dioecious**

Flowers that have only a male or a female part, with stamen and carpals on separate plants.

**Figure 9.10**    Sample of angiosperms. Angiosperrms are flowering plants, which reproduce by forming fruits which are often dispersed. From *BSCS Biology: An Ecological Approach*, 9ᵗʰ Edition by BSCS.

**Pollination**

Movement of pollen from one plant to another.

**Pollen grains**

The male gametophyte.

**Ovule**

The female gametophyte.

**Endosperm**

The nutritive tissue found inside the seeds of flowering plants.

This tube is directed chemically (usually with calcium) into the ovary. Two sperms then enter the ovary containing an egg. This enables one male sperm to fertilize a female egg in the ovule. The fertilized egg develops into an embryo forming a new plant. The second sperm fertilizes another structure, the central cell (2N) of the embryonic sac, which forms the nutritious endosperm. The endosperm provides food for the embryo as it germinates. The endosperm is triploid because it contains a 3N arrangement of nuclear material. The ovule develops into the hard external layer of a seed. This process is called **double fertilization** because two sperms fertilize two separate structures to form a seed.

Flowers function to attract other organisms, such as birds and insects to spread pollen or fruit seeds, carrying them farther from parent plants. Nectar, which is the sugary attractant to insects and birds, bring them to flowers. Sticky pollen grains attach to birds and insects, getting a ride to farther distances. Sometimes wind carries pollen across distances.

When flowers are fertilized, they develop into **fruits**. Fruits are defined as a matured or ripened ovary-containing seeds. The ovule of a carpel develops into the seeds of a fruit and the ovary develops into the fleshy part of the fruit. Fruits attract animals to eat them. Fruits are colorful, tasty, and good sources of energy for animals. Animals spread

**Parts of a Flower**

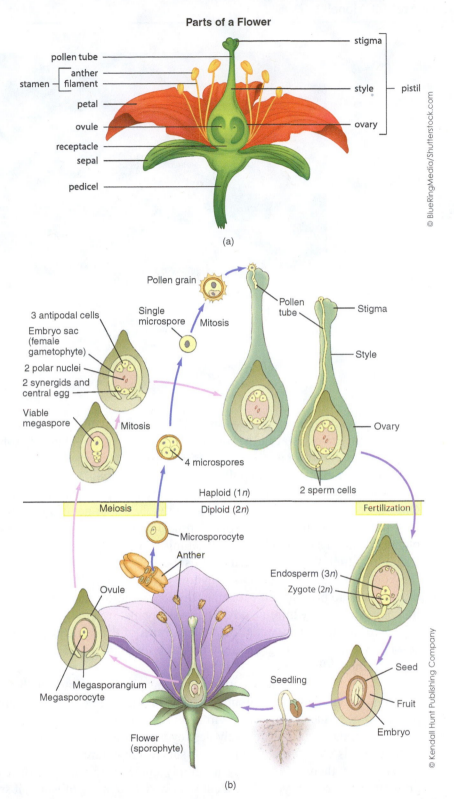

pollen tube

stamen — anther
filament

petal

ovule

receptacle

sepal

pedicel

stigma

style

pistil

ovary

© BlueRingMedia/Shutterstock.com

(a)

Pollen grain

3 antipodal cells

Embryo sac
(female
gametophyte)

2 polar nuclei

2 synergids and
central egg

Single
microspore    Mitosis

Viable
megaspore    Mitosis

4 microspores

Pollen
tube

Stigma

Style

Ovary

2 sperm cells

Haploid (1*n*)

Meiosis

Diploid (2*n*)

Fertilization

Microsporocyte

Anther

Ovule

Megasporangium

Megasporocyte

Endosperm (3*n*)

Zygote (2*n*)

Seed

Fruit

Embryo

Seedling

Flower
(sporophyte)

© Kendall Hunt Publishing Company

(b)

**Figure 9.11** a. Flower structure: male and female reproductive structures. The anther and filament make up the stamen of the male, which produces pollen. The sigma, style, and ovary together comprise the female structure called the carpel. The ovary of a female produces eggs. Pollination is facilitated via wind, insects, and birds. b. Pollination: Pollen lands on the stigma of the female and grows downward through the style. The pollen tube's growth is a different, more complex, process as compared with the moss gametophyte's simplicity. In mosses egg and sperm diffuse travel through water and land on female structures. Reproduction in flowering plants is a competition between pollen grains forming tunnels and reaching eggs within the ovary. c. Pollination is facilitated via wind, insects, and birds.

(c)

**Figure 9.11**    (*Continued*)

seeds by excreting them as they travel. This process, along with pollination, increases a plant's ability to grow at farther distances from its parent. Thus, bryophytes may have a reason to be jealous of angiosperms, if they could think about it.

Insect, animal, and wind pollination prevent inbreeding in several ways. First, wind and insect pollination transfer pollen to places at some distance from the parent plants. Once a tree roots, it is there for life. Its only chance to get away from family members is at the pollination or seed dispersal times. Outbreeding increases the biodiversity of plants by adding to the chances of getting as many genetic combinations as possible through mating with less related organisms.

For this reason, many plants with both male and female reproductive structures have mechanisms to prevent self-breeding. In fact, while some monoecious flowers self-pollinate, it occurs very infrequently. Stigmas may mature after pollen grains fall onto their surface or a stigma may not allow pollen grains of the same genotype to develop on its surface. Outbreeding is genetically desirable in plants, preventing the genetic defects found in inbreeding.

Asexual reproduction is common among plants, during which all offspring have the same genotype as the parents. Clearly, the drawback is genetic uniformity in new populations. In the event of an environmental change for which a particular genotype is susceptible, the results more likely destroy a population. Genetic diversity among plant populations is vital in maintaining their survival in nature.

However, there is efficiency in simply cutting off a branch of a willow tree and growing a new one. If you place a cut willow stem in a jug of water, it sprouts new roots easily. Asexual reproduction also occurs in strawberries, which form runners along the ground to sprout new plants. In dandelions, unfertilized seeds form new plants, which is very effective in colonizing new areas. Mosses in our story also have asexual reproduction by regenerating certain parts, called **gemmae**, of broken off pieces of moss.

However, danger remains for any plant population that is solely asexually reproduced. When a disease wipes out their singular genotype, as occurred for the American elm tree, they experience a widespread destruction from infection – the elm by the fungus *Ophiostoma ulmi*.

### COCONUT OIL – FRIEND OR FOE?

Coconuts are an example of a fruit that helps to spread offspring to new areas. Coconuts contain high amounts of fats attractive to animals, which consume them and spread their contents throughout the land. This spreads plant populations across larger surfaces, diminishing competition.

Whether coconut oil is good or bad for human health has been debated in recent studies. There is a clear answer – coconut oil has a very high amount of saturated fat. As described in Chapter 2, these oils are bad for the heart. The good fats – monounsaturated fat – are found only in small proportions, unfortunately. Coconut oil contains more than 92% saturated fat, and about 6% monounsaturated fat and 2% polyunsaturated fat.

Recall from Chapter 2 that saturated fats contribute to increased levels of "bad" cholesterol (LDLs, or low density lipoproteins) in the body. In contrast to coconut oil is olive oil, which contains only 14% saturated fat and 74% monounsaturated fat with 11% polyunsaturated fat. Olive oil is much better for the heart because of its high amount of monounsaturated fat, which is linked to "good" HDL cholesterol.

Recent research shows that coconut oil's saturated fats are medium-chain triglycerides (MCTs), which are linked to improving weight loss. However, these studies are only found in animal models and do not stand up at this point to recommendations for improving human health. Human studies do not support the benefit of MCTs over other saturated fat in risks for heart disease and obesity.

Often pointed out are the high rates of good heart health in Pacific Island and Asian populations. They have diets naturally high in coconut oil. Thus the relationship is drawn that their health is due to the coconut oil. However, consider that these populations have a more active lifestyle, eat primarily vegetarian foods and have little access to fatty meats.

Making a comparison to island diets is not scientific because of the many intervening variables affecting health. Island populations are so unlike those found in the developed nations because of their very different lifestyles. Why is there a movement afoot to show coconuts as healthy? Is it to bolster trade with nations or companies with products containing coconuts? Evidence clearly indicates that coconuts are bad for our heart health.

## Monocots and Dicots

In their initial development embryos in a seed produce their first leaf, called a cotyledon. Angiosperms produce seeds with two varieties: some contain an embryo with one seed leaf, called a monocot, and some contain an embryo with two seed leaves, called a dicot. Monocots include corn, rice, and lily plants, and dicot examples include maple trees, oaks, and peanut plants.

Monocots and dicots differ in some important structures . Dicots have a large vertical root called a taproot that burrows downward, anchoring the plant, while monocots have fibrous root system, which give increased exposure to water in soils. Some desert plants have fibrous root systems over 100 feet in diameter to obtain scarce water. Taproots are stronger and more secure for dicots, allowing them to grow much larger. Dandelions have a nasty taproot, preventing them from being pulled easily from garden

**Cotyledon**

The first leaf developed during the initial development of embryos in a seed.

**Monocot**

Angiosperms-produced seeds that contain embryo with one seed leaf.

**Dicot**

Angiosperms-produced seeds that contain an embryo with two seed leaves.

**Taproot**

Large vertical root that burrows downward, anchoring the plant.

**Fibrous root**

A root system made up of numerous branching roots and give increased exposure to water in soils.

and lawns. Stems of dicots are also different, ordered in vessels systems around a ring; but monocots have scattered vascular tissue. Monocot leaves have parallel vessels while dicots have branching vessels.

## Plant Tissues

Plant tissues of three types carry out their life functions. **Dermal tissue**, the epidermis, covers plants in a single layer of cells surrounding the entire organism, especially young plants. It functions in the same way that skin does, to cover and protect the human body. Dermal tissue specializes depending on the area in which it is located. In roots, dermal tissue develops into root hairs to increase surface area and thus absorption. Most absorption occurs in the root tips due to these hairs. In leaves, dermal tissue forms a waxy layer called a cuticle that helps the plant retain water.

On the underside of leaves, stomata or openings allow gas exchange (also discussed in Chapter 4). Stomata are surrounded by **guard cells**, which control their opening and closing (Figure 9.12). Based on water pressure within guard cells, stomata open when guard cells fill with water and close when they have less pressure. When guard cells take in water, they become turgid, and their cellulose fibers radiate outward. This causes guard cells to buckle, creating an opening space between them. When guard cells lose water, they become flaccid again, closing the hole. Closing minimizes water loss via evaporation from water contained within xylem of plants. However, stomata must remain open to obtain needed carbon dioxide for carbon fixation and oxygen for cell respiration. Water loss is a constant threat to a plant's survival, with a mature, temperate tree losing upward of 100 gallons of water each day. Moss gametophytes in our story do not have stomata; instead they obtain needed gases through diffusion in their thin "leaves."

**Vascular tissue**

Tissues that transports water, minerals and food throughout a plant.

Vascular tissue, or xylem and phloem, transports water, minerals, and food throughout a plant. Plants have no heart, unlike animals which use it to pump blood for transport. Instead, plants use physics principles to transport materials (not circulate them around) from one spot to another. Xylem conducts sap, which moves water and minerals

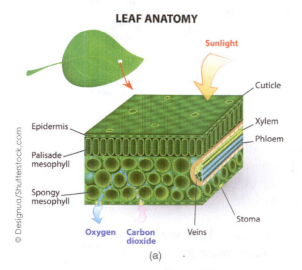

**LEAF ANATOMY**

Sunlight

Cuticle
Xylem
Phloem

Epidermis
Palisade mesophyll
Spongy mesophyll

Oxygen    Carbon dioxide    Veins    Stoma

© Designua/Shutterstock.com

(a)

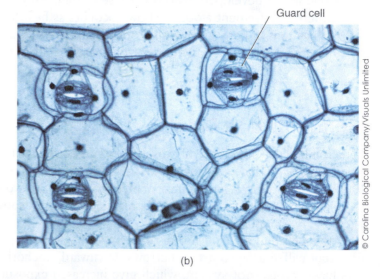

Guard cell

© Carolina Biological Company/Visuals Unlimited

(b)

**Figure 9.12** a. A general structure of a leaf. The mesophyll layer of a leaf contains the vascular bundle, which transports food (phloem) and water and minerals (xylem) through vessels. b. Stomata are pores on the underside of plant leaves. They are surrounded by guard cells which regulate their opening and closing.

upward, from roots to shoots of plants. These vessels contain cells that are dead at maturity, called tracheids and vessel elements (see Figure 9.15). Their cell walls make up the vessel walls of xylem. Tracheids are long, thin cells with tapered ends and thick lignin partitions. Vessel elements are wider and shorter, with thinner walls. Both contribute to the vessel tubes of xylem. Gymnosperms contain only tracheids, which make their vascular systems less efficient. Angiosperms have both tracheids and vessel elements, enabling faster delivery of materials.

Phloem conducts sap downward from leaves, where food is produced through photosynthesis, to all parts of a plant. (see Xylem and Phloem in Figure 9.12a) Phloem sap contains sugars and dissolved organic materials, which are needed for a plant's life functions. Sieve-tube members (Figure 9.15), which are cells that transport sap through phloem vessels, are alive at maturity, unlike xylem cells. They do, however, lack a nucleus, ribosomes, and vacuoles. In angiosperms, walls between sieve-tube members, called sieve plates, have pores to allow the flow of fluids. Phloem is also composed of companion cells (Figure 9.15), which lie next to sieve-tube cells. They are connected to sieve-tube cells through plasmodesmata, or gap-like cell junctions, allowing organelles to serve sieve-tubes. Together, these types of cells conduct sap through phloem in angiosperms and gymnosperms. Mosses in our story lack these vital vessels.

Plants use several types of cells composed of ground tissue that support their structure and store and produce food. Ground tissue includes parenchyma cells, also called the typical plant cell, which carries out most of the metabolism in plants (Figure 9.13a). For example, in leaves parenchymal cells produce sugars via photosynthesis; in stems and roots they have plastids which store starch; and in fruits, they make up the fleshy part. These cells photosynthesize, produce ATP, repair damaged cells, and make hormones. They carry out most of a plant's activities. Parenchymal cells are also precursors to other more specialized cells, changing into them at certain points in a plant's development. Whenever you see a live plant cell, it is likely a parenchyma cell. In fact, most of a plant is actually dead material; roughly 98% of most very old trees are dead cells.

Sclerenchyma cells are stringy and elongated, with thick cell walls (Figure 9.13b). Plant cell walls have embedded lignin, which provides support. Their irregular arrangement allows plants flexibility so that they can twist and bend. The stringy appearance of asparagus stalks after cooking them is exemplary of this cell type. Sclerenchyma is not alive at maturity, unlike other ground tissue. Sclerenchymal cells have lignin cell walls. Lignin comprises the woody parts of a plant and cannot be digested directly by animals.

**Tracheids**

Elongated cells found in the xylem of vascular plants that conduct the transport of water and mineral salts.

**Vessel element**

A cell type found in xylem.

**Sieve-tube members**

Cells that transport sap through phloem vessels, are alive at maturity, unlike xylem cells.

**Companion cells**

Are specialized parenchyma cells found in the phloem of flowering.

**Ground tissue**

Tissues that are neither vascular nor dermal and support a plant's structure and store and produce food.

**Parenchyma cells**

The typical plant cell that carry out most of the metabolism in plants.

**Sclerenchyma**

Stringy and elongated cells with thick cell walls.

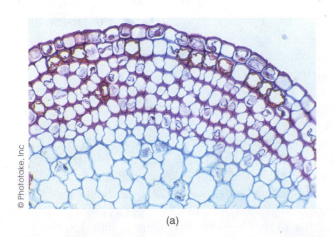

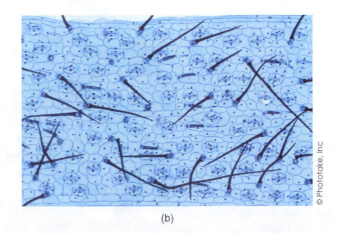

(a)       (b)

**Figure 9.13** Selected cells of the vascular system. a. Parenchyma. b. Sclerenchyma.

**Meristem**

A formative plant tissue responsible for growth whose cells divide to form plant tissues and organs.

**Germinate**

To begin to grow.

**Imbibition**

The process in which germination starts with the massive influx of water into the seed.

**Apical meristem**

Meristems that are found at the tips of roots and in shoot buds to begin primary growth.

**Lateral meristem**

A type of meristem that is found along the sides of stems and roots which gives rise to secondary growth.

**Root cap**

A section of tissue at the tip of a plant root.

**Zone of cell division**

One of the zones of development in which mitosis occurs in a slow but protected manner.

**Zone of elongation**

One of the zones of development in which cells elongate.

**Zone of differentiation**

One of the zones of development in which cells become one of the three types of plant tissues.

## WHY DO VEGGIES CAUSE GAS?

Does eating vegetables cause gas? Most vegetables, including beans and legumes, contain lignin components, such as sclerenchyma, that cannot be digested by intestinal enzymes in humans and most animals. Thus, it takes bacteria in the large intestines in humans, for example, to break down these materials, forming gases: Hydrogen ($H_2$), carbon dioxide ($CO_2$), and methane ($CH_4$). Flatulence develops from the buildup of these gases. The telltale sign of gas pains is a shooting sensation alleviated by movement, as abdominal muscle contractions move gas bubbles.

Many bean varieties contain raffinose oligosaccharides, a substance that is broken down by bacteria, causing significant amounts of flatulence. The presence of raffinose oligosaccharides leads to increased bacterial action and thus more gas. Oddly, the environment in animal large intestines is very similar to early Earth conditions, discussed in Chapter 7. It is anaerobic with the same gases permeating the region. Sulfur gas gives flatulence its odorous quality, probably quite similar to early Earth's atmosphere. Our first ancestral bacteria, the archaebacteria, resembled microbial life in our intestines today, and developed from similar conditions.

# Plant Growth

Plant growth occurs in all regions, emanating from the meristems, which are undifferentiated or unspecialized cells. Seeds remain dormant until a stimulus – usually wearing away of the seed coat through digestion, fire or water – causes a seed to germinate or grow. Germination starts with the massive influx of water into the seed. This process, called imbibition, expands the seed so that it ruptures. In the process, metabolic changes take place within the developing embryo to cause its rapid growth and resulting in a new sporophyte's root and shoot system. Enzymes digest endosperm nutrients, mostly starch, for an embryo to use. Primary growth thus starts right away, within meristems, elongating roots and stems. At the end of primary growth, cells differentiate into the three plant tissues, forming leaves and branches.

Meristems have the potential to divide into any type of plant cell. Mitosis occurs in meristems at very high rates. Apical meristems are found at the tips of roots and in shoot buds to begin primary growth (Figure 9.14). Primary growth is growth in length and gives rise to the three types of plant tissues. Lateral meristems are found along the sides of stems and roots which gives rise to secondary growth. Secondary growth is growth in width, thickening plants when they divide.

Primary growth in plants pushes roots through their soils. A root cap protects root meristems as they grow, secreting a polysaccharide that also lubricates the soil. In roots, there are three zones of development: a zone of cell division, in which mitosis occurs emanating from a quiescent center in the apical meristem, which contains cells that divide in a slow but protected manner; a zone of elongation, in which cells elongate over 10 times to push root tips through the soil; and a zone of differentiation, in which cells become one of the three types of plant tissues. Here, a protoderm gives rise to the epidermis; a procambium layer becomes xylem and phloem; and ground meristem emerges as ground tissue.

Primary shoot growth is very similar to root growth: first there is mitotic growth, then cell elongation and finally differentiation. Shoots contain vascular bundles that

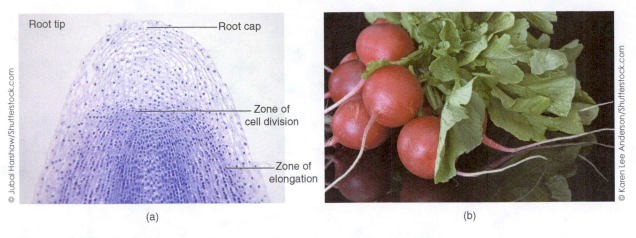

**Figure 9.14**    Root tip structure.

also grow along with them. Stems hold leaves and grow to maximize their sunlight intake. Plant hormones, discussed in the next section, lead these plant growth directions. Stems grow to develop into wood, our most valued nonedible plant product. Stem growth occurs in various plant areas, sometimes underground, as in potatoes and sometimes across the ground, as runners in strawberries. Stem and leaf growth alongside root development is seen in radishes in Figure 9.14b.

Secondary growth is also known as plant thickening. It occurs in all living plant tissues, emanating from two lateral meristems: the vascular cambium produces xylem and phloem, and the cork cambium produces thickened outer coverings of stems and roots. Vascular cambium occurs between xylem and phloem, growing new xylem cells continually toward the inside of the vascular cambium and new phloem cells toward the outside of the vascular cambium. Multiple layers of xylem form sapwood, which is poor at conducting water. However, layers closer to the vascular cambium are good transporters of water.

Cork cambium contributes to the girth of a stem and the external hard layer of bark. It is found outside of the phloem, adding new but dead tissue to the exterior of a stem. This new material is called cork (phellam) which is dead upon production and serves to protect inner layers of stems. Beneath the sapwood layer, at the very center of a tree, the heartwood is composed of dead parenchyma cells, vessel elements, and tracheids. Heartwood provides a support column for a plant but it is not active in a plant's life functions, such as transport of materials (Figure 9.15).

The age of a tree may be calculated by counting the number of rings of xylem produced each year: the thicker the ring, the better the growth in a given year. Cambium layers are cylinders of cells that remain young forever, making new plant tissues for growth. Plants have such longevity because meristems remain able to differentiate perpetually. This is the fountain of youth for trees.

## Transport of Water and Nutrients in Plants

The shock and awe that water and nutrients can move over one hundred feet up an oak tree sparks the simple question: How? Roots absorb water and minerals from the soil, as discussed earlier in this chapter. Roots also exchange gases with the soil, putting out carbon dioxide and taking in oxygen (to enable root cells to carry out cell respiration). Water and minerals are transported upward through xylem from roots because of an upward force or pull. This is called transpirational pull. Loss of water from leaves by evaporation through stomata as discussed earlier, is called transpiration and creates an

**Secondary growth**

Growth in vascular plants emanating from two lateral meristems resulting in wider branches and stems.

**Vascular cambium**

One of the lateral meristems that produces xylem and phloem.

**Cork cambium**

A tissue found in a plant's stem and is responsible for thickening stems and roots.

**Transpirational pull**

The process in which water and minerals are transported upwards through xylem from roots because of an upward force or pull.

**Transpiration**

Loss of water from leaves by evaporation through stomata.

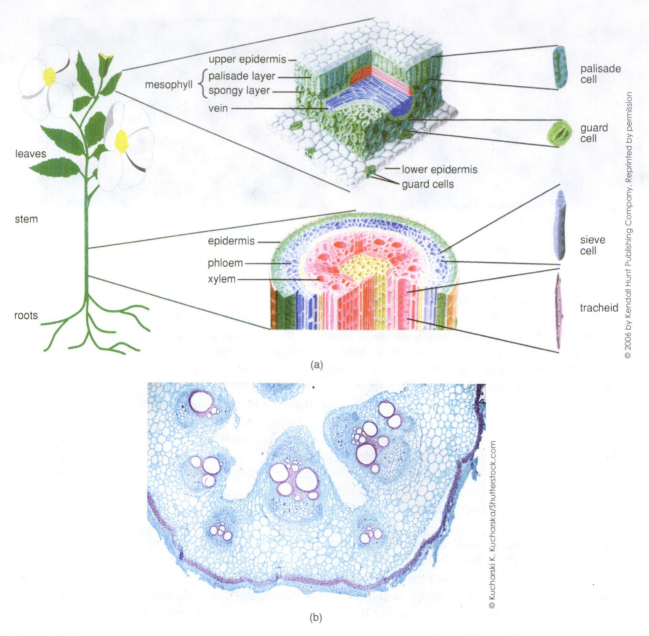

(b)

**Figure 9.15**    The cross section of a stem. a. Xylem and phloem occur together in vascular bundles circling the outer part of the stem. From *Biological Perspectives*, 3rd ed by BSCS. b. This is a pumpkin stem section.

upward force. Water leaving through stomata must be replaced as it evaporates, creating a suction-pull from roots to shoots (Figure 9.16a).

The tendency of water to leave an area is called its **water potential**. Water moves from an area with a higher water potential to an area with a lower water potential. It is measured in megapascals, MPa. The more water that leaves via stomata, the greater the water potential pulls driving water up a plant (Figure 9.16b). Water leaving stomata will lead to regions in xylem that are unoccupied. Adhesive (water–xylem wall attraction) and cohesive forces (water–water attraction) create a line of water from roots to shoots, with pressure from water potential and adhesion/cohesion resisting the force of gravity. Each time a water molecule leaves via transpiration, it must be replaced by a molecule beneath it in the line. This creates a pull from the top down in a plant, suctioning water and minerals into roots.

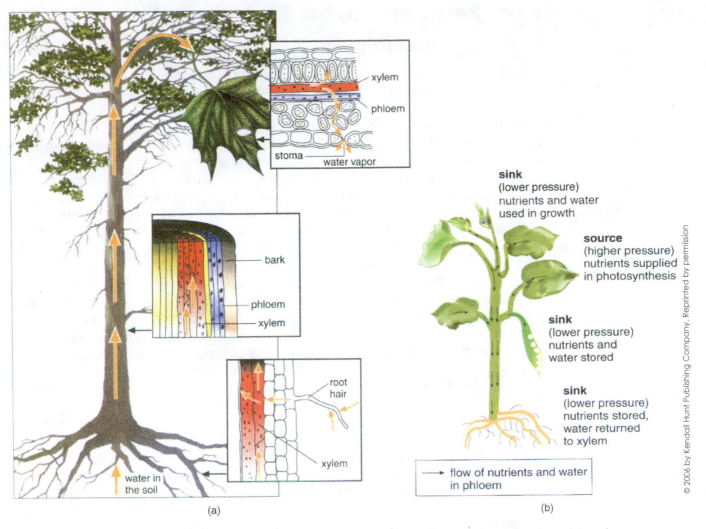

(a)

(b)

**Figure 9.16** a. Transpirational pull in a tree. Water moves through the xylem of a plant driven by concentration differences between soil and stomata. b. Plant nutrients move from source (high pressure) to sink (low pressure), bringing food down the plant. From *Biological Perspectives*, 3rd ed by BSCS.

In xylem, negative water potential exists in leaves as transpiration proceeds, causing a bulk transport of water and minerals up a plant. Transpiration may lose up to 200 liters of water per hour. Each hour, all of this water needs to be replaced. Xylem sap may rise up to 300 feet against the force of gravity. At night, transpiration is low, but roots keep taking in water causing water to continue to flow up xylem. Water droplets (dew) on leaves in the morning are formed from this excess water in plants overnight, causing guttation or droplet formation on leaves. Mosses in our story, which depend on diffusion, avoid this entire process.

In phloem, sucrose moves from source (leaves) to sink (nonphotosynthetic parts of plants) down a concentration and gravity gradient. Transpiration is a necessary evil: plants lose 90% of their water through stomata transpiration. However, it is required to allow gas exchange for photosynthesis and cell respiration. Transpiration also drives the movement of water and minerals up a plant and serves in evaporative cooling of plants; since the hottest molecules of water evaporate in transpiration, leaving a plant. Plant temperatures are lowered between 15 and 20°C in evaporative cooling, preventing high temperatures from denaturing enzymes in plants. In desert plants, the rate of transpiration is less because their enzymes can tolerate heat better, and water loss is the greater threat.

# Plant Responses to the Environment

## Hormones and Tropisms

While plants do not have motility, they do respond in many ways to environmental stimuli. To accomplish this, plants have hormones. Plant hormones are chemicals produced by a one tissue and carried to other tissues to cause a response in them. There are five types of plant hormones: auxins, abscisic acid, gibberellins, ethylene, and cytokinins. Each hormone performs functions to help a plant respond to cues in its environment.

**Phototropism**

A tropism in which the growth of a plant is toward sunlight.

The most important cue from the environment is sunlight. Phototropism is defined as the growth of a plant toward sunlight (see Figure 9.17). It occurs when **auxins**, a type of plant hormone, migrate from the light side to the darker side of a shoot tip. Auxin hormones cause elongation of cells on the darker surface, bending the shoot in the direction of sunlight (Figure 9.19). In general, auxins stimulate plant growth by cell division and elongation in root and vessel formation. Plants also use their hormones to carry out internal cycles throughout a single day or even through a century. Sunflowers turn toward the sun every day they are in bloom, like clockwork. They have an internal clock directing this activity, through the work of hormones. Yearly, the century plant, *Agave Americana* which lives 10–30 years, blooms only once at the end of its life and then dies. It is called a century plant by mistake, but probably due to its long period of time to bloom. Hormones guide these rhythmic workings, often referred to as an organism's biological clock.

**Geotropism**

The growth of shoots upward and roots downward in response to gravity, results also from plant chemicals.

Geotropism, or the growth of shoots upward and roots downward in response to gravity, results also from plant chemicals. Regardless of the orientation of a plant, whether upside down or not, roots grow downward and shoots grow upward in response to gravitational pull. **Abscisic acid**, another plant hormone, directs movement of roots downward, for example. Abscisic acid also serves plants by inhibiting their activity during stressful periods. During harsh conditions of dryness or the onset of winter, abscisic acid inhibits seed germination and stomata opening. Conservation of energy during these conditions is often critical to a plant's survival. Wasting energy on flower production during a spring ice storm might kill a plant.

**Leaf abscission**

Loss of leaves.

Most plants lose leaves in response to cold weather and freezing water. Plants do not detect the cold or freezing water, however. It would be bad if a cold spell could trick a plant into dormancy. Instead, each year, plants detect decreased sunlight during the changing seasons, causing them to lose their leaves. A loss of leaves is called leaf abscission. It enables plants to conserve water that would otherwise be lost through transpiration. While it is wasteful to make and lose leaves, plants are thus able to survive harsh conditions.

**Thigmotropism**

Any plant growth response to touch.

Any response to touch, called thigmotropism results in growth changes in plants. For example, climbing plants such as beans or ivy adhere to surfaces in response to their

**Figure 9.17**    Phototropism: A bean plant bends toward the sun. It is a way that plants "move" to obtain the necessary resources in the environment.

**Figure 9.18** A field of red cabbage, *Brassica oleracea*. Red cabbage contains the red pigment anthocyanin, which acts as a pH indicator. It turns red in acids and blue-green to yellow in alkaline solutions.

physical contact. **Gibberellins** are hormones that stimulate growth of plants in many ways, including their response to touch. Gibberillins work in several capacities: elongation of cells in roots and shoots, stimulation of mitosis in apical meristems, inducing seed germination, blooming of flowers, and enlargement of fruit size. These hormones are analogous to human growth hormone, and can cause enormous sizes in plants. Note the healthy, growing red cabbage, fed gibberillins by farmers, in Figure 9.18.

When fruits ripen together, as seen in a set of bananas, ethylene is at play. **Ethylene** is a gas that causes fruit and vegetable ripening in almost every part of a plant. Ethylene gas permeates through a pile of fruit leading to rapid ripening in all of them. Some fruits – strawberries, for example – are resistant to ethylene gas.

Cell division and growth are also initiated by **cytokinins** in almost every tissue in plants. Cytokinins are plant hormones that work in concert with auxins to stimulate and sustain growth in plants throughout their lifetimes (Figure 9.19). Cytokinins also stimulate new branches from lateral buds and seed germination.

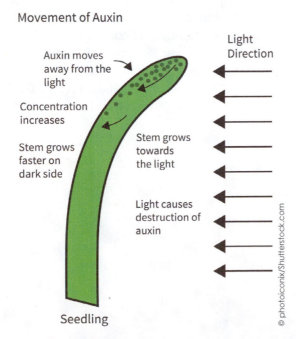

**Figure 9.19** How auxin works to bend plants toward light.

## Plant Defenses

In a plant's defense, it cannot move and cannot directly attack it stalker. It does, however, have a host of strategies, depending upon the plant, to combat herbivores. Some plants are carnivorous, such as the Venus fly trap, described earlier in the text. It works by trapping insects, a frequent friend (pollinator, as you recall) and enemy (herbivore) of plants, in a chamber, secreting enzymes to digest the fly's body. However, this example is not only a defense, but satisfies the fly trap's nitrogen requirement for the synthesis of proteins, a component of amino groups.

Other defenses manifest as chemicals in over 3,000 plant species. Consider *cannabis sativa*, the plant from which marijuana is derived. It contains a hallucinogen, THC or delta-9-tetrahydrocannabinol, which deters animals from continued eating the plant by causing them to be disoriented. Obviously, this chemical gives the characteristics of marijuana's effects. However, they occur due to plant defenses instead of human needs.

Many mechanical defenses also help plants to keep herbivores at bay. Raspberries and blackberries have thorny branches to deter animals from casually eating them. Waxy leaves and saps on stems have been shown in studies to cause insects to glide right off. For example, thick cuticles on the prickly pear cactus prevent water from forming droplets, within which fungi spores develop. Mimicry such as that of the Passiflora plants (see Chapter 7) is a mechanical defense mechanism as well. The yellow spots on their leaves resemble the eggs of the Heliconius butterfly, causing butterflies to lay their eggs elsewhere. While not always fail-safe, plant deterrents help them to survive herbivores.

## Summary

Plants play an integral role in our ecosystem and in human society. They emerged from oceans roughly 475 million years ago as adaptations of green algae. While on land, plants diversified into bryophytes and tracheophytes, with unique characteristics in each division. Bryophytes, including mosses, have simple structures and small sizes, producing spores but lacking a vascular system. Tracheophytes are diverse, usually tall, and always more complex than bryophytes, with seedless and seed varieties. Some tracheophytes, including gymnosperms and angiosperms, use seeds to reproduce. Gymnosperms include conifers that comprise many of our evergreen forests. There are many species of angiosperms, all of which use flowers and fruits in reproduction. Plants, unable to move as adults, respond to their environmental changes through the use of hormones. Plants protect themselves by chemical and mechanical style defenses, using trickery at times and toxic substances at other times. Plants remain a vital part of our ecosystem.

## CHECK OUT

**Summary: Key Points**

- Plants photosynthesize using sunlight to produce food for themselves and for the animals eating them.
- Plants evolved from green algae by developing roots, shoots, and methods such as seed formation to prevent them from drying out.
- Lower plants without a vessel system are bryophytes, such as mosses.
- Higher plants or tracheophytes developed transport vessels to help them to grow taller, which aides them in obtaining sunlight and dispersing their seeds.
- Plants are able to asexually reproduce, but attain genetic diversity through sexual methods.
- Dermal plant tissue covers plants, vascular tissue forms vessels for transport, and ground tissue is used for daily life functions of plants.
- Plants use hormones such as auxins to grow in response to light as well as develop new tissue.

## KEY TERMS

| | |
|---|---|
| alternation of generations | phototropism |
| angiosperm | pollen grains |
| anther | pollination |
| apical meristem | prothallus |
| carpal | root cap |
| companion cells | root system |
| cork cambium | sclerenchyma |
| cotyledon | secondary growth |
| dicot | seed |
| dioecious | shoot system |
| endosperm | sieve-tube members |
| fibrous root | sporophyte |
| filament | stamen |
| gametophytes | stigma |
| geotropism | style |
| germinate | tap root |
| ground tissue | thigmotropism |
| gymnosperm | tracheids |
| imbibition | tracheophyte |
| lateral meristem | transpiration |
| leaf abscission | transpirational pull |
| meristem | xylem |
| monocot | vascular cambium |
| monoecious | vascular tissue |
| ovary | vessel element |
| ovule | zone of cell division |
| parenchyma cells | zone of elongation |
| phloem | zone of differentiation |

# Multiple Choice Questions

1.  Which division of plants is LEAST likely to provide wood for human societal use?

    a.  angiosperm
    b.  bryophyte
    c.  gymnosperm
    d.  conifer

2.  Which group of organisms most likely aided in the first transition of primitive plants from aquatic to land environments?

    a.  animalia
    b.  sarcodina
    c.  fungi
    d.  monera

3.  The shoot of a plant is composed of all of the following EXCEPT:

    a.  a cotyledon
    b.  an endosperm
    c.  a leaf
    d.  a quiescent center

4.  The ferns belong to the _____ division of plants:

    a.  bryophyte
    b.  seed tracheophyte
    c.  seedless tracheophyte
    d.  gymnosperm

5.  In a cluster of mosses growing stalks that produce diploid grains, a stalk is called:

    a.  a gametophyte
    b.  a prothallus
    c.  a sporophyte
    d.  a quiescent center

6.  Which represents a logical order, from early to later, in the evolution of plants?

    a.  green algae → bryophytes → gymnosperm → angiosperm
    b.  angiosperm → gymnosperm → green algae → bryophyte
    c.  gymnosperm → angiosperm → bryophyte → green algae
    d.  bryophyte → gymnosperm → angiosperm → green algae

7.  The anther within an angiosperm is analogous to:

    a.  testes in humans
    b.  ovaries in humans
    c.  buds in yeast
    d.  daughters in prokaryotes

8. Which term includes all of the others?
   a. tracheophytes
   b. ferns
   c. gymnosperms
   d. conifers

9. In question #8, which process helps these organisms to obtain genetic dispersal and variation in their populations?
   a. photosynthesis
   b. pollination
   c. asexual reproduction
   d. species specificity

10. Which plant hormone is most responsible for growth of a group of geranium plants toward light on a window sill?
    a. auxins
    b. ethylene
    c. gibberillins
    d. abscisic acid

## Short Answers

1. Describe two ways in which plants benefit humans. List two ways in which plants are harmed by humans. Be sure to list and describe each.

2. Define the following terms: gymnosperm and angiosperm. List one way each of the terms differ from the other in relation to their 1) morphology; 2) diversity; and 3) reproduction methods.

3. A plant bears cones, contains pollen, and grows to over 50 feet. Make use of the characteristics of plants in this chapter to classify this organism. Why did you place it in its group?

4. A plant becomes wilted, loses its leaves, and enters a dormant state. Which plant hormone is likely involved in this process? What factors might determine whether a plant enters dormancy? In nature, what is the most important factor causing leaf abscission? Why?

5. During sexual reproduction in angiosperms, seeds have unique properties that help plants survive on land. List three characteristics of seeds that make them more efficient than spores. How are these characteristics helpful for seed plants in their survival on land?

6. List and draw the male and female reproductive structures in flowers. How are they different from each other? How is their arrangement important in limiting inbreeding?

7. Explain the process of transporting water within a vascular system. Use the following terms in your answer: xylem, water potential, transpirational pull, adhesion, stomata.

8. Transpiration is considered a necessary evil in plants. Explain why this is so.

9. What type of plant tissue is a parenchyma cell? How is it important in a plant's functioning?

10. Draw a diagram showing the underside of a leaf, with stomata open. Be sure to label guard cells and stomata in the diagram. Indicate the direction of water flow when stomata are open.

## Biology and Society Corner: Discussion Questions

1. Plants play an important role in our society, comprising a large portion of our trade, both import and exports, for our economy. Research the importance of heartwood in developing products for human use. Name one region where timber production led to negative ecological consequences. Could it have been prevented?

2. The movement to limit bad fats and increase good fats in our diets has led to many nutritional claims. Choose a type of plant oil that you consume, either through cooking with it or as a component of your dietary intake. Research the nutritional chemistry of the oil and make a recommendation on its benefits or drawbacks for your own health.

3. Ethylene gas naturally ripens fruits. Industrial calcium carbide is used to ripen fruits but contains arsenic and phosphorous, potential human health hazards. Research the use of calcium carbide to determine the consequences of its use. Should you or your loved ones avoid food ripened with calcium carbide? What are the current laws in the United States regarding its use?

4. Colony Collapse Disorder (CCD) is a disease affecting insects that live in communal colonies, most in the hymenoptera order. Many honey bees have died off as a result of CCD. Explain how this disease may affect genetic diversity and reproduction in plants. How may CCD have eventual impacts on human society?

5. Our opening story in this chapter showed the importance of bryophytes in flood control. Wetlands are areas designated with certain flora and animal life that make them unique. Research the geographical areas in your local neighborhood to determine where wetlands are designated. What plants and animals are found in wetlands? Research those wetlands and make recommendations to preserve those regions. Are there any societal forces endangering those wetlands?

Figure – Concept Map of Chapter 9 Big Ideas

# Moving on Land and in the Sea: Animal Diversity

**10**

A beaver family of four

Little John shares the work of the beaver

Beaver(s) are usually working hard, here shown is a Beaver family lodge built by hard work

This beaver prepares wood material for the dam

© Kendall Hunt Publishing Company

Beavers hard work changes the environment–this dam is impressive

## CHECK IN

**From reading this chapter, you will be able to:**

- Explain how animals play their roles in the Earth's ecosystem and in relation with human society.
- Trace the evolution of animals, from simpler to more complex organisms, explaining the four ways to classify animals
- Describe and Define vertebrate, invertebrate, Cambrian explosion, endoskeleton, exoskeleton, exotherm, endotherm, and chordate and use them appropriately to explain the four ways to classify animal phyla.

## The Case of the Homey Homeotherm

The stone patio had taken months to build. The farmer worked each day, meticulously placing each flat rock into the soil. He obtained the rocks from the river bed right next to the area on which he was building the patio. The stone patio had to be solid, so that the river would not wash it away in a storm. The rocks were heavy and required two people to lift them.

His only help was his sister's 11-year-old son, little John visiting from the city. Little John was not ambitious, seemed to argue about helping out, and was listless as his uncle worked. Little John had gotten into some real trouble with the law, hanging out back in the city with the neighboring kids. They sent Little John to get away from a bad neighborhood element. His sister thought that Little John would benefit by staying the summer in the countryside and helping with chores; but little John hated being there.

Digging was difficult, but the farmer laid each piece with care, solidly into the dirt. It was a beautiful patio which would be his grandest accomplishment. "Stones last forever, Little John," explained the farmer. But Little John did not care – he wanted to go home, back to his friends. The farmer was very unhappy with his nephew and gave up on him. "He'll wind up in trouble before he even grows up," he told his sister – "little John is a hopeless case."

A new neighbor moved in next door and made the situation all the more difficult. "There goes the neighborhood," remarked Little John, "beavers make terrible neighbors." After the beaver arrived, in less than a month's time the beaver family, a couple with six beaver babies, was creating a swamp around the stone patio. Quickly, the beavers used their front teeth to gnaw and fell trees from the farmer's surrounding forest. Beaver teeth grow continuously through their lives, sharpening as they chew and strong enough to build a three-foot dam across the river. "If this keeps up, the beaver dam will be 10 feet high and my stone patio will be underwater," thought the farmer, "I am going to have to kill the beavers."

Soon the beavers built a two-room lodge using sticks and mud. The beaver couple shared all of the work, from gathering plants and berries for food to felling trees and using their tails to pack their lodge with mud.

Little John became intrigued with the beaver lifestyle, reading about their ways. He found out that beavers are herbivores, monogamous (one spouse) through their lives, and have excellent hearing and smell to compensate for poor eyesight. They are the largest in the order Rodentia, with two major species: one in North America, *Castor Canadensis,* and the other in Eurasia, *Castor fiber*. They maintain their properties well, putting a great deal of effort into their territory. They use scent glands to mark their borders and an alarm call by flapping their tails against the water to let other family members know an invader is present. Beavers are family oriented and live in their lodge together with kin. They recognize kin by anal gland scents, helping them to get along as a family unit. They are homeotherms, meaning they maintain a stable internal body temperature, making their lodges important in keeping them warm.

European exploration of North America was in part based on trapping beavers for their skin to make clothing and their glands, which had perfume and medicinal purposes. So successful were the hunters that they were hunted to near extinction by the early 1800s. Before European settlement, the North American beaver population was at up to 150 million. There are now between 6 and 12 million beavers left in North America, a 90% decline from their peak. In Europe, their numbers also dwindled to the point of extinction at some points in history. Beavers are being reintroduced into many areas of Europe to repopulate them. After 200 years, beavers returned to New York City, making dams in the Bronx River.

The beaver children watched and learned from their parents all summer long. Surprisingly, Little John watched too, as the beavers worked and worked, building up a home for themselves. Little John began to help his Uncle around the farm a good bit more.

"I think I want a home of my own one day, like the beavers," said Little John, "don't kill them." At that moment, the farmer realized that Little John was not hopeless. The beavers set a good example for making a respectable life. "The beavers can stay as long as they like," responded the farmer to Little John.

**Homeotherm**

Organisms that maintain a stable internal body temperature.

---

## CHECK UP SECTION

This story indicates the complexity of behavior that may be seen in the animal kingdom. A beaver's place within our ecosystem is interesting – second only to humans, beavers modify their environment more than any other organism, building dams and creating wetlands and aquatic systems.

Study the biology of the *Castor* genus to determine its relationship with the ecosystem and human society. Research the island of Tierra del Fuego, in southern Chile, which experienced beavers as an invasive species in the past century. Explain how they are coping with the beaver population explosion. What factors are leading to this problem? What suggestions would you make to help protect their communities?

---

# Unity and Diversity of Animals

When most of us think about animals, we imagine mammals such as the beaver of our story. In reality, over 90% of animals are invertebrates – meaning that they do not have a backbone – and 75% of all animals are arthropods, of which most are insects (see Figure 10.1). In fact, 25% of all known species are beetles, the most diverse group of animals. Animals having a backbone, called vertebrates, include the more complex organisms such as humans, whales, beavers, and frogs. More than 1 million species of animals have so far been discovered. The branch of biology that is dedicated to the study of animals and their characteristics is zoology.

**Invertebrates**

Animals that lack a backbone.

**Vertebrates**

Animals having a backbone.

**Zoology**

The branch of biology that is dedicated to the study of animals and their characteristics.

**Diversity of Life on Earth**

| some major groups of organisms | insects | other animals | higher plants | fungi | protists | bacteria | viruses | total identified | total estimated to exist |
|---|---|---|---|---|---|---|---|---|---|
| number of species identified (approximate) | 950,000 | 281,000 | 270,000 | 72,000 | 40,000 | 4,000 | 4,000 | 1,750,000 | 4–112 million (working number = 14 million) |

**Figure 10.1**    This chart shows the diversity of animal species

Animal fossils emerged over 600 million years ago, probably from a single protist, with rapid diversification following in a short time frame – so short that it represents only 1% of life's history on Earth. These rapid increases in animal species occurred during a period in geologic time called the **Cambrian period,** and the burst of diversification is known as the Cambrian explosion. Since this time period, through roughly 530 million years, many species of animals developed and changed. The result today is roughly 30 different animal phyla.

While there are many differences in structures and lifestyles among the animals, they have amazing similarity. All animals have common methods of development, with similar structures and functions of organs and organ systems.

This chapter tours only nine of these phyla, which include almost all of the known animal species. Their adaptations in each animal phylum represent a branch taken in evolution to adjust to environmental changes. A sea urchin developed its structures to be a successful sea urchin and a beaver developed its ways of life to be successful in rivers.

Features unique to each phylum show how animals changed to suit their unique needs. No one phylum is better than another, and humans are not the "highest" species in development. An organism's evolutionary success is determined by its survival on Earth, which remains to be seen. Extant organisms are those that exist today and extinct organisms are those that have died off. Any species still existing in today's environment are thus far successful, because they is still here despite a competitive and harsh world.

Animals' sheer numbers and omnipresence on Earth attest to their ecological importance. The beaver in our story modifies its environment greatly, showing how even one family of animals may have significant ecological impacts. An excellent example of biodiversity and the importance of each species can be found simply by scooping a spade of dirt from one's garden: if you examine that sample carefully you might find earthworms, beetles, nematodes, spiders – all living within a small cube of land. Even a tiny area of the ocean might have lobsters, crayfish, eels, and sharks living around a sponge population.

Animals trump all other species in terms of their visible, unique differences, from beavers and parrots to monkeys, rams and the butterfly (see Figure 10.2). This vast diversity is attributed to their complex needs to live on land. All animals need to obtain food from other organisms, as they cannot carry out photosynthesis. They must therefore adapt ways to obtain food from their respective environments.

Thus, all animals are multicellular, heterotrophic organisms that have motility. They need to be able to move to obtain a mate, find food, and defend themselves. To meet these needs requires a complex body organization: obtaining food, a mate, and a proper environment suitable for living is no easy task. Animals often have specialized cells and

**Cambrian explosion**

A evolutionary event during which rapid diversification of multicellular animal life occurred.

**Extant**

Are organisms that exist today.

**Extinct**

Are organisms that have died off.

**Figure 10.2** Animal diversity: Each of these animals share common characteristics of life, but carry out different ways of performing those life functions. Birds and butterflies are aquatic while camels and rams live on land.

tissues that enable them to carry out their unique life functions. In most phyla, cells are coordinated, often in the form of muscle and nerve tissue, to allow motility and respond to stimuli. Animal cells lack a cell wall because structure and height given by walls is less important in animals than in plants. Instead, it is more important that animals are able to move about on land and in the water, without the encumbrances of a rigid cell wall. Their body plan differs among the animal phyla, but the same goals apply – eating, mating, and defense against predators.

# Four Ways to Classify Animals

While all animals evolved from a common ancestor – a single-celled, protist – each species has certain features. These may be classified according to a simple system. There are four key ways to divide animals into their groups (refer to Figure 10.5).

## Specialized Cells

**Specialized cells**

Cells that carry out a particular function.

First, animals may be either composed of specialized cells forming tissues or contain no specialized cells. The simplest body plan is found in the phylum Porifera or sponges, which are merely colonies of cells living in association with each other. Sponges do not have specialized cells; sponge cells do not coordinate activities but remain together, operating as solitary units of life. They are aggregates of cells which work well as a colony but they do not comprise a whole, unified organism. Sponges will be discussed as our first phylum. All other animal phyla are characterized by complex tissues, operating together in a coordinated way. Humans, for example, discussed in the next unit of this text, have a specialized, complex set of tissues – muscles, nerves, and bones – to enable movement and other life functions (see Figure 10.3).

## Symmetry

**Radial symmetry**

Symmetry that describes any organism that is structured so that when a line is drawn down the middle of it, at any orientation, both sides are identical.

Second, animals develop a shape that has either radial symmetry or bilateral symmetry (see Figure 10.4). Radial symmetry describes any organism that is structured so that when a line is drawn down the middle of it, at any orientation both sides are identical. Animals with radial symmetry include mostly slow moving or floating organisms such as sponges and sea anemones. Animals with bilateral symmetry, meaning that they are roughly identical upon surface observation when a line is drawn down their middle, include faster moving organisms. These include most of the more complex species, such as frogs, fish, and humans. They respond better to stimuli than less complex organisms, giving them improved means of hunting, mating, or escaping from predators.

**Bilateral symmetry**

The property of being roughly identical upon surface observation when a line is drawn down their middle.

## Molting

**Molt**

To shed the outer covering.

Third, some animals molt, or shed their external exoskeleton or outer covering as they grow, forming a new one to fit their new size. Molting organisms include spiders, lobsters, crayfish, and insects. While it appears a waste of energy to molt off an exoskeleton, this process enables growth while at the same time protecting organisms from predators.

**Exoskeleton**

A rigid outer covering of an animal.

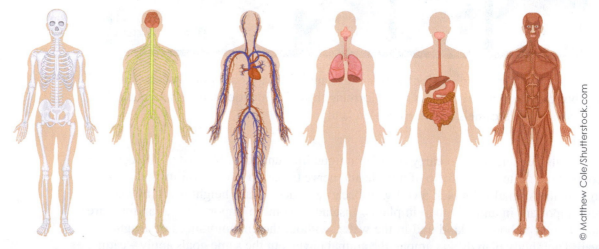

© Matthew Cole/Shutterstock.com

**Figure 10.3**   Humans, like other animals, have specialized structures and exhibit great complexity.

(a)           (b)

**Figure 10.4** All animals (except sponges) have either radial or bilateral symmetry. a. This butterfly exhibits bilateral symmetry. b. The sea urchins exhibit radial symmetry.

All other animals grow continuously as they add mass to their bodies. Earthworms, grasshoppers, squids, humans, and dogs are all animals that grow continuously. They sometimes have an endoskeleton or internal skeleton as a support system to give them structure.

## Body Cavity Formation

Fourth, animals are divided by the manner in which their body cavities form. Bilateral animals mature either by first forming a mouth or by first forming an anus during their embryo development. Organisms forming their mouth first are called protostomes, which include animals with simpler body plans such as flatworms, roundworm, and insects. Those animals forming their body cavity from the back, or anus region, are called deuterostomes. These include starfish, monkeys, and humans, which have the more complex body plans among animals. The developmental stages of these two groups are best observed as embryos; it is more difficult to see this development as adults. Body cavity development separates animals based on their evolutionary lineages. Deuterostomes are much more closely related with each other than with protostomes and vice versa.

## The Major Phyla

Animals consist of nine major, separate evolutionary lineages, as shown in Figure 10.5. The rest of this chapter examines each phylum in greater detail, showing the changed characteristics for each group in relation to their respective environments.

To start, beavers in our story are classified as mammals, the final phylum to be discussed. Beavers have specialized cells and bilateral symmetry; they do not molt but grow continuously, and form their anus first during development, making them deuterostomes.

## Porifera: The Scattered Sponges

Unlike the clever and hard-working beaver, which is clearly an animal, sponges may appear almost nonliving or plant-like to the casual eye. Sponges were once placed in the subkingdom Parazoa, which means "besides the animals," due to their evolutionary and physical differences from the other animal groups. They are sessile as adults and appear to sit in one spot for most of their life cycle. They do move as juveniles to colonize new areas and are heterotrophic. However, while they are the least animal-like of any members of the animal phyla, they are indeed animals.

**Endoskeleton**

Internal skeleton that acts as a support system to vertebrates.

**Protostomes**

Organisms that form their mouth first.

**Deuterostomes**

Animals belonging to the group Deuterostomia; in which the body cavity first forms from the back, or anus region.

**Sessile**

Immobile.

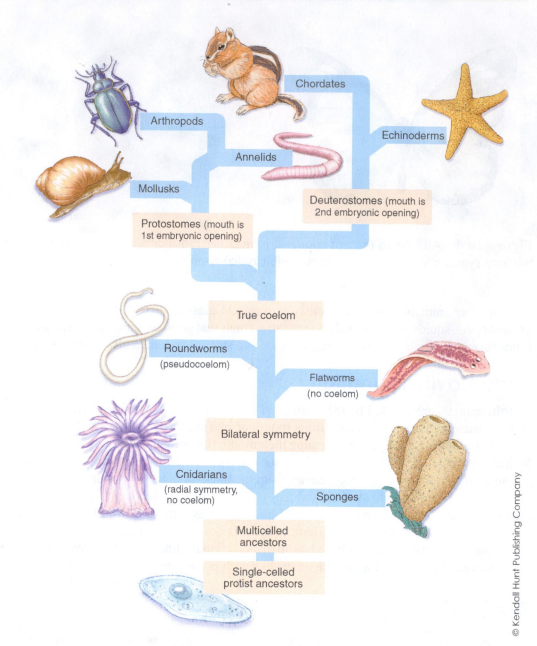

**Figure 10.5**   Phylogeny of animals: classification scheme based on the four characteristics of animals showing all nine phyla branched.

**Porifera**

A phylum of aquatic invertebrates that comprise of sponges.

**Epidermal cells**

A type of sponge cell that covers and protects sponges.

**Collar cells**

A type of sponge cell that has beating flagella move water through the internal cavity of the sponge.

**Amoebocytes**

A type of sponge cell that transports food through the sponge body.

Sponges, in the phylum Porifera, represent a large group of organisms that exist as colonies, aggregates of unspecialized cells. These cells lack coordination, are sessile for most of their life cycle, and have no symmetry of body plan, as described earlier in the chapter. Sponge cells are hollow tubes that contain pores that filter out food from the water passing through their bodies.

Sponge cells are not, however, haphazardly arranged. Sponges organize into sets of colonies, based on their particular species type. While sponges have no specialized tissues, they contain three types of cells to carry out life functions: epidermal cells cover and protect sponges; choanocytes or collar cells, which have beating flagella that move water through the internal cavity of the sponge; and amoebocytes, which transport food through the sponge body (see Figure 10.6). As microscopic food such as algae and bacteria travels through the sponge cavity, it becomes trapped by a sticky, gelatinous mucous on the surface of collar cells. Digestion occurs separately in each sponge cell. Sponges lack a transport system, so movement of digested food via amoebocytes is their only circulation method.

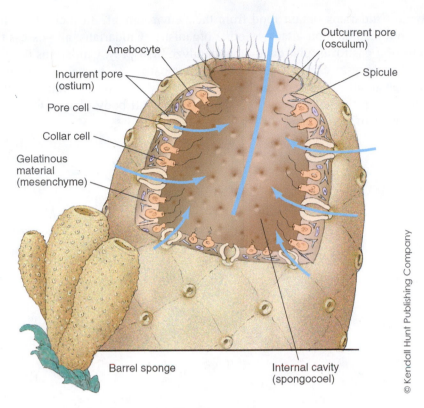

Ameboycte

Incurrent pore
(ostium)

Pore cell

Collar cell

Gelatinous
material
(mesenchyme)

Outcurrent pore
(osculum)

Spicule

Barrel sponge

Internal cavity
(spongocoel)

© Kendall Hunt Publishing Company

**Figure 10.6**     Sponge body cells.

Reproduction in sponges occurs either asexually or sexually. Fragmentation, a simple breaking off of a piece of the sponge, is usually the asexual route for reproduction. Alternatively, sponges are hermaphrodites, meaning that they have both male and female reproductive parts. However, only one sex is active at any one time, preventing self-fertilization. When a male-acting sponge produces sperm, it swims to the female part of a female-acting sponge. In the next phase, sponge larvae are free-swimming, the only stage at which they are motile. During this period, larvae float to find a new home to grow into future colonies.

Sponges, comprising 5,000 different species, were used in a number of ways by humans in the past. The hollow interiors of many sponges are absorbent and soft, so they were used as shock absorbers in army helmets during medieval times, and as cleaning and painting products in more recent memory. In the past, the absorbent quality of sponges made them useful in house cleaning activities. Our modern household sponges do not derive from living organisms, however, because sponges are scarce due to overfishing. Instead, kitchen and bathroom sponges are now made of manufactured materials.

## Cnidarians: Creatures with an Open Cavity

Cnidarians include jellyfish, sea anemones, hydras, and corals, which all contain an open body cavity. This cavity is called a coelenteron, or hollow cavity open to the outside environment, in which digestion occurs. Those containing such a cavity are called coelenterates. When we think of Cnidarians, we think of jellyfish, which are really not fish. Instead they belong to the phylum Cnidarian.

Cnidarians are among the most poisonous of all animals. If you have been stung by a jellyfish on the beach, you are familiar with its venom. Their gastrovascular cavity or "hollow gut," also known as a coelenteron, has only one opening. Within this cavity poisons and enzymes are secreted to carry out external digestion.

**Hermaphrodite**

A person or animal having both male and female reproductive parts.

**Cnidarian**

An aquatic invertebrate that comprises coelenterates.

**Coelenterons**

The open body cavity present in Cnidarians and opens to the outside environment, in which digestion occurs.

**Cnidocysts**

Stinging cells present in Cnidarians.

**Nematocysts**

Barbed threads found in tentacles of Cnidarians.

How do Cnidarians obtain food from their environment? In their sting, they also have poison to help in their attack on other organisms. Cnidarians, all possess tentacles at the ends of a gastrovascular cavity that paralyze their prey. Cnidarians have stinging cells called cnidocysts, each containing a set of nematocysts, barbed threads that thrust outward when another organism touches them. Usually a poison accompanies the barbed thread, engulfing its prey. Figure 10.7 shows the overall body plan of Cnidarians and varied examples of Cnidarians.

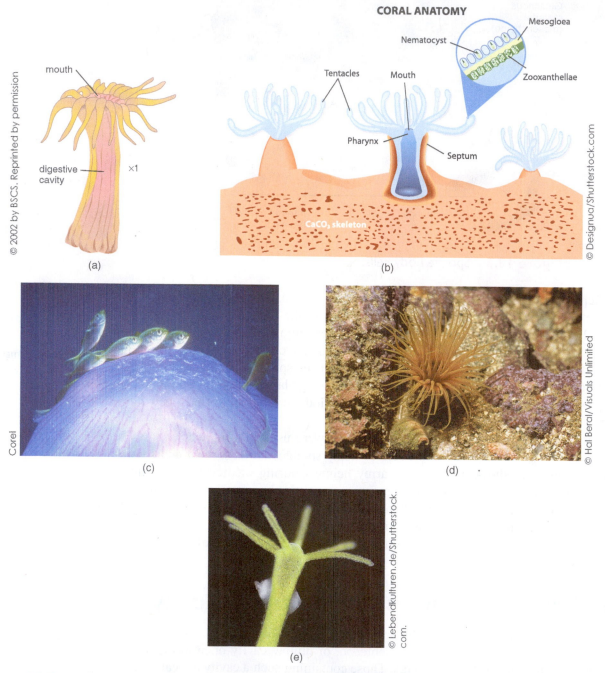

**Figure 10.7**    What do cnidarians look like? a. Radial symmetry of the cnidarian body plan; b. Cnidarian use stinging cells with nematocysts to attack prey. When cnidarian stinging cells are stimulated, they discharge toxic substances and nematocysts which paralyze prey; c. A jellyfish from the Red Sea; d. Burrowing sea anemone, *Pachycerianthus*; e. *Hydra*. a. From *BSCS: An Ecological Approach*, 9th Edition by BSCS.

Cnidarians are more coordinated than sponges. They contain a nerve network, which is a set of nerve cells to help them respond to stimuli. Cnidarians sense the outside world using their nerve networks. The nerve network does not have a centralized area (or brain) to process information, however. It acts only to respond to external stimuli. This system benefits Cnidarians evolutionarily, to enable them to be effective heterotrophs, moving and responding to stimuli to obtain food and defend themselves.

Cnidarians all have radial symmetry, digestion in their open body cavity, and consist of two layers, an ectoderm or outer layer and an endoderm, an inner layer. In between the two layers is a gelatinous filling called a mesoglea or "middle jelly" layer. These layers surround their gastrovascular cavity.

Cnidarians have both sexual and asexual reproduction, like the sponges. The life cycle of Cnidarians has two stages: the polyp and the medusa stage. During the polyp stage, cnidarians are sessile, and in the medusa stage they have movement. Polyp or medusa stages may last almost throughout a Cnidarian's life cycle or may comprise only a short period. Jellyfish are able to move through most of their lives in a medusa form, but sea anemones move very little and remain in a polyp form for most of their lives (Figure 10.8).

## Jellyfish

The "cup animals" which have a central cavity making them appear as cup-like, comprise mostly jellyfish. They have the characteristic stingers on their end tentacles. They are ominous in movies and after we get stung. They range in size from 2 cm in diameter to over 15 m, including tentacles, in the case of the Lion's Mane Jellyfish.

Their nerve networks respond to organisms surrounding them, often leading to their discharge of poison from their cnidocysts. A jellyfish sting may lead to serious harm and even death in humans. Often, multiple bites from the same jellyfish occur because of the many tentacles that it possesses.

**Nerve network**

Set of nerve cells that help Cnidarians respond to stimuli.

**Ectoderm**

The outermost layer of a Cnidarian.

**Endoderm**

The innermost layer of a Cnidarian.

**Mesoglea**

The gelatinous filling found in between the two cell layers in the bodies of Cnidarians and sponges.

**Polyp stage**

The stage in which Cnidarians are sessile.

**Medusa stage**

Cnidarians in their free swimming stage.

**Jellyfish**

Free-swimming marine creatures that have a central cavity making them appear as cup-like.

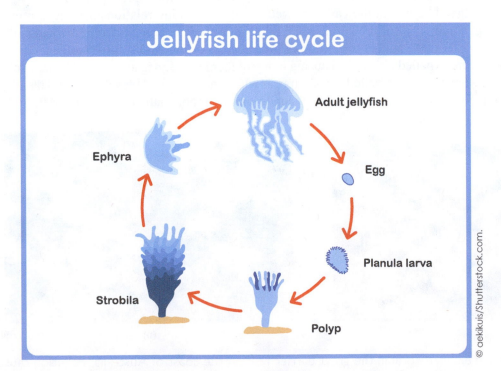

**Figure 10.8** Cnidarian life cycle: Obelia, a marine colonial organism.

## Sea anemones

Sea anemones appear as flower-like tubes, with their colors depending upon the pigments contained within them. They are not pretty flowers, however; they are carnivorous animals that sting their prey. Their tentacles move other organisms into their open body cavity, where they digest the prey. A sea anemone body cavity is divided into vertical chambers, each allowing digestion to occur separately. Sea anemones are often used in studies of fertilization because it is relatively easy to stimulate their production of gametes.

## Hydras

One of the most studied classes of cnidarians contains the hydra. Hydras are medusa-like, with a set of tentacles on the outside of their coelenterate opening. Hydras have a gland cell that secretes digestive enzymes into their body opening. Another digestive cell, the nutritive cell, uses a flagellum to mix food. Their pseudopods (false feet) extend outward from the hydra to absorb the digested nutrients. Hydras reproduce in part through an asexual process called budding. In budding, a new smaller hydra grows from the parent and falls off to start a new life. Hydras also have complex movement, such as gliding or somersaulting, coordinated by a nerve network guiding simple muscular movement.

## Corals

Corals live mostly in the polyp phase within large colonies composed of limestone skeletons. Their appearance is pretty, but they have tentacles, like other cnidarians, that sting and capture prey. Corals are cnidarians that secrete calcium carbonate (limestone) as their outer covering. This hard exterior gives corals their characteristic appearance. As they die, limestone layers build up, forming a complex structure. As discussed in another chapter some of the largest, most diverse ecosystems are composed of coral reefs. Beneath the sea, the 2,000-km long Great Barrier Reef, off the coast of Australia, is a complex of corals. It is so large that it may be seen from outer space.

Algae live within the cavities of corals in a symbiotic relationship. Algae provide oxygen and food for the coral, while corals afford algae a protected home and carbon dioxide to carry out photosynthesis. As temperatures increase in the changing climate, algae are expelled by corals, causing many of them to die. Algae give corals their colors and when they are expelled, corals appear white and are called bleached. Coral bleaching is a telltale sign that environmental conditions are problematic (see Figure 10.9).

### Sidebar glossary

**Sea anemone**

Water-dwelling animals that are brightly colored and fix themselves onto rocks.

**Hydra**

Freshwater organisms with a set of tentacles on the outside of their coelenterate opening.

**Budding**

A form of asexual reproduction in which new organisms develop from a bud as a result of cell division at one specific site.

**Corals**

Marine invertebrates that live in large colonies composed of limestone skeletons.

**Calcium carbonate**

A naturally occurring chemical compound, making up by coral skeletons.

**Coral bleaching**

The loss of algae from corals, and resulting coral death.

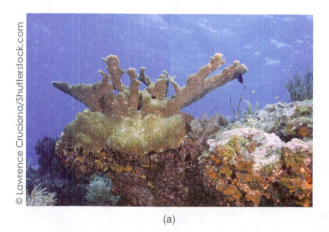
(a)

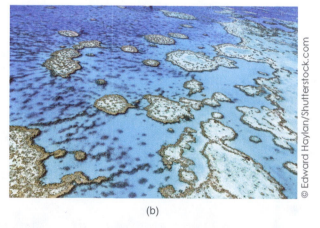
(b)

**Figure 10.9**    a. Bleached elkhorn coral in the great barrier reef off the coast of Australia. Pollution causes harm to coral reefs throughout the world. b. Bleaching at the Great Barrier Reef, Queensland, Australia.

Compared with mammals, such as the beaver in our story, cnidarians are quite simple. However, their simple design is adequate for their life functions. While sponges are collections of organisms, each cnidarian independently works to survive and respond to its environmental conditions.

# Worms

There are three phyla of worms: flatworms in phylum Platyhelminthes, segmented worms in phylum Annelida, and round worms in phylum Nematodes. All of these worm groups have bilateral symmetry, along with an ability to move forward. As a group, worms did not develop together from one branch of the evolutionary chain. Instead, they are less related to each other than to organisms in other animal phyla. Nematodes, for example, are more closely related to arthropods such as insects than to other worm classes. Annelids are closer to mollusks (clams and oysters) than to Nematodes. Annelids have a body cavity or coelom, surrounded by specialized tissues, and the other worms do not. All worm phyla have parasitic species, making them interesting in terms of human disease.

## Flatworms

Flatworms do not have a body cavity, but have a compact body plan, giving them their name. Flatworms lack a space or coelom between their organs and instead have a central gastrovascular cavity to propel fluids. They are the first animal phyla to develop a distinct head and tail end. Flatworms range in size from 1 mm to 20 m (65 feet). They are found in abundance, reaching over 20,000 species. They feed through a single mouth that also serves as the anus, where digested food is also expelled. Flatworms also contain clusters of photosensitive cells that detect light and movement. These appear in some species as **eyespots (**see Figure 10.10a).

When reproducing, they are either asexual or sexual in their processes. During asexual phases, they simply split in half through binary fission, leading to two new organisms. In a simple experiment, if a flatworm, the *Planaria* for example, is cut in half, it will regrow its lost parts, forming into two new organisms. *Planaria* are also hermaphroditic,

**Flatworms**

Any worm belonging to the phylum Platyhelminthes.

**Segmented worms**

Worms characterized by cylindrical bodies segmented both externally and internally.

**Round worms**

A nematode worm infesting the intestine of mammals.

**Gastrovascular cavity**

The primary organ of digestion found in Cnidaria and Platyhelminthes.

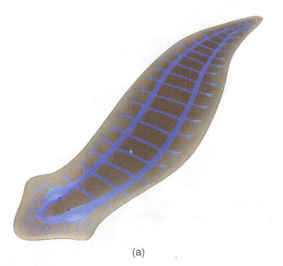

(a)

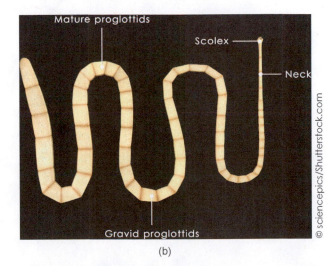

Mature proglottids

Scolex

Neck

Gravid proglottids

© sciencepics/Shutterstock.com

(b)

**Figure 10.10** a. Flatworms are a species in the phylum Platyhelminthes have well-developed organ systems. The *Planaria*'s body plan exemplifies this. Note its well-developed nervous system in the figure. Its many different systems interact to allow planarians to carry out their life functions. b. Tapeworms have long bodies and are able to produce thousands of eggs in animal intestines.

**Flame cells**

Specialized excretory cells found in certain invertebrates.

**Tape worms**

Parasitic flatworms that live in the intestines of people and animals.

containing both male and female gametes, used in their sexual phases of reproduction. *Planaria* contain flame cells that serve to remove their wastes. They also have a set of eyespots connected with nerve cords that enable them to respond to stimuli such as light.

There are over 5,000 species of parasitic flatworms called **tapeworms**, most causing disease in their host species. Blood flukes, also called schistosomes, cause intestinal pain and anemia in over 200 million people worldwide each year. Tapeworm larvae are found in uncooked meats, especially pork and fish. Larvae grow in size up to 2 m in human intestines, causing blockages and preventing nutrient absorption (see Figure 10.10b). This is why cooking meats is so important, preventing the spread of a number of worm-related diseases.

## Roundworms

There are 25,000 known roundworm species, the most abundant worms on Earth. They are found mostly in aquatic environments or wet soils. Nematodes or roundworms have a separate mouth and anus, as compared with flatworms, meaning that they do not eat and excrete through the same opening. Nematodes are almost always parasitic, living on the energy of their hosts, with 15,000 species responsible for human diseases. For example, *Trichinella* is a roundworm which infects human intestines, after which it burrows into muscle tissue (see Figure 10.11). Its related disease, called trichinosis, is potentially fatal and is caused by eating uncooked pork. Roundworms are spread by fecal contamination of food or soil.

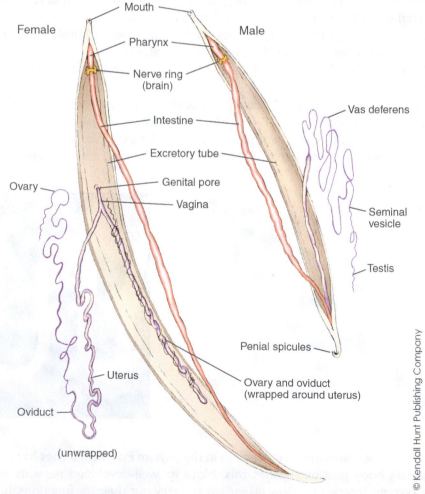

**Figure 10.11** Roundworms (*Trichinella*).

# Segmented Worms

Earthworms are the most common example used to describe segmented worms. They are rarely parasitic to humans. Annelids or segmented worm all have repeating chambers or units, which are called segments. Each segment has a digestive cavity, including a mouth and anus, which directs digestion through the worm. Each segment is identical to the others. The segments contain specialized organs that digest nutrients as they pass through and excrete wastes.

**Segmented worms** comprise 16,000 species and range in size from 1 mm to 3 m, in the case of the giant Australian earthworm. Earthworms are bulk-feeders, meaning that they pass food through their digestive systems as they burrow through the soil. They therefore recycle matter, passing digested material back into the soil. Earthworms mix and aerate the soil, helping other plants and animals to grow.

There are three groups of Annelids: marine, terrestrial, and leeches. Marine annelids or polychaetes combine bristles with their segments. They live on the seafloor and burrow through the soil to obtain food. Polychaetes use their tentacles to bring food into their mouths. Earthworms or oligochaetes are soil dwellers, for example, the common earthworm seen in your backyard. Earthworms also have a circulatory system, with aortic arches operating as simple hearts to pump blood. They contain nephridia, which are specialized tubes to excrete their wastes. A coelom or open body cavity also separates the organs of the annelid (see Figure 10.12). They are also hermaphroditic, able to reproduce sexually and asexually. In asexual reproduction, earthworms spilt at a special spot called their clitellum, which results in two new, identical offspring.

Leeches live in aquatic environments and have segmented bodies. About half of the leeches are blood suckers, using an anticoagulant chemical to keep blood flowing once they grab hold of a host. The other half of leeches act as predators, which feed on other animals.

# Mollusks

Mollusks include snails, clams, oysters, and squids, which are all soft-bodied animals most of which are protected by a hard outer shell. Some mollusks – for example, slugs and octopuses – have reduced shells or have lost their shells through evolution. However, all mollusks have the same three-point body plan: a muscular foot used for movement; a visceral mass containing the internal organs of the mollusk; and a mantle which secretes the outer shell (see Figure 10.13).

**Gastropods.** Slugs and snails are gastropods, mollusks with an enlarged foot to help them move. The slime of gastropods is used to defend against predators. For example, when a bird attacks a slug, its slime sticks to the bird's beak along with debris such as leaves and twigs. Some slime is toxic, harming predators that eat or touch them.

**Bivalves.** Clams, oysters, and mussels are bivalves: they have two shells hinged together. Bivalves are marine and freshwater organisms living in the mud underneath the water. When water passes through their gills, bivalves capture food particles making them filter feeders.

**Cephalopods.** Squids and octopuses comprise the majority of the third group of mollusks called cephalopods. Cephalopods are more mobile than other mollusks, with reduced or missing shells enabling greater flexibility. They have enlarged brains, giving their name "cephalo" which refers to their brain development; and cephalopod translates into head-foot. They also have well-developed sensory organs, enabling cephalopods to stalk and capture prey efficiently. Cephalopods are not, however, brilliant or even smart, as sometimes depicted by the media. An octopus, for example, is able to manipulate

---

**Segments**

The repeating chambers or units found in annelids.

**Polychaetes**

A marine annelid worm.

**Oligochaetes**

Aquatic and terrestrial worms.

**Aortic arches**

The simple hearts of segmented worms.

**Nephridia**

Excretory organs found in many invertebrates.

**Coelom**

An open body cavity that separates the organs of the annelid.

**Mollusk**

Invertebrates, chiefly marine, characterized by a soft unsegmented body and an external hard shell.

**Muscular foot**

One of the three-point body plans of mollusks, used for movement.

**Visceral mass**

One of three-point body plans of mollusks that contain the internal organs.

**Mantle**

One of the three-point body plans of mollusks that secretes the outer shell.

**Gastropods**

Mollusks with an enlarged foot.

**Bivalves**

The property of having two shells hinged together.

**Cephalopods**

The third group of mollusks characterised by a large head, eyes, and a ring for sucker-bearing tentacles.

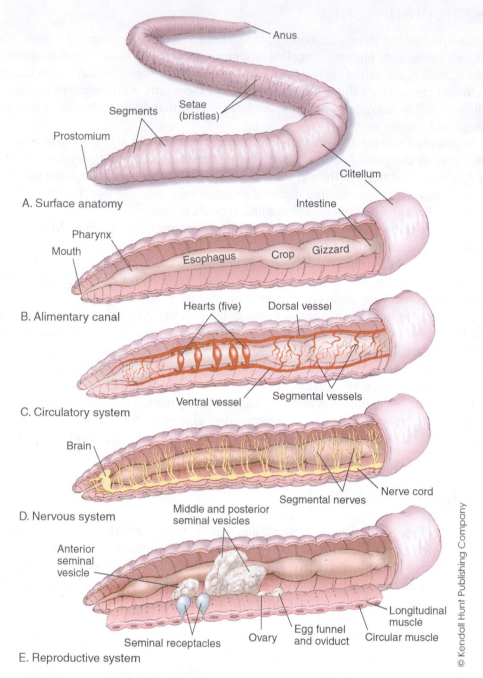

**Figure 10.12** Earthworm anatomy. The digestive system of the earthworm is complex, using several specialized chambers for the breakdown of foods. Its clitellum houses the earthworm's reproductive structures.

things very well – they are capable of untwisting lids of jars or prying to open clam shells – but intelligence is a difficult concept to demarcate. A cephalopod's brain activity is only thus far demonstrated in terms of its use of techniques to obtain prey and survive – not to plan and strategize in an abstract way. Its intelligence is overestimated by many news reports. While cephalopods have a distinct brain, it is difficult to measure their intelligence. Unlike the beaver in our story, which monitors and performs complex building and feeding activities, cephalopods do not exhibit planning behavior.

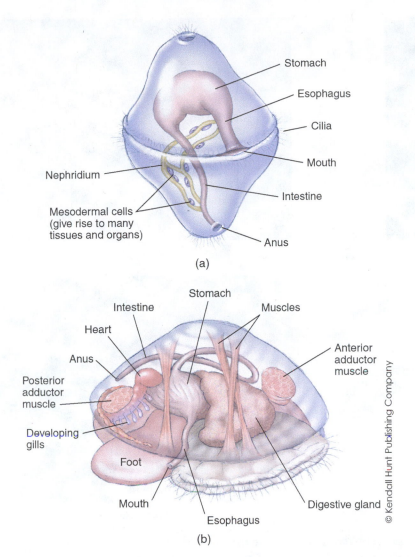

(a)

(b)

**Figure 10.13** Mollusk three-point body plan. Mollusks have bilateral symmetry and possess a muscular foot, a visceral mass containing its internal organs, and a mantle that secretes its protective shell.

# Arthropods

Once segments specialized, each with differing roles, such as mouth parts and antennae, arthropods emerged on life's scene. Arthropods are invertebrates with a specialized segmented body and a protective external skeleton, or exoskeleton, and joint appendages. Their body segments consist of a head, thorax (midsection), and abdomen.

Arthropods comprise the most numerous and diverse of all animal phyla, with over 950,000 known species. There are three major groups of arthropods including insects (flies and moths), arachnids (spiders and marine arthropods), and the crustaceans (lobsters and crayfish). There are over 90,000 species of flies alone, and arthropods total a population of billions ($10^{18}$) of organisms on Earth. Thus, they outnumber humans 150,000:1. If you sit in a forest or in a city subway, you are probably aware that arthropods are omnipresent – they are everywhere (see Figure 10.14).

All arthropod exoskeletons are composed of chitin, a protective polysaccharide that is the most abundant living protein on Earth. The hard exoskeleton protects arthropods.

**Arthropods**

Invertebrates with a specialized segmented body and a protective external skeleton, or exoskeleton, and jointed appendages.

**Insects**

Small invertebrates with a head, thorax, abdomen, six legs, and one or two pairs of wings.

**Crustaceans**

An arthropod characterized by having five sets of appendages.

**Arachnids**

An arthropod characterized by having eight legs.

© 2002 by BSCS. Reprinted by permission

© Hawk777/Shutterstock.com

(a)

(b)

(c)

(d)

**Figure 10.14**    What kinds of present-day arthropods exist? a. Spiders are arachnids. b. Lobsters are crustaceans. c. a millipede. d. dragonflies. a and b Corel. c. From *BSCS Biology: An Ecological Approach*, 9th Edition by BSCS.

When their body grows, an arthropod must shed its exoskeleton through molting. Until another exoskeleton grows back, the organism remains vulnerable to predators. While an exoskeleton helps them to defend against attack, it also preserves their internal water supply, especially important to those species that live in very dry areas.

All insects have wings so they are able to fly, avoiding their predators. Other arthropods, such as arachnids, have poison glands that produce toxins to paralyze their enemy. They then use their appendages to dismember their prey and eat the liquid contents. An arthropod's many specialized structures make them very resourceful in multiple situations. This is the reason they are able to live in so many areas.

## Arachnids

**Arachnids** are arthropods that have eight walking legs and live on land. They include spiders, the most numerous group, as well as ticks, mites, and scorpions. In addition to walking limbs, arachnids have a pair of feeding appendages to capture and kill prey.

Spiders construct often complex webs composed of silk to trap prey, including insects. Most spiders are harmless to humans. *Pholcus phalangioides* or cellar spider, commonly known as the daddy long-legs, is a gangly arachnid, but harmless to humans because its bite is unable to penetrate our protective skin layers (see Figure 10.15). However, some spiders, such as the black widow spider, cause a bite to humans that can be deadly (see Figure 10.16).

## Crustaceans

Almost all of the **crustaceans** are aquatic arthropods. Crustaceans, all have one feature in common: five sets of appendages with three sets used for feeding on prey and two sets to sense their environment.

Crustaceans are very different from each other in other ways, ranging from the small and sessile barnacle to the large, complex, and mobile lobster (see Figure 10.17b). Each species of crustacean possesses specialized appendages that serve unique functions. Lobsters and crabs have modified limbs that hold their eggs or new offspring.

Some crustaceans have value as a delicacy to humans: shrimp, crab, and lobster are expensive dishes in restaurants.

**Figure 10.15**   Daddy long-legs are harmless to humans. It looks gangly and leggy.

**Figure 10.16**   A female black Widow Spider *(Lactrodectus mactans)* sits in wait for prey on its web. Its venom is a neurotoxin, paralyzing its prey.

(a)

(b)

**Figure 10.17** Pillbugs and lobsters are examples of crustaceans. a. Pillbugs have little flesh to serve in human meals, but the lobster b. is a delicacy.

The only terrestrial crustacean is the pillbug, shown in Figure 10.17a, which is not at all tasty, with little meat for food appeal. Pillbugs are some of the oldest species of arthropods.

## Insects

Insects always have three pairs of appendages and one or two pairs of wings in addition to the other characteristics of arthropods. Insects comprise the most numerous group of arthropods, representing 60% of all animal species. Examples include beetles, flies, butterflies, and moths (see Figure 10.18).

**Figure 10.18** There are many examples of insects.

Egg

Larvae

Adult

Pupa

© kerstiny/Shutterstock.com

**Figure 10.19**    Metamorphosis of a moth.

Many insects grow through a process known as metamorphosis, in which they undergo a series of molts, changing to look more and more like their adult form in each stage. First, insects hatch as an egg to become larvae, which appear as caterpillars or maggots, for example. In this stage, larvae consume food to develop, molting into a pupa form, which is cocooned. In the cocoon, the pupa organs break down and adult organs develop rapidly. The adult emerges from the pupa cocoon, entering the world (see Figure 10.19). Caterpillars emerge as butterflies, and maggots become flies, while grubs develop into beetles.

The common housefly, *Musca domestica,* represents about 90% of all fly species (see Figure 10.20). Like many insects, houseflies carry pathogens, causing almost 100 different diseases in humans. Cholera, diphtheria, tuberculosis, and typhus are diseases carried by the common housefly. Insects spread diseases affecting about 250 million people a year and cause 2 million of their deaths. They also carry diseases that harm other animals. An outbreak of tularemia in Urbana's Meadowbrook Park in Illinois in 2013 killed a beaver population through fly and tick bites. Tularemia is caused by a bacterium spread by insect bites between animals.

**Metamorphosis**

A complete change of physical form.

**Egg**

The female reproductive cell in plants and animals.

**Larvae**

The active immature form of an insect.

**Pupa**

The stage between the larval and adult stage, in a cocoon.

**Adult**

Fully grown or developed.

© Yegor Lari/Shutterstock.com

**Figure 10.20**    A common housefly. Flies carry many microbes, some of which cause disease. Beaver fever and malaria are spread through the bite of mosquitoes.

# Echinoderms

Echinoderms are a group of marine animals with a spiny skin and an endoskeleton. Their hard endoskeleton is found beneath the skin, giving echinoderms their structure. The endoskeleton is made of calcium carbonate. The prefix "echin-" is Greek for spiny, which refers to the prickly plates covering their surfaces. The phylum includes sand dollars, sea cucumbers, sea urchins, and starfish, which are probably the most familiar to people. Echinoderms comprise 7,000 animal species, all characterized by adults with radial symmetry. They do not have body segments and instead have specialized structures within one larger unit.

Echinoderms have a water-vascular system, a set of internal channels that circulate water through their bodies, enabling gas exchange and waste removal. The water vascular system ends in small suction cup-shaped feet, called tube feet, which are used for holding prey. Echinoderms carry out complex movements but have no brain; instead, they have a nervous system composed of a central ring which branches into the appendages to sense their environment.

Echinoderms move using their water-vascular system, in which water fills the canals and pushes out through tube feet. Water enters starfish, shown in Figure 10.21, through a central canal and fills its arms. As water enters the arms, they move outward, and to contract them back into position, starfish use their muscular system. The system operates based on a hydraulic set up, with high-pressured water pushing outward to propel an echinoderm forward.

Food sources for echinoderms include algae, shrimp, sea urchins, sand dollars and mollusks, which starfish pry open to obtain their luscious interiors. The water-vascular system transports foods and wastes throughout an echinoderm body.

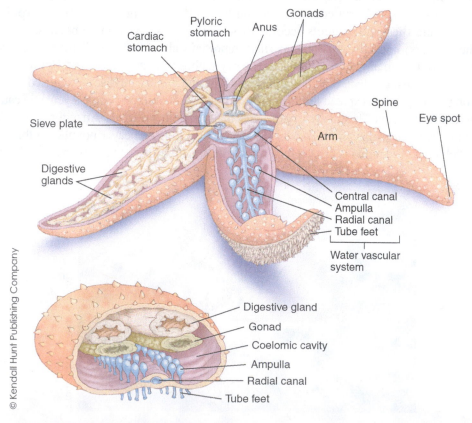

**Figure 10.21**    Starfish anatomy.

Chordates and echinoderms are the most closely related animal phyla. They are the only other animal phylum to contain an internal endoskeleton. While both phyla appear very different from each other, molecular and embryo evidence show their similarity. An echinoderm's embryo stage of development most closely resembles those found in chordates.

# Chordates

Fish, salamanders, reptiles, birds, mammals, and other common images of animals such as the beaver in our story are classified as chordates. Chordates are animals with a spinal cord or spinal cord-like structure (see Figure 10.22).

First, all chordates have a rod of tissue, called a notochord, extending from head to tail. In complex chordates, this tissue develops into a backbone, which is a set of nervous tissue surrounded by bones for protection. In simple chordates, the notochord is unprotected. Organisms with a backbone are classified into the subphylum **vertebrates**. Vertebrates have a backbone and an endoskeleton, shared in common only with echinoderms. Second, all chordates contain a nerve cord extending across their backside. Third, chordates have pharyngeal slits at some point in their development, which act as gills. Gills enable chordates to feed and breathe as water passes through them. Very often gills are only found in the embryonic stages of a chordates life, as in beavers and humans. A post-anal tail extends beyond their normal digestive tract at some point in development, as the fourth characteristic of chordates. Humans lose their tails after the embryo phase, with their tailbone or coccyx exhibited after birth as a vestige of this common phase.

## Subphyla: Lancelets and Tunicates

While vertebrates comprise one subphylum of chordates, two others exist. 1) *Lancelets* comprise about 20 species of small eel-like organisms. They resemble other chordates at both larval and adult stages; and 2) *Tunicates* comprise about 2,000 species, which appear as sessile organisms, with holes that pull in food and water. They most resemble chordates in their larval stages. As adults, tunicates are the size of our thumbs and appear as blobs of gelatinous masses. Both subphyla are filter feeders; they draw water through

**Chordates**

Are animals with a spinal cord or spinal cord-like structure.

**Notochord**

A tissue rod found in chordates that extends from head to tail.

**Backbone**

Set of nervous tissue surrounded by bones for protection.

**Nerve cord**

A dorsal tubular cord of nervous tissue present in chordates.

**Pharyngeal slits**

Openings in the pharynx that develop into gills in some chordates.

**Post-anal tail**

An extension of the spinal cord that extends beyond an animal's normal digestive tract at some point in development.

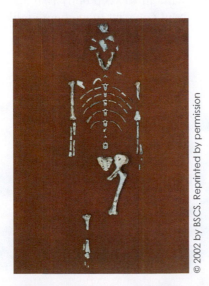

© 2002 by BSCS. Reprinted by permission

**Figure 10.22** Remains of early human ancestors show a distinct backbone, characterizing humans as vertebrates. From *BSCS Biology: An Ecological Approach*, 9th Edition by BSCS.

a system that filters out small, microscopic food particles in their feeding process. They do not contain a backbone and do not have a cranium; but they all have the other chordate characteristics.

# Vertebrates

## *Fish*

The first vertebrates were aquatic, evolving early in the Cambrian explosion, about 540 million years ago. They were jawless and lacked fins but obtained food through scavenging dead animals or sucking on prey. Without jaws, it was difficult for these early vertebrates to defend themselves or to obtain prey.

About 470 million years ago the fossil record indicates that the first jawed fish evolved in the sea. They had fins, which helped them to move quickly through water to obtain food. A typical fish has seven fins, which enable movement in almost any direction. Their mobility, coupled with the development of jaws and teeth, made fish excellent predators. This evolutionary advantage still serves fish well.

The evolution of fins and jaws was a major change in vertebrates. It resulted in rapid speciation into the many classes of vertebrates that we see in our environment. There are three major categories of fish. Cartilaginous fishes are those that have a skeleton composed of the flexible but solid connective tissue, cartilage. Cartilage is the same tissue found in our nose and in the discs between our backbones – it allows multiple movements due to its structure. There are about 880 species of cartilaginous fish including sharks and rays, as commonly known examples (see Figure 10.23). Their flexibility allows cartilaginous fish species to maneuver quickly in the water. Sharks have a keen sense of smell but poor eyesight. They are also able to sense small vibrations in the water, which help them to easily detect prey.

The second category of fish is the bony fishes that have skeletons composed of bone. Calcium phosphates strengthen bones, making them rigid much like those found in our human skeletons. Over 97% of fishes are a ray-finned type of bony fish, including 27,000 species. These have skeletal rays emanating from their central backbones. Examples of ray-finned fish include those we commonly see: guppies, bass, goldfish, and trout.

Bony-fish groups contain an internal swim bladder, which acts as a precursor to lungs. The swim bladder fills with air to keep bony fish buoyant. This buoyancy allows bony fish to remain afloat without constant movement, unlike cartilaginous fish, which

**Cartilaginous fishes**

Are fishes that have a skeleton composed of the flexible but solid connective tissue, cartilage.

**Bony fishes**

Type of fishes that have skeletons composed of bone.

**Ray-finned fishes**

Fishes characterized by skeletal rays emanating from their central backbones.

**Swim bladder**

Organ that is present in many bony fishes and helps them maintain buoyancy.

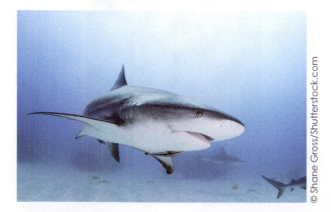

© Shane Gross/Shutterstock.com

**Figure 10.23**   Cartilaginous fish: a Shark. Their skeletons are made mostly of cartilage, which is softer than bone.

must keep moving to avoid sinking. Their buoyancy also helps them to conserve a great deal of energy. Bony fish have both an excellent sense of smell and eyesight to enable them to capture prey and avoid predators (Figure 10.24).

Almost all fish species reproduce by laying eggs that are fertilized externally as well as male and female gametes that are released into surrounding waters. There are a few exceptions, with many shark species having internal fertilization. Some fish, including guppies, carry their eggs within their mothers until a live birth.

The third, smaller group of fishes is the lobe-finned fishes. Lobe-finned fish species have a sturdy pelvis and two solid, muscular fins on the underside of their bodies. These resemble appendages, which later developed into limbs in terrestrial vertebrates, described in the next section. They have two primitive lungs developed from gills. Their lungs are developed, with a complex series of air sacs that facilitate gas exchange, much more efficient than that occurring in the gills of other groups. Lobe-finned fishes contain only eight species, including six species of lungfish (see Figure 10.25) and two

**Lobe-finned fishes**

A smaller group of fishes having a developed pelvis, primitive lungs and muscular fins — precursors to life on land.

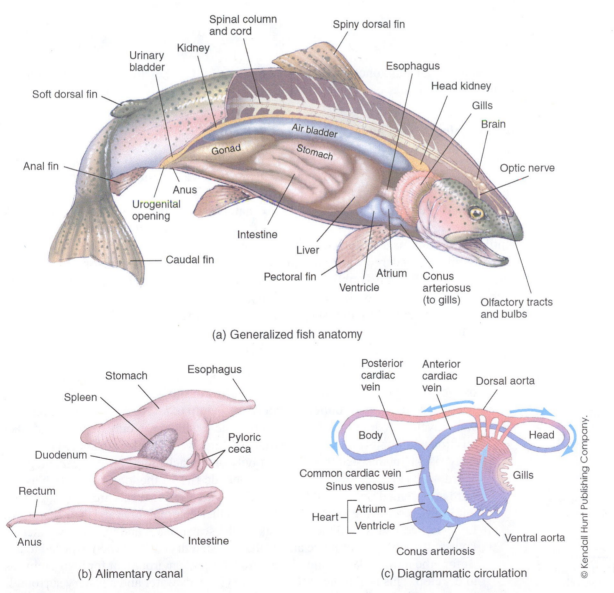

(a) Generalized fish anatomy

(b) Alimentary canal

(c) Diagrammatic circulation

© Kendall Hunt Publishing Company.

**Figure 10.24**    Bony fish body anatomy.

**Figure 10.25**   African Lungfish have a pairs of lungs on either side of their throat. They can survive long periods if their habitat dries up. This is one of only six living lungfish species.

species of coelacanth, once thought to be extinct. Lungfishes live in coastal wetlands, gulping air when they emerge from the water.

## Amphibians, the First on Dry Land

**Amphibians**

Vertebrates that live a portion of their lives in water and another portion on land.

The Greek "-amphibios" translates into "living a double life." It is apropos because amphibians, including frogs, toads, and salamanders, live a portion of their lives in water and another portion on land. Amphibians were the first vertebrates to develop features that enabled life on land. There are approximately 6,000 species of amphibians, comprising about 12% of all vertebrate species. All terrestrial vertebrates – amphibians, reptiles, and mammals – evolved from a common ancestor, the lobe-finned fishes.

Amphibians undergo a metamorphosis, in which their eggs develop into a larval stage called a tadpole, resembling a fish, with gills and no limbs. Then tadpoles develop limbs, air-breathing lungs (losing their gills), and external eardrums. Their movement to land is complete in this adult stage, except that their reproduction is linked to the sea because frog eggs lack a shell and would dry out on land (see Figure 10.26).

There were four vital adaptations for moving onto land:

**Lungs**

Pair of organs that people and animals use for breathing air.

- First, lungs developed in the lobe-finned fish to enable them to breathe air when oxygen concentrations were too low in warm coastal waters.
- Second, a **sturdy backbone** to support the weight of an organism moving on land was developed. Interlocking vertebral bones, which not only support but also cushion and flexibly move a terrestrial creature, were a major advance. The structured backbone resisted the pull of gravity and attached to four moveable legs.

**Limbs**

An arm or a leg of a person or animal.

- Third, limbs evolved from the sturdy underside fins of the lobe-finned fish as the third important development for moving on land. Lungfish fins are homologous to our limbs, resembling them in terms of both their internal form and their functions. Lobe-finned fishes used them to walk on the sea surfaces and in shallow waters.

**Egg**

The female reproductive cell in plants and animals.

- Fourth, the ability to produce an egg that resisted drying out when exposed to air. Before the move to land, animals required a liquid environment for their embryos to develop. This is still the case – an embryo needs water and a watery surrounding during its development

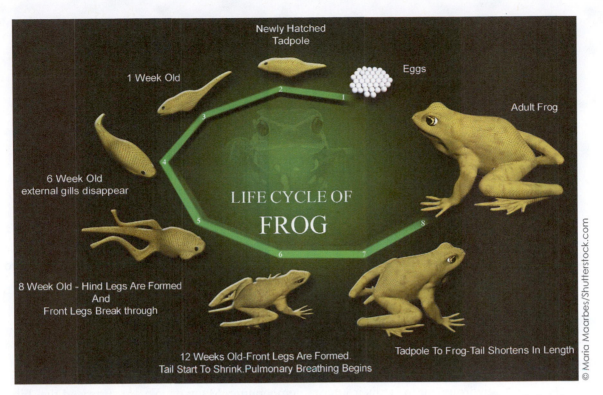

Newly Hatched
Tadpole

Eggs

1 Week Old

Adult Frog

6 Week Old
external gills disappear

LIFE CYCLE OF
FROG

8 Week Old - Hind Legs Are Formed
And
Front Legs Break through

12 Weeks Old-Front Legs Are Formed.
Tail Start To Shrink.Pulmonary Breathing Begins

Tadpole To Frog-Tail Shortens In Length

© Maria Maarbes/Shutterstock.com

**Figure 10.26**  Life cycle of an amphibian (*Rana arborea*). Amphibians undergo growth and developmental changes at certain points in their lives.

Amphibians still fertilize and lay their eggs in a watery ecosystem, called external fertilization because it occurs outside of their bodies. Other terrestrial animals, such as reptiles and mammals developed an amniotic sac to surround developing embryos. In these cases, embryos may exist within a mother or a waterproof eggshell, which both enable development to occur in a wet world.

Thus, terrestrial animals are divided into two groups, amphibians, which are non-amniotes and develop their eggs in the absence of an amniotic sac, and amniotes, such as reptiles, birds, and mammals, which develop their eggs encapsulated within an amniotic sac.

## Vertebrates: Reptiles, More Efficient on Land

Once the development of amniotes took hold on land, reptiles, which include snakes, lizards, turtles, crocodiles, and alligators, as well as birds (often called the feathered reptiles) and dinosaurs emerged on land. Reptiles are a group of amniotes that share certain common features.

These features are specially adapted to living on dry land. First, they developed waterproof skin in the form of scales, which prevent water loss. Second, they have internal fertilization, with egg and sperm combining inside their bodies. This prevents the need, as seen in amphibians, for a watery ecosystem to have sexual reproduction. Third, the **reptile egg** is the most important feature distinguishing them from other terrestrial animal groups. It is a self-contained pond, with an amniotic sac that surrounds a developing embryo with its own water and food supply. Requiring no water from an external source, reptile eggs form within a hard shell made of calcium salts (see Figure 10.27). Reptile **lungs,** the fourth feature helping their adaptation to land, are more efficient and better adapted lungs than amphibian lungs. Reptile lungs have a greater surface area and better exchanges of gases than amphibian lungs. Amphibians use lungs to breathe only

**External fertilization**

The fertilization process that occurs outside the bodies of animals.

**Non-amniotes**

Terrestrial animals that develop their eggs in the absence of an amniotic sac.

**Amniotes**

Terrestrial animals that develop their eggs encapsulated within an amniotic sac.

**Reptiles**

Cold-blooded vertebrates that crawl or creep.

**Scales**

Dermal or epidermal structures that form the external covering of reptiles, fishes, and certain mammals.

**Internal fertilization**

The fertilization process that occurs inside the bodies of animals.

(a)                                    (b)

**Figure 10.27** Adaptations of reptiles to inhabit land: note the eggs and scales in the examples. a. The Horned Adder snake is protected by its scales. b. This baby crocodile, hatching, has been protected by its egg.

**Ectotherm**

Organisms that rely on their environment to set their internal temperature.

**Endotherm**

Organisms that generate heat produced internally by cell respiration to maintain a stable internal body temperature.

as adults, and even then gas exchange mostly occurs through their skin. Reptiles use their lungs to breathe at all stages of their life cycles.

All reptile species except for birds are ectotherms, meaning that they rely on their environment to set their internal temperature. Many people are surprised that birds are reptiles and not part of mammals because they are endotherms, meaning that they generate heat produced internally by cell respiration to maintain a stable internal body temperature. Endotherms are also called homeotherms, as the title of our story describes the beaver. Birds have all of the characteristics of reptiles, except that they are homeotherms and have feathers for insulation.

Dinosaurs are a branch of reptiles that died off 65 million years ago as discussed in another chapter. They were the most dominant terrestrial animal roaming Earth during the Mesozoic era, from 250 million years ago to their extinction. The fossil record shows transitional species linking modern day reptiles to dinosaurs.

### ARE DINOSAURS REALLY RELATED TO BIRDS?

Dinosaurs are extinct because they are no longer roaming the Earth in the forms that once existed. However, molecular evidence shows that dinosaur DNA is very similar to that of birds. Scientists agree that the birds are the direct descendants of dinosaurs. Feathers evolved in birds and in reptiles, with some species of dinosaurs exhibiting feathers well before the evolution of birds. Back in the 1860s, paleontologists found feathers on dinosaur species linking over twenty species with bird-like feathers.

Fossils of dinosaurs and birds, along with molecular DNA evidence of both groups indicate that they share a close relationship. Birds are a type of dinosaur, and both are classified as reptiles. Birds branched off at a point in evolutionary history, with dinosaurs more like birds than any other reptile, fish, or amphibian species. Birds probably evolved from a group of two-legged dinosaurs known as theropods.

Studies of the *Tyrannosaurus rex* also show a close genetic relationship with birds. Collagen fibers, that are strands of proteins found in the soft tissues of animals, were studied to compare their make-up between the species. While it is obvious that dinosaurs no longer roam the planet, their related genes remain – mostly in the birds.

## Vertebrates: Birds, the Other Reptile

Birds were previously classified as mammals, in the class Aves, but are now recognized as reptile homeotherms. There are about 10,000 species of birds, almost all of which have flight. Those that are flightless – ostriches and penguins – probably evolved from birds that did fly. The evolution of flight was a dramatic shift in life's history. Being able to escape predators, find and out maneuver prey from the air, and soar high to reach new regions to colonize, make flight a great development.

All of the anatomical features of birds contribute to their ability to fly (see Figure 10.28). Birds have no teeth, which would otherwise add weight to their skulls and drag them down head first during flight. Instead, they chew or grind their food in a compartment next to the stomach called a gizzard. Their bones are honeycomb in structure, which gives their skeletons strength but is very light. Large spans of bones form wide wingspans without the weight of a heavy skeleton beneath them. A frigate bird has a wingspan of more than 2 m (6.6 feet) but only a weight of 113 g (4 ounces). Females contain only one ovary instead of the two found in other mammals, to reduce their weight. Birds have strong breast muscles, which we commonly called white meat, which expend a great deal of energy to create wing motions for flight. Feathers are light but serve to insulate birds all around their bodies. They are made of the same chemicals as reptile scales, showing their relatedness. The shape and size of the wings of birds are aerodynamic – they enable a "lift" based on wind currents in some bird species, such as eagles and hawks, soaring to great heights. These movements ended the story in Chapter 4, as the old lady watched the great tree's eagles soar, showing the beauty of such flight ability. Other birds, such as the hummingbird, require constant flapping of

**Birds**

Warm-blooded vertebrates characterised by feathers, wings, beak with no teeth, scaly legs, and typically by being able to fly.

©Panaiotidi/Shutterstock.com

**Figure 10.28**    The bird skeleton is lightweight, highly adapted for flight. Note the inside of its bones on the left, with many open spaces giving it a light weight.

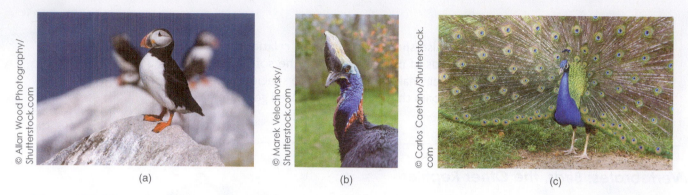

(a)      (b)      (c)

**Figure 10.29** All these modern birds have similar characteristics such as feathers and light bones, but each differs from the other markedly in its appearance. a. The Atlantic puffin is a protected species in Maine. b. The Southern Cassowary male is brightly colored to attract females. C. The male Peacock has a majestic tail to attract females.

wings to keep them afloat. All of these adaptations make birds a unique branch of reptiles. Figure 10.29 gives a sample of birds and their features.

Unlike other reptiles, birds are endotherms, able to use cell respiration to generate heat and maintain a stable internal body temperature. Flight requires a great deal of energy and its movement provides ample heat for maintenance of its body temperature.

## Vertebrates: Mammals, Homeotherms That Thrive on Land and in the Sea

**Mammals**

Are homeotherms that produce milk to feed their young.

Before birds appeared, mammals evolved from reptiles as small, nocturnal creatures. Mammals are homeotherms that produce milk to feed their young; they have hair for insulation and protection. The first mammals branched from reptiles about 200 million years ago. Mammals lived side-by-side with dinosaurs for 130 million years. After the dinosaur extinction, mammals lost a major predator and thrived. This resulted in rapid speciation and development of the numerous species of mammals we know of today. There are 5,300 species of mammals and most of them are terrestrial. There are 80 species of aquatic mammals, such as whales and dolphins, and 1,000 species of winged mammals, such as bats.

### DO OSTRICHES REALLY HIDE THEIR HEADS IN THE SAND?

Ostriches do not hide their heads in the sand or anywhere. Instead, what appears to be hiding in the ground is actually their moving their heads closer to their bodies to appear as a ball to predators.

Pliny the Elder (AD 23-79), a Roman historian and naturalist wrote of the ostrich: "[they must] imagine, when they have thrust their head and neck into a bush, that the whole of their body is concealed." giving rise to this myth. Ostriches are not cowards; they have very strong legs that they readily use to defend themselves and their young. They avoid predators to prevent unnecessary conflicts by maintaining a ball-shaped posture.

Hiding one's head in the sand is akin to being unaware of one's surroundings or happenings. However, ostriches have acute hearing and vision and are able to detect predators from long distances. They are also able to move at speeds up to 31 mph, two advantages that compensate for their being flightless.

Ostriches emerged on the evolutionary tree over 120 million years ago. Their unique strategies enabled them to survive for those many years. They are actually the fastest runner of all two-legged animals. They have adapted strong wings to hit back predators and defend themselves and their young.

There are three groups of mammals: eutherians, monotremes, and marsupials. Over 95% of mammals are eutherians, which are those developing their embryos internally and nourishing them using a placenta. A placenta is an organ inside the mother's body that provides food and removes the waste of a developing organism. Eutherians include most of our known mammals: cows, horses, dogs, cats, apes, and humans.

Marsupials have a short pregnancy, giving birth to small and not completely developed offspring. They are still embryonic and require further development within a protected region in its mother, most often a pouch. These offspring complete their development attached to their mother, nursing on their nipples, obtaining milk. Marsupials include kangaroos and koalas; they live in regions in Australia, New Zealand, and North and South America. Australia is the habitat for most species of marsupials, with little competition from eutherians. Thus, in Australia marsupials occupy ecological areas that are usually reserved for eutherians.

The only mammals to lay eggs are the monotremes, which include only two species: the duck-billed platypus and the spiny anteater. Both organisms nourish their young with milk after they are born. They live primarily in Australia and New Zealand, along rivers and streams. They use leaves and warm nests to incubate their eggs.

## Human Evolution

During the rapid speciation of mammals after the dinosaur extinction, some mammals evolved into primates. Primates evolved about 55 million years ago, with new features to make them more competitive. Primates are adapted to living in trees, with binocular vision to allow three-dimensional vision for jumping, shoulder and arm joints to rotate, and digits – fingers and toes – to grasp. These features of primates enabled them to move about more efficiently in an arboreal environment.

The primates include **prosomials**, including lemurs and tarsiers; anthropoids or monkeys; and hominids, which include the apes and humans. Apes such as orangutans, gorillas, and chimpanzees are most closely related to humans. Chimpanzees and humans share 99% of our genetic material in common (see Figure 10.30). The 1% of gene differences between us accounts for many of the features that make us human.

Primate evolution may be traced as a wide bush, with many evolutionary branches emerging from their common ancestors. Newly evolved features included the development of wider dental arches and stronger teeth for powerful chewing; and bipedalism (ability to walk on two legs), which allowed for faster movement on land. Bipedalism evolved first, about 4 million years ago, enabling our ancestors to leave trees and walk on land, opening up a new set of habitats including the grasslands (see Figure 10.31). Walking on two legs uses less energy than on four legs.

Several groups of bipedal hominids existed between 4 million and 1 million years ago, known as the genus *Australopithecus*. There were at least two species including

### Eutherians

Mammals that develop their embryos internally and nourish them using a placenta.

### Placenta

An organ inside the mother's body that provides food and removes the waste of a developing organism.

### Marsupials

A type of mammal in which young ones are born immature and continue to develop in a pouch.

### Monotremes

Primitive mammals that lay eggs.

### Anthropoids

A higher primate, including monkeys.

### Hominids

A primate belonging to the family Hominidae.

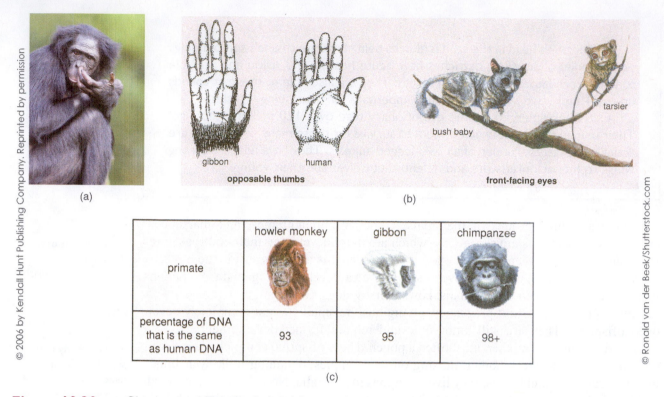

**Figure 10.30**  a. Chimpanzee *(Pan troglodytes)* have opposable thumbs and big toes. They are 99% genetically similar to humans. b. Primate characteristics. From *Biological* Perspectives, 3rd ed by BSCS. c. Comparing genetic similarities across selected primates. From *Biological* Perspectives, 3rd ed by BSCS.

*Australopithecus africanus* and *Australopithecus robustus*. They had a small brain size, their bodies were the size of chimpanzees, roughly 1 m in height (about 3 feet tall), and they weighed about 30 kg (60 pounds).

Up until about 1.5 million years ago, *Australopithicus* lived in Africa. With the development of smaller teeth, larger brains, and the ability to use tools, our modern genus *Homo* emerged. "*Homo* means human" in Latin, referring to the similarities between those early species and modern humans. Several groups of Homo evolved from Australopithecus including the group *Homo erectus* and *Homo ergaster*. *H. erectus* left Africa, migrating to regions in Asia and Europe, while *H. ergaster* remained in Africa. *H. erectus* were

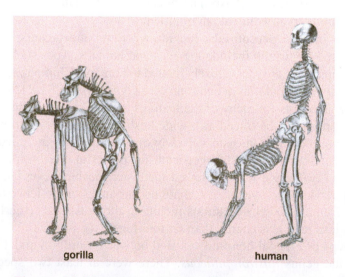

**Figure 10.31**  Adaptations for bipedal motion

successful hunters who shared food and worked cooperatively together forming communities. They had fire and lived in caves, protecting them from a host of environmental dangers. *H. erectus* lived from between 1.5 million years ago until 300,00 years ago.

Modern *Homo sapiens* or humans probably evolved from *H. ergaster. H. ergaster* evolved into *Homo heidelbergensis* and branched into either *Homo neanderthaensis* or *Homo sapiens*. Brain size and height and weight doubled as compared with earlier *Homo* forms. *H. erectus* evolved into another group that lived alongside them called *Homo floresiensis* about 200,000 years ago (see Figure 10.32).

Human evolution begins roughly 100,000 years ago based on fossil and molecular genetic evidence. *Homo sapiens* left Africa after their evolution, as a small group of only 100 people, diverging into Europe, Asia, and Australia, according to mitochondrial DNA evidence. They had an advanced sized brain capacity and this helped them to compete with the other groups of *Homos* that they encountered. About 15,000 years ago they crossed over into North America via Alaska, colonizing the New World.

Humans encountered three major groups of Homos in their journeys around the Earth. 1) *Homo neanderthalensis* or Neanderthals, our closest ancestor, lived from 150,000 to 35,000 years ago at the same time as modern humans. Neanderthals used fires, lived in caves, buried their dead, and lived in social groups. It is possible that we interbred with Neanderthals, but their lineage dies off, with ours emerging successfully into humans today. *Homo floresiensis* lived as small "hobbit-like" creatures, only about the size of their ancestors. They were more advanced in their social activity, using tools and living in groups. Their ancestors, members of *H. erectus,* lived alongside them for almost 200,000 years.

Once humans invaded regions of the Earth, all of its Homo relatives eventually became extinct. Neanderthals died off 30,000 years ago, *H. erectus* about 27,000 years ago, and *H. floresiensis* went extinct 12,000 years ago. Most anthropologists concur that humans exterminated all of the other *Homo* species, leaving us as the only surviving group in the continued evolution of our genus.

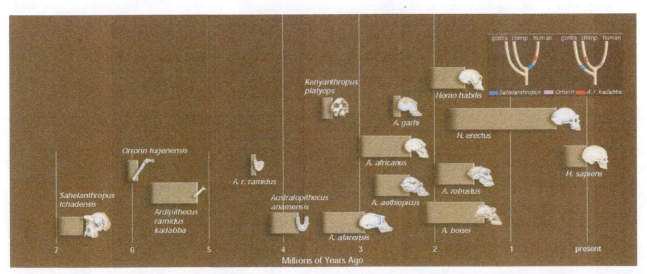

Hominid evolution. The fossil record of hominids shows that several species of hominids existed at the same time during the later stages of human evolution. New discoveries cause scientists to reconsider which of the early species led to modern humans. The branching diagrams in the upper right of the diagram show different hypotheses about how recently discovered fossils, Sehelanthropus, Orrorin, and Ardipithecus, might be related to humans. Although scientists are not certain about the evolutionary pathway that led to modern humans, new discoveries will undoubtedly lead to new ideas.

**Figure 10.32**   Hominid timeline.

# Summary

Animals emerged from a single protist roughly 600 million years ago. After the Cambrian explosion, divergence into 30 different phyla of animals reflects the many adaptations to changing environments encountered since that time period. Animals are broadly divided into invertebrates and vertebrates, depending on whether they contain a backbone. There are four characteristics which are used to classify animals: cell specialization, symmetry, molting, and direction of development. Porifera are the simplest animals, as sponges with aggregates of unspecialized cells; cnidarians are more complex, including the jelly-fish and corals; worms have segments with specialized structures, often causing diseases such as the intestinal tapeworm; mollusks are soft-bodied animals usually protected by a shell, such as oysters and clams; arthropods have segments, nonrepeating, and each with their own specialization, including insects and spiders; echinoderms, including starfish and sand dollars, have an endoskeleton; chordates developed a more complex nervous system, enabling more effective responses to stimuli. Chordates include the most complex animals, such as fish, frogs, reptiles, birds, beavers, and humans. Chordate's move to land included a number of adaptations such as lungs, scales, and shelled eggs to complete their move to land. Human evolution occurred relatively recently in earth's history: during the past 100,000 years. Our triumph over other *Homo* genus's has resulted in our lone existence among this group.

## CHECK OUT

### Summary: Key Points

- Animals interact with one another and with the environment to effect changes, as seen in the parasitic nature of the roundworms, the aerating effects of earthworms in soils and the vast ecological changes produced by the activity of the beaver.
- Animals increased in complexity as evolution progressed, adding special adaptations to exploit more environments and new environmental conditions.
- Animals are classified based upon their cell specialization, body symmetry, ability to molt, and direction of gut formation.
- Porifera are sponges with asymmetry and no specialization.
- Cnidarians have radial symmetry but all other animals have bilateral symmetry.
- Roundworms and arthropods are the only animal phyla that molt.
- Flatworms, segmented worms, mollusks, echinoderms, and chordates grow continuously.
- Echinoderms and chordates are the only animal phyla that develop from the back to the front.
- Human evolution of the *Homo* genus occurred only in the past 100,000 years, emerging from the *Australopithecus* genus.

## KEY TERMS

adult
amniotes, nonamniotes
amoebocytes
amphibians
aortic arches
anthropoids
arachnids
arthropods
backbone
bilateral symmetry
birds
bivalves
bony fishes
budding
Cambrian explosion
calcium carbonate
cartilaginous fishes
cephalopods
chordates
cnidarian
cnidocysts
collar cells
coelenterons
coelom
coral bleaching
corals
crustaceans
deuterostomes
echinoderms
ectoderm
ectotherm
egg
endoderm
endoskeleton
endotherm
epidermal cells
eutherians
exoskeleton
extant
extinct
external fertilization
eye spots
flame cells
flatworms
gastropods

gastrovascular cavity
hermaphrodite
homeotherm
hominids
hydra
internal fertilization
invertebrates
insects
jellyfish
larvae
limbs
lobe-finned fishes
lungs
mammals
mantle
marsupials
medusa stage
mesoglea
metamorphosis
mollusk
molt
monotremes
muscular foot
nematocysts
nephridia
nerve cord
nerve network
notochord
oligochaetes
pharyngeal slits
placenta
polychaetes
polyp stage
porifera
post-anal tail
protostomes
prosomials
pupa
radial symmetry
ray-finned fishes
reptiles
round worms
scales
sea anemone
segments

segmented worms

sessile

specialized cells

central canal

swim bladder

tape worms

tube feet

vertebrates

visceral mass

water-vascular system

zoology

# Multiple Choice Questions

1. Which animal is among the first to exhibit the environmental effects of pollution?

   a. Insects
   b. Corals
   c. Segmented worms
   d. Clams

2. The Cambrian explosion resulted in:

   a. rapid animal speciation
   b. rapid animal extinction
   c. decreased parasitic diseases
   d. decreased protist–animal relationships

3. Which animal phylum has a distinct endoskeleton?

   a. Cnidarian
   b. Polychaete
   c. Arthropod
   d. Echinoderm

4. 75% of animal species belong to the _____ grouping of animals:

   a. invertebrate
   b. arthropod
   c. chordate
   d. vertebrate

5. In a cluster of sea anemones growing together, a line may be drawn in any direction and both sides of it are the same. This refers to its

   a. radial symmetry
   b. bilateral symmetry
   c. protostome development
   d. deuterostome development

6. Which represents a logical order, from early to later, in the evolution of chordates?

   a. fish→reptiles→frogs→mammals
   b. reptiles→mammals→fish→bryophyte
   c. fish→frogs→reptiles→mammals
   d. frogs→fish→reptiles→mammals

7. Which phylum contains organisms with asymmetry:

   a. Cnidarian
   b. Mollusk
   c. Porifera
   d. Echinoderm

8. Which serves as a heart for earthworms?

   a. Segment
   b. Nephridia

   **c.** Aortic arch

   **d.** Coelom

**9.** Which group of organisms BEST represents the most important link between land and sea adaptations?

   **a.** Cartilaginous fishes

   **b.** Bony fishes

   **c.** Ray-finned fishes

   **d.** Lobe-finned fishes

**10.** Which anatomical adaptation BEST facilitated the *Homo* move from trees to grasslands?

   **a.** Bipedalism

   **b.** Use of fire

   **c.** Increased size

   **d.** Larger teeth

## Short Answers

**1.** Describe two ways in which animals, such as the beaver in our story, are beneficial to humans. List two ways in which animals harm humans. Be sure to list and describe each.

**2.** Define the following terms: vertebrate and invertebrate. List one way the terms differ from each other in relation to their: 1) anatomy; 2) diversity among the animal phyla; and 3) behavior.

**3.** An animal is discovered by a group of zoologists on the coast of Antarctica. It has radial symmetry, is heterotrophic, contains repeating segments, does not molt and develops mouth first. Use the characteristics of animals in this chapter to classify this organism. Why did you place it in its group?

**4.** A mollusk contains three basic parts to its body plan. List these three parts and describe their functions. Which is most important in a mollusk's survival? Are there any parts missing in some species?

**5.** Is a cephalopod, such as an octopus, really smart? Why or why not?

6. List and draw the body structure of a Porifera. Use the following terms in your drawing: epidermal cell, amoebocyte, collar cell?

7. Explain the "double-life" of an amphibian. Be sure to draw and label the steps of its life cycle. How is the double-life of amphibians an adaptation to living on land?

8. For question #7, how does the size of a pond affect an amphibian's reproductive process? If a pond is too large, what are its advantages and disadvantages? If a pond is too small, what are its advantages and disadvantages?

9. List four adaptations reptiles developed to make their transition to life on land? How is each important in a reptile's functioning in a terrestrial world?

10. Describe the two theories explaining how Neanderthals became extinct. Which is most plausible? Why?

# Biology and Society Corner: Discussion Questions

1. The Cambrian explosion resulted in a diversification of animal species. Each species developed into a more complex set of organisms as compared with those that came earlier. Does this make humans, last to evolve, the "highest" organisms in the evolutionary chain? Why or why not?

2. Some cnidarians, including jellyfish species, are notorious for their sting, inhabiting beach waters and washing onto the sand. Choose a pollution threat to cnidarians which has emerged in the past 25 years. Research the effects of the pollutant and make a recommendation on its benefits or drawbacks for your own health.

3. Almost all of the 15,000 species of nematodes cause animal diseases. If a pharmaceutical company could eradicate all of these organisms, should we proceed to do it? Should you or your loved ones be concerned about the destruction of nematodes? Why or why not?

4. Amphibians are an indicator species, much like the moths in Chapter 7, first showing the effects of environmental changes due to their delicate nature. Many amphibian species are in decline in the United States and in Puerto Rico. The island of Puerto Rico provides a study unit to view amphibian dynamics. Research amphibian population decline in Puerto Rico, and list three reasons for it. How may amphibian deaths have eventual impacts on human society?

**5.** Our opening story in this chapter showed the importance of beavers in flood control and their effects on humans. Beavers are second only to humans in their effects on the environment around them. How did humans affect other *Homo* species as *Homo sapiens* emerged in Africa and as they explored other parts of the world?

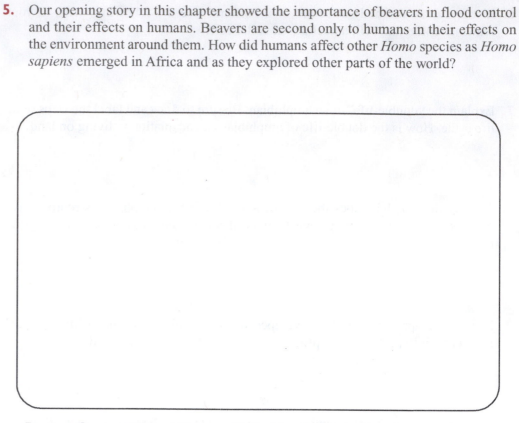

Figure – Concept Map of Chapter 10 Big Ideas

# UNIT 4
# The Dynamic Animal Body

# Animal Organization

## 11

## ESSENTIALS

Starfish

She throws her starfish back into the ocean

Many starfish wash up along the beach

Sabrina helps a starfish

Micrographs of the four tissue types of a Starfish

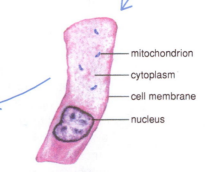

mitochondrion
cytoplasm
cell membrane
nucleus

**epithelial cell**
A Starfish, parts of its cell

## CHECK IN

**From reading this chapter, you will be able to:**

- Explain how knowledge about proper functioning of body tissues plays an important role in human health.
- Describe methods of studying human structures, and define anatomy, physiology, gross anatomy, histology, cytology, and developmental anatomy.
- Explain complementarity by giving an example connecting a structure's anatomy with its physiology.
- Explain how homeostasis is maintained using negative and positive feedback mechanisms of control, and define those terms.
- List, describe, and compare the four types of human body tissues.
- Locate different regions of the human torso using the language of anatomy, and organize and place different body structures within their appropriate organ systems.

# The Case of a Saved Star

A college freshman, Sabrina walked along the beach in South Texas, at Padre Island. There were parties and excitement with college students abounding, running into the water and celebrating a coveted Spring Break. Everyone was having a good time, except Sabrina. She walked alone along the beach, much as she did in most of her life. She wondered why nobody noticed the thousands of starfish washed up on the shore from the last night's storm surge. All of the starfish were dying on the shore, without seawater to nourish them.

Sabrina recalled the water chemistry lectures in class, which explained the need for a strict balance of ions in all living tissues. Her heart broke as she spied the drying out parts of each starfish, as they lay lifeless on the beach.

She saw their skin or epidermis, consisting of a thin cuticle made of epithelial or covering tissue cracking and dry. Bumps on the starfish, made mostly of calcium carbonate, jutted out in the form of spines. This was the starfish endoskeleton, made of connective tissue resembling bone, which maintains its structure and connects different parts of the animal. Sabrina looked more closely at the mouth of the starfish, trying to help it back into the water. When she touched its tube feet, they retracted backward. She recalled from biology class that each of the tube feet is sensitive to touch, connected to a set of nervous tissue forming a nerve net beneath the epidermis of the starfish. Nerves fire and cause motion in the arms of the starfish and in its muscles to close its mouth. The movements of their tube feet stimulate muscles within starfish arms to contract and grasp onto objects. Sabrina noted the weakness of the starfish muscles as they dried. Its central mouth could not close and its arms were unable to respond normally. She felt compelled to change this wretched situation.

A panic came over Sabrina as she realized that the animals needed to be returned, as quickly as possible, back into the sea. So, she started throwing the dying creatures back, one by one, into the ocean. It was an arduous task in the hot sun and she knew that she could not save all of the starfish.

A few college classmates ran by and called out to her, inquiring – "What are you doing?" Sabrina responded that she was returning the starfish to their homes. She said, "They'll die in the sand. Why don't you come here and help me bring them back?" Many laughs came thereafter from her crowd of "friends," with one of her worst critics exclaiming, "Why don't you come swim with us or would you rather be with the starfish? You know you can't save them all . . . it's hopeless."

Sabrina saddened for just a moment. Then, she picked up a single starfish and held it up to her peers and said, as she threw it back into the ocean, "At least I made a difference for this one!" Sabrina's classmates listened to her words and experienced a change of heart. One by one they joined her in returning the starfish to the ocean.

*Adapted from The Starfish Story: http://www.ordinary people change the world.com/ articles/the-starfish-story.aspx*

## CHECK UP SECTION

However, all animals including humans are composed of the same four types of tissues – epithelial, connective, nervous, and muscle – in its organization.

In the last chapter, the overview of animal diversity showed many different types of organisms.

Like all human systems, the starfish body is organized to perform at its best within the right conditions, with its four tissues working together. For example, the Starfish is structured to function with the right amount of water and temperature in its surroundings to prevent it from drying out.

Sabrina was inspirational in her attempts to save the life of the dying starfish. What societal checks do we have in place to maintain the proper balance and health of our tissues? Choose one lifestyle choice that disrupts this balance in society. Do you live your life to maintain the health of all of your tissues?

## Orientation to the Human Body

This story describes the four tissue types that compose our internal structures. The complexity of our systems and their capacity to work together to enable life functions is the focus of this chapter. The desiccation or drying out of Starfish tissues creates an imbalance in life functions of the animal, the basis of disease. Disease occurs when an organism's systems do not maintain balance while working together. Celiac disease, for example, impedes proper absorption of food for energy from foods consumed. In addition to describing the workings of the human organism, in the next chapters of this unit we will also consider some of the diseases associated within each organ system of the body.

So far, the text described many forms of living systems, from the elegant *Stentor* and fruiting bodies of slime molds to the giant redwoods and other ancient trees. However, it is a focus on human systems in this next unit that explains the many happenings within our bodies. A major goal of this text is to show how human biology and human society interact with other living systems.

In this unit, we will discuss the body systems separately, but in fact they interact with each other constantly to perform body functions efficiently and maintain a steady state. The relationship between different systems is shown in Figure 11.1. When one system fails to function properly, this affects the workings of other systems. For example, when a weak heart is unable to carry sufficient blood through the body, it also fails to push enough excess water out of the kidneys. As a result, swelling in the legs and abdomen are common signs of a weakening heart. Water balance is vital for the survival of all organisms, as shown in Sabrina's starfish's plight with dehydration in the story. All of the body systems in humans work in concert to carry out multiple, simultaneous functions to keep organisms alive. Figure 11.1 shows the relationship of the urinary system (which maintains water balance) with other interdependent systems.

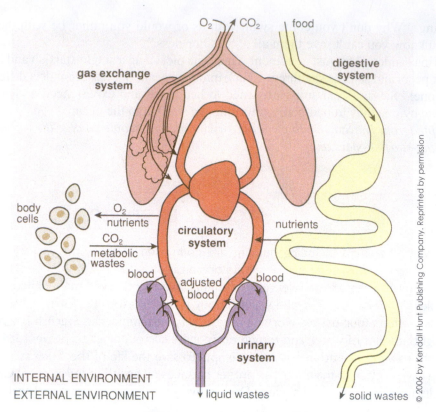

**Figure 11.1** The body's systems are interdependent upon each other. Human digestion is accomplished only with the help of other organ systems. For example, without the heart, food from digestion would not be brought to needed parts of the body. Without the kidneys, wastes from food would build up, and be fatal. From *Biological Perspectives*, 3rd ed by BSCS.

**Anatomy**

The study of the structure of body parts.

**Gross anatomy**

The study of body parts that can be seen without use of microscopy.

**Microscopic anatomy**

The study of structures too small to be seen with the naked eye

**Cytology**

The study of cell parts.

**Histology**

The study of tissues.

The drying out of the starfish in our story is a great introduction to the study of the tissue types that make up the organ systems that carry out the life functions of the human body. Outside of the seawater, starfish tissues quickly experience breakdown and a loss of structure. Environmental conditions need to be just right for those tissues to maintain their structure. The study of the structure of body parts is known as anatomy. Anatomy describes what a structure, such as an organ or tissue, looks like. There are several subdivisions of anatomy, depending upon the focus of study. Figure 11.2 illustrates that the branches of anatomy study life at different levels of its organization.

First, many medical techniques and surgeries work on structures visible to the naked eye, in what is termed gross anatomy. **Gross anatomy** refers to the study of body parts that can be seen without the use of microscopy. Some of these medical procedures include setting broken bones, treating skin wounds, or massaging muscles. Starfish tissues viewed by Sabrina in our story are also examples of gross anatomy.

Second, all study of structures too small to be seen with the naked eye is called microscopic anatomy. The use of a dissecting microscope or electron microscope includes study of cells and cell structures. The study of cell parts is called cytology, which researches cells and their organelles such as mitochondria and the nucleus. The study of tissues, or groups of cells performing the same overall functions is called histology. Histology studies groups of cells that are also too small to be seen casually without a microscope. Both cytology and histology are branches of microscopic anatomy.

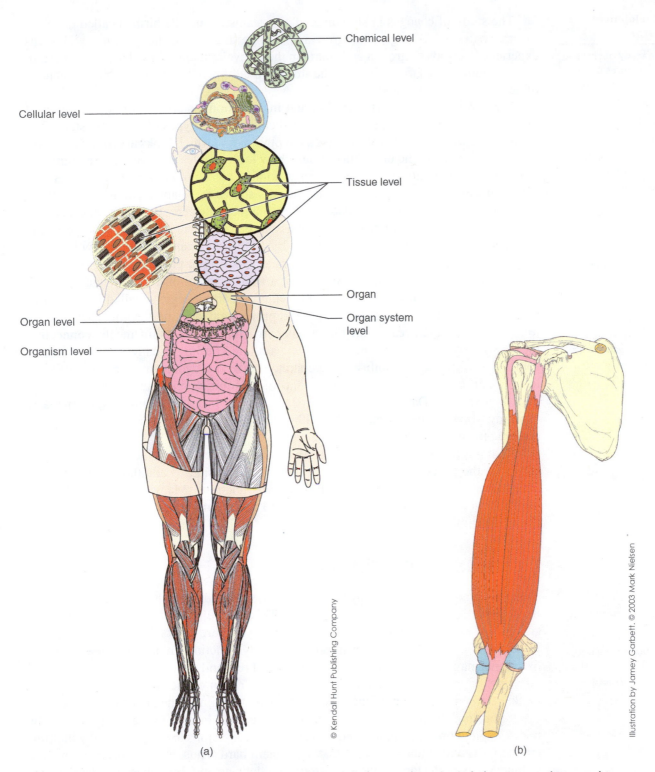

Chemical level

Cellular level

Tissue level

Organ

Organ system level

Organ level

Organism level

(a)

(b)

© Kendall Hunt Publishing Company

Illustration by Jamey Garbett. © 2003 Mark Nielsen

**Figure 11.2**   a. Hierarchy of body structure. Chemicals form cells, which form tissues that combine to form organs. Related organs form organ systems comprising a whole organism. b. Studying life's changes. The different branches of anatomy studying life at different levels of its organization. Gross anatomy is the study of those parts able to be seen with the naked eye. In this figure, the orientation of the upper limb muscles is studied by gross anatomy.

**Developmental anatomy**

The study of changes in structures of an organism since its birth.

**Embryology**

The study of anatomy before birth; looks at structures of developing embryos and fetuses

**Physiology**

The study of the function of body parts.

**Disease**

An imbalance in the proper working of a tissue, organ, or organ system.

**Senescence**

The process of aging.

**Complementarity**

The state in which the function of any body part always depends upon its structure.

The study of changes in structures of an organism since its birth is called developmental anatomy. All organisms grow and change throughout their lifetimes. Humans experience periods of growth and spurts in their development. These lead to their maturity into adulthood. Embryology, or the study of anatomy before birth, looks at structures of developing embryos and fetuses.

How an organism's parts function and malfunction, as seen in the dehydration of tissues in the starfish case, is a focus of medical and other types of scientific study. The study of the function of body parts is called physiology, which looks at how an anatomical structure works. The orientation of an elbow joint enables it to work in a certain way to bend and allow movement of the arm. A kidney cell's unique structure enables it to conserve water. The long strands of nerve cells shown in Chapter 1 transmit electrical messages to allow thought; we also saw, how plaques interfere with such transmissions to cause Alzheimer's disease. The starfish tissues began ceasing to function properly and weakened the organism in the dry, hot sand. All of these examples show how the anatomy of a body part helps to determine its proper working or physiology.

An imbalance in the proper working of a tissue, organ, or organ system is known as disease. All of our bones are joined together at certain regions called joints. Joints allow bones to move and at the same time protect the structures within them. Many joints are moveable because they have other structures, such as ligaments, connecting them together. When a joint malfunctions, pain and swelling usually limit motion and sometimes result in immobility. An overstretching of one's joints may result in damage to ligaments or other tissues surrounding the joint, requiring physical therapy and/or surgery as treatment. Damage to knees constitutes over 60% of joint injuries, particularly in young women active in sports. A healthy knee joint is shown in Figure 11.3. Note the multiple ligaments holding the bones in place.

Anatomy and physiology both study life at different stages. At the end of life, senescence, or the process of aging, accompanies the extension of adulthood into old age. Senescence is characterized by the loss of cell functions. Many age-related illnesses limit functionality in older people. In Chapter 1, aging of the brain led to our character Hans' development of Alzheimer's disease.

# Complementarity

The function (physiology) of any body part always depends upon its structure (anatomy). In other words, function always follows form. This is called complementarity because the anatomy of a structure complements the way that structure works. When complementarity fails to work properly, the result is disease or dysfunction. In our story, the Starfish's tissues normally function to process nutrients and respond to stimuli. Its tissues are held intact by an endoskeleton and muscles. The dehydration of the support structure led to weakened muscle and a nonfunctioning mouth, both an imbalance in its processes.

If you consider almost any anatomical part, you will be able to see complementarity. For example, bones in humans and in Starfish endoskeletons are impregnated with mineral deposits and strands of fibers that make them hard. Thus, in humans bones are able to function as protectors, with ribs surrounding delicate and thin lungs, and the flat sternum (breast bone) covering the vital beating heart. In each case, a living, hardened bone supports the structures that it surrounds. In particular, bones meet within joints in animals to enable protection but also movement. Consider the knee joint in Figure 11.3, which consists of two bones joined together with ligaments and tendons acting as straps. Muscles strengthen the joint and movement is possible with this unique arrangement. Without this unique structure, both the goals of movement and protection would not be possible.

The anatomy of a structure also fits its physiology to meet chemical needs in the body. Consider fish gills, which are composed of very thin tissues, with filaments only

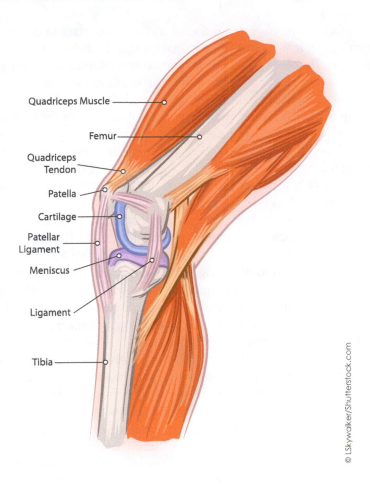

Quadriceps Muscle

Femur

Quadriceps
Tendon

Patella

Cartilage

Patellar
Ligament

Meniscus

Ligament

Tibia

© LSkywalker/Shutterstock.com

**Figure 11.3** The human knee is the body's largest and most complex joint. It is easy to damage and resembles two matchsticks stuck together with many ligaments and tendons. The muscle around it helps to add strength.

a few cell layers thick. The surface area of fish gills is large, reaching many square feet when dissected and spread out evenly. The thin construction of gills allows gas exchange across the moist membranes. In addition, blood moving through the gills flows in the opposite direction as the water current. This countercurrent exchange, as it is called, maximizes the rate of exchange between fish gills and the gases in the water.

Diffusion of gases, as discussed in Chapter 3, occurs only across short distances and through thin tissues. Oxygen gas is needed for proper cell functions, and carbon dioxide gas is a waste product. Both are transported into and out of the human body through the lungs. Gases must be continually exchanged in the lungs to enable cell respiration and energy uptake by cells, as described in Chapter 4. After all, a cell can only live a maximum of 4 minutes without oxygen before it dies! An imbalance in lung physiology occurs in the presence of pulmonary diseases such as asthma, emphysema, and lung cancer. In these diseases, the membranes for gas exchange no longer operate efficiently, preventing sufficient gas exchange. Breathing diseases will be discussed in greater detail in other chapters.

Studying how body parts work and their related diseases requires clinical analysis. Generally, medical practitioners first use observation and descriptions of symptoms to diagnose illness. They might observe a wound that does not heal and take pictures, measuring its progression. Physical manipulation is also used to either help treat symptoms or to diagnose the cause of a disease or injury. For example, when a joint is dislocated, often the best treatment is to place the bones back into their proper position.

**Observation**

The act of obtaining information from a primary source.

**Manipulation**

Manual movement of anatomical parts to either help treat symptoms or diagnose the cause of a disease or injury.

**Palpation**

The act of feeling with one's hand.

**Auscultation**

Listening to sounds produced within the body.

**X-rays**

A form of EM radiation that visualizes dense structures within the body

**CT scan**

Computerized axial tomography that produces detailed images of internal organs.

**MRI**

A technique that uses radio waves and magnetic field to generate detailed images of tissues and organs.

**Ultrasound**

A technique that emits high frequency sound waves and creates images based on the echo received back from the body part.

**Homeostasis**

Maintaining a steady set of environmental conditions.

Palpation, or feeling structures with one's hands, is important in determining diagnosis. In Sabrina's case, it was clear upon palpation that the muscle tissues of the Starfish were weakening. However, palpation in humans and animals requires experience to detect abnormalities. For example, when a lump or swelling in the testicles is found, it is indicative of testicular cancer, a disease affecting mostly young men between ages 17 and 34 years. Palpation of the lump requires professional analysis for a clear diagnosis. Auscultation, listening using a stethoscope, gives indications about heart health. Murmurs or turbulent blood flow sounds are detected using this procedure, often indicating a valve leak in the heart.

Some tests, such as blood tests and medical images, indirectly give data on a person's health. There are several types of medical imaging (see Figure 11.4). X-rays visualize dense structures within the body. The X-rays that are absorbed appear lighter and show a thicker structures such as a tumor. Bones are imaged well using X-ray imaging. In order to visualize softer tissues, other techniques were developed in the past half century. CT (or computerized axial tomography) scans use multiple sections of X-rays to image body regions. This method eliminated the need for many exploratory surgeries and led to a three-dimensional view of internal structures. However, CT scans have been recently associated with increased risks for developing cancer due to their use of a large amount of radiation. Even one CT scan increases the risk of getting cancer by 400 times. MRI (or magnetic resonance imaging) maps the body part's hydrogen atoms within water of soft tissues. The magnetic spin of hydrogen within water creates waves and an image to study. This is a generally harmless test that studies soft tissues and provides a three-dimensional image. Ultrasound technology, developed during World War II, emits high-frequency sound waves and creates images based on the echo received back from the body part. Ultrasound technology explores surface anatomy of parts within the body. It may look at the surface of internal abdominal organs or the structure of a developing fetus. Ultrasound has a low penetration of its sound waves into body structures, preventing it from showing deeper images.

# Homeostasis Is Vital for Carrying Out Life Functions

As discussed in Chapter 1, maintaining a steady set of environmental conditions is known as homeostasis. Body temperature, acid–base levels, and chemical concentrations in living systems must be in balance to keep organisms functioning properly. Internal conditions often vary within narrow limits; they do not exist in a fixed state but fluctuate

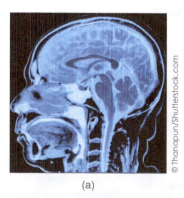
(a)

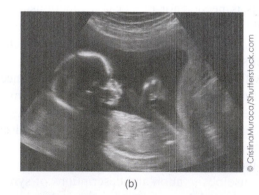
(b)

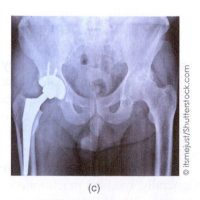
(c)

**Figure 11.4**  Images enable doctors to see inside regions without using exploratory surgery. a. MRI. b. Ultrasound. c. X-ray.

within a range of values. Body temperature is generally considered normal in humans at 98.6°F, but this value represents an average. Many people exhibit temperatures that are normally above or below this value. Some people run "hot," at 99.1°F and there is no disease pattern associated with their higher than normal temperature. Of course, when temperatures rise too high within an organism, chemicals such as enzymes do not work optimally or fail to function. Homeostatic processes keep conditions within a range that is acceptable for proper functioning.

As seen in our story, sometimes the ordered structure experiences failure, creating an imbalance in proper functioning. A failure to maintain homeostasis is commonly called **disease**. In our Starfish example in the opening story, homeostasis was disrupted by a change of environment from seawater to dry land. This led to drying out of tissues and changes in the water chemistry of the muscle tissue, weakening it. Usually homeostasis is maintained, in part by regulating the water chemistry of tissues in organisms.

The general components of a system that maintains homeostasis are shown in Figure 11.5. The system first receives information either from internal cues or from the external world. These cues are the stimulus, which is detected by a receptor, a special protein that monitors the environment. Receptors send messages, based on their stimulation, to a **control center** (in humans this is usually the brain) to cause or affect a response according to some set point value at which the organism should be maintained. The response is carried out by an effector, which is often a muscle that moves or a gland that sends out chemicals to carry out the response.

## Negative Feedback

Homeostasis is most often maintained in living systems by negative feedback mechanisms. During negative feedback, the response counters the effects of the original stimulus. For example, if body temperature rises above 98.6°F, receptors detect and send messages to the skin to sweat and lose heat and to blood vessels in the skin to expand or vasodilate to bring blood closer to skin surfaces and lose further heat. The opposite effect occurs when body temperature decreases, with messages directing the skin to sweat less and shiver to create heat and the capillaries to vasoconstrict (or narrow) to lessen blood flow (and thus lose heat) from the skin's surface. Negative feedback elicits an effect opposite that of the original stimulus. In the case of body temperature, the set point is maintained when temperature changes are detected. The mechanisms of controlling body temperature through negative feedback are important for every day health (Figure 11.6).

**Stimulus**

Something that causes an organ or cell to react.

**Receptor**

A special protein that monitors or received information from the environment.

**Control center**

An operational center for a group of related activities.

**Set point**

The normal value at which a variable physiological state stabilizes.

**Effector**

A muscle that moves or a gland that sends out chemicals to carry out the response.

**Negative feedback**

A key mechanism that regulates the physiological functions in living organisms.

**Vasodilate**

Widening of blood vessels.

**Vasoconstrict**

Narrowing of blood vessels.

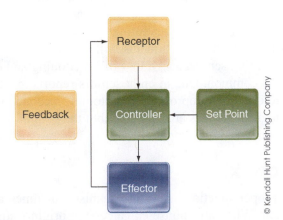

© Kendall Hunt Publishing Company

**Figure 11.5**    General components of a homeostasis system. This diagram represents a negative feedback system.

**Figure 11.6**    A negative feedback system. Body temperature is one of the internal systems that is maintained within a narrow range via a negative feedback system.

**Insulin**

A hormone released from the pancreas, which causes body cells to uptake glucose from the blood to be used by cells.

**Glucagon**

A hormone produced by pancreas, which causes the liver to convert stored glycogen into glucose, sent into the blood and thus available for cells to use.

Another example of negative feedback is the regulation of chemical levels within living organisms. In humans, blood sugar (or glucose) must be maintained around a set point of 90 mg/mL of blood (see Figure 11.7). When blood sugar levels increase, after eating a sugary food such as a donut, the hormone insulin is released from the pancreas. Insulin causes body cells to uptake glucose from the blood to be used by cells. This action decreases glucose levels in the blood. When sugar levels decline, as occurs in between meals, the pancreas produces the hormone glucagon, which causes the liver to convert stored glycogen into glucose, sent into the blood and thus available for cells to use. This prevents the damaging effects of low sugar levels in blood.

The set point of 90 mg/mL of blood glucose varies within a range of roughly 30 units. Many homeostasis mechanisms allow some flexibility around their set points, but often there is a point at which damage occurs. Movement beyond the sugar range in humans may lead to damage of body structures. Diabetes results from an inability to control blood sugar levels.

Negative feedback returns organisms to their optimal points of functioning. Sabrina's Starfish would not properly function within land conditions for very long. Thus, it is expected that it would eventually reset its water chemistry balance once returned to the ocean environment.

Disruptions in negative feedback, preventing returns to normal set points in the body, lead to disease. Disease and its treatments will be a focus of each chapter within this unit. For example, during senescence, a loss of functioning and of proper negative feedback mechanisms is common. One in eight seniors reports mental deterioration at some point in their life due to an improperly functioning feedback mechanism. In the aging process, physiological declines are expected, but the goal of biology and medical science is to find treatments to help people.

## Positive Feedback

**Positive feedback**

A key regulatory mechanism that enhances the original stimulus.

In order to maintain proper functioning of an organism, at times an original stimulus must be exaggerated. When a response enhances the original stimulus, it is called positive feedback. These events are uncommon in everyday functioning, but examples include unusual events such as blood clotting and childbirth.

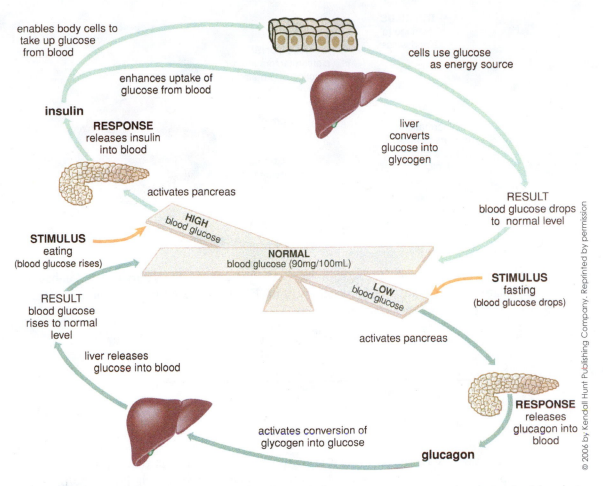

**Figure 11.7**    Blood sugar regulation. Insulin and glucagon restore sugar levels in the blood via a negative feedback system. From *Biological Perspectives*, 3rd ed by BSCS.

During blood clotting, special fragments of cells, called platelets, recognize a break in a blood-vessel wall. When platelets attach to the broken region, they release clotting proteins within them. This causes a clot to start forming to prevent blood loss. Platelets release chemicals that attract more platelets to the site of blood vessel breakage. The original stimulus (the start of the clot by platelets) is enhanced by chemicals released from platelets. The exaggeration of the original stimulus is necessary to get the job done – to form a clot big enough to patch the hole in the blood vessel – before ending the positive feedback mechanism. This process is depicted in Figure 11.8.

Another example of positive feedback occurs during labor contractions. During labor contractions, an odd stimulus for the body – a newly formed baby – elicits a drive to return the body back to normal. Normalcy in this case is a body without a baby inside its uterus. Thus, the goal of positive feedback during labor contractions is to give birth to the baby.

When the baby's head hits receptors on the cervix, which detect pressure, nerve messages are sent to the mother's brain to cause the release of the hormone oxytocin. Oxytocin enhances the stimulation of muscle contraction in the uterus, which further pushes on the baby. Oxytocin therefore causes more pressure on cervical receptors, stimulating more messages sent to the brain to release more oxytocin. It is a self-feeding system, in which more and more contractions stimulate more and more production of

**Platelets**

Special fragments of cells that recognize a break in a blood vessel wall during clotting of blood.

**Oxytocin**

A hormone released by the pituitary gland that enhances the stimulation of muscle contraction in the uterus.

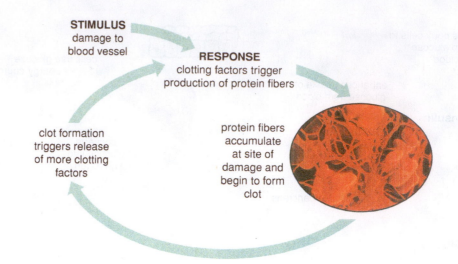

**STIMULUS**
damage to
blood vessel

**RESPONSE**
clotting factors trigger
production of protein fibers

clot formation
triggers release
of more clotting
factors

protein fibers
accumulate
at site of
damage and
begin to form
clot

**Figure 11.8**   Positive feedback: blood clotting within vessels is initiated and enhanced by platelets via a positive feedback system. David Phillips/ Visuals Unlimited; Basics about the Body's Organization.

## PROSTAGLANDINS AND PAIN RELIEF: "YOU LEFT ME, JUST WHEN I NEEDED YOU MOST…"

Prostaglandins have long been recognized as the hormones that increase pain perception. They have several functions, ranging from causing increased pain and enhancing uterine contractions during childbirth to causing menstrual cramps; they also play an important role in immunity and inflammation. During pregnancy, prostaglandins decrease, inhibiting pain and uterine contractions in the expecting mother. This is necessary because contractions would end a pregnancy. However, just at a time when an expecting mother could use pain relief, her prostaglandin levels shoot up during labor. These increase the force and frequency of labor contractions. However, pain is felt ever more acutely during this time because prostaglandins enhance sensations of pain. The concomitant benefit to increased prostaglandins during childbirth is that a mother, who feels great pain, is more likely to push her baby out than if there were little pain. Of course, this is cold comfort for the expecting mother who has enjoyed diminished pain throughout her pregnancy, with reduced levels of prostaglandins. The pain relief left her, just when she needed it most; but it is evolutionarily beneficial because the end result is a more likely successful childbirth. The processes of childbirth are shown in Figure 11.9.

Prostaglandins and pain have a continual association in our bodies at all times. Aspirin is used because it inhibits the effects of prostaglandins in the body, decreasing perceptions of pain. The Nobel Prize in Physiology or Medicine for 1982 was awarded to Bengt Samuelsson and Sune Bergstroem of Sweden and John Vane of England for their efforts in clarifying the role of prostaglandins in the human body.

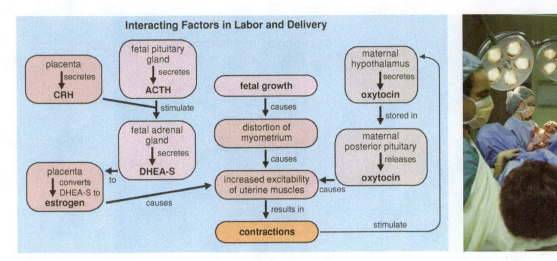

**Figure 11.9** Child birth process: during labor, uterine contractions are stimulated by prostaglandins and oxytocin, both hormones that enhance contractions. It is an example of positive feedback. From *Biological Perspectives*, 3rd ed by BSCS.

the hormone oxytocin that causes the contractions. As shown in Figure 11.9, oxytocin, alongside other factors, brings child birth to fruition. This feedback loop continues until child birth, which is the end result of the positive feedback mechanism. Giving birth returns the female's body to its normal state, one without a developing fetus.

## Systems of Homeostasis: Interplay between Endocrine and Nervous Controls

In each of the examples given in this section, homeostasis is maintained by two systems: 1) the endocrine system, which produces internal chemicals called hormones that cause a response in another organ or tissue; and 2) the nervous system, which transmits messages from one part of the body to another. Hormones such as insulin, glucagon, and oxytocin are made by glands and are only a small subset of chemicals that regulate an organism's life processes. Nerves form a network of cells throughout the bodies of most animals to rapidly communicate between different parts. While nerve messaging is rapid and almost instant, as can be observed when we touch a hot iron, endocrine responses require slower mechanisms of transport. Figure 11.10 shows the glands of the endocrine system and the branches and cells of the nervous system.

Hormones require diffusion to move in between cells of the body. They may take minutes or hours to work, unlike the milliseconds taken by ionic signals of the nervous system. Together, however, the endocrine and nervous systems work efficiently to maintain homeostasis in organisms. These two systems will be discussed in greater detail in other chapters.

**Endocrine system**

Glands and parts of glands that produce internal chemicals called hormones that cause a response in another organ or tissue.

**Nervous system**

Network of nerve cells that transmits messages from one part of the body to another.

## Discovery of Homeostasis

In Chapter 1, Charles Darwin was shown to be influenced by capitalism, which led to the development of his ideas about evolution. As a close parallel, Sir Walter Bradford Cannon (1871–1945), the first discoverer of homeostasis, developed his views in part based on his attraction to communist economic theory. Communism is defined as the economic model in which there is no individual ownership and land, property, and so

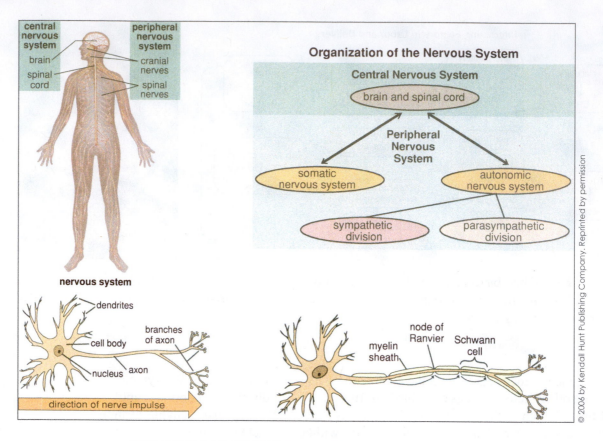

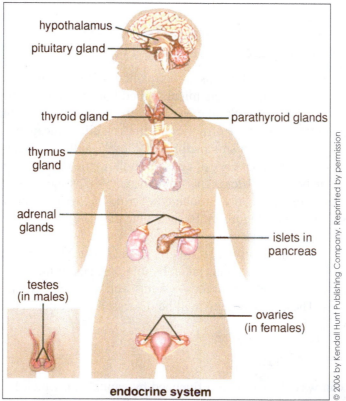

**Figure 11.10**   Hormonal and nerve systems interacting. Both systems communicate with the body's organs, tissues, and cells to integrate different systems. Hormones diffuse through the bloodstream to effect changes in body cells. Nerves send messages to all parts of the organisms. From *Biological Perspectives*, 3rd ed by BSCS.

on are held collectively. The government controls the means of production in a state or nation that is purely communist. In short, there is a central controller in communism as well as in homeostasis.

Communism is the opposite of capitalism, which led to the development of evolutionary thought. In capitalism, as you recall, individuals own the means of production and the fittest businesses survive. In communism, governmental control of the economy maintains a steady state in economic happenings. For instance, during the great depression, communist Russia was effectually immune from the economic downturn. While the communist government controlled which jobs people had, how they would be educated, their access to goods and services, capitalist countries allowed individuals the freedom to fail when the economy languished.

Cannon saw the benefits of communism while he watched his friends and family fail in business. He saw a benefit in communism as a way to stabilize the economy, through redistributing wealth and giving all citizens a small but stable portion.

Cannon was a Harvard physiologist and World War I medical doctor. While working on soldiers with injuries incurred during World War I, Cannon noticed that damage to nervous tissue in the brain or hormonal organs such as the pancreas could create major disruptions and death in his patients. Similarly, he noticed that removal of certain organs, such as the hypothalamus of the brain or the pancreas, resulted in an imbalance in homeostasis or disease. He determined that certain regions of the body of his patients were under a centralized control system. Making an analogy with the centralization of banks and centralized control of jobs and trade in communist economic theory, he hypothesized that the brain served much like the central government in communism, regulating many processes in the human body. He also determined that the nervous and endocrine systems were the two major conduits through which human homeostasis occur. The link between his social and economic leanings in developing Cannon's ideas on homeostasis is clear. Our purpose in relating the development of Cannon's hypothesis about the existence of a control system like homeostasis to his affinity for a particular economic view is not to debate the efficacy of true capitalism, true communism, or its hybrids, but it is to denote the importance of one's upbringing and one's society in influencing scientific thinking. Evolution and homeostasis are just a couple of examples of how society influences the direction of biological thought and of much scientific advancement.

# The Major Types of Tissues

As discussed in Chapter 1, groups of cells performing similar functions and with similar structures are called tissues. There are four types of tissues in animal systems: muscle, epithelial, nervous, and connective. (This can be remembered by an acronym using the first letter of each tissue: MENC). Each tissue has a general function:

1) **Muscle tissue** is composed of cells that are able to contract. Muscles either move materials (as in digestion) or bones and body parts (as in walking or breathing) in a variety of directions via their contractions. Heart, skeletal muscles, and organ muscle such as the stomach contain muscle tissue.

2) **Epithelial tissue** is made of cells that either covers other tissues or cells that produce hormones or other materials for export. The most obvious epithelial tissue is skin, a surface tissue, but epithelial tissue is found in many regions of the body to cover and support areas. In addition to skin, glands such as sweat glands and salivary glands, and linings of the throat, lungs, blood vessels, and digestive tract are composed of epithelial tissue.

**Muscle tissue**

A type of tissue that is composed of cells that are able to contract.

**Epithelial tissue**

Tissue made of cells that either covers other tissues or cells that produce hormones or other materials for export.

**Nervous tissue**

An excitable tissue specialized to send, store, and receive ionic impulses

**3)** Nervous tissue is an excitable tissue, specialized to send, store, and receive ionic impulses much in a way that electricity moves along copper wires. The brain and spinal cord are composed primarily of nervous tissue.

**4)** Connective tissue binds and supports different parts of the animal body. It is sometimes referred to as the misfit group of tissue because so many varied types of tissues belong to it. Blood, bone, ligaments, tendons, fat, and cartilage all belong to the connective tissue grouping.

**Connective tissue**

Tissue that binds and supports different parts of the body.

Histology studies each of the four types of tissues as well as a large number of different, specialized subgroups. Subgroups of tissues are structured in unique ways to help them to carry out their specified functions. The form of cells within each tissue type directs its functions, as expected in complementarity. For example, a group of fat cells have large storage vacuoles specialized to store fat. This enables the primary function of fat storage and insulation. The anatomy of a tissue always fits with its physiology.

There are blood vessels (connective tissue) permeating all of the Starfish tissue, enabling nutrients to arrive at cells. The digestive muscle bands (muscle tissue) through their arms propel Starfish. Along their entire digestive canal, absorption cell layers line the tract (epithelial tissue), which bring nutrients into the body. Nerves (nervous tissue) transect all of the other tissues to direct their activities, in part controlling the rate of processing of waste materials. Sabrina's Starfish case uniquely illustrates how internal tissues can be exposed for observation as they die away. Luckily, the return of the Starfish's four tissues into the ocean water usually results in continued, normal functioning. The four tissue types are shown in the Starfish anatomy in Figure 11.11.

# Epithelial

**Apical surface**

The free side of all epithelial tissues.

**Basement membrane**

A thin extracellular membrane underlying the epithelium of many organs.

**Simple**

An epithelial tissue that is only one cell layer in thickness.

**Stratified**

An epithelial tissue that is two or more layers thick.

**Squamous**

Flat-shaped cells.

**Cuboidal**

Composed of cubical elements.

Sweating, absorbing nutrients, walking on a hot floor, or making oils are all, in part, due to the functions of epithelial tissue. Epithelial tissue looks like sheets of cells arranged in different patterns. All epithelial tissues have a free side called the apical surface. Cells are anchored on their other side onto a basement membrane. Connective tissue, which nourishes epithelial cells, is always found on the other side of the basement membrane. The structure of epithelial cells is shown in Figure 11.12.

Note that epithelial cells are **avascular**, meaning that they do not contain blood vessels of their own to nourish them. Blood vessels within connective tissue provide nutrients and remove wastes for epithelial cells.

The system for naming epithelial tissue is based on two factors: number of layers of cells and the shape of the cell. When epithelial tissue is only one cell layer in thickness, it is called simple. When epithelial tissue is two or more layers thick, it is called stratified. Epithelial tissue has two names: its first name gives the number of layers the tissue has, simple or stratified; its second name indicates the shape of the epithelial cells, usually the shape of those cells nearest to the apical layer. The three shapes of cells are squamous, or flattened and squashed in appearance; cuboidal, or square, cube-like in shape; and columnar, or shaped like a brick, column-like. The epithelial cell shapes are shown in Figure 11.13. Epithelial cells found along the surface of Sabrina's starfish's skin, simple columnar in classification, protected the animal.

Each of these cell shapes is tightly held together and serves three functions: 1) to cover other tissues, organs, and systems, 2) to transport materials, and 3) to secrete products.

**l)** **Protection**: In their role of covering other structures, epithelial tissue serves as an important first line of defense in protection. Tight junctions and desmosomes hold epithelial cells closely, preventing leakage and penetration by enemies such

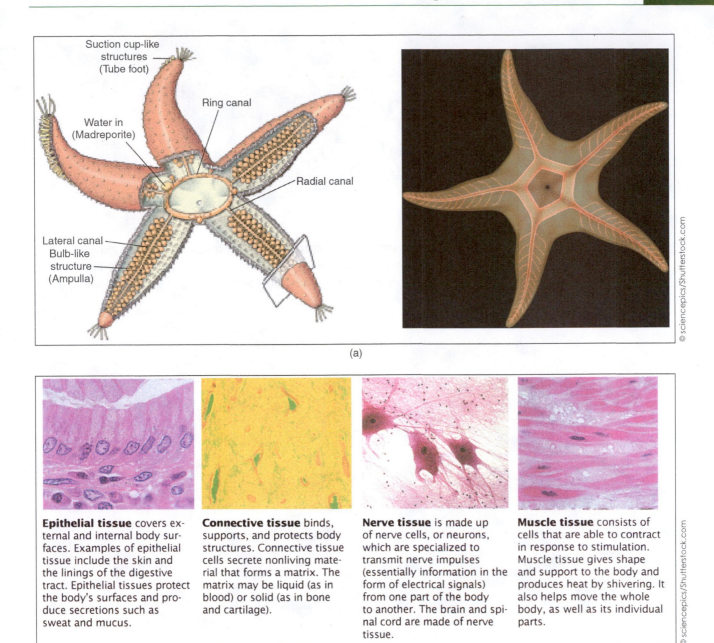

Suction cup-like structures (Tube foot)

Ring canal

Water in (Madreporite)

Radial canal

Lateral canal
Bulb-like structure (Ampulla)

© sciencepics/Shutterstock.com

(a)

**Epithelial tissue** covers external and internal body surfaces. Examples of epithelial tissue include the skin and the linings of the digestive tract. Epithelial tissues protect the body's surfaces and produce secretions such as sweat and mucus.

**Connective tissue** binds, supports, and protects body structures. Connective tissue cells secrete nonliving material that forms a matrix. The matrix may be liquid (as in blood) or solid (as in bone and cartilage).

**Nerve tissue** is made up of nerve cells, or neurons, which are specialized to transmit nerve impulses (essentially information in the form of electrical signals) from one part of the body to another. The brain and spinal cord are made of nerve tissue.

**Muscle tissue** consists of cells that are able to contract in response to stimulation. Muscle tissue gives shape and support to the body and produces heat by shivering. It also helps move the whole body, as well as its individual parts.

© sciencepics/Shutterstock.com

(b)

**Figure 11.11**    a. The body plan of a starfish. The four tissue types found within all animals including muscles and nerves in the arms of the starfish, the epidermal covering the skin and its underlying exoskeleton functioning as the support structures, made of connective tissue. b. The four type of tissues (micrographs).

as viruses and bacteria. In the skin, for example, multiple layers of cells prevent entrance by many microbes that inhabit our skin's surface. Internally, the stomach compartment has a very low pH, with acidic contents that would destroy stomach cells. Epithelial cells line the stomach to prevent such a breach from causing damage, such as ulcers.

2) **Transport**: Transport within the body works through epithelial cells acting as a gatekeeper to allow certain materials through its layers and not others. For example, in the kidneys, epithelial cells regulate in which molecules are eliminated

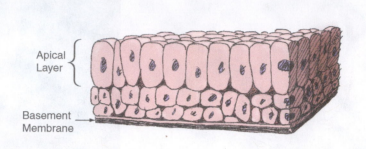

Apical
Layer

Basement
Membrane

**Figure 11.12**    General structure of epithelial tissue: apical surface, basement membrane, stratified layers, connective tissue is on the other side of the basement membrane of epithelial tissue. Adapted from *Anatomy I and Physiology Lecture Manual* by John Erickson and c. Michael French.

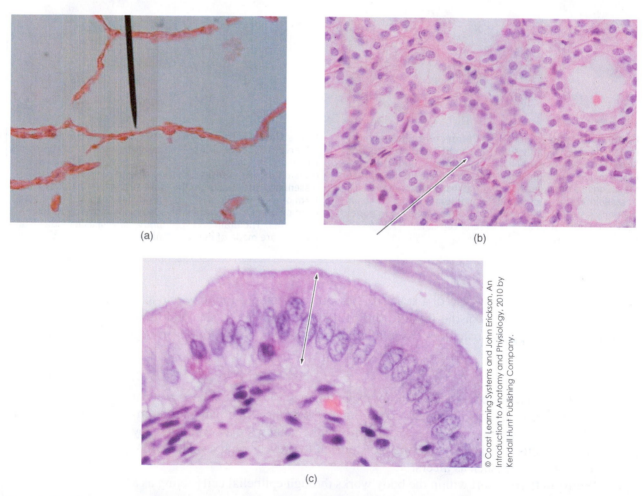

(a)

(b)

(c)

**Figure 11.13**    Examples of a. transport (air sac of lung) through the squamous shape; b. secretion (tubules of glands) from the cuboidal shape, and c. absorption (digestive tract) by the columnar shape.

as urine and which are reabsorbed into the bloodstream for use. Often based on size and shape of the molecule, epithelial cells filter out materials flowing through the blood.

**3) Secretion**: Glands are specialized groups of epithelial cells that secrete substances either out of the body or into body compartments. Materials that are secreted include hormones, saliva and sweat, milk, mucus, and earwax.

## Simple Epithelial Tissues

A number of simple epithelial types of tissues are found throughout the body. All of the simple forms are used for transport because the single layer of cells is useful as a screen to filter and absorb materials selectively. Multiple layers of stratified tissues would prevent the ease of transport afforded by a simple design. Would a thick tissue with many layers make sense surrounding an air sac? No. **Simple squamous** epithelial tissue, for example, is found in areas in which exchange of materials occurs. In air sacs of the lungs, where oxygen and carbon dioxide are rapidly moved through a thin membrane, a simple layer of epithelial cells facilitates gas exchange. Nutrients and wastes are moved between blood and the atmosphere via capillaries. When thickened, as in emphysema cases, transfer of gases is impeded and diseased lungs have poor oxygen exchange. Simple squamous epithelial tissue is also found in capillary walls, as it is thin enough to allow exchange of nutrients and wastes.

In another example, within the digestive system **simple columnar** epithelial tissue also acts as a filter to absorb smaller, digested products. Simple columnar cells are larger than squamous cells and thus involved in larger scale absorption processes. Large amounts of nutrients are needed for larger animals to survive and columnar cells are able to accomplish this.

## Stratified Epithelial Tissue

Owing to multiple layers of stratified cells, some tissues are best suited for protection. With many layers, damage to one or even several cell layers does not compromise the integrity of the tissue. Tissues and organs beneath stratified epithelial cells are well protected. To illustrate, skin is composed of stratified squamous epithelial tissue, numbering more than 100 layers in thickness in some areas, such as the soles of feet. Multiple layers of flattened cells protect the blood vessels, nerves, and bones that lie beneath and within the skin. Their layers form from a deeper, lower layer, called the basal layer. The basal layer of cells lies on a basement membrane that anchors it. The basal cell layer is usually cuboidal or columnar in shape and is mitotic. Stratified squamous epithelial tissues are also found in linings of the esophagus, mouth, and vagina.

Some stratified epithelial tissue forms glands. Glands are collections of epithelial cells that secrete a product such as hormones. When glands are active, their products are made in greater amounts. There are two types of glands: endocrine glands, such as the adrenal glands atop the kidneys and the pineal gland in the brain, which have no ducts into the bloodstream and produce hormones; and exocrine glands, which have ducts emptying their contents into the bloodstream, such as saliva, milk, mucus, and sweat. Exocrine glands serve in many parts of the body, such as the salivary glands to lubricate and provide nourishment.

Some stratified tissues form specialized structures. Transitional epithelium, for example, is used in areas of the body that require stretching such as the bladder, which holds urine. Its form fits its function. Transitional epithelial tissue looks at times cuboidal and at other times squamous in shape. When the bladder fills with urine, its cells

**Basal layer**

The lowest layer of epithelial layers.

**Glands**

Collections of epithelial cells that secrete a product such as hormones.

**Transitional epithelium**

A type of tissue that consists of multiple layers of epithelial cells, which looks at times cuboidal and at other times squamous in shape.

flatten and become squamous in shape. This shape change allows the bladder to be flexible and distend when urine enters it. Cells that accommodate fluid changes, such as transitional epithelial, are well suited for the urinary system in humans.

## Connective Tissue: An Overview

Connective tissue is the most abundant type of tissue in most animals. There are many types of connective tissue; these different types hold together and support many parts of the body. Connective tissue supports many functions of movement, such as walking; oxygen is transported by blood and used by muscles during walking; cartilage covers the ends of long bones to cushion them as the leg muscles move them; ligaments and tendons, also connective tissues connect bones and muscles together to allow the joint to move during running.

Construction of connective tissues uses the same general plan used for all connective tissue types, with its general structure given in Figure 11.14. An extracellular matrix made of noncell substances (polysaccharides and proteins) is found in every connective tissue type. The extracellular matrix may be solid (bone), semi-solid (cartilage), or liquid (blood plasma). Within the extracellular matrix, cells are found suspended such as: mature cells of the tissue, defense cells, or macrophages, and fibroblasts, which build new tissue. Connective tissue cells produce the extracellular matrix. Holding the cells together within its matrix, fibers (collagen, reticular, elastic) are embedded in connective tissue, giving

**Extracellular matrix**

Is a collection of proteins and carbohydrates found in every connective tissue type.

**Fibers**

Threadlike structure embedded in connective tissue, giving strength and support.

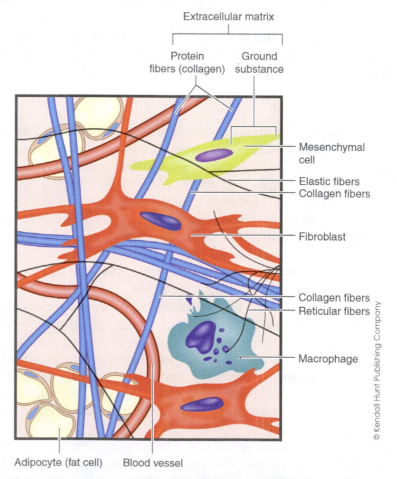

**Figure 11.14** Connective tissue structure. The fibers and ground substance of the extracellular matrix are infused with cells of many types.

strength and support. Fibers are nonliving and provide a site onto which cells may anchor. The fiber types each have a unique function: collagen or white fibers add strength, elastic fibers allow pulling on the tissues and reticular fibers join tissues together, also adding strength. As you may deduce, elastic fibers are found in tissues that need to withstand a great deal of tension or pull, such as in the ear or along some blood vessels. Collagen fibers are found in tissues that need added strength, such as bones and tendons. The construction of connective tissues in terms of cells and fibers determines their unique functions.

## Types of Connective Tissue

We will look first at Connective tissue proper (CTP), a set of tissue types that act as package materials in the body. CTP may be loosely or densely packed, with varying types shown in Figure 11.15. All connective tissues emerge from embryonic connective tissues, called mesenchyme.

The first type of loosely held connective tissue is areolar connective tissue, which cushions organs and other tissues. Areolar tissue mirrors the function of packaging peanuts in shipped boxes. When humans move in one direction, organs often slide past each other in the other directions and are cushioned by areolar tissue to prevent damage. The second loose CTP type is reticular connective tissue, which traps foreign invaders such

**CTP (connective tissue proper)**

A set of tissue types that act as package materials in the body.

**Mesenchyme**

Embryonic connective tissue that gives rise to all the connective tissues.

**Areolar**

packaging type tissue.

**Reticular**

A connective tissue that traps foreign invaders such as bacteria.

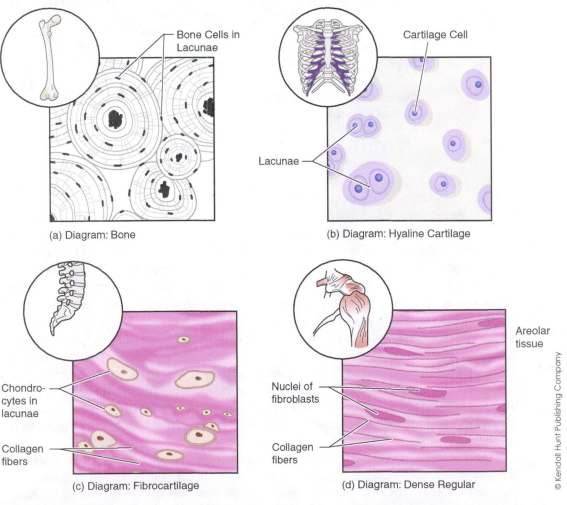

(a) Diagram: Bone

Bone Cells in Lacunae

(b) Diagram: Hyaline Cartilage

Cartilage Cell

Lacunae

(c) Diagram: Fibrocartilage

Chondro-cytes in lacunae

Collagen fibers

(d) Diagram: Dense Regular

Nuclei of fibroblasts

Collagen fibers

Areolar tissue

© Kendall Hunt Publishing Company

**Figure 11.15**    Types of connective tissues: each connects, supports, and anchors materials within the body.

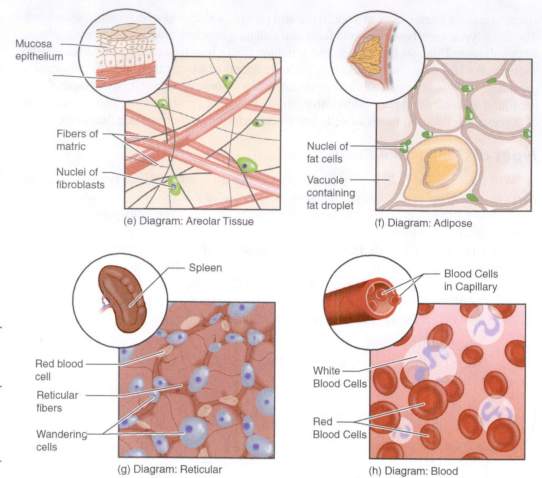

(e) Diagram: Areolar Tissue

Mucosa epithelium

Fibers of matric

Nuclei of fibroblasts

(f) Diagram: Adipose

Nuclei of fat cells

Vacuole containing fat droplet

(g) Diagram: Reticular

Spleen

Red blood cell

Reticular fibers

Wandering cells

(h) Diagram: Blood

Blood Cells in Capillary

White Blood Cells

Red Blood Cells

**Figure 11.15** (*continued*)

**Adipose**

A body tissue used for fat storage.

**Dense regular**

Connective tissue composed mostly of collagen fibers.

**Tendon**

Strong fibrous tissue that anchors muscles to bones.

**Ligament**

Band of tissue that anchor bones to bones.

**Dense irregular**

Connective tissue composed of irregularly arranged fibers.

**Special connective tissue**

A unique connective tissue that has either a rigid or a liquid extracellular matrix.

**Blood**

A red fluid that connects different parts of the body by providing nourishment and removing wastes.

as bacteria. Reticular connective tissue appears as a spider's web, with large amounts of reticular fibers strewn about. Reticular tissue is found in the lymphatic system, such as lymph nodes and the spleen, which function in immunity. Adipose tissue is also able to store energy in the form of fat, stored in adipose tissue. Adipose tissue cells contain very large fat vacuoles, sometimes comprising over 90% of the space in adipose tissue cells.

Dense CTP often contains fibers that give strength and flexibility to its tissues. Dense regular connective tissue is tightly packed CTP that is composed mostly of collagen fibers. These tissues include tendons, which anchor muscles to bones, and ligaments, which anchor bones to bones. Dense regular connective tissue has a high amount of tensile (pulling) strength. Alternatively, dense irregular connective tissues are composed of irregularly arranged fibers; because they crisscross in varied directions, these tissues are also able to withstand pulling forces from many directions. Dense irregular tissues are found in the dermis (lower part) of skin or in capsules of joints to allow pulling and movement along several planes. The skin, for example, is able to be pulled from several directions.

The second type of connective tissue, called special connective tissue, is unique because it can have either a rigid or a liquid extracellular matrix. Blood and bone are the most familiar type of connective tissue. Blood connects different parts of the body by providing nourishment and removing wastes in all parts of the body. However, the matrices of blood and bone are very different from each other. Blood is formed within the bones in marrow cavities, but has a liquid matrix. Its cells, such as red blood cells, travel in vessels and along with dissolved proteins within a watery matrix.

Bone has a solid form, with calcium salts embedded within fibers of its extracellular matrix. Bones connect together and form our skeletal system. The cellular arrangement of bone serves as support and protection for other body parts, with concentric rings or pillars of calcium salts. Bone forms a hard substance and is able to withstand tremendous compressive (pushing) and even tensile pressures. In fact, bone has the same tensile strength as steel of the same size and shape.

Cartilage is a dense connective tissue that provides cushioning support in vertebrates. It is a flexible, hard gel that is less hard than bone (a solid) but harder than tendon (a softer strap-like tissue). Cartilage is strong but has enough softness to act as a cushion between bones and joints; it also provides support in ears, noses, and fetal skeletons. In some organisms, such as the shark, it forms the entire skeletal structure.

While there are three types of cartilage, all forms have the same general structure. Cartilage is made of large amounts of collagen and elastin fibers along with proteins embedded in between the fibers. Each cell within cartilage is called a chondrocyte and is found inside a lacuna "lagoon" in which the chondrocyte sits.

Hyaline cartilage is composed of large amounts of collagen fibers, giving it strength. Hyaline cartilage is found in the nose, embryonic skeleton, trachea, and larynx (voice box), and forms the ends of long bones. It serves to cushion the ends of bones and enable joints to move without damaging associated bones.

Elastic cartilage is composed of large amounts of elastic fibers, which is able to withstand pulling forces. If you ever pull your ear, it works well to resist the force placed upon it. Elastic cartilage is found in the external ear and the epiglottis, which covers our windpipes. Elastic cartilage is poor in resisting compression. Place your ear in between your thumb and a finger and push it together. It folds quickly, without much resistance. Elastic fibers are very weak in maintaining shape against compressive forces. It has an extracellular matrix that contains randomly scattered elastic fibers surrounding its lacuna.

Fibrocartilage contains elastic and collagen fibers in its extracellular matrix. It is found in between the vertebrae of the backbones and in the pubic symphysis, which is a tuft of material that connects the hip bones. Fibrocartilage is able to resist both tensile and compressive forces effectively. For example, while our backbone cushions the forces placed upon it while walking or running, it is flexible enough to allow bending and stretching. It has an extracellular matrix with relatively parallel fibers surrounding its lacuna.

## Muscle

Muscles function to move things – bones, ligaments, tendons, internal materials within the intestines. Muscles contract when nerves stimulate them. Muscles have long, strap-like fibers, which work together to hold onto the body parts, particularly bones, that they move.

There are three types of muscle tissues (see Figure 11.16):

1) Skeletal muscle, which is long, and contains alternating patterns of proteins called striations, or stripes. Skeletal muscles are under voluntary control; thought is required to move them. They are found attached to bones of the skeleton and move the bones. Some skeletal muscles move no bones, such as the tongue. However, their purpose is to provide movement of the skeleton. Sabrina's starfish's arms and mouth are composed of skeletal muscle. It is therefore under voluntary control, allowing conscious regulation of the rate of movement.

---

**Bone**

Is the substance that has a solid form, with calcium salts embedded within fibers of its extracellular matrix.

**Cartilage**

A dense connective tissue that provides cushioning support in vertebrates.

**Chondrocyte**

A cell within cartilage.

**Lacuna**

An open space containing a chondrocyte in cartilage.

**Hyaline cartilage**

A type of cartilage that is composed of large amounts of collagen fibers, giving it strength.

**Elastic cartilage**

A type of cartilage that is composed of large amounts of elastin fiber, which is able withstand pulling forces.

**Fibrocartilage**

Cartilage that contains elastic and collagen fibers in its extracellular matrix.

**Skeletal muscle**

Long muscles that are found attached to bones of the skeleton and move the bones.

**Striations**

Alternating patterns of proteins in the skeletal muscle.

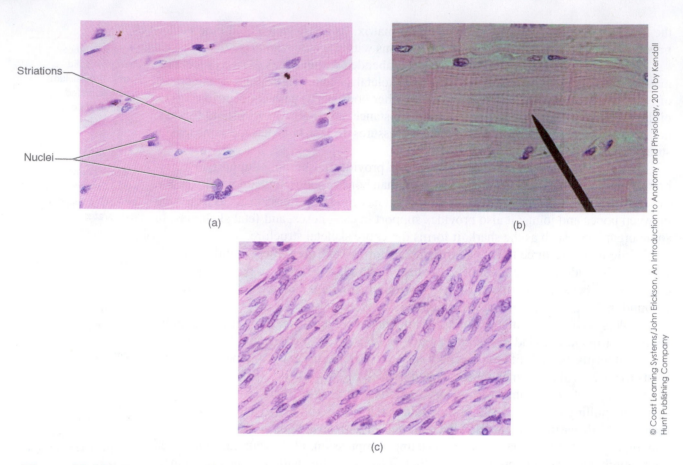

Striations

Nuclei

(a)

(b)

(c)

**Figure 11.16** Three types of muscle tissue. a. Skeletal muscle is found on the bones. b. Cardiac muscle is found in the heart. c. Smooth muscle is found in the organs.

**Cardiac muscle**

The muscle found only in the heart and beats spontaneously to pump blood throughout the body.

**Smooth muscle**

Muscle tissue that provides support and propels movement of food through the organs in which it is found

**Neuron**

Are special cells that store and transmit information in animals.

**Cell body**

The central part of the neuron that contains the machinery of the cell, with organelles and a nucleus that directs nerve functions.

2) Cardiac muscle is found only in the heart and beats spontaneously to pump blood throughout the body. Cardiac muscle is striated and involuntary, meaning that one does not need to think about this muscle to contract it. Cardiac muscles' cells are long and contain many branches. To add strength to withstand the pumping pressures placed upon them during heart beats, cardiac muscles have intercalated discs. Intercalated discs are specialized desmosomes that bind cardiac cells together.

3) Smooth muscle is not striated and is involuntary. Its cells are spindly in shape and are found on or within organs, such as the stomach or small intestines. They provide support and propel movement of food through the organs in which they are found. Smooth muscles are weaker than cardiac and skeletal muscles, lacking striations and support, but they function well in organs. Muscles along the digestive tract of Sabrina's Starfish are composed of smooth muscle. It is thus not controlled consciously and involuntarily holds in wastes as well as digestive tissues.

# Nervous

**Nerve cells,** or neurons, are special cells that store and transmit information in animals. They are found in the brain and spinal cord of vertebrates. Neurons are composed of a cell body that contains the machinery of the cell, with organelles and a nucleus that

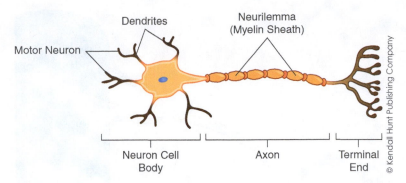

© Kendall Hunt Publishing Company

**Figure 11.17**   Giant multipolar neuron. A neuron conducts signals along its long anatomy. There are many dendrites that receive information from other neurons, sending messages to the cell body.

directs nerve functions. Two types of projections emanate from the cell body: dendrites, which receive signals, and an axon, which transmits information to other cells away from the cell body. The parts of the neuron are shown in the example in Figure 11.17.

Almost 90% of nervous tissue is composed of helper nerve cells called **glial cells**, also called neuroglia. Neuroglia cells do not carry or store information but instead assist in nourishing or supporting neurons. These are vital for nervous system functioning. Albert Einstein had more neuroglia than the average person, perhaps contributing to his higher level of intelligence.

In vertebrates, all of the neurons within the brain and spinal cord comprise the central nervous system, or **CNS**. Brain and spinal nerves process information as they enter and travel within the CNS. Those nerves found outside of the CNS are classified as the peripheral nervous system, or **PNS**. PNS nerves detect stimuli, and send messages to and from the CNS using peripheral nerves. Peripheral nerves send messages to muscles or glands to elicit a response. In Sabrina's Starfish, nerves within the nerve network connect beneath its epidermis along its radial nerve in each arm. Starfish nerves do not connect to have a brain, but are joined to a central ring to allow some degree of central control. PNS and CNS subdivisions in humans are shown in Figure 11.18.

Neurons communicate with one another but also with muscle cells to produce their movements. Nerve transmission may be compared with electricity, in that both transmit a current of ionic charges. In the case of nerve impulses, sodium and potassium ions move along neurons instead of electrons as found in electricity. Thus, transmissions excite cells that they move along.

# The Language of Anatomy

## Animal Organization

Animals are ordered systems, composed of groups of cells in many arrangements, including tissues. These systems are organized and work together to maintain life functions. The goal of all life is to separate itself from its environment. This separation allows organisms to keep conditions, such as pH and temperature, appropriate through using homeostasis for maintaining life functions.

As discussed in Chapter 1, life is ordered into a hierarchy of organized structures. Tissues and organs work together in the form of organ systems to carry out vital functions such as breathing and obtaining nutrition.

**Dendrite**

A long thread-like structure of the nerve cell that receives signals from other cells

**Axon**

A long thread-like structure of the nerve cell that transmits information to other cells away from the cell body.

**Neuroglia**

Helper nerve cells present in the nerve tissue.

**CNS (central nervous system)**

The part of the nervous system that consists of the brain and spinal cord.

**PNS (peripheral nervous system)**

The portion of the nervous system situated outside the brain and spinal cord.

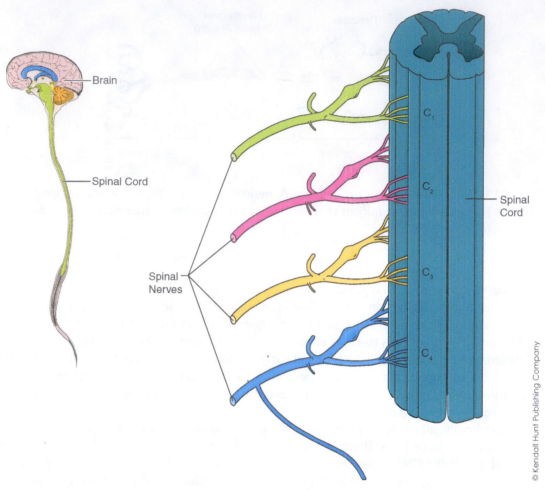

**Figure 11.18**   Divisions of the PNS (peripheral nervous system and CNS (central nervous system). The CNS consists of the brain and spinal cord nerves while all other nerves are considered part of the PNS. Illustration by Jamey Garbett.

When studying the anatomy and physiology of the human body, a specific language of anatomy is used. Medical terminology and references are described in the next section to demonstrate how to effectively communicate with in the medical community. One cannot merely state that there is a spot on one's liver; it is too general a statement. Instead, to be specific, in medicine, it is more specifically described as a lesion in the left quadrant of the liver, 1 cm in diameter. The metric system is used to quantify the size of the lesion. There is always a term for the specific location of the lesion on the human body to give detail to its description.

**Body landmarks (surface regions)**

Terms for the specific location of the lesions on the human body to give detail to their descriptions.

## Surface Regions

In medicine, there are numerous body landmarks or **surface regions** with specific names to describe the locales. These body landmarks are used regularly in medical communication. Many of the regions in Figure 11.19 reappear in this unit on human biology.

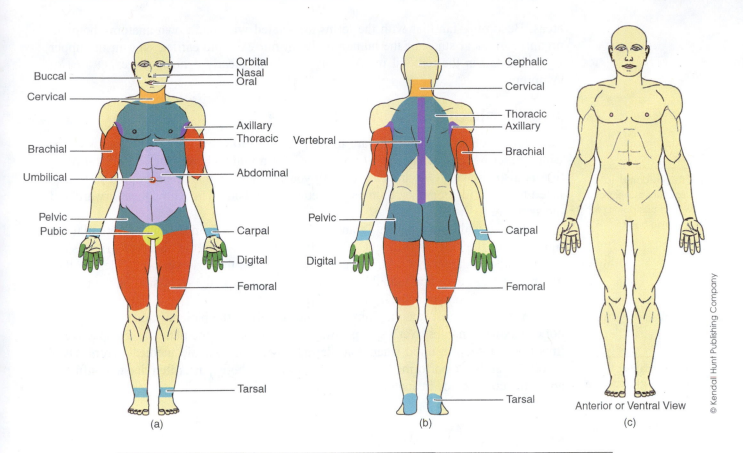

© Kendall Hunt Publishing Company

**Anatomical Landmarks**

| | |
|---|---|
| Abdominal: Anterior trunk between ribs and pelvis | Nasal: area of the nose |
| | Oral: mouth |
| Axillary: armpit | Orbital: area of the eye |
| Brachial: arm | Pelvic: area of the pelvis |
| Buccal: area of the cheek | Pubic: area of the genitalia |
| Carpal: wrist | Tarsal: ankle |
| Cephalic: head | Thoracic: chest |
| Cervical: region of the neck | Umbilical: navel |
| Digital: fingers and toes | Vertebral: area of the spine |
| Femoral: thigh | |

**Figure 11.19**   a. Surface regions of the human body. Each area of the body has a name, which often corresponds to structures found within that region. For example, the femoral region contains the femur (bone), the femoral nerve, the femoral artery, and the femoral vein. b. Table of anatomical landmarks. c. Anatomical position.

The bones, muscles, nerves, and other structures will often have the same name when they appear within one of the surface regions. For instance, the femur is the primary bone in the upper leg area or femoral region, which contains a femoral nerve, a femoral vein, and a femoral artery. Each of these structures includes the term "femoral" to indicate the region of the body where it is found. Anatomy is made much easier when terminology is remembered and ordered.

When studying the surface regions in Figure 11.19, find the anterior and posterior body landmarks on yourself and on a human torso, palpating the structures in those areas. Becoming familiar with the terms associated with one's own anatomy helps to organize medical study of the human body. In our example earlier, pain in the upper leg is located in the femoral region. This descriptor clearly denotes the exact area of symptoms.

## Anatomical Position

**Anatomical position**

The position that describes a specific way of positioning for a human body.

The regions of the human body are usually referred to using the anatomical position, which describes a specific way of positioning for a human body. The anatomical position is also shown in Figure 11.19(c). As you can see, in this positioning a person is faced forward, feet together, thumbs pointed to the outside. The palms face forward and the arms are to the side.

When considering a patient in anatomical position, it is important to note that the right side of the body is the right side of the patient and not the observer's right side. The left side of the patient is his or her left side and not the observer's left side. In other words, the right acromial (shoulder) region of a patient is the right side of the patient even though it is on your (the observer's) left side.

An observer must imagine himself or herself from the patient body's perspective, not one's own. The observer's side is always the opposite of the patient's. Perspective is important to keep in mind when considering directions to avoid medical errors. Over 2,700 surgeries are performed each year on the wrong body part, often due to confusion about the correct side in anatomical position.

**Directional terms**

Are words that describe a location on the human body.

**Proximal**

Situated close to a point of attachment.

**Distal**

Situated away from the point of attachment.

**Anterior**

At the front of or situated before.

**Posterior**

Backside.

**Lateral**

Of or relating to the side.

**Medial**

Situated in the middle.

**Superficial**

A surface marking.

**Deep**

A surface marking that is considered superficial and away from a surface.

## Directional Terms

Anatomical position is a starting point in describing directions and regions on the human body. However, direction is always a relative term. Directional terms are words that describe a location on the human body. Two locations should be given when using directional terms to show position. Consider the statement, "The United States is south…" "South of what?" should be the next question. A second location would show how the United States is positioned in relation to other areas. The United States is south of Canada but north of Mexico.

In anatomy, and all medical communication, this rule also applies. For example, the statement that the pollex (thumb) is distal makes no sense, except in relation to another anatomical structure. Using the terms found in Figure 11.20, complete the sentence: the pollex is distal _____. If you wrote in "distal to the arm (brachial region) or to the forearm (antebrachial region)," you were correct.

Directional terms are used to describe one location on the body in relation to its position with another location. The terms proximal and distal, used in our example, describe structures on limbs (arms and legs) only. Whenever a structure is farther from the point of attachment of a limb, it is considered distal or more distant from the point of attachment of the limb. When a structure is closer to the point of origin, it is a proximal location. The brachial (arm) region is thus proximal to the antebrachial (forearm) region. And the vice versa approach is also true: the antebrachial (forearm) region is thus distal to the brachial (arm) region.

Figure 11.20 depicts the general directions of the terms on a human body. The following directional terms are opposites: anterior (ventral) and posterior (dorsal); **superior** (cephalic) and **inferior** (caudal); lateral and medial; and superficial and deep. *Anterior* refers to the belly side of a human, and *posterior* refers to its back side. A four-legged animal's anal direction is called its posterior and its head is referred to as its anterior, but in humans the classification is different. In humans, any body part toward its head

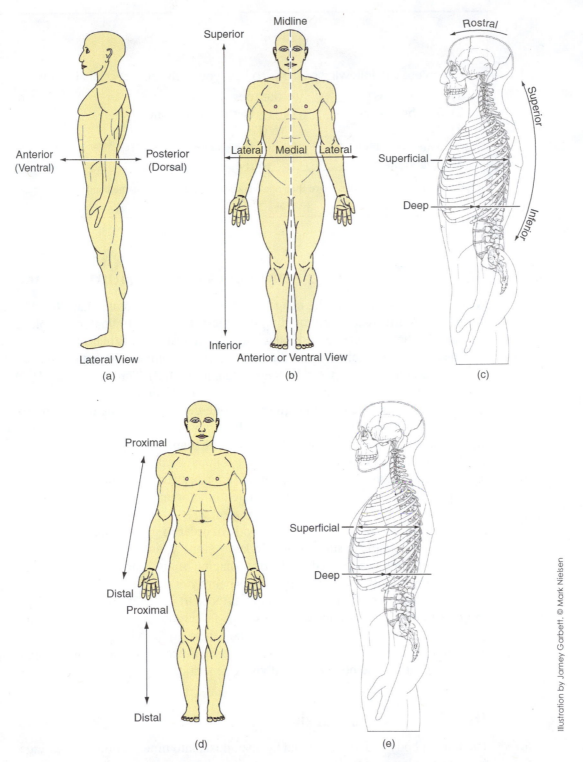

Lateral View
(a)

Midline
Superior
Lateral Medial Lateral
Inferior
Anterior or Ventral View
(b)

Anterior
(Ventral)
Posterior
(Dorsal)

Rostral
Superior
Superficial
Deep
Inferior
(c)

Proximal
Distal
Proximal
Distal
(d)

Superficial
Deep
(e)

Illustration by Jamey Garbett. © Mark Nielsen

**Figure 11.20** a–e. Direction terms are used in medicine and to indicate location in animals.

is said to be superior and anything away is said to be inferior. Any body part away from the midline along the middle of a human is termed lateral and toward this line is considered medial. A surface marking is considered superficial and away from a surface is called deep. For example, Sabrina's starfish's endoskeleton is deep to its epidermis but its epidermis is superficial to its endoskeleton.

---

**DIRECTIONAL TERMS PRACTICE**

Complete the following exercise to help you with the directional terms:

1) The knee is _____ to the toes.
2) The breastbone is _____ to the collar bone.
3) The elbow is _____ to the wrist.
4) The stomach is _____ to the spleen.
5) The kidneys are _____ to the spine.

\* Note that there may be more than one answer to the above fill-in-the blanks.

---

# Body Planes: Imaginary Lines on the Human Body

**Sagittal plane**

A vertical plane that divides the left and right side of an organism.

**Frontal plane**

The plane that divides the front (anterior) and back (posterior) regions of the body.

**Oblique plane**

A plane running at an angle to the organ or organisms.

While a plane conjures up the image of a flying aircraft, in mathematics and anatomy, it refers to an imaginary line dividing up different regions. In anatomy, a plane shows the cut and the perspective with which to view a structure. When observing different regions in the body, there are four different planes anatomists use that can move through the body to divide it into parts: 1). Sagittal; 2). Frontal; 3). Transverse; and 4). Oblique. These planes are shown in Figure 11.21. A sagittal plane divides the left and right side of an organism. When the sagittal plane runs along the middle of the organism, it is called a **midsagittal plane**. When the sagittal plane runs along the side of the middle of the organism, it is called a **parasagittal plane**. The plane that divides the front (anterior) and back (posterior) regions of the body is known as the frontal plane. The plane that divides the top (superior) from the bottom (inferior) part of the body is called the transverse plane. When a plane runs at an angle to the organ or organisms considered, it is called an oblique plane (not shown in Figure 11.21).

Planes divide different parts of the body in order to give perspective in surgery and in study. For example, a surgeon should know along which plane an incision should be made during a medical procedure. During imaging of body structures, it is vital to know the correct perspective with which a part is being viewed.

If you recall the movie "Ghost Ship," at the start of the movie, a metal rope breaks and cuts the passengers into two pieces, tops falling down to the ground. Which plane was sliced? A transverse plane made the whole movie quite a horror! Figure 11.21 shows three of the four planes along a human specimen. Select an organ from the human torso and practice using the body planes to visualize sections.

# The Abdominopelvic Regions

The human body is further divided by anatomists into nine sections, appearing as a tic-tac-toe grid along the body. These nine regions delineating specific parts of the abdominal and pelvic area (abdominopelvic for short) are shown in Figure 11.22: Right and left hypochondriac, epigastric, right and left lumbar, umbilical, right and left iliac (inguinal), and hypogastric. Figure 11.22 also shows the nine regions with the organs of the human body found within those regions. As practice, use Figure 11.22 to organize a list of the organs found within each region. For example, the rectum is located within the hypogastric region of the body. It is at the end of the digestive canal and beneath most of the digestive organs.

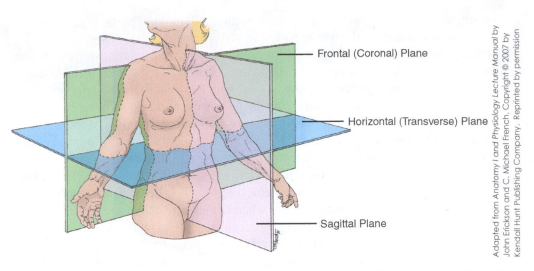

Frontal (Coronal) Plane

Horizontal (Transverse) Plane

Sagittal Plane

**Figure 11.21**  The four body planes: 1) Sagittal; 2) Frontal; 3) Transverse; 4) Oblique (not shown and rarely used, but divides at an angle). Each divides body parts in two different sections to study. Adapted from *Anatomy I and Physiology Lecture Manual* by John Erickson and C. Michael French.

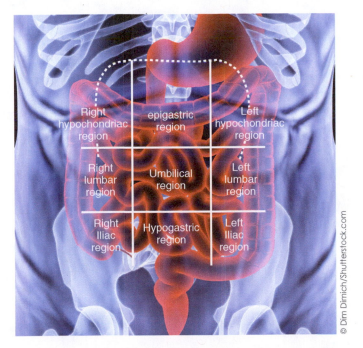

**Figure 11.22**  Abdominopelvic regions: The abdomen and pelvis is together divided into nine specific areas.

## Organ Systems

This unit will treat the major organ systems of the body. It is important to note that, while we look at them individually here, they do not operate in isolation. As mentioned previously, they act as an integrated whole, to keep an organism alive and functioning. This is no easy task and requires a careful interplay between the organs of each system with one another to maintain homeostasis.

You might think of the body as a machine, with many moving parts. If just one has a malfunction or imbalance, the whole body can suffer from disease.

As you study the organ systems in the coming chapters, their relationships with one another will become clear. The optimal pH and ion conditions, regulated by the kidneys (renal system) enable proper nerve transmissions, which use ions. Healthy nerves allow for clear brain functioning and control of our skeleton's movements. Moving one's skeleton helps the heart's health by relieving it from some work in pumping blood because muscles also push blood along the body. A healthy heart pumps blood through the kidneys efficiently, and does not do damage by exerting excess pressure.

Inspect the organ systems in Figure 11.23 and identify the organs in the list below. Assign each organ to the proper organ system(s) to which it belongs above. In which organ system are the kidneys, which fail first when humans are dehydrated, like the Starfish in the story, classified? Yes, in the urinary organ system; but all organ systems work together to the extent that, when one fails, others compensate. The heart races faster when kidney failure occurs to compensate for the buildup of wastes. Intravenous fluids can quickly help to return a dehydrating person's water chemistry back into balance. As discussed at the start of this chapter, organ systems are interdependent, working together closely for proper body functioning.

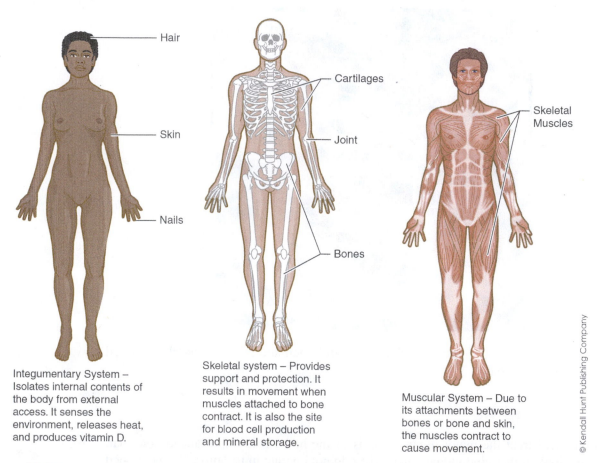

Hair

Skin

Nails

Cartilages

Joint

Bones

Skeletal Muscles

Integumentary System – Isolates internal contents of the body from external access. It senses the environment, releases heat, and produces vitamin D.

Skeletal system – Provides support and protection. It results in movement when muscles attached to bone contract. It is also the site for blood cell production and mineral storage.

Muscular System – Due to its attachments between bones or bone and skin, the muscles contract to cause movement.

© Kendall Hunt Publishing Company

**Figure 11.23** An overview of the twelve organ systems of the human body. Organ systems of the human body

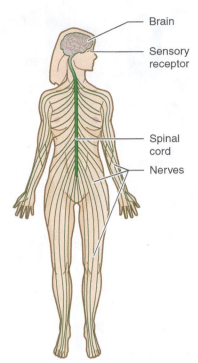

- Brain
- Sensory receptor
- Spinal cord
- Nerves

Nervous System – Provides rapid communication, sensing of the environment, analysis and decision making, and stiumulation of muscle.

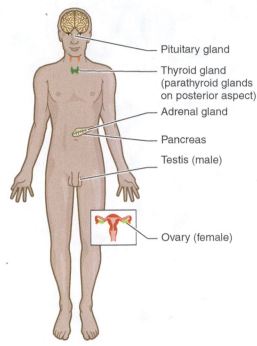

- Pituitary gland
- Thyroid gland (parathyroid glands on posterior aspect)
- Adrenal gland
- Pancreas
- Testis (male)
- Ovary (female)

Endocrine system – Provides slow sustained communication between cells through chemical messages (hormones)

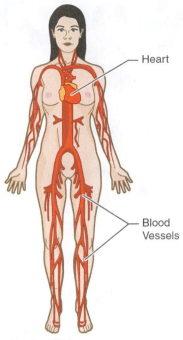

- Heart
- Blood Vessels

Cardiovascular system – Consists of a pump (the heart) and vessels through which substances can be transported in the blood

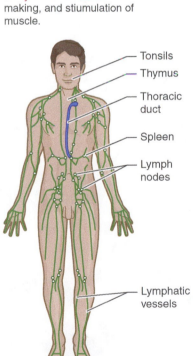

- Tonsils
- Thymus
- Thoracic duct
- Spleen
- Lymph nodes
- Lymphatic vessels

Lymphatic System – Collects and cleans excess tissue fluid and returns it to the bloodstream. It is responsible for immunity.

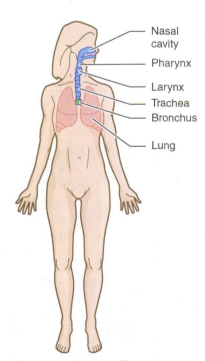

- Nasal cavity
- Pharynx
- Larynx
- Trachea
- Bronchus
- Lung

Respiratory System – The location where atmospheric gases can exchange with blood. It also assists with controlling blood pH.

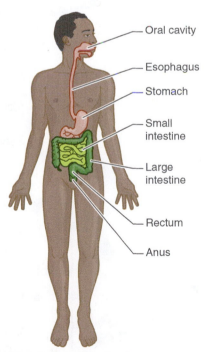

- Oral cavity
- Esophagus
- Stomach
- Small intestine
- Large intestine
- Rectum
- Anus

Digestive System – Breaks down nutrients to their simplest components then absorbs those nutrients into the bloodstream.

**Figure 11.23**   (*continued*)

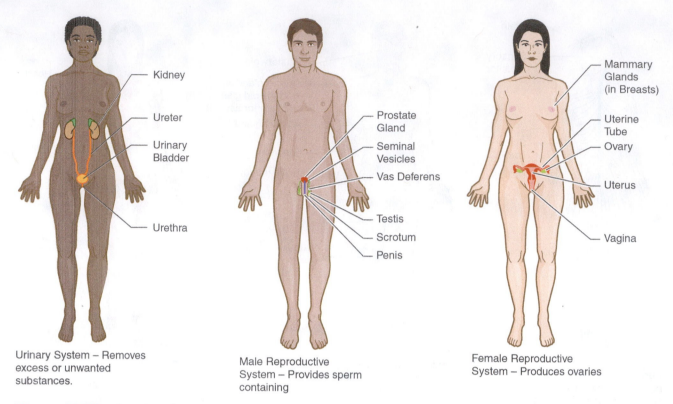

Urinary System – Removes excess or unwanted substances.

Male Reproductive System – Provides sperm containing

Female Reproductive System – Produces ovaries

**Figure 11.23**    *(continued)*

a)    Identify the following structures using the human torso in Figure 11.23:

|                  |                  |
|------------------|------------------|
|                  | Ovaries          |
| Brain            | Pancreas         |
| Cartilages       | Rectum           |
|                  | Small intestines |
| Esophagus        | Skin             |
|                  | Spinal cord      |
| Heart            | Spleen           |
|                  | Stomach          |
| Kidneys          | Testes           |
| Large intestines | Ureters          |
|                  | Urinary bladder  |

b)    Assign each of the above organs to the following systems:

Digestive:
Urinary:
Cardiovascular:
Endocrine:
Reproductive:
Respiratory:
Lymphatic:
Nervous:
Integumentary:

# Summary

Anatomy, the study of structure, and physiology, the study of function of body parts, together describe the shape, size, orientation, and workings of structures, and processes in the human body. The anatomy of a body part always fits its physiology – in other words, *its form fits its function*. The structures work together to maintain balance, or homeostasis. Most body systems maintain homeostasis through negative-feedback mechanisms, which return the system—body temperature or sugar availability, for example – back to normal conditions. The four tissue types of which organs are composed work together in concert to add diversity and specialization to life's processes. A language of anatomy is used in medicine to precisely communicate about these body structures and living processes.

---

## CHECK OUT

### Summary: Key Points

- Understanding the role of tissues, organs, and organ systems helps individuals make better, more informed decisions about their health.
- Complementarity of structures leads to their unique roles in living systems.
- A variety of methods are used to study animal body structures, with different strategies to help identify and treat diseases.
- Homeostasis, the maintenance of stable internal environments, adds stability to living systems.
- The economic theory advocating communism parallels homeostasis, both using a central control center. This relationship helped scientists to discover the role and processes of homeostasis in organisms.
- The four types of tissues are structured to carry out unique roles within the body.
- The four types of tissues work together to carry out life's processes.
- Body structures are placed in humans in expected and specific locations.
- Body structures are identified, described, and measured using the specific language of anatomy.

---

## KEY TERMS

| | |
|---|---|
| adipose | cartilage |
| anatomical position | cell body |
| anatomy | complementarity |
| anterior | connective tissue |
| apical surface | CTP (connective tissue proper) |
| areolar | CT scan |
| auscultation | cuboidal |
| axon | cytology |
| basement membrane | deep |
| blood | dendrite |
| body landmarks (surface regions) | dense irregular |
| bone | dense regular |
| cardiac muscle | developmental anatomy |

directional terms
disease
distal
elastic cartilage
embryology
endocrine system
epithelial tissue
extracellular matrix
fibers
fibrocartilage
frontal plane
glands
glucagon
gross anatomy
histology
homeostasis
hyaline cartilage
insulin
lateral
ligament
medial
mesenchyme
microscopic anatomy
muscle tissue
MRI
negative feedback
nervous system
nervous tissue
neuroglia (glial cells)

neuron
oblique plane
observation
oxytocin
palpation
physiology
platelets
positive feedback
posterior
proximal
receptor
reticular
sagittal plane
set point
simple
skeletal muscle
smooth muscle
squamous
stratified
striations
superficial
tendon
transitional epithelium
ultrasound
vasoconstrict
vasodilate
ventral
x-rays

# Multiple Choice Questions

1. Which played a role in causing problems for the Starfish, in the opening story?
   a. Water chemistry balance
   b. Homeostasis disruption
   c. Changed environmental conditions
   d. All of the above are true

2. This study of tissues within the gums of teeth is classified as:
   a. physiology
   b. senescence
   c. embryology
   d. histology

3. The anatomy of a tooth fits its physiology, enabling it to tear food into small parts. Which term best describes the relationship between chewing and tooth anatomy?
   a. Histology
   b. Complementarity
   c. Cytology
   d. Dentarity

4. "Listening" is most closely associated with:
   a. auscultation
   b. observation
   c. palpation
   d. CT scan

5. A set of cells comes together and send signals out to add more cells to an area. This refers to:
   a. positive feedback
   b. bilateral feedback
   c. negative feedback
   d. organ feedback

6. Which represents a logical order, from low sugar levels to normal sugar levels, in the process of negative feedback in glucose regulation?
   a. insulin release → uptake by cells → kidney → increased glucose
   b. insulin release → liver activated → glucose → increased glycogen
   c. glucagon release → insulin release → liver → increased glucose
   d. glucagon release → liver activated → glycogen → increased glucose

7. Which social or economic development is most closely associated with the discovery of homeostasis?
   a. Alliance systems
   b. Communism
   c. Capitalism
   d. Democracy

8. Which tissue both carries and stores messages?
   a. Muscle
   b. Epithelial
   c. Nervous
   d. Connective

9. Which group of tissues is BEST described by the terms "apical," "basement membrane," and "stratified?"
   a. Muscle
   b. Epithelial
   c. Nervous
   d. Connective

10. Which anatomical structure is found in the digestive system and in the right iliac region of the body?
    a. Appendix
    b. Brain
    c. Kidney
    d. Stomach

## Short Answers

1. List the four types of tissues found in humans, as described in this chapter. Research the condition of dehydration to explain how a failure of some of the tissues or organs may lead to disease.

2. Explain how the return of Starfish to freshwater might not help to Sabrina's starfish in the story. Use the terms "osmosis" and "diffusion" in your explanation. Why?

3. Define the following terms: smooth muscle and striated muscle. List one way the two terms differ from each other in relation to their a. anatomy; b. location in the human body; and c. role in controlling digestion.

4. Homeostasis is a process that uses different mechanisms within the body. Describe one way positive feedback maintains homeostasis. Give an example. Describe one way negative feedback maintains homeostasis. Give an example.

5. Any epithelial tissue contains three basic parts to its tissue plan. List and draw the basic structure of a stratified cuboidal epithelial tissue. Be sure to include the labels: apical surface, connective tissue and basement membrane. Which of these terms is most important to its nourishment and survival?

6. Is a cardiac muscle under voluntary or involuntary control? Why? Which two organ systems are most responsible for maintaining homeostasis within the body? Which acts more quickly? Which works more slowly? Give an example and name another system that works with these two to maintain homeostasis.

7. A patient's axillary lymph node is enlarged. Which areas of the body would you check to determine this diagnosis?

8. A poison is given a drug to destroy fat vacuoles. Which type of connective tissue would be most affected? Why?

9. An old man was fond of saying, "A fool and his cartilage is soon parted…" Describe the problems associated with overworking an area of the body. Choose one of the three types of cartilage and explain how its loss leads to illness.

10. Describe the anatomical position. How is it important in medical treatments and procedures?

## Biology and Society Corner: Discussion Questions

1. In the opening story of the chapter, Sabrina's starfish could not properly function because it did not have the proper nutrients on dry land? Our human bodies are also uniquely structured and its environment should be protected, especially those of children. To what extent should parents be held accountable for their children's (a) use of helmets during cycling; or ( b) use of life jackets during swimming; and (c) the diets parents provide for their children?

2. The human body is not structured to resist all the forces placed upon it during athletic competitions. High school and college football are both notorious for player injuries, especially head (nervous) and cartilage (connective tissue). What precautions

should be taken to prevent such injuries? Should there be limits to participation in these sports? To what extent are players participating "at their own risk"? Should they be compensated or even insured when placing themselves at anatomical risk during these sporting events?

**3.** How is the study of the organization and functions of living systems important to our society? How did a knowledge of water chemistry help Sabrina realize the need to return dying Starfish back into the ocean in the story? Should you or your loved ones be concerned about the role of water in their diets to support different tissues? Would such a knowledge have helped Sabrina's friends better appreciate the plight of the Starfish?

**4.** Cartilage deterioration and damage is responsible for many types of replacement surgeries in the United States. However, a statistic showing that knee replacements and hip surgery, performed on 775,000 Americans last year (both treating cartilage problems) offer no more pain relief after surgery than experienced by patients before surgery. Construct an argument in favor of and one against, the use of such surgeries.

**5.** In the above question #4, medical decisions are based on science and data from a patient. Explain how a parasagittal CT scan of the knee, to determine a need for surgery, might have drawbacks. Are treatments sometimes worse than cures for patients? Give an example of a case in which this might be true.

**6.** Which two organ systems are most responsible for maintaining homeostasis within the body? Which acts more quickly? Which works more slowly? Give an example and name another system that works with these two to maintain homeostasis.

Figure – Concept Map of Chapter 11 Big Ideas

# Nutrition and Digestion

# 12

Paulo is very chatty at parties

A person suffering from anorexia

Is this a healthy food choice?

Obesity is an international epidemic

Energy is all around us

© Kendall Hunt Publishing Company

## CHECK IN

**From reading this chapter, you will be able to:**

- Explain how nutrition and digestion disorders interface with societal factors.
- Define and describe the eating disorders anorexia nervosa and bulimia, and explore characteristics of and factors contributing to eating disorders and the obesity epidemic.
- List and describe the six types of nutrients, both micronutrients and macronutrients.
- Explain the prevailing theories about how weight is gained and lost.
- Describe mechanical and chemical digestion, listing the structures in the alimentary canal and describing their functions in the digestive process.
- Locate and describe the diseases associated with the alimentary canal.

## The Case of the Sweet Breath Date

"I know that I talk too much and should keep my mouth shut, but I am a social creature – how is that bad?" asked Paulo, as he described the latest events at the dorm party last night. "I think the two new members of our floor are going to get romantic." Paulo said. But Tommy did not want to hear what amounts to a whole lot of gossip from his friend. Tommy said to Paulo, "Just stop talking about people – you are getting yourself in trouble."

Paulo was a nice person; always willing to help you study or greet you with a smile whenever a bad day came along. But he had one major flaw – he was a gossip. He told everyone everything, and it caused many bad feelings among their friends. The sad part was that Paulo was totally unaware that he was doing anything wrong. He said whatever came to his mind – but you at least knew where you stood with Paulo. He always said the truth, at least as he saw it. Paulo particularly liked to diagnose people using his pre-medial knowledge, and at times offended them.

"Paulo, you are going to be a physician's assistant. You will have to keep confidentiality when it comes to your patients. Sometimes you can be very insensitive." explained Tommy. "I want you to be on your best behavior when I introduce you to my new girlfriend, Jenny, tomorrow night." Paulo was hurt; of course he knew how to be discreet. He was a pre-health student and knew a great deal about medicine and talking to patients. "I don't need you to tell me that – I know how to impress the ladies. Trust me." bragged Paulo, allaying Tommy's concerns.

It was time for the big night out, at a college party in the pub downtown for the spring semi-formal. Everyone was going to be dressed to the hilt and it would be a blast. Paulo was actually a lot of fun at parties – always had something to say or a funny story – and kept things lively. Paulo was his best friend, and Tommy planned to introduce Paulo to his new girlfriend, Jenny. They all planned to meet before the party at the First Street Café, where Paulo could get more closely acquainted with Jenny.

Jenny was the person Tommy most wanted – and needed – to impress. Tommy was sure that the pub party would be a great chance at showing Jenny that he and his friends were fun and that she could relax around him. Tommy was, however, a bit worried about his relationship with Jenny. When they went out to dinner, she never ate more than a few bites of food. It was a difficult situation. Tommy was concerned that Jenny was nervous around him and that she might even have an eating disorder. Tommy pondered, "Maybe she will feel a little support speaking with his close friend and getting to know him better this way; it might really help to relax Jenny if his friends hit it off with her."

Jenny came into the cafe looking terrific in a new black dress, and Tommy wanted to show that he had class too. He brought her around the cafe, introduced her to his many friends and then came Paulo sitting at a table. "Paulo, this is Jenny, my new friend who I was telling you about." "It is a pleasure to meet you, Jenny," said Paulo in a very gentlemanly manner. But as Paulo approached Jenny, he noticed her breath smelled and inappropriately asked her, "Do you have ketone breath?"

Paulo had just left his student physician's assistant clinicals, and smelled the breath of a patient who had anorexia nervosa. He noticed that Jenny had the same breath. Paulo explained to the group, "In this condition, a person does not intake enough calories. The body breaks down oxaloacetic acid in the Krebs cycle to get the needed energy. It changes it into glucose, and then into the needed calories." Tommy looked perplexed at Paulo but he went on, "You see, the Krebs cycle shuts down and acetyl-CoA cannot get into it. So, it changes into ketones, which are organic acids".

"Don't you see, Jenny…" explained Paulo, "your breath is sweet just like ketones."

Tommy was obviously horrified at Paulo for being so rude.

---

## CHECK UP SECTION

This story describes the process of ketosis, which arises from a lack of calories. The way food is processed and stored is associated with a number of diseases, including anorexia as well as obesity. Many Americans suffer from these illnesses.

List the symptoms of anorexia nervosa. Research its biological and psychological symptoms and develop a plan for treating a person with this illness.

Why are cases of anorexia so prevalent in the United States society today? How do aspects of our culture contribute to the development of anorexia nervosa and related illnesses?

---

# Eating Disorders

This chapter first discusses the social issues facing people in relation to the digestive system. It explores the causes and possible resolutions to anorexia, bulimia, and obesity, disorders found emanating from both psychological and physiological processes of nutrition practices and the digestive system. In the next part of the chapter, the role of nutrients in food processing in the body is connected with weight gain and loss. Finally, an overview of the parts of the digestive system is given, alongside its associated disorders.

## Anorexia and Bulimia

Eating disorders, as depicted in our story of Jenny, are omnipresent. It is estimated that about 8 million Americans – almost 3% of the U.S. population – have an eating disorder. Of those, 7 million are women and one million are men. Anorexia nervosa is an eating disorder characterized by a loss of appetite for food and a fear or refusal to maintain normal body weight. One in 200 women suffers from anorexia nervosa. About 3% of females have another eating disorder, related to anorexia called bulimia. Bulimia is characterized by purging food to maintain lower body weight. Roughly 10–15% of those suffering with an eating disorder are male. Anorexia nervosa and bulimia are the two most common eating disorders in the U.S. population.

Weight is a sensitive issue for people to discuss. It is laden with psychological issues based on how a person views herself or himself. In the story, Paulo's insensitivity to others shows lack of awareness and knowledge about how to address the disease. While Paulo seems to understand the biology of ketosis, he does not appreciate the psychological complexity of eating disorders in society.

**Anorexia nervosa**

An eating disorder characterized by a loss of appetite for food and a fear or refusal to maintain normal body weight.

**Bulimia**

An eating disorder characterized by abnormal and constant craving for food alongside purging.

In a study by the National Association of Anorexia Nervosa and Associated Disorders, upward of 10% of anorexics die within 10 years of onset of the disease, and 20% are dead after 20 years. Often, the person with an eating disorder feels shame, guilt, and denial. Sadly, as a result, only 1 in 10 people with eating disorders ever receive treatment. This contributes to low recovery rates of between 30 and 40%. Actresses Justine Bateman, Tracy Gold, Jane Fonda, and actor Sam Attwater each suffered publicly with eating disorders (see Figure 12.1).

Anorexia nervosa and bulimia are often derived from an individual's desire to maintain control over his or her life through food. However, the origins of the diseases are complex and in part stem from society's focus on body image. We are not sure if Jenny in our story really had anorexia or bulimia, but eating disorders constitute a rising threat to young adults – especially females, aged 15–24 years for whom the death rate from anorexia nervosa is 12 times higher than for all other causes of death combined.

## The Obesity Epidemic

While Jenny showed a few signs of anorexia nervosa, many Americans exhibit symptoms of a disease at the other end of the spectrum that is growing at an alarming rate – they are overweight or obese. The obesity epidemic occurring in developed nations contributes to a host of illnesses and premature death. Heart disease, stroke, cancers, and diabetes are only a few diseases with strong associations with obesity.

**Obesity**

The state of being overweight with a BMI greater than 30.

© Featureflash/Shutterstock.com

**Figure 12.1** Actor Sam Attwater suffered with bulimia at an earlier time in his life.

The headlines are not lying when they report obesity as an epidemic in the United States: "A chubbier America!"; "Obesity nation…"; and "More donuts, less carrots." In 2010, the Centers for Disease Control reported that 69.2% of Americans are either obese or overweight. There are many reasons for the change in society, but it is a relatively recent phenomenon, accelerating in the past few decades. In 1914, this statistic was very different: only 5% of Americans were overweight or obese. In 2014, over 70% of Americans fall into an overweight or obese category.

How is obesity determined? Obesity is defined by a body mass index (BMI), which gives a number value to a person's body mass based on his or her height and weight. The National Heart, Lung and Blood Institute (NHLBI), a division of the National Institutes for Health (NIH) uses the BMI as a baseline for measuring the health of both males and females. A BMI from 25 to 29.9 is considered overweight, and a value of 30 or more is considered

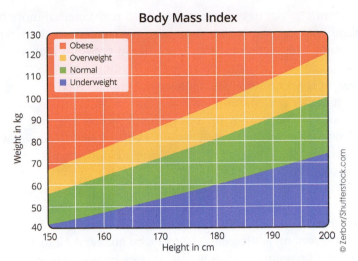

**Body Mass Index**

Legend:
- Obese
- Overweight
- Normal
- Underweight

Weight in kg (y-axis): 40, 50, 60, 70, 80, 90, 100, 110, 120, 130
Height in cm (x-axis): 150, 160, 170, 180, 190, 200

© Zerbor/Shutterstock.com

**Figure 12.2** Body mass index (BMI). This chart uses weight and height to calculate a number to evaluate one's overall body mass health.

obese. A BMI below 18.5 is considered underweight. While only a small portion of the population are underweight, a century ago, more than 40% of military recruits were rejected based on low weight. Being underweight is associated with poor nutrition and development. Use the chart in the accompanying Figure 12.2 to determine your BMI category, based on your height (with shoes) and weight.

Obesity is associated with a number of diseases, increasing the risk for high blood pressure, heart disease, type II diabetes, stroke, sleep disorders such as sleep apnea, certain cancers, and arthritis. These illnesses are based on known physiology about how the body works. For example, chronic high blood pressure, which can worsen with weight gain is based on physical laws. Each pound of fat a person gains requires 1,000 feet of extra blood vessels to support it (see Figure 12.3a). As a result,

© Pikul Noorod/Shutterstock.com

**Figure 12.3** a. "A Pound of Fat" – 1000 feet of blood vessels are built, required to support a pound of fat gained in a person.

the heart has to work harder to overcome the added resistance of more blood vessels. The more pounds gained, the higher the blood pressure from the heart required to push blood through body vessels. Thus, one's blood pressure increases as one gains weight.

Gaining weight is not an automatic indicator of higher blood pressure. Some people may be healthy at higher weights, and some may have high blood pressure and be within normal weight ranges. There are many unexplained factors affecting how diseases are developed and who responds to treatments. Of course, obesity does not need to be a permanent condition – if a pound of fat is lost, most blood vessels are reabsorbed, and blood pressure generally decreases.

While obesity is an increasing problem for adults, a particular worry is the rising obesity rate among children and adolescents. The Centers for Disease Control and Prevention also reported that 12.4% of younger children aged 2 to 5 are obese, 17% of those aged 6 to 11. Among adolescents, 17.6% who are aged 12 to 19 are classified as obese, and many will remain obese as adults. It is estimated that 30% of children who are obese will become obese adults. Increasing obesity rates are reported throughout the industrialized world.

## Why Is Obesity Rising?

Increasing obesity may be a result of a variety of factors. First, as mentioned in Chapter 6, obesity genes have an effect on people's weights, with 50–70% of the trait due to the effects of genes. While genetics plays a role, changes in lifestyles and in the types and quantities of foods that we eat compared to diets in the past may influence obesity prevalence. Although shuffled, the same genes are in our population as existed 100 years ago when the obesity rate was only 5%. What are the primary factors contributing to the modern increases in average weight?

First, in industrialized nations, such as the United States and western Europe, trends in types of work have changed. Society transformed into a service and white collar labor force. Whereas a century ago more than half of the population was involved in farm work, today most people have white collar jobs. Over 90% of the labor force in the United States has a sedentary office job. Many of the once active jobs, for example, construction and factory jobs, are now assisted with or replaced by machinery. All of this resulted in diminished physical requirements at the job.

Second, nutritional choices have changed since 100 years ago. In the past, food was not processed, and diets were dependent on whole foods such as grains, fruits, and vegetables, with small quantities of meats and cheeses (both high in calories). Today, more than 70% of people eat some form of fast foods every week. Fast foods – French fries and burgers being popular choices – are high in sugar and energy, adding calories without the same nutritional value as whole foods. Check the nutritional values of each of these foods. Note that preparation of foods matters – a plain baked potato is healthy in comparison with its French fried version – with unprocessed foods containing less fat, salt, and calories. One plain potato contains only 68 calories; it has a low saturated fat level, is low in cholesterol and salt, and has ample vitamin $B_6$, potassium, and vitamin C. French fries contain concentrated amounts of simple sugar and are high in saturated fat and salt, with few of the nutrients of a plain potato.

There are also many ingredients in foods, for example, high fructose corn syrup, that were not present a century ago. The role of HFCS (high fructose corn syrup) in foods was discussed in Chapter 2. The average American consumes about 62 pounds of HFCS per year and is another example of changed diets over the past century. Food is also more readily available, in particular meats, cheese, and processed products, often served in large portions, adding calories, fat, and sugar. Even one-third of homeless people in the United States are obese. Obesity is now ranked as the second preventable cause of death in the United States, second only to smoking. Causes of the obesity epidemic are complex. While greater access to food products for the public is a valued change in society, knowledge about nutrition (and digestion) will help people to make better decisions about the foods they choose.

# Nutrients

The goal of eating is to obtain nutrients for the healthy functioning of our bodies. Nutrients are the substances that the body uses to obtain energy and to maintain the body's activities, such as growth, repair, and reproduction. Figure 12.4 shows foods containing the six classes of nutrients: water, vitamins, minerals, carbohydrates, lipids (fats), and proteins.

Macromolecules (carbohydrates, lipids, and proteins) are the nutrients that possess energy within their bonds, and are referred to as macronutrients because our bodies require them in relatively large quantities. An overview of the macromolecules was given in Chapter 2. Vitamins and minerals are known as micronutrients, because they are needed in small quantities. Micronutrients include vitamins, minerals, and water. Micronutrients do not have long bonds from which the body can derive energy. However, micronutrients are vital for processes that unleash energy from bonds in macronutrients. Vitamins and minerals are associated with processes that break food down into useable energy. Water is also an essential micronutrient in which all cell reactions occur. All of the micronutrients are needed in living systems.

# The Micronutrients

**Vitamins.** There are 13 vitamins that are essential to human life processes. While vitamins and minerals are required only in very small amounts, they are needed to facilitate chemical reactions in almost every living process. The 13 vitamins are classified into two categories. Fat-soluble vitamins include D, A, E, and K, which accumulate in fatty tissues in the body. Fat-soluble vitamins may become toxic in large doses. Water-soluble vitamins include all of the other vitamins such as the B vitamins and vitamins C and K. Water-soluble vitamins can be taken in large doses and do not become toxic because they are eliminated through the urine.

Vitamins A, C, and E (you can remember these as "ACE") are antioxidants, which eliminate molecules with extra electrons, called free radicals, thus preventing damage to body structures. Free radicals cause cancer and heart disease, so vitamins A, C, and E are associated with better health and fitness. A lack of essential micronutrients, as occurs in cases of malnutrition and eating disorders, as described in our story, can be dangerous. Nerve and heart-muscle impairment occur with disruptions in mineral balances. The major vitamins and minerals and their primary function in the body are given in Figure 12.5.

---

**Nutrients**

The substances that the body uses to obtain energy and to maintain the body's activities, such as growth, repair, and reproduction

**Macronutrients**

Macromolecules that possess energy within their bonds.

**Fat-soluble vitamins**

Includes D, A, E, and K, which accumulate in fatty tissues in the body.

**Micronutrients**

Chemical substance required in small quantities, namely vitamins and minerals.

**Water-soluble vitamins**

Includes B vitamins and vitamins C and K, can be taken in large doses and do not become toxic because they are eliminated through the urine.

**Antioxidants**

The substances that eliminate molecules with extra electrons, thus preventing damage to body structures.

**Free radicals**

Molecules with extra electrons that cause damage to body structures.

**Figure 12.4** You are what you eat. The six nutrients in our foods

| Group | Example Foods | Major Nutrients Supplied in Significant Amounts | |
| --- | --- | --- | --- |
| | | **By All in Group** | **By Only Some in Group** |
| Fruits  | Apples, bananas, dates, oranges, tomatoes | Carbohydrate<br>Water | Vitamins: A, C, folic acid<br>Minerals: iron, potassium<br>Fiberw |
| Vegetables  | Broccoli, cabbage, green beans, lettuce, potatoes | Carbohydrate<br>Water | Vitamins: A, C, E, K, and<br>B vitamins except B$_{12}$<br>Minerals: calcium, magnesium,<br>iodine, manganese, phosphorus<br>Fiber |
| Grain products (preferably whole grain; otherwise, enriched or fortified)  | Breads, rolls, bagels, cereals (dry and cooked); pasta, rice, other grains; tortillas, pancakes, waffles; crackers; popcorn | Carbohydrate<br>Protein<br>Vitamins: thiamin (B1), niacin | Water<br>Fiber<br>Minerals: iron, magnesium selenium |
| Milk products | Milk, yogurt, cheese, ice cream, ice milk, frozen yogurt | Protein<br>Fat<br>Vitamins: riboflavin, B$_{12}$<br>Minerals: calcium, phosphorus<br>Water | Carbohydrate<br>Vitamins: A, D |
| Meats and meat alternates 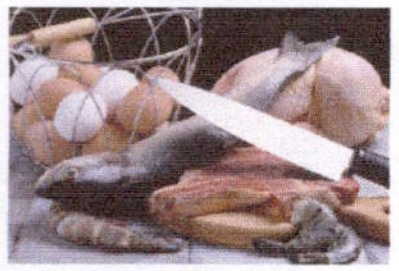 | Meat, fish, poultry; eggs; seeds; nuts, nut butters; soybeans, tofu; other legumes (peas and beans) | Protein<br>Vitamins: niacin, B$_6$<br>Minerals: iron, zinc | Carbohydrate<br>Fat<br>Vitamins: B$_{12}$, thiamin (B$_1$)<br>Water<br>Fiber |

Source: Christian, Janet, and Janet Greger. *Nutrition for Living*, 3rd ed. San Francisco: Benjamin Cummings, 1991.

**Figure 12.5** The major vitamins and their functions.

| Vitamin | Significance | Sources | Daily Requirement | Effects of Deficiency | Effects of Excess |
|---|---|---|---|---|---|
| **Fat-Soluble Vitamins** | | | | | |
| A | Maintains epithelia; required for synthesis of visual pigments | Leafy green and yellow vegetables | 1 mg | Retarded growth, night blindness, deterioration of epithelial membranes | Liver damage, skin peeling, central nervous system effects of nausea, anorexia |
| D | Required for normal bone growth, calcium and phosphorus absorption at gut, and retention at kidneys | Synthesized in skin exposed to sunlight | Nonea | Rickets, skeletal deterioration | Calcium deposits in many tissues disrupting functions |
| E (tocopherois) | Prevents breakdown of vitamin A and fatty acids | Meat, milk, vegetables | 12 mg | Anemia; other problems suspected | None reported |
| K | Essential for liver synthesis of prothrombin and other clotting factors | Vegetables; production by intestinal bacteria | 0.7–0.14 mg | Bleeding disorders | Liver dysfunction, jaundice |
| **Water-Soluble Vitamins** | | | | | |
| B₁ (thiamine) | Coenzyme in decarboxylation reactions | Milk, meat, bread | 1.9 mg | Muscle weakness, central nervous system and cardiovascular problems, including heart disease; called beriberi | Hypotension |
| B₂ (riboflavin) | Part of FMN and FAD | Milk, meat | 1.5 mg | Epithelial and mucosal deterioration | Itching, tingling sensations |

aUnless there is poor exposure to sunlight for extended periods; alternative sources are provided in fortified milk

FMN = Flavin mononucleotide

NAD = Nicotinamide adenine dinucleotide

FAD = Flavin adenine dinucleotide

(continued)

**Figure 12.5** The major vitamins and their functions.     *(Continued)*

| Vitamin | Significance | Sources | Daily Requirement | Effects of Deficiency | Effects of Excess |
|---|---|---|---|---|---|
| **Water-Soluble Vitamins** *(continued)* | | | | | |
| Niacin (nicotinic acid) | Part of NAD | Meat, bread, potatoes | 14.6 mg | Central nervous system, gastrointestinal, epithelial, and mucosal deterioration; called pellagra | Itching, burning sensations, vasodilation, death after large dose |
| B₆ (pyridoxine) | Coenzyme in amino acid and lipid metabolism | Meat | 1.42 mg | Retarded growth, anemia, convulsions, epithelial changes | Central nervous system alterations, perhaps fatal |
| Folacin (folic acid) | Coenzyme in amino acid and nucleic acid metabolism | Vegetables, cereal, bread | 0.1 mg | Retarded growth, anemia, gastrointestinal disorders | Few noted except in massive doses |
| B₁₂ (cobalamin) | Coenzyme in nucleic acid metabolism | Milk, meat | 4.5 mg | Impaired iron absorption causing pernicious anemia | Polycythemia |
| Biotin | Coenzyme in decarboxylation reactions | Eggs, meat, vegetables | 0.1–0.2 mg | Fatigue muscular pain, nausea, dermatitis | None reported |
| Pantothenic acid | Part of acetyl-CoA | Milk, meat | 4.7 mg | Retarded growth, central nervous system disturbances | None reported |
| C (ascorbic acid) | Coenzyme; delivers hydrogen ions | Citrus fruits | 60 mg | Epithelial and mucosal deterioration; called scurvy | Kidney stones |

Source: Coast Learning Systems/John Erickson, *An Introduction to Anatomy and Physiology*, 2010 by Kendall Hunt Publishing Company.

### DOES VITAMIN C PREVENT COLDS?

Upper respiratory tract infections, such as the common cold, invade mucous membranes through the nose and mouth. They are caused by a virus that is quite delicate but, when it enters the body, leads to mucous and inflammation symptoms of the common cold.

The Mayo Clinic conducted a series of studies on the effect of taking vitamin C on the duration and prevention of colds. They discovered that there were no differences between those people taking vitamin C and those taking a placebo (sugar pill).

Linus Pauling, a famous chemist of the 1930s, who discovered aspects of atomic structure, claimed that vitamin C boosts immunity. In Linus Pauling's book "Vitamin C and the Common Cold," he suggested that 1,000 mg of vitamin C per day would ward off the common cold. However, over 11,000 research studies have failed to find data to support his claim about the effects of vitamin C on prevention of colds.

It is true that vitamin C is an antioxidant, and foods containing vitamin C, such as fruits and vegetables like oranges, broccoli, have large amounts of antioxidants. Thus, diets high in these vitamins tend to improve one's overall health. A large glass of orange juice contains about 100 mg of vitamin C, and it is certainly better than many soft drinks. The relationship between vitamin C and health may emerge, however, from the association of vitamin C with other healthy foods. Figure 12.6 shows the innocent tablets of vitamin C, which stimulated all of this research and debate.

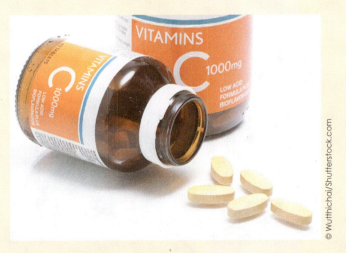

© Wutthichai/Shutterstock.com

**Figure 12.6**   Vitamin C tablets, are they helpful or useless?

# Minerals

They are inorganic substances that form ions in the body, which help to perform many functions. Calcium strengthens bones and teeth, iron is central to the transport of oxygen in the blood, and iodine is needed for thyroid hormone manufacture, as some examples. Minerals in the body are shown in Figure 12.7.

**Minerals**

Are inorganic substances that form ions in the body, which help to perform many functions.

**Figure 12.7** Minerals in the body and their functions

| Mineral | Significance | Total Body Content | Primary Route | Recommended Daily Intake |
|---|---|---|---|---|
| **Bulk Minerals** | | | | |
| Sodium | Major cation in body fluids; essential for normal membrane function | 110 g, primarily in body fluids | Urine, sweat, feces | 1.1–3.3 g |
| Potassium | Major cation in cytoplasm; essential for normal membrane function | 140 g, primarily in cytoplasm | Urine | 1.9–5.6 g |
| Chloride | Major anion in body fluids | 89 g, primarily in body fluids | Urine, sweat | 1.7–5.1 g |
| Calcium | Essential for normal muscle and nerve function, structural support of bones | 1.36 kg, primarily in skeleton | Urine, feces | 0.8–1.2 g |
| Phosphorus | As phosphate in high-energy compounds, nucleic acids, and structural support of bones | 744 g, primarily in skeleton | Urine, feces | 0.8–1.2 g |
| Magnesium | Cofactor of enzymes; required for normal membrane functions | 29 g, 17 g in skeleton and the rest in cytoplasm and body fluids | Urine | 0.3–0.4 g |
| **Trace Minerals** | | | | |
| Iron | Component of hemoglobin, myoglobin, and cytochromes | 3.9 g, 1.6 stored (ferritin or hemosiderin) | Urine (traces) | 10–18 mg |
| Zinc | Cofactor of enzymes systems, notably carbonic anhydrase | 2 g | Urine, hair (traces) | 15 mg |
| Copper | Required for hemoglobin synthesis, as cofactor | 127 mg | Urine, feces (traces) | 2–3 mg |
| Manganese | Cofactor for some enzymes | 11 mg | Feces, urine (traces) | 2.5–5 mg |

Source: Coast Learning Systems/John Erickson, *An Introduction to Anatomy and Physiology*, 2010 by Kendall Hunt Publishing Company.

Diets high in vitamins and minerals help carry out life functions and are associated with good health. Over 50% of Americans take vitamin and mineral supplements. But is there really a need for these additions to our diets? Most nutritionists agree that adequate amounts of vitamins and minerals, especially antioxidants, are a key to a healthy lifestyle. However, most also agree that supplements are not necessary for most adults. Few adults suffer from diseases associated with lack of micronutrients, such as goiter (thyroid malfunction due to lack of iodine) and rickets (bone weakness due to lack of calcium), which are found more frequently in the developing world. A diet properly balanced with whole foods provides an adequate supply of micronutrients, without the need for regular vitamin or mineral supplements.

## Water

The longest period of time a person has ever survived without the intake of water is 17 days, by a prisoner sentenced to death in Italy in the early 1900s. Water is an essential nutrient, comprising between 60 and 80% of the volume of cells, and is the medium in which all cell reactions take place.

Water is needed in almost every function carried out by organisms. All chemical reactions occur in a watery environment. Water is the medium that transports nutrients and wastes through the body. It lubricates joints, the brain, spinal cord, and eyes. Body temperature is regulated, as discussed in the last chapter, in part by sweating, which requires a large amount of water. A human can sweat up to 8–12 L (2–3 gallons) of sweat in a day. The need for water is continual to replenish it.

Water must be taken in the right amounts. The USDA recommends that an average adult person, requiring 2,000 calories of energy per day, needs between 2 and 3 L (roughly a half gallon) of water. Intake also depends upon age and different health conditions. Pregnant women require at least 1 additional liter of fluids per day. Water is found in many foods, such as milk, juice, and even many types of bread.

Excessive intake of water may lead to disruption in mineral balance (Figure 12.8). In a recent radio game show, called "Hold Your Wee for a Wii," a caller was dared to drink several liters without urinating. The caller, 28-year-old mother of three, Jennifer Strange, was found dead after trying to win one of Nintendo's Wii game consoles.

**Water**

An essential nutrient, comprising between 60% and 80% of the volume of cells, and is the medium in which all cell reactions take place

**Figure 12.8**    Person drinking water on radio talk show.

Ms. Strange may have suffered from water intoxication caused by a decreased concentration of sodium called hyponatremia. In cases of hyponatremia, sodium levels become so diluted that they do not function properly in nerve transmission, and heart and brain activity are disrupted. The radio show was under investigation for endangerment of health by the FCC.

**Food Plate (MyPlate guided diet)**

A nutrition guide modeled for healthy eating in the United States.

**Food pyramid**

A pyramid-shaped graphic representation that represents the optimal number of serving to be taken each day.

# Macronutrients

Macronutrients hold, within their bonds, the energy that drives life's processes. Macronutrients including proteins, lipids, and carbohydrates each store energy. The National Academy of Sciences has recommended intakes for our diets to include all of the macronutrients in accordance with a MyPlate guided diet. The MyPlate nutrition guide emphasizes only sparing amounts of dairy and fats, with guidelines shown in Figure 12.9a. It also emphasizes less meat and dairy and promotes more fruits and vegetables to be

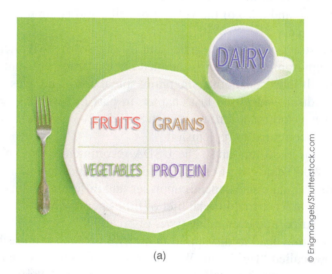

© Enigmangels/Shutterstock.com

(a)

| Food Group | Serving Size | Recommended Number of Servings per Day |
|---|---|---|
| Fats, oils, sweets | | use sparingly |
| Milk, yogurt, cheese | 1 C milk or yogurt; 1½ C cheese; 2 C cottage cheese | 2-3 |
| Meat, poultry, fish, dried beans, eggs, nuts | 2-3 oz extra lean meat, poultry, or fish; 1½ C cooked dried beans; 6 T peanut butter; 2 eggs | 2-3 |
| Vegetables | 1 C raw; ½ C cooked; ¾ C juice | 3-5 |
| Fruits | 1 medium size; ½ C canned; ¾ C juice | 2-4 |
| Bread, cereal pasta | 1 slice bread; 1 oz dry cereal; ½ C cooked cereal, rice, or pasta | 6-11 |

fats, oils, sweets

milk, yogurt, cheese

meat, poultry, fish, dried beans, eggs, nuts

vegetables

fruits

bread, cereal, pasta

(b)

**Figure 12.9** a. Guides to eating right by the National Academy of Sciences: the "MyPlate" Food Guide. b. The old USDA Food Pyramid gives guidelines for obtaining recommended daily allowances for vitamins and minerals. USDA.

consumed every day. The food plate is in greater agreement with most nutrition studies, and replaced older food guides, such as the Food Pyramid, which emphasized higher intakes of carbohydrates (Figure 12.9b).

Let's review how macronutrients drive life's processes.

# Proteins

Proteins compose many structures and functional chemicals in the human body, as you may recall from Chapter 2. Muscles are composed of almost 50% protein; bones and other body structures are also constructed with proteins. The Food and Nutrition Board of the National Academy of Sciences' Institute of Medicine report *Dietary References Intakes for Energy, Carbohydrate, Fiber, Fat, Protein, and Amino Acids (Macronutrients)* recommends daily protein intake of 10–35% of total calories. A slight increase in protein intake is recommended for athletes engaged in muscle-building programs. However, proteins alone do not build up muscles; muscle building requires resistance training, and proteins only act as raw materials.

How do proteins act as building materials? First, as described in Chapter 2, protein molecules are composed of long chains of amino acids, held together by peptide bonds. Amino acids may be compared to bricks in a house, which the body rearranges to build other structures. These "bricks" are reordered based on what foods we eat and the structure we require.

Second, proteins form specific shapes to enable then to perform multiple functions in the body. Proteins, in the form of the hemoglobin, carry oxygen gas in the blood. Proteins form enzymes that break down foods in our digestive system, a topic that will be discussed later in this chapter. Proteins also work as enzymes to break and form bonds. Enzymes are used in almost every chemical reaction within our bodies, from changing fat into energy to storing water and building new cells.

Finally, proteins are found in many types of foods. Animal sources of protein include meats such as chicken, fish, and beef, and dairy products such as cheeses and milk. Vegetable sources include mostly seeds, nuts, beans, and grains, which are used by plants for growing their new offspring.

All animals require each of the 20 amino acids that compose proteins. Only eight of the amino acids are considered essential, meaning that they must be taken in by diet. The essential amino acids are not able to be synthesized by the body. The other 12 nonessential amino acids may be produced from other forms of amino acids and thus are not required in a diet to survive. Many combinations of foods have all of the amino acids required for survival. In other words, the components of a nutritious meal complement each other by providing all of the essential amino acids. Meals such as rice and beans have amino acids in each food that together constitute a full set of proteins. Some foods, such as a slice of beef or an egg, contain all of the needed amino acids and are said to be complete protein meals. Vegetarians must be particularly careful to choose foods that provide the needed sources of proteins. Vegetarian diets are great alternatives to traditional cooking, as long as one chooses the right set of foods to yield a complete protein meal. Vegetarian diets are beneficial in that they are often healthy, low in fats and are able to provide all of the needed nutrients.

**Essential amino acids**

Amino acids that are not synthesized by the body.

**Nonessential amino acids**

Amino acids made by the human body and thus are not required in a diet to survive.

# Lipids

Lipids, including fats, waxes, and oils, are the focus of many weight management and nutrition studies. Lipids contain more than twice the energy per gram of other macronutrients and are therefore cited as a culprit in the obesity epidemic. A gram of fat has

9 calories, compared to 4 calories per gram for carbohydrates and protein. Fat stores a great deal of energy. A person's stored energy in the form of fat lasts them roughly 4 to 5 weeks when deprived of food.

A normal intake of fat is about 20–35% of one's total daily intake. Lipids are necessary for life functions, ranging from the cushioning provided by adipose tissue to maintaining the integrity of cell membranes within the fluid mosaic structure. However, excess lipid intake is related to greater risks for heart disease, stroke, and many other illnesses. Alternatively, when there is a lack of calories (as occurs in eating disorders such as anorexia nervosa) the body taps into its fat energy reserves. The breakdown of fats produces ketones, an acid responsible for the ketone breath alluded to in the story.

Lipids are composed of long-chained molecules of fatty acids and glycerol, which contain many bonds, all holding energy. Foods are, in part considered "good" or "bad" based on their fat content, with some examples shown in Figure 12.10. As classified in Chapter 2, saturated fats, with multiple single bonds, are associated with heart disease, and unsaturated fats, containing double bonds, with better health.

Trans-fats, or those lipids created by adding hydrogen to oils to make them solids, such as margarine, have the strongest links to heart disease and premature dying. Both saturated fats and trans-fats increase levels of LDL (bad) cholesterol in the blood, a substance that is thought to build up on artery walls. This leads to higher risks for stroke and heart attack.

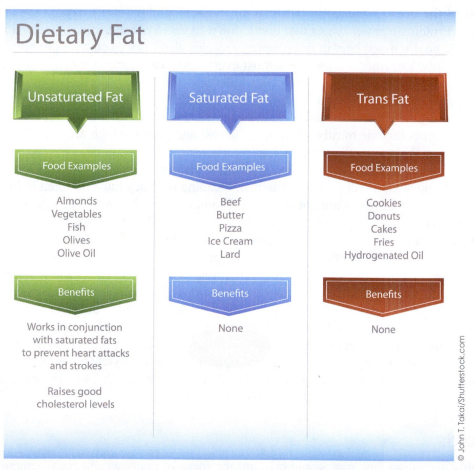

© John T. Takai/Shutterstock.com

**Figure 12.10**    The different kinds of fats in our foods. Trans-fats and saturated fats are most harmful to a person's health. Unsaturated fats are linked to healthy hearts.

## ARE FATS REALLY THE CULPRIT IN HEART DISEASE?

In addition, recent studies published in the Journal *Nature Medicine*, and the *New England Journal of Medicine*, link the dangers of red meat to bacteria in human guts. In the first study, l-carnitine, found in high proportions in processed red meat, provoked bacteria to produce trimethylamine N-oxide (TMAO). TMAO is an organic substance that alters cholesterol processing in the liver. In the second study, lecithin, also found in high amounts in eggs and red meats, led to the formation of TMAO by gut bacteria.

TMAO is associated with heart disease and strokes. The effects of TMAO have been demonstrated in a Cleveland Clinic study of over 4,000 people, in which those having higher levels of TMAO were more likely to have a heart attack or stroke over a three-year period. The findings show that other chemicals, l-carnitine and lecithin, and perhaps not saturated or trans-fats, might be the cause of heart disease and premature death — not the fats in foods directly.

It is possible that those foods containing saturated fats also contain substances causing bacteria to form TMAO. In the Cleveland Clinic study, those subjects eating two eggs per day had increased levels of TMAO, but those subjects adding an antibiotic to kill gut bacteria did not have increased levels of TMAO and did not suffer the same risks as those not taking the antibiotic. Use of antibiotics to kill gut bacteria and improve health is not recommended due to the need for bacteria in our digestive system (discussed later in this chapter). Those wishing to avoid the negative effects of TMAO on their health should increase their vegetable and fiber intake, and limit red meat and eggs.

## Carbohydrates

Access to quick energy in the body is provided by carbohydrates. Carbohydrates are composed of single molecules called simple sugars or monosaccharides, as discussed in Chapter 2. Carbohydrates are transformed into sugar and then into energy, in processes described in Chapter 4. They provide structure in cell membranes, enable growth and fuel movement. The Food and Nutrition Board suggests that 45–65% of our diets should draw from carbohydrates.

Just as all fats are not the same, all carbohydrates do not have the same contribution to rises in sugar levels in the blood. The glycemic index gives a numerical value to carbohydrates depending upon their spike in blood-sugar levels. Foods with a high glycemic index more rapidly increase sugar levels than those with a low glycemic index. For example, a donut with a high proportion of simple sugars has a high glycemic index because its sugars are quickly released into the bloodstream. Whole grain breads, which have longer chained carbohydrates, must be broken down to release simple sugars into the blood stream and thus have a lower glycemic index.

Foods with lower glycemic indices are advised because they take longer to process and do not spike levels of sugars and thus insulin in the body. Spikes in sugar levels contribute to increased risks of diabetes, or excess sugar in the blood (discussed in Chapter 14). Diets containing low glycemic index foods include fruits, vegetables, whole grain cereals and breads, whole grain rice, wheat. Processed foods such as white breads, candy, cookies, cakes, and donuts have a high glycemic index. It is best to

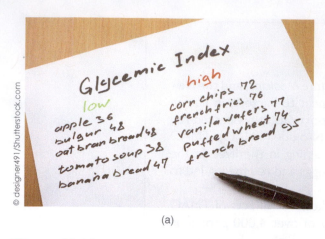

(a)

(b)

**Figure 12.11**    a. Choose foods with a low glycemic index. Foods with a lower glycemic index give less of a sugar rush after eating. b. Fruits and whole grain cereals have a lower glycemic index.

choose foods with a low glycemic index, along the lines of those examples given in Figure 12.11.

The storage form of glucose in our bodies is glycogen, found in the liver. When energy is needed, glycogen is broken down into simple sugars and transformed into energy by cellular respiration. Large amounts of water are stored along with glycogen; in fact up to four pounds of water are associated with every one pound of glycogen. This is the reason why a diet almost always *initially* works well, as glycogen is used up and water is excreted through the kidneys; water weight is lost but not fat. As a diet enters the fat-usage stage, weight loss slows dramatically because fat burning is a slow process in contrast to the water-weight loss associated with glycogen. If a person stops their diet, the weight is rapidly gained back because glycogen restores in the liver along with the water weight.

All carbohydrates are not created alike and therefore diets addressing these differences are essential for good health. Whole, unprocessed foods that contain higher amounts of larger, complex carbohydrates such as fruits, vegetables, and whole grains will limit risks for disease. Our physiology is based upon over 90,000 years of adaptations to diets of hunter and gatherers. They did not have processed foods, and our bodies are therefore not adapted to properly process these kinds of foods.

## IS THE NEOLITHIC (CAVEMAN) DIET A GOOD GUIDE TO EATING RIGHT?

Proponents of the Neolithic or "Caveman" diet propose that eating like a caveman or cavewoman is the right way to obtain nutrients and keep within a normal weight range. They argue that *Homo sapiens* developed digestive processes over the past 150,000 years based on the hunter/gatherer lifestyle that characterized early humans. These processes thus resulted from over 180 million years of mammalian evolution, 65 million years of primate evolution, 5 years of hominid evolution, and 2 million years of our genus *Homo*.

The idea of the Neolithic diet is that it took a long time to develop our digestion and, in an evolutionarily short period of time, our diets have made radical changes.

Some nutritionists argue that to more closely parallel the diets to which we are adapted, people should follow the prehistoric nutrition system called the Neolithic diet. This diet includes the kinds of foods that were eaten by prehistoric humans. They claim that the MyPlate guidelines conflict with the Neolithic diet; and thus they collide with a diet type that our metabolism requires because of many years of evolution.

Prehistoric diets likely included mostly whole grains, fruits, nuts, and vegetables, as recommended in the previous sections. In cave-era times, there were few opportunities to eat meat and fish because they are more difficult to capture and hunt. Thus, human digestion was adapted to greater proportions of plant foods. Obviously, proteins and fats were therefore less a part of the Neolithic diet, far less than most Americans consume today.

The Neolithic diet most likely contained not only a large amount of fresh fruits and vegetables, but also many whole-grain (unrefined) starches such as rice, tubers, acorns, and grasses. Early humans likely ate small game meats such as frogs, birds, snakes, fish, and even insects to obtain most of their fats and proteins. It would have been extremely rare (and probably coveted) to find refined sugars in the form of honey.

On the other side, our modern diet is laden with refined sugars and high-fat foods, never encountered by our ancestors and thus not well adapted by our genes. Americans regularly consume too much refined sugars and fats, and not enough micronutrients from lower calorie foods such as fruits and vegetables.

Neolithic diet supporters contend that modern society is at odds with genetic predispositions of digestion emanating from the past. First, they cite our activity levels. We are less active today than compared with our ancestors and even people from 100 years ago. This, they cite, is contributing to the modern obesity epidemic. Preagricultural people led more active lifestyles with caloric needs of about 3,000 calories per day. Nomadic lives required hunting of game and gathering of vegetables, as well as fighting the elements and caring for young. However, they consumed fewer calories from fat. Corn-fed stock today contains approximately 29% fat as compared with wild game of our ancestors, which had only 4% fat.

Critics of the modern American diet often cite the prevalence of pizza. Pizza has become rapidly integrated into our diets within the past three decades. Perhaps consider when was the last time you ate pizza? It may be alarming to you that the components of pizza are drizzled cheese and oils. An average slice of pepperoni pizza contains 298 calories, with 37% fat, 47% carbohydrates, and only 14% protein. A serving of deer meat, for which our bodies are more adapted, contains only 32 calories per ounce and has 18% fat, 0% carbohydrates, and 82% protein (Figure 12.12). A mismatch between modern day food choices and our physiologic origins are possible contributors to the current obesity epidemic, according to proponents of the Neolithic diet.

**Neolithic diet**

The prehistoric nutrition system.

CAVEMAN

PALEO
DIET
PYRAMID

NUTS & BERRIES

VEGETABLES & FRUITS

WILD MEAT & HEALTHY FAT

© happyexplorer/Shutterstock.com

(a)

© PathDoc/Shutterstock.com

(b)

**Figure 12.12** a. Caveman diet is filled with fruits and vegetables; and b. Modern person eating pizza.

## How Is Weight Gained and Lost?: Food, Energy, Metabolism, and Weight

### Energy Is measured in Calories

**Calories**

The amount of energy required to raise 1 gram of water by °1 Celsius.

Food energy is measured in small units commonly known as calories **(cal)**, which are defined as the amount of energy required to raise 1 gram of water by 1°C. However, food calories are actually kilocalories (kcal) of energy, each of which equals 1,000 calories. We use kcal to measure food energy quantities because food contains much more energy than single calories. Thus, a yogurt with 100 calories on its label is really 100 kcal or 100,000 calories.

As we have said before, macronutrients provide different amounts of energy. One gram of lipids provides 9 calories of energy, while 1 g of carbohydrate and 1 g of protein each contain only 4 calories of energy. However, a diet low in fat may not lead to less intake of energy. Intake of energy is based upon the total calories in one's diet.

A simple rule applies to gaining and losing weight: calories in vs. calories out. The body may be compared to a meticulous accountant – it measures how many calories are coming into the body and how much is leaving it. If a person takes in more calories than are expended in a day, then that person will gain weight. If a person takes in fewer calories than are expended in a day, then the person will lose weight. To gain a pound of weight, a person needs to take in 3,600 more calories than he or she uses. Thus, to lose a pound of weight, a person must take in 3,600 calories less than is used.

## Basal Metabolic Rate

Of course, weight gain and loss depend upon how well those calories in one's diet are burned off. The energy needs of an individual are calculated based on a base level of energy needs, doing nothing but survive in a sitting position within a single day. This energy requirement is known as the basal metabolic rate (BMR), for a person who is at rest, with no energy requiring digestion and only minimal energy needed for movement. The BMR for a human is roughly 1 calorie per hour per gram of body weight. This translates into a need for 1,700 calories per day for a person weighing 170 pounds. In reality, calorie needs are about 50% higher for most people, who are involved in some physical activity such as walking and conducting daily activities. The same 170-pound person would therefore need about 2,550 calories per day to maintain their weight.

Basal metabolic rate differs for different species of animals. Generally, the smaller animals have faster heartbeats and a faster metabolic rate, requiring more energy. For example, a tiny shrew has a heart rate of more than 500 beats per minute (humans average only 80 beats) and a BMR 35 times higher than humans. An elephant, large in size and stature, has a heart rate closer to 40 beats per minute and has a BMR only 4% that of a shrew.

Different people have differing BMRs based on their genetics, muscle composition, and hormonal factors. The greater the number of muscle cells, which use more energy than an average cell, the higher one's BMR. Some people burn calories better than others and are said to have a faster metabolism, or total set of cellular reactions. Some have a slower metabolism.

A recent set of research studies reports that BMR may not be the best measure of overall health. Instead, a person's waist-to-height ratio is more important in predicting heart health and longevity than BMR. It is recommended that the waist-to-hip ratio should not exceed 0.5. In other words, a person's waist circumference should not be more than half of his or her height in inches or centimeters. Thus, a person at a height of 6 feet, or 72 inches, should have a waist 36 inches or smaller to be considered healthy. Abdominal fat, indicated by a larger waist size, is associated with stroke, heart diseases, and various cancers. People with an apple shape, or those carrying fat around their midsections, are thus at greater risk for developing disease than those with a pear shape, who carry extra weight around their hips. Figure 12.13 depicts the two body types.

Some studies on fat distribution point to body fat percentage as the best measure of a person's risk for disease. In these calculations, folds of fat are measured at various locations on a person's body. Measurements are taken at the waist, back of the upper arm and at the back regions. To be considered healthy, a male should have a body fat percentage less than 15%, and females should have not more than 24% body fat. Women require more body fat due to hormonal and childbearing influences. Young adults who are healthy should range between 12 and 15% total body fat for men and between 20% and 25% for women.

**Basal metabolic rate (BMR)**

The minimal rate of energy used by an organism at complete rest.

**Metabolism**

Chemical processes occurring in a living organism that are necessary for life maintenance.

**Apple shape**

A body shape that is characterized by excess body fat in the abdominal region.

**Pear shape**

A body shape characterized by extra weight around the hips.

**Alimentary canal**

The tube and associated organs of digestion, which includes all of the parts of the digestive system that contribute to the breakdown of food.

**Anal canal**

Terminal part of the large intestine.

**Mouth**

An opening in the lower part of the face.

**Pharynx**

A tube that starts behind the nose and mouth connecting to the esophagus.

**Esophagus**

Food tube.

**Stomach**

An internal organ sac that holds and digests food before entering the small intestines.

**Peristalsis**

The involuntary muscular contractions of the digestive tract by which contents are forced onward.

**Rectum**

The final part of the large intestine.

**Accessory organs**

Associate organs that produce, store and/or release chemicals to carry out the processes of the breakdown of food.

**Salivary glands**

The gland that secretes saliva.

**Liver**

A large glandular organ found in the abdomen of vertebrates.

**Pancreas**

A diffuse gland located near the stomach.

© PixelJoy/Shutterstock.com

**Figure 12.13**   Apple vs. pear shapes. It is better to be a pear than an apple. Abdominal fat located around the organs in the abdomen is linked to cardiovascular disease and increased cancer risks.

The obesity epidemic described in the earlier section reflects a change in diets to fast foods and prepared meals as one factor causing increasing weight. These foods have higher numbers of calories and thus increase the amount of energy taken in, on average, by people.

Of course, there are many factors leading to the increasing average weight in our populace. As average weights go up, the appeal of a thin body grows. Eating habits and therefore weight have strong social and psychological components, a theme of our story of Paulo and Jenny. First, to fully consider these factors, let's review how the energy is released from the macromolecules by our digestive system.

## The Digestive System: How Humans Break Down and Absorb Food

### The Alimentary Canal: A Tour of the Digestive System

The alimentary canal is the tube and associated organs of digestion, which includes all of the parts of the digestive system that contribute to the breakdown of food. It is composed of about 24 feet of tubes and as such, it is referred to as a "tube-within-a-tube" arrangement within our tube-like bodies. The structure of the digestive system in humans is given in Figure 12.14.

The alimentary canal comprises a series of tubes from mouth to anus including a mouth chamber surrounded by a tongue and teeth; a pharynx or back of the throat; a food tube called the esophagus; which empties into a flexible compartment, the stomach; which brings food into the **small intestines** where most of it is absorbed; and empties into the **large intestines**, after which it is eliminated through the anal canal and rectum. Food is propelled through the alimentary canal through rhythmic contractions of its muscular walls, called peristalsis.

The alimentary canal is connected with associated organs, called accessory organs, which produce, store, and/or release chemicals to carry out the processes of the breakdown of food. Accessory organs include the salivary glands of the mouth, which secrete saliva; a **gallbladder**, which stores bile, a juice released within the canal; the liver and pancreas, which both make enzymes to break down food.

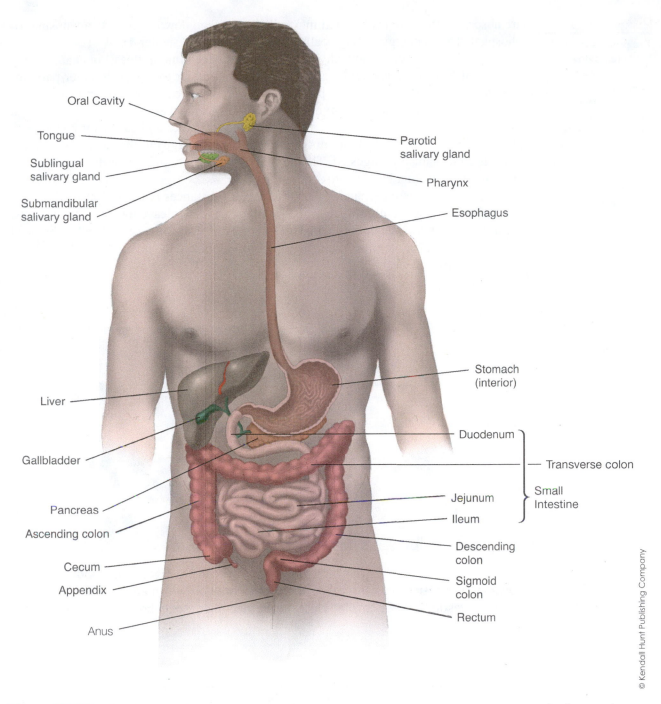

Oral Cavity
Tongue
Sublingual salivary gland
Submandibular salivary gland
Liver
Gallbladder
Pancreas
Ascending colon
Cecum
Appendix
Anus

Parotid salivary gland
Pharynx
Esophagus
Stomach (interior)
Duodenum
Transverse colon
Jejunum
Ileum
Small Intestine
Descending colon
Sigmoid colon
Rectum

**Figure 12.14**   The human digestive system. The alimentary canal in humans is composed of several compartments that mix and digest food.

# Digestion

The alimentary canal, also called the digestive tract, carries out the process of digestion. Digestion is defined as the mechanical and chemical breakdown of food. It readies food for absorption into the body. Digestion does not provide instant energy because the food is not broken down into ATP. Instead, digestion breaks down food particles into a size that is able to be absorbed by the alimentary canal at certain points. After food particles

**Digition**

The process in which food breaks down mechanically and chemically. Mechanical digestion changes only the size of food particles, making them smaller and easier to digest, while chemical digestion changes the structure of the substances being digested.

are absorbed, they are transported into cells and broken down to release adenosine triphosphate (ATP) energy through cellular respiration (see Chapter 4).

Digestion occurs mechanically and chemically. **Mechanical digestion** changes only the size of food particles, making them smaller and easier to digest. Smaller particles have a greater surface area on which digestive enzymes can work to break the particles down. This process is shown in Figure 12.15. Mechanical digestion first occurs in the mouth. Through chewing food, particles become smaller and easier for enzymes to attach to. After the mouth, food moves through the esophagus to the stomach, where it is further mechanically digested. Through the churning of muscles on the walls of the stomach, food further breaks into smaller bits.

**Chemical digestion** changes the structure of substances being digested. Enzymes are produced in the mouth, stomach, and small intestines to cleave food at certain points, breaking them into smaller compounds. The smaller substances are able to be absorbed by the lining cells of the small intestines. Let's give a look at the processes and structures along the alimentary canal.

**Figure 12.15** Surface area increases after mechanical digestion. Food is broken down into smaller pieces, increasing the surface area on which enzymes can act to aid in food digestion. These smaller pieces of Belgian cheese will be easier to digest than a large piece.

## DOES DIGESTION CAUSE CRAMPING IF YOU DON'T WAIT TO SWIM AFTER YOU EAT?

There is no need to wait to swim after you eat, as digestion does not interfere with swimming. Muscle cramps while swimming do not occur as a result of eating a meal. No measurable changes in peristalsis occur while a person swims.

This myth might have originated because the parasympathetic nervous system becomes activated during and after eating. The parasympathetic nerves activate digestion and slow messages to other body parts. However, no evidence has shown muscle dysfunction or heart problems during exercise simultaneous with digestion.

## Mouth

The first part of digestion, called ingestion, requires food to be brought into your stomach. Ingestion typically takes about one minute, the time to chew and swallow food. It involves the mouth, tongue, teeth, and esophagus. Figure 12.16 shows the process of swallowing and the rhythmic motion of muscles in the throat that propel food into the stomach. Figure 12.16 also gives an overview of the digestive organs alongside their enzymes produced and the foods absorbed in those different regions. **Peristalsis** is defined as this set of digestive muscle contractions, which begins with smooth muscle bringing food into the stomach. Because the muscles carrying out peristalsis act involuntarily once food is swallowed to move the bolus (ball) of food through the alimentary canal, a person could drink and eat upside down.

Food, and sometimes the smell or thought of food, stimulates the production of saliva in our mouths. When food enters the mouth, it stimulates a nervous reflex, leading to saliva release by the salivary glands. Over 1 L of saliva is produced by the salivary glands per day and released into the mouth. The salivary glands of the mouth are identified in Figure 12.17.

While mechanical digestion is accomplished by the tongue and teeth, breaking food into smaller bits, saliva contains the enzyme salivary amylase which chemically breaks down starch into smaller polysaccharides. About 20% of the starch ingested into the alimentary canal is broken down by salivary amylase. Digestion of carbohydrates is completed later in the small intestines.

Buffers within the saliva also neutralize the acids in the mouth and limit damage to the teeth. A flora of bacteria lives in our oral cavities; they produce acids that cause dental caries (cavities) and gum disease. Certain bacteria, such as *Streptococcus mutans* and *Streptococcus sobrinus,* are particularly good producers of acid when sweet foods enter the mouth cavity. These bacteria are closely linked to tooth decay, but their populations are greatly reduced in the presence of fluoride toothpaste.

**Ingestion**

Consumption of a substance by living organisms.

**Bolus**

A rounded mass of food.

**Salivary amylase**

An enzyme present in the saliva that chemically breaks down starch into smaller polysaccharides.

**Enamel**

The strong covering protecting the teeth.

**Dentin**

The bony tissue of a tooth.

**Cementum**

A glue-like substance holds the ligament that connects the dentin to the underlying bones of the face.

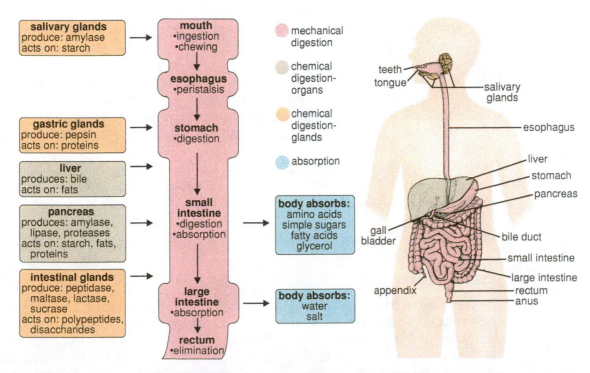

**Figure 12.16**    Ingestion of food. Food moves quickly from the mouth (via the esophagus) to the stomach. It then takes 24–48 hours to move through the rest of the digestive tract, with enzymes released and absorption of nutrients in different regions. BSCS by Doug Sokel. Corel.

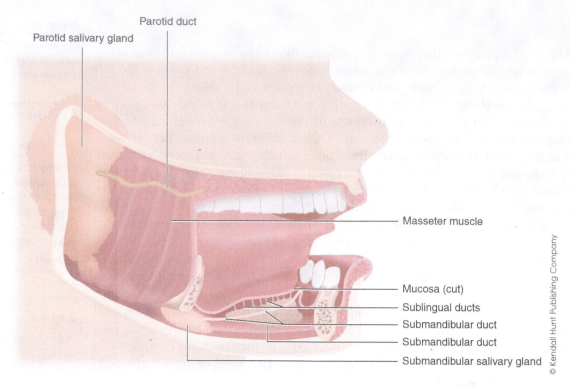

**Figure 12.17** Salivary glands in the mouth release saliva for digestion of starches. Saliva also buffers the mouth, limiting damage to teeth from its often caustic acidity.

**Pulp cavity**

A cavity within the dentin containing blood vessels and nerves.

**Incisors**

The four front teeth evolved and adapted for cutting and tearing.

**Canines**

The front teeth on the side, long and narrow, evolved to tear and pull foods.

**Pre-molars**

Teeth situated between canine and molar teeth and suited for grinding and chewing foods.

**Molars**

A grinding tooth found at the back of the mouth and suited for grinding and chewing foods.

**Epiglottis**

The flap of elastic cartilage covering the trachea.

Teeth, as shown in Figure 12.18, are protected by a strong covering called enamel, composed of calcium salts. These calcium salts are not replaced, so wearing of enamel results in bacteria's access to tissues beneath. The bony tissue of a tooth, called dentin, lies below the enamel layer. Dentin is connected to the underlying bones of the face via a ligament held together with glue-like substance called cementum.

A cavity within the dentin contains blood vessels and nerves. This cavity is known as the pulp cavity or root canal, a name that scares many people from the dentist's chair. In order to deaden the nerve and remove infected tissue from the pulp cavity, a "root canal" is performed, which is often associated with pain at the dentist. Teeth are surrounded by the gingiva or gums, composed mostly of connective tissue providing a blood supply to the region. Gum disease, also known as periodontal disease, is a leading cause of tooth loss.

The front teeth, called incisors and canines, are long and narrow; both types evolved to tear and pull foods. Premolars and molars at the back of the mouth are suited for grinding and chewing foods, such as vegetables and fruits. A look at an animal's teeth indicates its diet. For example, meat eaters such as cats and bats have sharp teeth while those eating plants have flat, wide teeth.

## Esophagus

As the tongue brings food from the mouth it first enters the pharynx. The bolus of food passes the **pharynx** at the back of the throat, and the pharynx opens into two passageways: the trachea and the esophagus. The trachea, also known as the windpipe, connects to the lungs, and the **esophagus** travels to the stomach. The trachea is covered with a flap of elastic cartilage called the epiglottis, which prevents food from entering into the trachea.

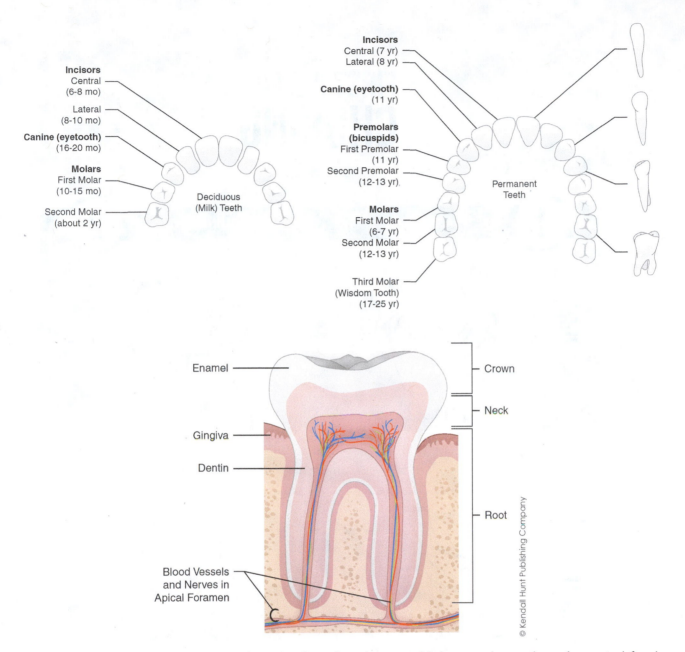

**Figure 12.18**    Tooth anatomy and types of teeth in humans. Molars are best adapted to grind foods such as vegetables and fruits. Canine and incisors are better adapted for pulling and tearing foods such as meats.

When food arrives at the pharynx, the voice box or larynx moves quickly upward, pushing the epiglottis over the trachea. When a person is choking, the bolus of food becomes lodged in between the epiglottis and the larynx. Choking victims can be helped by the Heimlich maneuver, which forces air out of the trachea to dislodge the food blockage. The steps of the Heimlich maneuver are given in Figure 12.19.

The esophagus, also known as the food tube, is a muscular tube that propels food down into the stomach. The superior part of the esophagus is composed of striated muscle, enabling voluntary movement of food downward. The inferior portion of the esophagus is composed of smooth muscle, which is not under conscious control.

**Larynx**

The part of throat containing the vocal cords.

**Trachea**

A tube-like portion of the respiratory tract that connects to the lungs.

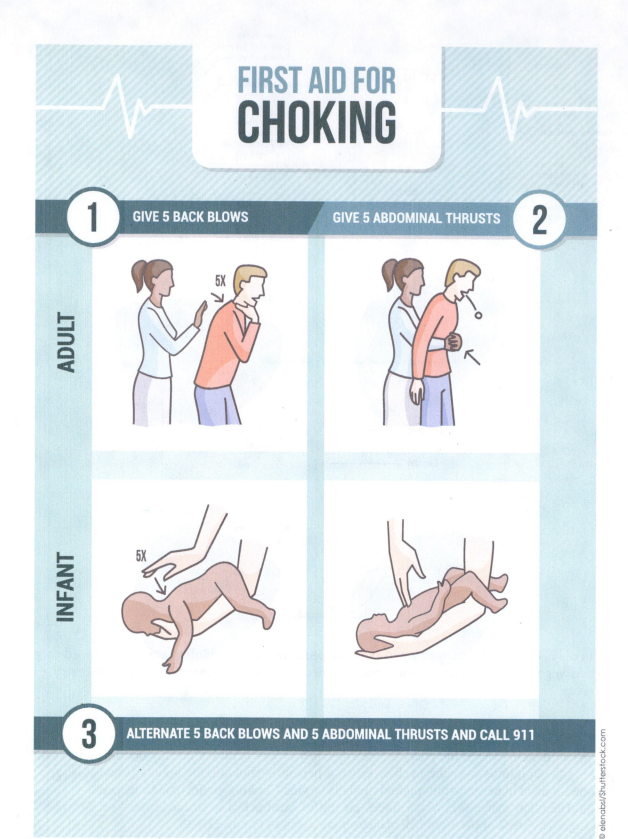

**Figure 12.19**   The Heimlich Maneuver is vital in saving choking victims.

## Stomach

No digestion occurs in the esophagus, through which food is transported quickly into the stomach, a flexible and muscular J-shaped organ. Proteins and fats are not broken down up to this point. The walls of the stomach are quite muscular with three layers and a compartment within is able to hold up to 4 L (one gallon) of food and fluids. Food remains in the stomach for up to 2 hours as it is digested. Figure 12.20 shows the structure of the stomach.

The top region of the stomach, where it connects to the esophagus, is guarded by the cardiac sphincter or gastro-esophageal sphincter. The cardiac sphincter opens and closes as a circular muscle. It prevents stomach fluids, which are acidic, from back flow into the esophagus. In the stomach, folds called rugae churn food to mechanically break it down. The bottom region of the stomach, which empties into the small intestines, is controlled by the pyloric sphincter.

When food enters the stomach through the gastro-esophageal sphincter, stomach-lining cells found within gastric pits in the stomach wall produce hydrochloric acid (HCl) and pepsinogen. Pepsinogen is an inactive enzyme and unable to break down nutrients. After it is released into the stomach cavity, pepsinogen is quickly transformed into an active enzyme called pepsin. Pepsin requires the acidity created by HCl in the stomach cavity to become active and digest proteins. When a pepsinogen molecule is released into the stomach, HCl in the stomach cavity cleaves the tail off the molecule, revealing an active site that enables pepsin to break down protein. This action occurs in open areas in Figure 12.21.

The pH of the stomach ranges between 2 and 3. Acids break down plant materials such as fiber, and structures that otherwise are not digested such as cellulose and bind proteins. This process enables food to more easily move through the digestive tract. The acids of the stomach also serve to destroy the many bacteria in our foods. Even cooked, bacteria – which number more than 100 million per cubic centimeter – enter into the stomach. Most bacteria die when entering such an acidic condition. However,

**Cardiac sphincter**

The muscle surrounding the opening between the stomach and esophagus.

**Pyloric sphincter**

Muscle fibers around the stomach opening between it and the duodenum.

**Pepsinogen**

The inactive precursor to pepsin.

**Hydrochloric acid (HCl)**

An aqueous solution of hydrogen chloride.

**Pepsin**

An enzyme produced in the stomach.

**Rugae**

Series of folds produced by folding the wall of an organ.

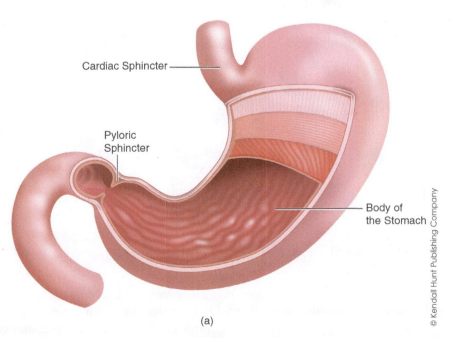

Cardiac Sphincter

Pyloric Sphincter

Body of the Stomach

© Kendall Hunt Publishing Company

(a)

**Figure 12.20** Stomach anatomy. a. The stomach muscle layers help propel food (by peristalsis) from the gastro-esophageal sphincter to the pyloric sphincter. b. Micrograph of stomach.

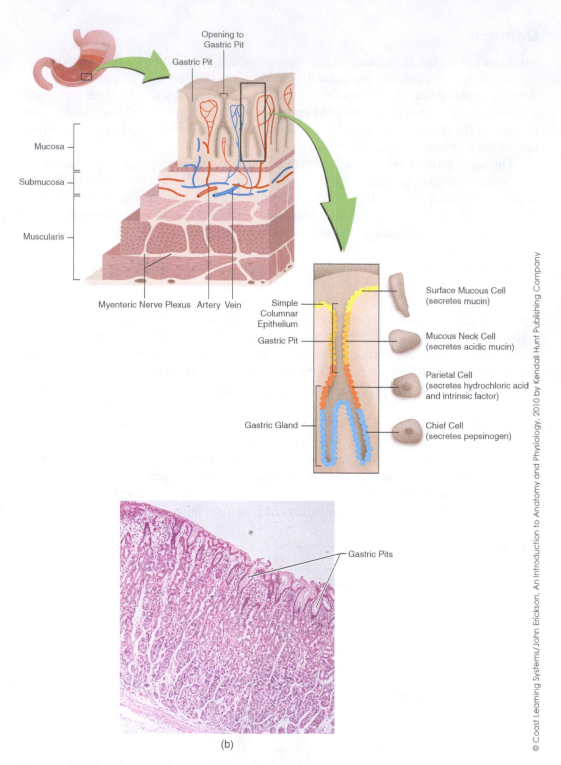

**Figure 12.20**   (*continue*)

some food-borne illnesses, such as *Staphylococcus aureus* and *Salmonella* are resistant to the acidic environment of the stomach. Acidity from the stomach has powerful effects. In our story treating eating disorders, many bulimics suffer damage to their oral cavity from stomach acids. In bulimia, which involves purging of food, stomach acids regurgitate to damage teeth and gums. Acidity from the stomach also damages the linings of the esophagus in its victims.

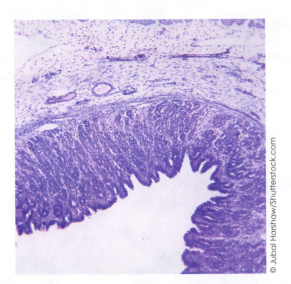

**Figure 12.21**    Cells in the pits of the stomach secrete pepsinogen and HCl. Pepsinogen is inactive when produced but is transformed into the active enzyme pepsin. The enzyme action of pepsin on proteins breaks these substrates down into amino acids.

## DOES PUTTING HOT FOOD IN THE REFRIGERATOR RUIN THE FOOD?

Hot foods should be immediately placed into the refrigerator to avoid foodborne illnesses. There is little truth to leaving out a dish of food to "bring out its flavors." The same flavor will develop, perhaps more slowly, in the refrigerator. Moreover, the increased risk of dangerous bacteria multiplying when food is left at room temperatures far outweighs any minimal loss of flavor from refrigerating cooked foods immediately.

The cooling down phase from hot food (57°C or 135°F) to room temperature (15°C or 41°F) is known as the danger zone because certain types of bacteria, *Salmonella, Pseudomonas* and *E. coli*, for example, grow well at 37°C, and *Staphylococcus* (often found on skin surfaces) thrives at 25°C.

Some bacteria, such as *Pseudomonas*, are also able to grow at refrigerator temperatures, but the cold slows their growth drastically. Placing hot food in the refrigerator does require more energy since it takes more electricity to bring a meal from a higher temperature than from room temperature, but the tradeoff in added safety is well worth the extra expense.

## Small Intestine

The acidic ball of food, called the chyme, passes into the small intestines through the pyloric sphincter. The small intestines are narrow, thin tubes roughly 20 feet in length. The name "small" emanates from its narrow width, but it is actually a very long tube winding around abdominal organs. Peristalsis moves food slowly through the small intestines, taking up to 8 hours. The liver and pancreas are connected to the small intestines via a common duct.

**Chyme**

Acidic ball of food.

**Small intestine**

The portion of intestine that lies between the colon and stomach.

**Villi**

Small folds or projections lining the walls of the small intestine.

**Microvilli**

Smaller villi.

The walls of the small intestines are covered in villi, which are small folds or projections. On these small projections are thousands more smaller villi called microvilli. Together, villi and microvilli increase the surface area of the small intestines to aid in digestion and absorption of nutrients. Figure 12.22 successively shows the smaller and smaller folds of the small intestines.

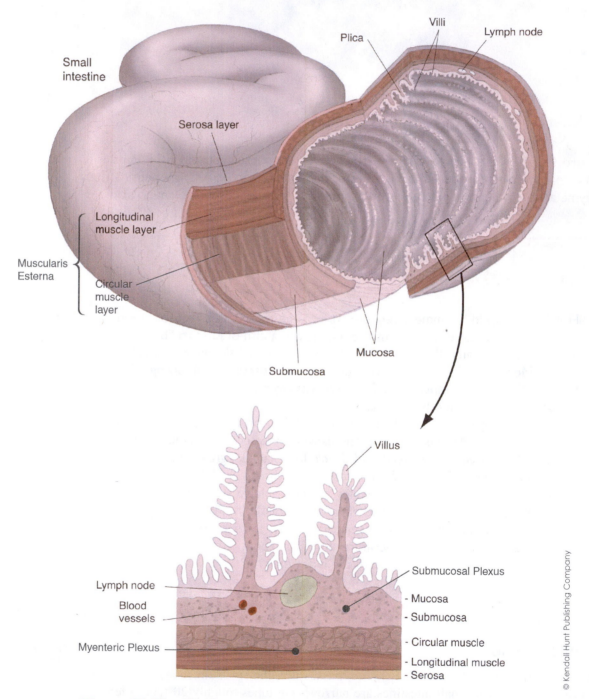

© Kendall Hunt Publishing Company

**Figure 12.22** The layers of the digestive system. Folds of the small intestines form villi and microvilli. Villi contain capillaries and lymph vessels that absorb and transport nutrients to the rest of the body.

In Celiac disease, a malfunction of villi creates cramping and intestinal discomfort. There are many forms and levels of severity of Celiac disease. An inability to digest gluten is blamed on increasing cases. The causative effect of gluten on intestinal efficiency is still being researched.

Complete digestion of all of the macronutrients occurs in the small intestines. Intestinal cells produce enzymes that complete the digestion of all of the macronutrients. Food becomes small enough to be absorbed through the cell membranes of intestinal lining cells.

How is this done? As chyme enters into the top portion of the small intestines, called the duodenum, it stimulates the release of enzymes. Some enzymes are made by the intestinal lining cells and others are made by the pancreas and the liver. As food moves through the small intestines, it enters a middle region called the jejunum in which digestion continues. Accessory organs, discussed below, aid in the digestion of the macronutrients. After this point, food is small enough to be absorbed, as shown in Figure 12.23. It reaches the end region of the small intestines, called the ileum. Absorption occurs in the intestinal cells of the ileum. By the end of digestion in the small intestines, all of the macronutrients are absorbed into the body.

**Major Accessory Organs: Pancreas.** Acidic chyme entering from the stomach stimulates intestinal cells to make the hormone secretin. Secretin causes the **pancreas** to release bicarbonate ($HCO_3^-$) and pancreatic juice. Bicarbonate is a buffer and neutralizes the acidic chyme entering the intestines. For enzymes to work in the small intestines, a basic pH of roughly 8 is required. Pancreatic juice also contains enzymes that digest all of the macronutrients: proteins, carbohydrates, and lipids are broken down into their smallest units – amino acids, simple sugars (monosaccharides), fatty acids, and glycerol – to be absorbed by the intestines.

**Intestinal cells**

Cells lining the GI tract.

**Duodenum**

The top portion of the small intestine.

**Jejunum**

The second part of the small intestine.

**Ileum**

The third portion of the small intestine.

**Bicarbonate ($HCO_3^-$)**

A buffer that neutralizes the acidic chyme entering the intestines.

**Pancreatic juice**

A secretion of the pancreas that contains enzymes that digest all of the macromolecules.

**Secretin**

A digestive hormone secreted by the duodenum.

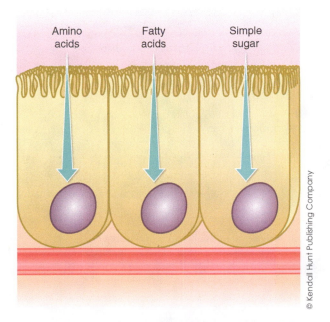

© Kendall Hunt Publishing Company

**Figure 12.23** Digestion of the three macronutrients in the small intestines: Almost all nutrient absorption occurs across the lining cells of the small intestines. Villi help to increase surface area for absorption in the small intestines.

**Major Accessory Organs: Liver.** When fats arrive at the duodenum, intestinal cells make the hormone cholecystokinin (CCK), which slows peristalsis. This gives the small intestine more time with fatty foods to complete their digestion. CCK also stimulates the **liver** to produce bile, a salt that emulsifies fats. Emulsification breaks fat globules into smaller globules, helping their digestion by enzymes. Bile is not an enzyme and serves to mechanically break down fat by pulling apart the hydrophobic and hydrophilic regions. The way bile works is depicted in Figure 12.24a.

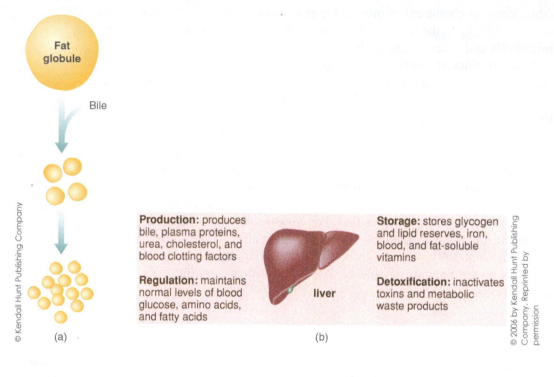

**Production:** produces bile, plasma proteins, urea, cholesterol, and blood clotting factors

**Regulation:** maintains normal levels of blood glucose, amino acids, and fatty acids

**Storage:** stores glycogen and lipid reserves, iron, blood, and fat-soluble vitamins

**Detoxification:** inactivates toxins and metabolic waste products

(a)  (b)

© 2006 by Kendall Hunt Publishing Company. Reprinted by permission

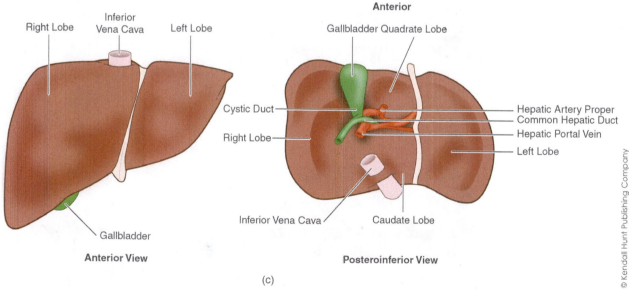

(c)

**Figure 12.24** a. Fat emulsification. The way in which bile works. Bile breaks down larger globules of fat into smaller ones. When fat is in the small intestines, bile is released from the gall bladder to break it down. b. Functions of the liver. From *Biological Perspectives*, 3rd ed by BSCS. c. The liver.

Like the pancreas, the liver is an accessory organ, and both aid in the digestion of macronutrients. Bile is made in the liver but is stored in the gall bladder, a small organ on the underside of the liver. Bile is released only when fat arrives at the duodenum. The liver has a number of functions within the body in addition to its role in digestion:

**Gall bladder**

A small organ on the underside of the liver.

1) Detoxification of harmful substances such as alcohol and drugs
2) Storage and packaging of fat into cholesterol: HDLs and LDLs
3) Deamination of proteins by removing amino groups and forming urea in the urine
4) Transformation of amino acids into carbohydrates
5) Regulation of glucose levels in the blood by storing glycogen
6) Production of plasma proteins, clotting factors, and bile; and storage of fat-soluble vitamins
7) Inactivation of hormones

While the liver is only an accessory organ in the digestive system, it plays a major role in our life functions. The digestive and accessory organs work together to accomplish the complex task of digestion. An overview of the principal digestive enzymes and their regions of activity in the body are given in Figure 12.25.

**Summary of Digestion**

| Type of Food Ingested | Enzymes Involved | | | Resulting Nutrients |
|---|---|---|---|---|
| | In Mouth | In Stomach | In Small Intestine | |
| Carbohydrates | salivary amylase | | pancreatic amylase | simple sugars |
| Proteins | | pepsin | pancreatic proteases | amino acids |
| Lipids | | | pancreatic lipase | fatty acids; glycerol |

**Figure 12.25** Principal digestive enzymes. Enzymes digest macromolecules in different regions of the alimentary canal. Ed Reschke.

### WHEN COMBINING CERTAIN FOODS, CAN DIGESTION BE SLOWED, AIDING WEIGHT LOSS?

Some diets argue that eating separate macromolecule meals (proteins, lipids, or nucleic acids) or combining certain foods with one another, helps people to lose weight. They claim that some combinations of foods poison the body and lead to weight gain and other ailments. It is true that certain foods, such as vitamin C, help the ileum to absorb more iron from foods. However, the small intestines make enzymes for all of the macronutrients at the same time. The macronutrients are digested independent of one another, with combinations irrelevant to digestion or absorption. As long as a macronutrient is broken down into small enough particles, it is absorbed in the small intestines. There is no truth to this myth.

### Large Intestine

The final phase of food processing occurs in the **large intestines**, also known as the **colon**. The large intestines are much wider (3 inches) than the small intestines (1 inch) but they are shorter, at only 3–6 feet in length. The large intestines form a loop around the abdominopelvic cavity, surrounding the small intestines. They attach to the small intestines at the ileum and receive remaining wastes. They then ascend, travel across the body, and descend into the rectum. Feces or final waste products are expelled through the anus. The human large intestine is shown in Figure 12.26.

It takes between 12 and 24 hours for food to travel through the large intestines. In the 1980s, the Mayo clinic measured the time it takes for digestion in 21 healthy people. It estimated that the transit time for food moving throughout the alimentary canal is 33 hours for men and 47 hours for women.

The large intestines are excellent recyclers of the micronutrients: they absorb over 90% of the water needed for the body, along with dissolved minerals, salts, and vitamins from feces. The colon concentrates wastes, drying them out as water and these vital micronutrients are absorbed. The last part of the colon, the rectum acts a storage compartment for our final waste products.

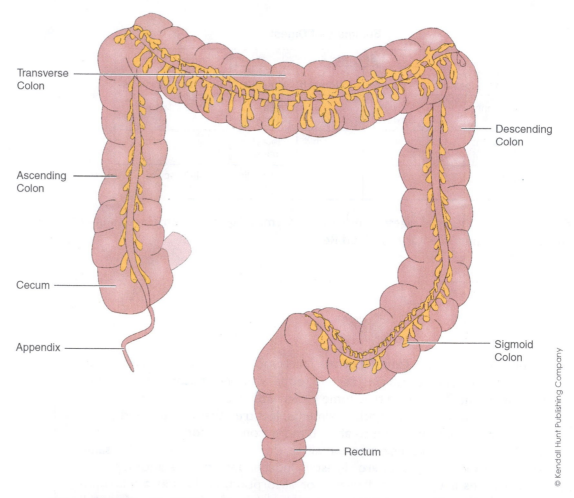

Transverse Colon

Descending Colon

Ascending Colon

Cecum

Appendix

Sigmoid Colon

Rectum

**Figure 12.26**   The human large intestine. A wide tube, which houses bacteria and is responsible for most of the water absorption in the human body. It is also known as the colon and carries food to the rectum and anus.

The composition of feces is about 95% bacteria by dry weight. A flora of bacteria normally resides in the large intestines. These bacteria are useful to humans. They guard against other, more harmful bacteria that would otherwise take hold of the area. Large intestinal bacteria provide nutrients such as vitamin K and a type of vitamin B (biotin) as they break down wastes. Anorexia nervosa, as pointed to in our story, can have drastic effects on heart and nerve functions, as minerals such as sodium and potassium are not absorbed in adequate amounts. This disruption of mineral balance is a frequent cause of death in anorexics.

The large intestines provide an oxygen-free environment. They house bacteria that break down wastes to produce nonoxygen gases as by-products: methane, hydrogen, sulfur, carbon dioxide, and water (also the composition of any flatulence – intestinal gas). The gases found in the large intestines of humans are very similar those found on early Earth. Thus, the bacteria within our intestines also resemble those found on Earth over 4 billion years ago. Often, environmental cleanliness is measured by tracking certain bacteria in soils and water, namely coliform and *E. coli*, normal intestinal bacteria.

## DO VEGETABLES CAUSE GAS?

Bacteria break down plant and animal materials that are undigested in humans. Undigested feces move *en masse* into the large intestines, where bacteria transform feces into gas vapors. Vegetables, particularly beans and legumes, have high amounts of these materials, such as cellulose, which only bacteria can consume. Raffinose oligosaccharides, found in high doses in beans, cabbage, and Brussels sprouts produce larger volumes of gas as they are broken down by bacteria. The gases cause the commonly associated intestinal pains. Abdominal muscles contract to move the gas bubbles through the intestines more quickly.

*The Role of Fiber.* Diets high in fiber, long chained carbohydrates, help to cleanse the digestive system. Fiber, found in grains, vegetables, and fruits, is not digested or absorbed in the alimentary canal. It adds bulk and speeds the movement of food through the digestive tract. In chronic constipation cases, a long term treatment plan is to increase fiber intake to prevent constipation by softening feces and making it easier to move through the many convolutions of the digestive tract.

Diet changes are always the best first strategy to improve one's digestion. Laxatives are also used to soften stools. Laxatives often contain magnesium salts, which increase the salt concentration in the feces. Since water follows solute, salt draws more water out of the body and into the large intestines. The added water softens feces, making it easier to expel.

Fiber is also related to lower risks for colon cancer and heart disease. Some forms of fiber, which dissolve in water called soluble fiber, attach to cholesterol in the digestive tract and thus eliminate fat from the body. Diets high in soluble fiber are associated with lower risks of heart disease and stroke. Insoluble fiber, which does not dissolve in water, serves as roughage to cleanse the intestines.

**Soluble fiber**

Fibers that dissolve in water.

**Insoluble fiber**

Fibers that do not dissolve in water and serve as roughage to cleanse the intestines.

## Common Diseases of the Digestive System

### *Heartburn*

*Heartburn*. Heartburn occurs when pain along the esophagus results from acids of the stomach leaking out of the upper regions through the gastro-esophageal spinchter. Leakage from either sphincter in the stomach leads to damage in adjoining organs. Indigestion due to leakage of acidic stomach contents into the esophagus causes damage to linings and may result in serious discomfort for its sufferers.

It may occur when someone eats too much or too quickly, propelling food upward into the esophagus. Antacids can alleviate the discomfort of indigestion. However, antacids interfere with digestion because the stomach functions best when its acidity is normal.

**Gastro-esophageal reflux disease (GERD)**

A chronic condition caused when acids continue to escape into the esophagus.

When acids continue to escape into the esophagus, a chronic condition called gastro-esophageal reflux disease (GERD)can develop and continual damage to the esophageal lining may lead to Barrett's esophagus, characterized by a change into precancerous cells. In 10% of cases of Barrett's esophagus, GERD leads to esophageal cancer, a very dangerous disease claiming over 95% of its victims (discussed in chapter 2). GERD is treated by stronger medications to reduce acidity in the stomach and by surgery to repair the sphincter.

## Ulcers and Stomach Cancer

*Ulcers*. A species of bacteria called *Heliobacter pylori* thrive in acidic conditions and can cause forms of stomach (peptic) ulcers. Ulcers are open sores in the lining of mucous membranes that can develop into infections. It is estimated that one-third of all people have *H. pylori* in their bodies. In 1985 Barry Marshall, a young physician and scientist, demonstrated that *H. pylori* caused ulcerations in the stomach lining by drinking a vial of *H. pylori* to prove that it was the cause of, rather than a result of, ulcers. Indeed, he developed numerous irritations after drinking the vial. However, *H. pylori* are only active in some people, for reasons that are not fully understood.

*Stomach Cancer*. Peptic ulcers are a risk factor for developing stomach cancer. Stomach cancer has a low survival rate because it is often found too late and metastasizes to other parts of the body. As we saw in Chapter 5, cancer when found early, is usually curable. Some cancers, such as stomach cancer, are not screened regularly or found early enough. Stomach cancer occurs at a high incidence in Japan, where adults are regularly screened for the disease. An open or bleeding peptic ulcer is obvious in Figure 12.27.

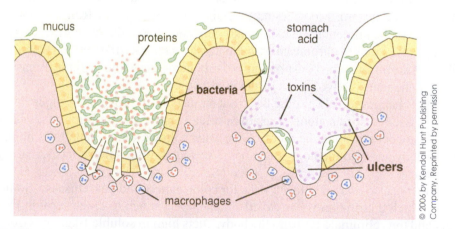

© 2006 by Kendall Hunt Publishing Company. Reprinted by permission

**Figure 12.27**  Ulcers are not caused by stress alone but by some forms of bacteria. From *Biological Perspectives*, 3rd ed by BSCS.

# Colon Cancer

*Colon Cancer.* Diets high in insoluble fiber are related to lower risks for colon cancer, the third highest cause of cancer deaths in the United States. Colon cancer develops from a small polyp, which grows larger and becomes cancerous over time. A colonoscopy places a flexible tube into the colon to detect polyps and colon cancer (Figure 12.28). If removed early, colon cancer has a 90% survival rate. Polyps are removed during a routine colonoscopy. Surgery may be required upon detecting colon cancer. A controversial limit to screen for colon cancer recommends an end to screening after age 75 because a person is statistically likely to die from other causes by that age. Is this a fair recommendation by the American Medical Association? Are seniors being deprived of screening tests that could save their lives in the name of statistics?

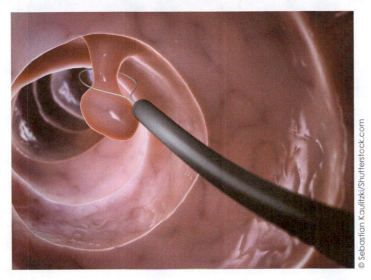

© Sebastian Kaulitzki/Shutterstock.com

**Figure 12.28**   A colonoscopy examination checks the topography of the large intestines to detect signs of colon cancer. In this image, a polyp (small growth of cells) is removed within the colon.

## DOES BLACKENED (OR SEARED) MEATS CAUSE CANCER?

Whenever a meat is smoked or grilled (especially when it is blackened), it has been shown to develop chemicals that are cancer causing (carcinogenic). High temperatures convert normal chemicals in meats into heterocyclic amines (HCA's), which have been shown to cause a variety of cancers. The smoke emanating from grilling contains polycyclic aromatic hydrocarbons (PAHs). PAHs are known carcinogens that attach to meats as they are grilled or smoked. Salami, smoked fish, bacon, hamburgers, and hot dogs contain PAH levels associated with cancer.

However, limited research has been conducted on human subjects to make these claims. Animal studies act as a model for possible links between grilled and smoked meats and cancer (Figure 12.29). One recommendation is to remove the skin on smoked or grilled meats, which is to likely contain the majority of the PAH chemicals. As in our opening story, food is a source of health and disease.

**Figure 12.29**   Grilling meats and PAHs, does charred meat cause cancer?

# Summary

Eating disorders and obesity are illnesses influenced by societal factors centering on people's focus on body image and food. While food is needed to survive, with micronutrients and macronutrients required in the right amounts, food is also a source of disease. Eating too many calories leads to obesity and eating too few calories, to a point of being seriously underweight, also leads to health problems. A person's healthy body weight may be determined using the BMI. Nutrients are broken down and absorbed through digestive processes. Digestion and its related diseases occur throughout the alimentary canal and its accessory organs. Eating disorders and obesity, both leading causes of health hazards, are a result of disruption in normal nutrition and digestion.

---

## CHECK OUT

### Summary: Key Points

- Psychological and societal factors both influence what foods a person chooses.
- Changes in diet and exercise in the past century have led to both the obesity epidemic and eating disorders in our society.
- Nutrients, both micronutrients (water, minerals, and vitamins) and macronutrients (proteins, lipids, and carbohydrates), are essential for the proper workings of the human body.
- Calories taken in vs. calories used by the body determines whether a person gains, maintains, or loses weight.
- While mechanical digestion physically breaks down food into smaller pieces, chemical digestion changes food by rearranging bonds, into smaller substances.
- The alimentary canal transports food through the mouth, pharynx, esophagus, stomach, small intestines, large intestines, rectum, and anus.
- Digestive cancers, such as stomach, esophageal and colon cancer, and ulcers, along with eating disorders, impact proper digestive processes.

## KEY TERMS

accessory organs

alimentary canal

anal canal

anorexia nervosa

antioxidants

apple shape

basal metabolic rate (BMR)

bicarbonate ($HCO_3^-$)bolus

bulimia

calories

canines

cardiac sphincter

cementum

cholecystokinin (CCK)

chyme

dentin

digestion, mechanical, chemical

duodenum

enamel

epiglottis

esophagus

essential amino acids

fat-soluble vitamins

food plate

food pyramid

free radicals

gall bladder

gastro-esophageal sphincter, or pyloric sphincter

gastro-esophageal reflux disease (GERD)

hydrochloric acid (HCl)

ileum

incisors

ingestion

insoluble fiber

intestinal cells

jejunum

large intestines or colon

larynx

liver

macronutrients

metabolism

micronutrients

microvilli

minerals

molars

mouth

Neolithic diet

nonessential amino acids

nutrients

obesity

pancreas

pancreatic juice

pear shape

pepsin

pepsinogen

peristalsis

pharynx

premolars

pulp cavity

rectum

rugae

salivary amylase

salivary glands

secretin

small intestines

soluble fiber

stomach

trachea

villi

water

water-soluble vitamins

# Multiple Choice Questions

1. Which disease is MOST influenced by the role food plays in body image society?
   a. Colon cancer
   b. Herpes
   c. Ulcers
   d. Anorexia nervosa

2. A characteristic of modern society that has resulted in changes in the rates of obesity over the past century is:
   a. less physical activity in jobs.
   b. more varied food choices.
   c. decreases in HFCS use in foods today.
   d. decreased productivity by workers.

3. Which term is best associated with metabolism?
   a. BMR
   b. BMI
   c. BMO
   d. BBB

4. Vitamin A belongs to the_____ group of nutrients.
   a. smallest
   b. largest
   c. macronutrient
   d. micronutrient

5. When dieting, the most weight is lost during the first few weeks. This weight loss is attributed to loss of:
   a. lipids
   b. proteins
   c. minerals
   d. water

6. Which represents a logical order, from start to finish, in the movement of food along the alimentary canal?
   a. esophagus → pyloric sphincter → stomach → large intestines
   b. pyloric sphincter → esophagus → large intestines → stomach
   c. esophagus → large intestines → stomach → pyloric sphincter
   d. esophagus → stomach → pyloric sphincter → large intestines

7. Which is NOT a function of the liver?
   a. Production of bile
   b. Storage of glycogen
   c. Absorption of nutrients
   d. Elimination of wastes

8. Mechanical digestion is carried out by:
   a. teeth
   b. pepsin
   c. pancreatic juice
   d. amylase

9. Which organ does NOT produce enzymes that digest nutrients?
   a. Liver
   b. Stomach
   c. Pancreas
   d. Small intestines

10. *Heliobacter pylori* is most closely associated with:
    a. tooth decay
    b. acid reflux disease
    c. colon cancer
    d. ulcers

## Short Answers

1. Describe two ways in which society influences the foods people choose to eat. Define and describe one digestive disease that is influenced by the food in one's diet.

2. Define the following terms: anorexia nervosa and bulimia. List one way the terms differ from each other in relation to their a. symptoms; and b. prevalence in the U.S. population. Explain how the terms are similar.

3. Explain how BMI is used in determining healthy weight. Is it always a reasonable measure of a person's health? Why or why not?

4. Which micronutrient is the most important in maintaining ion balances within the body? Which organs are most affected by micronutrient imbalance?

5. Mechanical digestion plays an important role in breaking down foods. Define mechanical digestion and give an example of it within the alimentary canal. Is a human able to survive using only mechanical digestion?

6. Colon cancer is a major killer. List two ways a person may limit their risks of dying from colon cancer.

7. Trace the movement of sugar water through the alimentary canal. Be sure to list 1) each region it is digested and 2) each region it is absorbed.

8. For question #7, how would the answer change if the food were a large slice of fatty bacon?

9. Explain how GERD is affected by both diet and anatomy.

10. Describe a plan for best improving weight loss in a client. Be sure to include the terms BMI, BMR, muscle mass, insoluble fiber, and water.

## Biology and Society Corner: Discussion Questions

1. Fast foods, such as McDonald's and Kentucky Fried Chicken, are becoming more common in many countries in Western Europe, including Greece. Research and explain the changes in diet and obesity rates in Greece over the past 25 years.

2. The growing obesity epidemic is particularly concerning among children. What changes (if any) would you suggest to parents to help their children eat healthy. Do you think it is ethical to mandate your suggestions in, for example, school lunches?

3. Roughly half of Americans take vitamin supplements. It is a multimillion-dollar industry. Yet, most nutritionists agree that the majority of these vitamin takers do not need them. Should the government limit their consumption? Why or why not?

4. Barry Marshall, to show that *H. pylori* causes (and is not a result of) ulcers in the stomach, went to extreme measures, drinking a vial of the bacteria. Do you think it was ethical for him to put his life in danger with an untested hypothesis? Would an animal model have been a good enough evidence for the research question? Why or why not?

5. This chapter focuses on many diseases associated with digestive illnesses. The survival rate for many digestive cancers, such as colon cancer, is high when caught early in its onset. Would you recommend mandatory screenings at certain ages? Do you agree with limiting screening after certain ages, as for colon cancer, as recommended by the American Medical Association?

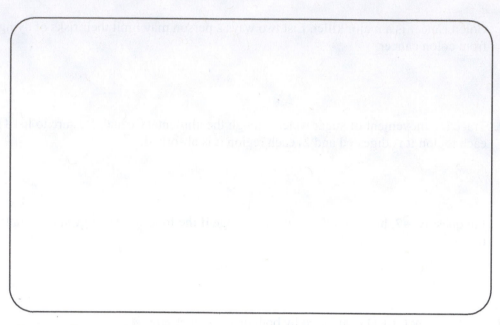

Figure – Concept Map of Chapter 12 Big Ideas

# The Heart-Lung Machine: Circulation and Respiration

## 13

## ESSENTIALS

A father and daughter

A father is alone

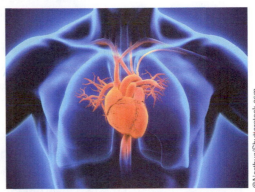

A heart and lungs

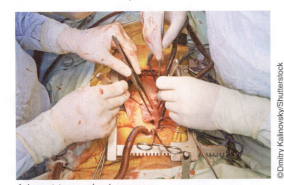

A heart transplant surgery

A car accident

## CHECK IN

**From reading this chapter, you will be able to:**

- Explain how medical treatment advances and prevention help improve heart and lung health in society.
- Describe the functions and components of the blood.
- Trace the movement of blood within the heart and blood vessels and the electrical activity that corresponds to an ECG.
- Connect the role of high blood pressure with cardiovascular diseases in society.
- Trace the movement of air through the respiratory system.
- Explain how gases are exchanged within the lungs
- Describe the cardiovascular and respiratory diseases and their current treatments.

## The Case of his Daughter's Heart

2014: Charles' heart beat ever so gently, with a regularity that reminded him every moment that Peggy was still with him – in him – all the days of his life.

Flashback 15 years: It would be her first trip away from home. Peggy looked forward to her weekend camping trip, high in the mountains of the Adirondacks. Charles' daughter, Peggy was a lively and driven teen but she wanted to be a little more independent. It would be her first trip away from home and the first time she spent a night away from her father. Charles had a strange feeling about the day. He felt worse than usual, as if life would soon end. He was quite ill, after all, and struggled to walk around the house; even a trip across the room had become a breathless chore.

Charles' heart was weak, with a condition called congestive heart failure. He had a series of heart attacks, which weakened his heart muscle to about 15% of its normal strength. The doctors checked his left ventricle ejection fraction, which is the amount of blood pumped out of the heart by its strongest muscle, the left ventricle. Charles would need a new heart but it would be hard to find a matching one. The doctors leveled with Charles, informing him that if he did not find an organ within the next year, he would die.

Charles felt a horrid chill after hearing the doctor's news that he would soon die – an ominous loneliness that he and his Peggy would part. He looked through old pictures of his daughter, tracing each event as she grew up through the years. Their years together would stay with them forever.

Charles was both a father and a mother to Peggy. After her mother died, when Peggy was only 8 years old, they had only each other. Charles had not ever spent a night away from his daughter since she lost her mother. They had a closeness about which they never spoke, but was deeply felt between them. Some people told Charles that he was overprotective.

Nevertheless, Charles was pleased that Peggy travelled with her friends to the Adirondacks. Peggy might soon be alone, without Charles to care for her, if biology had its way. So, Charles knew that he had to let go. The trip would be good for Peggy and for him. "She needed to be with other people; to have a good time," Charles thought. "Peggy always gave a part of herself to everyone she met."

This is what made the phone call so ironic: "Your daughter is in a coma. She had a car accident on the Northway. I am very sorry sir," spoke the voice on the phone. Charles was numb and his horror was immense.

At the hospital, doctors emphasized that his daughter would die shortly. She was an organ donor and she had offered her heart to you in her donation paperwork.

Charles could not do it: "Was it ethical?"; "How macabre."; "Who could take their own daughter's heart?" thought Charles. But it was the right decision or Charles too would soon die.

The doctor explained that it was his daughter's wish that he live on through her. It was a gift of the deepest love. Charles took his daughter's heart.

---

## CHECK UP SECTION

Organ donation is a key to survival for many people facing serious illness, like Charles in the story. But obtaining a donor for a needed transplant can be difficult. About 6,500 people die each year while waiting for a transplant.

Many organs are able to be transplanted after a person dies, while some organs may be transplanted while a person is alive. Research the organs used in both deceased and live organ transplants. Explain the risks associated with organ transplants.

Owing to the shortage of organs, should the government require all people to donate their organs upon death? Why or why not?

---

## Blood: Life's Force

Early religions viewed blood as a life force, with special healing and magical powers. Many religions integrated these views into their beliefs and customs, from sacrifices and blood bonding to specific uses of foods containing blood. The Mayan Indians, for example, believed that blood sacrifices needed to be offered to keep the cosmos in balance. Figure 13.1 depicts an image of their ceremonies.

**Figure 13.1**  Mayan Indian sacrifices. Blood sacrifices were used in many rituals throughout human history. In this figure, a Mayan Lord runs a rope through the tongue of its subject.

Blood has a mystique, perceived by the ancients, that also intrigues us today. This chapter answers some of the ancients' questions: How does blood heal us? Why do we die when we lose too much blood? How is blood able to carry our nourishment? Science discovered the chemical components of blood to better understanding its role within our bodies. In this chapter, we will first look at how blood functions. We will then study how the heart and lungs work together to transport gases, nutrients, and wastes.

## What is Blood?

**Blood**

A viscous fluid composed mostly of water, with dissolved salts, proteins and cells within the liquid.

Blood is a viscous fluid, composed mostly of water, with dissolved salts, proteins, and cells within the liquid. The blood comprises almost 8% of the body weight of most adults. Males, on average have between 5 and 6 L (1.5 gallons) of blood and females have 5–6 L. Males have slightly more red blood cells than females, to enable more oxygen transport to muscles. When blood travels within its vessels, it is dark red with less oxygen than when it is exposed to the air, appearing bright red due to the enhanced oxygen content. Blood is carried throughout the body to perform many tasks: healing, waste removal, temperature regulation, nutrient transport, and acid–base balance. Blood is slightly basic, with a pH of about 7.4. Its pH must remain within strict limits for the body to function properly.

**Plasma**

The straw-colored liquid that makes up 55 percent of the blood.

The composition of blood is mostly water but contains cells and solutes within it. Roughly 55% of the blood is made of a straw-colored liquid, called the plasma. Blood plasma is 90% water and contains more than 100 dissolved solutes, which include calcium, potassium and urea. Dissolved solutes are often included in blood tests to diagnose a person's health. A high amount of urea indicates kidney malfunction, for example, because this solute should be removed from the blood through the kidneys. The rest of the blood is composed of the formed elements, which are those cells and cell fragments "formed" by the body. About 45% of the blood is composed of the formed elements. When blood is centrifuged, as shown in Figure 13.2, heavier materials such as the formed elements, move to lower parts of the test tube.

**Formed elements**

The cells and cell fragments formed within the blood and have a definite shape.

The formed elements of the blood include white blood cells, red blood cells, and platelets:

**White blood cells**

Large blood cells that help the body fight infections.

1)  White blood cells (also called leukocytes) play a role in protection from disease and illness. When a pathogen or disease-causing agent, such as a bacterium or virus enters, white blood cells are the first line of defense for the body to guard itself. As you might recall from Chapter 8, there are many microbes surrounding us. They are kept at bay by our body's defenses.

**Pathogen**

Any disease causing organism.

Each white blood cell plays a different role in the defense our bodies, which will be elaborated upon in Chapter 15. There are five types of white blood cells: neutrophils, lymphocytes, monocytes, eosinophils, and basophils. However, all of them patrol the blood vessels to search for signs of an invasion. Normal white blood cell numbers range between 4,800 and 10,800 cells per cubic millimeter. When these readings increase, infection or other illnesses are indicated.

**Red blood cells**

Blood cells that contain hemoglobin and carry oxygen to and from the tissues.

2)  Red blood cells (also called erythrocytes) carry oxygen and carbon dioxide gases. They are the most numerous of blood cells, comprising over 95%. Red blood cells are flexible, biconcave cells, fairly small in size at 7.5 μm in diameter. They are quickly made and swiftly disposed of, lasting only about 120 days. (Other cells can last a lifetime, such as nerves and heart cells). Red blood cells are destroyed in the spleen and liver, as discussed in Chapter 12. Red blood cells are filled with hemoglobin molecules, each able to transport oxygen and carbon dioxide gases.

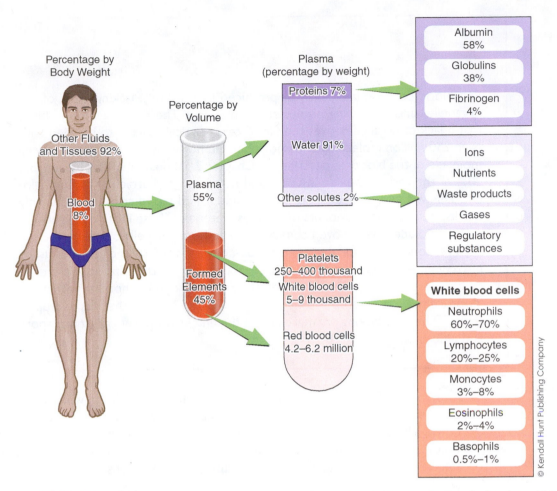

**Figure 13.2**    The components of human body: white and red blood cells, and platelets. When centrifuged, the lighter layer of watery plasma sifts to the top of the tube and the denser red blood cells migrate to the bottom. A thin layer of white blood cells forms in the middle of the two other layers.

Porphyria, as you may recall from our story in Chapter 6, is an example of a blood disorder in which abnormal hemoglobin causes symptoms similar to that of a vampire. Red blood cells are unable to properly transport gases in porphyria.

There are other types of red blood cell disorders, each of which may have serious health consequences. When hemoglobin is abnormal, it affects a red blood cell's oxygen-carrying capacity. In thalassemia, for example, a faulty or absent hemoglobin chain makes the molecule fragile and less able to carry oxygen. A more serious disease, sickle-cell anemia, was described in Chapter 5. Here, the red blood cell forms a sickle shape under conditions in which there is less oxygen, such as during physical exertion. Then, sickle-shaped cells jam up within blood vessels, causing clots, often affecting organs.

When blood lacks normal oxygen-carrying capacity, it is called anemia. The most common form of anemia is iron deficiency anemia. Iron holds oxygen within a hemoglobin molecule within its heme group of the hemoglobin molecule. When iron is lacking, often due to low amounts in the diet, blood is unable to carry sufficient oxygen.

**Thalassemia**

The condition in which a faulty or absent hemoglobin chain makes the molecule fragile and less able to carry oxygen.

**Sickle cell anemia**

A serious condition that affects the RBCs

**Anemia**

The condition in which blood lacks normal oxygen carrying capacity.

**Iron deficiency anemia**

A condition characterized by lack of healthy RBCs in blood.

## CRITICAL REASONING: WHY ARE RED BLOOD CELLS MADE SO CHEAPLY?

Red blood cells are inexpensively produced by the body, lacking most of the organelles and containing no nucleus at maturity. The first reason for this is that a lack of organelles gives red blood cells physical space – to be filled with over 250 million molecules of hemoglobin. Hemoglobin is the oxygen carrying molecule of the blood, as described in Chapter 2. Almost all of the oxygen and 20% of the carbon dioxide in the blood is transported attached to hemoglobin. This will be discussed later in the chapter. Red blood cells are like supply trucks, carrying large amounts of gaseous materials. These include oxygen and carbon dioxide gases, moved constantly to and from all cells of the body.

A second reason for the way a red blood cell is constructed lies in its usage of oxygen. As a supply truck, it would be a problem if the supplies were used by those transporting them. Similarly, the presence of mitochondria within red blood cells would use up the very oxygen needed by cells that depend upon them. Thus, cells lacking organelles serve the unique purposes of fast transport and efficiency.

**Platelets**

Are chips of cells that form clots within vessels to prevent blood loss.

**Clotting factors**

Are proteins that undergo a series of chemical reactions that halt bleeding.

**Thrombin**

An important enzyme in blood that facilitates clotting of blood by converting fibrinogen to fibrin.

**Protime**

A blood test that measures the rate at which blood clots.

**Hemophilia**

A condition of uncontrolled bleeding when blood clots occur too slowly.

**Thrombosis**

Clots forming in the wrong places (an unbroken vessel).

Another form is pernicious anemia, which presents itself when a patient lacks vitamin B-12 in her or his diet. Sometimes people with anemia feel run down. They lack energy due to the low levels of oxygen in their blood. Many times, pernicious anemia develops during pregnancy. Pernicious anemia is treated with vitamin supplements or vitamin B-12 injections.

3) Platelets are chips of cells that form clots within vessels to prevent blood loss. Within them, platelets contain granules filled with clotting factors. Clotting factors are proteins that undergo a series of chemical reactions that halt bleeding.

Platelets work by latching onto roughened surfaces where a break or tear occurs within the body. They stick to the collagen fibers along the damaged walls of vessels. Platelets first work by causing vasoconstriction of blood vessels, which lessens blood flow to the area. This slows bleeding. Platelets and damaged cells release chemicals that activate a chain of over 200 clotting factors and other proteins. These factors lead to the forming of a clot.

An important enzyme at the end of this series of chemical reactions is thrombin. Thrombin mediates the last steps of the reactions, and is measured by a blood test to determine the rate at which blood clots. This rate is commonly called its protime, or time to clot: it is the time it takes for the reactions to form thrombin. Thrombin causes the formation of fibrin threads, which trap platelets and red blood cells within a clot, as shown in Figure 13.3. Also note that calcium ions are important in mediating the end steps of clotting.

When blood clots too slowly, a condition of uncontrolled bleeding occurs, called hemophilia. Hemophilia is a disorder of certain clotting factors within platelets. There are three types of hemophilia: A, B, and C; each lacking a different clotting factor. In hemophilia, a prolonged bleeding in skin and within body organs occurs. Genetically engineered clotting factors are an effective treatment for hemophilia, eliminating the need for blood transfusions.

Sometimes clots form in the wrong places. In an unbroken vessel, such a clot is called a **thrombus** or thrombosis. A thrombosis in the heart (coronary)

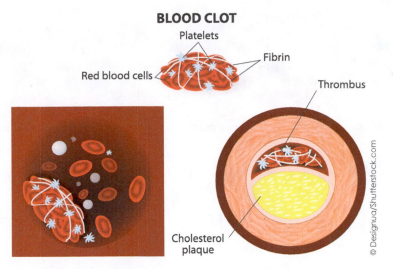

**Figure 13.3** A blood clot also called a thrombus. Over 200 steps lead to the formation of a clot. Red blood cells get caught in a mesh of fibrin strands during the formation of a clot.

arteries is known as a heart attack. A thrombosis in the brain causes a stroke. In many cases, after sitting too long, in an airplane or after an operation, blood pools within the legs and may form a deep vein thrombosis.

The danger of a thrombus is in its ability to break apart and travel to other parts of the body, causing obstructions. A floating thrombus is dangerous and known as an embolus. When an embolus lodges in the lungs, a common pathway for its travel from the legs, it is a pulmonary embolism. Treatments for thrombosis and embolism cases include blood thinners such as warfarin and heparin. These are two strong blood thinners used to treat and prevent clots.

Blood cells are produced in the bone marrow, a compartment within bones that stores stem cells. Stem cells are specialized cells that are able to develop into many types of cells, given particular conditions. For example, under some conditions stem cells become skin and in other conditions, stem cells become a heart muscle. Stem cells are also referred to as **pleuripotential stem cells**, because they have multiple potentials to become any types of cell depending upon the environment. Blood cells form from stem cells within the bone marrow. This is a reason why damage to the bone marrow cavity, as in cancer treatments, often leads to anemia (low red blood cell production) and infection (low white blood cell production) (Figure 13.4).

## Why Blood?

Why is blood, and its movement within the body, laden with historical and religious connotations in society? Because blood performs so many tasks that satisfy our immediate needs, vital to our survival. Without oxygen, a cell dies in less than 4 minutes, as mentioned in a previous chapter. In our story, Charles suffers from a weak heart that cannot pump the blood sufficiently. Let's give a look at the primary roles for blood in the body. The functions of the blood include:

1) **Transport**: Blood contains nourishment for every cell in the form of nutrients (macromolecules) and oxygen and removes unwanted wastes, including

**Heart attack (myocardial infarction)**

The condition in which heart muscle is damaged from the sudden blockade of coronary artery by blood clot.

**Stroke**

The sudden diminution of brain cells due to lack of oxygen caused by obstruction or rupture of a blood vessel of brain.

**Deep vein thrombosis (DVT)**

The condition that occurs when a blood clot forms in one of body's large veins, most commonly in legs.

**Embolus**

A floating thrombus.

**Pulmonary embolism**

The condition in which an embolus lodges in the lungs.

**Bone marrow**

A compartment within bones that stores stem cells.

**Stem cells (Pleuripotential)**

Are specialized cells that are able to develop into many types of cells, given particular conditions.

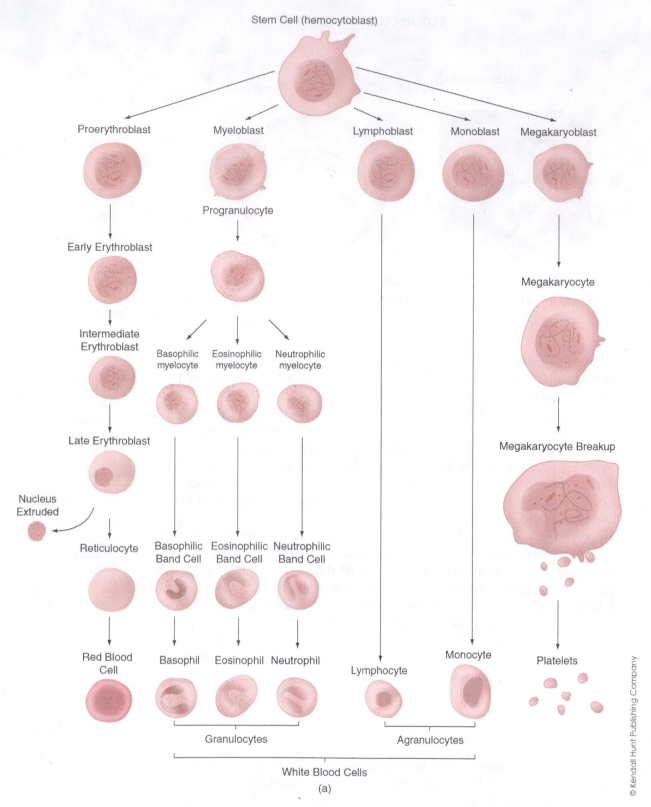

Stem Cell (hemocytoblast)

Proerythroblast     Myeloblast     Lymphoblast     Monoblast     Megakaryoblast

Progranulocyte

Early Erythroblast

Megakaryocyte

Intermediate Erythroblast

Basophilic myelocyte     Eosinophilic myelocyte     Neutrophilic myelocyte

Late Erythroblast

Megakaryocyte Breakup

Nucleus Extruded

Reticulocyte     Basophilic Band Cell     Eosinophilic Band Cell     Neutrophilic Band Cell

Red Blood Cell     Basophil     Eosinophil     Neutrophil     Lymphocyte     Monocyte     Platelets

Granulocytes     Agranulocytes

White Blood Cells

(a)

**Figure 13.4** a. Cells of the blood and their origins. All cells emerge from pleuripotential stem cells. b. White blood cells: a close up look.

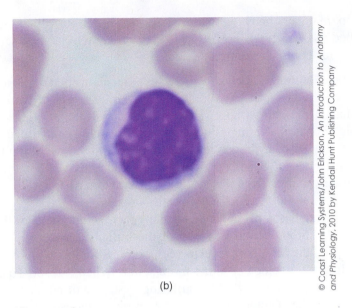

(b)

**Figure 13.4** *(Continued)*

carbon dioxide gas. Every cell needs a continual supply of nutrients and a constant removal of wastes. Nutrients are brought through the blood as dissolved particles, as described in Chapter 12. After macromolecules are digested and absorbed into the alimentary canal, they are transported, used, and removed as wastes through the blood.

As nutrients are processed in metabolism, wastes accumulate and are toxic if not removed by the blood. Many wastes contain nitrogen, such as urea, and are removed immediately as they are formed. Nitrogen wastes build up in serious cases of kidney failure, called uremia. If uremia is left untreated, death is certain. Carbon dioxide gas is also a waste product of metabolism, but is removed through the lungs. Carbon dioxide is a product of cellular respiration, described in Chapter 4.

Hormones are also moved through the body. These chemical messengers stimulate cells to use nutrients for growth, development, and reproduction. Hormones help tissues and organs communicate through a chemical message system, described in the next Chapter 14.

2) **Defense**: The blood is used for the protection of the body. White blood cells within the blood carry out phagocytosis to removed microbes and harmful substances, such as cancer cells and pathogens. Antibodies, or specialized proteins, disable, destroy, and reveal harmful organisms that have invaded the body. Blood serves a role in defending the body from infection and disease in ways that will be discussed in Chapter 15.

In defense of body damage, clotting factors within the blood form a clot to prevent blood loss. An abrasion on the skin, for example, is healed by a series of steps started by clotting factors. This process was described earlier in the chapter, leading to a solid clot along the wall of a blood vessel.

3) **Temperature Regulation**: Blood is also able to store heat as it moves through the body. When blood vessels expand or contract, they change the amount of blood and heat reaching body surfaces. This regulates heat arriving at different areas. When blood vessels expand (or vasodilate), more blood goes to the skin, causing heat loss. When blood vessels constrict (or vasoconstrict), it causes less to blood movement to the skin and thus heat conservation. Blood movement

helps to maintain body temperature, a vital goal for all homeotherms. As described in Chapter 11, homeostasis keeps body temperature set around 37°C (98.6 °F) for optimal enzyme functions.

4) **Acid–Base Balance**: Blood contains a set of reactions that regulate its pH called the carbonic acid–bicarbonate buffering system. The pH of blood cannot vary more than .1 pH unit without serious health consequences. Thus, the carbonic acid–bicarbonate buffering system absorbs and releases hydrogen ions. In this way, the number of hydrogen ions is regulated and the pH of the blood is buffered or maintained. Usually, blood pH is held strictly at 7.4 in healthy people. Even small variation from this set point, a symptom of congestive heart disease described in the story, may lead to death.

Stabilizing blood pH depends on many factors, but includes the role of carbonic acid. Carbonic acid is a part of the buffering system that prevents the movement of the acid–base level to veer too far from its set point. When it gives up a hydrogen ion, blood becomes more acidic. When it absorbs a hydrogen ion, blood becomes more basic. The buffering system will be discussed in greater detail later in the chapter, in relation to respiratory gases.

**Carbonic acid-bicarbonate buffering system**

A set of reactions that regulate the pH of blood.

# Cardiovascular System: Heart and Vessels

## Heart

Blood is propelled through the body of vertebrates using a specialized muscular organ called the heart. The heart in humans is a fist-sized organ, weighing about one pound. It is very strong, with cardiac tissue beating about 70 times each minute, functioning as the most reliable pump that has ever been developed. In fact, the heart beats 1 billion times in a person's lifetime. The heart muscle, called myocardium, is thick and flexible.

Some organisms, such as the earthworm use a series of small hearts to pump blood. Mammals and birds have a four-chambered heart to transport blood through a series of connected blood vessels. Together, the heart and blood vessels are called the cardiovascular system.

**Heart**

A specialized muscular organ that propels blood through the body of vertebrates.

**Cardiovascular system**

The system comprising the heart and blood vessels

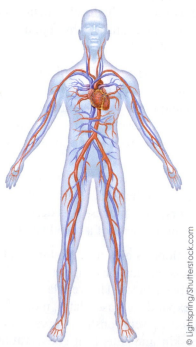

© Lightspring/Shutterstock.com

**Figure 13.5** The heart and it blood vessels: arteries are colored red and veins are colored blue.

The human cardiovascular system acts as a heart–lung machine, with blood vessel connections between the two organs. The heart pumps blood and the lungs trade different gases with the blood. Lungs exchange gases to remove carbon dioxide wastes from it and infuse needed oxygen. In our story, Charles' breathing was also affected by his weakened heart. It was unable to pump blood through the lungs adequately, making him short of breath. Let's trace the movement of blood between the heart, lungs, and body to study how heart functioning affects one's overall health.

## Movement of Blood in the Heart and Vessels

The center of the cardiovascular system is the heart. The heart consists of four chambers: **right** and **left** atrium and **right** and **left** ventricle; and a number of vessels connecting the heart with the body and the lungs. In humans, the chambers and blood vessels are lined with simple squamous tissue, which appears smooth to the touch, called endothelium. The chambers and vessels of the heart are shown in Figure 13.6.

When blood is received from the body, it travels through the **superior** and **inferior** vena cava into the **right atrium** of the heart. Let's trace the movement of blood through the heart and the rest of the body using Figure 13.7. The right atrium, a small chamber with little pumping ability, moves blood through the tricuspid valve and into the **right ventricle**. The right ventricle pumps blood out of the heart through the pulmonary arteries to the lungs. In the lungs, blood exchanges its gases.

The connection between the heart and lungs is known as the pulmonary circuit. So far, the blood in the vessels mentioned has been **deoxygenated**. Deoxygenated blood has a lessened amount of oxygen within it after exchanging with body cells. As such, the lungs are said to **oxygenate** blood within it, adding oxygen for the body. Carbon dioxide, built up by body processes, leaves the blood through the lungs. Oxygenated blood then travels back into the left side of the heart, through the pulmonary vein and into the **left atrium**.

When blood returns from the lungs through the pulmonary veins, it completes its travel through heart on the left side. Blood makes its way through the bicuspid or mitral valve into the **left ventricle**. The left ventricle is a very powerful, thick muscle which pumps blood out through the **aorta** to the rest of the body. The vessel connection between the heart and body cells is called the systemic circuit. The systemic circuit connects the heart with the whole of the body systems. When the aorta forks, it sends blood to the body in two ways: to the head and arms in one direction and to the lower body and legs in another direction.

Some vessels branch off and resend blood toward the heart in what is called the coronary circuit. The coronary circuit supplies the heart walls with oxygenated blood. Why does a blood supply need to return to the heart, when there is ample blood within the heart? The reason is: blood within the heart is unavailable to it for use by its cells. Blood within the heart travels through it but the heart's endothelial lining prevents the transfer of nutrients. Instead, the coronary circuit is comprised of **coronary arteries** which branch into capillaries along the surface of the heart, allowing exchange of oxygen and nutrients.

In our story Charles had congestive heart disease, in which the heart is too weak to pump blood efficiently. Charles had shortness of breath in this condition, because blood could not sufficiently move out of the lungs through the left side of the heart. Thus, his pulmonary vein backed up, leading to fluid buildup in his lungs. This presents as breathing difficulty during exertion. Symptoms may also include fluid accumulation in the legs and abdomen, because the left side of the heart is weak. When the heart cannot sufficiently pump out fluid from the body, it accumulates in lower regions.

**Atrium**

An entry chamber of the heart from which blood is passed to the ventricles.

**Ventricle**

A chamber of heart that receives blood from the atrium.

**Endothelium**

The squamous tissue lining the chambers of heart and blood vessels.

**Vena cava**

A large vein that carries deoxygenated blood into the heart.

**Tricuspid valve**

A heart valve between the right atrium and ventricle and keeps blood from flowing back into the atrium.

**Pulmonary artery**

The artery that carries blood from the right ventricle to the lungs.

**Lungs**

A pair of breathing organs.

**Pulmonary circuit**

The connection between the heart and lungs.

**Deoxygenated blood**

Blood that lacks oxygen.

**Pulmonary vein**

A vein that carries oxygenated blood from lungs to the heart's left atrium.

**Mitral (bicuspid) valve**

A heart valve located between the left atrium and left ventricle.

**Systemic circuit**

The vessel connection between the heart and body cells.

**Coronary circuit**

The system in which some vessels branch off and resend blood toward the heart.

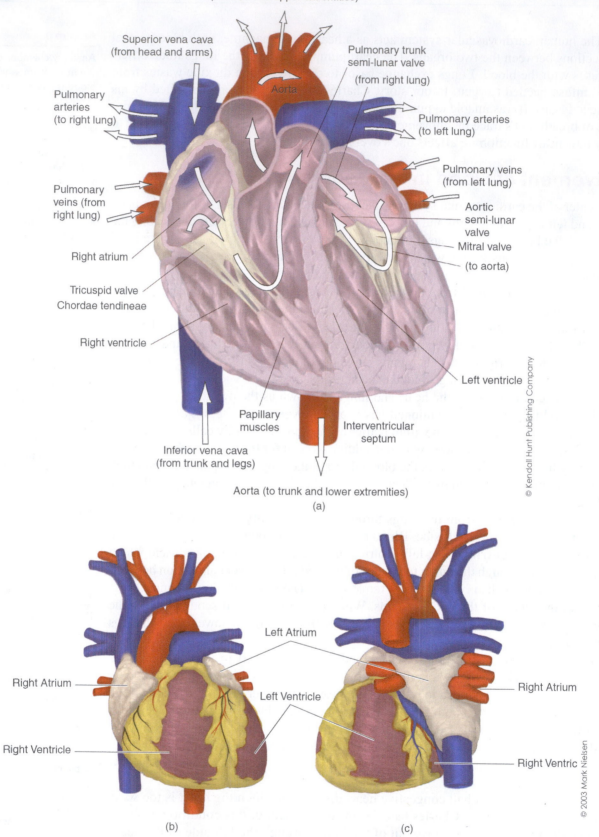

**Figure 13.6** a. The heart and its chambers. The four chambers of the heart pump blood through the lungs and the body. b. Anterior view of external heart. Illustration by Jamey Garbett. c. Posterior view of external heart. Illustration by Jamey Garbett.

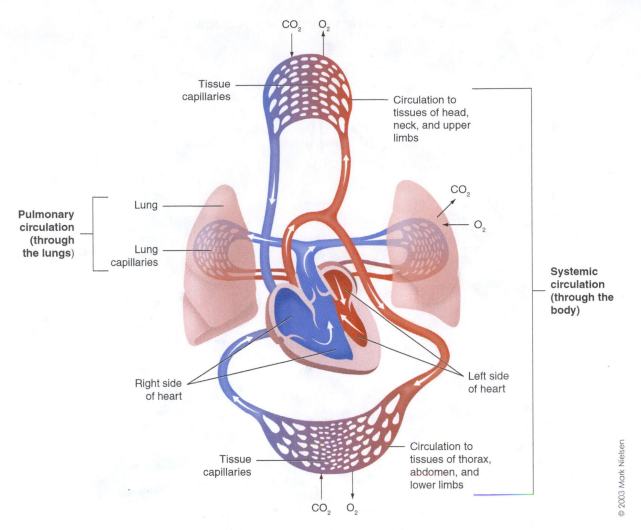

**Figure 13.7**   Cardiovascular system: tracing the traveling of blood through the heart, lungs, blood vessels. The systemic and pulmonary circuits are linked by the heart. Illustration by Jamey Garbett.

# Heart Beats: Electricity Activity

The heart beats because of a control center, called the Sinoatrial (SA) node, which regularly sends an electrical message initiating a heart muscle contraction. Cells of the SA node are found in the wall of the right atrium and send out a message to cause the heart to beat. They are said to be autorythmic, meaning that they beat independent of the nervous system (on their own).

   Each time the heart beats, the SA node sends its message, causing a wave-like electrical dispatch leading to both atria beating at the same time. This message hits another region, called the atrioventricular (AV) node. The AV node sends another electrical message down a set of nerves in the heart called the Bundle of His and into the left and right ventricles along Purkinje fibers, causing the ventricles to contract. The rhythmic contractions of the heart keep blood flowing through all of the circuits in a regular manner.  An electrocardiogram (ECG) traces the conduction of electricity through the heart per heartbeat, as shown in Figure 13.8. An ECG detects an abnormal rhythm, called an arrhythmia. A common type of arrhythmia, atrial fibrillation is found in over 2% of the population.

**Autorythmic**

Cardiac muscle cells that beat independent of the nervous system.

**Atrioventricular (AV) node**

Small mass of neuro muscular fibers located at the base of the interatrial septum.

**Sinoatrial (SA) node**

The center that controls the heart beats.

**Purkinje fibers**

Specialized heart muscle fibers that carry electrical impulses controlling the contraction of ventricles.

## ECG and electrical activity of the myocardium

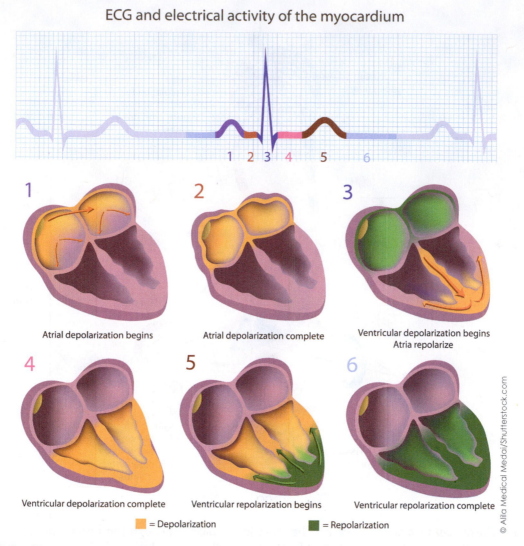

**1** Atrial depolarization begins

**2** Atrial depolarization complete

**3** Ventricular depolarization begins
Atria repolarize

**4** Ventricular depolarization complete

**5** Ventricular repolarization begins

**6** Ventricular repolarization complete

■ = Depolarization    ■ = Repolarization

© Alila Medical Medaj/Shutterstock.com

**Figure 13.8**  Electrical conduction in the heart and a normal ECG. Depolarization occurs in step-like waves along the muscle of the heart.

**Bundle of His**

A collection of heart muscle cells that transmit electrical impulses from the AV node to the interventricular septum and ventricles.

**Arrhythmia**

A condition in which the heart beats in an abnormal rhythm.

# Diseases of the Cardiovascular System
# Heart Attack: Myocardial Infarction

As a person ages, though, the heart parts wear and weaken, leading to cardiac disease. Heart or cardiac disease is the leading cause of death in the United States and the developed nations. The most common cause of cardiac disease is a heart attack or myocardial infarction (MI). An MI leads to death of myocardial tissue and therefore poorer functioning of the heart. One third of MIs lead to instant death because it disrupts the electrical beating of the heart just described. The heart becomes functionless in these cases, with no pumping of blood. Many victims can be saved with a shock to bring the beating back into a normal rhythm.

The health of survivors of MI is determined by the amount of damage to the myocardium. The greater the myocardial damage, the poorer the outcome for the patient. The amount of damage to heart muscle after an MI is measured by troponin levels, which are muscle proteins of the heart which show up after tissue is destroyed. The higher the troponin levels, the greater the extent of damage to heart muscle.

# Arteriosclerosis

In our story, Charles was an end-stage cardiac patient, with a heart weakened enough to require a transplant. There are many effective treatments for heart disease to prevent or slow progression to the degree of congestive heart failure that Charles experienced.

A major cause of heart attack and heart damage is a buildup of plaques in coronary arteries called arteriosclerosis (also referred to as atherosclerosis). Plaque buildup is strongly associated with cholesterol and bad fats in the blood, such as LDL described in Chapter 12. When blood flows through a plaque laden area on a coronary artery, blood flow becomes more turbulent. This disruption from a smooth flow causes platelets to initiate clotting. Platelets recognize rough surfaces, in part by detecting turbulent blood flow. Thus, wavy walls of coronary arteries caused by plaques constitute a high risk for MI.

When plaques are detected, diet changes and/or medicines may be recommended to lower a patient's cholesterol. If the condition is more serious, with plaques blocking more than 60% of arteries, surgery may be required. A less invasive surgery is angioplasty, in which a tube called a catheter is inserted into the coronary artery to the area of plaque buildup. The tube expands, sometimes placing a hollow metal tube in place, called a stent to hold the area open. Sometimes, a replacement of the diseased artery is required, in open heart surgery called coronary artery bypass graft (CABG). The surgery takes a health artery or vein from another area on the body, such as the chest or leg, and uses it to replace the diseased artery.

**Arteriosclerosis**

A chronic condition characterized by abnormal thickening of vessel walls.

**Artery**

A vessel that carries oxygenated blood away from the heart to cells, tissues, and organs.

**Angioplasty**

A surgical procedure to widen obstructed arteries or veins.

**Coronary artery**

An artery supplying blood to the heart.

**Coronary artery bypass graft (CABG)**

A type of surgery that improves blood flow in the heart.

---

## WE ARE WHAT WE TAKE INTO US!

Choosing nutritious foods to eat is a daunting task. There are so many ingredients that are hard to pronounce and chemicals that are unfamiliar, in processed foods. This is a reason why people have a hard time choosing the right foods. To start, eating foods that are not processed, such as fresh vegetables and fruits or meats that you cook yourself, is healthier than those that contain many of the chemicals added in the processing of meals. Fats in food contribute to heart disease, a leading cause of death.

In considering foods and other substances entering your body, consider their components. Figure 13.9 shows the many substances found within cigarettes and yet more than 20% of Americans smoke. Tobacco companies used research by the Tobacco Research Council showing that there is no basis for its link to lung cancer for years. How is this so when so many dangerous chemicals make up cigarettes? Some statistics were used for years to give a pass to smoking and its link to disease.

What is inhaled, such as chemical in the air alos affects our health. For example, a recent study points to increased risks of heart attack when jogging along roads with high traffic. It is surmised that the pollutants inhaled by joggers stimulate changes in the heart to elicit an attack. We are what we eat as well as the air we breathe into our bodies.

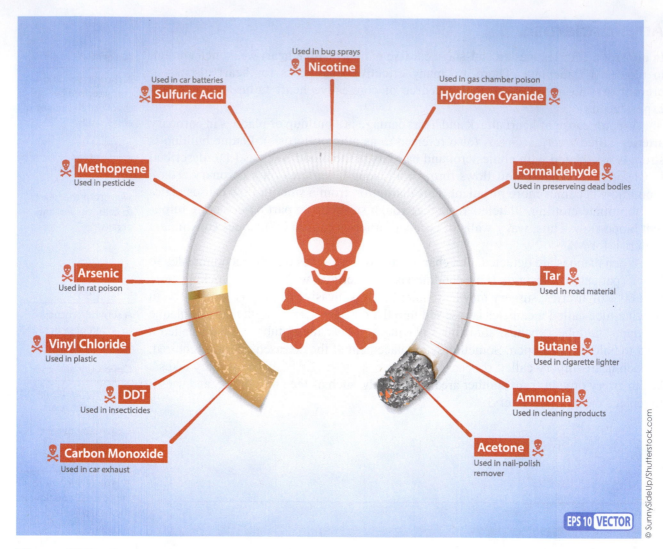

Used in bug sprays
☠ **Nicotine**

Used in car batteries
☠ **Sulfuric Acid**

Used in gas chamber poison
**Hydrogen Cyanide** ☠

☠ **Methoprene**
Used in pesticide

**Formaldehyde** ☠
Used in preserveing dead bodies

☠ **Arsenic**
Used in rat poison

**Tar** ☠
Used in road material

☠ **Vinyl Chloride**
Used in plastic

**Butane** ☠
Used in cigarette lighter

☠ **DDT**
Used in insecticides

**Ammonia** ☠
Used in cleaning products

☠ **Carbon Monoxide**
Used in car exhaust

**Acetone** ☠
Used in nail-polish remover

EPS 10 VECTOR

© SunnySideUp/Shutterstock.com

**Figure 13.9** Cigarettes and their many harmful chemical components. The harmful substances found in cigarettes are linked to many chronic diseases.

---

**Semilunar valves**

A valve of heart that prevents backflow into vessels.

**Regurgitation**

The most common valve problem.

**Murmur**

An abnormal sound made by blood during the heartbeat cycle.

# Heart Valve Disease

Another wear-and-tear problem in the heart emanates from the valves. **Heart Valves** prevent backflow of blood into chambers of the heart. There are four valves in the heart, as shown in Figure 13.6: the **mitral** and **tricuspid**, which prevent backflow into the atria; and the semilunar valves, which prevent backflow into vessels.

The most common valve problem is backflow or regurgitation, detected by auscultation as a murmur, described in Chapter 11. Most often, the mitral valve, due to high pressure from the left ventricle, is the culprit. When the backflow is bad enough, the heart repumps the same blood over and over, weakening the myocardium. This enlarges the heart and creates a smaller chamber within. Smaller chambers are not able to pump as much blood as a normal-sized chamber. This results in less efficient heart pumping ability. In these cases, surgical replacement with a cadaver, pig, or artificial valve is recommended. Murmurs are very common and usually do not require intervention. In fact, over 80% of people over age 65 have a leaky valve.

# Cardiovascular Disease: Treatment Progress

Treatment and surgery for cardiovascular disease has advanced greatly over the past 25 years. While it still afflicts over 30 million people each year, the death rate and quality of life for cardiovascular patients has improved greatly. In the past half-century, medical developments have cut deaths from cardiovascular disease by 80% for strokes and 70% for heart attack, between 1950 and 2000. Many heart diseases link to a buildup of plaques of fat in vessels. Figure 13.10 shows the development of atherosclerosis or clogging of the vessels. Medical procedures and drugs have helped patients with atherosclerosis.

# Blood Vessels

Blood is carried to and from the heart by different types of blood vessels. The cardio-vascular system consists of three sets of vessels: arteries, veins, and capillaries. Let's compare the three types, with a look at their differing structures and functions:

1) **Arteries** are the set of vessels that transport blood away from the heart, to all parts of the body. (You can remember that the word "artery" starts with "A" and "away" also starts with "A".) Arteries contain oxygenated blood, obtaining oxygen from the lungs and transporting it to body cells. The only exception to this rule is in the pulmonary arteries, which carry blood away from the heart to the lungs to become oxygenated. Blood within the pulmonary arteries is deoxygenated, the only artery to have this.

   There is a great deal of pressure within arteries because they are nearer to the heart, the source of that pressure. Walls of arteries are thick to withstand the extra pressure, containing smooth muscle. The aorta is the largest and thickest artery in the human body, receiving the highest pressure blood because it is so near to the heart. Smooth muscle allows for changes in the diameter of arteries, in vasoconstriction and vasodilation. Thick arteries branch into smaller ones,

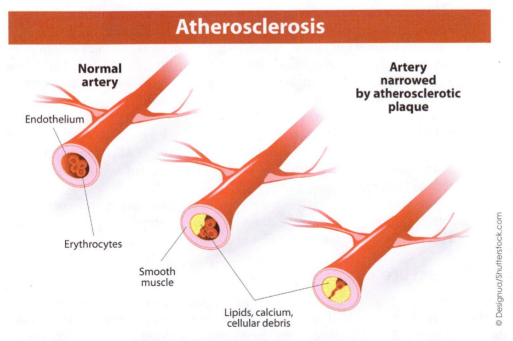

**Atherosclerosis**

Normal artery
Endothelium
Erythrocytes
Smooth muscle
Lipids, calcium, cellular debris

Artery narrowed by atherosclerotic plaque

© Designua/Shutterstock.com

**Figure 13.10** Atherosclerosis and heart disease. The development of plaques in the blood vessels is a major cause of stroke and heart attack.

**Arteriole**

A small branch of artery leading to a capillary.

**Capillary**

A tiny blood vessel that connects arteries and veins.

**Capillary bed**

The whole system of capillaries of the body.

**Venules**

Small veins connecting capillaries with larger systemic veins.

**Vein**

A blood vessel that carries deoxygenated blood back to the heart after it has picked up wastes and carbon dioxide from body cells.

**Varicose veins**

The condition in which valves are incompetent within the legs leading to the formation of blood pools.

eventually into arteries only a few cells thick, called arterioles. Arterioles lead into all of the organs and tissues of the body until they branch into capillaries.

2) Capillaries connect arteries and veins and are much smaller than both vessels, with a 10 μm diameter the size of a red blood cell. The walls of capillaries are thin enough (only one cell layer thick) to allow exchange of materials. These walls are porous, with holes sometimes large enough for whole cells to move through. In capillaries, blood exchanges wastes and nutrients through diffusion with body cells.

Free trade between capillaries and body cells is rapid and extensive, with 50,000 miles of capillaries in an average human adult. Capillaries form a capillary bed, in which the many capillaries branch out, slowing the movement of materials within them. Capillary walls create friction against the blood moving through them. This slows the speed of blood, enabling exchange to occur more completely.

Capillaries are the first vessels to carry deoxygenated and nutrient-poor blood back to the heart. Capillaries converge into larger vessels called venules, which combine to form into veins.

3) Veins carry deoxygenated blood back to the heart after it has picked up wastes and carbon dioxide from body cells. The only exception to this rule is in the pulmonary veins, which carry oxygenated blood toward the heart from the lungs. Pulmonary veins are oxygenated because they have just picked up oxygen from the lungs. The pulmonary veins are the only veins to carry oxygenated blood. The vena cava, discussed earlier in the chapter, is the largest vein in the human body, returning blood from all parts of the body back to the heart.

Vein structure is similar to arteries, both lined with an endothelial layer and both contain thick walls compared to the single layer of capillaries. Owing to their distance from the heart, veins have less pressure than arteries, about one-tenth that of arteries. Veins therefore also have walls that are thinner than arteries, with a lessened pressure not requiring thick walls. Walls of veins therefore have much less smooth muscle than arteries.

Recall that capillaries, with their porous walls and extra resistance, slow the blood and reduces its pressure. With less pressure, veins could lack fast enough movement of blood through it. In fact, over 60% of blood remains within veins at any one time. However, veins are adapted against the backflow of blood and blood pooling. They contain valves, which are one way doors interspersed throughout the vein system.

When valves are incompetent within the legs, blood pools in a condition called varicose veins. These may be painful and are associated with extended periods of standing, such as for nurses, cashiers, and teachers or due to extra pressure caused by pregnancy. Varicose veins are treated by laser surgery or injections that fade or remove them.

The three vessel types each have different functions with structures to enable them to perform them. A comparison of the structure of arteries, veins, and capillaries is given in Figure 13.11.

## Blood Pressure

High blood pressure is a root cause for many of the diseases of the cardiovascular system. **Blood pressure** is defined as the amount of force per unit area on blood vessel walls. It is measured in millimeters of mercury using a blood pressure cuff, usually against the brachial (upper arm) artery. It is a measure of artery pressure within the body.

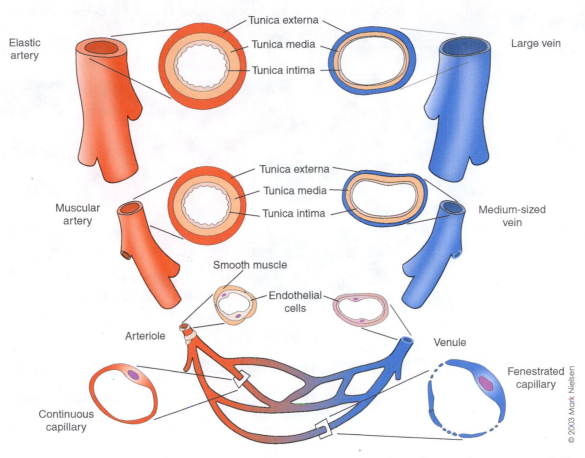

**Figure 13.11**   Comparison of the structure of arteries, veins, and capillaries. Arteries are thick-walled while veins have large lumen. Capillaries are adapted for exchange, with walls only one cell layer thick. Illustration by Jamey Garbett.

When ventricles contract, there is a greater force exerted against the walls of arteries. This is known as a person's systolic pressure. When ventricles fill, less force is exerted against artery walls. This is known as the diastolic pressure. Two numbers are given in a blood pressure reading. The higher number or systolic pressure is the top number and the lower number, or diastolic number, is the bottom number. Thus, a reading of 120/80 indicates that there is 120 mmHg of pressure against a person's artery when their ventricles are contracting and 80 mmHg of pressure against that person's artery when their ventricles are relaxing. A normal adult blood pressure reading should not exceed 120/80 mmHg.

High blood pressure is defined as a chronic elevation of pressure above the normal 120/80, for a consistent period of time. This period ranges from a month to several months. Pressures between 120/80 and 140/90 mmHg are considered borderline cases. These individuals are considered prehypertensive and should be monitored more closely. Roughly three quarters of prehypertensive adults will develop high blood pressure. Hypotension occurs when blood pressure is below 100/60 mmHg and is (when not caused by disease) related to longevity. Of course, high blood pressure, along with other factors such as diet, family history, and older ages, leads to cardiovascular disease. Figure 13.12 shows the multiple factors causing cardiovascular disease.

High blood pressure itself may be caused by several factors: stress, anxiety, smoking, excess weight, or it may simply be inherited. Excess salt in one's diet is also related

**High blood pressure**

A chronic elevation of pressure above the normal 120/80, for a consistent period of time.

**Table 13.1**    Blood Pressure Ranges. Normal Blood Pressure is Below 120/80 mmHg.

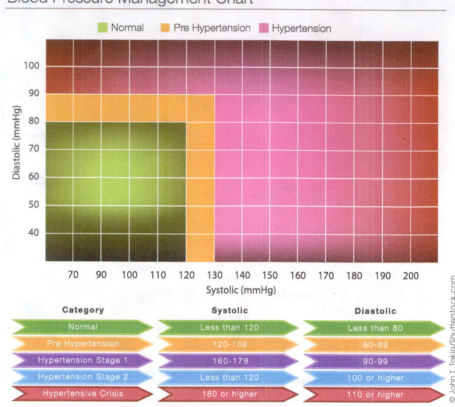

| Category | Systolic | Diastolic |
|---|---|---|
| Normal | Less than 120 | Less than 80 |
| Pre Hypertension | 120-139 | 80-89 |
| Hypertension Stage 1 | 160-179 | 90-99 |
| Hypertension Stage 2 | Less than 120 | 100 or higher |
| Hypertensive Crisis | 180 or higher | 110 or higher |

© John T. Takia/Shutterstock.com

to high blood pressure. As described in Chapter 2, water follows solutes. When added salt enters the bloodstream, water follows, increasing its pressure.

Regardless of the cause, high blood pressure, over time, damages vital organs, and the linings of arteries. For example, damage to vessels may result in plaque buildup, stroke, and heart attack. This damage is irreversible but, if high blood pressure is treated before damage occurs, its effects are limited.

© travellight/Shutterstock.com

**Figure 13.12**    High blood pressure and heart disease are linked to many factors. Blood pressure is a major contributor to cardiovascular disease.

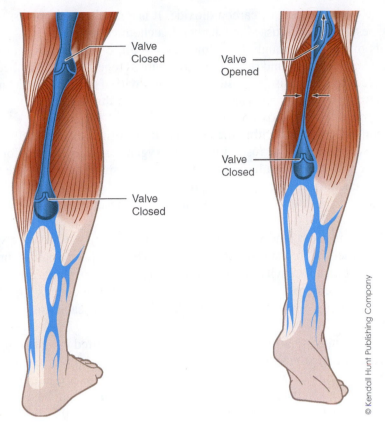

**Figure 13.13** Muscular pump: skeletal muscles contract, moving blood back to the heart. One-way valves allow the blood to move toward the heart but not backward. This reduces the workload of the heart with blood movement without a need for the heart's pumping.

Regular exercise helps to reduce the risk of high blood pressure. It reduces weight and therefore the number of blood vessels in each pound. Each blood vessel gives resistance, so weight loss reduces resistance from blood vessels. When exercising, muscles are used to breathe and to move. When skeletal muscles contract during exercise, for example, it propels blood through veins back to the heart. This movement of blood is termed the muscular pump. The muscular pump helps pump blood through the vessel system, reducing the workload on the heart. Thus, each time a person exercises, muscles help to move blood. Its action is shown in Figure 13.13.

During breathing, other muscles contract in the chest and abdominal cavity. This muscle movement is termed the respiratory pump, which forces blood back to the heart as well through the veins. Both the respiratory and muscular pumps reduce strain on the heart.

# The Respiratory System

## What Is Respiration?

The most pressing need for body cells is the uptake of oxygen and the release of carbon dioxide gas. Our bodies are in constant need of these two processes, accomplished by respiration. Respiration is defined as the taking up of oxygen gas from the environment

**Muscular pump**

A collection of skeletal muscles that aid the heart in blood circulation.

**Respiratory pump**

The movement of blood when other muscles contract in the chest and abdominal cavity.

**Respiration**

The process of taking up of oxygen gas from the environment and the release of the waste gas, carbon dioxide.

and the release of the waste gas, carbon dioxide. It is a mechanical process and differs from cellular respiration, discussed in Chapter 4, a chemical process.

During its movement through the pulmonary circuit, blood exchanges gases in the lungs. Therefore, circulation and respiration are coupled together, as described earlier in this chapter. Needed oxygen is obtained by the blood within the lungs and carbon dioxide waste is removed into the air. The heart pumps blood through lungs, which gives it oxygen and removes carbon dioxide from it.

Respiration uses muscles within the chest cavity to bring air into the lungs for gas exchange. However, in cellular respiration, the oxygen taken in is used for obtaining energy. Cellular respiration results in the breakdown of sugar from the use of oxygen through several chemical processes. Both processes are related; each involving oxygen and carbon dioxide transport. The subsequent sections describe the mechanical act of respiration.

Gas exchange in most land vertebrates occurs within lungs, which are specialized organs with branched and moist respiratory surfaces. Humans have a pair of lungs, which both rest within the chest cavity. Gases are exchanged in two places in the body: 1) between the external environment and the air sacs or alveoli of the lungs; and 2) between oxygenated blood and body cells. The movement of gases in each area is depicted in Figure 13.14.

The process of moving air into and out of the lungs is called mechanical breathing. The act of breathing involves the taking of air into the lungs, known as inspiration and expelling air to the outside world, called expiration. The lungs are like balloons, with very little pumping capacity on their own. They inflate and deflate based on the pressures surrounding them in the chest cavities. During inspiration, the ribs move upward

**Air sacs (alveoli)**

Tiny sacs within the lungs where exchange of oxygen and carbon dioxide takes place.

**Mechanical breathing**

The process of moving air into and out of the lungs

**Inspiration**

The process of taking of air into the lungs.

**Expiration**

The process of expelling air to the outside world.

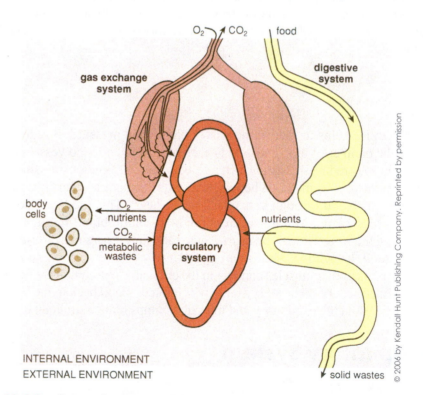

**Figure 13.14**    Gas exchange in cells occur in air sacs of the lungs and between body cells and the circulatory system. From *Biological Perspectives*, 3$^{rd}$ed by BSCS.

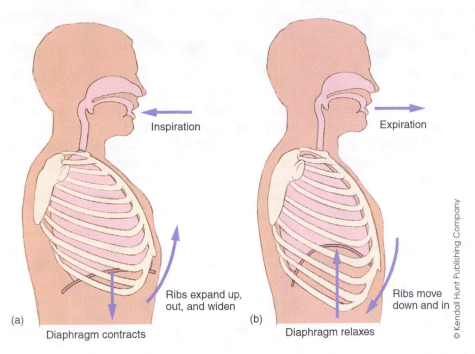

Inspiration

Expiration

Ribs expand up, out, and widen

Ribs move down and in

(a)

(b)

Diaphragm contracts

Diaphragm relaxes

© Kendall Hunt Publishing Company

**Figure 13.15**   Mechanical breathing: inspiration and expiration are accomplished by movements of muscles of ribs and the diaphragm.

and outward, and a sheet of muscle at the bottom of the chest moves downward. These movements open up the chest cavity, forcing air into the lungs. In expiration, the opposite occurs, with the ribs moving downward and inward and the diaphragm moving upward, decreasing the size of the chest cavity. The smaller size forces air out of the lungs during expiration. The process of mechanical breathing is given in Figure 13.15.

Humans take 12–20 breaths per minute, on average. Count the number of times you breathe in 15 seconds and multiple that by four. This will give your number of breaths per minute. The more carbon dioxide in the blood, the greater the number of breaths a person needs to take per minute. Rapid breathing blows off carbon dioxide and can be a sign of diminished oxygen in the blood. Air can also be moved in and out of the lungs very rapidly, reaching over 100 mph, in some coughing fits. This speed is important in the ability of a cough to dislodge food from one's respiratory passageways.

## Anatomy of the Respiratory System

Humans use a respiratory system to uptake oxygen from the environment and eliminate carbon dioxide. The respiratory system includes all of the organs used to move air into the body. The system resides within the thoracic (chest) cavity. Air moves from the **nose** and **mouth** through the pharynx, **larynx**, **trachea** (windpipe), and into the bronchial tubes within the **lungs**.

Let's trace the route air takes as it moves through the respiratory system. As air is inhaled, or brought into the body through the nose, it is filtered by small hairs called **cilia**. Dust, microbes and other particles larger than 4 μm are removed as air moves through the convoluted nasal cavities.

**Respiratory system**

The system by which oxygen is taken into the body from the environment and carbon dioxide is eliminated.

**Pharynx**

A tube that starts behind the nose and mouth connecting to the trachea.

**Bronchial tubes**

Tubes that let air in and out of the lungs.

## ASBESTOS: THE WONDER MATERIAL . . .?

Some particles are so small that they cannot be caught by our nasal canals. Asbestos is a material that is smaller than 4 µm, able to enter into our lungs. It is a wonder substance — able to be molded into any size or shape and is therefore used in insulation, siding, and fire resistant materials — but also has a deadly link to lung cancer. Even a brief exposure to asbestos in the air can greatly increase one's chances of getting a type of lung cancer called mesothelioma (Figure 13.16).

It might surprise you that evidence of the ill effects of asbestos on human health has been around for a long time. Some governments were quick to outlaw or limit its use. Even Nazi Germany limited the use of asbestos in the 1930s, based on damaging data about the substance. At the same time, the United States unfortunately continued using asbestos up until the 1980s in many facets of building. Government agencies later admitted that the use of asbestos, due to its ability to enter and irritate the lungs, should be banned.

**Larynx**

Voice box.

**Vocal cords**

Two elastic cords stretching across the upper end in the larynx.

**Trachea**

A tube-like portion of the respiratory tract that connects to the lungs.

Let's trace the movement of air as it flows through the body. Figure 13.17 shows the anatomy of the human respiratory system. First, when air moves to the back of the throat or pharynx, it shares the area with food until it reaches the larynx or voice box. The larynx sits at the start of the trachea, creating sounds as air rushes through its folds. It is composed of a set of cartilage structures. At the top of the larynx, two elastic cords stretch across the upper end, called the vocal cords. When the vocal cords tighten, the pitch of a sound created increases and when it loosens, the pitch decreases.

As air rushes over the larynx, it travels through the trachea, or windpipe. The epiglottis, discussed in Chapter 12, remains in an open position, allowing air into the trachea. The trachea is about 4 or 5 inches long, held open by rings of hyaline cartilage. It branches into two **primary bronchi**, which enter into the two lungs, branching further into smaller **secondary bronchi** and finally within the lungs into **bronchioles**. There

© Tom Grundy/Shutterstock.com

**Figure 13.16** Asbestos fibers are linked to lung diseases, including lung cancer, particularly mesothelioma. This photo shows natural asbestos fibers, mined from rocks.

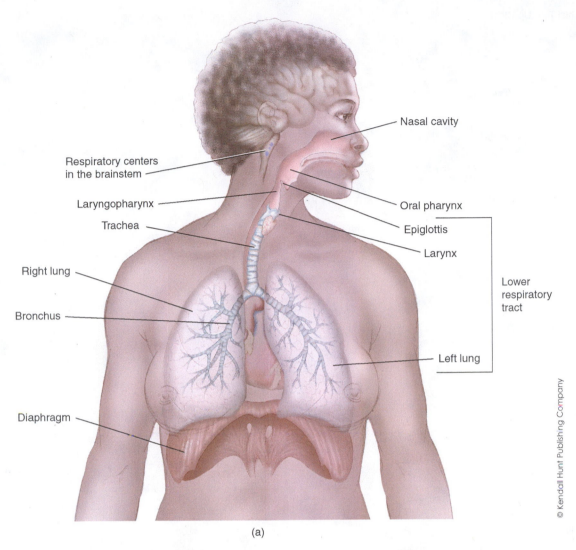

Nasal cavity

Respiratory centers
in the brainstem

Laryngopharynx

Trachea

Right lung

Bronchus

Diaphragm

Oral pharynx

Epiglottis

Larynx

Lower
respiratory
tract

Left lung

(a)

**Figure 13.17**   Human respiratory system a. An overview of major respiratory organs. b. The bronchial tree.

are hundreds of thousands of tiny bronchioles in the lungs, each ending in an air sac or **alveolus**. There are about 300 million alveoli in a human lung. Each is composed of squamous epithelial cells, thin enough to allow for easy transport of gases.

## Exchange in the Lungs

Blood enters the lungs from the heart. As blood moves into the lungs, through the pulmonary artery, the vessels branch into smaller and smaller sizes until they become capillaries, reaching the lung's alveoli. The alveoli are delicate sacs surrounded by a network of vessels, called **pulmonary capillaries**. This network completely surrounds the air sac. Alveoli are a perfect place for gas exchange, because transport of gases occurs over a short distance between vessels surrounding the sac and the air in the sac.

Oxygen and carbon dioxide gases are traded between blood in the lungs and the atmosphere, through diffusion. Diffusion, as you may recall from Chapter 3, occurs when particles move from higher to lower concentrations.

Bronchial Tree Up-Close

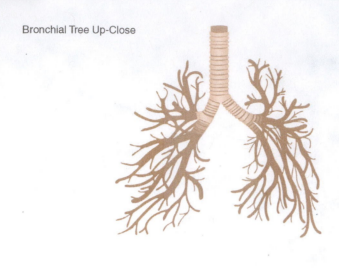

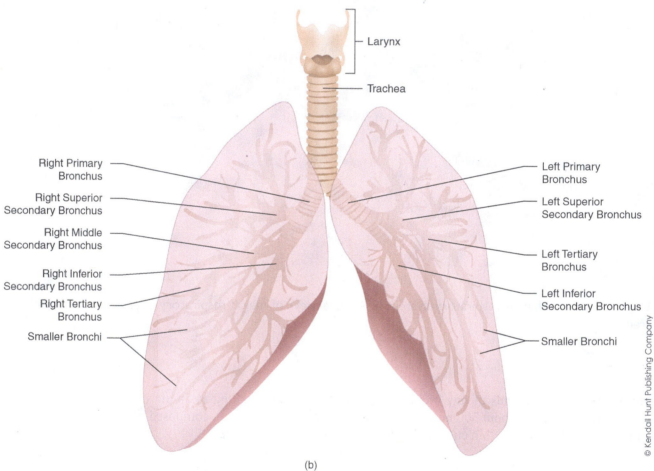

Larynx

Trachea

Right Primary Bronchus

Right Superior Secondary Bronchus

Right Middle Secondary Bronchus

Right Inferior Secondary Bronchus

Right Tertiary Bronchus

Smaller Bronchi

Left Primary Bronchus

Left Superior Secondary Bronchus

Left Tertiary Bronchus

Left Inferior Secondary Bronchus

Smaller Bronchi

(b)

**Figure 13.17** (*Continued*)

Gases are no exception to this rule, except that they move according to partial pressures of the gases. They are composed of particles in differing concentrations and pressures within the lungs. Oxygen gas, for example is in a high concentration within an air sac and a lower concentration within pulmonary capillaries surrounding the air sac. Thus, oxygen gas moves (from a higher to lower partial pressure) into pulmonary capillaries. The opposite occurs for carbon dioxide gas. Respiring cells cause higher pressures of carbon dioxide to accumulate within pulmonary capillaries, creating a diffusion pressure out of them and into the air sac. Thus, carbon dioxide gas moves from pulmonary capillaries into the alveoli and out of the body as waste. The movement of both oxygen and carbon dioxide gases within the lungs are shown in Figure 13.18.

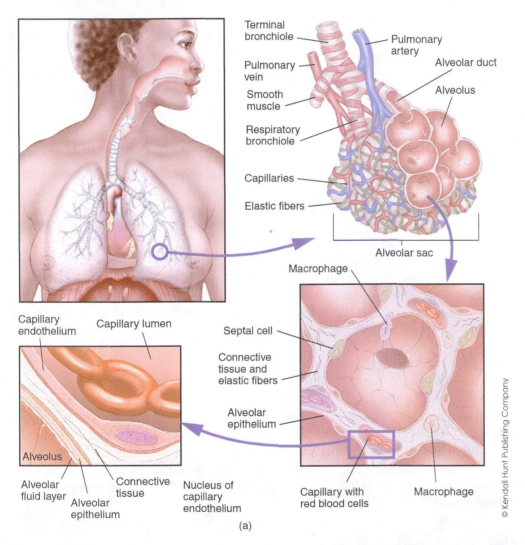

© Kendall Hunt Publishing Company

(a)

**Figure 13.18**  a. Terminal bronchioles. Alveoli gas exchange. Oxygen and carbon dioxide gases are transferred between the air and the blood within the air sacs of the lungs. The accompanying figure shows the air sac (alveolus). Carbon dioxide and oxygen gases are exchanged across the alveolus membrane. Oxygen enters red blood cells where it attached to hemoglobin within the capillaries. b. Exchange of gases in air sacs and in tissue cells.

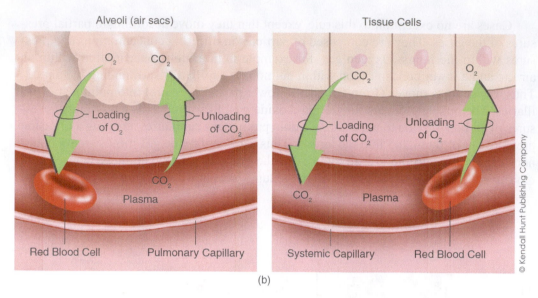

**Figure 13.18**   *(Continued)*

## Lung Compliance

The ability of the lungs to properly inflate and exchange gases is termed lung compliance. Several factors affect a lung's compliance. First, the passages of the respiratory system, namely the bronchial tubes, determine the amount of air reaching the lungs. When the bronchial tubes are wide open, there is enough air to reach the lungs. However, in some conditions, such as asthma and in allergies, lung compliance is compromised because the diameter of the air passageways becomes smaller. This limits the amount of needed gases reaching and leaving the lungs.

Second, the lungs must inflate and deflate continuously to function properly. This ability is called its resilience. Lung resilience is limited when parts of the lung are not flexible, as occurs in emphysema, when lung tissue becomes stiff or fibrosis (filled with fibers).

Finally, in order for gas exchange to occur, the lungs are normally filled with fluid to keep its cells moist. Healthy lungs contain surfactant, a special chemical that helps to keep the alveoli open despite this fluid. Surfactant decreases the surface tension of fluid within the lungs. As you may recall in Chapter 2, surface tension causes a liquid to "stick" to itself. Surfactant interferes with the surface tension of water in the lungs, preventing alveoli from collapsing into each other. In premature births, usually more than one month early, surfactant-producing cells are too immature to make surfactant. This may result in troubled breathing, in a condition called infant respiratory distress syndrome (IRDS).

**Resilience**

The ability of the lungs to inflate and deflate continuously to function properly.

**Surfactant**

A special chemical that helps to keep the alveoli open by reducing the surface tension of fluid within the lungs.

## Gas Transport in Blood

Oxygen is carried within the blood in several ways. Only a small portion of oxygen is dissolved within the blood in its transport. Instead, over 98% of the oxygen carried within the human body is carried on the hemoglobin molecule in red cells. As discussed earlier in the chapter, hemoglobin is a small molecule composed of four polypeptide chains, each holding a heme group with iron at its center.

Hemoglobin holds on to oxygen at its iron core, bound in a form called oxyhemoglobin. Each hemoglobin molecule is able to carry up to four oxygen molecules. With over 250 million hemoglobin molecules per single red blood cell, 1 billion oxygen molecules are carried per red blood cell! The ability of blood to move large amounts of oxygen in a short period of time is enormous.

Hemoglobin is not permanently affixed onto an oxygen molecule. Oxygen moves off of hemoglobin based on the pressure difference it encounters as it moves through the body. When there is a high pressure difference, oxygen moves quickly away from hemoglobin. When there is less of a need, oxygen is slower to move. Figure 13.14 shows the sites at which oxygen moves off of hemoglobin in the body.

Hemoglobin is much like a parent who gives money only when it is most needed, saving some for a rainy day. Hemoglobin conserves some oxygen, at some points to give to tissues when they need it most, as in cases of strenuous exercise. At these times, there is a decreasing in hemoglobin's affinity for oxygen. Hemoglobin becomes less able to hold onto oxygen during exercise and when a need for oxygen presents itself.

Carbon dioxide is also transported through the body to be eliminated as a waste product. About 20% of carbon dioxide in the blood is bound to hemoglobin, in the form of carbaminohemoglobin. About 7% is dissolved and over 70% is carried as the bicarbonate ($HCO_3^-$) ion. When carbon dioxide dissolves in the blood, it rapidly forms into carbonic acid. Carbonic acid ($H_2CO_3$) is a weak acid, and rapidly breaks into bicarbonate ($HCO_3^-$) and hydrogen ions:

$$CO_2 + H_2O \rightarrow H_2CO_3 \rightarrow HCO_3^- + H^+$$

This equation is reversible, meaning that it also moves in the opposite direction depending upon the amount of bicarbonate and carbon dioxide present. If there is a large amount of bicarbonate in a system, it will drive the reaction to the left, producing more carbon dioxide. When carbon dioxide builds up in the blood, the reaction shifts to the right and blood becomes acidic.

# Diseases of the Respiratory System

## Respiratory Acidosis

During respiratory acidosis, when the lungs and heart do not sufficiently transport needed gases within the body, acidic blood develops. A buildup of carbonic acid and hydrogen ions lowers the pH of the blood in this situation. As you may recall from earlier in the chapter, pH of the blood must be held within stringent conditions. Veering too far from the set point may result in serious health consequences. The end stages of ketosis, lung and heart disease often result in acidosis of the blood and death. In our story, Charles was saved by his daughter's heart transplant. Otherwise, he faced death by respiratory acidosis. His blood would not have sufficiently pumped out carbon dioxide from his lungs, and would have made his blood acidic by his final weeks. As a result, changes in the pH would have ceased his life functions. Hemoglobin gives off more oxygen when blood is acidic. This helps the situation but acidic blood is a serious health hazard (Figure 13.19).

**Oxyhemoglobin**

A bright red complex of oxygen and hemoglobin present in oxygenated blood.

**Carbaminohemoglobin**

One of the forms in which carbon dioxide exists in blood.

**Respiratory acidosis**

A condition that occurs when the lungs and heart do not sufficiently transport needed gases within the body, leading to the development of acidic blood.

**Figure 13.19** A patient is breathing poorly, using oxygen tanks to supplement low oxygen levels in the blood.

## The Bends

Usually carbon dioxide and oxygen are the only gases transported within the blood. However, during deep-sea diving, nitrogen gas may accumulate in a diver's blood in a condition known as the bends. When a diver ascends too quickly to the surface, nitrogen gas bubbles come out of solution too quickly. This leads to air pockets that interfere with proper blood flow. As a result, blockages act like clots, to prevent blood flow to needed areas. Placing a victim in a compression chamber in order to slowly equalize the pressure treats the bends. The deep-sea diver in Figure 13.20 needs to be very careful to ascend from the water slowly to prevent the bends.

## Carbon Monoxide Poisoning

In cases of suffocation due to car exhaust, carbon monoxide is the culprit. Carbon monoxide poisoning occurs when carbon monoxide (not carbon dioxide) binds to hemoglobin, replacing oxygen. This occurs because carbon monoxide (CO) more easily binds to hemoglobin than oxygen. CO is 200 times stickier to hemoglobin than oxygen. Thus,

**Figure 13.20** Deep-sea diver. This diver is exploring a wreck.

**Figure 13.21** High oxygen hyperbaric chamber. A treatment for the bends, which slowly brings nitrogen gas out of solution and prevents it from forming large bubbles in blood vessels.

when carbon monoxide poisoning occurs, treatment with 100% oxygen or a hyperbaric compression chamber works best. The hyperbaric chamber, shown in Figure 13.21 works to force it off of the affected hemoglobin molecules.

## Altitude Sickness

When altitude affects a person's ability to breath, it is called altitude sickness. Usually after a person travels from a low altitude to one that is over 8,000 feet, there are some health consequences. There is a decreased pressure of oxygen at such heights. Therefore, with air density too low for sufficient oxygen levels, the body needs to acclimatize to lower oxygen conditions. Acclimatization occurs when more red blood cells are formed and when the lungs develop more capacity to hold more air to compensate for the new conditions. Athletes often train in higher altitudes to naturally acclimatize, giving them an advantage when they return to normal altitudes.

Blood is able to carry oxygen based upon the number of red blood cells that it has. In blood doping, red blood cells are added to an athlete's blood. One way is to store her or his blood and then inject it before a competition. This increases the number of red blood cells and hemoglobin a person holds. Blood doping, banned by the Olympics today, enhances an athlete's performance and is considered unethical. Lance Armstrong, an Olympic medal winner, lost his awards due to allegations of blood doping.

Blood doping also has negative health consequences. It adds red blood cells to the blood, thus increasing the thickness or viscosity of it. Thicker blood may initiate clots and increase risks for heart attack and stroke. Olympic cyclist Lance Armstrong, in Figure 13.22 was accused of using blood doping to improve his performance.

**Altitude sickness**

The condition in which the altitude affects a person's ability to breath.

## Lung Cancer

The leading cause of death in the United States from cancer is a result cancer of the lungs. One-third of all cancer deaths are due to lung cancer. In lung cancer, a bleeding mass or growth blocks the normal passage and exchange of gases in the lungs. It is often a painful disease, with a poor prognosis; survival is less than 20%, 5 years after diagnosis.

**Figure 13.22** Lance Armstrong Cycling

What are the causes of lung cancer? Smoking is cited as the number one cause of lung cancer, while a distant by second asbestos exposure. However, what percentage of people actually gets lung cancer if they smoke? You might guess 20% or even 50% . . . but the actual number is much lower. Only 1% or 1 in 100 smokers ever gets lung cancer in his or her lifetime.

So, is smoking really linked to lung cancer? Consider the alternate data: on any lung cancer floor in the hospital, over 90% of lung cancer patients were smokers. Most lung cancers are related to smoking, given this set of data. Tobacco companies in the 1930s through to the 1980s manipulated these statistics to give the public the impression that lung cancer was not caused by smoking. Smoking is now accepted as the leading cause of lung cancer, among other illnesses including other respiratory diseases, cancers, stroke and heart disease. The lungs of a smoker are shown in Figure 13.23.

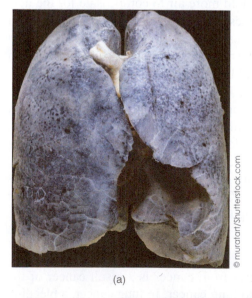
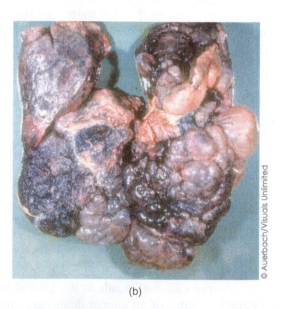

(a)      (b)

**Figure 13.23** Smoking damages the tissues and conducting tubes of the lungs. a. The accompanying images show evidence of a lung tumor linked to smoking. b. Smoking destroys healthy lung tissues.

## Chronic Obstructive Pulmonary Disease

In chronic obstructive pulmonary disease (COPD), limitations in sufficient breathing occur. One form of COPD is emphysema, which manifests as deteriorated alveoli. In emphysema, lungs lose their elasticity and air sacs are hardened, unable to properly exchange gases. Sufferers therefore are required to breathe more frequently and with more exertion. It can be an exhausting disease, with great effort expended to improve lung compliance. Emphysema is treated with increased oxygen and lung reduction surgery.

Asthma, another form of COPD, is an inflammation of the respiratory passageways. Gas exchange is harmed because the smaller size of the respiratory passageways limits the amount of air moving in and out of the lungs. Asthma is treated with steroids to relax muscles around the respiratory passages. An asthma sufferer is shown in the photo in Figure 13.24.

## Controls of Heart and Lung Actions

The heart and lungs are controlled by hormones and nerves that direct them to work faster or slower. When certain nerves are activated, called the sympathetic nerves, they cause the heart to beat faster and the force of its contraction to be stronger. An opposing set of nerves, called the parasympathetic nerves, slow the heart down. Hormones are also released to speed up and slow down the cardiovascular system. Epinephrine and thyroxin both speed up the heart while acetylcholine slows it down. At times the needs of the body require changes in the rate at which the cardiovascular system works. When a need to run away from a scary situation arises, the heart rate must change to accommodate changing conditions. The role of nerves and hormones in regulating body functions will be discussed in more detail in Chapter 14.

The rate of breathing is determined by a number of factors within our internal environment. The respiratory system is controlled by hormones and nerves much like the cardiovascular system. Our brains, particularly a region called the medulla oblongata, detect carbon dioxide levels. When the levels are too high, the medulla sends nerve messages to the ribs and diaphragm to stimulate more breathing. This eliminates the increased carbon dioxide detected in the blood and simultaneously adds oxygen for cellular use.

**Emphysema**

A condition in which lungs lose their elasticity and air sacs are hardened, unable to properly exchange gases.

**Asthma**

A respiratory condition characterized by inflammation of the respiratory passageways.

**Medulla oblongata**

The inner part of the brain.

© Antonio Guillem/Shutterstock.com

**Figure 13.24**    Asthma sufferer

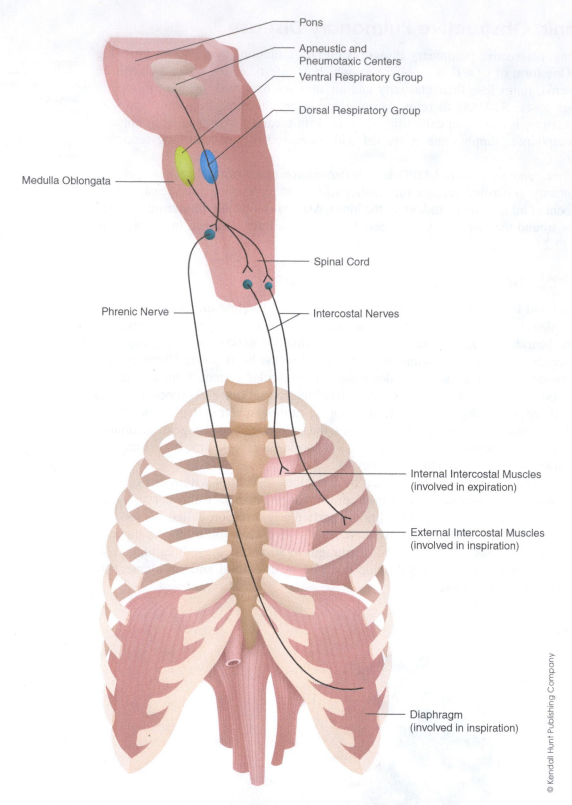

Pons

Apneustic and
Pneumotaxic Centers

Ventral Respiratory Group

Dorsal Respiratory Group

Medulla Oblongata

Spinal Cord

Phrenic Nerve

Intercostal Nerves

Internal Intercostal Muscles
(involved in expiration)

External Intercostal Muscles
(involved in inspiration)

Diaphragm
(involved in inspiration)

© Kendall Hunt Publishing Company

**Figure 13.25**    Respiratory controls of breathing. The medulla oblongata of the brain detects carbon dioxide levels. The brainstem is linked via nerves to communicate blood gas information to other parts of the body.

Other factors also change one's rate of breathing. Consider a situation in which a person gets excited, as before an exam; her or his heart rate and breathing rate increases so that more oxygen can be transported around the body. This excitement is then brought back to normal breathing levels after the situation subsides. Some nerves slow respiration, as occurs when the Pons of the brain is stimulated. Emotion, as detected in the hypothalamus of the brain, affects respiration. The controls of breathing are shown in Figure 13.25.

## ARE SCARING A PERSON, EATING SUGAR, BREATHING INTO A BAG OR DRINKING A GLASS OF WATER REALLY THE BEST WAYS TO STOP HICCUPS?

It is only a myth that scaring a person, eating sugar, breathing into a bag and drinking a glass of water stops hiccups. A look at nerve anatomy indicates the best cure. As shown in this chapter, nerves are sent from the brain to the heart and lungs to cause their movements. The phrenic nerve, sent from the fourth cervical plexus in the neck sends impulses to the diaphragm muscle below the lungs to stimulate breathing. When the diaphragm contracts, it changes pressure in the thoracic cavity, which we know from the chapter readings, creates a gasp of air to move inward.

When pressure or damage to the phrenic nerve occurs at any point along its pathway, hiccups may occur. Hiccups are rapid gasps of air, producing a sound in the larynx. Imagine yourself hiccupping for 68 years: Charles Osborne, an Iowa farmer hiccupped from 1922 until his death in 1991. Mr. Osborne hiccupped every few seconds, with the only relief occurring when he slept. He had the longest attack of hiccups in recorded history.

The best cure for hiccups was discovered by accident. When a 60-year old man was admitted to the hospital with acute pancreatitis, the man developed hiccups lasting two days. By accident, a new treatment for chronic hiccups was discovered. After doctors attempted a variety of treatments, they found no success. However, during a routine rectal examination the hiccups quickly stopped. While his hiccups resumed again within a few hours, the rectal examination was repeated, and the hiccups did not return. The rectal examination, was surmised, massaged the phrenic nerve to stop its firing and its stimulation of the diaphragm muscles. This cure ironically came a year too late in 1992 . . . Charles Osborne did not live to see his cure.

## Summary

The circulatory and respiratory systems are linked together with a common goal: Transport of needed materials throughout the body. The heart and lungs are linked by a set of blood vessels known as the pulmonary circuit. Blood is the fluid used to accomplish transport as well as protection, temperature and pH regulation. As blood moves through the pulmonary circuit, it forms oxygenated blood ready for the body to use. Blood moves through different types of vessels – arteries, capillaries, and veins – as it makes its way to needed cells and back to the heart. The respiratory organs function together to transport to and from the blood. A lung's ability to function is measured by its compliance. There are many diseases related to the cardiovascular and respiratory systems, with emerging treatments, such as organ transplants, improving survival, and quality of life for its sufferers.

# CHECK OUT

## Summary: Key Points

- Medical treatments, such as organ transplants described in the story and prevention of disease such as diet and exercise have improved life expectancy and quality of life for cardiovascular and respiratory illnesses.
- Blood is composed of plasma, which is mostly water and dissolved solutes and the formed elements, which are cells and cell fragments in the blood.
- Blood serves to transport materials within the body; it also regulates body temperature, internal pH, and protects the body from pathogens.
- Blood returns to the heart via the vena cava; is pumped from the right side to the lungs, back to the left side of the heart, and is propelled to all parts of the body through the pumping activity of the left ventricle, all based on electrical messages sent from the sinoatrial (SA) node.
- High blood pressure damages heart vessel linings and other organs, increasing risks for stroke and heart attack.
- When air enters the nose and mouth, it is conducted through the pharynx, larynx, trachea, bronchi, and into the air sacs of the lungs.
- The alveoli exchange oxygen and carbon dioxide gases, based on the partial pressure of each gas, between air in its sacs and pulmonary capillaries surrounding them.
- Numerous treatments for cardiovascular and respiratory diseases have helped improve survival and the quality of life of many people including: angioplasty, heart bypasses, surgery, steroids, oxygen, and hyperbaric compression chambers.

## KEY TERMS

| | |
|---|---|
| air sacs (alveoli) | capillary |
| altitude sickness | capillary bed |
| anemia | carbaminohemoglobin |
| angioplasty | carbonic acid–bicarbonate buffering |
| arrhythmia | system |
| arteriosclerosis | carbon monoxide poisoning |
| arteriole | cardiovascular system |
| artery | clotting factors |
| atrioventricular (AV) node | coronary artery |
| atrium | coronary artery bypass graft (CABG) |
| asthma | coronary circuit |
| autorythmic | deep vein thrombosis (DVT) |
| the bends | deoxygenated blood |
| blood | embolus |
| bone marrow | emphysema |
| bronchial tubes | endothelium |
| Bundle of His | expiration |

formed elements

heart

heart attack (myocardial infarction)

hemophilia

high blood pressure

inspiration

iron deficiency anemia

larynx

lungs

mechanical breathing

medulla oblongata

mitral (bicuspid) valve

murmur

muscular pump

oxyhemoglobin

pathogen

pharynx

plasma

platelets

protime

pulmonary artery

pulmonary circuit

pulmonary embolism

pulmonary vein

Purkinje fibers

red blood cells

regurgitation

resilience

respiration

respiratory acidosis

respiratory pump

respiratory system

semilunar valves

sickle cell anemia

sinatrial (SA) node

stem cells, pleuripotential

stroke

surfactant

systemic circuit

thalassemia

thrombin

thrombosis

trachea

tricuspid valve

varicose veins

vein

vena cava

ventricle

venules

vocal cords

white blood cells

# Multiple Choice Questions

1.  Which treatment uses a tube to open clogged coronary arteries?

    a.  Angioplasty
    b.  Bypass grafting
    c.  Lung-reduction surgery
    d.  Hyperbaric compression chambers

2.  Which term does NOT fit with the others?

    a.  Plasma
    b.  Erythrocyte
    c.  White blood cell
    d.  Platelet

3.  A disease in the _____ would MOST affect blood cell production?

    a.  blood
    b.  bone marrow
    c.  heart
    d.  lungs

4.  Carbonic acid functions to:

    a.  protect blood from pathogens.
    b.  buffer blood from pH changes.
    c.  transport carbonated liquids.
    d.  exchange carbon dioxide and oxygen gases.

5.  Which transmits electrical signals through the left ventricles, during a heartbeat?

    a.  Sinoatrial (SA) node
    b.  Atrioventricular (AV) node
    c.  Mitral valve
    d.  Purkinje fibers

6.  Which represents a logical order, from start to finish, in the movement of blood through the heart and lungs?

    a.  right atrium→pulmonary artery→lung→left ventricle
    b.  pulmonary artery→right atrium→left ventricle→lung
    c.  left ventricle→right atrium→pulmonary vein→lung
    d.  right atrium→pulmonary vein→lung→left ventricle

7.  Which traps particles, larger than 4 μm, during respiration?

    a.  Epiglottis
    b.  Nasal cavities
    c.  Trachea
    d.  Echinoderm

8. Which is NOT a part of the region that exchanges gases within the respiratory system?
   a. Alveoli
   b. Trachea
   c. Pulmonary capillaries
   d. Air sacs

9. If pressure of oxygen within an alveolus is 40 mmHg and within pulmonary capillaries is 35 mmHg, which is expected to occur?
   a. Oxygen will move into the capillaries
   b. Carbon dioxide will move into the capillaries
   c. Oxygen will move out of the capillaries
   d. There is no net movement of gases in the alveolus under these conditions

10. Which increases a person's risk for developing high blood pressure?
    a. Smoking
    b. Being overweight
    c. Genetics
    d. All of the above

## Short Answers

1. Describe two treatments in which cardiovascular life expectancy have been improved upon since the 1950s.

2. Define the following terms: vein and artery. List one way each of the terms differ from each other in relation to their 1) anatomy; 2) function in blood transport; and 3) placement between the heart and lungs.

3. The formed elements make up an important part of the blood. Choose one of the formed elements and 1) explain how it is used in the body and 2) what diseases are linked to the malfunctioning of that formed element.

4. Draw a sketch of the heart, using arrows to trace the movement of blood through the sketch. Which is vessel is the most likely to be damaged by high pressure from the heart?

5. How does the muscular pump aid in the prevention and treatment of high blood pressure?

6. Describe the anatomical changes that occur during inhalation and exhalation. Use the following terms in your description: diaphragm, ribs, chest cavity, lungs, inflate, deflate.

7. Define lung compliance. Explain three factors that contribute to lung compliance.

8. For question #7, which factor is most affected in infant respiratory distress syndrome?

9. Where does sound get produced in the respiratory system? How do the vocal cords create higher pitched sounds?

10. Describe three ways that carbon dioxide is transported in the blood. How is carbon dioxide buildup dangerous to a person's health? Be sure to discuss the role of respiratory acidosis in your answer.

## Biology and Society Corner: Discussion Questions

1. Organ donation is a voluntary option exercised by a small percentage of people in the United States. Over 10% of people will die each year while waiting for an organ transplant. As shown in our story, however, it can be life-saving. Should the government mandate all citizens to donate their organs upon their death? Why or Why not?

2. In the above question, consider that there is a black market for organs, traded illegally throughout the world. The average price paid for a kidney was $150,000 USD. It is estimated that 11,000 organs were traded illegally in 2010. Would mandatory organ donation, if implemented, affect patient healthcare? What is the likelihood that people will be allowed to die (or be killed) in the hospital to obtain their organs, because they are so valuable? What safeguards could be implemented, if any, to prevent such as problem?

3. Smoking is a health hazard, yet more than 20% of adults in the United States smoke. Smoking is also on the rise in other nations, especially in the developing countries. The Tobacco Institute and Council for Tobacco Research continues to present data to question the link between smoking a lung cancer. Should the government intervene to make laws restricting or banning the use of tobacco products, considering that it is harmful to us? Why or why not?

4. While air quality has improved in cities over the past 20 years, air pollution is still a risk factor affecting heart and lung health. For example, studies show that jogging near a polluted highway increases the risk of heart attack by more than four times. Children who grow up in a city are also more likely to develop asthma than those in rural areas. How might this information impact whether you or your family's choice in where to live? Would you rather live in a city or in a rural area, based on this information?

5. Blood doping has been banned by the Olympics because it gives unfair advantage to players. Lance Armstrong, Olympic gold medal winner in cycling, has been stripped of his medals due to his alleged use of blood doping. Construct an argument against the banning of blood doping, defending Lance Armstrong. What other factors, besides blood doping, give one athlete an advantage over another? Should these be considered in your argument?

Figure – Concept Map of Chapter 13 Big Ideas

# Regulation: Nervous, Musculoskeletal, and Endocrine Systems

**14**

## ESSENTIALS

Lifting large objects, especially the wrong way, can cause injury

Pain from the back and neck can emerge in other areas of the body. A burning and painful arm is caused by discs compressing on the nerves of the neck going down the arm

Neck pain is a common ailment but may indicate anatomical causes

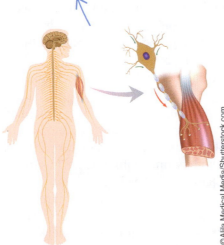

Nerves go to muscles, transmit pain, and conduct messages

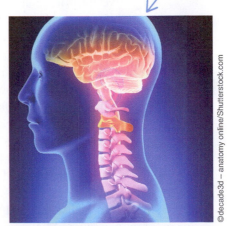

The discs in the neck: herniated discs usually cause pain

© Kendall Hunt Publishing Company

# The Case of the Burning Arm

The pain was intense – it shot like an electric shock, moving through his living tissues – along his arm, with a burn all across his upper back and neck. "I should never have lifted that heavy rock . . . . these rocks will rip the very flesh off of my body," sighed Richard.

It had been weeks since Richard worked to build his stone wall on the upper part of his land. The farmers built beautiful stone walls, and Richard prided himself on continuing in their tradition. It was his hobby and Richard needed one; he was so energetic and healthy that the diversion of stone wall building was always a great feeling of accomplishment.

However, this last time was different. Richard knew that he had taken on too much for his body to bear. The rocks were larger and heavier and Richard was not accustomed to such strenuous lifting. He was a determined fellow, however, and he would build a new and larger wall than he had ever constructed. At the very moment that Richard lifted the corner rock on his land, he knew that something had gone terribly wrong. He felt a burn down his arm and a weakness in its muscles.

Richard waited in pain for days until he decided to go to his doctor. While Richard explained to the doctor that he must have pulled a muscle in his shoulder, his doctor knew the diagnosis right away. "Richard, I think you have a herniated disc in your neck. We will order an MRI to check it out." Richard was perplexed. "If the pain is in my left arm, what does the neck have to do with it?" he asked. The doctor explained that a slipped or ruptured disc was pressing on the nerves coming out of his neck that stimulate his arm muscles. Nerves leave the neck as a group of nerves called the cervical plexus, giving messages to muscles throughout the arm.

The prognosis was as confusing as the cause of the symptoms. The doctor explained that in most cases, for roughly 80%, the slipped disc retracts back into place and no further treatment is needed. When the disc does not stop causing pain, drugs for pain and steroid shots can be given to reduce swelling and help buy time until it heals. In serious cases, neck surgery is needed. It all depends on Richard's pain level.

Richard certainly did not want surgery but he was in pain. The doctor recommended that Richard wait a while, perhaps months, to see if the pain gets better. Surgery would be a last resort and was dangerous because they would need to go into his neck. Surgeons would clip or remove the disc and they would be very close to all of his nerves, including his spinal cord. There was even a chance of total paralysis. Richard was given an anti-inflammatory and sent home. Richard was in severe pain.

The most difficult part of this problem for Richard was that he did not know what would happen to him because of this malady. He had read that over 90% of people experience regular back and neck pain and get better, but he was not. He could not work for the past two months and, in this condition, Richard could barely move his arm.

"How surprising that it was all simply because of this small disc, a piece of cartilage that cushions the vertebrae of the back," Richard thought to himself. How could something so small and moveable still be causing him pain after merely moving rocks? What a disaster; Richard needed to do something; he needed to get back to work and earn a living. Would Richard get better, face surgery, or live a life of pain?

Richard repeated to himself, "I should never have lifted that heavy rock . . . ."

---

## CHECK UP SECTION

Neck and back pain afflict many people. They are sometimes driven to alternative medicine, often when traditional medical approaches fail.

Research the types of alternative medical approaches to treating pain. Be sure to describe the techniques and philosophy used by chiropractors and acupuncturists to treat pain. How does it differ from the traditional medical community?

Most medical doctors do not work with nor recognize alternative medicine practitioners. Would you recommend that Richard try an alternative medicine practitioner, such as a chiropractor? Why or why not?

---

# The Nervous System

## Regulation

Whether shooting pain down the arm, as shown in our story, or a change in blood pressure described in other chapters, the changes that occur in the body require regulation. Regulation, or control over functions of the body, is accomplished by a combination of the nervous (nerves), muscular (muscles), skeletal (bones), and endocrine (hormones) systems working together. Each plays a role in responding to changes and adapting to those changes. These systems work toward the common goal: to maintain homeostasis. In this chapter, we will survey the four systems to study how they regulate processes in the human body.

**Regulation**

Control over functions of the body.

## Pain

Pain is a response to some malfunction, sensed by the nervous system. It indicates that something is wrong – a cancer or an inflamed nerve, for example – which manifests as a symptom. Symptoms are studied by doctors to determine the cause and course of treatment. The philosophy that the medical community uses to approach each case is based on previous outcomes and treatments of other patients. Many times a diagnosis changes or a prognosis (expected outcome) is uncertain in medicine. For example, in the story, sometimes a herniated disc retracts and sometimes it does not. Some people (5% of cases) have MRIs that show herniated discs but do not have any symptoms and experience no pain. Some people have extreme back pain and have almost no disease in their medical imaging.

Medicine treats the person and the case, but not the image. Each patient reacts differently to treatments and diseases in their outcome. Sometimes it is difficult to predict who will do well and who will not, although there are indications based again, on prior cases. Medical treatments always have uncertainty in their outcomes, for these reasons.

Our story described the suffering of Richard, like many people, who lives in pain every day because of a disorder of the musculoskeletal or nervous systems. Sometimes pain is at the site of its cause and sometimes it is referred pain, sent to another site away from its cause. In Richard's case, the neck problem sent shooting pain down his arms instead of his neck, a common symptom of herniated discs. The pain was referred to another area of the body besides the neck.

More than 14 million people in the United States live with chronic pain. Many of these cases involve back and neck problems, because nerves emanate from the back in many areas, each of which sense pain. Others are diagnosed with diseases such as fibromyalgia (inflammation of the nerves), arthritis (inflammation of the joints), and endometriosis (inflammation in the reproductive system). Pain is personal – no one can see it or feel it but the sufferer – but pain is real for people and is it felt.

Should a person turn to alternative medicines, such as chiropractic help and acupuncture? Medical treatments, such as surgery, physical therapy, and drugs, seek to repair anatomical problems. Alternative medicines do not intervene in injuries, and instead rely on holistic strategies for pain relief. There are many studies supporting the use of alternative medicines. However, the medical community does not officially endorse nor work with these practitioners. More research needs to be done and more collaboration with traditional medicine to improve outcomes for alternative medicine pain patients.

Many famous people suffer with chronic pain due to back injuries, such as George Clooney and Jennifer Grey shown in Figure 14.1. Clooney fell while filming the thriller Syriana in 2005, which resulted in a tear in his dura mater that surrounds and contains the fluid of the spinal cord. After multiple surgeries to repair the tear, he continues to suffer from pain, and considered "ending it all," at points due to the severity of his pain. Grey was in a car accident in 1987, causing her neck injury and pain. She treated it using

**Referred pain**

Pain that is not at the site of its cause.

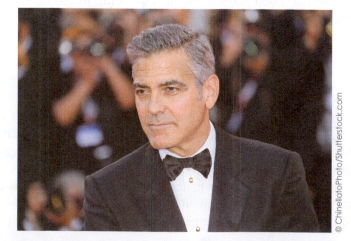

© ChinellatoPhoto/Shutterstock.com

**Figure 14.1**     George Clooney (Venice Film Festival, 2012).

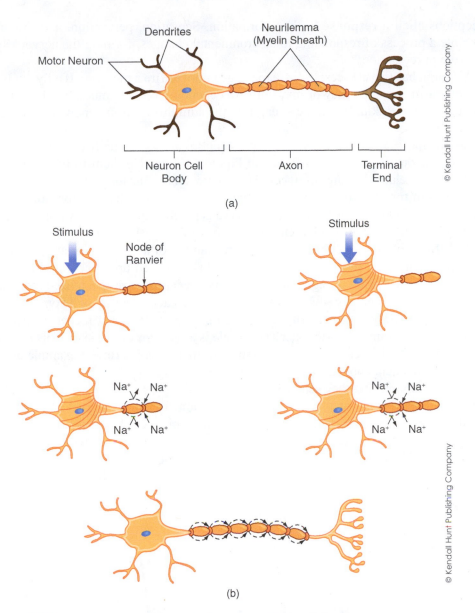

**Figure 14.2** a. Neuron (nerve cell) Structure: The cell body contains most of a neuron's organelles. b. Impulses are received by dendrites and travel along the axon to terminal branches. They jump from node to node bypassing the myelin sheath, a neuron's insulation. Impulses reach terminal branches ready to travel to another nerve cell, muscle, or gland.

ice packs and Advil. Grey eventually had surgery to place a metal plate into her neck to stabilize her vertebrae, which eased her pain.

## Nerves

Almost all animals except the sponges, including vertebrates and invertebrates, have a nervous system and nerve cells or neurons. Neurons were described in Chapter 11, and their general structure is shown again in Figure 14.2. Neurons are specialized cells that receive information in the form of a sensation, which brings information to the brain and spinal cord. That information is processed in the brain and is then called a **perception**.

**Neuron**

A nerve cell.

**Sensation**

Information received by the neurons.

Perceptions elicit a **response** to the information. Sensation, perception, and response constitute a process of responding to environmental changes, forming the nervous system's role in regulation.

A neuron has two choices: to fire a message or not to fire a message. If a fly lands on a person's arm, a neuron may or may not fire; depending on how much the fly disrupts the neurons on the skin. When the strength of the stimulus is strong enough, a message by a nerve is sent.

Neuron message signals are called nerve impulses and are actually a flow of charged ions. Long axons in neurons, as shown in Figure 14.2, enable them to transmit nerve impulses over relatively long distances. The sciatic nerve, our longest, is about 1 m (3 feet) in length traveling down the legs. Sciatica is pain down the leg along the sciatic nerve, a common ailment when the sciatic nerve is pinched or touched by bone or cartilage. Let's review the neuron's structure in Figure 14.2.

Richard's pain described in our story is an example of a stimulus, or any change in the environment that causes a response. A stimulus is picked up by neurons. The start of a nerve response process begins with receptors. **Receptors** are specialized structures that sense stimuli. There are several types of receptors, each specific to the type of stimulus it receives. Each also performs a special function, with select receptors given in Figure 14.3. For example, a Meissner's corpuscle is a receptor on the skin. This receptor fires when mechanical changes (e.g., pressure) from a stimulus (in the example above, the fly) are sufficient enough to warrant a receptor's firing.

When receptors fire, they send a nerve impulse along a sensory neuron (also called afferent neurons). Sensory neurons bring information from the external environment, toward the brain and spinal cord. Within the brain and spinal cord, interneurons organize and connect those messages. Motor neurons (also called efferent neurons) bring nerve

### Nerve impulse

Neuron message signals that are actually a flow of charged ions.

### Stimulus

Any change in the environment that causes a response.

### Sensory neuron

Neurons that bring information from the external environment, toward the brain and spinal cord.

### Interneuron

A neuron that transmits impulses between other neurons.

### Motor neuron

A nerve cell that brings nerve impulses from the brain and spinal cord to a muscle.

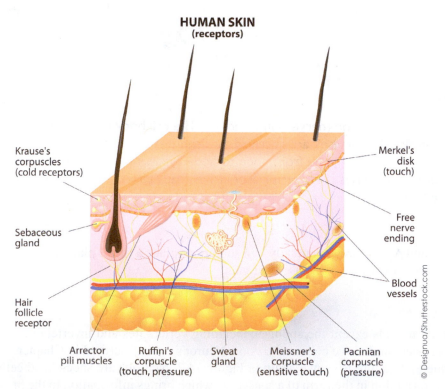

**HUMAN SKIN**
**(receptors)**

Krause's corpuscles (cold receptors)

Sebaceous gland

Hair follicle receptor

Merkel's disk (touch)

Free nerve ending

Blood vessels

Arrector pili muscles

Ruffini's corpuscle (touch, pressure)

Sweat gland

Meissner's corpuscle (sensitive touch)

Pacinian corpuscle (pressure)

© Designua/Shutterstock.com

**Figure 14.3**   Types of Sense Receptors (aka corpuscles) on the skin.

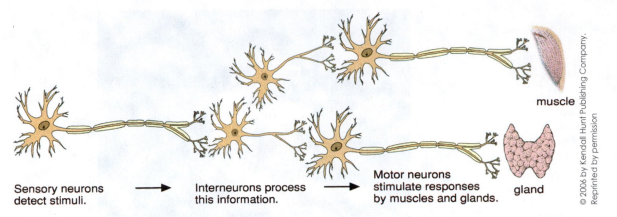

Sensory neurons detect stimuli. → Interneurons process this information. → Motor neurons stimulate responses by muscles and glands.    muscle    gland

**Figure 14.4**    Neurons Work Together in the Nervous System. Sensory neurons, interneurons, and motor neurons. Note that sensory and motor neurons have a myelin sheath that allows for rapid transmission of impulses along axons. From *Biological Perspectives*, 3rd ed by BSCS.

impulses away from the brain and spinal cord to elicit a response by an animal. In our example of the fly sitting on an arm, the stimulus or mechanical changes brought out by the fly need to be strong enough to cause a receptor to fire. The message would then be sent via a sensory neuron to the brain to perceive the fly's presence. After it is integrated using interneurons, the brain is likely to send a motor neuron message to the arm muscles, directing them to swat the fly. The passage of nerve impulses along the three different types of neurons is shown in Figure 14.4.

## THE DOLLAR BILL DROP: CAN A PERSON CATCH A DOLLAR BILL WHEN IT IS DROPPED IN BETWEEN FINGERS?

Place a dollar bill in-between another person's thumb and middle fingers, with their fingers extending around the face picture on the bill. Drop the dollar bill at a random time. Keep the catcher's hand stationary. This prevents the catcher from lunging forward toward the bill.

You will see that the catcher's nervous system is always too slow to catch the bill. The dollar bill drop is a measure of a person's reaction time. Reaction time is the time it takes for a person to react to a stimulus. In this case, the stimulus is the dropping of the dollar bill.

The reaction to the dollar bill dropping takes too long for the sensory neuron, interneuron, and motor neuron to work together to elicit a response in time. The nerve message is sent from the brain, after seeing the dollar drop. There are too many nerve cells and too many gaps in-between the nerve cells for a person to consistently catch a bill.

Through only random chance, roughly 1 in 1,000 people, some will inevitably luck out and catch the bill – but usually the catcher is too slow. Figure 14.5 shows that it is a clever bar trick that can win some money in betting – but also a black eye!

**Central nervous system (CNS)**

The part of the nervous system consisting of the brain and the spinal cord.

**Brain**

A part of central nervous system that functions as the command center of the body.

**Spinal cord**

A long cord of nerve tissues that connect the brain to the other parts of the body.

**Meninges**

A series of protective membranes that surround the spinal cord nerves.

**Peripheral nervous system (PNS)**

The portion of the CNS that is outside the brain and the spinal cord.

**Somatic nervous system**

Part of the PNS that controls the voluntary movements in the body.

**Autonomic nervous system**

System of involuntary nerves.

**Figure 14.5**    Catching a Dollar Bill Is Not Easy!

# Organization of the Nervous System

The example above shows that the branches of the nervous system have differing roles. The different branches of the nervous system and their interactions are clarified in Figure 14.6. The brain and the spinal cord function as the center of the nervous system and are together called the central nervous system (CNS). The brain is the command center of the body, interpreting stimuli and capable of higher-order thought. This will be discussed later in this chapter. The spinal cord nerves are surrounded by a series of protective membranes called the meninges, within the vertebrae (neck and back bones). The spinal neurons conduct impulses up and down the spinal cord.

All of the other nerves outside of the CNS are part of the peripheral nervous system (PNS). Peripheral nerves may be classified within the somatic nervous system, which is under voluntary control, such as those motor neurons, directing muscles to swat a fly; or they may be part of the autonomic nervous system, which are involuntary nerves.

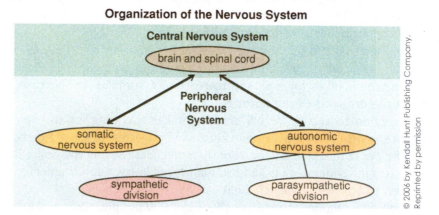

**Figure 14.6**    Organization of the Nervous System: the CNS (brain and spinal cord) and PNS (all other nerves). Receptors of the PNS send nerve impulses to the CNS. The brain sends messages via motor neurons to effectors (muscles or glands). From *Biological Perspectives*, 3rd ed by BSCS.

Autonomic nerves are those that control involuntary activities, such as smooth muscles of the alimentary canal or heart muscle, discussed in Chapters 12 and 13.

Autonomic nerves may be stimulated during times of excitement or fear, in a set of neurons comprising the sympathetic nervous system. Sympathetic nerves increase heart rate, stimulate muscles, and raise blood pressure. They also slow digestion because energy is directed instead to the muscles and the heart. Sympathetic nerves carry out the "fight or flight response," which is characterized by an energetic reaction to fearful or exciting stimuli. A student getting handed an examination paper or confrontation with a friend, both stimulate sympathetic nerves and a sympathetic nervous system response.

The opposing set of nerves is termed the parasympathetic nervous system, which are stimulated when the body calms down, under relaxing conditions. The parasympathetic nervous system acts in opposition to sympathetic responses. They instead slow down the heart rate, relax the muscles, lower blood pressure, and speed up digestion. When you had a frightening event, did you feel that you would never calm down? Parasympathetic nerves brought your body back to normal; back to homeostasis. In Richard's case in our story, relaxation is a technique used to improve pain, which we might recommend to him despite the stresses of a herniated disc.

## Do Nerves Use Electricity?

Regulation by the nervous system is fast and efficient. Neurons fire, and a flow of ionic charges move along them. In our story, Richard felt the effects of nerve impulses because a disc compressed his cervical neurons, sending pain and heat sensations down his arm along their axon paths. While it felt like electricity to Richard; what really is a nerve impulse?

A nerve impulse is similar to an electric current because both consist of a flow of charged particles. Anyone who has touched a live wire knows the feeling that Richard felt down his arm: burning pain and numbness due to the flow of charged particles along the stimulated nerve.

However, a nerve impulse differs from electricity in several ways. First, a nerve impulse is actually a wave of positively charged ions, namely sodium and potassium ions. Electricity is the flow of negatively charged electrons within copper wire. Impulses are much slower than electricity, moving only 2 m/s; while electricity rapidly travels millions of meters in 1 s. The strength of a nerve impulse also stays the same, but electricity weakens over a distance. This is why powerhouses cannot be too far from the homes they supply; the electrical strength would be too weak.

## Nerve Impulses

Neurons are said to be "at rest" when they are not carrying a nerve impulse. However, nerve cells are really active, even when they are not firing impulses. They conduct cellular activities, mostly within their cyton. Across the membrane of the neuron, a difference in the numbers and types of ions exists. Recall from Chapter 3 that the sodium–potassium pump actively transports sodium out of and potassium into the cell. The sodium–potassium works continually to create a difference in charge across the neuron membrane. This difference in charge is called the neuron's resting potential. Its inside is more negatively charged than the outside because more $Na^+$ is pumped out than $K^+$ is pumped into the cell. The resting potential of a neuron is $-70$ millivolts (mV) because its inside is relatively more negative (due to the positive sodium ions pumped outside) compared with the outside. Figure 14.7 gives the cutaway of an axon to show the resting potential of a neuron.

**Sympathetic nervous system**

A part of the nervous system that increases heart rate, stimulates muscles, and raises blood pressure.

**Parasympathetic nervous system**

Opposing set of nerves that are stimulated when the body calms down, under relaxing conditions.

**Resting potential**

The potential of a cell that does not exhibit the activity resulting from a stimulus.

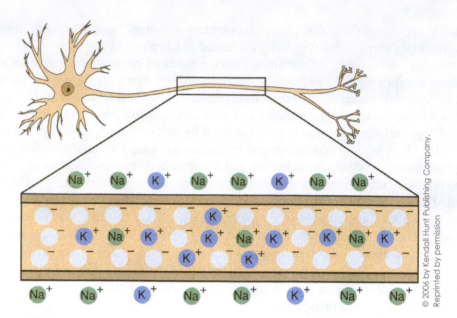

**Figure 14.7** Resting Membrane Potential. The sodium–potassium pump works to maintain a charge difference across the cell membrane. This action (by the presence of negatively charged proteins within the cell) causes the inside of a nerve to be more negative by −70 mV, called its resting potential. From *Biological Perspectives*, 3rd ed by BSCS.

**Critical threshold potential**

The critical level (-55mV) at which the entire neuron fires a nerve impulse across its membrane.

**Action potential**

A change in the electric potential across the plasma membrane that occurs when a cell is stimulated.

**Myelin sheath**

Pads of insulation that prevent the action potential from weakening.

**Synapse**

The gap or region separating neurons from other cells or each other.

**Neurotransmitter**

Special chemicals that carry a nerve impulse to new cells.

A stimulus may cause a disruption of that resting potential. Much like an earthquake, a stimulus causes openings, in this case in the neuron's membrane, to leak sodium ions into the cell. The stimulus must be strong enough to disrupt the resting potential. When the potential rises to −55 mV, called the critical threshold potential, the entire neuron fires a nerve impulse across its membrane. A fly, for example, would disrupt the resting potential if it bites or walks on a person's arm. It requires enough detectable force to hit the −55 mV threshold. The stimulus then opens up channels in the dendrite's membrane, allowing $Na^+$ ions to flow into the cell. Some $K^+$ ions leak out at roughly the same time. This forms a wave of ion flow called an action potential. The flow of sodium and potassium ions creates a moving charge in Figure 14.8, which is an action potential.

The action potential travels through a neuron, down the axon, which is covered in pads of insulation called the myelin sheath, which prevents the action potential from weakening. The action potential continues until it reaches the terminal branches at the end of the axon. Here, another shake-up occurs, with sacs along the terminal branches sent off to cross a gap to reach another neuron or muscle cell. The gap or region separating the neurons from the other cells is called its synapse. A picture of the synapse is given in Figure 14.9. Nerve impulses travel across synapses and require extra time, slowing their transmission. These slow-downs cause people to fail to "catch" the dollar bill described earlier.

# Neurotransmitters

Sacs traveling through the synapse are filled with neurotransmitters, which are special chemicals that carry a nerve impulse to new cells. Roughly 25 types of neurotransmitters have been identified, each of which serves an important role in nervous system regulation. Table 14.1 gives selected neurotransmitters, their actions, and structure. These are the most important to know for a working knowledge of human biology and health study.

Neurotransmitters are important in memory, mood, and pain perception. Acetylcholine is a neurotransmitter that is associated with good memory skills. It is found in

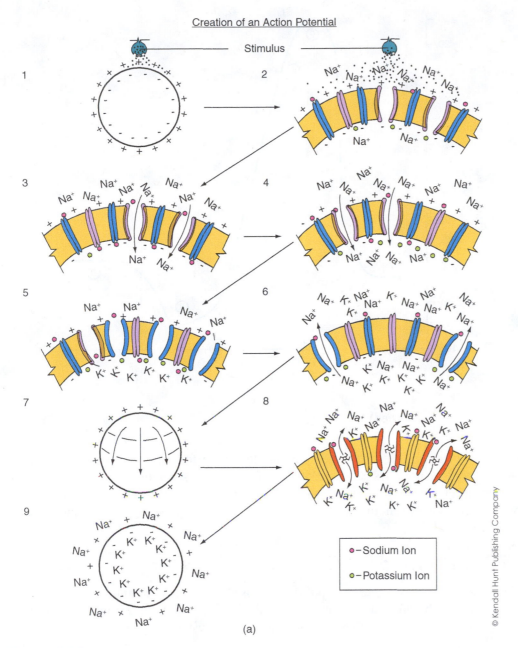

(a)

**Figure 14.8**    a. Formation of an action potential. Nerve impulse transmission along the axon: ions flow into and out of the neuron forming a wave-like current, called the action potential. b. Action potential changes in membrane polarity associated with the action potential corresponding to chemical movements in part (a).

© Kendall Hunt Publishing Company

decreased amounts in Alzheimer's patients, as you may recall from Chapter 1. Several other neurotransmitters influence mood and even personality. Serotonin and endorphins, for example, are both types of neurotransmitters that improve mood and inhibit pain and depressive feelings. They are found in increased amounts in happier people and decreased in depressed people. There are ways to improve mood and decrease pain. Some manufactured drugs, such as some antidepressants, work by increasing the amount of these neurotransmitters in the blood. Prozac, for example, blocks the normal reabsorption of

**Serotonin**

A type of neurotransmitter that improves mood and inhibits pain and depressive feelings.

**Endorphins**

A type of neurotransmitter that improves mood and inhibits pain and depressive feelings.

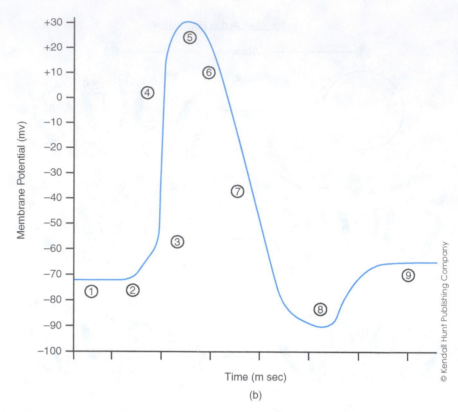

(b)

**Figure 14.8**   *(continued)*

serotonin, increasing it within the body and lessening feelings of depression. A more natural way is to use physical exercise and training. Both serotonin and endorphins also increase after physical activity, such as regular aerobic exercise. Medical doctors are increasingly recommending exercise for patients with depressive mood disorders, for these reasons.

## Special Senses

While the mechanics of nerve impulse transmission explains communication and regulation within the body, a look at our special senses – smell, taste, touch, vision, and hearing – gives us a better view of how our nervous system helps us to respond to the

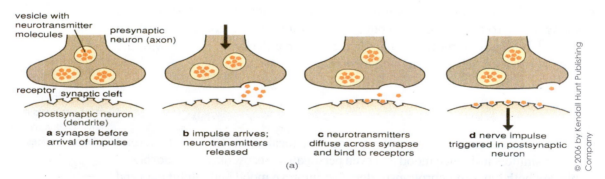

**a** synapse before arrival of impulse

**b** impulse arrives; neurotransmitters released

**c** neurotransmitters diffuse across synapse and bind to receptors

**d** nerve impulse triggered in postsynaptic neuron

(a)

**Figure 14.9**   a. The Synapse and Traveling Nerve Impulse. Nerve impulses must travel across synapses, which are gaps between nerves through which ionic charges are carried. Neurotransmitters (nerve chemicals) carry the charge across the synapse. From *Biological Perspectives*, 3rd ed by BSCS. Reprinted by permission. b. Steps for transmission of action potential across synapse.

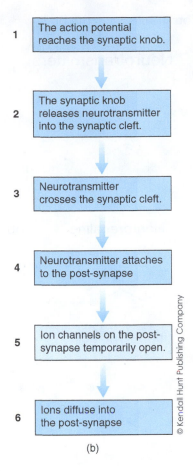

(b)

**Figure 14.9**   *(continued)*

world around us. Some people have a more keen sense of smell than others. It is also reported that failure in one of the senses, such as blindness, results in development of another sense, such as improved hearing.

Different organisms also have heightened senses: dogs are able to hear pitches much higher than humans; dogs also have a sense of smell 40 times more powerful than humans, able to detect drugs and track suspects; ants are able to see very little but are able to sense chemicals that are one-billionth of a gram in weight; and nocturnal owls have excellent eyesight and depth perception, able to detect their prey very easily. A photo of an owl, seriously focused on its prey, is given in Figure 14.10.

Even different humans have differing sense abilities. Children are able to sense more frequencies of music than adults, whose hearing is diminished as a result of age. If a group of school kids set their cell phones to ring at a certain range of pitches, their teacher will not hear it! Our differences are based on our individual physiology. What one person is able to hear may not have access to another due to the differing sensitivity of their senses.

## Gustation

The ability to taste or gustation starts by chemoreceptors in the mouth, tongue, and cheeks. Specialized receptors are called taste buds, which fire when chemicals from food attach to them. Gustatory receptors are thus chemoreceptors because they

**Chemoreceptor**

A sensory cell that is stimulated by chemicals.

**Table 14.1**   Selected Neurotransmitters in the Body

## Neurotransmitters

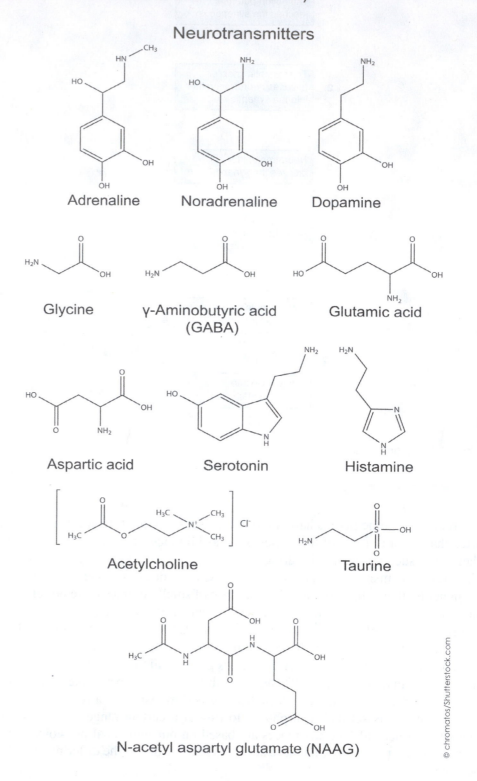

Adrenaline    Noradrenaline    Dopamine

Glycine    γ-Aminobutyric acid (GABA)    Glutamic acid

Aspartic acid    Serotonin    Histamine

Acetylcholine    Taurine

N-acetyl aspartyl glutamate (NAAG)

© chromatos/Shutterstock.com

are stimulated by chemicals. Gustation is stimulated most strongly by taste buds on the tongue.

There are roughly 10,000 taste buds in the human mouth and on the tongue. The sides of the tongue pick up sour tastes and the front senses sweet. The front and sides overlap with other taste buds to sense sour chemicals and the back of the tongue pick

**Figure 14.10**   An Owl Watches Its Prey before Attacking. Visual clues help many organisms locate and obtain food.

up bitter sensations. Different regions of the tongue detect different tastes in foods, as shown in Figure 14.11. Studies report that men and women, on average, differ in their taste preferences, with men preferring salty treats more often and women preferring sweets.

Taste has strong nerve responses that remain in one's memory. Do you know someone who has had food poisoning? They are likely to have lost interest in whatever food they ate at the time, due to the association of that food with the experience. Taste is a strong force in society. Cultures are built around food, and more than half of a hunter-gatherer's time was spent obtaining it. Our modern culture also centers on eating, a need for survival. When taste is associated with satiety from hunger, the food becomes more desirable. Many Europeans, as in Germany, eat pig brains for breakfast. While Americans may consider it distasteful, pig brains probably satisfied hunger in these nation's people.

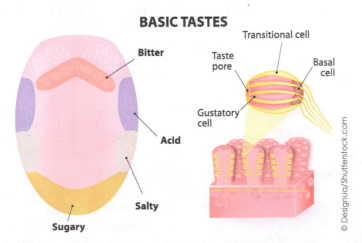

**Figure 14.11**   The Tastes of the Tongue: taste buds for different stimuli are in different regions of the tongue. While more receptors for sour (acidy) substances are located mostly on the sides of the tongue, as shown in this figure, receptors of all types are found strewn throughout the mouth.

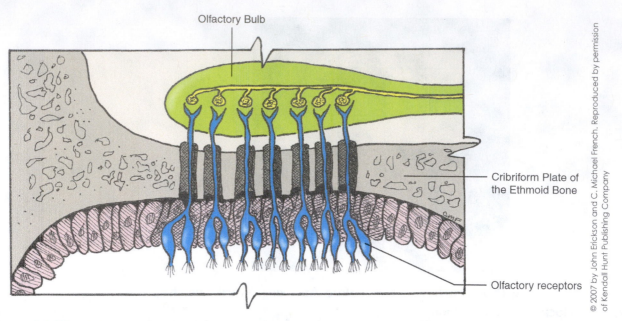

**Figure 14.12**   Olfactory (smell) Receptors Are Linked to the Brain by Running Nerves across the Nasal Bones. There are small holes in these bones that allow the passage of nerves to the brain. The olfactory bulbs in the brain interpret smells. Adapted from *Anatomy I and Physiology Lecture Manual* by John Erickson and C. Michael French.

## Olfaction

**Olfaction**

The sense of smell.

A keen sense of smell or olfaction is great for culinary students and chefs. Try cooking without a good sense of smell and try eating with your nose plugged. You taste is compromised and you cooking skills diminish greatly. Figure 14.12 points out the relationship between the brain and the olfactory receptors in our noses.

Olfactory receptors are located in the superior portion of the nasal cavity, covering an area of about 5 cm². Smells are a result of chemicals diffusing through the air. Thus, olfactory receptors are also chemoreceptors. When chemicals attach to olfactory receptors, nerve impulses are sent to a special region of the brain, the olfactory bulbs. Olfactory bulbs are located at the front part of the brain. Nerves travel through a thin bone separating the brain from the nasal cavity, called the cribriform plate, covered in holes for which nerves may travel. Damage to the frontal parts of the brain may inhibit smell. A famous chef sued for damages after a car accident damaged his sense of smell. He received millions of dollars because his cooking suffered with the loss of his olfactory senses.

### ARE NETI POTS SAFE?

A young boy was admitted to the hospital in central Texas in August 2007. He had fever, felt poorly, had transient pains, and lost his sense of smell. Instead, he smelled a burning odor despite there being none in his surroundings. He had been swimming in a lake during his days at summer camp. Doctors could not identify the cause of his symptoms until a sample of his cerebrospinal fluid showed that he had amoebas in his CNS. The boy received aggressive treatment for the amoeba but he died five days after being admitted to the hospital.

The story above reveals a case in which an amoeba, a single-celled protist discussed in Chapter 8, caused a fatal infection of a child's brain: amoebic meningoencephalitis. When the amoeba, *Naegleriafowleri*, enters the brain through the nasal passages, it travels through the cribriform plate. It first attacks the olfactory bulbs of the brain. Thus, the first symptom is loss of smell or a sense that something is burning or rotting. Olfactory tissue is instead being attacked and in a sense "burned" away by the amoeba. Later symptoms include a loss of balance, seizures, and confusion and hallucinations. The entire course of the disease takes under two weeks, almost always resulting in death. It is difficult to treat and even detect early, over 98% of people die within this short period. From 1995 to 2004, the amoeba has killed 23 people only in the United States, according to the U.S. Centers for Disease Control and Prevention.

While this is a rare illness, it has been associated with swimming in warm water lakes and ponds. More recently, two cases are thought to have arisen from water wells and municipal water supplies. *Naegleria fowleri* was found in the home water supply of both victims. Neti pot, a device used to irrigate (clean) nasal systems, may be the cause. Water is poured from the neti pot through one nostril and then it flows out of the other. While it is an ancient technique, it is now being recommended by the medical community for allergy, cold, and sinusitis treatment. The use of sterile water is therefore recommended when using a neti pot.

# Vision

Human use of vision is well developed as compared with many other animal species. It is our sharpest sense, used for hunting, reading, writing, and fine movements. Our eyes are developed to take in light rays and convert them into a nerve impulse to be registered within the brain. Let's trace the movement of light as it travels from the outside world and into our brains, using Figure 14.13.

The outermost tunic (covering) of the eye is called the sclera. When light travels into the eye, it first hits the cornea, which covers the anterior (front) chamber of the eye. Light moves through the cornea and into the front chamber, known as the aqueous humor. The aqueous humor contains fluid through which light passes until it is focused by the lens. The colored iris surrounds the opening to the lens, called the pupil. The lens of the eye is very hard in structure but flattens to focus the rays of light passing through it. The ciliary body surrounds the lens, which regulates the lens' shape. After passing through the lens, light moves as a focused beam through the eye's posterior chamber called the vitreous humor. The inside of the eyeball is lined with a thin, tanned coat called the retina.

Light strikes the retina, which contains receptor cells called photo-pigments. Photo-pigments come in two forms: rods and cones, which send impulses to the brain when they are stimulated by light waves. Rods are sensitive to low levels of light, allowing sight at night. Cones are more sensitive to light and used to see during the daytime. There are three types of cones: red-, green-, and blue-sensitive. Using a combination of cones enables us to see colors. (Both cone and color start with "C" to help you remember). Some people (usually males) are color-blind because their green or red cones do not function properly. It is a sex-linked trait, as you may recall from Chapter 6. Recall that waves are energy, and light waves transfer that energy to ionic impulses because of photoreceptors.

**Cornea**

Transparent part of the eye.

**Aqueous humor**

The clear fluid present between the cornea and lens of the eyes.

**Iris**

The colored part found around the pupil of the eye.

**Pupil**

The opening in the center of iris.

**Lens**

A very hard structure of the eye but flattens to focus the rays of light passing through it.

**Ciliary body**

A part of the eye located between the choroid and iris.

**Vitreous humor**

Posterior chamber of the eyes.

**Photo pigments**

Special pigments found in the retinal rods and cones.

**Rods**

One form of photo pigment that sends impulses to the brain that give black-and-white perception.

**Cones**

A form of photo pigment that sends impulses to the brain that give color perception.

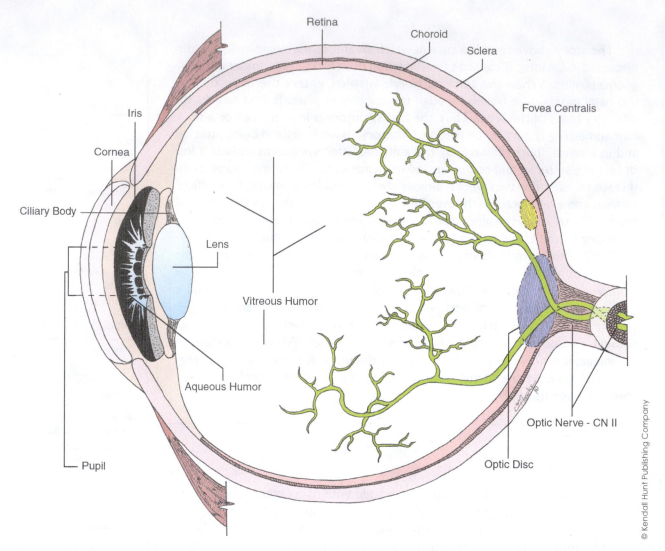

**Figure 14.13** Eyeball Anatomy. Light's movement through the structures of the eyeball cause vision. Light passes through the eye to the photoreceptors in the retina along the backside. Nerve impulses are sent via the optic nerve to the brain to be processed.

## Changing Light into Nerve Impulses

**Rhodopsin**

One type of photo-pigment found in rods that responds to light by changing shape and generating a nerve impulse and sending it to the brain.

Rods and cones are shown in Figure 14.14, in which the process of forming a nerve impulse from light is described. One type of photo-pigment found in rods for example, called rhodopsin, responds to light by changing shape and generating a nerve impulse and sending it to the brain. Rhodopsin is composed of two parts: opsin, which is a gly-coprotein, and retinal, a derivative of vitamin A. Absorption of light by photo-pigments occurs as energy from light waves strikes them. Photo-pigments change their structure upon light absorption. The retinal part of rhodopsin converts from a bent or cis-retinal form to a trans-retinal form, which is straight in its shape. The shape change causes an earthquake-type response in the retina, stopping $Na^+$ from flowing into the adjacent nerve cells. These nerve cells are inhibitory, so that when they cease to function, other nerve cells are triggered.

**Bipolar cells**

A neuron that has two processes.

Thus, nerve impulses are sent from receptors within the retina. First, bipolar cells fire sending impulses to **ganglion cells**, and eventually, via the optic nerve to the brain.

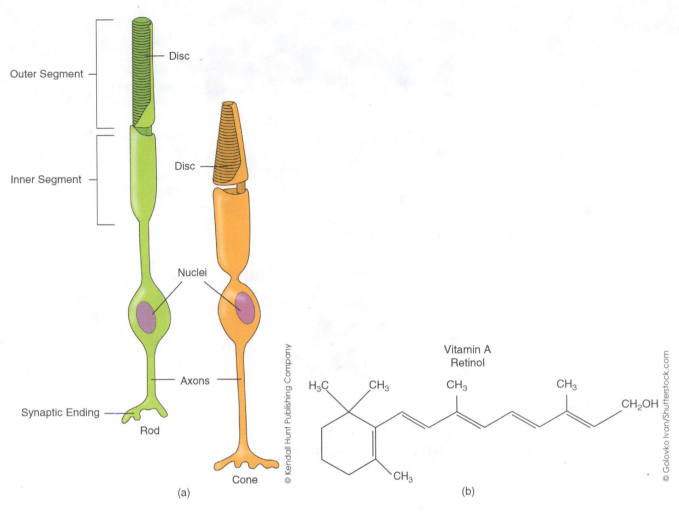

Outer Segment

Inner Segment

Disc

Disc

Nuclei

Axons

Synaptic Ending

Rod

Cone

(a)

© Kendall Hunt Publishing Company

Vitamin A
Retinol

$H_3C$    $CH_3$    $CH_3$    $CH_3$    $CH_2OH$

$CH_3$

(b)

© Golovko Ivan/Shutterstock.com

**Figure 14.14** a. Photoreceptors in the eye, rods, and cones contain photo-pigments stacked in discs along their top regions. b. Vitamin A (retinol) is important in regenerating retinal.

In the brain, visual messages are sensed and interpreted. The brain's functions will be discussed in the next sections; however, the movement of impulses through the retina is shown in Figure 14.15.

# Hearing

The human ear is adapted to transform sound wave energy into mechanical energy. The ear is a funnel through which waves are concentrated more and more, strengthening them. Have you ever seen old movies, with hearing aids that were simply funnels placed up against an ear? The principle applies to modern hearing aids as well: concentrating sound waves enables hearing. Different species have differing abilities to hear: Dogs hear high pitches well and some even hear echoed sound, in echolocation used by bats. However, they all hear using the same process, with sound waves moving through the ear structure shown in Figure 14.16.

First, sound waves are funneled in the outer ear, which consists of a pinna and an eardrum. The pinna acts as a funnel to concentrate sound to the eardrum, which vibrates. At this point, the traveling sound transforms into a physical entity: a vibration in the

**Outer ear**

Pinna and eardrum.

**Pinna**

Projecting part of the external ear.

**Ear drum**

The membrane separating the outer ear from the middle ear.

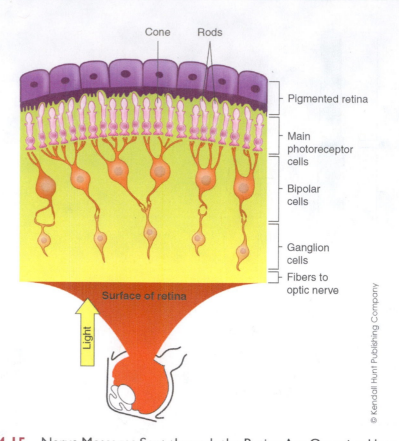

Cone    Rods

- Pigmented retina
- Main photoreceptor cells
- Bipolar cells
- Ganglion cells
- Fibers to optic nerve

Surface of retina

Light

© Kendall Hunt Publishing Company

**Figure 14.15**   Nerve Messages Sent through the Retina Are Organized by a Network of Neurons; bipolar cells and ganglion cells each help light's message to pass to the optic nerve and onto the brain.

**Middle ear**

Middle ear bones found inside the eardrum.

**Hammer**

A bone that is the outermost of the three small bones in the middle ear.

**Anvil**

A tiny bone in the middle ear.

**Stirrup**

The innermost bone of the middle ear.

**Semicircular canals**

Part of the inner ear filled with a fluid substance.

**Cochlea**

A spiral-shaped cavity of inner ear.

**Oval window**

An oval-shaped opening that is the start of the inner ear.

eardrum. Vibrations travel along the middle ear, which consists of three ear bones: the hammer (malleus), anvil (incus), and stirrup (stapes) or (HAS, to help you remember).

The oval window is the start of the inner ear, which consists of two parts: the semicircular canals and the cochlea. The semicircular canals are responsible for balance. As the stirrup vibrates, the oval window passes energy along the vibrations to the fluid within the cochlea. The oval window is much smaller than the eardrum. Why? Well, what is better to walk with in a muddy field: boots or high heels? Of course, high heels will sink much faster than boots. This is because the surface area of the high heels is so small that it concentrates the weight of a person to press down into the ground. The oval window thus concentrates sound strongly into the cochlea.

This concentration of sound waves transmits into the liquid of the cochlea, appearing as liquid waves. Waves in the cochlea bend tiny hairs sitting atop receptors, along its membranes. Each time a hair bends, receptors fire a nerve impulse up the auditory nerve and to the brain. When sound travels, it either has many waves, making it a higher pitched sound, or it has waves that are high, making it louder. As we age, hearing usually changes, decreasing in the range of sounds we are able to hear. There are numerous causes of hearing loss, from viral infections and arthritis to deterioration of the membranes in the cochlea. For example, when ear bones develop arthritis or joint disease, it can impair hearing. Hearing aids are useful but often do not return one's hearing successfully to its full capacity. Different parts of the membrane within the cochlea detect different pitched sounds, as identified in Figure 14.17.

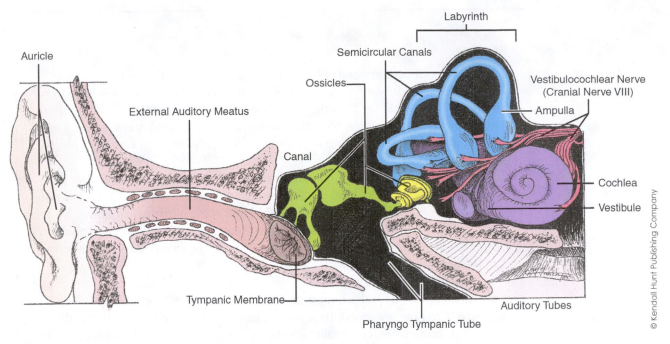

**Figure 14.16** Structure of the Human Ear; vibrations are transmitted through the ear to bend hairs within the cochlea, forming nerve impulses.

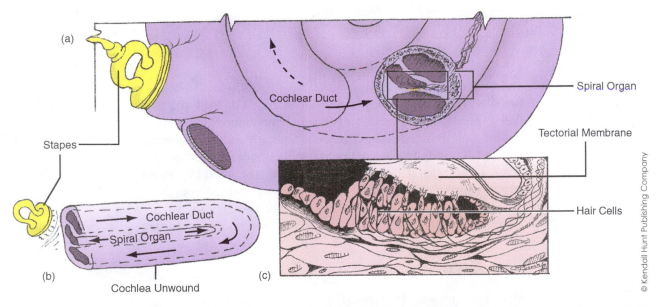

**Figure 14.17** a. Distinguishing Pitch Occurs within the Cochlea. Different parts of the membrane within the cochlea detect different pitches of sound waves. b. If unwound, the cochlea would form a U shape. c. The cochlea has a spiral shape, with fluid and hair cells that are attached to the tectorial membrane.

## Touch

Human touch, the last of the five senses, is studied as an important treatment in sickness and recovery. Massage therapy, acupuncture, and osteopathic medicine use touch to alleviate symptoms of pain and help in a patient's recovery. We began the chapter with a reflection on alternative medicines and whether they would help Richard in our story. All of those medicines focus on the efficacy of touch.

Many pieces of data are sensed by receptors other than through the five traditional senses. Many organisms, for example, are able to navigate based on the Earth's magnetic field. Birds, whales, eels, and sharks migrate using metal within their brains to align with the magnetic fields. Snakes are able to sense heat or infrared energy in other organisms, and jellyfish have sensory cells to sense gravity.

Other data are detected internally by the human body. For example, proprioceptors are used to help us balance based on gravity. Proprioceptors are found in our muscles, tendons, and joints and send information about body positioning to the brain. Damage to these areas can lead to balance problems. Some studies report that cholesterol-reducing drugs, called statins, may damage these regions.

The sensing of pain, described in our story of Richard, uses special receptors called nociceptors. Pain is perceived differently by different people. This is because it has many aspects to how it is perceived and felt. Sometimes there is a mind–body connection, allowing some people to relax and reduce their pain while not working for others. There is also a cultural component to pain, with some cultures showing more or less reaction to the same pains. Because so many people suffer from pain, whole areas of medicine and research are dedicated to pain management.

**Mechanoreceptors**

A sense organ responding to physical changes.

**Thermoreceptors**

A sense organ responding to temperature.

**Nociceptors**

A sense organ responding to pain.

**Cerebrum**

The largest region of the brain.

**Cerebellum**

The posterior part of the brain, involved in coordination.

**Dura mater**

The outer layer of meninges.

**Arachnoid**

The middle layer of meninges.

**Pia mater**

The inner layer of meninges.

Touch is classified into several categories, based on the types of receptors that are stimulated: pressure by mechanoreceptors, temperature by thermoreceptors, and pain by nociceptors, described earlier in the chapter. Most of these are found on our skin and are used to respond to internal or external cues. When we have pain, for example, in our elbow, it indicates that there is damage and we should be careful with it. Some drugs, such as morphine, block the nerve impulses transmitted from the site of pain to the brain. It is a cut in the communication between the two areas, a way to treat pain. Painkillers are addictive and a leading cause of substance abuse in our society.

## The Brain

All of the senses are interpreted by the brain, a complex organ capable of great thought, but only about 1300 g (3 pounds) in weight. It is gray in color and mushy in texture, resembling a ball of bumpy play dough, just a bit softer though. However, the appearance of the brain merely skims its great functions: the brain is very much a fantastic organ. When holding a brain, it is amazing that all of our emotions, thoughts, intelligence, and even spirituality arise from this small yet complex set of tissues.

The brain is composed of three regions: (1) the cerebrum, or the largest part of the brain, which is divided into two hemispheres, (2) the cerebellum, the posterior part of the brain, and (3) the **brain stem**, which consists of the midbrain, pons, and medulla oblongata.

The brain and spinal cord are both protected by three coverings, together called the **meninges**. The outer layer of the meninges is the dura mater ("tough mother"), made of strong fibrous tissue. The middle layer is known as the arachnoid (spidery) layer, which has vessels that appear spidery. The inner layer is called the pia mater ("delicate

<u>Cerebrospinal Fluid Structures and Flow</u>

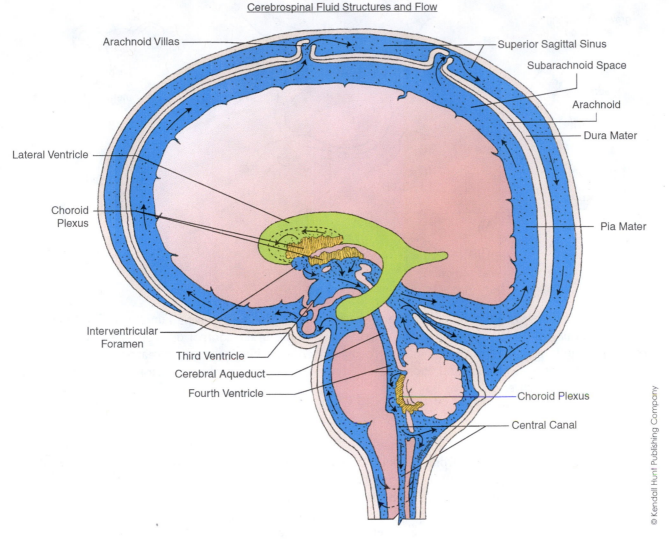

**Figure 14.18**    Brain Anatomy. The brain and its protective coverings, the meninges.

mother"), which contains capillaries that nourish underlying brain tissue. The brain and its coverings are shown in Figure 14.18.

Below the meninges, many structures are visible on the surface of the brain. The corpus callosum is an obvious attachment area, connecting the two hemispheres of the cerebrum. The rounded area underneath the corpus callosum is called the thalamus. It is like the grand central station of the brain, receiving all of the sense information from the body. The rest of the brain interprets that information. The region below the thalamus is the hypothalamus, which has many functions in the body (e.g. thirst, hunger, and temperature regulation). Attached to the hypothalamus is the pituitary gland. It regulates the hormonal system, which will be discussed in the last section of this chapter.

The regions of the brain that are responsible for basic survival are shown in Figure 14.19. At the front of the brain there are two olfactory bulbs, described earlier as important in sensing smell. The pons and medulla and midbrain and **cerebellum** are also visible, which comprise the central core or the brainstem (Figure 14.21). The brainstem carries out the basic functions of life: breathing, heart rate, and blood pressure are controlled by the medulla; and hearing, seeing, and pain perceptions are functions of the midbrain. The thalamus is shown transmitting messages to multiple parts of the brain.

**Corpus callosum**

An attachment area that connects the two hemispheres of the cerebrum.

**Hypothalamus**

The region below the thalamus.

**Thalamus**

The rounded area underneath the corpus callosum.

**Pons**

Part of the brainstem that helps relay impulses from cortex to cerebellum.

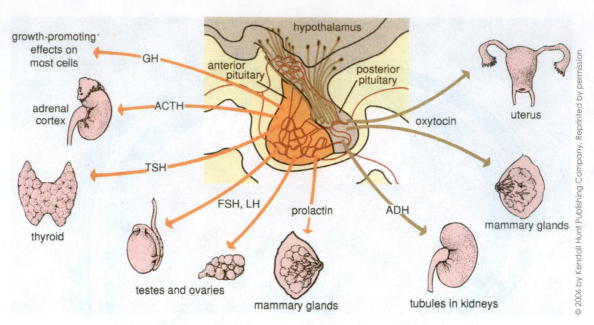

growth-promoting effects on most cells

GH

adrenal cortex

ACTH

TSH

thyroid

FSH, LH

testes and ovaries

prolactin

mammary glands

hypothalamus

anterior pituitary

posterior pituitary

oxytocin

uterus

ADH

mammary glands

tubules in kidneys

**Figure 14.19** Different Regions of the Brain Are Responsible for Basic Living Functions. The brainstem controls breathing and blood pressure. The major lobes of the brain control different life functions. The cerebrum is the largest part of the brain, controlling higher-level thinking, such as reasoning and language. The cerebellum controls balance and coordination, located as a sphere in the posterior region of the brain. From *Biological Perspectives*, 3rd ed by BSCS.

### Brainstem

The portion of brain that consists of pons, midbrain, and medulla oblongata.

### Midbrain

The short part of the brainstem above the pons.

### Central core

The foundational part of an organism that helps regulate basic life processes.

### Limbic system

A group of brain structures found on both sides of the thalamus.

### Amygdala

A section of brain associated with fear, panic, and aggression.

The thalamus is the "grand central station" of the brain because it is the first to receive information from the body. It quickly sends out nerve messages to be interpreted by other brain structures.

The cerebellum controls balance and coordination. A great basketball player usually has a more developed cerebellum to enhance motor (movement) skills. The cerebellum is located toward the posterior region of the brain.

We might recommend to Richard that there is a mind–body connection to pain perception that occurs within the central core. It is believed that smaller diameter nerve fibers transmit pain up to the brain, but large diameter fibers carry other nonpain information. Acupuncture claims to work by stimulating large diameter fibers. This blocks the sensations of small nerve fibers and pain messages.

The central core is our most primitive part of the brain, developed first evolutionarily as shown in part in Figure 14.19. The next set of structures to develop in the animal kingdom was the limbic system. In humans, the limbic system consists of the **thalamus**, which transmits messages to the **hypothalamus**, hippocampus, and amygdala, and other limbic structures. Each limbic structure controls behaviors related to emotional situations such as sexual drive and evaluating threatening situations. The inner portion of the brain holds the limbic system, as shown in Figure 14.20. The three most important limbic structures direct many of our basic needs and desires:

1) The **hypothalamus** is responsible for our basic drives: hunger, thirst, sex, and sleep. Sleep is a strange phenomenon. It is a needed function, but encompasses almost one-third of our lives. The longest a person has gone without sleep has been 264 hours, after which the person, a volunteer in an experiment, began hallucinations and showed physical problems such as heart rhythm effects. Mice kept awake indefinitely always die, but first develop heart disease, high blood pressure, and cholesterol problems.

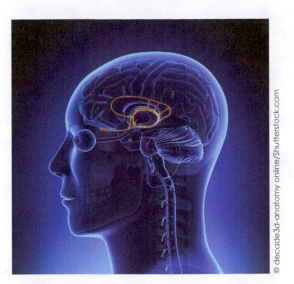

© decade3d-anatomy online/Shutterstock.com

**Figure 14.20**    The Limbic System (shown in red): important are emotional responses such as evaluating threatening situations, sexual behavior, and aggression. It evolved earlier than the cerebrum in animal evolutionary history.

**2)** The **hippocampus** is involved in short-term memory inputs, acting as a conduit to long-term memory storage in the cerebrum. When the hippocampus is damaged, a person cannot recall recent events. In the movie *50 First Dates*, the main character is afflicted with damage to her hippocampus, falling in love with the same person over and over.

   As we age, the hippocampus decreases in its size and ability to function. A 75-year-old loses more than 50% of their hippocampus functionality in tests comparing them with 17-year-olds. This is why learning a language, such as anatomy, at a younger age is easier than at older ages.

**3)** The **amygdala** is associated with fear, panic, and aggression. Recent studies show different activity levels in the amygdala of violent criminals as compared with nonviolent people. However, where are the nerve impulses most associated with our personality and temperament?

Our higher thoughts and actions – visual memory, speech, reasoning, and love – manifest within the neurons of the **cerebrum**. The cerebrum is the largest region of the brain, with two hemispheres (right and left) and four lobes: frontal, parietal, temporal, and occipital, shown in Figure 14.21. The regions of the brain are associated with different, higher-level activities. For example, the ability to understand language is accomplished by a small area between the parietal and temporal lobes called Wernicke's area. Different regions of the brain work together to accomplish complex tasks and thoughts.

Electroencephalography (EEG) experiments show that the frontal lobe is responsible for much of our personality, intelligence, and skeletal movements. Activity in the left frontal lobe is associated with social and friendly demeanor, while activity in the right frontal lobe is most active in people who are depressed, argumentative, and crabby. Is personality that simple? Of course not but, the other regions of the brain contribute to higher levels of thought: speech, sensation, and sensory integration occurring in the parietal lobe; hearing and visual sensing as well as language comprehension in the temporal lobe; and receiving and processing visual cues in the occipital lobe. The left and right sides of the brain control functions on their opposite sides. For example, the right side of the brain controls the left side of the body. Different regions of the brain also control different sensing and motor activities in the body. Figure 14.22 maps the right and left functions.

**Hippocampus**

Part of the brain involved in short term memory inputs, acting as a conduit to long term memory storage in the cerebrum.

**Frontal lobe**

The anterior part of the brain that is responsible for much of human personality, intelligence, and skeletal movements.

**Parietal lobe**

One of the four major lobes of the brain that contains an area concerned with higher levels of thought (speech, sensation, and sensory integration).

**Temporal lobe**

One of the four major lobes of the brain that contains an area concerned with hearing and visual sensing as well as language comprehension.

**Occipital lobe**

The posterior lobe of the brain.

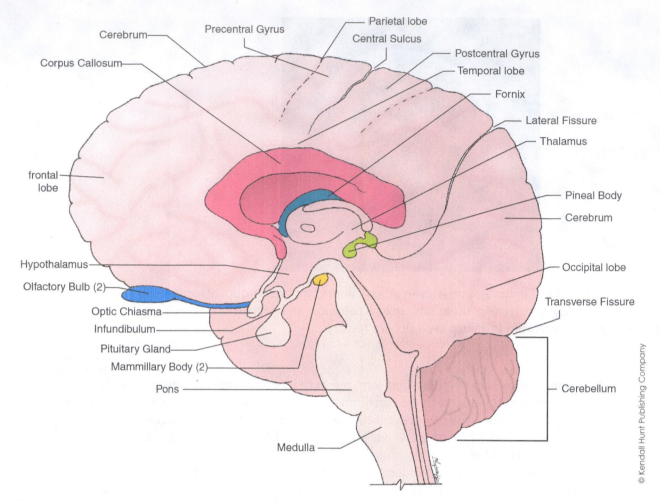

**Figure 14.21** General Structure of the Brain.

© Kendall Hunt Publishing Company

Brain function is complex, and each person is unique in his or her personality and abilities because of this complexity. Figure 14.22 shows that damage to one area of the brain affects particular abilities and normal functions. Consider the case of Phineas Gage. In 1848, Mr. Gage, a railroad worker, was injured while packing explosives. A three-and-a-half foot iron bar was driven through his face just beneath his left eye. Shockingly, he walked to the doctor and made what appeared to be a complete recovery. Soon, however, his personality changed, going from a friendly and hard-working man to a vulgar, hostile, and lazy vagrant. He drifted job to job and after he died 13 years later, his brain was preserved. It showed that he damaged portions of his frontal lobe, affecting his personality.

As a person ages, the size of her or his brain also shrinks. Most people lose about 10% in their lifetimes. The frontal lobe, associated with personality, decreases by over 50% by the age of 90 years. Altered levels of acetylcholine and decreased numbers of neurotransmitters in general are responsible for cognitive impairments in aging. However, there is a natural cell death, called apoptosis, which is a necessary process. Normal losses in cells are needed to make room for new ones and to grow and properly develop. In schizophrenia, normal apoptosis of dopamine-producing neurons does not occur. This leads to excesses in the neurotransmitter dopamine, causing fragmented thoughts and actions, classic symptoms of the disease.

**Apoptosis**

Cell death that occurs as a part of an organism's growth.

**Dopamine**

A neurotransmitter type that plays an important role in a number of different brain functions.

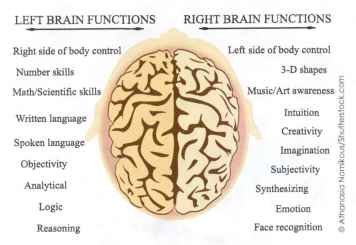

LEFT BRAIN FUNCTIONS    RIGHT BRAIN FUNCTIONS

Right side of body control    Left side of body control

Number skills    3-D shapes

Math/Scientific skills    Music/Art awareness

Written language    Intuition

    Creativity

Spoken language    Imagination

Objectivity    Subjectivity

Analytical    Synthesizing

Logic    Emotion

Reasoning    Face recognition

© Athanasia Nomikous/Shutterstock.com

**Figure 14.22**    Specialized Hemispheres of the Human Brain. The left and right sides of the brain each control opposite sides of the body. The left side controls the right side of the body and the right side of the brain controls the left side of the body. Each side of the brain has greater control of certain characteristics. In addition, there is a differing amount of control each body region has within the brain. The lips, for example, occupy a much larger part of the brain than the arm. Thus, the lips are more sensitive to feeling than an arm. Pinch your lips and then pinch your elbow. Which has greater sensation? Even if you pinch your elbow as hard as possible, it will be difficult to elicit a pain response.

# The Muscular System

## Characteristics of Muscles

Nerves stimulate muscles, attached to the bones of the skeleton, to move the body. Muscles are the workhorse of the body. They respond to directives from nerves to cause them to move. The three types of muscles were described in Chapter 11: skeletal, cardiac, and smooth. Each muscle type has a nerve that is attached to it to stimulate its movement. We will focus on the skeletal muscles in this chapter, as they respond to nerves.

Muscles make up over 50% of a human body's mass. Muscle cells are specialized to transform ATP energy into mechanical movements. They function in moving different bones of the skeleton, maintaining posture, and in stabilizing the joints. Muscles also generate a great deal of heat in their actions, maintaining body temperature. In Chapter 12, sphincters also guarded entrances and exits within the alimentary canal.

Muscle cells have the unique ability to both contract when stimulated by a nerve and also extend back into their original shape. The ability of a muscle cell to resume its original length after a contraction is called its elasticity.

When muscles contract, the direction in which they move is called their action. An action is based on two places: the location at which a muscle is attached to a bone that moves, called its insertion and the location at which it is attached, the origin. The origin of a muscle is usually another bone onto which the muscle is secured. A muscle action causes the movement of a bone in a certain direction or set of directions. These movements are always stimulated by a nerve, called the innervation.

Two sets of muscles frequently act in opposition to each other: while one is contracting, the other is relaxing during an action. The biceps of the arm, for example, contract at the same time that the triceps relax, both contributing to coordinated movement of the arm. An example of this antagonistic muscle movement is illustrated in Figure 14.23.

**Elasticity**

The ability of a muscle cell to resume its original length after a contraction.

**Insertion**

Movable end of a muscle.

**Origin**

The location (bone) at which muscles attach.

**Action**

The direction in which muscles move when they contract.

**Innervation**

The distribution of nerves to an muscle.

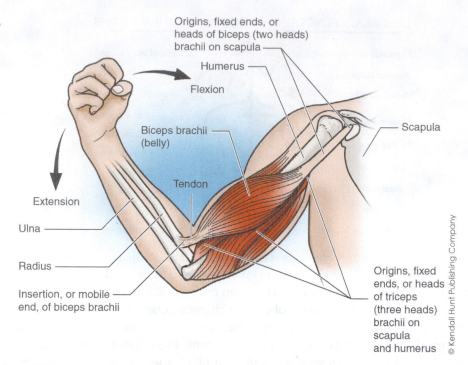

Origins, fixed ends, or
heads of biceps (two heads)
brachii on scapula

Humerus

Flexion

Biceps brachii
(belly)

Tendon

Extension

Ulna

Radius

Insertion, or mobile
end, of biceps brachii

Scapula

Origins, fixed
ends, or heads
of triceps
(three heads)
brachii on
scapula
and humerus

© Kendall Hunt Publishing Company

**Figure 14.23**    Muscles and Bones Work Together to Pull Bones and Rotate Limbs Around Joints. Muscle groups are antagonistic with each other, with one muscle often contracting, while another relaxes. Action of biceps and triceps work antagonistically to cause movement in the arm.

## Muscle Cell Organization

Whole muscles, such as the biceps, are actually composed of smaller strands of muscle fibers. Bundles of these muscle fibers are known as fascicles, which are further comprised many smaller muscle cells called muscle fibers. Muscle cells are made of thousands of intertwining myofibrils. Myofibrils are individually made of a series of contractile units called sarcomeres. The sarcomere is the fundamental contractile unit of muscles. They contract inwardly like an accordion, after they are stimulated by nerve impulses. When a muscle cell contracts, either all of its sarcomeres contract together, or none at all. This principle of muscle contraction is known as the all-or-none response. Richard's pain resulted from his nerves firing into muscles along his arm, neck, and back, making them contract and tense up. This causes pain. The structure of muscle fibers is given in Figure 14.24.

## Sliding Filament Theory

Over 80% of sarcomeres are made of two types of proteins: actin and myosin. This is why proteins are so important in building strong muscles. Actin and myosin slide past each other folding upon itself and expanding much like an accordion. When sarcomeres contract, the sliding filament theory explains the steps of the process. These steps may be traced using Figure 14.25:

1) A nerve impulse changes the ionic potential along the muscle cell membrane, called the sarcolemma.
2) Changes in the membrane potential of the sarcolemma cause calcium ions to flow into the muscle cell.

**Fascicle**

Bundle of muscle fibers.

**Muscle fibers**

The functional muscle cell.

**Myofibrils**

A rod-like protein structure in a muscle cell.

**Sarcomere**

A series of contractile units that make up the myofibrils.

**All-or-none response**

All sarcomere contract in a muscle fiber or none at all.

**Sliding filament theory**

The theory that explains muscle contraction.

**Sarcolemma**

A nerve's cell membrane.

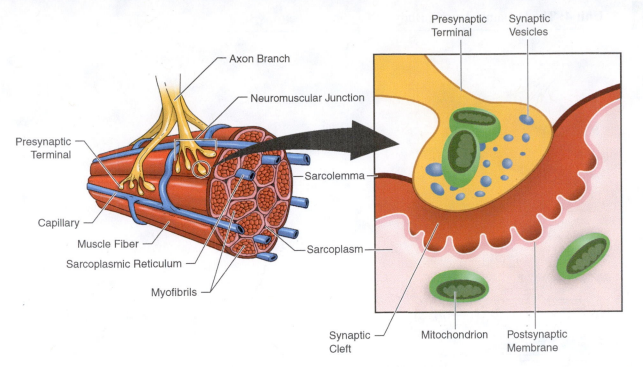

Figure 14.24 labels: Axon Branch, Neuromuscular Junction, Presynaptic Terminal, Capillary, Muscle Fiber, Sarcoplasmic Reticulum, Myofibrils, Sarcolemma, Sarcoplasm, Presynaptic Terminal, Synaptic Vesicles, Synaptic Cleft, Mitochondrion, Postsynaptic Membrane

© Kendall Hunt Publishing Company

**Figure 14.24** Muscle Anatomy. Muscles are composed of many fibers running alongside each other. Orientation of muscle fibers into parallel bundles causes movement of muscles along only a single plane.

3) Calcium ions combine with a protein, troponin (as you recall in Chapter 13, troponin is tested to see the extent of a heart attack), which changes shape and causes its attached tropomyosin to slide. Tropomyosin looks like ropes covering binding sites on actin. When these sites are revealed, they allow myosin heads to attach to them.

4) When myosin heads attach, ATP is used to cause the heads to swivel. This is known as the powerstroke because the sarcomere contracts inward using energy or power.

5) Relaxation occurs when another ATP molecule is used to release myosin heads from their attached positions.

## Rigor Mortis

Because relaxation takes energy, both contraction and relaxation during exercising a muscle are important. During rigor mortis, after a person dies their muscle stiffen. There are old horror stories, before embalming, which tell of people sitting up in their coffins after a few hours from their death. A particularly scary scene of rigor mortis is shown in Figure 14.26.

When the dead begin deteriorating, calcium ions flow through the holes formed and into the muscle cell. This changes troponin shape and starts the sliding of filaments. The problem arises during relaxation, which requires ATP energy. After death, there is no new energy and so the muscles remain stiffened for up to 12 hours.

## Fast vs. Slow Twitch Fibers

The time period comprising a contraction and relaxation is called a twitch. Some muscles twitch faster than others. Fast-twitch muscles contract up to 10 times faster than slow-twitch muscles (Figure 14.27). Some athletes have more of each type, contributing to their specialized skills in a sport. Long-distance runners, for example, have greater

**Troponin**

A protein found in all muscle.

**Tropomyosin**

A protein rope that plays an important role in muscle contraction.

**Powerstroke**

Movement of filaments using ATP during the contraction of muscle.

**Relaxation**

A state of freedom from skeletal muscle tension and anxiety.

**Rigor mortis**

Stiffening of the body that happens a few hours after death.

**Twitch**

The time period comprising a contraction and relaxation.

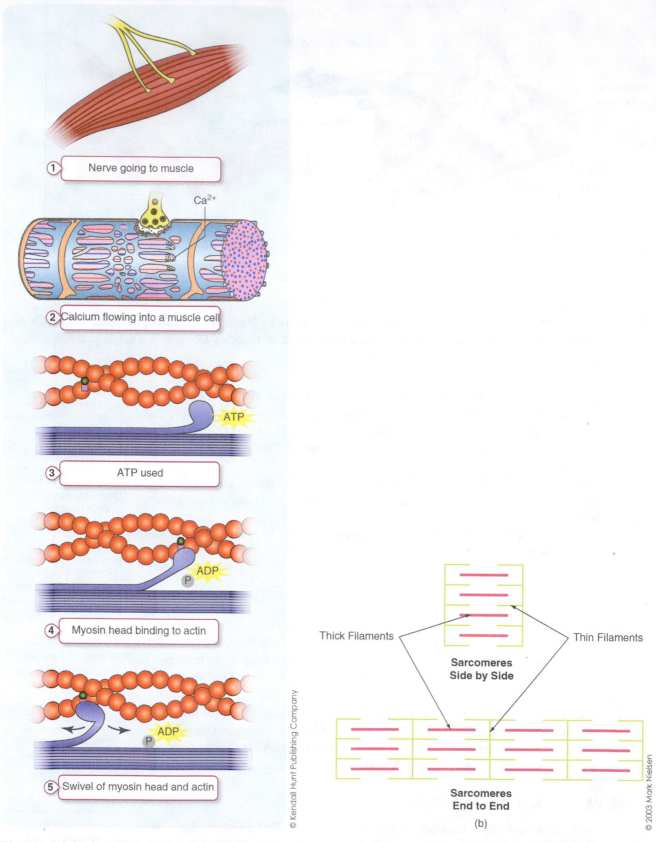

① Nerve going to muscle

Ca²⁺

② Calcium flowing into a muscle cell

ATP

③ ATP used

ADP
P

④ Myosin head binding to actin

ADP
P

⑤ Swivel of myosin head and actin

© Kendall Hunt Publishing Company

Thick Filaments

Thin Filaments

**Sarcomeres
Side by Side**

**Sarcomeres
End to End**

(b)

© 2003 Mark Nielsen

**Figure 14.25** a. Sliding Filament Theory: sarcomere contraction follows specific steps to cause muscle movement. When motor neurons send impulses to the muscle, calcium ions cause actin and myosin to attach. ATP is used in the powerstroke to move the muscle filaments. b. Sarcomeres. Illustration by Jamey Garbett.

**Figure 14.26** A Deceased Person Sitting Up in a Coffin Was Commonplace (before Embalming) in Days Gone By. Rigor mortis sets in within hours of death, causing muscle contractions, but there is no energy to relax the corpse. Movement results in rigid contractions found only in the dead.

proportions of slow-twitch fibers than sprinters. Sprinters have more than 70% classified as fast twitch. Each person may develop their muscle fibers, but will not change their proportion nor their inclination for a particular type of sport. Someone who is physiologically meant for long-distance running will probably never be a great sprinter and vice versa.

Exercise has many benefits to muscles. It increases the size of muscles called hypertrophy. It also increases the number of blood vessels that supply it as well as the number of mitochondria from which it derives its energy. Exercise does not, however, affect the number of muscle cells we have; it only improves what we are genetically given.

Of course, disuse of muscles leads to its rapid loss. Have you ever had your arm or leg in a cast or sling for a few weeks? It probably resulted in a very skinny arm with a lot of muscle loss. Humans lose about 5% of their muscle mass per day in those that are not used! There are over 200 muscles in the human body, with some superficial muscles shown in Figure 14.28.

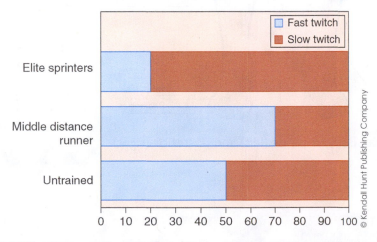

**Figure 14.27** Fast vs. Slow Twitch Percentages in Different Athletes. Fast-twitch muscles enable sprinters to move in quick spurts and slow-twitch muscles best suit long-distance runners to endure long periods of less-intense exercise.

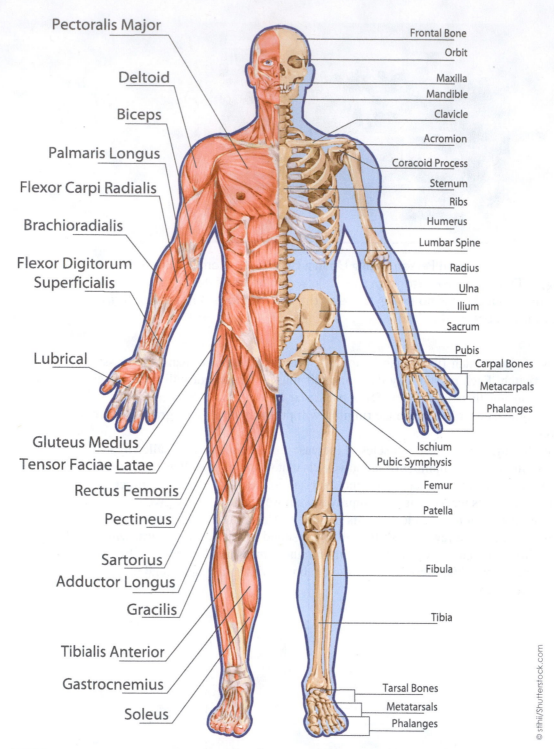

Pectoralis Major

Deltoid

Biceps

Palmaris Longus

Flexor Carpi Radialis

Brachioradialis

Flexor Digitorum
Superficialis

Lubrical

Gluteus Medius

Tensor Faciae Latae

Rectus Femoris

Pectineus

Sartorius

Adductor Longus

Gracilis

Tibialis Anterior

Gastrocnemius

Soleus

Frontal Bone

Orbit

Maxilla

Mandible

Clavicle

Acromion

Coracoid Process

Sternum

Ribs

Humerus

Lumbar Spine

Radius

Ulna

Ilium

Sacrum

Pubis

Carpal Bones

Metacarpals

Phalanges

Ischium

Pubic Symphysis

Femur

Patella

Fibula

Tibia

Tarsal Bones

Metatarsals

Phalanges

© stihii/Shutterstock.com

**Figure 14.28** Muscles of the Human Body: anterior (left) and posterior (right).

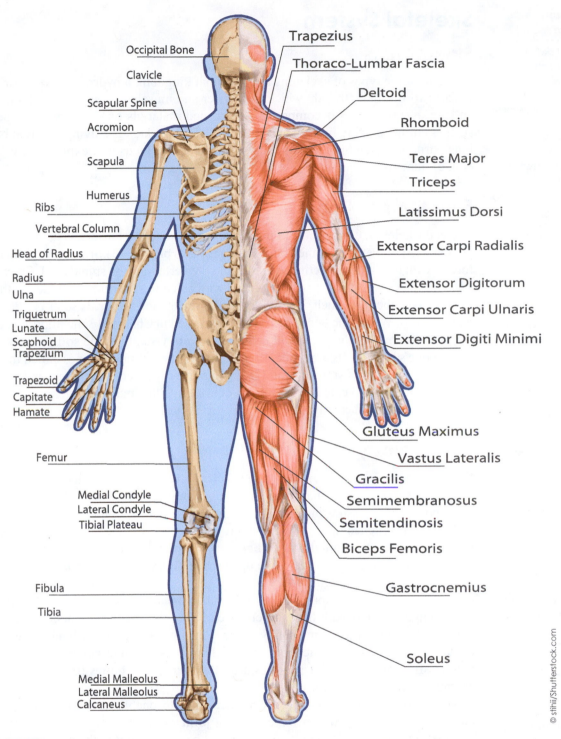

Occipital Bone
Clavicle
Scapular Spine
Acromion
Scapula
Humerus
Ribs
Vertebral Column
Head of Radius
Radius
Ulna
Triquetrum
Lunate
Scaphoid
Trapezium
Trapezoid
Capitate
Hamate
Femur
Medial Condyle
Lateral Condyle
Tibial Plateau
Fibula
Tibia
Medial Malleolus
Lateral Malleolus
Calcaneus

Trapezius
Thoraco-Lumbar Fascia
Deltoid
Rhomboid
Teres Major
Triceps
Latissimus Dorsi
Extensor Carpi Radialis
Extensor Digitorum
Extensor Carpi Ulnaris
Extensor Digiti Minimi
Gluteus Maximus
Vastus Lateralis
Gracilis
Semimembranosus
Semitendinosis
Biceps Femoris
Gastrocnemius
Soleus

© stihii/Shutterstock.com

**Figure 14.28**    (*continued*)

**Hydrostatic skeleton**

A type of skeleton found in earthworms and jellyfish, which use water and muscles for support and movement.

**Exoskeleton**

An external covering that is composed of hardened materials on the outside of the body.

**Endoskeleton**

A living, internal but hard inner structure found in animals and humans.

**Yellow marrow**

Hollow cavities within bones filled with fat.

**Morphology**

A particular structure or shape.

**Spongy bone**

Tissue found inside the bones that resemble a sponge.

**Trabeculae**

Small spindles that make up the spongy bone.

**Compact bone**

A portion of bone that is dense and contains a very little open space.

**Articulation**

The areas of connection between two bones.

**Surface markings**

The distinctive features found on human bones.

**Projection**

Sites for muscle attachment or joint connections.

**Depression**

Blood or nerve openings in human bones.

# Skeletal System

## Skeletons

There are three types of skeletons in the animal kingdom: a hydrostatic skeleton, found in earthworms and jellyfish, which uses water and muscles for support and movement; an exoskeleton, which is composed of hardened materials on the outside of the body, such as chitin covering insects; and an endoskeleton, which is a living, internal but hard inner structure found in animals such as humans. Our focus will explore the endoskeleton of humans.

## Functions of Bones

The endoskeleton, composed of our bones, works with the muscle system for (1) **support** and (2) **movement** of the body. The bones also provide (3) **protection** from outside forces, with the skull protecting the brain and the vertebrae, the spinal cord, for example. The bones are hard structures, mineralized with calcium salts and embedded with multiple layers of fibers. As such, its salts act as (4) **storage for minerals** such as calcium and phosphorous. Within the bone marrow, we discussed in Chapter 13 the importance of the skeletal system's role in (5) **blood cell production**. Usually in the soft or spongy parts of the bone's blood cells are made. In hollow cavities within the bones called the yellow marrow, (6) energy is stored as fat. Have you ever broken a chicken bone and pulled out its yellow marrow to eat? Many cultures consider yellow marrow a delicacy because it is filled with fat and energy.

## Morphology of Bones

While there are 206 bones in the human skeleton, bones are classified into four general categories based on their shape or morphology: **long**, **short**, **flat**, and **irregular**. Within these bones, their tissues include: (1) spongy bone and (2) compact bone. Bone tissue anatomy was touched upon in Chapter 11, if you wish to review. Spongy bone is composed of many small spindles or trabeculae of bone with many open spaces. Compact bone is dense and contains very little open space. Some examples from each bone classification are given in Figure 14.29.

Long bones (e.g., femur, fibula, and phalanges) are longer in length than in width. They contain a shaft with heads at either end. **Short bones** (e.g., tarsals and carpals) are cube shaped and contain spongy tissue. **Flat bones** (e.g., skull bones) are thin and contain two layers of compact bone sandwiching a spongy internal area. Bones that are not classified within these categories are called **irregular bones** (e.g., vertebrae). For examples, see Figure 14.29.

Many of the bones connect with each other to enable movement. The areas of connection between two bones are called a **joint** or articulation. There are numerous articulations of the skeletal system, which will be pointed out in the next sections.

Human bones also have distinctive features on their morphology called surface markings, which include either projections (sites for muscle attachment or joint connections) or depressions (blood or nerve openings). Each of the surface markings serves a purpose for human body functions. Projections often attach to a muscle or ligament, and depressions such as small holes (or foramen) act as opening for blood vessels and nerves. Some examples of surface bone markings are shown in Figure 14.30.

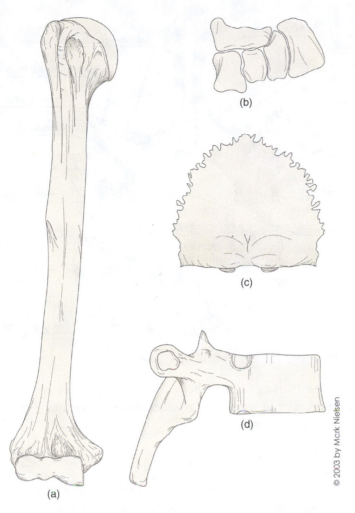

© 2003 by Mark Nielsen

**Figure 14.29** The Four Types of Bones: Long, Short, Flat, and Irregular. Each shape is suited for its function. A flat bone (e.g., skull bone) is often used to protect the underlying organs and tissues. a. long; b. short; c. flat; d. irregular. Illustration by Jamey Garbett.

# The Human Skeleton

The human bones are a part of the axial skeleton, which comprises the skull bones, ribs, sternum, and vertebrae, and the appendicular skeleton, which consists of bones of the limbs, the pelvic girdle, and the pectoral girdle (shoulder). Both skeletal categories are important to learn, as shown in Figure 14.31, which gives just a few of the major bones of the human skeleton.

The axial skeleton holds the center of gravity for the human body. It resists the many pressures placed upon it by gravity. Walking, running, and working in a garden require a strong resistance by the axis of the vertebrae and supporting bones of the axial skeleton. The axial skeleton contains skull bones, the ribs, and the bones of the vertebrae.

The skull is composed of two sets of bones: cranial bones (cranium), which enclose the brain, and the facial bones, which is attached to the muscles of the face. There are eight bones that make up the cranium, which are curved around the skull: The **frontal**

**Axial skeleton**

The portion of skeleton that consists of the skull bones, ribs, sternum, and vertebrae.

**Appendicular skeleton**

The portion of skeleton that consists of bones of the limbs, the pelvic girdle, and the pectoral girdle.

**Cranial bones**

The bone that encloses the brain.

**Facial bones**

The bones that attach to the muscles of the face.

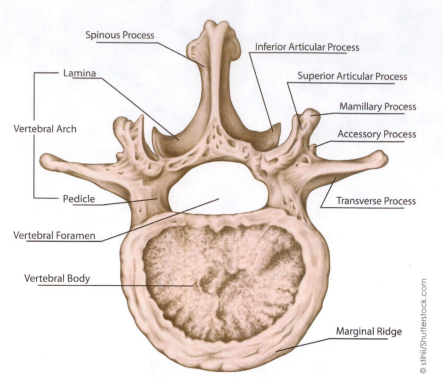

Spinous Process

Inferior Articular Process

Superior Articular Process

Mamillary Process

Accessory Process

Lamina

Vertebral Arch

Transverse Process

Pedicle

Vertebral Foramen

Vertebral Body

Marginal Ridge

© stihii/Shutterstock.com

**Figure 14.30**  Bone Markings: Every Bump, Groove, or Hole on Our Bones Has a Name and a Function. The figure shown is an example of the many surface regions of one of our vertebrae (back bones).

**Zygomatic bone**

Bone that forms an important part of the cheeks

**Vomer**

A small bone found inside the nose.

**Mandible**

Jawbone.

**Sacrum**

A large, wedge-shaped bone located between the two hip bones of the pelvis.

**Coccyx**

A small triangular-shaped bone located at the base of the spine.

bone make up the anterior part of the cranium; the two **parietal** bones are along the sides of the cranium; the two **temporal** bones are inferior to the parietal bones, placed along the lateral part of the skull; the two zygomatic bones form the cheeks; and the occipital bone is at the back of the head. Note that the names of the bones generally correspond with the names of the underlying parts of the brain. For example, the occipital bone covers the occipital region of the brain.

There are 14 facial bones, all paired except for the vomer bone (inside the nose) and the mandible (jaw). Palpate the following areas on your own body to discover these bones and surface markings using Figure 14.32.

1) Mastoid process: the roughened area behind your ears.
2) Temperomandibular joints: open and close your mouth to hear these click.
3) Superior orbital foramen: apply pressure in the middle of your eyebrow and feel for the indentation.
4) Nasal bones: feel the bridge of your nose.
5) Mandibular angle: feel the angle of your jaw.
6) External occipital protuberance: feel the back, inferior portion of your skull – it is the bump back there.
7) Zygomatic bone and arch: feel your cheekbone and its bridge toward the ear.
8) Hyoid: squeeze medially just below the mandible.

The axial skeleton consists of the **vertebrae** of the spine (24 single vertebrae bones plus two fused bones, the sacrum and coccyx) and the thorax (ribs and sternum). The vertebral column extends from the skull to the pelvis (hip bones) and is the main axial support. Of the 24 single vertebra of the spinal column, 7 are in the neck, which are called

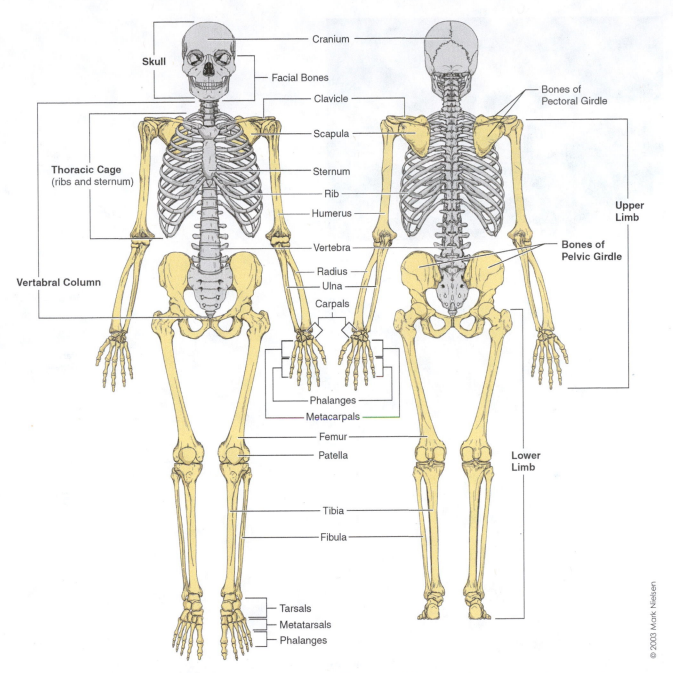

**Figure 14.31** Major Bones of the Human Skeleton. There are 206 bones found within humans. Illustration by Jamey Garbett.

cervical vertebrae, 12 are along the thorax, which are referred to as thoracic vertebrae, and 5 make up the lower back, known as lumbar vertebrae. You can remember this by when you eat: BREAKFAST at 7 a.m., LUNCH at 12 noon, and DINNER at 5 p.m. The vertebrae are shown in Figure 14.33.

**Cervical vertebrae**

The top seven vertebrae of the spinal column that form the neck.

**Thoracic vertebrae**

The 12 vertebrae along the thorax.

**Lumbar vertebrae**

The five vertebrae that make up the lower back

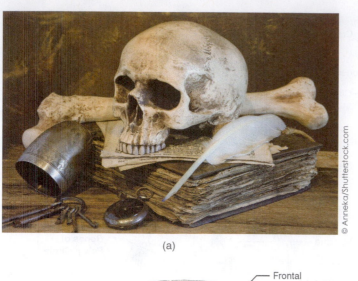

(a)

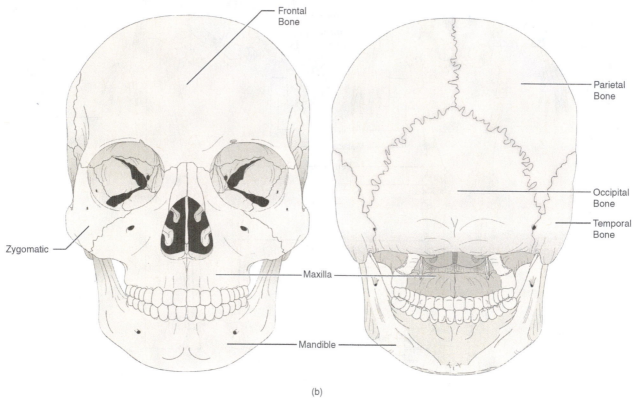

(b)

**Figure 14.32** Skull Bones Are Flat and Curved, Forming a Covering Around the Brain. a. b and c: Illustration by Jamey Garbett.

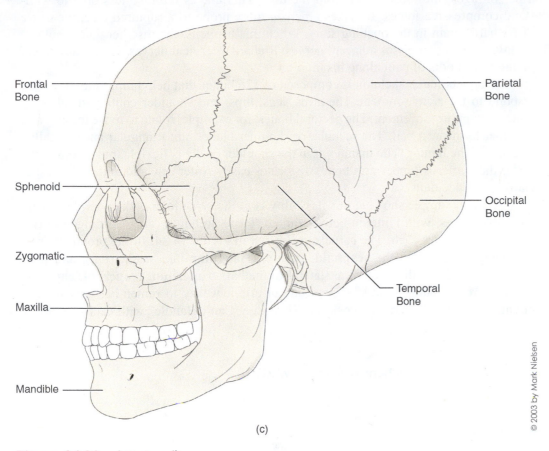

Frontal Bone

Sphenoid

Zygomatic

Maxilla

Mandible

Parietal Bone

Occipital Bone

Temporal Bone

© 2003 by Mark Nielsen

(c)

**Figure 14.32**    (*continued*)

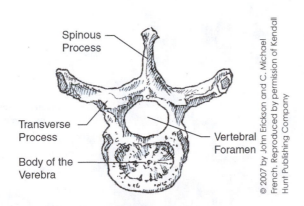

Spinous Process

Transverse Process

Body of the Verebra

Vertebral Foramen

© 2007 by John Erickson and C. Michael French. Reproduced by permission of Kendall Hunt Publishing Company

**Figure 14.33**    Vertebral Bones Are Irregularly Shaped and Contain Holes to Allow the Spinal Cord through It. Adapted from *Anatomy I and Physiology Lecture Manual* by John Erickson and C. Michael French.

**Intervertebral disc**

Pads of fibrocartilage that separate individual vertebrae.

**Humerus**

Upper arm bone.

**Carpals**

Any bone of the wrist.

Vertebrae are separated by pads of fibrocartilage called intervertebral discs. As a person ages, they lose water from the disc. This makes the disc less able to withstand compressive forces. It makes a person more prone to a ruptured disc, the cause of Richard's pain in the opening story. When this happens, the disc herniates (bulges) backward and compresses adjacent nerves. Richard's cervical nerves were compressed, giving him shoots of pain along his muscles.

The appendicular skeleton is composed of 126 bones that help humans to move and respond to their environment. The arms, legs, hips, and shoulder contain all of these bones to support movement. The pectoral bones, for example, rotate to move the arm and shoulder to throw a ball. The scapula, or shoulder blades, are triangular and are called "wings" of the human. The main joint in the shoulder girdle is flexible, but it comes at a price: the humerus (upper arm bone) dislocates easily (inferiorly and anteriorly), especially in car accidents.

Another example showing our flexibility is in the wrist joints. The wrist has many moveable bones, with joints in-between them. There are eight carpals arranged in two rows of four bones each. The carpals are shown in Figure 14.34. In the proximate row are (lateral to medial) the **scaphoid**, **lunate**, **triangular** (aka triquetrum), and **pisiform** bones; in the distal row are (lateral to medial) **trapezium**, **trapezoid**, **capitate**, and **hamate**. A naughty little saying, but an efficient way to remember these bones, might be: "Some Lovers Try Positions That They Can't Handle" with each first letter

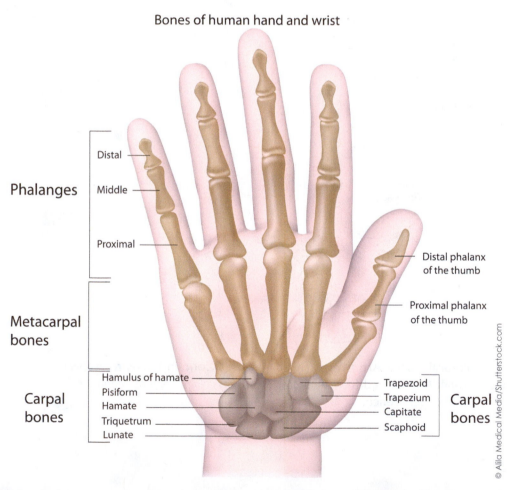

**Bones of human hand and wrist**

Phalanges
- Distal
- Middle
- Proximal

Metacarpal bones

Carpal bones
- Hamulus of hamate
- Pisiform
- Hamate
- Triquetrum
- Lunate

Distal phalanx of the thumb

Proximal phalanx of the thumb

Carpal bones
- Trapezoid
- Trapezium
- Capitate
- Scaphoid

© Alila Medical Media/Shutterstock.com

**Figure 14.34**    Carpal Bones in the Wrist Glide Past Each Other to Create Movement.

corresponding to a carpal, in that order. The carpals are held very closely together by ligaments, which restrict joint movement between them.

Carpal tunnel syndrome is an inflammation of the joints in-between the carpals due to, primarily, repetitive movements such as using a computer keyboard. The damage of the cartilage and/or ligaments between the carpals can require physical therapy, anti-inflammatory medications, or surgery.

## Bone Remodeling and Disease

Bones grow and change over a person's lifetime. Have you ever experienced growing pains at night, when most of our bone growth occurs? It can wake a person from even a deep sleep. Bones grow and remodel according to the forces placed upon them. This phenomenon is called Wolff's law. When a person exercises, for example, it stimulates bones to grow stronger in certain areas. Bones remodel and add material to areas that are more likely to break. For example, in the femur, the area most likely to buckle is the thickest, as shown in Figure 14.35. Euler's buckling equation, an engineering formula to show where a cylinder is most likely to fail, when applied to a femur, shows a thicker area in the region predicted to break. (Euler's equation from which this information is derived is: $P_{cr} = pi^2 E/L_2$, with $L$ equal to the length of the tube, $E$ the elasticity of the substance, and $P_{cr}$ the critical load amount to cause buckling.) This phenomenon is shown in Figure 14.35.

**Wolff's law**

The phenomenon which states that bones grow and remodel according to the forces placed upon them.

**Euler's Buckling equation**

An engineering formula to show where a cylinder is most likely to fail, when applied to a femur, shows a thicker area in the region predicted to break.

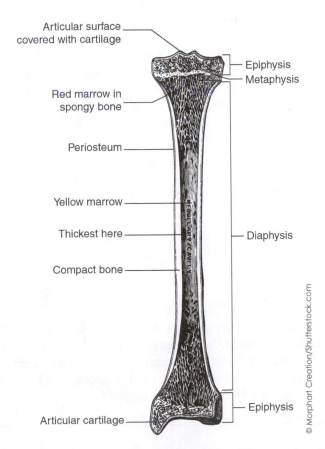

Articular surface covered with cartilage
Epiphysis
Metaphysis
Red marrow in spongy bone
Periosteum
Yellow marrow
Thickest here
Diaphysis
Compact bone
Epiphysis
Articular cartilage

© Morphart Creation/Shutterstock.com

**Figure 14.35** Long Cut of a Femur and Wolff's Law. The area in which a bone experiences the most stress is often the thickest. Bone remodeling, according to Wolff's law, adds bone to the area it is most needed.

Bones recycle and replace themselves – compact bones every 10 years and spongy every 3–5 years – so that we are made of all new bones in these periods. Replacement occurs slowly, and regular exercise helps to keep bones strong.

Bones have tremendous strength: Long bones have half as much compressive strength as steel of the same size and the same tensile (pulling) resistance as steel! The arrangement of the calcium salts and the fibers within bone gives it great hardness.

**Osteoarthritis**

A disease in which bones and their joints deteriorate, usually because the cushioning cartilage in between these bones wears out.

While fractures (breaks) in bones are the most common injury, other disorders involve a long-term problem in many people. Osteoarthritis is a disease in which bones and their joints deteriorate, usually because the cushioning cartilage in-between these bones wears out. It is not merely a disease of the elderly. A high proportion of runners experience arthritis in their knees because of chemicals, called metalloproteases, produced during their runs. These chemicals wear out their joints. Arthritis forms bone spurs or projections that inflame joints and cause pain. More than 85% of people will get osteoarthritis in their lifetime. The key to healthy joints is strengthening the ligaments, muscles, and tendons surrounding the bones. This gives the joints stability and limits its wear and tear.

**Osteoporosis**

Thinning and weakening of bones.

**Osteoclast**

Bone destroying type cell.

**Osteoblast**

Special bone building cells.

**Hydroxyapatite**

An essential component and major ingredient of normal bone.

Osteoporosis, a thinning and weakening of bones, occurs most often in the elderly, particularly in females. It is believed that decreased estrogen levels are associated with osteoporosis. Osteoporosis is due to a higher amount of osteoclast, or bone-destroying cell activity, and a lower amount of osteoblast, or bone-building cell activity.

Soda is especially bad for osteoporosis. Not only does soda replace other, more nourishing beverages such as milk or orange juice, which contain calcium to build strong bones and teeth, but phosphorous in many sodas causes a leaching effect of calcium out of the bones. Phosphorous is an essential component of bones in the form of calcium phosphate salts and hydroxyapatite ($(Ca_3(PO_4)_2 * (OH)_2)$). As can be observed from the equation, both calcium and phosphorous are needed in these bone-building materials. However, when phosphorous is added to our diets in large amounts via soda consumption, it combines with calcium in the blood and leaches it out. Research at Tuft's University studied a large number of both men and women, and discovered that women who drank three or more phosphorous-containing sodas had a 4% lower bone mass density than women who drank non-phosphorous-based soda or less soda. We are not entirely sure of the mechanism, but soda is linked to osteoporosis. Regular exercise and diets rich in calcium and vitamin D are the best strategy to fight the effects of thinning bones. Figure 14.36 shows osteoporotic bone and its weak trabeculae compared with normal bone.

### IS A BROKEN BONE STRONGER THAN THE ORIGINAL?

Because bones remodel according to the forces placed upon them, special bone-building cells called **osteoblasts** build new bone tissue when pressure is applied or when there is a break in the bone. Bones will overcompensate and build even more material in areas of breaks. Therefore, yes a broken bone is indeed often stronger and thicker than its original form.

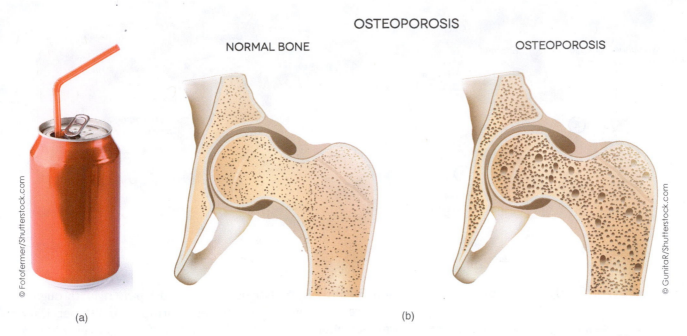

**Figure 14.36**   a. Soda b. Osteoporotic bone trabeculae are thin and fragile compared with normal bone.

# Endocrine System

## Glands and Basics

Controls over bone growth and remodeling are due to, in part, hormones. Hormones (from the Greek "to arouse") are chemical messengers that cause change or direct activity in another area of the body. Any cell that secretes a hormone is an endocrine cell. Hormones arouse the functions of another area of the body, called the target cells.

Some hormones, such as steroid (fat-based) pass easily through cell membranes. Other hormones must have receptors to enter a cell. Many target cells have specifically shaped receptors on them to attach to the shape of their respective hormones. This is called the receptor–hormone match. The methods by which hormones enter a cell and cause changes are illustrated in Figure 14.37.

The **endocrine system** includes all of the groups of cells, called endocrine glands, which produce hormones. It is much slower to act than the nervous system because its messengers are chemicals that diffuse through the circulatory system. Nerves fire like electricity, but hormones travel slowly like cargo ships.

There are seven major glands of the endocrine system. Figure 14.38 gives the major endocrine glands in the human body: the hypothalamus, the pituitary gland, the thyroid, the parathyroid, the pancreas, the adrenal glands, and the pineal gland. These glands produce an alphabet soup of hormones, with names given alongside their glands in Figure 14.38.

The **hypothalamus**, in the brain, is connected to and controls the activity of the pituitary gland. The pituitary gland is considered the master gland of the endocrine system because it sends messages to stimulate all of the other glands. We will take a look at some of these glands to study how they regulate life functions. An overview of the glands and the hormones they produce is in Figure 14.38.

**Hormones**

Chemical messengers that cause change or direct activity in another area of the body.

**Endocrine cell**

Any cell that secretes a hormone.

**Target cell**

Any cell having a specific receptor for an antibody, hormone, or antigen.

**Receptor-hormone match**

Match making between the receptors of target cells and their respective hormones.

**Pituitary gland**

The master gland of the endocrine system that sends messages to stimulate all of the other glands.

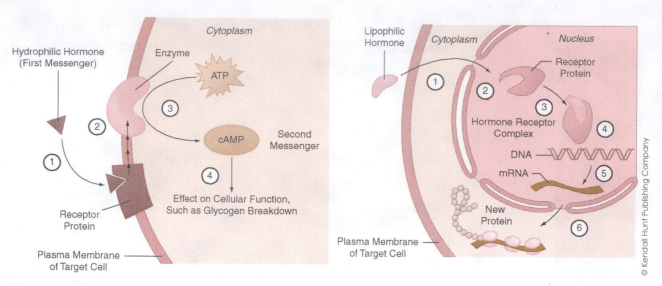

**Figure 14.37**    The Hormone Communication Process. Some hormones (steroids) pass right through a membrane, while other hormones (peptide) require a receptor protein. These hormones dock specifically to receptors, bringing them into the cell. Hormones often work within the nucleus of a target cell, eliciting a cellular response.

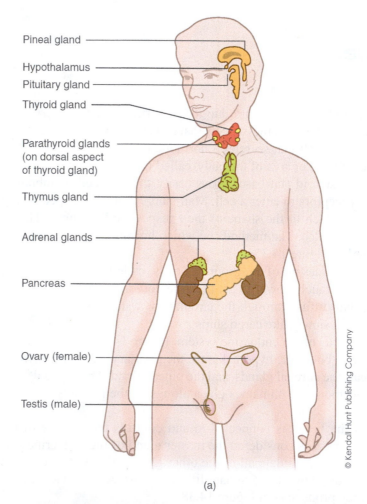

(a)

**Figure 14.38**    a. Major Glands of the Endocrine System and their Hormones. b. Functions of Endocrine Glands. From *Biological* Perspectives, 3rd ed by BSCS. c. Pituitary control over other glands. From *Biological* Perspectives, 3rd ed by BSCS.

**Peptide hormones** can influence only those cells that have receptors on the cell membrane to which they can bind. Once bound, they activate other proteins associated with the cell membrane. For example, some of these proteins might be enzymes that regulate essential steps in cellular reactions, such as the conversion of glucose into glycogen.

| Gland/Hormone | Primary Target | Principal Action |
|---|---|---|
| **Hypothalamus** | | |
| vasopressin (antidiuretic hormone) | kidneys | promotes reabsorption of water |
| oxytocin | mammary glands, uterus | stimulates milk release and uterine contractions (labor) |
| **Pituitary** (anterior lobe) | | |
| growth hormone (GH) | most cells | promotes growth; stimulates protein synthesis |
| adrenocorticotropic hormone (ACTH) | adrenal cortex | stimulates secretion of steroid hormones |
| follicle-stimulating hormone (FSH) | ovaries, testes | in females, stimulates egg formation and estrogen production; in males, helps stimulate sperm formation |
| luteinizing hormone (LH) | ovaries, testes | in females, stimulates ovulation; in males, promotes testosterone secretion |
| prolactin | mammary glands | in females, stimulates milk production and secretion |
| thyroid-stimulating hormone (TSH) | thyroid gland | stimulates secretion of thyroid hormones |
| **Parathyroids** | | |
| parathyroid hormone (parathormone) | bone, kidney, intestine | stimulates release of calcium from bone; elevates calcium levels in blood |
| **Thyroid** | | |
| calcitonin | bone, kidney, intestine | inhibits release of calcium from bone; decreases calcium levels in blood |
| thyroxine | most cells | regulates metabolism; has roles in growth and development |
| **Thymus** | | |
| thymosin | immune system | promotes maturation and activity of immune system |
| **Adrenal** (inner portion) | | |
| epinephrine | liver, muscle, adipose tissue | increases heart rate; increases blood sugar levels; prepares body for fight-or-flight response |
| norepinephrine | smooth muscle of blood vessels | constricts blood vessels, thus regulating blood pressure |
| **Pancreas** (endocrine tissues in islets) | | |
| glucagon | liver | raises blood sugar levels |
| insulin | muscle, adipose tissue | lowers blood sugar levels |

(b)

**Figure 14.38** (*continued*)

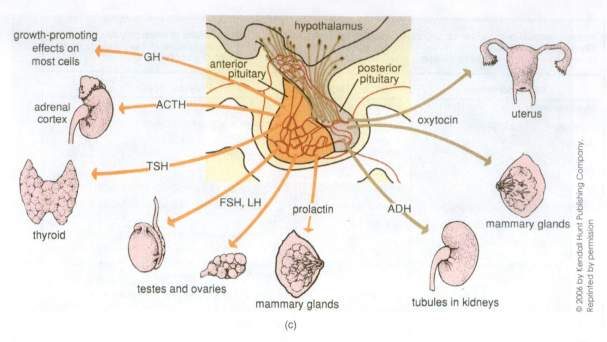

growth-promoting effects on most cells

GH

adrenal cortex

ACTH

TSH

thyroid

FSH, LH

testes and ovaries

hypothalamus

anterior pituitary

posterior pituitary

oxytocin

uterus

prolactin

ADH

mammary glands

mammary glands

tubules in kidneys

(c)

**Figure 14.38** *(continued)*

**Parathyroid hormone (PTH)**

Hormone produced by the parathyroid glands help in regulating the amount of phosphorous and calcium in the body.

**Calcitonin**

A hormone made by the thyroid directs calcium ion uptake by the bones.

**Insulin**

A hormone produced in the pancreas that regulates the amount of glucose in blood.

**Glucagon**

A hormone that raises blood sugar level.

**Beta cells of pancreas**

Cluster of cells found in the pancreas, which makes the insulin.

**Alpha cells of pancreas**

Cluster of cells found in the pancreas, which makes glucagon.

# Hormones Regulate Homeostasis

## Calcium and Bones

Almost every activity in the body involves hormones. Bones grow, as described in the previous section, when calcium levels are sufficient within the blood. Bones store 99% of calcium in the body, so that one hormone called parathyroid hormone (made by the parathyroid gland) directs it to give up calcium. Back in the blood, calcium is then available for a body's use. When calcitonin, another hormone made by the thyroid, directs calcium ion uptake by the bones, it can again be used for strengthening the bone tissue. The control of calcium in the blood is an example of how hormones regulate the aspects of our body. Blood calcium is kept at a balance of 9–11 mg/100 mL. The process of calcium homeostasis is traced in Figure 14.39.

## Blood Sugar and Diabetes

Normal levels of blood sugar (glucose) in the blood (80–120 mg of glucose/100 mL blood) are maintained by two hormones: insulin and glucagon. When carbohydrates are consumed in an animal's diet, special beta cells of the pancreas release insulin. Insulin causes the direct uptake of glucose into body cells. Insulin also stimulates the liver to store glucose in the form of glycogen.

As sugar levels decrease, alpha cells of the pancreas release glucagon, which stimulates the liver to release glucose from its stored glycogen reserves. This is a negative feedback mechanism, described in Chapter 11, which maintains the homeostasis of sugar to around 90 mg/100 mL blood consistently through a lifetime.

Diabetes is a disease in which glucose levels remain higher than normal in the blood and urine. It may lead to numerous health problems including nerve damage, heart disease, blindness, and death, if untreated. There are two types of diabetes: Type I and Type

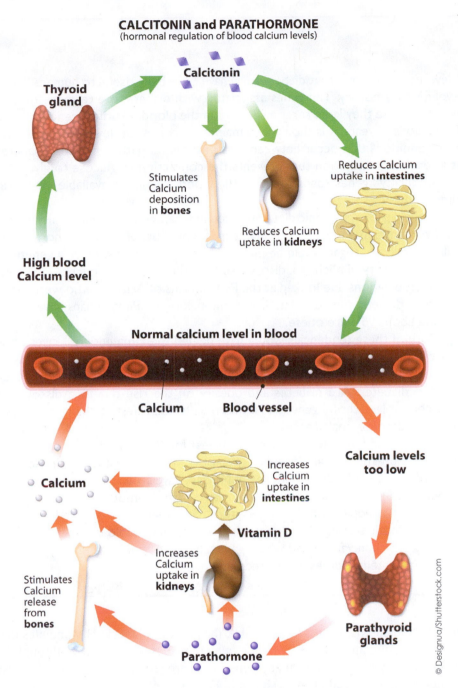

**CALCITONIN and PARATHORMONE**
(hormonal regulation of blood calcium levels)

**Calcitonin**

**Thyroid gland**

Stimulates Calcium deposition in **bones**

Reduces Calcium uptake in **kidneys**

Reduces Calcium uptake in **intestines**

**High blood Calcium level**

**Normal calcium level in blood**

**Calcium**          **Blood vessel**

**Calcium levels too low**

Increases Calcium uptake in **intestines**

**Vitamin D**

**Calcium**

Increases Calcium uptake in **kidneys**

Stimulates Calcium release from **bones**

**Parathyroid glands**

**Parathormone**

© Designua/Shutterstock.com

**Figure 14.39** Calcium Homeostasis Is Accomplished by the Action of Two Hormones: calcitonin and PTH (parathyroid hormone).

II. Type I diabetes is an autoimmune disease resulting from an attack on the pancreatic cells making insulin. Type II diabetes is a result of resistance to insulin by cells.

Diabetes is associated with diets high in refined sugars because excess sugars "wear out" insulin receptors. This wearing down process is known as down regulation. The receptor–hormone match wears out due to overuse.

## Metabolism

Regulation of the sum of chemical activities, called metabolism, is a large task. It involves several endocrine glands: When the hypothalamus activates the anterior pituitary to

**Diabetes Type I**

An autoimmune disease resulting from an attack on the pancreatic cells making insulin.

**Diabetes Type II**

A medical condition as a result of resistance to insulin by cells.

**Down regulation**

Decrease in the number of effective receptors on cell surfaces.

## IS DIABETES A GOOD ADAPTATION?

New research shows that diabetes may have once been useful to humans in our evolutionary history. The genes associated with diabetes are termed "thrifty" genes because they keep sugar levels high in the blood. "Thrifty" genes evolved for millions of years, it is thought, to maintain higher sugar levels during starvation conditions that occur between meals for hunter-gatherer tribes. Diabetes is a result of a mutation that prevents the conversion of glucose to glycogen. Thus, "thrifty" genes spare sugars in the blood, keeping it available for use in times of need.

Consider the Neolithic diet, discussed in Chapter 12, in which calories are hard to find. Someone with "thrifty" genes would benefit because normal circulating levels of sugar would be maintained longer. James V. Neel, a geneticist at the University of Michigan, discovered this "thrifty gene" sequence in several human populations. He looked at the Pima Indians of Arizona, who were more apt to be diabetic and store fat. About one-half of the Pima Indians have diabetes and about 95% are obese.

Both of these tendencies would have helped the Pima endure longer periods of starvation conditions in evolutionary history. In modern society, with food readily available, people with "thrifty" genes are more prone to obesity and diabetes. Are diabetes and obesity on the rise due to a dissonance between evolved thrifty genes and modern diets? Can lifestyle changes improve these modern-day maladies?

Studies show that things are not hopeless for those predisposed to diabetes: Pimas practicing traditional lifestyles of hunting and gathering, in isolated parts of the Sierra Madre mountains of Mexico, have significantly lower rates of diabetes (8%) and obesity (rare) as compared with the modern US Pima Indian population. If people with thrifty genes are more aware of their predisposition through genetic testing, they could alter their diets to avoid diabetes and obesity. A diet rich in variety and whole grains, fresh vegetables, and fruit and low in fat and protein sources is thus recommended.

---

**Thyroid stimulating hormone (TSH)**

Hormone produced by the pituitary gland and stimulates the thyroid gland.

**Thyroxin**

A hormone that increases metabolism throughout the human body.

**Hyperthyroidism**

An overactive thyroid, which results in too much thyroxine, causing nervousness, excess energy, sometimes enlarged eyes and irregular heart rates.

**Hypothyroidism**

An underactive thyroid, which results in weight gain, intolerance to cold and higher cholesterol.

**Adrenal glands**

Glands that sit atop both kidneys and control a variety of body functions.

**Adrenal cortex**

The exterior portion of the adrenal gland that secretes steroid hormones.

---

produce thyroid-stimulating hormone (TSH), the process starts. TSH stimulates the thyroid to manufacture and release thyroxine, which increases metabolism throughout the human body. If a problem arises at any point in the process, disease occurs. The process of thyroid-regulating metabolism is shown in Figure 14.40.

Hyperthyroidism is an overactive thyroid, which results in too much thyroxine, causing nervousness, excess energy, sometimes enlarged eyes (exophthalmos), and irregular heart rates. In the long term, it can lead to cardiovascular disease. Hypothyroidism results from an underactive thyroid and insufficient amounts of thyroxin, causing weight gain, intolerance to cold and higher cholesterol, and in the long term also cardiovascular disease. Both can be treated successfully with drug therapy, or in the case of hyperthyroidism, destruction of part of the thyroid.

## Control atop the Kidneys

The adrenal glands sit atop both the kidneys, controlling a variety of body functions. The adrenal cortex, the exterior portion of the gland, secretes steroid (fatty-based) hormones

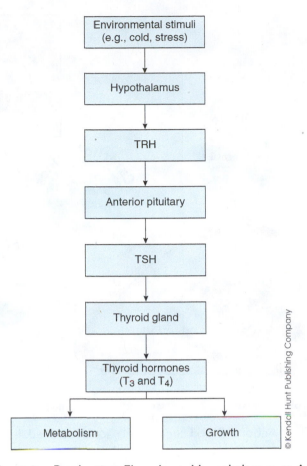

**Figure 14.40**    Thyroxine Production Flowchart: Hypothalamus -> Anterior Pituitary -> Thyroid. Thyroid hormones limit their own production by a negative feedback mechanism. Any problem in one process within the thyroxine production sequence may lead to a thyroid disorder.

including **mineralocorticoids** that control mineral and water balance; **glucocorticoids that** help regulate glucose levels; and **gonadocorticoids** that contribute to female sex drive. All of these activities are under the control of the pituitary gland that directs the activities of the adrenal cortex.

The adrenal medulla, the inner part of the gland, is like a knot of nerves, connected with the sympathetic nervous system described earlier in this chapter. When its nerves are activated, the cells secrete epinephrine and **norepinephrine**, both of which initiate a fight-or-flight response. Stimulation of the sympathetic nervous system increases the heart rate, blood pressure, and short-term available energy. It is the fastest acting of the endocrine glands because it is partially nervous tissue. The adrenal glands and their actions are shown in Figure 14.41.

## Pineal Gland

Have you ever wondered why babies are so sleepy? Why do you become more tired at night instead of during daylight? The pineal gland, found deep within the brain, produces melatonin. Melatonin is a hormone that makes a person sleepy. The more active the pineal gland is, the more melatonin is produced.

**Adrenal medulla**

The inner part of the adrenal gland.

**Epinephrine**

A hormone secreted by adrenal medulla.

**Pineal gland**

A small gland located deep within the brain.

**Melatonin**

A hormone produced by the pineal gland that makes a person sleepy.

# ADRENAL GLAND
(hormones)

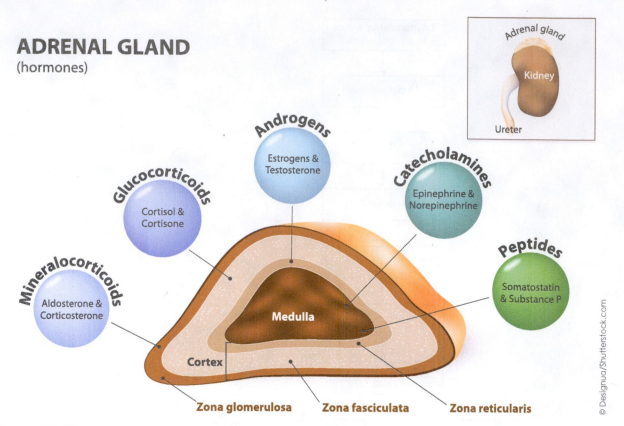

**Figure 14.41** Adrenal Glands and Their Actions. Blood sugar levels are controlled, in part, by the action of the adrenal glands.

Babies have very active pineal glands. Pineal glands also become activated in darkness, producing more melatonin. As we age, our pineal glands decrease in activity, with a person of age 80 requiring only 5 h of sleep compared with a growing child, who needs over 12 h. The chart in Figure 14.42 shows how sleep needs change with age.

## Reproduction

The gonads, ovaries, and testes are also endocrine glands. Each gland secretes hormones: male testes produce androgens, the class of male hormones responsible for male sex characteristics and sex drive; female ovaries produce estrogens, the class of female hormones responsible for female characteristics and sex drive. Both males and females have androgens and estrogens, just in different amounts. Males have more androgens and females have greater proportions of estrogens. Reproductive hormones will be discussed in greater detail in Chapter 16.

## Pheromones

**Pheromone**

A chemical that travels between different organisms to interconnect them.

Communication between organisms rather than within them occurs frequently in nature. Chemicals called pheromones travel between different organisms to interconnect them. Ants communicate with each other, and conduct complex wars and food recruitment, through use of pheromones. Human females, within the same dorms, have been shown to exhibit reproductive cycles in tandem with each other, due to pheromones. Consider the complex actions of ants, shown in Figure 14.43, all directed by pheromones.

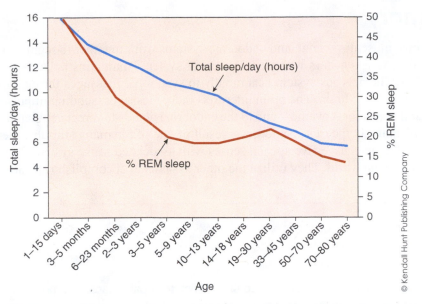

**Figure 14.42** The Amount of Sleep a Person Needs Decreases with Age. In the course of a lifetime, human sleeping time decreases by more than 50%. The pineal gland becomes less active as we age, causing less melatonin and fewer hours of sleep.

**Figure 14.43** Ants Move in a Line to Carry Out Activities, such as Food Procurement. Pheromones are used to communicate and coordinate activities in ants.

## Pain and Paracrine glands

Many tissues produce substances that diffuse from one tissue into another tissue. These chemicals are called paracrine regulators, which bind to receptors on neighboring cells to elicit a response. Prostaglandins are a type of paracrine regulator that affects inflammation and pain in tissues. Prostaglandins were discussed in Chapter 11 and its role in pain perception.

Aspirin and other pain medications inhibit the enzymes that help to produce prostaglandins, reducing pain and inflammation. These drugs can damage linings of the stomach and cause bleeding. In Richard's case, weighing the pros and cons of pain medication is a difficult task.

**Paracrine regulators**

Chemicals that bind to receptors on neighboring cells to elicit a response.

# Summary

The nervous, musculoskeletal, and endocrine systems work together to regulate life processes. Homeostasis occurs when the concert of systems maintains balance within the human body. The nervous system acts rapidly to send ionic messages through the body and ultimately to the brain. The brain interprets these messages, sending impulses to the muscles and bones for movement and to the endocrine glands to direct cellular activity in target cells. Pain, a type of sensation, is a result of one of the many stimuli experienced by humans. Humans have five special senses with associated sense systems to perceive and respond to the world. They utilize the other systems to accomplish regulation.

---

## CHECK OUT

### Summary: Key Points

- Medical treatments, such as surgery, physical therapy, and drugs, seek to repair anatomical problems, but alternative medicines do not intervene in injuries, and instead rely on holistic strategies for pain relief.
- The nervous system is organized into central and peripheral divisions, with several subgroups. Nerve impulses are a flow of positively charged ions, transmitted by neurotransmitters across synapses.
- The five senses: gustation, olfaction, vision, hearing, and sensation each begin by stimulation of special receptors sending nerve impulses to the brain for interpretation.
- The brain developed from the simple central core, then a more complex limbic system and then higher brain centers in the cerebrum.
- The sliding filament theory shows that muscle fibers slide past each other when stimulated by nerves, resulting in a muscle contraction.
- The bones of the human skeleton are divided into an axial and appendicular skeleton.
- Endocrine glands produce hormones that regulate various activities in the human body.

---

## KEY TERMS

| | |
|---|---|
| action | calcitonin |
| action potential | carpals |
| adrenal glands, cortex-, medulla-, | cerebellum |
| all-or-none response | cerebrum |
| amygdala | central core |
| anvil | central nervous system (CNS) |
| apoptosis | chemoreceptor |
| appendicular skeleton | ciliary body |
| aqueous humor | coccyx |
| arachnoid space | cochlea |
| articulation | compact bone |
| autonomic nervous system | cones |
| axial skeleton | cornea |
| bipolar cells | corpus callosum |
| brain | cranial bones |
| brainstem | critical threshold potential |

sympathetic nervous system
synapse
target cell
temporal lobe
thalamus
thermoreceptors
thyroid-stimulating hormone (TSH)
thyroxin
trabeculae

troponin
tropomyosin
twitch
vertebrae, cervical-, thoracic-, lumbar-
vitreous humor
vomer
Wolff's law
yellow marrow
zygomatic bone

# Multiple Choice Questions

**Reflection questions:**

1. It is hypothesized that acupuncture works by stimulating:
   a. sensory neurons
   b. motor neurons
   c. large diameter nerve fibers
   d. small diameter nerve fibers

2. Which term(s) does NOT fit with the sympathetic nervous system?
   a. Fight-or-flight
   b. Somatic
   c. Autonomic
   d. Peripheral

3. A nerve fires when it hits _____, which is defined as the _____ threshold potential.
   a. −55 mV; critical
   b. −70 mV; resting
   c. +70 mV; impulse
   d. 0 mV; resting

4. Changing from wave energy to mechanical energy, by movement of hairs, is accomplished in:
   a. vision
   b. hearing
   c. gustation
   d. olfaction.

5. Which type of receptor triggers gustation sensation?
   a. Thermoreceptor
   b. Mechanoreceptor
   c. Chemoreceptor
   d. Photoreceptor

6. Which represents a logical order, from oldest to most recent, in the development of the brain over evolutionary time?
   a. Cerebellum → medulla → hippocampus → cerebrum
   b. Medulla → cerebellum → hippocampus → cerebrum
   c. Hippocampus → medulla → cerebellum → cerebrum
   d. Cerebrum → medulla → hippocampus → cerebellum

7. Which swivels in the powerstroke step, during the sliding filament theory?
   a. Actin
   b. Troponin
   c. Myosin
   d. Tropomyosin

8. Which bone is NOT a part of the axial skeleton?
   a. Frontal
   b. Humerus
   c. Occipital
   d. Temporal

9. Which correctly MATCHES an endocrine gland with its hormone?
   a. Adrenal medulla –- thyroxin
   b. Thyroid – parathyroid hormone
   c. Epinephrine – adrenal cortex
   d. Pineal – thyroid-stimulating hormone

10. Which decreases as person ages and is associated with sleepiness?
   a. Melatonin
   b. Thyroxin
   c. Mineralocorticoids
   d. All of the above

## Short Answers

1. Medical treatments, especially in the treatment of pain, always have uncertainty in their outcomes. Give two reasons for this uncertainty in medicine.

2. Define the following terms: sympathetic and parasympathetic nervous system. List one way each of the terms that differ from each other in relation to their: a. organization in the nervous system; b. function; and c. relationship with each other.

3. What is the point of the corpus callosum in brain functioning?

4. Draw a sketch of the brain, using arrows to show five structures (or regions) of the cerebrum, central core and limbic system. What does the medulla oblongata regulate. Can a person live without one?

5. How does vitamin A help in the prevention and treatment of poor eyesight?

6. Pretend you are a light wave. Describe the pathway that you take after entering into the human eyeball. Name five structures that you pass through. What do you become, at the end of the vision process? Where is your final place in the body?

7. Define rigor mortis. Explain the role of ATP in this phenomenon.

8. For question #7, how do calcium ions affect rigor mortis? Why?

9. Describe the steps of calcium regulation in the body. Be sure to use the following terms in your explanation: calcitonin, PTH, thyroid, parathyroid, and osteoblasts?

10. Describe the symptoms of osteoarthritis and osteoporosis. How are the two conditions similar? How are they different? Which would you rather have?

## Biology and Society Corner: Discussion Questions

1. The research presented in this chapter shows that amygdala activity in the brain is associated with criminality. Should convicts, who are more prone to criminality, be treated differently (more leniently) because of their underlying predispositions? Why or Why not?

2. The use of night shifts is vital for many professional services: nursing, air traffic control, airline pilots, and military personnel, to name a few areas. Considering that the pineal gland acts to make us tired in the night, is it ethical to keep people awake every week on night shifts? To answer this, consider: What are the effects on the body of keeping someone on long-term night shifts? Does medical care quality suffer at night time hours? What could be done, if anything, to the problems associated with night shifts?

3. Bert's son, Bob, is short for his age. Bob is only 13 years old, but Bert wants his son to be taller than he was? He experienced a great deal of stress and bullying in school due to his height. The doctor predicts that Bob is projected to grow to 5 ft 4 in as an adult. Bob wants Bert to try Human Growth Hormone (HGH) to help him grow. Research the uses and side effects of HGH. Should Bob take HGH, given his case? Why or why not?

4. Creatine is used as a supplement to improve muscle mass building and endurance training. It is thought to add extra phosphate for ATP to aid in training. The International Olympic Committee, most professional sports leagues, and the National

Collegiate Athletic Association do not prohibit the use of creatine. Research the biology of creatine, its effect on training, and the side effects. On the basis of your research, would you take creatine in your own training? Would you recommend restrictions in creatine use by professional athletes? Why or why not?

5. Steroids mimic testosterone, a type of androgen hormone normally produced in the body. Steroids are banned in the Olympics and in professional sports. Steroid use builds muscle mass rapidly, giving energy and bulk to aid in muscle-building exercises. Construct an argument for and against the banning of steroids. Which factor gives you the most certainty in your argument?

Figure – Concept Map of Chapter 14 Big Ideas

# A War against the Enemy – Skin's Defenses and the Immune Attack

# 15

## ESSENTIALS

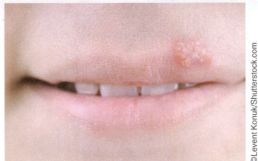
A college student has a cold sore

Students studying organic chemistry

College stress can be overwhelming

**ANTIBODY**

Antigen binding site — Antigens — Antigen binding site

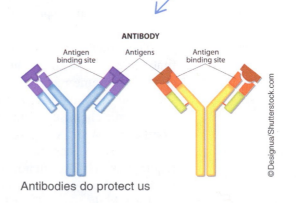
Antibodies do protect us

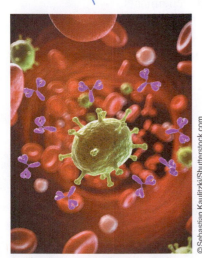
Antibodies attack viruses, such as herpes

## CHECK IN

**From reading this chapter, you will be able to:**

- explain how individuals with a contagious infection have been treated through history in society.
- list and describe the three lines of defense provided by the immune system.
- connect the structure and functions of skin with immune defense.
- describe and compare three general types of white blood cells of the immune system and connect them with the events of inflammation.
- compare and contrast cell-mediated and humoral immunity, connecting the events of each to defense against pathogens.
- explain the process of tissue regeneration and list, in a hierarchy, the tissues according to their ability to regenerate.
- list and explain the four types of acquired immunity.
- define and describe the functions of the lymphatic system.
- list the diseases occurring when the immune system malfunctions.

# The Case of the Recurring Chemistry Nightmare

We all met in the Commons, the name for our cafeteria at the college. The Commons was the center of social life, where we would eat all our meals, have coffee, and plan the parties for the week. It was also a place where you were on stage and where you were watched – studied, looked at, and analyzed – by your fellow students. It was a rumor mill, in short.

Rumors started with my smell. In organic chemistry laboratory, we used chemicals that left a scent on me. Whenever I went to the Commons after the lab session, I felt people noticed my odor. I remember the day when we produced amines in the lab; amines were one of the worst smelling functional groups of all the chemicals.

I had a dinner date that evening right after lab (no time to shower). I had looked forward to it for weeks. When the smell of amines sifted through the air in the Commons, I knew something was wrong. My date hurried home after eating very little of the dinner. We usually got along well, but we had not much to say to each other afterwards; and a rumor spread that I smelled. That smell rumor stuck with me through the years.

The larger problem of rumors for me spun out of the organic chemistry class, every time there was an exam. There were four exams in the class plus a final, each of which elicited the same recurring nightmare. But the nightmare was real and it cost me my social life. It should have been no big deal, except for the rumors.

Organic chemistry exams were stressful for me. It is true that I was strained when an exam came. While studying, I encountered reduction reactions, functional group movements, I had to keep track of electrons; but the worst were synthesis reactions: which required us to figure out how to make an organic compound from its constituents. I think the fear of these exams caused my recurring nightmare – herpes!

I would develop a fever blister on the same spot on my lips with every test. People watched me at the Commons, with my large lip blisters . . . rumors spread throughout campus that *I had contracted herpes*. It is true that the organic chemistry exam gave me herpes – but not the kind that was sexually transmitted, herpes simplex II. An organic chemistry exam cannot cause a sexually transmitted disease . . . although it can cause lots of other issues. Instead my cold cores were a result of the other form: herpes simplex type I, which is not sexually transmitted. Many people have it and it is obtained through touch or even when we are in our mother's womb. Stress can bring out the virus.

Herpes simplex I presents as a fever blister or cold sore at the edge of the lips. It is always within the body but rears its painful and ugly effects at certain times – during stress, illness, sunburn, for example – any weakening of the immune system. Usually the immune system – the white blood cells, antibodies, interferon chemicals – keeps the virus dormant. A herpes virus recognizes and attacks nerve cells at the skin of the lips, causing them to fire and give pain. Herpes, in me, remains quiet until my immune system weakens during organic chemistry exams.

Most people at the college did not know about the difference between the two types of herpes. I tried to explain it over and over that more than 80% of people have herpes type I. I don't think anyone was listening. Being on stage at the Commons led to unknown rumors about me; but is it all in my head? I am not sure of it, but am I being discriminated against because of my illness? I am hoping that this note may clear my name.

## CHECK UP SECTION

Was the watchfulness of the college community real or imaged by our character in the story? It is difficult to know. However, often throughout history, it is true that when illness spreads in a society, those who are deemed contagious are shunned. The plague, leprosy, and more recently AIDS sufferers have experienced discrimination due to their illness.

Research the types of illnesses spread in the past two centuries, in which its victims also felt social discrimination. Choose one of these illnesses and (1) describe how people suffered physically, (2) describe how ill people suffered due to society, and (3) make recommendations on how to prevent society-based discrimination due to illness in modern society.

## The Immune System's War

The story of herpes at the start of this chapter illustrates not only how our body is susceptible to pathogens for a long time but also how the society responds to those of us with these illnesses. The media and people are animated by stories of disease and contagion. But how do these stories affect our perceptions of disease in society?

In the news, a flu epidemic or a new strain of virus captures the headlines and public interest almost each day. Billions of pathogens including the viruses, bacteria, fungi, and parasitic protists (described in Chapter 8) continually inhabit the surfaces of our bodies and the items we touch. There are many agents of infection to talk about (and fear) from the news. A 2010 media report showed that 72% of shopping carts had fecal bacteria! But how susceptible are we?

The immune system acts as a very effective bulwark against all of these pathogens. The immune system has several layers in its line of defense. It should be thought of as a fight against invaders, using strategies and military operations similar to war. The battle is often bloody and pus filled, much like the results of a cold sore described in our story,

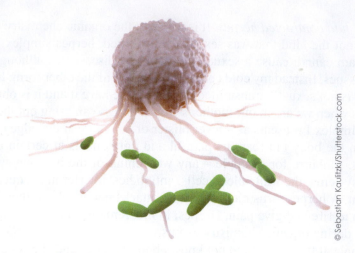

© Sebastian Kaulitzki/Shutterstock.com

**Figure 15.1** Immune Cells Are Actively Protecting Us in the Blood. The immune cell in the image is engulfing a number of enemy antigens. Defense of the body is much like a war, with tactics and troops (cells) defending their home turf.

or worse infections. Figure 15.1 paints the picture: infectious diseases surround us but are fought by the immune system traveling within our blood.

The human immune system includes the set of disease-fighting factors that protect against pathogens. These include three general lines of defense against the invading enemy. First, the **physical barriers** of the body provide protection by preventing pathogens from gaining entrance into the body. The skin and mucous membranes comprise most of this first line of defense.

After entering the body, pathogens encounter the second line of defense called **nonspecific immunity**. Most multicellular organisms have a nonspecific immune system. In this second line of defense, inflammation, fever, and chemical defenses operate to generally attack the invader. The name "nonspecific" denotes the fact that these processes do not target any particular intruders. For example, when *Staphylococcus aureus*, a common bacterium causing skin infections, occupies an area of the body, general inflammation and fever often occur. The immune system does not produce any substances aimed directly against *S. aureus*. Instead, it fights off the invaders by stimulating fever, inflammation, and phagocytosis, for example. These general strategies are employed by the immune system to kill any and all intruding pathogens without a specific targeting.

Immune cells do specifically attack certain pathogens in their last line of defense. They make cells and chemicals especially for certain invaders. The third line of defense is called **specific immunity**, during which the body employs methods to *specifically target* pathogens. It directs proteins called **antibodies** to bind with specific microbes. Antibodies stimulate white blood cells, produced and geared only for a particular pathogen. In our example of *S. aureus* infection, specific immunity produces specialized agents to target and destroy the bacterium. The three lines of immune defenses are delineated in Figure 15.2.

**Immune system**

A system that includes the set of disease-fighting factors that protect against pathogens.

# Physical Barriers: First Line of Defense

## Border Patrol: The Skin and Mucous Membranes

In our story, the skin is depicted as a site for attack from the inside, by the herpes virus. The skin usually functions as a defense from pathogens and other damaging agents on the outside of the body. The skin is the physical barrier, like "barbed wire" in trench warfare, which stops the battle before it begins.

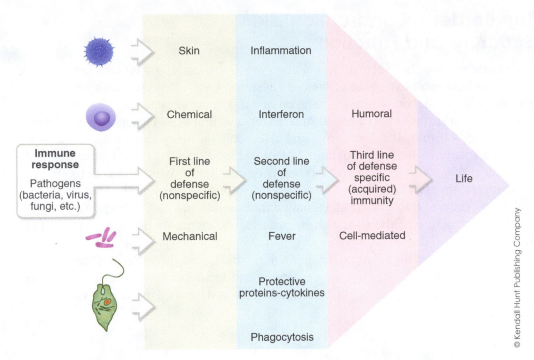

**Figure 15.2**    Three Lines of Defense against Pathogens: Physical Barriers, Nonspecific Immunity, and Specific Immunity.

Preventing war is the best form of defense, so pathogens that are kept out do not cause harm. Skin has multiple layers through which pathogens need to travel before gaining access to the body. The skin also has an acidic pH, roughly 3–4, which drives away most infectious microbes.

Saliva, tears, and mucous act as a border patrol. These secretions contain antibodies and other special chemicals that attack pathogens. Secretions are produced from modified areas of the skin, especially in regions most vulnerable to attack – our openings: eyes (tears), ears (wax), mouth (saliva), and the nose (mucous). The anus has resident bacterial defenders, discussed in Chapter 12. Openings are an easy conduit for pathogens to enter, but these secretions and bacteria both act as border patrol. Figure 15.3 shows some of the body's vulnerable orifices.

**Figure 15.3**    Orifices of the Human Body: Nose, Mouth, Eyes, Ears, and Anus. These are the vulnerable regions that may be breached during the first line of defense (nonspecific immunity) of the immune system. This one-week-old baby has a sensitive immune system, with orifices that may be easily breached.

# The Border's Construction: Skin Structure and Function

The skin is not simply a sheet of beauty upon which to gaze. It is a vibrant, active organ system and the most expansive of the body systems. Our college student's cold sore in this story shows that there is more going on in the skin than aesthetics.

Skin has a surface area of 15–20 square feet, a weight of 9–11 pounds, and a thickness of 0.5–4.0 mm. It is expansive and also contains many specialized structures, with one square inch of skin holding: 15 feet of blood vessels, 12 feet of nerves, 100 oil glands, 650 sweat glands, and thousands of circulating white blood cells. Within these layers, immune cells work to stave off infections such as herpes. First, the physical layers prevent microbes from gaining a footing into the human body. Figure 15.4 shows the structure of the skin with its dead cells continually being produced in its surface.

Human skin has three layers: the epidermis, which is on the surface of the skin; the dermis, which is directly below the epidermis; and the hypodermis, the deepest layer. The layers are held together tightly by cell junctions, as described in Chapter 3. However,

**Epidermis**

The outer layer of the skin.

**Dermis**

The middle layer of the skin, containing most of its organs and sense receptors.

**Hypodermis**

The deepest layer of skin.

Anatomy of the Epidermis

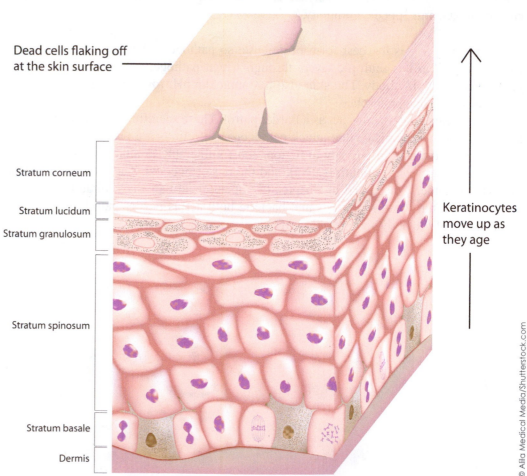

Dead cells flaking off at the skin surface

Stratum corneum

Stratum lucidum

Stratum granulosum

Stratum spinosum

Stratum basale

Dermis

Keratinocytes move up as they age

© Alila Medical Media/Shutterstock.com

**Figure 15.4** General Structure of the Skin Is Continually Renewing and Dead Cells Are Constantly Sloughing Off.

in some situations, such as wearing poorly fitting shoes while walking, a blister forms due to a separation of the three skin layers. Blisters from cold sores, as described in our story, break apart the layers of the skin due to the pressure from battle between herpes and the immune system.

## The Outside Border: Epidermis

The **epidermis** contains a series of five layers of flattened cells (Figure 15.4). Epidermal layers mostly contain dead cells filled with keratin, especially the outer layers. These cell layers become filled with granules of keratin protein, which is protective. From outside to inside, epidermal layers include the: stratum corneum, stratum lucidum, stratum granulosum, stratum spinosum, and stratum basale. You may remember them with a saying: "cells like getting sun burnt"; with the start of each word in the sentence also starting with the same letter as the layer names. Figure 15.5 gives a micrograph of the layers of the epidermis.

The outmost layer, the stratum corneum serves as the best protection from pathogens, with 20–80 cell layers. The cells in this layer are water repellant and easily flake off, making those inexpensive cells to lose or to protect us. The stratum basale is the only fully living cell layer. It is also mitotic and gives rise to the other epidermal layers. It gives rise to so many cell layers, and so continuously, that the skin would be six feet in diameter if stratum corneum cells did not flake off by the time we are 80 years old.

## The Inside Border: Dermis and Hypodermis

The **dermis** lies directly beneath the epidermis and contains most of the organs and living cells of the skin. It is separated from the epidermis by the dermal papillae, a wavy layer of the skin, which is also responsible for our fingerprints. Use Figure 15.6 as a guide to integumentary anatomy: the hair root and its hair follicle are used to detect insects and the arrector pili muscle helps hair to stand on end. Four enlarged receptors called corpuscles act to sense stimuli: the Pacinian corpuscle looks like an onion and senses deep pressure, Meissner's corpuscle is another receptor that senses light touch,

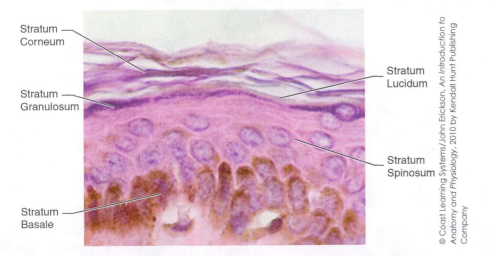

**Figure 15.5** Layers of the Epidermis. Each layer has specific functions and characteristics. The outer layer, the stratum corneum is thickest and visible to the eye.

© Coast Learning Systems/John Erickson, An Introduction to Anatomy and Physiology, 2010 by Kendall Hunt Publishing Company

**Keratin**

A protein that is the principal constituent of nails, hair, and skin tissues.

**Stratum corneum**

The outermost layer of the epidermis.

**Stratum granulosum**

A thin layer of granular cells in the epidermis located between stratum lucidum and stratum spinosum.

**Stratum lucidum**

A clear layer of dead skin cells in the epidermis located between stratum corneum and stratum granulosum.

**Stratum spinosum**

A layer in the epidermis located between stratum granulosum and stratum basale.

**Stratum basale**

The deepest layer of the epidermis, mitotic.

**Dermal papillae**

A wavy layer of the skin which is also responsible for human fingerprints.

**Arrector pili muscle**

Small muscles attached to hair follicles in skin.

**Hair follicle**

A structure from which hair grows.

**Hair root**

Part of hair embedded in a hair follicle.

**Meissner's corpuscle**

A receptor that senses light touch.

**Pacinian corpuscle**

A receptor that senses deep pressure.

**Ruffini's corpuscle**

A receptor that senses heat.

**Krause's corpuscle**

An bulbous cell that senses cold.

**Sweat glands**

A tubular gland that secretes sweat.

**Langerhans cell**

Special white blood cells that reside within the skin.

Ruffini's corpuscle (ending) senses heat, and Krause's corpuscle (bulb) senses the cold. Sweat glands and **sebaceous oil glands** produce their secretions, and blood vessels and nerves carry materials and cells through the skin to parts of the body.

Deep to the dermis, the **hypodermis** layer is primarily composed of adipose (fat) cells. Stored fat acts as a protection to absorb shock and cushion tissues it surrounds. Along the layers of the skin there are also special cells of the skin. Special white blood cells called Langerhans cells reside in the stratum spinosum are able to defend the layer if invaded. Keratinocytes produce keratin granules that infuse the layers of the epidermis to aid in protection. Melanocytes make the skin pigment called melanin, which gives the skin its darker tones and protects from ultraviolet light, as you might recall from Chapter 5. All of these cells work to provide the first line of defense in immunity.

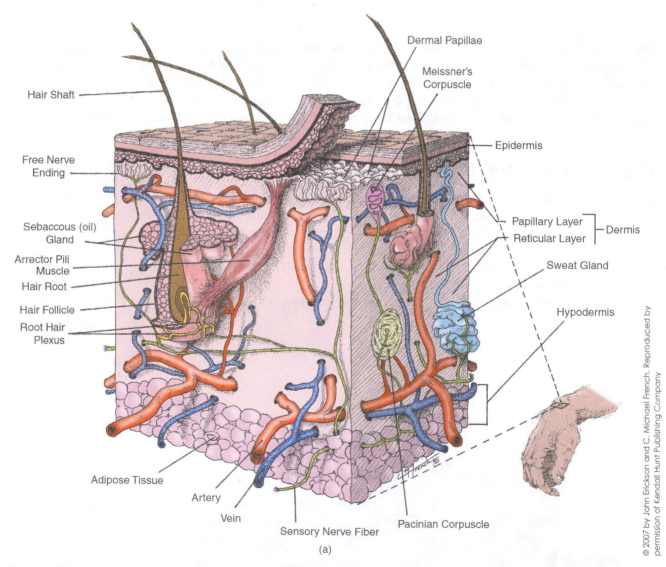

(a)

**Figure 15.6** a. The Human Integumentary System: Epidermis, Dermis, and Hypodermis Structure, along with Specialized Components. The dermis of the skin contains most of the organs and is responsible for many of its varied functions. Adapted from Anatomy I and Physiology Lecture Manual by John Erickson and C. Michael French. b. Sensory receptors of the skin.

SENSORY RECEPTORS IN SKIN

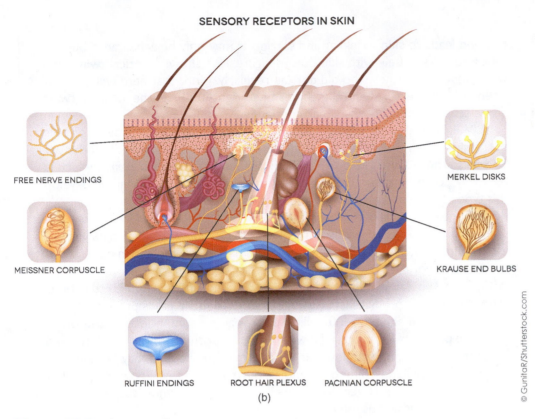

FREE NERVE ENDINGS

MEISSNER CORPUSCLE

MERKEL DISKS

KRAUSE END BULBS

RUFFINI ENDINGS     ROOT HAIR PLEXUS     PACINIAN CORPUSCLE

(b)

© GunitaR/Shutterstock.com

**Figure 15.6**    (continued)

**Melanocyte**

Cells that make the skin pigment melanin.

**Keratinocyte**

An epidermal cell that produces keratin granules.

---

### IS A TAN HEALTHY?

Skin color denoted social status since ancient society. In the time of Cleopatra, lighter skin tones were admired. However, by the 1920s, tanned skin became in style. A tan is a sign of good health and attractive in our society. Most of us know that it is associated with skin cancer but accept that aspects of darker skin tones are considered appealing. After all, celebrities show off their tanned bodies at the beach and the media is quick to give them attention.

Skin darkens when ultraviolet (UV) light strikes melanocytes in the skin. This stimulates melanocytes to make more melanin, a response to give protection. Melanin shields the nucleus of skin cells from the damaging effects of UV light. It was believed that tan skin protects from the effects of UV light and this is true. However, the way to get tanned skin is by exposure to the dangerous effects of UV light in the first place.

Thus, the more one is exposed to UV light, a source of radiation, the greater the damage to the integumentary system. UV light is clearly associated with an increased rate of aging of the skin. As you might recall from Chapter 11, collagen is a protein giving support to most areas of the body. Skin has an abundant amount of collagen ropes holding it together. When UV light strikes the collagen fibers, they form cross-linkages and become brittle. Collagen cross-linkages break apart, decreasing their amounts. Less and brittle

**Skin cancer**

A condition characterized by the abnormal growth of cells of the skin.

**Basal cell carcinoma**

A type of cancer, which is relatively common, but very rarely kills its victims.

**Melanoma**

The most dangerous form of skin cancer.

**Squamous cell carcinoma**

A type of cancer characterized by a flaky, reddened area.

collagen leads to skin wrinkling and sagging. It may only be appearance but UV light also links closely with skin cancer. Figure 15.7 shows identical twins, one who tanned and other who did not tan regularly. Note that here will be a difference in skin features and the effects of sunlight between the identical twins if one stops tanning but the other continues.

UV light is also associated with skin cancer, the abnormal growth of cells of the skin. The rate of skin cancer incidence has increased markedly in the last 80 years. There are three types of skin cancers: basal cell carcinoma, which is relatively common (almost one third of the US white population will develop this type of skin cancer) but very rarely kill its victims; squamous cell carcinoma, which appears as a flaky reddened area and is able to spread and kill; and melanoma, which is very dangerous and rapidly spreads (a few months) in its victims.

The incidence rates of malignant melanomas are on the rise in populations all over the world. This trend is alarming: they occurred for 1 person in 1,500 in 1930, to 1 in 250 in 1981, to 1 in 87 in 1996, to 1 in 79 in 1999. Projections from this pattern predict increases by roughly 7% more cases per year. Americans now spend almost $400 million each year on suntan lotions and other cosmetic products for tanning. However, a tan is nothing more than destroyed collagen and changing cells into cancer. Malignant melanoma may be detected by using the ABCD (asymmetry of moles, border irregularity, coloration differences, and diameter - larger than a pencil eraser makes a mole suspicious) rule for identifying suspicious moles.

## The Role of the Border: Skin Functions

Protection is not the only function afforded by the skin. The skin is considered an organ system because it is complex and performs numerous functions for the body. It is known as the integumentary system because it maintains the integrity of the body, covering it

**Figure 15.7**   If One of These Twins Stops Tanning, She Will Likely Have Less Wrinkles and a Decreased Risk of Skin Cancer in Years to Come.

but also managing many aspects of homeostasis. As shown in the previous section, the skin contains all of the tissue types as well as a variety of smaller organs. Together, these parts comprise the integumentary system and play a central role in:

1) **regulation of body temperature**: The skin contains blood vessels that are able to expand and contract, changing the amount of heat and blood to skin surfaces. When blood vessels dilate, more blood is brought to the surface, losing heat and lowering body temperatures. Humans also are capable of sweating, sometimes profusely. An average person sweats out 500 mL (a cup) per day but is able to sweat up to 12 liters (3 gallons)!

2) **excretion of wastes**: Nitrogen-containing wastes are removed from the kidneys, which will be discussed in Chapter 16. However, some nitrogen-containing wastes are excreted via sweat, with urea, uric acid, and ammonia components of sweat and all containing toxic nitrogen.

3) **blood regulation**: About 5% of blood remains in the integumentary system at any one time. The skin acts as a blood reservoir, sending blood to needed areas depending on a person's activities. For example, after eating or exercising, blood is sent to the digestive system or to the muscles, where it is needed.

4) **sensation of stimuli**: The corpuscles of the skin sense different stimuli. They are adapted to respond to the environment to send these messages to the brain, in ways described in Chapter 14.

5) **vitamin D synthesis**: When cholesterol present in the skin layers is exposed to UV light, it forms vitamin D, which is needed in the digestive tract to absorb calcium. When people are deprived of sunlight, their intake of calcium is impaired because of a lack of vitamin D. This problem is especially noticeable in the elderly population who are restricted to the indoors. Sunlight helps them to increase their levels of vitamin D and thus calcium needed for strong bones and teeth.

---

## HOLD THE SHOWER, PLEASE! . . . IS BATHING OR SHOWERING EVERY DAY RECOMMENDED?

Our first line of defense, the skin and its related membranes, are covered by a normal skin flora of bacteria and other microbes. Most of these bacteria are not harmful and in fact, fight off other more harmful strains of microbes.

Although most adults in the United States bathe or shower every day, it may not be necessary or recommended. First, compared with the past in which farming lifestyles required physical strength and sweating, modern society has led to more passive lifestyles and people get much less dirty. Second, bathing too frequently reduces the population of protective *Staphylococcus*-type beneficial bacteria on the skin surface. *Staphylococcus* occurs normally as a part of the skin microbiome, which includes over 1,000 species of bacteria covering our two square meters of skin surface. *Staphylococcus* makes up only 5% of the skin microbiome, with many other defending microbes protecting us. Protecting our skin by bathing and showering is vital in staving off infections. However, can we bathe too much?

Some studies claim that daily bathing may make people more susceptible to skin infections, including MRSA, mutant-resistant *S. aureus*. Figure 15.8 shows the superbug MRSA and an infection caused by its invasion of the skin.

Skin that is too frequently cleansed with soaps and shampoos removes protective oils from the surface. This causes dry skin, cracking, and openings through which microbes, such as the herpes virus in our story, may enter the body. Although bathing and showering are not ecofriendly and wastes valuable water, its effects on our health should also be considered. Is this a good case for being dirty?

**Jaundice**

A disease characterized by yellow coloration of the skin.

**Bilirubin**

A substance produced by the digestive system during the breakdown of RBCs.

## Malfunctions of the Border: Skin and Disease

Besides skin cancer, other skin changes also indicate illness. For example, the skin requires constant supply of blood for nutrients. When that supply is cut off, which occurs when lying in bed for too long, bed sores often develop. Bed sores, called decubitus ulcers, form dead tissue rather quickly when the blood supply is pinched by pressure from lying down.

Skin color is also important an indicator of a patient's health. When a yellow coloration of the skin is noted, as seen in jaundice, it indicates that the liver is not functioning properly. Bilirubin a substance produced by the digestive system should be broken down

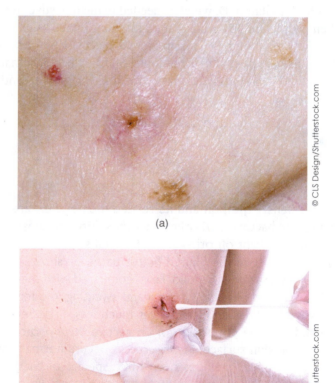

(a)

© CLS Design/Shutterstock.com

(b)

© Tibanna79/Shutterstock.com

**Figure 15.8** MRSA and Its Invasion of the Skin. a. MRSA first appears like a spider bite. b. As MRSA progresses, it creates an inflammation with pus oozing.

by the liver. When the liver malfunctions, bilirubin accumulates in the skin layers and results in the yellow coloration. When a liver disease occurs or when a baby's liver is temporarily immature upon its birth, jaundice is presented in a patient. Bronzing of the skin indicates that a person's adrenal glands are hypoactive, a symptom of Addison's disease.

## Internal Borders: Stomach and Respiratory Tract Defenses

When pathogens pass through the skin's protective borders, they encounter harsh conditions in the stomach and in the respiratory tract. The stomach, as you recall from Chapter 12, has a very acidic pH of between 2 and 3, killing most bacteria on foods. The respiratory tract is also lined with cells containing cilia, which beat upward to move pathogens out, as described in Chapter 13. When microbes do enter our lungs, specialized immune cells are able to phagocytize them in the next line of defense, nonspecific immunity. Let's take a closer look at the immune cells and their tactics during nonspecific immunity.

### IS THERE REALLY A 3-SECOND RULE FOR FOOD?

You should not pick up food after it falls to the floor, even if it is less than 3 seconds. When food touches the surface, either the floor or the other areas suspected of increased microbes, pathogens attach easily. Research from 2003 and 2006 studies showed that all of the foods falling to the ground have some level of contamination by microbes. Whether in 2 or in 10 seconds, all of the foods had significant amounts of *Escherichia coli* and *Salmonella* contamination.

Our surrounding surfaces have $10^8 = 100,000,000$ per square centimeter (about the size of your fingernails) of bacteria along with large numbers of protists, viruses, and fungi, all described in Chapter 8. Many of these microbes are associated with illnesses such as food poisoning and intestinal sicknesses. Stomach acidity and antibodies in our saliva fight off pathogens, but not always. That said, eating food from any floor or dirty surface is not recommended, even after an overview of the immune system defenses.

## Nonspecific Immunity: The Second Line of Defense

When the first line of defense fails, nonspecific immunity procedures take over the defense of the body. Immune cells called specific white blood cells (introduced in Chapter 13) conduct a military operation to (1) identify the invading pathogen, (2) recruit new immune cells to the infected area to help in the fight, and (3) attack and destroy the invaders. Tissue injury due to pathogens (but also physical trauma and harmful chemicals) causes necrosis or tissue death. When cells die as they are invaded by pathogens, they release chemical messengers to trigger white blood cells to respond to the invasion.

**White blood cells**

Blood cells that help body fight infections.

**Necrosis**

Death of tissue.

# The Start of Warfare: Inflammation

**Inflammation**

A series of events which identify, recruit and attack invading cells, causing swelling.

**Mast cell**

A type of immune cell.

**Histamines**

Chemicals that bring more blood to a site of infection by vasodilation of the vessels surrounding the area.

Do you recall stepping on a nail or getting a paper cut? Any disturbance of the border of our bodies, whether damage by a pathogen or due to a physical injury, elicits an inflammatory response. Inflammation is a series of events that identify, recruit, and attack invading cells. It results not only in the pathogen's destruction but also in redness, heat, swelling, and sometimes, pain. Figure 15.9 describes the steps in eliciting an inflammatory response to pathogens.

When an intruder, such as the herpes virus mentioned in the story, enters through the first lines of defense, damage to the cells occur. The start of warfare follows when damaged fibers are recognized by a type of immune cell called a mast cell. Circulating mast cells patrol the border looking for damaged fibers. When contacting out of place fibers, mast cells release two types of chemicals: histamines and heparin. In our story, the area of infection led to a cold sore, which felt inflamed and swollen due to this process.

Histamines are chemicals that bring more blood to a site of infection by vasodilation of the blood vessels surrounding the area. This causes more blood flow (and heat and swelling) in infected areas. More blood flow serves the purpose of increasing transport to the area. If more blood is directed to the site of infection, then more immune cells will arrive as well. Any military operation requires more troop transport to a region of attack, through roads and rails. Similarly, histamines open up the area to bring in more troops or in our example, immune cells.

Heparin is a chemical that acts as a blood thinner. It is also found in rat poison and medicines that slow blood clotting. When heparin is released into the site of infection in an

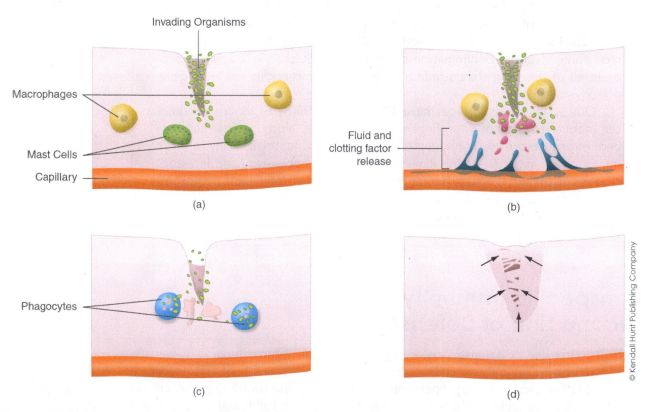

© Kendall Hunt Publishing Company

**Figure 15.9** (a) Nonspecific Immunity: Identifies, Recruits, and Attacks Pathogens. White blood cells (phagocytes) arrive at the site of infection and histamine and heparin are released simultaneously. (b) White blood cells (also called macrophages or phagocytes) attack pathogens at the site of infections. (c) White blood cells clean up the infection site by gobbling up dead cells and debris. (d) Mitosis replaces dead tissues.

area, the region is less able to clot blood. It might seem strange to slow clotting in an area of a cut, which would require a clot to stop its bleeding. However, the external area of the cut does continue to clot, in which there is little heparin. Instead, the site of battle, where the pathogen combats its immune cells, remains free from clots. This area allows free movement for immune cells to carry out their defense operations. Free mobility is vital in a battle, in which immune cells acting much like tanks are able to attack their enemy.

# The Tanks: Cells of the Immune System

## Neutrophils

White blood cells are phagocytes that attack in nonspecific immunity. They gobble up the invading pathogen by phagocytosis, as described in Chapter 3 and white blood cells shown in (Table 15.1). As seen in Figure 15.9, phagcytosis of pathogens is accomplished by cells that are the first to arrive at a site of infection, neutrophils. Neutrophils are white blood cells that quickly ingest pathogens. They push through the endothelial layers of capillaries and enter the site of battle. Neutrophils attempt to contain the invader before it has time to divide and colonize an area.

Neutrophils act like "fast tanks," with rapid speed in an area of infection (they arrive within 1–2 hours after an invasion). They are cheaply built and rapidly made – and they are self-destruct units, destroying themselves after they ingest pathogens. Neutrophils only live 3–5 days and are found in high numbers during bacterial infections. Neutrophils make up between 50% and 70% of white blood cells. High counts of neutrophils in the blood indicate an acute bacterial infection.

## Macrophages

Macrophages are the largest (from the term "macro") of the white blood cells, developed from monocytes (both shown in Table 15.1). They are expensive to build and slow to arrive at the site of infection. They are the slow moving but "powerful tanks" of the immune system. Although they are slow to arrive, they make a big impact, carrying out large-scale phagocytosis to engulf their enemies.

Macrophages engulf whole pathogens and display the pathogen's parts, called antigens, on their surface. Antigens are any molecule or cell part that initiates an immune response. The display of antigens is called macrophage presentation. Presentation of

**Neutrophils**

A type of WBC that are the most abundant in mammals and are first to arrive at an invasion.

**Macrophage**

The largest of the white blood cells.

**Antigens**

Any molecule or cell part that initiates an immune response.

**Macrophage-presentation**

The display of antigens by a white blood cell.

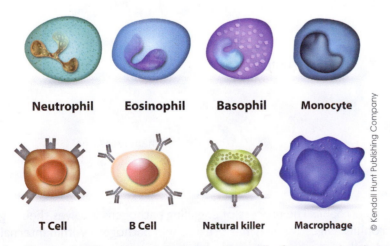

Neutrophil    Eosinophil    Basophil    Monocyte

T Cell    B Cell    Natural killer    Macrophage

© Kendall Hunt Publishing Company

**Table 15.1**    Types of White Blood Cells, the "heros" of immune battles

the pathogen helps other immune cells to identify the invader. For example, in our story, the narrator's herpes simplex I viruses would be ingested by macrophages and their antigens displayed on the surface. All other immune cells would "see" this as an example of what to search for and destroy as an enemy in the body. This process will be discussed further in the next section on specific immunity in this chapter.

Macrophages do not die upon phagocytosis of their enemy. They live for months or even years, circulating through the blood to ready for the next encounter. They are often engaged in heavy battle, phagocytizing invading cells. Macrophages make up only 5% of white blood cells.

Macrophages are often found in pus, a yellow fluid emerging from a site of infection. Pus indicates that a battle was serious and contains parts of damaged pathogens and macrophages. In our story, the character's blisters form from a heavy battle between herpes viruses and macrophages. The pus within cold sores is composed, in part of giant cells, which are enlarged macrophages. Giant cells become so large, filled with their enemy, hence their name. The process of phagocytosis by neutrophils is shown in Figure 15.10.

When defense against an intruder fails, the body walls off an area called an abscess. Abscesses can break apart and become dangerous when its infection spreads. The way to treat any abscess is to drain the pus within it and administer antibiotics. Pimples

**Pus**

A yellow fluid emerging from a site of infection.

**Giant cells**

Enlarged macrophages.

**Abscess**

Collection of pus that builds within the tissue of the body.

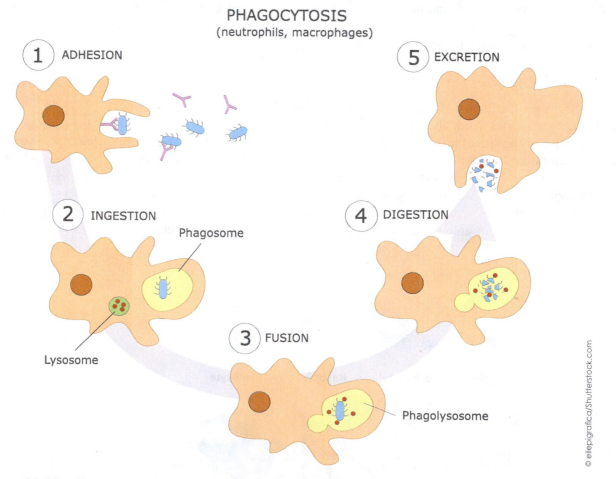

PHAGOCYTOSIS
(neutrophils, macrophages)

1 ADHESION

5 EXCRETION

2 INGESTION

Phagosome

4 DIGESTION

Lysosome

3 FUSION

Phagolysosome

© ellepigrafica/Shutterstock.com

**Figure 15.10** Phagocytosis by a Neutrophil. The process of engulfing pathogens involves their digestion by lysosomes in vacuoles within neutrophils. Digestive enzymes break down pathogens within internal vacuoles and eventually expel them from the cell.

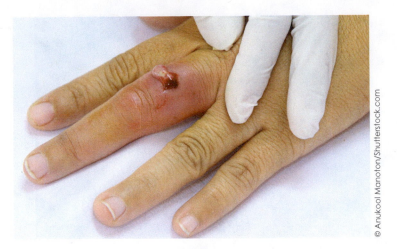

© Anukool Manoton/Shutterstock.com

**Figure 15.11** Abscesses and Pus. Abscesses are walled off infections and the body isolates it to fight it.

on the skin are examples of minor abscesses, in which *Staphylococcus* (discussed in Chapter 8) bacteria invade hair follicles. An example of a skin abscess is shown in Figure 15.11.

## Lymphocytes

Neutrophils and macrophages work in the nonspecific immunity division, gobbling up whatever pathogens they encounter. However, lymphocytes are white blood cells that work to specifically target invaders. Lymphocytes are thus classified as specific immunity. They are the "special forces-type tanks" of the immune system able to specifically target pathogens.

    Lymphocytes are the smallest of the white blood cells but contain the largest nucleus. They are the brains of the immune cell operation; their nucleus directs complex specific-immunity methods. They are in the last line of defense and are protected from the rigors of battle along the border of the body. Lymphocytes are expensive to build and remain in the body for many years, giving long-term immunity to illnesses.

    There are two types of lymphocytes: **B cells** and **T cells** (both shown in Table 15.1). Both types of white blood cells are produced in the bone marrow. T cells leave the bone marrow to mature in the thymus (hence the name "T" cell) and B cells remain, maturing in the bone marrow (standing for "B" cell). B cells, T cells, and antibodies, along with their attack strategies, will be discussed in the subsequent sections of this chapter.

## Chemical Warfare

Both neutrophils and macrophages secrete chemicals to attack the invading pathogens. The secreted chemicals, hydrogen peroxide ($H_2O_2$) and hypochlorous acid (both components of household bleach), are deadly to bacteria and fungi. When $H_2O_2$ is secreted from neutrophils during the height of its battle with a pathogen, it releases small spears along with the chemical in what is termed a respiratory burst. The spears cut into a pathogen's cell membrane creating damaging holes. Figure 15.12 will help you visualize the defense strategies of neutrophils.

    Chemical warfare also occurs when interferons are released from dying cells. Interferons are small proteins that bind with receptors on neighboring cells. It is released in small amounts by the dying cells and is the chemical weapons used in immunity.

**Lymphocyte**

White blood cells that work to specifically target invaders.

**Specific-immunity**

The third line of defense in the human immune system.

**Respiratory burst:**

The rapid release of hydrogen peroxide and superoxide radical from neutrophils.

**Interferon**

Small proteins which bind with receptors on neighboring cells.

© stevemart/Shutterstock.com

**Figure 15.12** $H_2O_2$ (Hydrogen Peroxide) Released from a Neutrophil. Many terms describe the vicious attack by the respiratory burst. In this image, peroxide shoots out of a neutrophil like a gas stream hitting the enemy antigen.

Interferons cause neighbor cells to produce antimicrobial proteins and also recruit more white blood cells to the site of invasion.

**Pyrogene**

A chemical that causes fever.

Some chemicals called pyrogenes cause fever, a natural and needed response by the body. Fever increases the average body temperature in response to disease or illness. Pathogens such as bacteria work best at body temperature. When pyrogenes activate, warmer body temperatures slow growth and inhibit their survival. Pyrogenes initiate a "slash-and-burn strategy" to defense, by raising temperatures to destroy pathogens and sometimes our own body cells as collateral damage. A very high fever (105°F) is dangerous because our own proteins may denature (unfold) and cease functioning.

Taking anti-inflammatory medicines and aspirin help to lower a fever. But do they also interfere with the military operations of the immune system? A study showed that those patients taking aspirin took longer to recover from chicken pox than those taking a placebo.

## Specific Immunity: The Third Line of Defense

The final line of defense operates carefully to match their substances with invaders. **Specific immunity** occurs when the immune system specifically targets certain types of pathogens. It directs the body to make materials and gear them for attack against a particular antigen. Specific immunity is the final and most precise phase of the immune response.

Macrophage presentation (shown in Figure 15.13), introduced in the previous section, starts specific immunity. The macrophage holds antigens from their engulfed invader, on the surface of its cell membrane. The macrophage acts much like a flag-carrier during a battle. The flag is the enemy antigen, shown on the macrophage membrane. Its specific shape informs all other immune cells to target specific pathogens. Immune cells can then search for and destroy only those cells. For example in our story, if the herpes virus antigens are presented, a specific, tailored immune response will be initiated.

There are two types of specific-immunity strategies:

**T-helper cell**

A specific type of T-cell that attaches to the macrophage to start specific-immunity.

1) **Cell-mediated Immunity**. Macrophages that present immediately attach to other immune cells. The attachment is mediated by the specific fit of the shape of the presented antigen and the receptors on these immune cells. A specific type of T cells, called a T-helper cell, attaches to the macrophage to start specific

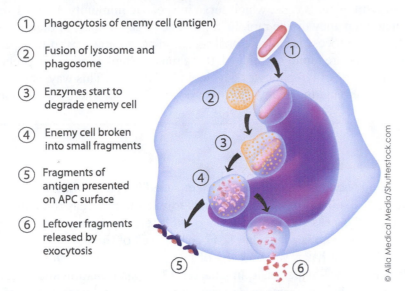

**Role of an Antigen-Presenting Cell**

1. Phagocytosis of enemy cell (antigen)
2. Fusion of lysosome and phagosome
3. Enzymes start to degrade enemy cell
4. Enemy cell broken into small fragments
5. Fragments of antigen presented on APC surface
6. Leftover fragments released by exocytosis

© Alila Medical Media/Shutterstock.com

**Figure 15.13**   A Macrophage "Presents" the Antigen That It Engulfs. It then places the antigen on its cell surface, which attract T-helper cells. T-helper cells activate other cells of the immune system to initiate both cell-mediated and humoral immunity.

immunity, as shown in Figure 15.15 and 15.17. T-helper cells stimulate other cells to target invaders.

The use of T cells in specific immunity is called cell-mediated immunity because *cells* drive the processes. In cell-mediated immunity, T-helper cells send out chemical messengers when binding to macrophages. T-helper cells stimulate other T cells, called cytotoxic T cells (or killer T cells) to kill specific invaders. T-helper cells are shown attaching in Figure 15.15. Cytotoxic T cells also attack cancer cells in our body. (see Figure 15.17 for an overview of cell-mediated immunity). Our immune system plays a role in fighting off many diseases – perhaps more than we know – which affects our health. Direct evidence for the role of the immune system in suppressing cancer is shown in Figure 15.14. In this figure, a T cell directly attacking a cancer cell is shown.

**Cell-mediated immunity**

An immune response that is based on on antigen-specific T lymphocytes.

**T-cell**

A type of lymphocyte that matures in the thymus.

**Cytotoxic T-cells**

Killer T-cells stimulated by T-helper cells to kill specific invaders.

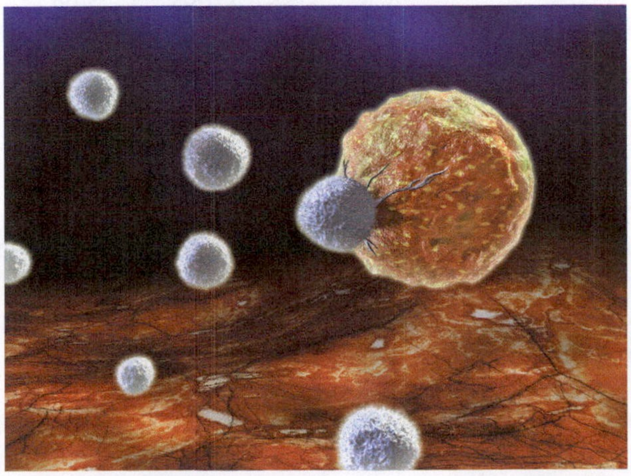

© Andrea Danti/Shutterstock.com

**Figure 15.14**   A Cytotoxic T Cell (shown in grey) Attacks Cancer Cell. This is direct evidence for the importance of the immune system in fighting and preventing cancer.

T-helper cells also cause other T-helper cells to divide more quickly, forming new ones. They also stimulate B cells, which serve in specific immunity. When T or B cells encounter an antigen, they divide rapidly to produce large numbers of their own type of immune cells. This process is called clonal selection because clones of the specifically needed T and B cells are rapidly produced. It is a massive tank production process, with many of the same type of lymphocytes developing all at once. This way, very few T and B cells need to be circulating at any one time, saving the body energy to produce them.

**Clonal selection**

The process by which T-cells or B-cells divide rapidly to produce large numbers of their own type of immune cells on encountering an antigen.

---

### MHC PROTEINS: KNOW THY ENEMY!

All cells of our body have antigens on them called **MHC (major histocompatibility) proteins**. These are indicators to tell immune cells the difference between invading cells and those of our own. It tells the immune system who thy enemy is and who is a friendly "self" cell (or cell of the body).

If a foreign cell has a different set of MHC proteins, the immune system will attack it; using T-helper cells. In Chapter 13, organ donation and rejection was discussed. MHC proteins in donated organs need to match those of the recipient. In our chapter 13 story, Charles was able to obtain his daughter's heart because they were genetically close enough in their MHC compatibility.

MHC proteins have also been shown to determine mating behavior between organisms. Mice will evaluate each other's MHCs to find a suitable mate. By smelling potential mates, they base their choice on how different a mate's MHC is from their own. Studies show that mice are choosy; selecting only those mates with different MHC proteins. This avoids inbreeding and its negative effects, described in Chapter 6.

---

**B-cell**

A type of white blood cell that produces antibodies.

**Memory cell**

Type of lymphocytes that continue to defend against pathogens long after they are gone.

**Humoral immunity**

A form of immunity where plasma cells and B lymphocytes produce antibodies.

**Plasma cell**

A type of lymphocyte produces antibodies.

2) **Humoral Immunity**. When T-helper cells combine with presenting macrophages, they send out chemical messengers. These messengers activate B-cell lymphocytes to become plasma cells. Plasma cells may also become memory cells, which are lymphocytes that continue to defend against pathogens. They may circulate in blood for years and even a lifetime to defend after exposure to an illness. Figure 15.15 describes how B cells develop to contribute to our immune response. How do these two cell types – plasma and memory cells – function to accomplish such a feat?

The answer is that they use antibodies to attack invaders. Humoral immunity occurs when plasma cells produce **antibodies** or proteins that bind with and attack invading antigens (antibodies are also called immunoglobulins). An overview of humoral immunity is given in Figure 15.17. They are the "fighter jets" of the immune system, specifically targeting their enemies. They also look like airplanes, with a set of wings and a cockpit. Unlike other methods, which harm many cell types, antibodies attack only those pathogens for which they are specifically made to fight against. They pursue certain targets much like military airplane searches for specific enemies.

The name humoral immunity originates from the Greek and Roman words for liquids in the body, also called "humors." Blood, phlegm, yellow bile, and black bile were the four humors of the ancients. Oddly, the ancients were on the right track – all of our humors including sweat, tears, milk, saliva, and blood – contain antibodies within them to defend us.

Antibodies specifically target pathogens and have a "Y" shape that, mentioned earlier, resembles a fighter plane. Antibodies have a specific shape on their "wings," onto

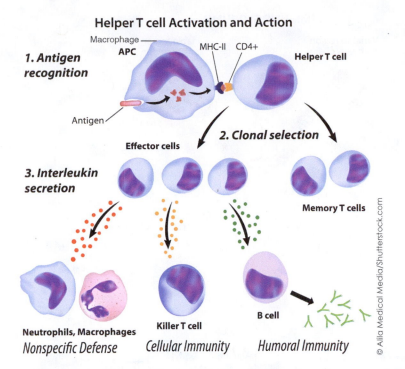

**Helper T cell Activation and Action**

**Figure 15.15** Humoral Immunity. T-helper cells stimulate B cells to specialize and make antibodies and memory cells. Humoral immunity uses antibodies to specifically attack invading pathogens.

which antigens attach. Memory cells circulate in the body long after an infection ends, producing antibodies whenever they are needed. Memory cells are antibody factories. When a new pathogen invades our bodies, it takes 7–9 days to start humoral immunity. When there are memory cells already in the blood for pathogens, it takes only 1–2 days to obtain enough antibodies to defend ourselves.

Antibodies are not really fighter jets – they are instead composed of four polypeptide chains, joined together by sulfur bridges. The wings of the airplane are made of variable regions, which vary from antibody type to antibody type, and **constant regions,** which are the same across antibodies. Constant regions allow the body to mass produce antibodies cheaply with the same parts. However, a variable region is necessary to give antibodies their specific fit with antigens. Antibodies have a three-dimensional shape, which attaches to specific antigens. The antibody structure and it functions are given in Figure 15.16.

When they attach to an invader, antibodies harm the pathogen by using a series of possible mechanisms. First, antibodies often coat the invading pathogen, to enable macrophages to attach more easily. This process is called opsonization, which also means "to make tasty." The four mechanisms of antibody action are to: (1) **neutralize** antigens, which cover their harmful parts (like muzzling a dangerous animal), (2) **agglutinate** (or clump) antigens together to help them be "seen" by white blood cells, (3) **precipitate** out antigens, which brings them out of dissolved form, also allowing immune cells to identify them, and (4) initiate a **complement** cascade of chemicals. Complement does not mean a positive remark (like you have a nice shirt on) but instead is a series of chemical reactions that blow holes in a pathogen's cell membrane. The goal of all four mechanisms of antibodies is pathogen cell lysis. In our story, the way in which the herpes cold sore virus is combatted is partially through the strategies employed by antibodies: they are clumped and neutralized to prevent their spread. Specific antibodies are made against herpes simplex I to conduct this immune-specific process.

**Variable region**

Regions that vary from antibody type to antibody type.

**Opsonization**

The process by which antibodies often coat the invading pathogen to enable macrophages to attach more easily.

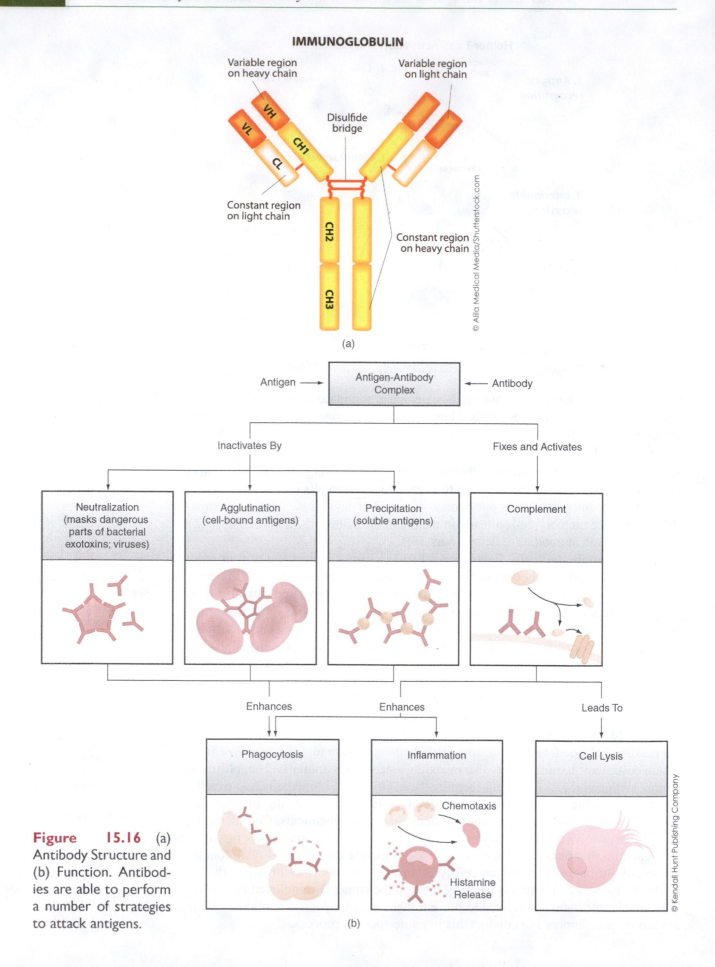

**Figure 15.16** (a) Antibody Structure and (b) Function. Antibodies are able to perform a number of strategies to attack antigens.

When an immune response ends, T-suppressor cells enter the field of battle. T-suppressor cells turn off the activities of the immune system. It acts to "demilitarize" the area, shutting down cells and chemicals that cause immune responses. T-suppressor cells end the immune response in an area and ready it for regeneration of new tissue to rebuild. Much like peace-keeping troops after a war, T-suppressor cells serve only to maintain order as rebuilding of the area takes place. Macrophages phagocytize final debris and white blood cells migrate out of the area. T-suppressor cells guide the final phase of the immune response to pathogens. Figure 15.17 shows the many interactions between cells of humoral and cell-mediated immunity.

**T-suppressor cell**

Type of immune cells "demilitarize" an immune response when an immune response ends.

**Fibrosis**

The scarring or thickening of connective tissue.

**Fibroblast**

Specialized cells that produce and maintain connective tissue.

**Regeneration**

Replacement of damaged tissues with an original tissue type.

# Rebuilding after the War: Regeneration of Tissues

The scars of battle lead to necrotic tissue and tissue injury. Regions affected are repaired and rebuilt after T-suppressor cells calm the immune response. There are two ways to rebuild damaged areas in the body: regeneration and fibrosis. Both require specialized cells called fibroblasts to lay down a network of collagen fibers on which to add replacement materials and cells. Then, capillaries invade the area, adding nutrients to bring further construction materials. However, regeneration and fibrosis differ in the kinds of cells reoccupying the damaged region.

Both types of responses begin after macrophages engulf pathogens that they encounter in the body, digest them, and display antigens from the pathogens on their surfaces. The presence of these antigens enables the macrophages to activate, or signal, specific helper T-cells.

**antigens:** specific molecules, such as the proteins that appear on the surface of pathogens, that elicit an immune response

**lymphocytes:** white blood cells that act in specific immune responses

**T-cells:** lymphocytes that mature in the thymus gland

**helper T-cells:** cells that help recruit more lymphocytes to defend the body

**lymphokines:** molecules produced by activated helper T-cells that stimulate other lymphocytes

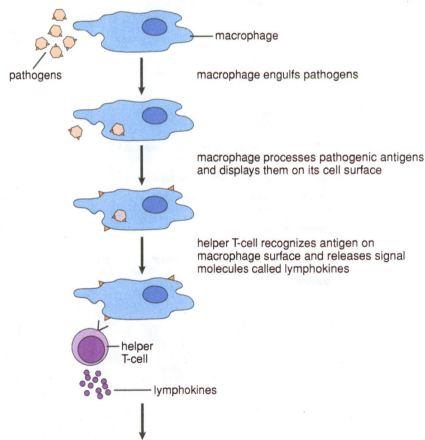

pathogens

macrophage

macrophage engulfs pathogens

macrophage processes pathogenic antigens and displays them on its cell surface

helper T-cell recognizes antigen on macrophage surface and releases signal molecules called lymphokines

helper T-cell

lymphokines

(a)

**Figure 15.17** An Overview of Specific Immunity. Macrophages, B cells, and T cells carry out immune functions in concert with each other. From *Biological Perspectives*, 3rd ed by BSCS.

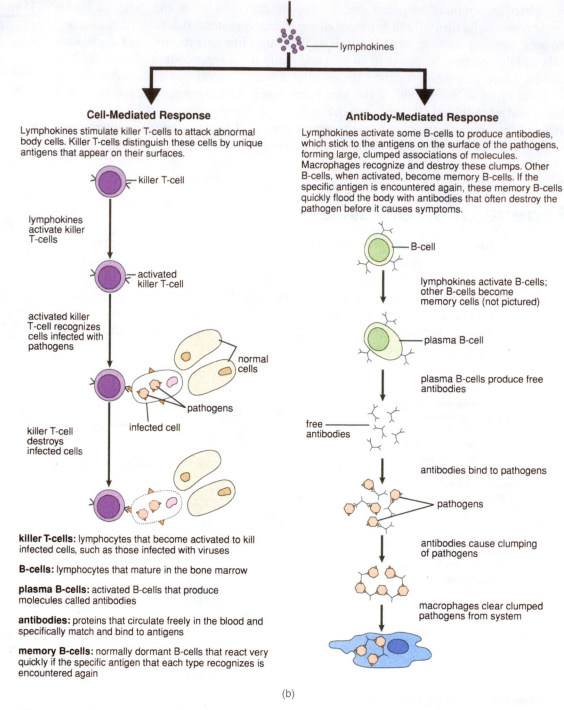

**Cell-Mediated Response**

Lymphokines stimulate killer T-cells to attack abnormal body cells. Killer T-cells distinguish these cells by unique antigens that appear on their surfaces.

killer T-cell

lymphokines activate killer T-cells

activated killer T-cell

activated killer T-cell recognizes cells infected with pathogens

normal cells

pathogens

infected cell

killer T-cell destroys infected cells

**killer T-cells:** lymphocytes that become activated to kill infected cells, such as those infected with viruses

**B-cells:** lymphocytes that mature in the bone marrow

**plasma B-cells:** activated B-cells that produce molecules called antibodies

**antibodies:** proteins that circulate freely in the blood and specifically match and bind to antigens

**memory B-cells:** normally dormant B-cells that react very quickly if the specific antigen that each type recognizes is encountered again

lymphokines

**Antibody-Mediated Response**

Lymphokines activate some B-cells to produce antibodies, which stick to the antigens on the surface of the pathogens, forming large, clumped associations of molecules. Macrophages recognize and destroy these clumps. Other B-cells, when activated, become memory B-cells. If the specific antigen is encountered again, these memory B-cells quickly flood the body with antibodies that often destroy the pathogen before it causes symptoms.

B-cell

lymphokines activate B-cells; other B-cells become memory cells (not pictured)

plasma B-cell

plasma B-cells produce free antibodies

free antibodies

antibodies bind to pathogens

pathogens

antibodies cause clumping of pathogens

macrophages clear clumped pathogens from system

(b)

**Figure 15.17** (continued)

When scar tissue, composed of fibrous connective tissue, replaces damaged regions, **fibrosis** occurs. Scar tissue does not carry out the functions of the original tissue replaced. For example, scar tissue that replaces heart muscle after a heart attack does not pump blood as its original cardiac cells would have.

Scar tissue can be harmful when it occupies areas of the body and gets in the way of normal functioning. In the example of scar tissue in the heart, it may occupy space in ventricles and limit the holding capacity of the heart. However, scar tissue does have benefits – it is infection resistant – which prevents further intrusion by pathogens in the areas it replaces. But, scar tissue is not desirable because it does not function as the original cells.

However, **regeneration** is the most desired strategy, which replaces areas with the *same type of tissue* that was destroyed. Regeneration occurs because cells that were lost are regrown or newly added with the same functioning tissue. This is optimal because the damaged area returns to its previously functioning state. Of course, regeneration depends on the extent of the damage to tissues and tissue type affected. When there is extensive damage to any tissue, regeneration is limited and areas are instead replaced with scar tissue. These incidences are obvious because they leave a scar, as with many surgical cuts and accidents.

Some tissues are less able, inherently because of their type, to regenerate. In heart damage, for example, cells do not readily regenerate and recovery is therefore limited. Some tissues easily regenerate, such as the skin that contains epithelial cells. Recall, when you last got a cut on your hand; it probably did not take more than a few days to note marked healing. But when a nerve gets damaged, as seen in our story of Richard in Chapter 14, its effects are seen for years and sometimes permanently. Figure 15.18 gives a hierarchy of regeneration based upon tissue type affected.

Epithelial and many connective tissues, such as bone and blood, readily regenerate given the proper nutrients and conditions. These are constructed of materials that are easier to build than those tissues that have poorer regeneration capacity. In our story, the cold sore victim heals quickly, after her epithelial regenerates. How do her other tissues fare?

Smooth muscles in organs and dense regular connective tissues, such as ligaments and tendons, have regularly arranged fibers. These are more difficult to lay down during replacement than the scattered cells of skin, for example. Cartilage, with no blood vessels to give them direct nourishment and to bring immune cells, is unable to regenerate and does worse than smooth muscles and regular connective tissue. In the story, none of these tissues were damaged, which is why healing is so quick from cold sores.

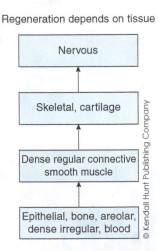

**Figure 15.18** Hierarchy of Regeneration of Tissues. The most regenerative tissues are at the bottom of the hierarchy (epithelial [skin] and connective [bone, areolar, and blood]), then dense regular connective (tendon and ligaments) and smooth muscle (e.g. on the stomach walls), then cartilage and skeletal muscle, and finally, the least regenerative tissues are on top (nerve and heart muscle).

Of course, the most difficult to regenerate are cardiac and nerve cells. In our story, the narrator's nerves were inflamed, causing pain associated with cold sores. However, nerve damage does not usually occur in herpes simplex I. When brain damage occurs, as in strokes, recovery requires retraining of remaining cells rather than regeneration of damaged nerves. Both cardiac and nerve cells are complex and difficult to build, in the first place. Thus, their regeneration is very limited.

Recent studies indicate, however, that some regeneration occurs in the heart and along nerves but very slowly. A peripheral nerve cell heals at a rate of one millimeter a month, cold comfort for Richard, our nerve pain sufferer in the story in Chapter 14.

## Preventing Future Attacks: Acquired Immunity

A vaccine is a medicine containing a weakened or dead pathogen or piece of a pathogen, which is injected into our bodies to start an immune response. Vaccines do not cause the disease but instead stimulate the body to produce antibodies to prevent from future attack. Vaccination is the series of medicines that are administered to prevent various illnesses and their spread through society.

Before vaccinations, the rate of death from infectious diseases were profoundly higher than today: In 1900, the infant mortality rate was upwards of 30% and today it is below 5%; in 1920, there were upwards of 200,000 cases of diphtheria and today only one; in 1952, polio paralyzed more than 20,000 people and today no one; and in 1985, there were 20,000 cases of children infected with *Haemophilus influenzae* type b, causing meningitis (12,000) and pneumonia (7,500) and in 2002, only 34 cases. Chicken pox, an infection that caused over 4 million cases per year, caused suffering with pustules and painful itchiness throughout the body. In 1995, a vaccine for chicken pox was introduced, cutting cases by over 75%. It saves people from more than 100,000 hospitalizations and some deaths every year. All of these advances are due to immunity derived from modern medicine: vaccines acting as aids to our immunity.

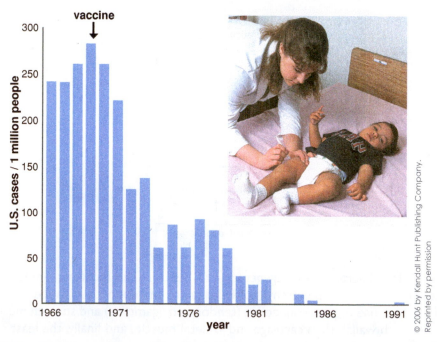

**Incidence of Rubella Before and After Vaccine**

**Table 15.2**   The effects of vaccines on Rubella incidence in the US. From *Biological Perspectives*, 3rd ed by BSCS.

**Acquired immunity** occurs when an organism obtains a prior-set of immune defenses to a pathogen *before* that pathogen enters the body. It is the immune system's military preparation for war before it happens.

There are four types of acquired immunity, depending upon the way in which the immunity was acquired. When a vaccine activates the immune system to produce antibodies in a response to it, it is termed active artificial **acquired immunity**. The chicken pox vaccine, in the example earlier, gives active acquired immunity to patients. Vaccinations are a part of artificial acquired immunity because a medical treatment (the vaccine) stimulates an immune response. It is not a natural process and is therefore called "artificial" immunity. Most vaccines carry out active acquired immunity, such as polio, measles, and influenza vaccines.

When a medicine gives immunity to a patient without stimulating their immune system, it is called passive artificial **acquired immunity**. In passive acquired immunity, medicines are composed of antibodies, which simply pass into the body. Rabies vaccination, the discovery described in the story in Chapter 8, uses passive acquired immunity to fight off rabies and prevent its infection. The patient's immune system does nothing, merely receiving the antibodies needed to combat the rhabdovirus.

Natural acquired immunity occurs without intervention from the medical profession – in other words, naturally and using the body's normal processes. In our story, the herpes infection leads to natural reactions but no long-term immunity. The story's narrator remains infected with herpes for life. Although this is often the case after an infection, sometimes immunity for the longer term may be acquired.

When the immune system defends itself and enables future defense against pathogens, it is termed active natural **acquired immunity**. The immune system actively produces memory cells and/or antibodies to guard against future attack. It engages in the battles described in this chapter, which ultimately leads to memory cells and circulating antibodies. Some natural immunity is long-term, such as chicken pox and measles, with memory cells lasting for years. Other illnesses, such as the gastrointestinal illness caused by the Norovirus, have an immunity lasting for only a few days.

Passive natural **acquired immunity** occurs when immunity is obtained naturally but without the work of the immune system. During nursing, for example, a child passively receives antibodies from mother's milk, giving immediate defense against pathogens. It is passive because the child's immune system does not activate to make antibodies; and it is natural because it is from a normal process, lactation.

## Active artificial

The condition that occurs when a vaccine activates the immune system to produce antibodies in a response to.

## Passive artificial

The condition that occurs when a medicine gives immunity to a patient without stimulating their immune system.

## Active natural

A type of immunization that occurs when the immune system defends itself and enables future defense against pathogens.

## Passive natural

The condition that occurs when immunity is obtained naturally but without the work of the immune system.

### DOES DRESSING WARM IN COLD WEATHER PREVENT THE COMMON COLD?

The rhinovirus is spread from person to person through contact or through airborne means. Chapter 8 showed that the common cold is caused by the rhinovirus. Staying warm does not physically prevent the rhinovirus from touching and attaching to our skin and mucous membranes. However, when a person takes care of their health, by eating right and dressing warm, the overall immune system stays stronger to fight off infections. Some studies show that cold weather may suppress the immune system. Suppressed immunity gives the rhinovirus better chances to gain a foothold in our bodies and invade cells. However, there is no direct link between keeping warm and preventing the common cold.

During cold weather, however, more people are driven indoors. Thus, the population density increases — as in the mall, in schools, or in dorms — which leads to a more rapid spread of the rhinovirus. Thus, during the winter, common colds (and other sicknesses such as influenza) occur at higher rates in the population.

**Lymph**

The fluid of the body that carries excess liquids and cells not normally transported by the circulatory system.

**Lymphatic system**

The series of vessels and their organs carrying lymph.

# Defense Stations: The Lymphatic System

Lymph is the fluid of the body that carries excess liquids and cells not normally transported by the circulatory system. The series of vessels and their organs carrying lymph is called the lymphatic system. The organs of the lymphatic system include: the spleen, lymph nodes, small intestinal areas, and thymus, depicted in Figure 15.19. Lymph nodes

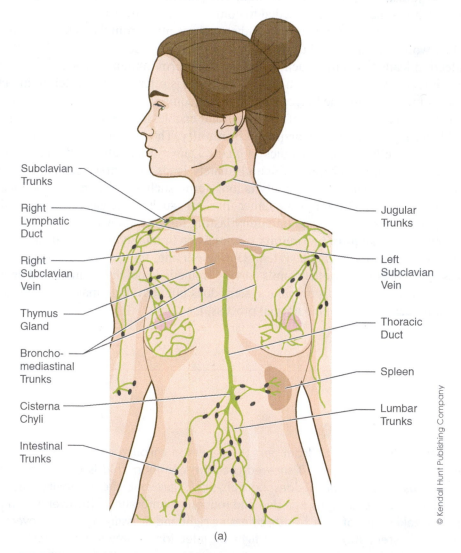

Subclavian Trunks

Right Lymphatic Duct

Right Subclavian Vein

Thymus Gland

Broncho-mediastinal Trunks

Cisterna Chyli

Intestinal Trunks

Jugular Trunks

Left Subclavian Vein

Thoracic Duct

Spleen

Lumbar Trunks

© Kendall Hunt Publishing Company

(a)

**Figure 15.19** (a) The Lymphatic System: Vessels and Organs of the System Are Found throughout the Human Body. (b) Lymph nodes are sites of immune defenses by white blood cells against invaders. White blood cells migrate between blood and lymph to defend areas of the body. From *Biological Perspectives*, 3ʳᵈ ed by BSCS.

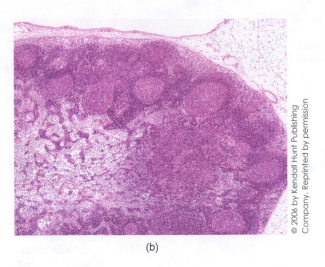

(b)

**Figure 15.19**  (*continued*)

act like "defense stations," strewn throughout the body and filled with defending white blood cells. Movement of lymph through lymphatic vessels does not occur with the pumping of a heart. Instead, muscle movements in the body push lymph fluids through toward the chest.

The lymphatic system plays a role in (1) immunity and (2) transport of fluids and nutrients.

**Immunity and Lymph**. White blood cells pack lymph nodes, as shown in Figure 15.19. These white blood cells phagocytize bacteria, viruses, and cancer cells removing them. Although lymph nodes serve as defense stations, white blood cells may defend against invaders at any point in the lymph vessels. Lymph nodes are armed at all times. During an illness, lymph nodes enlarge because heavy battle occurs at these defense stations. The tonsils, the largest lymph nodes in humans, enlarge to become swollen when the body fights a throat infection. In our story, herpes simplex I does not get noticed by white blood cells in lymph nodes, a reason there is little swelling of nodes during cold sore infections.

Lymph nodes serve to fight against cancer cells. When a cancer cell travels into a lymph node, white blood cells attempt to kill it before it can spread to other parts of the body. A failure at the lymph nodes leads to the spread of cancer.

When diagnosing stages at which cancer has spread, lymph nodes are removed to determine if cancer cells have entered the lymph. Classifying cancer into different stages is based on the size and location of the tumor as well as the extent of its spread. However, generally, if cancer cells remain localized and do not spread to lymph nodes it is classified as Stage I. If cancer enters the lymph nodes, cancer is at Stage II. This means that some of the cancer cells of the tumor invaded the blood, through which they travelled across lymph tissue. When cancer is at Stage III, it is found extensively in the lymph nodes. When it is classified as the highest Stage IV, it is discovered in distant areas of the body. The higher the stage, the more difficult a cancer is to treat. The stages, as applied to breast cancer, are given in Figure 15.20.

**Transport of Fluids and Nutrients**. As cells exchange materials with capillaries, diffusion occurs. When some fluid is lost to the interstitial (in-between cells) areas of the body, they are collected by the lymphatic system. When lymph vessels do not work properly, lymphedema may occur. For example, in deep vein thrombosis cases, described in Chapter 13, roughly 45% of patients experience damage to valves in their lymphatic vessels.

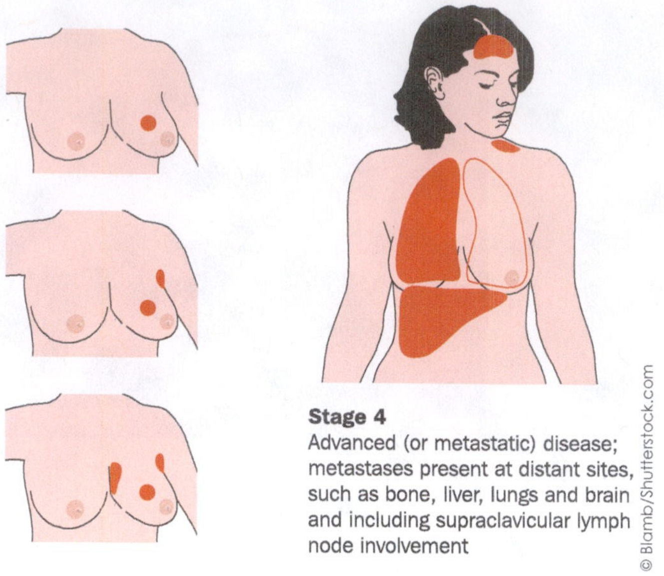

**Stage 1**
Early disease: tumour confined to the breast (node-negative)

**Stage 2**
Early disease: tumour spread to movable ipsilateral axillary node(s) (node-positive)

**Stage 3**
Locally advanced disease tumour spread to the superficial structures of the chest wall; involvement of ipsilateral internal mammary lymph nodes

**Stage 4**
Advanced (or metastatic) disease; metastases present at distant sites, such as bone, liver, lungs and brain and including supraclavicular lymph node involvement

© Blamb/Shutterstock.com

**Figure 15.20**    Breast Cancer, Stages I through IV. All cancer follows a similar classification scheme based on the extent to which it has travelled.

**Lacteal**

A lymphatic vessel of the small intestine that absorbs digested fats.

In an extreme case, mostly in the tropics, parasitic worms carried by mosquitoes cause infection leading to elephantiasis, or swelling of the extremities. In these cases, extremely large swellings of the arms and legs occur, as shown in Figure 15.21.

Lymph vessels also extend into the villi within the small intestines, in the form of lacteals. From lacteals, lymph vessels transport fats from the small intestines to the liver, where fats are further processed.

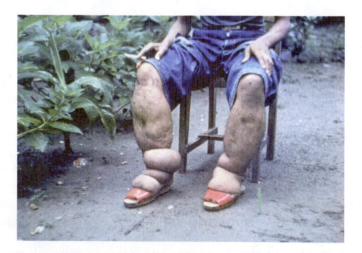

**Figure 15.21**    Lymphedema Patient: Leg Swelling. It occurs because lymph nodes that are damaged cannot transport excess fluids back to the blood vessels. As a result, swelling results from water retention in lymph vessels. CDC photo

# Malfunctions in our Immune Defenses

## Our Immune System Attacks its own Troops: Autoimmune Disease

When the immune system overreacts to its own cells, it is called an autoimmune disease. Antibodies attack their own cells during an **autoimmune attack**. Antibodies normally recognize foreign antigens, separating "self" from "nonself" in the way they identify structures. Recall that normally, antibodies attach to antigens and stimulate immune responses. However, when antibodies attach inappropriately to our own body cells, an immune response is initiated at the wrong time. Multiple sclerosis, rheumatoid arthritis, lupus, Graves' disease of the thyroid, and sarcodiosis all involve an autoimmune attack. Each is characterized by inflammation and tissue damage as a result of the immune system attacking normal body cells. Figure 15.22 depicts the deformations of joints in rheumatoid arthritis shows how disabling the disease can become.

**Autoimmune disease**

A disease in which the immune system overreacts to its own cells.

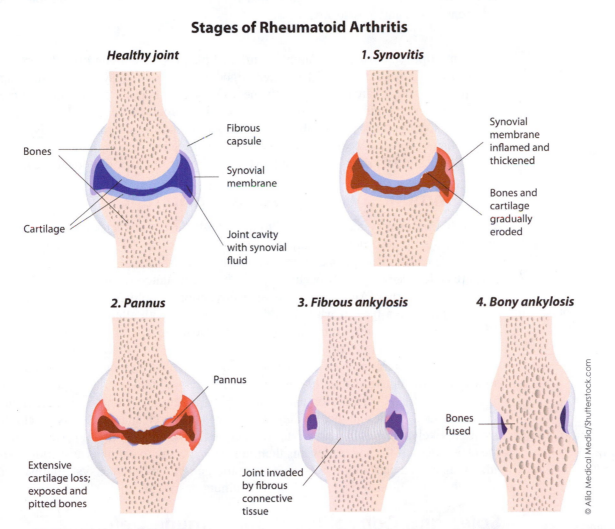

## Stages of Rheumatoid Arthritis

*Healthy joint*

Bones

Cartilage

Fibrous capsule

Synovial membrane

Joint cavity with synovial fluid

*1. Synovitis*

Synovial membrane inflamed and thickened

Bones and cartilage gradually eroded

*2. Pannus*

Pannus

Extensive cartilage loss; exposed and pitted bones

*3. Fibrous ankylosis*

Joint invaded by fibrous connective tissue

*4. Bony ankylosis*

Bones fused

© Alila Medical Media/Shutterstock.com

**Figure 15.22**   Rheumatoid Arthritis. Symptoms of joint damage include swelling, inflammation, and even dislocation of joints within affected areas.

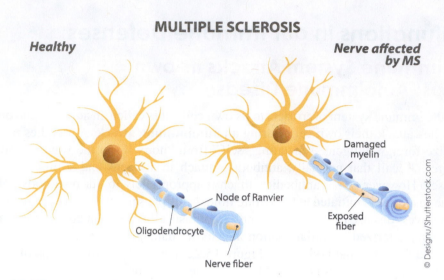

**MULTIPLE SCLEROSIS**

*Healthy*

*Nerve affected by MS*

Damaged myelin

Node of Ranvier

Oligodendrocyte

Exposed fiber

Nerve fiber

© Designu/Shutterstock.com

**Figure 15.23** Multiple Sclerosis (MS): Plaques on Oligodendrocytes Disrupt normal Nerve Transmissions. MS causes abnormal symptoms in patients such as double vision, odd sensations, and muscle weakness.

In multiple sclerosis, for example, hardened plaques form along oligodendrocytes. Figure 15.23 shows a normal and attacked oligodendrocyte. You may recall from Chapter 11, oligodendrocytes are neuroglia in the nervous system. Plaques prevent normal transmission of nerve impulses to the brain because this neuroglia is destroyed. Cell destruction leads to symptoms such as abnormal vision and sensations across the body. In later stages, there is muscle weakening and possible death. Some studies indicate that a virus may initiate multiple sclerosis, causing cytotoxic T cells to attack oligodendrocytes.

## Our Immune System Overreacts to Terror: Allergies

When the immune system overreacts to a "nonself" antigen, such as pollen, peanuts, or chemicals in latex gloves, an allergic reaction occurs. An allergy is an inappropriate immune reaction to antigens that are otherwise not harmful. Antigens that cause allergies are called allergens. They occur when the body encounters an antigen that should be recognized as normal. An allergy is like an overreaction, in politics, which leads to a war.

Basophils are a type of white blood cell associated with allergies (shown in Table 15.1). They send out histamines and heparin to cause the swelling, inflammation, and runny nose common in allergies.

Basophils determine whether or not a cell is one of its own or a foreign cell to be attacked. In the determination of "normal" and "foreign," childhood exposure is shown to be associated. When a person experiences pine pollen, for example, as a child, he or she is less likely to be allergic. Studies also show that very clean households for children are more likely to result in higher rates of allergies for them as adults. It is surmised that the cleanliness of the house prevents their normal exposure to antigens as a child. When adults raised in a clean household encounter allergens, their immune system mistakenly views the antigens as foreign, initiating an immune response.

## Spies and Corruption of our Immune Defenses

The most serious malfunction of the immune system occurs when it does not work efficiently, called immunodeficiency. There are numerous examples of diseases characterized by weak immune systems. Some of these are a result of viruses, which invade and

**Allergy**

An inappropriate immune reaction to antigens that are otherwise not harmful.

**Allergen**

Antigens that cause allergies.

**Basophil**

A type of white blood cell associated with allergies.

**Immunodeficiency**

The most serious malfunction of the immune system occurs when it does not work efficiently.

destroy our defenses. Viruses, as described in Chapter 8, act like spies, which take over a cell's defenses. They corrupt the normal processes of a host cell. A weakened immune system makes a cell susceptible to invaders.

In our story, stress weakened the immune system of the main character. It resulted in the eruption of herpes along the lips. Sometimes, the immune system is weakened not because of stress, but due to other illnesses such as lung disease, neutropenia, as well as many cancers. Other weakened immune systems result due to viral infection, as in HIV and AIDS.

When only one of the immune system strategies is compromised, all of its defense mechanisms are threatened to fail. The results can be devastating, with the many microbes ready to invade, as described in Chapter 8.

Consider AIDS, or Acquired Immune Deficiency Syndrome, the most widely known immunodeficiency. AIDS as viral illness was introduced in Chapter 8 as well, which described its lysogenic life cycle akin to a spy operation. The HIV virus invades a cell and remains dormant, like a spy, to attack a cell from the inside at any time.

The HIV virus attacks and destroys T-helper cells, which in turn limits humoral and cell-mediated immunity in its sufferers. This spy-like virus leaves only the first two lines of defense, the skin and nonspecific immunity, to fight off diseases. When T-helper cells fail to stimulate plasma cells, antibodies cannot be produced in sufficient amounts to fight off other infections. AIDS victims succumb to infectious illnesses such as pneumonia as well as cancers such as Kaposi's sarcoma.

HIV infection affects less than 1% of the U.S. population, representing about 1 million people. However, only 40,000 of those infected with HIV actually get symptoms of AIDS. In developing nations, the infection and death rates are much higher; with a lack of adequate access to the expensive AIDS cocktail of medicines. In sub-Saharan Africa in nations such as Namibia and South Africa, infection rates from HIV are 25% or more in the adult population. AIDS patients have new treatments to stave off the effects of weakened immunity, but a cure to save T-helper cells is still not realized. At this time there is no effective vaccination for HIV because the virus mutates so rapidly and there is no cure for AIDS. A map of HIV infection rates across the world is shown in Figure 15.24.

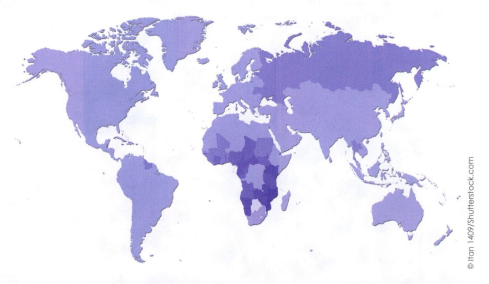

© Ifan 1409/Shutterstock.com

**Figure 15.24**  Map of HIV Infection Rates across the World. The highest rates are found in sub-Saharan and South Africa. The darker the shade purple, the higher percentage affected within that nation.

### CAN A PERSON ACQUIRE WARTS FROM TOADS OR FROGS?

Warts are acquired after a papillomavirus enters body cells and seizes control of its machinery. Viruses, as you recall from Chapter 8, are intracellular spies, which take control of cells. There is no evidence that toads or frogs act as vectors for human papillomavirus.

A wart is actually a benign tumor, which is defined as an abnormal cell growth that does not spread. A wart is not a malignant cancer and is not life endangering. However, warts are troublesome and do spread from person to person. Viruses are species-specific, meaning that they cannot spread from one species to another species (there are exceptions, discussed in Chapter 8). However, if a human papillomavirus enters into an open or delicate area of the skin of another human, the virus may spread.

Thus, you cannot generally get warts from other species, such as frogs. Figure 15.25 shows a toad kissing a woman with a cold sore. The toad cannot get a cold sore from the woman he is kissing. Also, while a toad may appear to have wart-like features, it does not mean that it has warts. Toads often have only roughened skin.

## Summary

The immune system defenses include: a protective border of skin and mucous membranes; a nonspecific immunity including white blood cells and chemicals; and a specific immunity that utilizes antibodies and specialized cells to specifically target invading pathogens. The skin or integumentary system serves several purposes in the human body. It maintains a protective border to defend against pathogens. When the skin defenses are breached, a series of nonspecific and specific immune responses take place. There are malfunctions of the immune system including allergic reactions, autoimmune diseases, and immunodeficiency such as AIDS.

**Figure 15.25**  A Toad Kissing a Woman with a Cold Sore.

## CHECK OUT

**Summary: Key Points**

- People who suffer from contagious diseases such as herpes, leprosy, and AIDS have been shunned by many in society. Understanding about the nature of these illnesses dispels many fears.
- The immune system has three lines of defense: the skin and mucous membranes; nonspecific immunity; and specific immunity.
- The skin and mucous membranes have multiple layers and secretions, which act as a strong border; skin serves other functions, such as blood storage and vitamin D synthesis.
- Neutrophils, macrophages, and lymphocytes not only ingest pathogens but also play a role in initiating inflammation and specific immune defenses.
- Cell-mediated immunity uses cells in hand-to-hand combat to battle pathogens. Humoral immunity uses antibodies to specifically target pathogens.
- Some tissues regenerate well, such as skin, bones, and blood; while others have little or no regeneration capability, such as nerves and cardiac tissues.
- The lymphatic system comprises the lymph nodes and other organs, serving as defense stations to battle invading pathogens.
- The immune system malfunctions when (1) it attacks itself, as in autoimmune disease; (2) it overreacts to antigens, as in allergies; and (3) it is weak, as in AIDS.

## KEY TERMS

abscess
acquired immunity, -active artificial, -passive artificial, -active natural, -passive natural
agglutination
allergy
allergen
antibody
antigens
arrector pili muscle
autoimmune disease
B cell
basal cell carcinoma
basophil
bilirubin
cell-mediated immunity
clonal selection
cytotoxic T cells
dermis
dermal papillae
epidermis
fibroblast
fibrosis
giant cells

hair follicle
hair root
histamines
humoral immunity
hypodermis
inflammation
interferon
immune system
immunodeficiency
keratin
keratinocyte
Krause's corpuscle
jaundice
lacteal
Langerhans cell
lymph
lymphatic system
lymphocyte
macrophage
macrophage-presentation
major histocompatibility proteins (MHC)
mast cell
melanocyte
melanoma

Meissner's corpuscle
memory cell
necrosis
neutralization
neutrophils
nonspecific immunity
opsonization
Pacinian corpuscle
plasma cell
pyrogene
pus
regeneration
respiratory burst
Ruffini's corpuscle
sebaceous glands

skin cancer
specific-immunity
squamous cell carcinoma
stratum basale
stratum corneum
stratum granulosum
stratum lucidum
stratum spinosum
sweat glands
T cell
T-helper cell
T-suppressor cell
variable region
white blood cells

# Multiple Choice Questions

**Reflection questions:**

1. Contagious diseases are difficult to cope with because of:
   a. pain
   b. recurrence
   c. fear of rejection in society
   d. all of the above

2. Which term(s) does NOT fit with nonspecific immunity?
   a. macrophage
   b. pyrogene
   c. neutrophil
   d. antibody

3. Which part of the skin acts as a border with many layers of defense?
   a. dermis
   b. epidermis
   c. hypodermis
   d. sudoriferous sweat gland

4. To begin specific immunity, antigens are presented by:
   a. macrophages
   b. neutrophils
   c. antibodies
   d. fibroblasts

5. Which directly leads to the formation of antibodies?
   a. humoral immunity
   b. nonspecific immunity
   c. cell-mediated immunity
   d. autoimmunity

6. Which represents a logical order, from MOST to LEAST able to regenerate, for tissues of the human body?
   a. epithelial → cartilage → ligaments → nerve
   b. epithelial → ligaments → cartilage → nerve
   c. ligaments → epithelial → nerve → cartilage
   d. nerve → cartilage → ligaments → epithelial

7. A mother nurses her child, giving her baby vital antibodies through the milk. This is an example of:
   a. natural active acquired immunity
   b. natural passive acquired immunity
   c. artificial active acquired immunity
   d. artificial passive acquired immunity

8. Which is NOT a function of the lymphatic system?
    a. transport of fats
    b. transport of carbohydrates
    c. transport of excess fluids
    d. immune defenses

9. Which correctly MATCHES an immune system disease with its classification?
    a. allergy – thyroxin
    b. autoimmune – allergen
    c. immunodeficiency – rheumatoid arthritis
    d. multiple sclerosis – autoimmune

10. White blood cells are found in all of the following places EXCEPT:
    a. lymph nodes
    b. lymph vessels
    c. dermis
    d. stratum corneum

## Short Answer

1. Contagious diseases are often recurring. Give one reason for why they recur and one reason for why they are socially taboo in some societies.

2. Define the following terms: heparin and histamine. List one way each of the terms differ from each other in relation to their (a) function, (b) role in inflammation, and (c) relationship with each other.

3. List the three lines of defense of the immune system. Which uses antibodies to target pathogens? Which uses lymphocytes to defend against pathogens?

4. Draw a sketch of the inflammatory response, using arrows to show the role of the two chemicals in inflammation. Be sure to include the white blood cells involved in inflammation. How does a mast cell begin the inflammatory response?

5. How do memory cells play a role in natural active acquired immunity? Do they play a role in natural passive acquired immunity? Why or why not?

6. Pretend you are an antibody. Describe the four pathways that you take to destroy a pathogen. Which pathway most directly damages your enemy?

7. Sketch a hierarchy of tissues, in terms of their ability to regenerate from damage to them. Explain the role of mitosis in tissue regeneration.

8. For question #7, explain the role of scar tissue in tissue repair? Give the pros and cons of scar tissue to answer this question.

9. Describe the steps of macrophage presentation in the immune response. Be sure to use the following terms in your explanation: antigen, pathogen, T-helper cell, plasma cell, and antibody. How do lymph nodes play a role in this process?

10. Describe the symptoms of multiple sclerosis. How does the disease progress? What does the latest research show is the cause? How is the disease classified?

## Biology and Society Corner: Discussion Questions

1. Is taking anti-inflammatory and fever reducer drugs good or bad for fighting pathogens? Why or why not? Discuss the role of society in its expectations of doctors to provide relief from illnesses.

2. The following statement was made in a courtroom in a medical malpractice suit: "This doctor is often accused of treating the condition and its symptoms and not the whole patient" as a reason for the malpractice. This case involved the mistreatment of a person with HIV infection, who is shunned in some societies. Without knowing the details of the case, construct an argument for the plaintiff, the patient. Construct a counterargument in favor of the defendant, the medical doctor.

3. The treatments for AIDS and HIV infected individuals are very expensive, tallying several thousand dollars a month. People in developing nations die at much higher rates than those infected in the United States. Research this disparity and plan a strategy to lessen the gap between survival rates between these two groups. List three specific ways that you would attempt to accomplish your task. List the drawbacks to each of your strategies.

4. Allergies are on the rise in populations all over the world. Some scientists blame changing climate and others claim that there is simply better diagnosis of allergies compared with the past. Research the biology of allergies and its changing rates in

the population. Based on your research, which factor do you think is most important linking to the rise in allergies?

5. Jim Carey, a prominent U.S. movie actor, has campaigned against the negative effects of vaccines on childhood health. Based on the information in this chapter, construct an argument against stopping the use of vaccination. Which factor gives you the most certainty in your argument?

Figure – Concept Map of Chapter 15 Big Ideas

# Urogenital Functions in Maintaining Continuity

# 16

## ESSENTIALS

A normal childbirth

©Olesia Bilkei/Shutterstock.com

Stages of Childbirth

Fetal expulsion

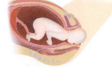

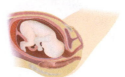

Cervical dilatation

Delivery of placenta

©Alila Medical Media/Shutterstock.com

Portrait of a Victorian lady, Mrs. Eddy

©Dimitris_k/Shutterstock.com

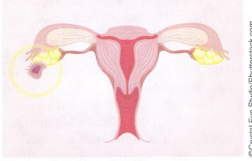

An embryo (center of circle) develops in the wrong place (within the fallopian tube) called an ectopic pregnancy

©Crystal Eye Studio/Shutterstock.com

## CHECK IN

**From reading this chapter, you will be able to:**

- explain how developments in medical technology play an important role in changing women's health.
- define excretion and describe how it is accomplished in human kidneys.
- explain a urinalysis, analyze its results and connect it to urinary diseases.
- compare asexual and sexual reproduction and external and internal fertilization.
- trace the movement of sperm and egg from production to fertilization and identify the organs in the reproductive system.
- describe the changes in structure during embryology.
- identify and describe the diseases of the reproductive system.

## The Case of the Stone Baby

The autopsy was attended by over 20 persons. In the year of our Lord, 1851, Dr. William H. H. Parkhurst, the friend and attending physician of the deceased, presented a short history of an unusual and rare case. The widow of Amos Eddy of Herkimer County, New York died after a long and unusual set of symptoms. After the autopsy was complete, Dr. Parkhurst shuffled his papers and began: "I present to you the most unusual case I have ever encountered . . . Mrs. Eddy, my former patient, it appears has been pregnant for 50 years!"

The audience whispered loudly in a show of disbelief. "It is not possible," one doctor called out, "for a mother to go beyond 45 weeks gestation and live." "Aha" replied Dr. Parkhurst, "if you let me explain the case, you will see how it was and is possible."

He went on, "Ladies and gentlemen, I met Mrs. Eddy in 1842, 40 years after her first signs of pregnancy. She described her marriage to Mr. Eddy in 1795 and her days as a new bride for their first happy years together. Then, this woman became pregnant with child. Her early months passed with the usual symptoms of pregnancy: the catamenial secretion ceased; a recurrent nausea occurred and normal feelings of pregnancy through to the last eight and one half months gestation."

Dr. Parkhurst paused, sadly in memory of his patient, and then continued, "But in the last month of her pregnancy, Mrs. Eddy, while preparing supper with a large pot suspended over the fire, received a shock when it gave way due to the weight of the meal. This stimulated labor contractions about two hours later. Mr. Eddy readied the horse and buggy to bring her to the doctor. Birth was thought to be imminent. But the labor pains subsided and Mr. Eddy kept his horse and buggy ready for the next month . . . but no baby came. More weeks went by and Mrs. Mrs. Eddy felt the pains of bloat and labor contractions but no doctor could deliver a baby for her."

"Mrs. Eddy's health deteriorated." The doctor explained that months passed and Mrs. Eddy did not feel well and was near exhaustion and death. "But after about one year and one half, Mrs. Eddy's health began to slowly improve. Her abdomen remained swollen except for the reabsorption of adipose." The doctor, however, described her condition as chronic pregnancy, explaining that "forty years after her pregnancy, I still saw her undergo periods of labor contraction. I tried to deliver the baby, and it would have been a 40-year-old newborn."

The audience was dismayed at the images of the newly deceased Mrs. Eddy. Dr. Parkhurst began the autopsy to verify his suspicions. He described how he came to believe in Mrs. Eddy's pregnancy: "I had suspected Mrs. Eddy of this unusual condition but I dared not tell her, so many years after the start of her pregnancy. I became aware of another lithopedian (calcified fetus) when reading about a similar case in Cooperstown, New York."

The autopsy then confirmed the case of the stone baby. Everyone in the audience was astonished to see the opened pelvic area showing a baby, six pounds in weight and facing his mother's spine. The fetus was fully grown, wrapped in a bony and cartilaginous casing. One leg, one foot and one elbow were fused together. The baby was almost completely ossified – it was now a stone baby!

The problem that the stone baby had, which caused the abnormal condition, was then discovered. There were no connections to the uterus by way of an umbilical cord. Instead, the baby's nourishment was obtained by mesenteric arteries. It had been grown in the fallopian tubes of its mother instead of the uterine wall, which would be normal. Growth of an embryo in any other place (abdominal cavity; fallopian tube) is known as an ectopic pregnancy. Often it is painful and if left without surgical intervention, as in days of the past, it may be deadly for the expecting mother.

Dr. Parkhurst made his final remarks: "Believe it or not … the baby turned to stone."

*Based on a true story from *Transactions of the American Association of Obstetricians and Gynecologists* for the Year 1888, Volume 1, p. 305. The Association, 1888.

---

## CHECK UP SECTION

The story above depicts the true case of the result of an ectopic pregnancy in the early 1800s in Herkimer, New York. Ectopic pregnancy occurs when a fetus develops outside of the uterus. Its extreme results were due to a lack of knowledge about ectopic pregnancy. There were only 300 documented cases of lithopedians diagnosed in history.

Research the types of problems that occur during an ectopic pregnancy. Be sure to describe the techniques and methods now used to help women with ectopic pregnancies.

A battery of prenatal tests is soon becoming available to detect over 1,500 diseases before a fetus is born. How might the emergence of these tests affect decisions to terminate a pregnancy? Would you consider using the technology to find out, for example, if your newborn is likely to have a disease?

---

## The Urinary System

Within the pelvic region described in our story, a lithopedian formed within Mrs. Eddy. The story shows some of the changes in this area of the body related to pregnancy. The reproductive system includes all of the pelvic structures and it products related to the forming of new organisms. In humans, the pelvic region is also home to the urinary system, which is responsible for eliminating wastes.

The two systems are studied together because they are located in the same region. They comprise what is called the urogenital system, shown in Figure 16.1. The urinary and reproductive systems will be discussed in separate sections of this chapter. Although found close together, they each work to accomplish distinct goals – the urinary system produces urine and the reproductive system produces babies – but both are the focus of this chapter. Both maintain continuity: reproduction helps the species endure and the urinary system stabilizes water and ion balance.

**Reproductive system**

The system that includes all of the pelvic structures and it products related to forming new organisms.

**Urinary system**

The system that is responsible for eliminating wastes.

**Urogenital system**

The system comprising of the reproductive organs and the urinary system.

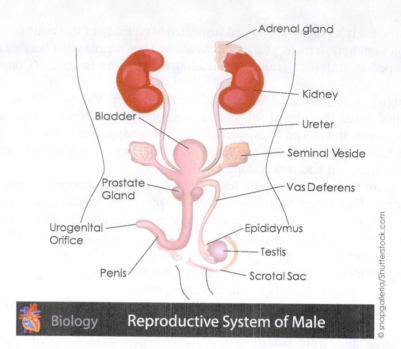

Biology    Reproductive System of Male

© snapgalleria/Shutterstock.com

**Figure 16.1**    Sketch of the Male Urogenital System. The reproductive and urinary systems lie mostly within the pelvic region.

# Regulating Water Balance

**Osmoregulation**

The process by which organisms control their fluid intake along with dissolved solute balances.

First, let's look at the way organisms regulate their water and ion levels. All organisms, whether on land or in the sea, control the amount of water they take in and lose using a process known as osmoregulation. Osmoregulation is the process by which organisms control their fluid intake along with dissolved solute balances. Organisms have adapted structures in order to work with their environments to maintain water balance.

In freshwater systems, for example, water is in excess and solutes within organisms need to be brought inward. Some organisms have special structures within cells to help control water balance. The *Paramecium* described in Chapter 8, for example, has contractile vacuoles, which displace water as it enters. In another example, freshwater fish have gills that take up salt from their surroundings and excrete large amounts of dilute water. In salt water, the opposite situation happens. Saltwater organisms conserve water and prevent salts from entering into them. Salt water fish, for example, drink large amounts of water to compensate for the solutes and produce small amounts of very concentrated urine to rid the solutes. This helps them to conserve water. Some saltwater fish gills also excrete salt back into the environment.

**Excretion**

The process of eliminating wastes from organisms.

The removal of wastes from organisms is known as excretion. The main waste products are: **urea**, which contains nitrogen; carbon dioxide (eliminated through the lungs); salts; and water. There are four organs responsible for excretion of these waste products in humans (Figure 16.2): (1) the liver processes nitrogen wastes by combining them to form into urea and removes bile from red blood cell breakdown via intestines and feces, detailed in Chapter 12; (2) the lungs remove carbon dioxide waste and some water vapor, in processes described in Chapter 13; (3) sweat glands, shown in the previous Chapter 15, remove water and salts in the form of sweat; but (4) the kidneys are the main organ of excretion in vertebrates.

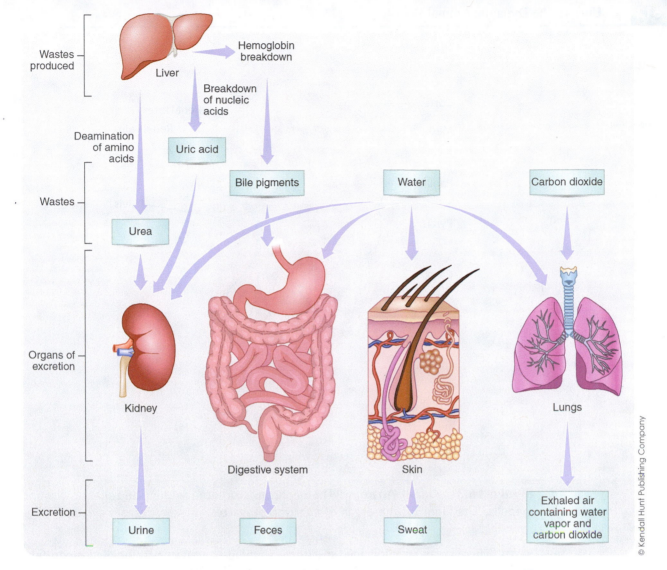

**Figure 16.2** Organs of Excretion: Humans dispose of their metabolic wastes through four major organs: the kidneys, lungs, liver, and sweat glands.

Image labels:
Wastes produced — Liver — Hemoglobin breakdown
Breakdown of nucleic acids
Deamination of amino acids — Uric acid
Wastes — Bile pigments — Water — Carbon dioxide
Urea
Organs of excretion — Kidney — Digestive system — Skin — Lungs
Excretion — Urine — Feces — Sweat — Exhaled air containing water vapor and carbon dioxide

© Kendall Hunt Publishing Company

# Kidneys

Water is a scarce commodity on land, where animals need to conserve water. Water balance and water scarcity in the environment are topics discussed in Chapter 18. Terrestrial organisms require systems adapted for their lives on land. Animals, therefore, must remove the nitrogen wastes produced during metabolism while also conserving water. This is a conundrum – wastes accumulate in the body's fluids and yet they cannot be simply removed because the land organism must conserve its fluids.

To solve this dilemma, terrestrial vertebrates developed urinary systems adapted for limited water conditions. In humans, there are two organs called the kidneys, which filter blood, remove wastes, and at the same time, conserve needed materials including water. The kidneys are bean shaped and fist-sized, lying in the posterior region of the lumbar area on either side of the spine.

Any structure or function referring to the kidneys is known as "renal." So, the name renal arteries indicate that they are entering the kidneys. The renal capsule covers the kidney, affording it some protection. The outer portion of the kidney is known as the renal cortex and the inner portion is the renal medulla. The medulla empties into the renal

**Kidney**

Organ that filters blood, removes wastes, and at the same time conserves needed materials including water.

**Renal capsule**

A fibrous layer surrounding the kidney, affording it some protection.

**Renal cortex**

The outer portion of the kidney.

**Renal medulla**

The inner portion of the kidney.

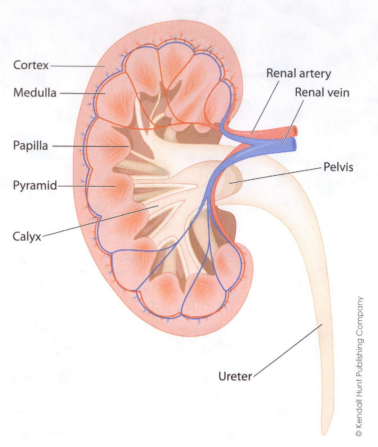

Cortex

Medulla

Papilla

Pyramid

Calyx

Renal artery

Renal vein

Pelvis

Ureter

© Kendall Hunt Publishing Company

**Figure 16.3**    Kidney Anatomy. The nephron is situated within the renal cortex and medulla. The kidney is the organ of filtration of wastes in the excretory system.

pelvis. There are pyramid-shaped structures oriented around its pelvis called **medullary or renal pyramids**. The renal pelvis removes liquid from the kidney (Figure 16.3).

Let's trace the flow of urine from the kidney to the outside of the body (Figure 16.4). Each kidney is connected to a tube called a ureter, which transports urine from the renal pelvis to a holding tank called the urinary bladder. Urine is stored in the urinary bladder until it is removed from the body through a final tube, the urethra. The urinary system connects with the reproductive system within the urethra in males. It remains separate from the reproductive system in females. We will explore these systems more closely in the next sections.

However, pregnancy places pressure on the urinary system. Many pregnant women experience bladder leak and swellings, for example. In our story, Mrs. Eddy's fetus grew to six pounds. As you may see in Figure 16.4, this placed pressure on her urinary bladder and other organs. Obviously, in normal cases, pregnancy is a temporary condition and the urinary system fortunately returns to normal functioning. This was not the fortune of Mrs. Eddy, however.

## Functions of the Kidneys

The kidneys perform five functions for the body, besides excretion; they: (1) regulate blood volume and the pH of the blood; (2) remove toxic nitrogen- and sulfur-containing compounds; (3) make erythropoietin, a hormone that regulates red blood cell formation; (4) make renin, a hormone that regulates blood pressure; and (5) maintain $Ca^{++}$ ion balance, which activates vitamin D in the body.

**Renal pelvis**

A hollow funnel that removes liquid from the kidney.

**Ureter**

A tube connected to each kidney responsible for transporting urine from the renal pelvis to the urinary bladder.

**Urinary bladder**

An organ that holds the urine.

**Urethra**

The duct by which urine is removed from the body.

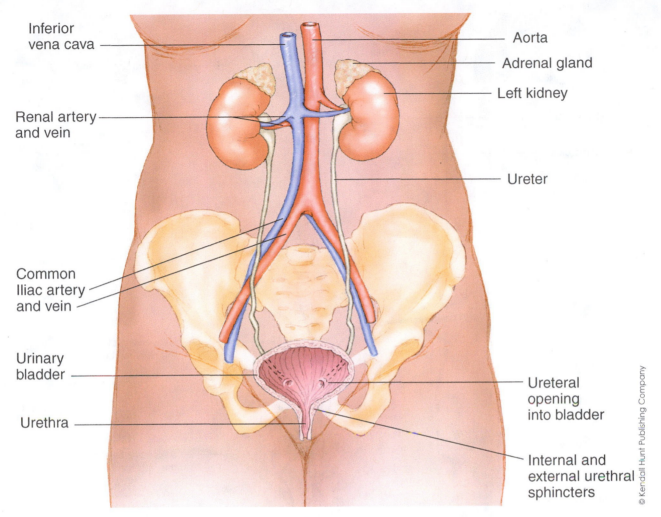

Inferior vena cava

Renal artery and vein

Common Iliac artery and vein

Urinary bladder

Urethra

Aorta

Adrenal gland

Left kidney

Ureter

Ureteral opening into bladder

Internal and external urethral sphincters

© Kendall Hunt Publishing Company

**Figure 16.4** Urinary System Anatomy. The kidneys produce urine, which passes through the ureters and into the bladder, in which it is temporarily stored. At urination, the bladder releases urine through the urethra to the outside.

Although excretion of wastes and water conservation are key goals of the kidneys, they have an intricate connectedness with other systems of the body. For example, if blood volume is not properly regulated by the kidney, the heart will have problems functioning. Too much liquid in the blood might be too difficult for the heart to pump. The kidneys are often discussed at the end of a unit in human anatomy and physiology. This is because they link to all other systems of the body through their multiple functions.

## Special Cells of the Kidneys: Nephrons

The kidneys filter up to 2,000 liters of blood per day. A person's urine output varies as well day-to-day, with 0.5–2.0 liters of urine excreted daily, depending upon the amount of liquids taken in. Kidneys are composed of specialized cells called nephrons, the functional unit of the kidneys. Each kidney contains 1 million nephrons, all of which produce urine.

As shown in Figure 16.5, nephrons are composed of (1) long tubes and (2) blood vessels surrounding those tubes. The blood vessels at the start of a nephron form a ball of capillaries called the glomerulus. The glomerulus is surrounded by Bowman's capsule, (the glomerular capsule) which receives substances from the glomerulus.

**Nephron**

The functional unit of the kidney.

**Glomerulus**

A ball of blood vessels at the start of a nephron.

**Bowman's capsule**

A cup-like sac surrounding the glomerulus.

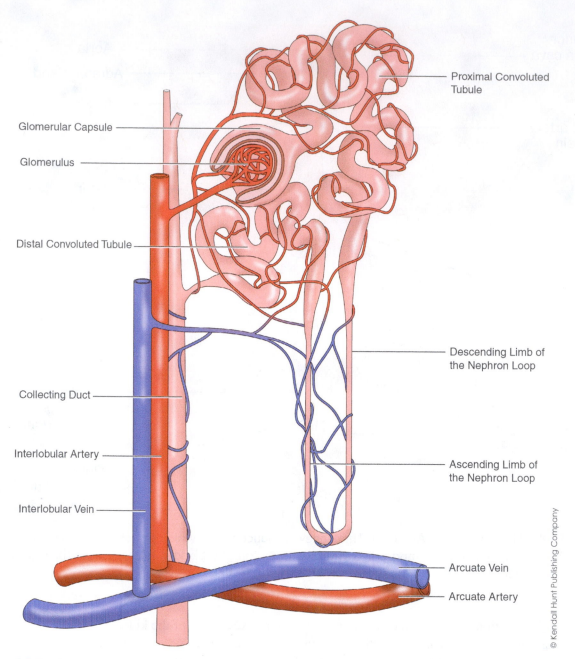

Proximal Convoluted Tubule

Glomerular Capsule

Glomerulus

Distal Convoluted Tubule

Collecting Duct

Interlobular Artery

Interlobular Vein

Descending Limb of the Nephron Loop

Ascending Limb of the Nephron Loop

Arcuate Vein

Arcuate Artery

© Kendall Hunt Publishing Company

**Figure 16.5** Nephron Structure. There are three steps to urine formation in the kidneys: filtration, reabsorption, and secretion. Much reabsorption of water from the collecting ducts concentrates urine by conserving water.

**Loop of Henle**

A large tube in the nephron that descends and then ascends.

**Collecting duct**

A collecting tube that receives urine from several nephrons.

These capillaries contain blood from the body. Vessels then leave the glomerulus (along with its blood) moving through and surrounding the nephron tubes. A large tube in the nephron, descending and then ascending, is called the Loop of Henle. The Loop of Henle transports its contents into the collecting duct, after which it is removed from the kidneys. The glomerulus is found in the cortex of the kidneys, while the Loop of Henle and collecting duct are located in the renal medulla.

The nephrons work by filtering out some substances from the blood. Those substances enter nephron tubes. Materials that remain in nephron tubes eventually become urine. Those materials remaining in the blood are returned to the body. In this way, the kidney controls which materials are kept and which are removed from the blood.

# The Kidney has a Three-Step Process to Make Urine

Blood enters the kidneys so it can pick up urea from the liver, and ions, salts, and water from body cells. The kidneys carry out a three-step process to produce urine, using the structures shown in Figure 16.5.

1)  Filtration. When blood enters the kidneys, it is first filtered in the glomerulus. The glomerulus has tremendous pressure exerted on its capillary walls. When blood flows through the glomerulus, small substances such as water and salts pass through the filter and large substances remain in the blood. Filtration is strictly based on size. Large proteins should not be found in the urine, for example, because normally they are too large to pass through the glomerulus. Holes in the capillaries of the glomerulus determine whether or not a substance will pass through it. Substances that pass through and into the nephron tubes are called **filtrate**. Water, salts, glucose, urea, and small amino acids are normally found in the filtrate. Red blood cells, white blood cells, and large proteins should not be found in the filtrate.

    The kidneys concentrate liquids from the blood as they flow through. High pressure in the glomerulus sends filtrate into Bowman's capsule. Materials within Bowman's capsule move along the nephron through the Loop of Henle and the collecting duct.

**Filtration**

The process by which substances in blood are separated out by pressure through kidneys.

2)  Reabsorption. As indicated, filtration is unselective and removes all small items while keeping large ones in the blood. Some materials that were filtered out need to be returned back into the blood by the process of **reabsorption**.

    Over 2,000 liters of blood pass through the kidneys each day, and yet only 1.5 liters of urine are produced. This means that 1998.5 liters are reabsorbed by the kidneys every day. How does this much liquid return to the blood and body? It begins by the nephrons actively transporting $Na^+$ ions out of the nephron tubes by active transport (out of the ascending limb of the Loop of Henle in Figure 16.5). This creates a hypertonic environment outside of the tubes. The blood vessels that surround the nephron thus also become hypertonic. As described in Chapter 3, water follows solute. So, water travels back into the blood from the tubes to be reabsorbed by the body.

    The filtrate in the nephron is modified all along its path through the nephron. Valuable items are reabsorbed; they include water, macromolecules (such as carbohydrates and amino acids), and ions (such as potassium and sodium) which all serve important roles in the body. They are reabsorbed at certain points along the tubes of the nephron due to the osmotic differences across the nephron.

**Reabsorption**

The process by which some materials that were filtered out are returned back into the blood.

3)  Secretion. Some materials that are not filtered from the blood still need to be removed. The removal of unwanted or unneeded substances from the blood is called **secretion**. Substances are actively and passively transported out of the blood (and into the nephron tubes) by secretion. For example, to maintain pH balance, $H^+$ ions are actively transported into the nephron tubes (and out of blood vessels) to increase the pH of the blood.

**Secretion**

The removal of unwanted or unneeded substances from the blood.

One point in the nephron, the collecting duct, is particularly permeable to water but not to salt. Here, a large amount of water diffuses into the collecting duct and is removed from the body. To recap, the environment around the collecting duct is hypertonic, drawing water from the tubes back into the blood. Active transport of ions by the nephron creates a hypertonic environment surrounding tubules. This pulls most of the water out of the tubules and back into the blood. Animals adapted to conserve water usually have long Loops of Henle. Cats, for example, have very long Loops of Henle, concentrating their urine. This is the reason why cat urine is so smelly: the solutes remaining in the filtrate cause odor and there is very little water within its urine. Some animals are so efficient at concentrating their urine, such as kangaroo rats, that they never need to drink any water. They obtain all of their fluids from only the foods that they eat.

In summary, as filtrate moves along the tubules of the nephron, it becomes more and more concentrated, losing water to a hypertonic environment. Because regions surrounding the tubules are high in solutes due to active transport, water follows solute out of the nephron. Concentrated filtrate is produced and water is conserved for the body. Excretion by the kidney is an excellent adaptation to water conservation by land-dwelling animals. The movement of water, ions, and urea along a nephron is shown in Figure 16.6.

## Urine Indicates a Person's Health

The end result of kidney function is to conserve water and needed substances but also to excrete wastes and harmful products. This produces healthy urine with normal substances found within it. The kidneys are very effective and over 99% of water is returned to the body, along with glucose and amino acids. Urine normally contains any wastes and unwanted materials made by the body or taken into the body. Drugs and toxic chemicals are removed efficiently by the kidneys. A urinalysis tests urine for normal and abnormal substances in the urine. Normal values for various substances, including whole cells (leukocytes) and a battery of ions in the urine, are given in Figure 16.7.

The kidneys are excellent processors, able to efficiently remove wastes and toxic substances from the blood. As such, most drugs (legal and illegal) ingested are found during drug testing through a urinalysis. Metabolized products of Ecstasy, cocaine, and marijuana are easily detected in drug screenings. In the case of the fat-soluble drug,

**Urinalysis**

An analysis that tests urine for normal and abnormal substances.

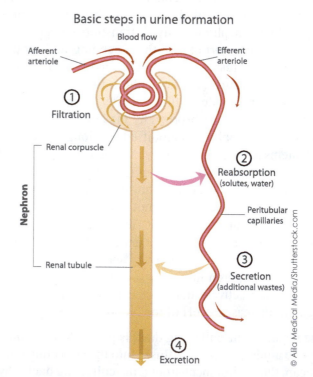

**Figure 16.6** How Do Nephrons Work? Movements of materials along the nephron. Water passes through the nephron tubules, becoming more and more concentrated. Active transport of Na+ ions out of the thick filament of the ascending segment of the Loop of Henle drives water out of the nephron tubules (passively) and back into the blood. The three-step process is shown here.

**Figure 16.7** Urinalysis. A test strip is used to show a patient's urine results. The values are compared with a normal reading. The figure shows abnormal constituents that may be found in urine and their associated pathology.

| Substance Found | Potential Causes for its Presence |
|---|---|
| Glucose | Pathological: diabetes mellitus<br>Nonpathological: excessive sugar intake |
| Blood | Pathological: bleeding in the urinary tract as a result of bacterial infection or kidney stone; chronic sympathetic stimulation |
| Protein | Pathological: kidney disease—glomerulonephritis, hypertension<br>Nonpathological: excessive exercise, pregnancy |
| Pus, White Blood Cells | Pathological: urinary tract infection |
| Bilirubin | Pathological: if excessive, liver malfunction |
| Ketones | Nonpathological excessive breakdown of lipids |

© Kendall Hunt Publishing Company

marijuana, it remains in body cells much longer than most drugs. Marijuana can be detected for a month or longer after its last use. Drug testing in the workplace is a common practice, which screens employees for illegal drugs (Figure 16.8).

## Excretion is Expensive

However, kidney processes come at a high price – they are very expensive. Active transport taking place in the kidneys conserve water but take a great deal of ATP energy. It is, therefore, a costly tax on the body. However, excretion is a necessary strategy to employ in order for organisms to live on land.

Excretion is essential because it removes nitrogen wastes. For all organisms to stay alive, nitrogen wastes must be found in low concentration. Nitrogenous wastes are toxic and interfere with normal life functions. Have you ever sniffed ammonia, a cleaning product? It is actually very dangerous and in an unventilated room may lead to disorientation and health problems.

© new photo/Shutterstock.com

**Figure 16.8** Drug Testing at the Workplace. Should people be required to take drug tests as a condition for their employment?

**Ammonia** (NH₃) is the first by-product made from the metabolism of proteins and nucleic acids. Ammonia is converted into urea in the liver by combining it with carbon dioxide, as described in Chapter 12. This process takes a great deal of energy, but ammonia is so toxic that it would kill humans if it remained in the system for more than a few minutes. Urea is able to remain in the body longer, without causing as much harm. Our kidneys are the human adaptation to solving the problem of conserving water while also removing nitrogen wastes.

Some organisms do not have the problem of processing nitrogen wastes. Aquatic organisms, such as fish and aquatic snails, are able to directly excrete ammonia into their watery environment. They do not require the energy needed for processing nitrogen wastes.

Other organisms go a step further than humans, transforming ammonia into the pasty white precipitate, **uric acid**. It is a more energy-consuming process than forming urea or simply excreting ammonia. Birds, insects, terrestrial snails, and some reptiles produce uric acid as a waste product. Some nitrogenous waste products from different organisms are seen in Figure 16.9.

## Why Uric Acid?

But why would an organism go through the extra work of making uric acids? Why do birds, for example, bother producing uric acid, instead of other substances? Ammonia made by fish is less expensive to process. We know that uric acid is more expensive to make but there must be some evolutionary benefit. In other words, why did different systems of excretion develop differently in organisms?

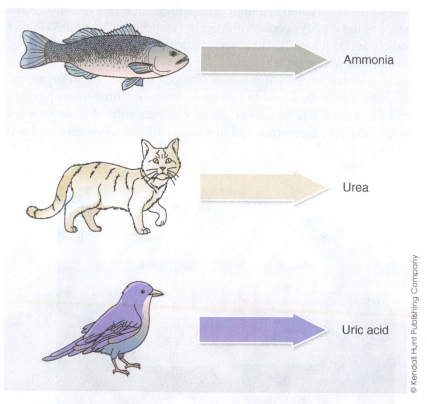

© Kendall Hunt Publishing Company

**Figure 16.9** Types of Excretory Products in: Fish (ammonia), Birds (uric acid), and Cats (urea). Nitrogenous wastes are excreted differently by each animal, with humans producing a less toxic form, urea. Humans, just like cats, excrete small amounts of ammonia and uric acid alongside mostly urea.

Let's consider bird droppings, which are mostly uric acid. You may recall your first ice cream cone, vanilla, topped while you were not looking with uric acid . . . and it does not taste like chicken. Then why would a bird excrete it? We stated earlier that uric acid is a precipitate, meaning that it is in semi-solid form. This is different from urea, the human substance of excretion, which dissolves in water.

How are birds different from humans? One way is that they lay eggs in which their embryos develop. Human embryos develop in a watery amniotic sac. What would happen in a bird's egg, if birds instead excreted urea? Urea, dissolved in water, would diffuse throughout a bird's egg. It would therefore travel into the bird embryo and its toxicity would kill it. Instead, uric acid precipitates out at the corner of an egg and does not harm the chick embryo. Birds would end as a living unit if urea were its excretory product. In humans, the wastes of a fetus are removed by blood vessels connections with its mother. Thus, we do not need to form uric acid as our main excretory product.

## Controlling Kidney Functions

As discussed in Chapter 14, hormones play an important role in regulating processes in the body. The three hormones that help control our urine output are: aldosterone, angiotensin II, and antidiuretic hormone (ADH). Each of these works to conserve water in our bodies. All three hormones work by reabsorbing water into the capillaries surrounding the nephron tubules. This increases water levels within the blood. Recall from Chapter 14 that ADH is produced in the hypothalamus of the brain and stored in the pituitary gland just below it. When the hypothalamus detects that the blood has become too salty (too much solute; too little water), it produces ADH. This increases the amount of water in the blood. By ADH increasing the permeability of the collecting duct (and other tubules) to water, it then moves freely back into the capillaries and is therefore conserved. This decreases urine volume output and increases blood pressure.

## Malfunctions of the Kidneys

Sometimes the kidneys fail to work properly, resulting in impaired kidney function. The rate at which a kidney filters blood is measured to determine kidney impairment. If the filtration rate drops to 50% or less, a person is classified as being in kidney failure. The top two causes of kidney failure are high blood pressure and diabetes. Both of these diseases damage delicate blood vessels in the kidneys.

During impaired kidney function, improper filtering of the blood often leads to rising pH values and solute concentration changes. In total kidney failure, ionic balances are more disrupted and toxic substances accumulate within days leading to death. For

**Aldosterone**

A hormone produced by the cortex of adrenal gland.

**Angiotensin**

A hormone that promotes aldosterone secretion in blood and causes blood pressure to rise.

**Antidiuretic hormone (ADH)**

A hormone released by the pituitary gland that helps in water retention in the body.

**Impaired kidney function**

The failure of kidneys to function properly.

**Kidney failure**

A medical condition in which the filtration rate falls to 50 percent.

### DOES DRINKING ALCOHOL MAKES YOU URINATE?

Alcohol is a depressant, slowing down all of the processes in humans. It also slows the hormones and nerves in the brain. The hypothalamus and the hormones controlling the kidney also slow in their actions. When ADH acts more slowly, for example, more urine output results. This is why "breaking the seal" is not a myth after drinking alcohol. Alcohol stimulates a person's need to urinate and it remains in effect for many hours.

**Dialysis**

A treatment for kidney impairment.

Mrs. Eddy in our story, the pressure and changes associated with her stone baby probably damaged her kidneys and led to her poor initial health in the first years after fertilization.

The treatment for kidney impairment is dialysis, which filters the blood and restores normal ion concentrations. There are two types of dialysis (Figure 16.10 show hemodialysis): **hemodialysis**, which carries a patient's blood through semi-permeable tubes that filter it; and **peritoneal dialysis**, which pumps dialysis solution into a patient's abdominal cavity to exchange nutrients and wastes. Hemodialysis is performed three times each week in the hospital and peritoneal dialysis is done at home each night.

When kidney failure is permanent, dialysis must be continued lifelong or until a kidney transplant is performed. A person may live healthy life with only one kidney. However, for a transplant to be successful the proper MHC match must be found. The National Kidney Foundation estimates that each day, 17 people die from kidney failure. Organs from living donors have a better chance at success than those from dead donors. Organ donation was discussed in Chapter 13, showing the many sides of the matter.

Some advocate allowing people to sell their kidneys for money to alleviate the donor shortage. Others find the sale (and endangerment of health of the donor) of organs an unethical temptation, baiting people with money. It is possible that genetically modified organs may one day be produced from other organisms, as described in Chapter 5.

# Reproduction: An Introduction

## Types: Sexual and Asexual

When we think about reproduction as a system, it conjures up images of two people mating; perhaps in a sexy scene from a movie. **Reproduction** is defined as the making of new organisms from existing ones. It is the final system treated in this unit, and yet, it is not necessary for an individual's survival.

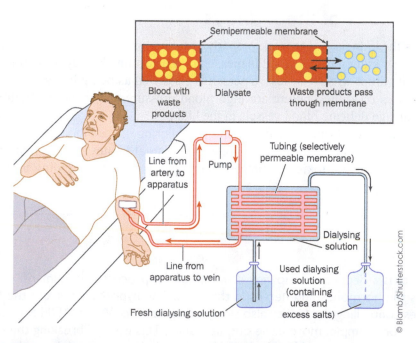

**Figure 16.10** How Kidney Dialysis Works: dialysis helps patients to cleanse their blood of nitrogenous and other wastes.

An organism may live a long span without reproducing. In fact, studies indicate the greater the number of birds in a clutch (eggs in a nest), the shorter a mother bird's lifespan. Offspring take energy and sap a parent's strength in the process. However, reproduction is necessary for the survival of the species.

Therefore, reproduction is a luxury and not a necessity for an individual to survive. Then, why have sex? As discussed in Chapter 6, finding a partner and engaging in sex are costly. It causes battles between males and costs energy to attract females. The answer, biologically speaking, is to improve the genetic quality of the species (how scientific!). Sexual reproduction, during which two individuals contribute genetic material to their offspring, produces two genetically different organisms (Figure 16.11). This results in genetic diversity among the offspring. Most animals and plants carry out **sexual reproduction**.

An evolutionary reason for sex is to increase genetic variety in a species, as you may recall from Chapter 6. This is termed genetic variation in a population. If there are more types of organisms (e.g., some resistant to a killer bacteria and others resistant to a killer fungus), some will survive environmental changes and some will die. If all of the organisms were the same, then whole species would be more likely to become extinct when the environment changes.

Organisms benefit by reproducing sexually, adding variety to their genetic make-up. Some organisms, such as earthworms (Figure 16.12), are able to both contain both male and female organs, enabling them to reproduce sexually with more partners. This increases their chances for genetic transfer of materials.

Nonetheless, at their own risks, some organisms simply do not generally engage in sexual reproduction. Many organisms, including all prokaryotes, and some plants, protists, and animals resort mostly to reproduction without using a partner. They carry out asexual reproduction. This is the process by which a single individual produces new offspring, without genetic material contributed from a partner. All offspring reproduced asexually are genetically identical to their parents. The benefit of asexual reproduction is that it is fast and easy. No partner needs to be solicited and large numbers of new offspring are produced. Asexual reproduction would have also prevented Mrs. Eddy's condition in our story of the stone baby.

There are several forms of asexual reproduction in plants and animals including: (1) budding, whereby a new organism grows directly from the body of its parent. A bud then falls off its parent to develop into a new adult. The *Hydra* described in Chapter 10 reproduces by budding; (2) fragmentation, in which a piece of a parent breaks off and

**Sexual reproduction**

The process in which two individuals contribute genetic material to their offspring.

**Genetic variation**

An evolutionary reason for sex is to increase genetic variety in a species.

**Asexual reproduction**

The process in which a single individual produces new offspring, without genetic material contributed from a partner.

**Budding**

A type of asexual reproduction whereby a new, smaller organism grows directly from the body of its parent.

**Fragmentation**

The process in which a piece of a parent breaks off and forms a new organism.

**Figure 16.11**    Two Humans in a Sexy Scene.

**Figure 16.12**    Earthworms Are Hermaphroditic, Meaning that They Reproduce Sexually and Asexually.

**Parthenogenesis**

A virgin birth.

**Fertilization**

The process of combining male gametes with female gametes.

forms a new organism. In willow trees, by cutting a branch off at any point and placing it in water, the branch grows roots ready for planting in new soil. The new willow will grow, genetically identical to its parent, into a new willow tree (both processes are shown in Figure 16.13); and (3) parthenogenesis, or a virgin birth (Figure 16.14) results from a female's egg developing into a new organism without being fertilized by sperm. In species of the order hymenoptera, which include ants, bees, and wasps, the queens are capable of giving birth to unfertilized eggs under certain conditions. These eggs develop into males that are haploid. When these haploid males mate with females, their offspring are genetically very similar: 75% identical in gene composition between sisters, to be exact. This makes them the most genetically identical organisms on earth (besides identical twins and asexually produced organisms). The results of parthenogenesis on societal organization of these organisms will be explored in Chapter 20.

## External and Internal Fertilization

The process of forming egg and sperm cells during meiosis was described in Chapter 6. How do these gametes unite, during fertilization, to form a new organism, a zygote? There are two strategies to accomplish fertilization. The first to evolve was **external** fertilization, during which sperm and eggs unite outside of the male and female.

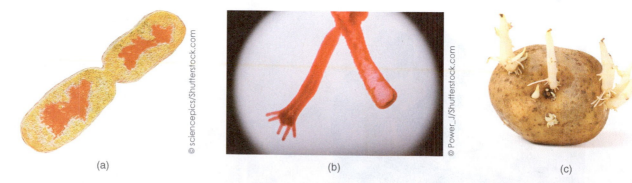

(a)                    (b)                    (c)

**Figure 16.13**    a. Binary Fission. b. Budding Fission. c. Fragmentation (when a piece of this potato is removed and grown into a new plant).

**Figure 16.14** Parthenogenesis: a queen bee reproduces sexually to create females but gives a "virgin birth" to produce males. This image shows a queen bee surrounded by her workers.

Chapter 10 described an example of external fertilization in frogs, which use amplexus. In amplexus, male frogs grab onto females from behind and wait until they release eggs, then releasing their own sperm. This process may go on for days or weeks, leading to female frog deaths in extreme cases of males holding on too long.

During external fertilization, egg and sperm enter a watery world to travel through to unite. This is a major drawback of external fertilization: it must occur in a watery environment to be successful. Most land animals avoid this strategy because watery areas are often ephemeral and dry out quickly.

Thus, terrestrial animals evolved another strategy: internal fertilization. **Internal fertilization** instead deposits sperm directly into the female's reproductive system. The act of placing a male structure into a female's reproductive tract is called copulation. Copulation solves the problem of dry land and prevents desiccation of sperm and eggs. Internally, a fluid environment protects the gametes. A mammal's reproductive tract is lined with immune defenses and a stable, wet environment.

Fertilization is more successful internally than externally. Then, why do so many fish, frogs, and toads survive, which reproduce externally? To compensate for the adverse conditions experienced by gametes during external fertilization (Figure 16.15), organisms compensate with large numbers of gametes. This way, at least some will make it to fertilization. Also, these organisms have cues to synchronize their mating strategies: water temperature, courtship rituals, chemicals secreted (pheromones), and even phases of the moon – help dictate the optimal times to mate for external fertilizers.

**Amplexus**

Mating behavior seen in frogs and toads.

**Copulation**

The act of placing a male structure into a female's reproductive tract.

# Male Reproductive System

## Male Structures

In humans, reproduction is sexual and internal. Although there are obvious differences between human external male and female anatomy, the inside structures of each are unique to their functions. Let's study male reproductive anatomy to explore the way a sperm cell travels to a female and becomes a new human being.

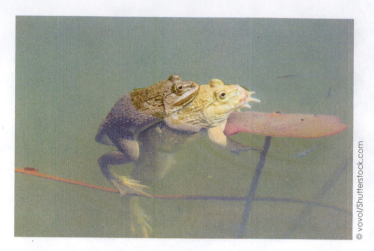

**Figure 16.15**    Two Frogs Carry Out External Fertilization According to Environmental Cues. A male frog squeezes a female until her eggs are released. His sperm then mix as eggs stream from the female.

**Penis**

The male reproductive organ.

**Scrotum**

A sac beneath the penis.

The external anatomy of the male (Figure 16.16) consists of two structures: a penis and scrotum. The scrotum is a sac beneath the penis, the male reproductive organ. Muscles (corpus cavernosum, depicted in figure) within the penis become engorged during sexual arousal to enable copulation. There are three cylinders of muscles within the penis. During arousal, their spongy tissues fill with blood. Viagra, the sexual enhancement drug, works by vasodilation of blood vessels in the penis. This causes erection, independent of sexual desires during normal copulation.

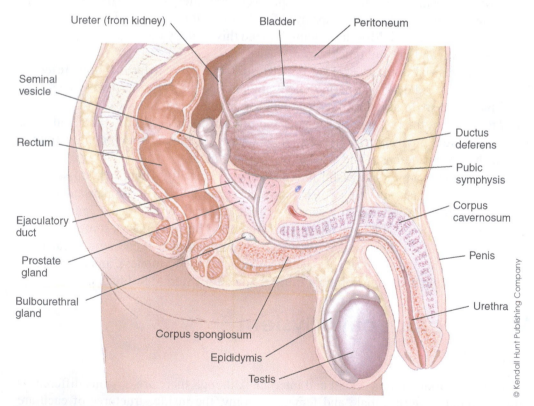

**Figure 16.16**    The Parts of the Male Anatomy. The structures produce and transport sperm and fluids (together called semen) into a woman.

Within the scrotum, there are two testes (or testicles), each suspended by a thick spermatic cord. The testes make both sperm and androgen hormones (Chapter 14). Interstitial cells lie in between the tubules in the testes. They produce androgens, which develop the male reproductive system. Hormones of reproduction will be discussed later in this chapter.

Sperm are produced by spermatogonia cells in the testes through a process known as spermatogenesis. Some organisms produce more sperm and deliver it more quickly to females than others.

Spermatogonia undergo meiosis, as described in Chapter 6, producing 100 million sperm each day. Starting at puberty, a male will produce sperm until the end of his life. Spermatogonia line highly coiled structures within the testes (Figure 16.17) called seminiferous tubules. Sperm have a long tail called a **flagellum** and an acrosome "head" filled with enzymes that digest their way into a female egg cell. Sperm, developed in the seminiferous tubules, are now ready for release.

## Tracing a Sperm's Travel

Let's trace the movement of sperm as it makes its journey through the male reproductive system, using Figure 16.16 and 16.17. Sperm, once produced in the testes, move out of seminiferous tubules. Seminiferous tubules converge together to form the epididymis, located mostly on the back side of each testicle. It is an extended and coiled tube, roughly 18-feet long. Sperm mature and become motile in the epididymis, feeding mostly on glycogen along with other macromolecules.

During ejaculation, or release of sperm from males during copulation, sperm enter into the vas deferens. The vas deferens carries sperm from the epididymis through the urethra, a duct passing out of the penis.

## Making Semen

In its trip, other fluids are added to sperm to produce semen. Semen is composed of sperm plus fluids from glands. These fluids help sperm to make their journey through the female reproductive system. There are three glands that add substances to make semen (Figure 16.17). The prostate gland, just below the urinary bladder, secretes a milky and basic fluid to buffer the effects of the acidic female environment sperm will first encounter. Prostate fluids make up about 30% of the volume of the semen. They also contain nutrients and enzymes used by sperm. The bulk of semen volume, about 60% arises from seminal vesicles. Seminal vesicles contribute fructose and amino acids as food sources for sperm. Their secretions also contain prostaglandins (Chapter 11), which stimulate small contractions in females. This enables sperm to more easily travel within a female's reproductive tract. A small amount of fluid (just a few drops) is donated to the semen at the end of ejaculation by bulbourethral glands. They are located at the base of the penis adding mucous and sugars to aid lubrication.

At the moment of ejaculation, about 280 million sperm are released in the semen. Sperm only comprise 1%–5% of the total volume of semen. The role of semen is to help sperm travel to the female egg by protecting and nourishing it. Sperm donation, used commonly in fertility clinics, assists in reproductive technologies for infertile couples and single women who want to have children (Figure 16.18).

**Testes**

Organs that produce sperma.

**Interstitial cells**

Cells that produce androgens.

**Spermatogenesis**

The process by which sperm are produced by spermatogonia cells in the testes.

**Acrosome**

A caplike structure covering the top end of the sperm.

**Seminiferous tubules**

Highly coiled structures within the testes.

**Epididymis**

An elongated organ that stores sperm and transports them from the testes.

**Semen**

Male reproductive fluid.

**Prostate gland**

A gland located just below the urinary bladder that secretes a milky and basic fluid to buffer the effects of the acidic female environment sperm will first encounter.

**Seminal vesicle**

A gland situated behind the bladder and above the prostate gland in males.

**Bulbourethral gland**

A small gland located at the base of the male reproductive organ and donates small amount of fluid to the semen at the end of ejaculation.

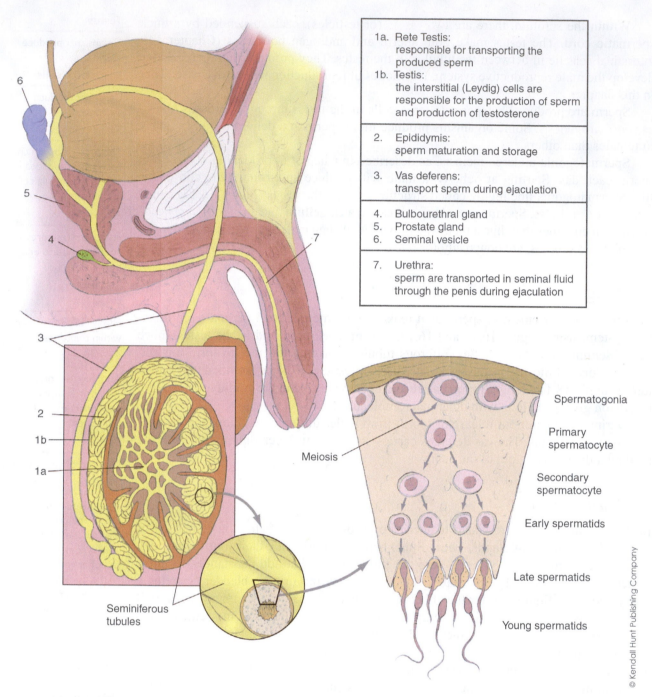

| | |
|---|---|
| 1a. | Rete Testis: responsible for transporting the produced sperm |
| 1b. | Testis: the interstitial (Leydig) cells are responsible for the production of sperm and production of testosterone |
| 2. | Epididymis: sperm maturation and storage |
| 3. | Vas deferens: transport sperm during ejaculation |
| 4. | Bulbourethral gland |
| 5. | Prostate gland |
| 6. | Seminal vesicle |
| 7. | Urethra: sperm are transported in seminal fluid through the penis during ejaculation |

© Kendall Hunt Publishing Company

**Figure 16.17**    Sperm Formation. Sperm occurs as haploid cells, with a long flagellum and a head containing the haploid nucleus.

# Female Reproduction

## Female Structures

Females carry out the same process of making sex cells as males – meiosis. Genetically eggs are the same as sperm, as haploid (N) cells. But the reproductive processes and structures in females are so much more complex.

© Yurlick/Shutterstock.com

**Figure 16.18** Sperm Donation Advertisements Seek to Recruit Sperm Donors, Paid Significantly Less than Egg Donors.

Gametes form in females by oogenesis, a process which takes place in cells of the ovaries. Ovaries are the female reproductive structures, two of each, which make and release eggs into the reproductive tracts. An ovary is roughly the size of a large almond. Cells of the ovaries, called oogonia carry out meiosis to produce eggs. An egg develops within the follicle of an ovary (Figure 16.19).

The egg develops surrounded by follicle cells. These cells form a protective layer to nourish the egg until it is released. At birth, a female has over 1 million follicles. In adulthood, oogonia occur in each of over 40,000 follicles in the ovaries. Only 400 follicles (and their eggs) will ever mature in any woman's lifetime, one each month. The remaining eggs and their follicles disintegrate with age, but modern technology has made egg donation possible. Eggs when fully developed are ready for release. They are very expensive and donation requires more risks to the donor (Figure 16.20).

**Oogenesis**

A process which takes place in cells of the ovaries.

**Ovary**

Female reproductive structure.

**Follicle**

A small ovarian sac that contains a maturing ovum.

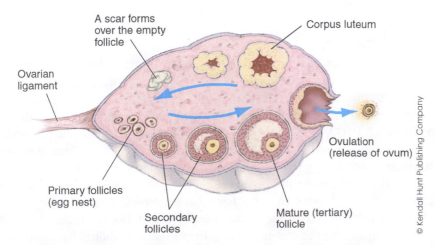

© Kendall Hunt Publishing Company

**Figure 16.19** The Event of Ovulation. The stages of development of an egg in the ovary. The final step, ovulation releases the egg for fertilization.

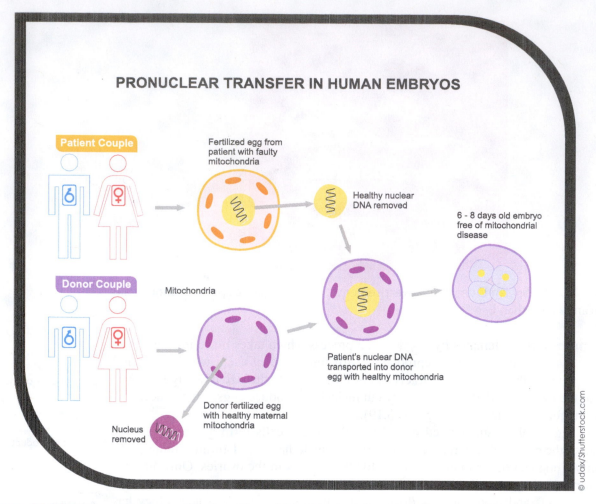

**PRONUCLEAR TRANSFER IN HUMAN EMBRYOS**

Patient Couple

Fertilized egg from patient with faulty mitochondria

Healthy nuclear DNA removed

6 - 8 days old embryo free of mitochondrial disease

Donor Couple

Mitochondria

Patient's nuclear DNA transported into donor egg with healthy mitochondria

Donor fertilized egg with healthy maternal mitochondria

Nucleus removed

© udaix/Shutterstock.com

**Figure 16.20** Donor Eggs Are Used to Treat Mitochondrial Diseases. Pronuclear transfer refers to the transfer of a nucleus from one egg to another. Egg donation is paid well but carries risks by invasive procedures used to obtain eggs within ovaries.

**Ovulation**

The process of producing and discharging eggs from ovary.

**Corpus luteum**

Hardened mass of follicle.

**Oviduct (fallopian tube)**

Tube through which an ovum passes from an ovary.

**Uterus**

A thick, muscular organ in which a fetus develops.

**Endometrium**

A membrane that lines the womb.

## Tracing an Egg's Travel

Let's trace the travels of an egg through the female reproductive system (Figure 16.21). Release of an egg from a follicle is known as ovulation. The follicle becomes a hardened mass known as the corpus luteum. The corpus luteum develops into an endocrine gland, secreting estrogen and progesterone, hormones used during pregnancy (see the corpus luteum in Figure 16.19). As the egg is released, it moves past finger-like projections of the fimbriae and into the oviduct. The oviduct is also known as the **fallopian tubes**. It is the tube through which an egg passes on its way to the uterus.

The uterus is a thick, muscular organ (also called the womb) with walls well vascularized to supply it with nutrients. It is here that a fetus develops. In our story, Mrs. Eddy's lithopedian developed in the wrong place – the fallopian tubes. Her embryo should have implanted into the uterus and developed there. It should have grown in the part of the wall with a rich blood supply, called the endometrium. A traveling egg is fertilized by a sperm, optimally in the oviduct. As a zygote, it moves along to implant in the endometrium, if pregnancy is successfully started.

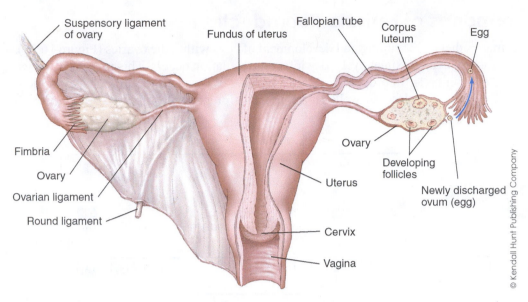

**Figure 16.21** Female Reproductive Structures. An egg moves from the ovaries through the fallopian tube and when fertilized, it finds a home in the uterus.

## External Structures: Outside the Cervix

On the external side, the female reproductive system consists of tissues arising from the same embryonic layers as the male penis and scrotum. The vulva are external female genital structures: the clitoris and labia surround the opening of the vagina, or birth canal. The clitoris is a site of external female stimulation. Just inside, the vagina receives semen to begin a sperm's journey through the female reproductive tract. The lowest portion of the uterus, separating it from the vagina, is the circular opening called the cervix (Figure 16.22).

**Vagina**

A muscular tube leading from the outside to the cervix of the uterus in female mammals.

**Clitoris**

A site of external female stimulation.

Clitoris

Labia minora

Labia majora

Urethral opening

Vestibule

Vaginal entrance

Anus

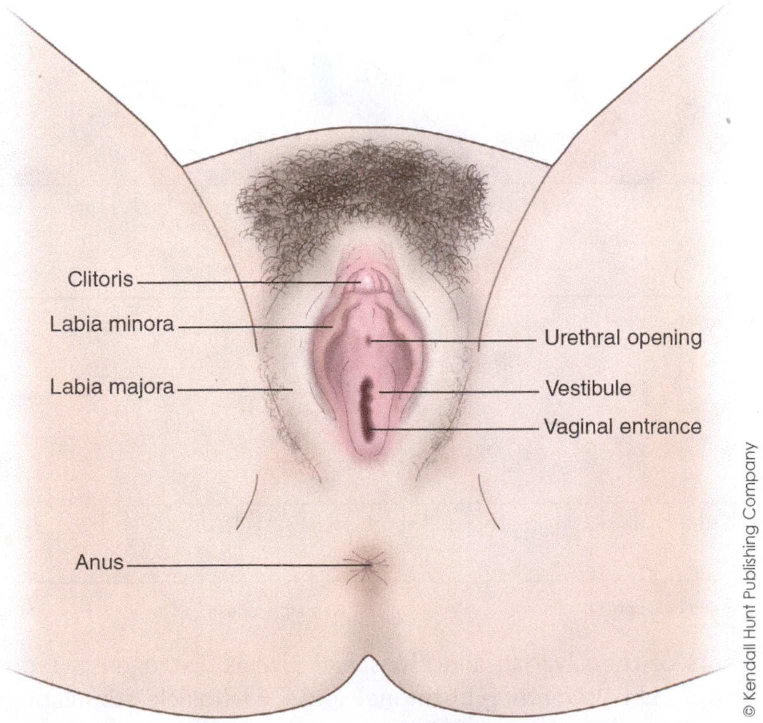

**Figure 16.22** External Female Anatomy. The external labia surround the vaginal opening. The clitoris is an excitatory center in sexual foreplay.

# Hormones of Female Reproduction

Hormones dictate the timing and development of eggs within the ovaries (Figure 16.23). Each month, at approximately 28-day intervals, ovulation occurs. This process is known as the ovarian cycle. In order to ready the reproductive system for a pregnancy, a menstrual cycle takes place as well. Hormones also prepare the uterus for an embryo's

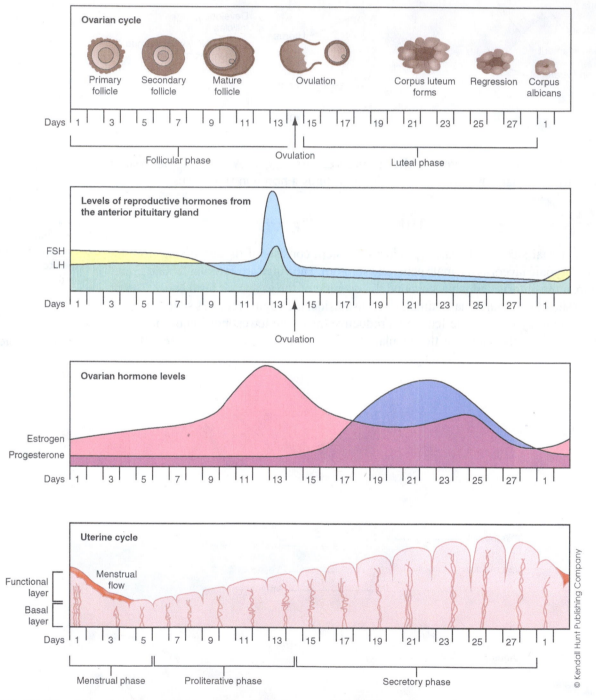

**Figure 16.23**   Hormones and Events of the Ovarian and Menstrual Cycles. Estrogen and progesterone prepare the uterus for egg implantation. LH (luteinizing hormone) and FSH (follicle-stimulating hormone) stimulate the follicles in ovaries to development.

implantation during the menstrual cycle. In our story, chronicling Mrs. Eddy's pregnancy, "the catamenial secretion . . . " described refers to menses or menstruation. This process takes place in regular intervals, but ceases during pregnancy:

1) Menstruation. When bleeding first appears, the endometrial linings are shed in what has traditionally been called a "period" or **menstruation**. Menstruation lasts between 3 and 5 days. At the end of menstruation, levels of estrogen decrease. Low amounts of estrogen cause the pituitary gland to release follicle-stimulating hormone (FSH). FSH as the name indicates develops eggs in their follicles.

2) Proliferation. Follicles produce estrogen once again, which develops the uterine wall. The uterus grows and develops during the proliferation phase, just before pregnancy occurs. The endometrium thickens and increases in its blood supply. In this phase, the egg travels to the uterus and when fertilized, it is ready to implant.

3) **Ovulation**. About half way through the menstrual cycle (14 days), high levels of estrogen stimulate the pituitary to release a spike of luteinizing hormone (LH) into the bloodstream. This sharp rise causes the bursting of a follicle and its release of an egg in ovulation. Just after ovulation, the corpus luteum fills with fluid and secretes estrogen and progesterone, both of which thicken the endometrium. This readies the uterus for another implantation by an embryo.

If pregnancy occurs, a hormone human chorionic gonadotropin (hCG) prevents the breakdown of the corpus luteum. This allows the continued release of estrogen and progesterone. Both hormones maintain the endometrium enabling implantation and pregnancy. The cycling of menses stops to allow development of an embryo.

However, if pregnancy does not occur, the corpus luteum continues to emit its hormones for about 12 days. This prevents another follicle from developing so that two eggs are not moving through the tract at any one time. When the corpus luteum does finally break down, there is a decline in the amount of estrogen. The endometrial lining then sheds and menstruation once again occurs. Also, the pituitary again secretes FSH and LH, stimulating another ovulation, starting the process over.

## Menarche and Menopause

Reproductive cycling continues from menarche, the age at first menstruation during puberty, until menopause at which time ovulation and menstruation cease. Menopause usually occurs between ages 45 and 55 but there is no exact moment. The onset of menopause is related to the number of viable eggs remaining in the ovaries. Of course, estrogen levels drop in postmenopausal women. Humans are the only species to lose their fertility at a time in their lives. It is likely a result of our longevity that has made this an event – our life expectancy throughout most of human history was well below 40 years of age!

Menarche is taking place at younger ages in our society. In the early 1900s, menarche was on average at age 15–16, but more recent studies indicate average ages reaching 11 and 13 years of age. The graph in Figure 16.24 shows the changing ages at first menarche from the mid-1800s to 2000 in France. U.S. data parallel those found in France. The causes are likely many-fold: changes in diet, easier access to foods, increased intakes of artificial hormones in milk, for example. However, the exact cause is unknown. Additionally, the longer estrogen is in the system, the greater the risks for reproductive cancers and even heart disease. Early menarche may be a health problem in the longer term.

**Menstrual cycle**

The process of ovulation in women and other female primates.

**Menstruation**

Discharge of blood from the uterus.

**Proliferation**

Rapid growth of cells by producing new parts.

**Menarche**

Beginning of menstruation.

**Menopause**

The time when ovulation and menstruation cease.

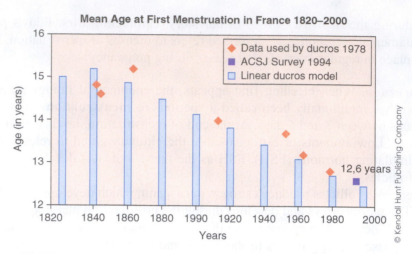

**Figure 16.24**    Graph of Age at First Menarche and Year Since 1820 in France.

# What Happens After an Egg Meets Sperm?

## *Fertilization*

With between 150 and 350 million sperm in the vagina, only about 100 sperm will survive their travels to the oviduct, where eggs are fertilized. It takes about 30 minutes to make it to the site of fertilization. Only one sperm is able to successfully penetrate an egg cell. A layer of proteins called the zona pellucida surrounds and protects the egg.

As the sperm approaches the egg, calcium ions are released from the zona pellucida, which increase sperm speed. They become even more competitive, riled up by the calcium in what is termed sperm activation. This ensures that the competition is fierce and that the best or most powerful sperm wins. Sperm compete in several ways, for example by producing a mate plug to prevent other males from penetration of a female after he has had intercourse with her.

The acrosome of the successful sperm's head fuses with the egg's protein layers surrounding the egg. When the acrosome's enzyme contents are released, they digest through the protein layer. This opens a channel pathway through which the sperm travels to enter into the egg. When this occurs, it stimulates changes in the coat surrounding the egg to prevent a second sperm penetration (Figure 16.25).

Once within the egg, a sperm disintegrates except for its nucleus. The sperm's entrance into the egg causes the final step in meiosis for the egg. The egg divides once again in Meiosis II to become a mature egg, an ovum and releases from it a second polar body. When the haploid sperm nucleus fuses with the egg's nucleus, a new diploid cell is produced, the zygote.

# Embryology

A zygote looks as close to a human as a football. How does this spherical ball become a specialized human being? The study of the embryo and the stages it undergoes is known as embryology.

The first step in embryology is mitosis, called cleavage. The more cells a zygote has, the less likely it is to die if something goes wrong in any one of them. Thus, very rapidly after fertilization, in about 30 hours, a zygote undergoes cleavage. As the zygote develops, it divides mitotically, over and over to create a ball of cells as shown in

---

**Zona pellucida**

A layer of proteins that surrounds and protects the ovum.

**Sperm activation**

The process by which sperm are additionally mobilized by calcium and even more able to fertilize the egg.

**Ovum**

A mature egg.

**Zygote**

A diploid cell that is produced when the haploid sperm nucleus fuses with the egg's haploid nucleus.

**Embryology**

The study of the embryo and the stages it undergoes.

**Cleavage**

Division of cells in the early embryo stage.

**Figure 16.25**    Sperm Enters the Female Egg by Burrowing in through Protective Layers. Sperm contacts, penetrates, and fuses its genetic material with an egg's nucleus. Millions of sperm attempt to fertilize one egg. It is an intense competition.

Figure 16.26. First 2, then 4, then 8 cells divide to become a morula, or a solid ball of cells that resemble a raspberry. During cleavage and the forming of a morula, there is a great increase in the number of cells but not the overall size. Its energy is placed into mitosis and not growth.

After the sixth day of continual mitosis, a morula forms a hollow cavity that contains a mass of cells about 1,000 in number. The newly formed structure is called a blastula. The blastula's inner cells are called its inner cell mass. The inner cell mass

**Morula**

A solid ball of cells formed by cleavage of a fertilized ovum.

**Blastula**

A hollow ball of cells at the early stage of development.

**Inner cell mass**

Inner cells of the blastula.

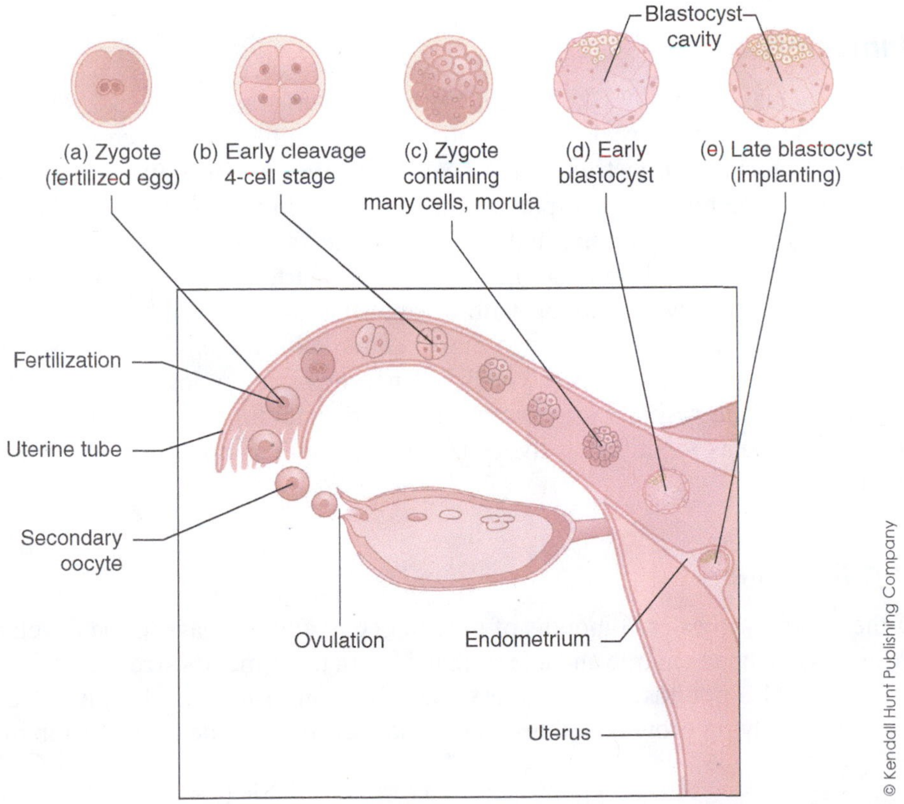

(a) Zygote
(fertilized egg)

(b) Early cleavage
4-cell stage

(c) Zygote
containing
many cells, morula

(d) Early
blastocyst

Blastocyst
cavity

(e) Late blastocyst
(implanting)

Fertilization

Uterine tube

Secondary
oocyte

Ovulation

Endometrium

Uterus

© Kendall Hunt Publishing Company

**Figure 16.26**    Cleavage in a Zygote. Cell division forms a ball of cells, the morula after only 4 days from fertilization.

**Amnion**

The fluid sac that protects a fetus.

**Chorion**

Fetal membrane that nourishes the fetus and becomes a part of its placenta.

**Gastrulation**

A developmental stage in which three distinct germ layers are formed.

**Ectoderm**

Outer layer of the gastrula, which develops into the skin and nervous system.

**Mesoderm**

The middle layer of gastrula that develops into the body's organs and muscles.

**Endoderm**

Inner layer of gastrula forming the digestive and respiratory tracts.

**Gastrula**

The entire mass of cells in an embryo developing after the blastula stage.

**Notochord**

A flexible rod of nerve tissue that develops in all chordates.

**Neurulation**

The process by which the ectoderm folds to become the brain and spinal cord.

**Trimester**

A normal pregnancy divided roughly into three parts.

**Gestation**

The process of development in the womb.

**Embryo**

An unborn offspring in the early stages of development.

eventually develops into the embryo and then into a human being. The outer 1,000 cells become structures that support the embryo: the amnion, which is the fluid sac that protects a fetus; a chorion, which nourishes the fetus and becomes a part of its placenta.

In the next phase of development, gastrulation results in the formation of three distinct germ layers (Figure 16.27). When cells of the blastula migrate inward, they form three germ layers. Each layer is specialized to become a different part of the body. The ectoderm is the outer layer of the gastrula, which develops into the skin and nervous system. The mesoderm is the middle layer and develops into the muscles and skeleton. The inner layer, the endoderm, becomes the lining of the digestive and respiratory tracts, along with associated organs: the liver and pancreas. The entire mass of cells, along with their germ layers, is defined as the gastrula.

During the third week, neural tube development occurs, with nerves running along the backside of the embryo. They form a notochord, which is a flexible rod of nerve tissue that develops in all chordates (described in Chapter 10). The ectoderm folds to become our brain and spinal cord in a process known as neurulation.

## Stages of Pregnancy

A normal pregnancy lasts 40 weeks and is divided roughly into three parts called trimesters. The process of development in the womb is termed gestation. Each trimester of gestation is characterized by different changes (Figure 16.28). During the first 8 weeks, when the zygote is implanted in the uterus, it is called an embryo. After this point, organs and other major structures form, demarcating a point of higher development. The embryo after 8 weeks is then called a **fetus**.

### First Trimester

In the first 6–7 days, the embryo still travels to the uterus and implants only on the seventh day. In the second week, gastrulation occurs, forming the three germ layers. Organs begin to form after gastrulation. A heart beats after the third week of gestation, and at the end of the fourth week, pharyngeal arches form in the embryo. These will later become the pharynx, larynx, and features of the face and neck. By the fourth week, all of the major organs, including the eyes and heart, are formed. In the second month, arms and legs develop, along with organs such as the liver, gall bladder, and pancreas. However, very little growth in mass occurs, with the embryo still weighing only about 1 g. In the third month, the embryo's nervous system develops. This allows it to have reflexes and responses to stimuli, including suckling. A mother's ability to make milk also begins simultaneously. Her mammary glands are stimulated to ready for milk production.

### Second Trimester

During the second trimester, 3 months of growth occur, adding mass to the developing fetus. Muscles and bone enlarge and a fetus may kick in this time. Its size increases from only 1 to 600 g (1.3 pounds). There is less specialization in months 4, 5, and 6 and a focus is increasingly on growth. A fetus still cannot survive outside of the womb in this period.

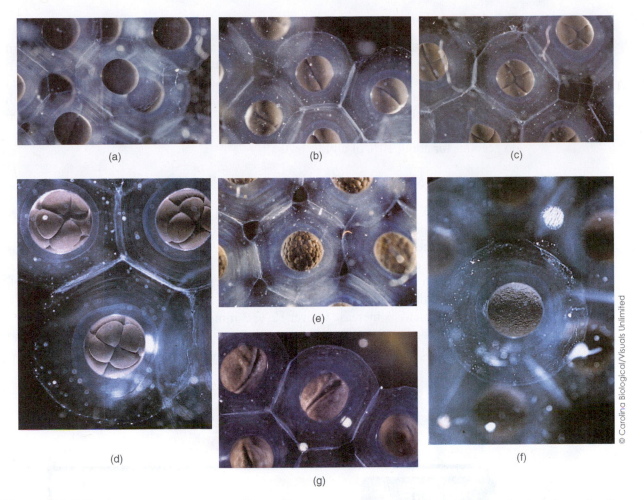

(a)    (b)    (c)

(d)    (e)

(g)    (f)

© Carolina Biological/Visuals Unlimited

**Figure 16.27** Early Embryo Development. Getting to gastrulation through inward migration of cells and specialization into three germ layers, the ectoderm, mesoderm, and endoderm.

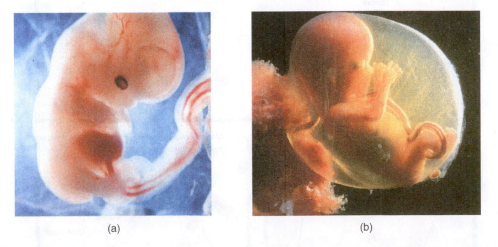

(a)    (b)

**Figure 16.28** Later Stages of Human Development. a. 6-week-old human embryo (2.5 cm), 1 inch long. b. 16-week-old human fetus (15 cm), 6 inches long. Lennart Nilsson/Bonnier Alba AB./*A Child is Born*, Dell Publishing Group.

### *Third Trimester*

Size increases markedly during the third trimester and a fetus's weight increases from 1.3 pounds to an average of 7.1 pounds at the ninth month. After 24 weeks, modern technology can keep babies alive, but preterm births have risks associated. The earliest preterm baby kept alive was after only 21 weeks gestation, but this is an anomaly. Birth defects and trouble breathing (especially before 34 weeks, when certain lung cells are matured) are at higher risk for premature infants. Premature births should be avoided but do account for about 10% of all live births.

## Birth and After

**Parturition**

The contractions and dilation of the cervix during child birth.

**Labor**

The opening of the cervix and uterus contractions leading to the birth of the baby.

Contractions and dilation of the cervix begin normal birth processes, also known as parturition. Hormones, such as prostaglandins and oxytocin, stimulate uterine contractions. These movements cause the cervix to open and allow the fetus space. The contractions are known as labor. The baby is delivered when the cervix is dilated large enough, usually about 10 cm in width. Widening of the cervix is accompanied by effacement, which is a thinning of the tissues surrounding the cervix.

When contractions increase to every 2 or 3 minutes, birth usually occurs. Delivery happens when the baby's head passes through the vagina. The umbilical cord is cut and clamped, and the baby is induced to breathe on its own. After the birth, a brief relaxation

## FETUS DELIVERY

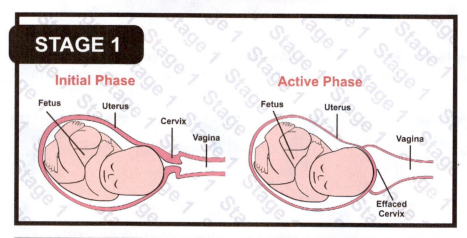

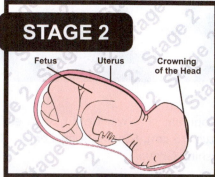

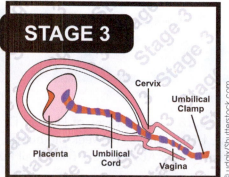

© udaix/Shutterstock.com

**Figure 16.29** Phases of Birth: Stage 1: Stage 2: Birth, and Stage 3: After-birth.

period is followed by smaller contractions that propel the placenta out of the uterus. This is commonly called the after-birth. The phases of birth are shown in Figure 16.29.

When a baby begins suckling on its mother's breasts, the effect is to stimulate milk production, or lactation. Suckling causes prolactin to be released from the pituitary gland and induces mammary cells to make milk, in the process called **lactation**. This is one of the few examples when one organism (a baby) causes the production of hormones in another organism (the mother). The positive feedback mechanism for lactation comes from mammary glands shown in Figure 16.30.

In the first few days, a yellowish fluid called colostrum is released. Colostrum is high in proteins and antibodies, which helps develop an infant's immune system in its early days. Colostrum slowly dissipates over the first week and milk production increases. Milk is high in fat and energy, giving needed nutrients to newborns.

Lactation decreases a woman's chances at fertility (though not completely) because suckling prevents LH from being released by the pituitary. LH plays a role in ovulation, without which there is no future pregnancy. This period of infertility lasts only about 6 months and is called "lactational amenorrhea." Nursing is highly recommended by the American Medical Association (AMA) as a natural way to give both immunity and nutrients to newborns during their first year of life. Recent studies report fewer infections and better cognitive skills among children who were nursed compared to bottle-fed at critical points in their development. There is debate on how long and where to breast-feed, but its value is supported by the research.

**After-birth**

A brief relaxation period followed by smaller contractions that propel the placenta out of the uterus after child birth.

**Lactation**

Formation of milk by mammary glands.

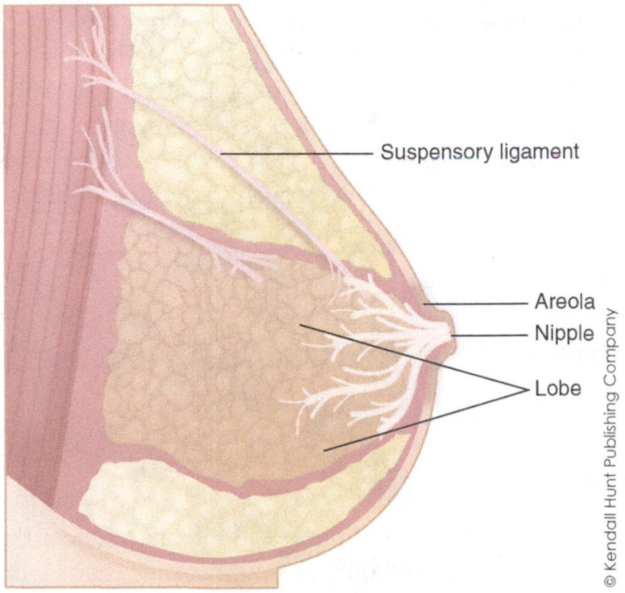

**Figure 16.30**  Mammary Gland Milk Production and Suckling. Note that the mechanism is an example of positive feedback control. There is hormonal–nerve communication between the hypothalamus, oxytocin, and nerves in the nipples.

## SHOULD GRANDDAD STOP BEING SCREENED FOR PROSTATE CANCER?

The AMA now recommends that screening for prostate cancer only prevents one death for every 1,000 men screened over 75 years of age. The AMA based its decision on a meta-analysis (grouping of studies) which shows no statistical significance in survival rates between those screened and those unscreened for prostate cancer after 75 years of age. The medical community cites side effects from false positives and cancer treatments as linked to prostate testing. Therefore, screening men after 75 is no longer recommended.

Thus, the elderly men you may know, some in your own family, are being told by their doctors to stop screening for prostate cancer. However, suppose you have prostate cancer and early detection could make you the person whose life the test saves? Is a healthy 75-year-old ready to die, doing nothing about his risks? Foregoing screening after certain ages is still controversial.

The test to screen for prostate cancer is called a prostate-specific antigen (PSA) test. It measures the amount of proteins given off by prostate cancer cells. Consistent levels above 2.5–4.0 generally call for a cancer screening called a biopsy. In a biopsy, cells from the prostate are taken and analyzed under the microscope for cancer. This test should be done along with a digital examination to check if the prostate is rough or smooth in morphology. If it is rough, it is more likely to be cancerous. Many factors such as symptoms – pain – and a person's family history are considered in determining risks.

Also, perhaps symptoms, patient family history, and other factors should be considered before simply not testing people after 75. Many people are healthy at after age 75. This is probably a reason many men still insist on getting tested, regardless of medical advice. Screening catches cancers early and prevents some deaths; but should one go against the doctor's advice?

# Malfunctions of the Reproduction System

## Male Cancers

### Prostate Cancer

Reproductive cancers in men are major killers, with prostate cancer the second leading cause of cancer deaths in men, second only to lung cancer. About 1 in 36 men will die from prostate cancer, but it is a very common disease. If a man lives long enough, his chances of getting prostate cancer is 100%! However, most men die with prostate cancer not because of prostate cancer. This is because most prostate cancers metastasize very slowly. Thus, there is controversy on determining how aggressive prostate cancer treatments should be and at what ages screenings should be ceased.

### Testicular Cancer

A reproductive cancer of young adults occurs as testicular cancer. It is the most common cancer in young men ages 17–34. It presents as a lump or swelling and may or may not be painful. Many forms are aggressive and travel through the pelvic region to stage

IV cancer within months. Thus, monthly self-screening is recommended to check for changes occurring on the testicles. Early detection is the key to survival of testicular cancer, like most other cancers.

### Penile Cancer

Penile cancer is relatively rare but is associated with not being circumcised. Its incidence is less than 0.01%, but it can be deadly. There is controversy regarding circumcision or the removal of foreskin from the penis. Although uncircumcised males have greater chances of contracting venereal diseases, there are low risks associated with circumcision as a surgical technique. In one recent case, an infection from instruments used during circumcision, lead to an infant's death.

## Inflammations in Male Organs

A common ailment of athletes is a result of physical trauma to the external male anatomy. In sports, testicles are sometimes twisted, resulting in a swelling. Orchitis is the swelling of the testicles and scrotal sac. Its severity depends upon the damage to the testes. Sometimes, the testes are twisted and blood supply is cut off, requiring surgery. Usually, testes can be manipulated to resupply the testes with blood. When the epididymis is inflamed, either due to physical trauma or due to a microbe causing venereal disease such as chlamydia, it is called epididymitis. It may damage the epididymis to cause infertility in males.

## Infertility

Some of the diseases described result in infertility, or an inability to conceive a child after one year. Infertility affects about 10% of couples. Its causes are evenly split between malfunctions of male and female reproductive systems. There are numerous effective treatments for infertility.

A low amount of semen does not indicate infertility. Fertility is determined by sperm health in males and reproductive health in females. Infertility in males occurs when sperm are abnormal or low in numbers to prevent pregnancy. Male infertility is often a result of physical trauma, hormonal deficiencies, or scarring of tissues along reproductive linings. Sometimes damage occurs in males and females as a result of sexually transmitted diseases. Microbes such as the herpes virus or chlamydia bacteria cause damage in reproductive structures.

Assisted reproductive technology to improve fertility is used widely. They include a variety of techniques, but the most common form is *in vitro* fertilization (IVF) (Figure 16.31). In IVF, female eggs are collected and combined with sperm in a petri dish. The resulting zygote (or fertilized egg, as you recall from other chapters) develops into an eight-cell stage embryo and is then inserted into the female's uterus. The zygote implants into the uterine wall, allowing development to occur.

Other techniques allow for sex selection by sorting sperm for those containing Y chromosomes and then carrying out *in vitro* fertilization. This method is called sperm sorting and uses a florescent dye, which latches onto DNA in sperm. Because X chromosomes are about 3% larger than Y chromosomes, each sperm type is separated using a small electric shock. The success rate for sperm sorting is about 70%–90%.

To achieve 100% success rates to obtain the desired sex of a child, pre-implantation genetic diagnosis (PGD) is used. In this procedure, females are given hormones to develop their eggs. Their eggs are then surgically removed and used in IVF to produce several embryos. A single cell from each embryo is removed and their chromosomes are analyzed to determine sex. The desired XX or XY embryo is then placed into the

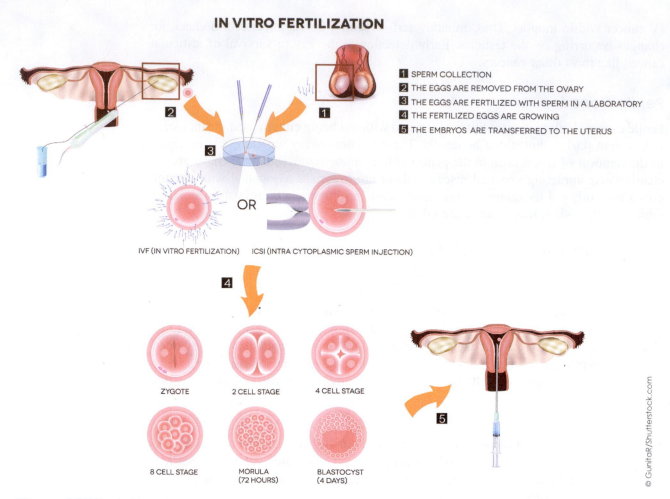

**IN VITRO FERTILIZATION**

1 SPERM COLLECTION
2 THE EGGS ARE REMOVED FROM THE OVARY
3 THE EGGS ARE FERTILIZED WITH SPERM IN A LABORATORY
4 THE FERTILIZED EGGS ARE GROWING
5 THE EMBRYOS ARE TRANSFERRED TO THE UTERUS

IVF (IN VITRO FERTILIZATION)   ICSI (INTRA CYTOPLASMIC SPERM INJECTION)

ZYGOTE   2 CELL STAGE   4 CELL STAGE

8 CELL STAGE   MORULA (72 HOURS)   BLASTOCYST (4 DAYS)

© GunitaR/Shutterstock.com

**Figure 16.31** *In Vitro* Fertilization (IVF) Involves Several Steps, from Egg Removal to Fertilization and Zygotes Growth to Implantation in a Uterus. IVF helps thousands of couples conceive despite infertility.

mother's uterus. PGD (cost of $12,000) is more invasive and costly than sperm sorting (cost of $1,500), but chances are far greater of obtaining the desired sex of a child.

# Contraception

Of course, the opposite of infertility is the desire to prevent pregnancy. There are a number of contraception methods, which either prevent ovulation, fertilization, or implantation. Birth control pills work by interfering with ovulation. The pills contain estrogen and progesterone, which prevent FSH and LH from stimulating follicles and ovulation. Because eggs are not released, pregnancy is prevented. Condoms, a diaphragm, and abstinence prevent fertilization. Each as varying degrees of success, but abstinence is 100% if used. Condoms come in at 89-99% when used correctly. Some devices may be implanted into the uterus, preventing implantation. The intrauterine device (IUD) is inserted into the uterus by a doctor and has 99% effectiveness. However, there have been multiple cases of infection and discomfort in IUD use. See Table 16.1 for an overview of birth control methods.

# Female Cancers

Women's cancers are very common, in part because of the continual new production of cells during reproductive processes. As shown in previous sections, these cycles require hormones, secretions, and new cells to be produced continually. When mitosis occurs at

**Table 16.1** Methods of contraception and their efficacy.

**Common Birth Control Methods**

| Method | Description / Explanation | User | Failure Rate* | Disadvantages and Side Effects |
|---|---|---|---|---|
| contraceptive patch | patch worn on abdomen or buttocks provides transdermal delivery of estrogen and progesterone; one patch per week for three weeks; period occurs during fourth week | female | 1–2 | may cause skin irritation at patch site; may cause irregular periods and headaches, nausea, and breast discomfort similar to side effects of oral contraceptives |
| diaphragm | flexible wire circle covered with latex and placed over cervix prevents sperm from entering uterus; generally used with spermicide | female | 21 | must be fitted by physician; may slip during intercourse; must remain in place for several hours after intercourse |
| intrauterine device (IUD) | small T-shaped plastic or copper device placed in uterus prevents implantation of fertilized egg; effective for 10 years | female | <1 | must be inserted by physician; can cause pelvic inflammatory disease, cramps, and increased menstrual flow; is dangerous if pregnancy results |
| male condom | thin sheath made of latex placed over penis to prevent sperm from entering vagina; often coated with spermicide | male | 11 | decreases sensitivity in male; latex can deteriorate over time; only latex condoms help protect against AIDS |
| female condom | lubricated polyurethane sheath shaped like male condom; placed into vagina to prevent entry of sperm | female | 21 | not as effective as latex condom; no protection against AIDS |
| natural family planning (rhythm method) | abstinence from sexual intercourse around the time of ovulation | male/ female | 20 | difficult to predict when ovulation will occur |
| oral contraceptive (combined pill) | synthetic estrogen and/or synthetic progesterone taken throughout menstrual cycle to prevent ovulation | female | 1–2 | must be prescribed by physician; can cause heart disease, hypertension, and stroke; may increase risk of infertility |
| oral contraceptive (progestin only minipill) | synthetic progesterone taken throughout menstrual cycle to prevent ovulation | female | 2 | Must be prescribed by physician; may cause irregular bleeding, weight gain, breast tenderness |
| spermicides | foams, creams, jellies, or suppositories containing chemicals that kill sperm; placed in vagina before sexual intercourse | female | 20–50 | can cause allergic reactions |
| Surgical sterilization (female) | oviducts in female are blocked, preventing egg from reaching sperm | female | <1 | must be performed by physician; usually not reversible |
| Surgical sterilization (male) | vasa deferentia in male surgically cut and tied off, preventing sperm from joining semen | male | <1 | must be performed by physician; may not be reversible |

* Failure rate is the percentage of women who become pregnant during the first year of birth control use. The data come from clinical trials submitted to the Food and Drug Administration during product reviews. The failure rate for the rhythm method is estimated from the published literature. For comparison, about 85 out of 100 sexually active women who wish to become pregnant would become pregnant within one year.

high rates, its chances for error also increase. In female processes, errors in cell division result in types of cancers.

The most common form of cancer in females is breast cancer. It affects one in nine women and begins as a lump or swelling in the mammary regions. The survival rate from breast cancer, if detected early, is over 90%. However, it requires regular mammograms to screen for abnormal growths. Ovarian cancer is abnormal growth in the ovaries, a deadly cancer because it is often not caught early. It symptoms include bloating and abdominal discomfort, but it is difficult to detect. As a result, ovarian cancer has a survival rate of only 40% after five years.

There is a high genetic link to both breast and ovarian cancer. The *BRCA 1* and *BRCA 2* genes are strongly associated with these cancers. Angelina Jolie underwent a controversial procedure after finding out that she tested positive for both genes, giving her an 89% chance of getting breast cancer. She elected to have both breasts removed in a double mastectomy. This improves her chances to over 95% of not getting the cancer. It is controversial because surgery has risks and financial costs. Indeed, some health insurance companies will not pay for the pre-emptive surgery had by Ms. Jolie (Figure 16.32).

The debate about pre-emptive surgery and early screening for breast and ovarian cancer continues to make headlines. The U.S. Preventive Services Task Force, as well as the American Cancer Society, recently announced that screening for the BRCA genes is not recommended and may lead to false positives and undue stress. It asks doctors to not request screening of women unless the patient has a family history of reproductive cancers.

From our knowledge of genetics and pedigrees, consider that many cancers are recessive in origin and skip generations. Also, pedigrees are quite incomplete for most of us. Do we really know what great aunt Bertha died from in 1936? Because one in three people will die from a cancer-related illness, cancer is probably in *everyone's* family; not just some people's.

The CA-125 blood test specifically screens for ovarian cancer. Its use has also been placed into question, when in study of over 78,000 women screening was shown not to help survival from the disease. In addition, about 10% had false positives resulting

© Joe Seer/Shutterstock.com

**Figure 16.32** Angelina Jolie Was Predisposed to Cancer, Harboring the Mutated BRCA Genes. She had a mastectomy to decrease her risks of breast cancer. This pre-emptive surgery is elective and debated.

in unnecessary ovary removal. However, when used with other methods of screening, fewer errors are likely to be made. A trans-vaginal ultrasound for example detects more ovarian cancer correctly. Thus, I recommend the CA-125 and BRCA gene tests be considered by all women after age 40.

# Summary

The urinary and reproductive systems reside within the same pelvic region in close proximity, together referred to as the urogenital system. The two systems separately accomplish their goals: the reproductive system produces new offspring and the urinary system conserves water and removes wastes. The nephron is the functional unit of the kidney, a human's main organ of excretion. It concentrates urine as it flows through, removing wastes and selectively bringing back needed solutes and water into the body. Reproduction may be sexual or asexual, externally or internally fertilized, each method with pros and cons. As sperm move through the reproductive tract, fluids are added to form semen, which is injected into females during copulation. Reproductive diseases include cancers, inflammation, sexually transmitted diseases, and infertility.

---

## CHECK OUT

### Summary: Key Points

- Medical technology advanced, along with society, to help newborns and pregnancy; but it has also led to ethical challenges when deciding about fetal health.
- Excretion eliminates wastes while conserving water and other needed solutes and cells.
- Urinalysis is used to detect disease as well as illegal drugs.
- Sexual reproduction and internal fertilization, used by humans, leads to fewer but more genetically diverse offspring than asexual and external reproduction.
- Sperm move from the testes and epididymis through the vas deferens and into the female reproductive tract during copulation.
- Diseases of reproduction include a number of cancers, inflammation, sexually transmitted diseases, and infertility.

---

## KEY TERMS

| | |
|---|---|
| acrosome | bulbourethral gland |
| aldosterone | chorion |
| after-birth | cleavage |
| amnion | clitoris |
| amplexus | collecting duct |
| angiotensin II | copulation |
| antidiuretic hormone (ADH) | corpus luteum |
| blastula | dialysis, hemo-, peritoneal-, |
| Bowman's capsule | ectoderm |
| budding | embryo |

embryology
endoderm
endometrium
epididymis
excretion
fertilization, -internal, -external,
filtration
follicle
fragmentation
gastrula
gastrulation
genetic variation
gestation
glomerulus
impaired kidney function
inner cell mass
interstitial cells
kidney
kidney failure
labor
lactation
Loop of Henle
menarche
menopause
menstrual cycle
menstruation
mesoderm
morula
nephron
neurulation
notochord
oogenesis
osmoregulation
ovarian cycle
ovary

oviduct (fallopian tube)
ovulation
ovum
parturition
parthenogenesis
penis
proliferation
prostate gland
reabsorption
renal capsule
renal cortex
renal medulla
renal pelvis
reproduction, sexual-, asexual-,
reproductive system
scrotum
semen
secretion
seminal vesicle
seminiferous tubules
sperm activation
spermatogenesis
testes
trimester
urinalysis
urinary bladder
urinary system
ureter
urethra
urogenital system
uterus
vagina
zona pellucida
zygote

# Multiple Choice Questions

**Reflection questions:**

1. Ectopic pregnancies are treated differently today than 150 years ago because:
   a. new technologies are able to identify ectopic pregnancies
   b. surgery is now available to remove ectopic pregnancies
   c. termination of dangerous pregnancies are now acceptable in society
   d. all of the above

2. Which is NOT a function of the kidneys?
   a. pH regulation of the blood
   b. activation of vitamin D by calcium absorption
   c. removal of nitrogen containing compounds
   d. body temperature regulation

3. Which process occurs in the glomerulus?
   a. filtration
   b. secretion
   c. reabsorption
   d. activation

4. Trace the movement of urine from its production to end. Which is correct?
   a. kidney to bladder to ureter
   b. ureter to bladder to urethra
   c. glomerulus to ureter to renal pelvis
   d. renal pyramid to urethra to bladder

5. Which should NOT be found in normal urine?
   a. white blood cells
   b. sodium ions
   c. potassium ions
   d. water

6. Which represents a logical order, from production to ejaculation, of a sperm's movement through the male reproductive tract?
   a. epididymis → vas deferens → testes → urethra
   b. testes → vas deferens → epididymis → urethra
   c. testes → epididymis → vas deferens → urethra
   d. urethra → vas deferens → epididymis → testes

7. A mother gives birth but afterwards another set of contractions occurs, leading to:
   a. abnormal growth removal
   b. after birth of the placenta
   c. umbilical cord severing
   d. birth of the uterus

8. In which structure do sperm mature and become motile?
   a. epididymis
   b. testes
   c. vas deferens
   d. urethra

9. Which correctly MATCHES an organism's method of reproduction with its characteristic?
   a. internal fertilization – many offspring
   b. asexual reproduction – many offspring
   c. external fertilization – few gametes
   d. sexual reproduction – few gametes

10. All of the following is true about breast cancer EXCEPT:
    a. it is the leading cause of female cancer deaths
    b. it is linked to genetics
    c. it is treatable best at early stages
    d. it does not spread past mammary tissues

## Short Answer

1. Pregnancy occurs in three trimesters. Discuss changes in the fetus and mother during these three trimesters. Name one medical technique used today (and unavailable a century ago) that aids pregnancy.

2. Define the following terms: oviduct and vas deferens. List one way each of the terms differ from each other in relation to their (1) function; (2) role in transport of gametes; and (3) relationship with each other in reproductive anatomy.

3. List the three germ layers of an embryo and identify the tissues that they will become. In which phase of embryology are the germ layers found?

4. Draw a sketch of the nephron using arrows to show the role of ions and water in excretion. How does excretion occur in the nephron of kidneys? Be sure to include the three steps of excretion in your discussion.

5. In kidney impairment, what three processes are measured to determine the extent of kidney disease? What substances might be found in the urine that indicates kidney disease?

6. List the three hormones that control urine formation in the kidneys. Describe how alcohol affects kidney function. Connect kidney function with these hormones in your answer.

7. Explain the role of hormones during the ovarian cycle in females. What is the function of ovulation?

8. For question #7, explain how the menstrual cycle works in tandem with female hormones?

9. Describe the benefits of external fertilization to frogs and other land animals. What are its drawbacks?

10. Describe the symptoms of ovarian cancer. How may the disease be screened? What is the latest research about screening? What are the survival rates for ovarian cancer compared with breast cancer?

## Biology and Society Corner: Discussion Questions

1. Androgen insensitivity syndrome (AIS) is a disorder in which androgens (male sex hormones) are dysfunctional in males. Although sufferers are genetically male (XY), they appear as females with external genitalia and undescended testes.

   Maria Patino, a former Olympic athlete was found to have AIS. Maria considered herself a woman and the news of having AIS was a shock. She was banned from female competition due to her condition. Research AIS and this case: Was the decision by the Olympics committee justified? Why or why not? Did her condition give her unfair advantage over other women in her competition?

2. Suppose you have a long lost uncle, who has died and will leave you his fortune of 3.5 million dollars. There is one catch – you must have a son within two years. You cannot have a daughter. The uncle was oddly eccentric.

Would you (and your partner) consider using sex selection techniques described in this chapter to conceive a boy? Defend your answer.

If you answered yes, which technique would you use, sperm selection or PGD (pre-implantation genetic diagnosis)?

3.  Would you donate your kidney to your _____ if you had an acceptable MHC match? What factors influenced your decision?

    a.  brother
    b.  father
    c.  mother
    d.  friend
    e.  stranger

4.  The 1,500 prenatal tests being developed will soon come to the market. It will tell expecting parents about many aspects of the health of their fetus. Would you consider termination of a pregnancy if you find it is likely to be born:

    a)  and suffer from spinal muscular atrophy and die within nine months of birth;
    b)  and have cystic fibrosis, expected to suffer lung problems and die by age 20;
    c)  and have Huntington's disease, in which the person is healthy until age 40 and then suffers progressive muscular weakness and dies by age 50.

    *Many younger readers draw the line against termination here because they think at 40, one has lived a full life already!

    d)  will get Alzheimer's disease at age 60 or suffer manic depression its whole life.
    e)  will have dwarfism
    f)  no termination is acceptable

5.  Workforce screenings use urinalysis to test for drugs in their employees. Should drug testing be allowed under the law or is it a violation of privacy? What about for certain jobs? Which jobs, if any or all, would you suggest for mandatory drug testing?

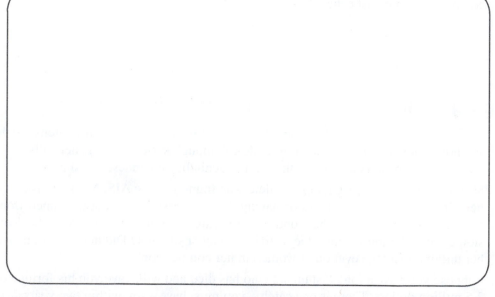

Figure – Concept Map of Chapter 16 Big Ideas

# UNIT 5
## A Small Hole Sinks a Big Ship – Our Fragile Ecosystem

# Population Dynamics and Communities that Form

## 17

## ESSENTIALS

Cheryl is a model

Her prince has arrived!

The cane toad (*Bufo marinus*) has taken over large parts of Australia's ecosystems. It is considered an invasive species

Toads are everywhere, as an invasive species that has a high rate of reproduction because they lack natural predators to keep their population numbers in check

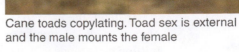

Cane toads copylating. Toad sex is external and the male mounts the female

© Kendall Hunt Publishing Company

## CHECK IN

**From reading this chapter, you will be able to:**

- Explain how invasive species and changes in their populations affect human society and the environment.
- Describe the characteristics of populations, its demographics, and how populations are studied.
- Define and describe invasive species, ecology, population ecology, logistic model of growth, exponential model of growth, carrying capacity, fertility rate, age structure diagram, survivorship curve, ecological footprint, density-dependent factor, density-independent factor, niche, habitat, resource partitioning, biotic factor, abiotic factor, predation, herbivory, parasitism, commensalism, mutualism, and competition.
- Apply models of population growth in human and nonhuman populations.
- Compare the two types of life histories and apply them to real examples using survivorship curves.
- Describe the roles organisms play within their community.
- List the types of population interactions within a community and give real examples.

## The Case of the Terrible Toads

Plop, Plop – into her drink! "What just fell into my drink?!" Cheryl called out. Cheryl, the partier did not like toads or frogs or anything slimy or warty. Cheryl was a model, tall with flowing blonde hair; and as such, she had no use for amphibians. She left the United States last night for a great holiday get away to an island off the coast of Australia.

"How sheik," Cheryl mused as her plane landed. On her vacation, Cheryl expected a beach hotel with hot days in the sand and hotter parties at night. At the airport, the taxi driver picked her up, quickly asking her, "Do you want to go out with me tonight?" Cheryl replied coldly, "I am sorry. I am here to meet my prince." The driver was disappointed. However, Cheryl had an agenda. She hoped to meet celebrities on the island and start her acting career, now that she finished her days at the university. This trip was her emancipation from school.

As they drove in from the airport, down through the desert, and to the beach, Cheryl noticed a strange sight along the beautiful beachfront – there were warty creatures. They were hanging from the trees, jumping all over the roads, and hopping in unison like dense mats.

The road could no longer be seen and the driver simply ran the animals over, unfazed by their presence. Cheryl was aghast, asking the driver: "Driver, what are these things?" "The cane toad, of course, and they are here to stay." he replied serenely. Cheryl demanded to him, "Turn back, we are leaving this place!" However, it was too late; the plane had gone and would not be back to the island for one week.

Cheryl did not do her homework about the island or cane toads before booking her trip. The cane toad had been introduced to Australia in 1935, with the hope that it would prey on the destructive cane beetles. About 3,000 cane toads were released into the wild. The experiment was, however, a failure in controlling the beetle populations.

Instead, the cane toad became a much larger problem. Within a few years, millions of cane toads swarmed Australia and continue today to be a pest organism. There are 200 million cane toads in Australia, and the government has identified it as a key threat to the environment and other organisms.

Cane toads, scientifically named *Bufo marinus*, appear ugly to many, including Cheryl. They are large, chubby, and have dry, warty skin. They breed easily, having frequent copulation. *B. marinus* is also toxic, affecting the heart muscle of those organisms that consume it, including humans. Cane toads cause the death of many types of native species to Australia. Many pets, cats and dogs especially, eat the toad and die from its venom. Cane toad venom is secreted as a milky liquid from its parotid salivary glands located over its shoulders.

*Bufo marinus* has no natural predators to keep their population in check. They are native species in South America and the Southern United States, but their natural predators keep their numbers manageable in those areas. Those organisms eaten by cane toads are also being depleted, such as a number of insects. Species consuming those insects are also endangered, with little food remaining after the toads enter into an area. Thus, interactions between the different organisms in Australia have been shifted out of balance. The cane toad is therefore a dangerous invasive species to Australia.

Cheryl saw (and was touched by) things from the sky, from the ground, and from the water. All over, the toad creatures rubbed up against her in ways she had never seen before. The cane toad became, very quickly, Cheryl's worst nightmare.

That evening, the driver met Cheryl in the local bar. Cheryl sat with her drink, a cane toad and the driver at a table. Cheryl looked at the driver, the toad looked at Cheryl, and the driver looked at the toad. Cheryl looked angry. The driver addressed Cheryl the best way he could: "Here's your Prince. Pucker up, Cheryl…"

---

### CHECK UP SECTION

The cane toad in the story is considered an invasive species, which is any species not native to a region but grows rapidly due to lack of natural predators or parasites to keep them in check.

Over 4,500 invasive species have invaded the United States. Research the types of invasive species impacting the area in which you live or study. Be sure to describe the history of how the invasive species entered into the area and the techniques of eradicating the invasive species.

What changes have occurred in (a) society and (b) in the environment; as a result of the invasive species? How does it differ from organisms native to the region?

---

# Ecology is based on Studying Populations

## Order in a Population

In our story, invasive cane toads impact the environment and other organisms. They appear as a nuisance and repulse Cheryl on her vacation. However, their biological impacts on the island might be a bit more complex than just a mere bother to Cheryl.

How much land area will *B. marinus* take over? What can be done to slow its population growth? What organisms are eaten by the cane toad? How will the toads impact other species, such as freshwater turtles and crocodiles? All of these questions are answered through ecology, the study of the interactions between organisms and their environments. The term ecology derives from the Greek words "oikos," which means home and "logos," which is translated into "the study of." Together, ecology means the study of our home on this Earth.

**Ecology**

The study of the interactions between organisms and their environments.

**Population**

Organisms of the same species inhabiting a specific area.

**Ecosystem**

The interaction of the environment with a community of organisms.

**Biosphere**

All of the Earth's ecosystems.

This unit studies nature and the environment and is divided into three chapters. In this chapter, we will explore how biology occurs within populations. The environment is organized into different groupings (Figure 17.1). A population is a group of organisms of the same species living in an area. A population of organisms, such as *B. marinus*, for example, grows and is structured in ways that are studied by ecologists. In the latter part of this chapter, the **community** (aka biocenoses) is studied. A community is a set of populations interacting with each other.

In Chapter 18, we will study how the environment interacts with those communities, in a grouping called an ecosystem. All of the Earth's ecosystems are collectively known as the biosphere, which will be the focus of Chapter 19, the last chapter in this unit. The components of the biosphere as well as threats to its health will be explored. Figure 17.1 shows the hierarchy of environmental organization from population, community, and ecosystem to biosphere.

Ecology is based on the dynamics of populations – a population is the basic unit of study in ecology. The ways a group of species grows, shrinks, and breeds, for example, show the dynamics occurring within a population.

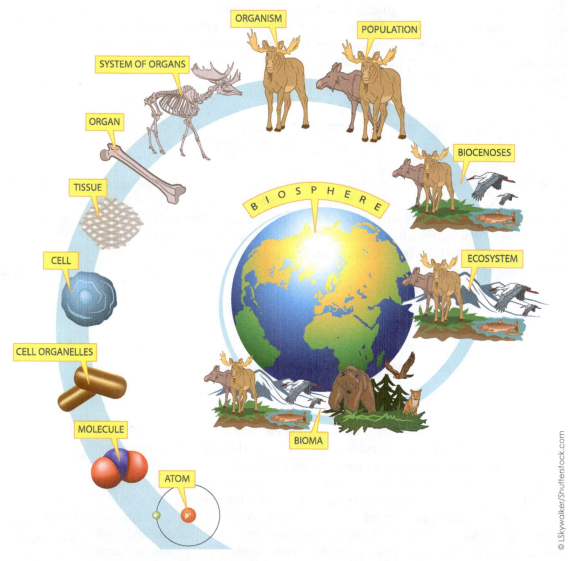

© LSkywalker/Shutterstock.com

**Figure 17.1** Hierarchy of environmental organization: individual, population, community, ecosystem, and biosphere of organisms such as the cane toad. The toad's role within each of these organizations should be studied to help combat its invasiveness.

Population ecology studies a population of organisms and how it interacts with its environment. It explores population patterns of growth and interactions with other species. *B. marinus* is treated through this chapter as a unit of study to explore: how its role in biology shapes the environment and other species around it. We will compare cane toad population growth to look at human populations. This chapter especially probes human population growth in its role underlying many ecological issues today.

## Population Demographics

The rate at which a population grows and shrinks is measured in several ways. Trees are tagged to measure their arrangement in a forest and Americans have a census every 10 years. Both of these are methods to quantify and study a population. The data collected by ecologists about the statistics of a population of species is known as its demographics. Population demographics tell us a great deal about how a population is structured in terms of age, size, and density.

Population demographics help scientists predict how populations will change over time and affect other organisms. A population size gives the number of organisms in a population, and a population density reveals the number of organisms per area of land in an ecosystem. For example, the total number of cane toads in Australia is over 200 million, but on the island in our story, there are fewer toads totally but they have a greater population density. Higher population density could have more impacts on the environment than sheer numbers. *B. marinus* patterns should be studied in greater detail to determine the answers.

A population distribution shows the arrangement of organisms of a population across a particular region. Most cane toads are clumped on the northeastern areas of Australia.

Whether a population grows or shrinks depends on four factors: the number of births (new born additions); the number of deaths (those leaving permanently); the number of immigrants (new organisms from other areas); and the number of emigrants (organisms leaving the area). Population growth occurs when more individuals are entering than leaving a population. It can be calculated by the following equation: Growth rate = (Births + Immigrants) – (Deaths + Emigrants). In Mexico, for example, a rapidly growing population is a result of higher birth rates than death rates. Death rates declined much more than birth rates in the 20th century, leading to a need for emigration to stave off even larger population increases.

In the case of finding ways to decrease the cane toad population, natural predators for the cane toad would increase its death rate. So far, this strategy has not shown much success because the cane toad is toxic to many predators, such as the crocodile that swallows a cane toad whole, but ingests enough toxin to kill itself. In another approach, by decreasing its birth rate, growth would slow as well.

Some studies look to introduce sterile males into populations to prevent births. This strategy has had limited success. Changing climate would drive populations of cane toads out of areas by making conditions unfavorable. However, limiting immigration and increasing emigration sounds easier than it is – there is difficultly for humans to accomplish desired environmental change, especially without harmful consequences.

## Population as a Unit of Study

A population is the primary focus of ecologists because *it is the unit of study* of ecology and evolution. An individual is not as important in studying trends associated with changes in gene flow and in the environment. While a single person may respond one way to a changing factor, such as increased sunlight, her or his reaction is not so important to ecologists. If a person moves to Florida and becomes tanned, possibly dying from skin cancer after 30 years of exposure, an ecologist cannot make strong ecological conclusions. A single data point cannot drive research findings. If, however, offspring of

**Population ecology**

The study of a population of organisms and how it interacts with its environment.

**Population growth**

The increase in the number of individuals inhabiting a place.

**Demographics**

The data collected by ecologists about the statistics of a population of species.

**Population size**

A measurement of population that gives the raw number of organisms in a population.

**Population density**

A measurement of population that reveals the number of organisms per area of land in an ecosystem.

**Immigrants**

New organisms moving in from other areas.

**Emigrants**

Organisms leaving an area.

**Births**

New born additions.

**Deaths**

The end of life; those leaving permanently.

populations who move to Florida die off more rapidly as a group 30 years after their emigration, ecologists can draw stronger conclusions about environmental effects. Groups of organisms (populations) are studied in ecology to make generalizations.

Population demographics are not measurements of an individual: births, deaths, immigration, and emigration are all population terms. They are used to view how whole groups change and form patterns within ecosystems. Ecology requires that ecologists are both inductive and deductive, terms that harken back from Chapter 1. These are each methods of finding the answers to scientific questions. Ecologists make predictions and form models based on information obtained from their measurements of groups. Ecology forms hypotheses to test about certain questions in deduction: a study introducing sterile males into a population of cane toads, for example. Induction looks at the many factors about cane toads – their toxins, fertility, and relationships with other organisms – to form ideas on how to solve the toad infestation. To illustrate, perhaps a genetically modified hardy reptile could increase *B. marinus* death rates. Our story would then have a happier ending. There are many ecological approaches that could be taken. Both induction and deduction are vital in studying population ecology.

Individuals were the focus of the previous unit in this text, on anatomy and physiology. It studied the individual and its components: the parts of the digestive system, the function of the liver, and the ways the heart works and malfunctions. It studied the organism down. However, ecology is a type of macrobiology, which studies the organism up. It looks at the organization of populations. It shows the movement of genes through a population and not simply the genes of an individual. The ways populations form patterns in their arrangement and the responses to other organisms and nonliving factors are an ecologist's field. The chapters in this unit represent a shift from micro- to macrobiology.

## Population Growth

As described in Chapter 1, populations tend to increase in size and overreproduce, unless checked by predators. Charles Darwin, in Chapter 1, describes population expansion as a driving force in evolution, as you may recall. Populations grow too large, and competition for scarce resources leads to a survival of those best adapted. Populations grow until they exhaust their available resources.

An exponential model of population growth depicts a population increase that is unlimited. This model assumes that there are no predators, unlimited resources, and no pathogens. A population will grow, under these conditions, to its biotic potential or maximum possible growth under ideal conditions. Population increases for *B. marinus* in Australia have reached almost its biotic potential. Our story shows, like the cane toad, many invasive species often grow unchecked by predators. They enjoy conditions of full resource availability. Their growth is exponential (Figure 17.2a) and continues until changing conditions limit them.

Populations can never continue to grow unchecked. Even prokaryotes, which grow at the biotic potential most of their lives, become checked with changing environmental conditions and competition between themselves. The main limiting factor to population growth in cane toads is one another. When some cane toads are killed by human methods, it stimulates even more growth because the competition diminishes between toads themselves.

In natural ecosystems, populations are limited by a scarcity in resources. This results in a logistic model of population growth. In this model, a population first grows slowly, during a lag period, as population gains in size and colonizes an area (Figure 17.2b). Then, a time of rapid and unchecked growth, called the exponential period, occurs when resources are unlimited and the numbers in the population explode exponentially. After this phase, growth slows as limiting factors, such as food, spaces, light, and water, become scarcer. These are called density-dependent factors because they become

**Biotic potential**

Maximum possible growth achieved by organisms under ideal conditions.

**Exponential model of population growth**

A model that depicts the increase of population growth at a constant rate.

**Logistic model of population growth**

A model that depicts the decrease of population growth rate with the increasing number of individuals.

**Exponential period**

A time of rapid and unchecked growth.

**Density-dependent factors**

Factors that limit the population size, whose effects are dependent on the number of individuals of a population.

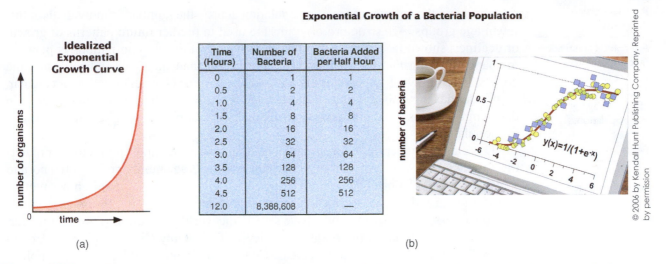

**Figure 17.2**    a. Exponential growth of *B. marinus* (cane toad). The cane toad population has expanded since 1935 without natural predators. Bacterial growth data, in the chart above, mirror an exponential pattern of growth. b. Logistic model of growth for deer, normally kept in check by limited resources and predators, shows an S-shaped curve of growth. Deer are usually kept in check by many factors including predators. From *Biological Perspectives*, 3rd ed by BSCS.

limiting only after populations reach higher densities. This period continues till a population reaches its carrying capacity or K, defined as the maximum number of individuals an environment is able to sustain in the long term. Most organisms follow this pattern of growth over time. When deer are introduced into a new area, they increase and level off in their rates of growth in a logistic pattern. The logistic model of population growth appears as an S-shape. Figure 17.2b shows this pattern for deer populations in Australia.

**Carrying capacity (K)**

The maximum number of individuals an environment is able to sustain in the long term.

## Human Population Structure

Human population has surpassed 7 billion people and is expected to reach 9 billion by 2050. Each year, 80 million new people are added to the total world population, with birth rates higher than death rates (Figure 17.3). During the past 1,000 years, human world population expanded exponentially. Most ecologists predict the emergence of logistic S-shaped growth in the 21st century.

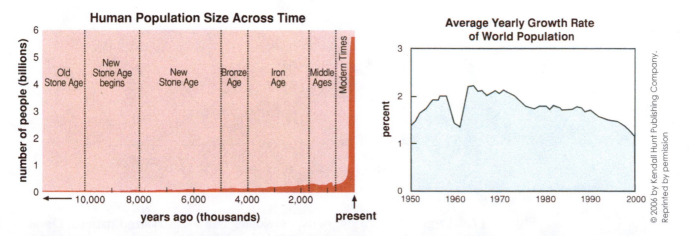

**Figure 17.3**    Human population growth. World population grew rapidly since the outbreaks of the plague in Europe and Asia. Humans are experiencing exponential population growth today. From *Biological Perspectives*, 3rd ed by BSCS.

**Age structure diagram**

A graphical illustration that are used to predict future patterns of growth or declines of various age groups in a population.

**Fertility rate**

Is the average number of children born to females in a population.

Studying the **age structure** of a population depicts the number of individuals at different age groups. Age structure diagrams are used to predict future patterns of growth or declines, shown in Figure 17.4. The age structure of developing nations is pyramid in shape, showing that there are many more young than old. This indicates that the population is growing. In developing nations, the age structure diagram is rectangular, with individuals evenly distributed at each age level. This indicates that populations in developing nations are stable. Age structure diagrams help us to predict rates of growth for populations.

Of course, number of children actually being born is an important factor in predicting population growth. Fertility rate is defined as the average number of children born to females in a population. The world fertility rate declined from 6.5 children per mother in 1950 to 2.5 today. The declines in fertility rate, due to education programs, contraception access, and other methods, have slowed the world population growth. However, fertility rate is still above the replacement level of 2.1. Only China, with rates around 1.6, will experience population decreases due to their one child per couple policy.

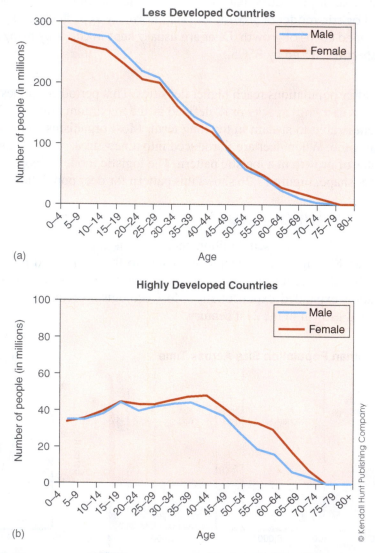

© Kendall Hunt Publishing Company

**Figure 17.4**  Age structure diagrams: developing and industrialized nations. Developing nations show a greater proportion of younger people and thus are predicted to have higher rates of growth.

The growth curve for past and current world population is shown in Figure 17.3, a result of the effect of high fertility rates in the world.

In 1996, the US population was 235 million. At the time, it was estimated that the carrying capacity of the United States was 250 million people. Today, we have over 310 million people, overshooting carrying-capacity estimates. Were the original estimates wrong? Were ecologists just naysays?

Factors changed, which could not be predicted at the time, which enlarged the United States carrying capacity substantially. There have been significant advances in farming and improvements in production and preservation of foods. The use of genetically modified organisms to increase food supplies, expansion into new habitats to obtain new resources, and new crop methods has increased our carrying capacity to 500 million, according to some estimates.

The US population grows each year by 3.3 million people, making us the fastest growing industrialized nation in the world, according to the US Census Data. The US population, given current rates and density-dependent factors, will reach 500 million in 2050. With new estimates of a carrying capacity, ecologists fear the effects of reaching these levels. Resource limitations are eventually expected to limit population growth. Of course, the wealthiest people consume most of the world's resources (Figure 17.5).

In every population, starvation, disease, and violence are the results of reaching carrying capacity. On the other hand, Often, density-independent factors decrease populations. These are the factors that increase death rate regardless of density. Landslide, earthquakes, and floods destroy life by directly killing organisms or reducing the size of their resources.

Both density-independent and density-dependent factors interact to affect population size. It is likely that cane toad populations will experience diseases, new predators, and natural disasters such as desertification, given enough time. If the numbers of toads increase, there are more chances for transmission of viruses and parasites. With greater density, disease is shown to spread more quickly. A population's increase is also seeds of its own destruction – for humans as well as cane toads.

A population's use of resources determines how quickly it will reach the carrying capacity. The ecological footprint of an organism or a population is defined as the amount of resources used: land, fuel, water, food, and other items. Some nations such as Sweden and New Zealand have small ecological footprints, while the others like the United States, Japan, and England have large ones. The ecological footprint of an American is 24 acres

**Density-independent factors**

Factors that limit population size, whose effects do not depend on the number of individuals of a population.

**Ecological footprint**

A population is defined as the amount of resources used.

© Christos Georghiou/Shutterstock.com

**Figure 17.5** The wealthiest people in the world use more resources than others. The wealthiest 16% in the world consume 80% of the world's resources. The cartoon depicts the United States as overusing the world's energy resources.

**Figure 17.6** Our ecological footprint varies for each nation in the world. The ecological footprint of an American is 24 acres worth of resources, but an ecological footprint from an Egyptian is 4 acres and from Bangladeshi 1.5 acres.

worth of resources, but one from an Egyptian is 4 acres and Bangladesh is 1.5. The map in Figure 17.6 gives a visualization for our resource use on the planet.

The United States, for example, represents only 5% of the world's population, but consumes 30% of the world's natural resources. The richest 16% of people consume 80% of the world's resources. This disparity between nations and the unsustainable levels of resource management have dire long-term predictions.

However, some ecologists and economists have a more positive outlook. They surmise that the carrying capacity can forever be increased by human innovation, as done in the past. As cited earlier in this section, this positive argument relies on the power of the human mind to solve the population explosion issue. Possible solutions include harvesting ocean algae, mass producing farm fish, and reducing population on Earth by traveling to Mars and the moon. All of these are plausible, but each possibility is only extrapolation at this time.

> "Once it was necessary that the people should multiply and be fruitful if the race was to survive. But now to preserve the race it is necessary that people hold back the power of propagation," by Helen Keller, deaf and blind author and lecturer.

**Life history**

Series of changes an organism undergoes during its lifetime.

**Opportunistic life history, r-selected strategy**

type of life history when parents have many young and invest very little in each.

## Survivorship Curves and Life History Strategies

The life history of organisms in a population is the set of inherited characteristics of an individual that tell us how it lives. An organism's life history, also called its **life strategy**, includes its fertility rate, breeding patterns, life span, and age at first reproduction.

There are two types of life histories, each representing opposite ends of a spectrum. The first type, used by dandelions and flies, is termed an opportunistic life history. It is also called an r-selected strategy. In this method of living, there are large numbers of young per breeding event, very little or no care of the young, shorter periods of development, and a higher mortality in early life. Usually, those organisms with opportunistic life histories live a short time and cannot care for their young. Their strategy is to have as many offspring as possible, putting their success in quantity of children and not quality of caring for them. Most flies live only a few weeks. This means that a successful life history does not include a long life raising young. It instead focuses on rapid development of young to adulthood so they do not need care. The cane toads in our story had an opportunistic life history, putting little effort into raising young. Opportunistic organisms often exhibit population booms like the one depicted in our opening story.

The second type of life history, called an equilibrial life history, occurs when parents invest in extended care to their young, live a long time, and have few offspring. It is also called a K-selected strategy. Their efforts go into quality of care and not quantity of young. Organisms such as elephants, coconut palm, and humans exhibit an equilibrial life history. They have few offspring and invest heavily in each. A whale and human usually produce only one newborn at a time. A coconut palm waits a long period of time before producing limited numbers of coconuts.

The two life history types are given in Figure 17.7.

Each species evolved a particular life history to optimize the survival of its members. While there are maximum life spans in every species, not all species will reach this age. In humans, the oldest documented case was Jeanne Calment, who lived up to the age of 122 years, 164 days. Our life history is K-selected and we have longevity, on average long exceeding the time needed to care for our young. However, humans are limited by genetics and may live for only so long.

**Equilibrial life history, K-selected strategy**

A type of life history that occurs when parents invest in extended care to their young, live a long time and have few offspring.

(a)                                           (b)

**Figure 17.7**  Opportunistic and equilibrial life histories. a. r-selected species such as quinoa produce many tiny seeds in one growing season. b. The coconut palm grows slowly and produces few seeds in its entire lifetime.

### A SUPERCENTENARIAN LIVES FOR 122 YEARS . . .

Jeanne Calment was born in Aries, France, on February 21, 1875. She is the oldest centenarian in history, dying at age 122 in 1997. Longevity ran in her family: her mother died at age 86 and her father died at age 94. These were very old ages in the 1800s, when medical treatments were limited.

She married her cousin but he died of food poisoning in 1942 at age 47. They had only one daughter who died of pneumonia in 1934. She then raised her grandson who died in 1963 from injuries in a car accident. Jeanne never worked a job and attributed her long life to not letting herself get stressed and to eating a diet rich in olives.

Did Jeanne Calment take care of her health? She smoked until age 119 and ate two pounds of chocolate every week. She rode her bicycle until age 100 and although going blind and hard of hearing in her last few years, she remained mentally alert and capable. She appeared humorous: When asked on her 120th birthday what kind of future she expected, she answered, "A very short one." and on her 110th birthday, she commented, "I've waited 110 years to be famous. I count on taking advantage of it. Why did Jeanne Calment live for so long . . .?

The maximum lifespan in humans is between 100 and 120 years. Jeanne Calment is an extreme example of survivorship. Human life expectancy is the age at which 50% of people in one's age group have died. To illustrate, life expectancy for men in the United States is 78 years. This means that by age 78 a man has lost half of his cohorts. Life expectancy for women is 82 years. A survivorship curve plots the number of survivors in a population over time.

Ecologists have classified three types of survivorship curves. Type I curves show most individuals surviving until the end of life, when death occurs in high proportions. Humans and large animals, which carry out equilibrial life histories, exhibit this type

**Survivorship curve**

A graph that gives number of survivors in a population over time.

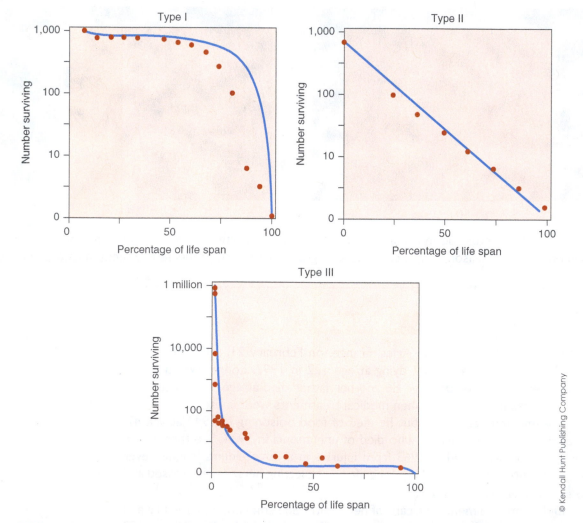

**Figure 17.8**   Three types of survivorship curves show the numbers of individuals surviving throughout their lifespans.

of survivorship curve. Most individuals live long enough to care for their young. Equal rates of death at every age occur in Type II survivorship curves. This distribution is seen in organisms, such as those in the wild, lizards, birds, and small mammals, which have equal chances at predation and environmental dangers at all ages. Those with Type III curves die off very early in life, depicted by an inward bulge in the graph. Organisms with a Type III curve include frogs and marine animals that give off large numbers of fertilized eggs into the wild. There is a high chance of death due to environmental dangers, as the eggs are defenseless and unprotected. Figure 17.8 depicts the three types of survivorship curves.

## Characteristics of Communities

### Roles

The second part of this chapter explores the features of a community. It examines the interactions of organisms within a community of populations. A forest, with maple trees and white pines, reptiles, amphibians, and humans constitute a community. It is a set of different populations living together. A community may also be something unseen – such as the microbiome of bacteria residing on your skin – which contain millions of species not visible to humans (Figure 17.9).

A biological community therefore includes all of the populations of organisms living in an area at a particular time, and their relationships with each other. The role an organism plays in the community is it ecological niche. Its niche displays how an organism interacts with all of the features of its environment. The space an organism occupies, including all of the factors with which an organism interacts, is known as its habitat. A habitat is the area in which an organism lives. Some factors in a habitat are nonliving, called abiotic factors. These include soil, sunlight, temperature, and rainfall. Other factors comprise the living things, called biotic factors. These include plants, animals, and microorganisms. Organisms use both biotic and abiotic factors to interact with other members of the community and the environment.

An organism's ecological niche may be limited by the resources it is actually able to use. An organism's fundamental niche is the area and resources that it is theoretically able to utilize. Its realized niche is the area and resources actually able to be used by a population. Consider the barnacles on the Scottish seacoast, as an example. They consist

**Ecological niche**

The role an organism plays in its environment.

**Habitat**

The space an organism occupies, including all of the factors with which an organism interacts.

**Abiotic factors**

The non-living factors in a habitat.

**Biotic factors**

Factors that comprise the living things in a habitat.

**Fundamental niche**

The area and resources that an organism is theoretically able to utilize.

**Realized niche**

The area and resources that an organism is actually able to use.

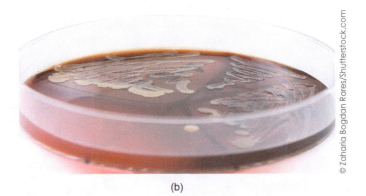

(a)                                                    (b)

**Figure 17.9** Communities come in big and small sizes. a. Forest community. b. Microbiome community taken from human skin.

**Figure 17.10**   The *Balanus* and *Chthamalus* genera along the Scottish seacoasts occupying different realized niches.

of two different genera: the *Balanus* is best adapted to exploit resources at lower portions of the coast and the *Chthamalus* genus is better suited for upper parts of the shore (Figure 17.10).

# Interactions within Communities

## Competition

**Competition**

The activity that occurs when organisms strive for the same limited resources.

**Competitive exclusion principle**

A principle, which states that organisms will compete with each other in an area until one goes extinct.

In the example above, the fundamental niche of both genera of barnacles is the entire Scottish coast. Under ideal conditions, with the other genera not around, each are able to use resources in the upper and lower regions of the coast. However, in reality, their realized niches matter more. Realized niches are exploited because of competition between the two genera of barnacles.

Competition occurs when organisms strive for the same limited resource. Competition reduces the survival of both organisms. They spend their energy competing with one another. Russian scientist G.F. Gause, in the 1930s developed the competitive exclusion principle, which states that organisms will compete with each other in an area until one goes extinct. He studied two species of paramecium, *P. caudatum* and *P. aurelia*. When grown separately in test tubes, each species thrived, using resources. When grown in the same test tube, *P. aurelia* drove *P. caudatum* into extinction (Figure 17.11).

Species do evolve to coexist with each other. All birds consume berries, but different species are adapted for different sizes, shapes, and types of berries. Competition does not always need to cause species extinction in an area. When two competitors coexist

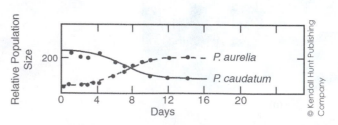

**Figure 17.11**   Competitive exclusion principle: graph of *P. caudatum* and *P. aurelia*

in the same area, they use resources in different ways, in a process known as resource partitioning. Resources are subdivided into different categories, and each category is separately used by competitors.

Resource partitioning is often accomplished by character displacement. In character displacement, organisms evolve characteristics to help them to partition resources. Sometimes new traits cause partitioning based on location, as described in our examples of barnacles in Scotland. Sometimes different resource use is temporal, or based on time of the day spent in a habitat. Bats, for example, hunt for prey at night, limiting their competition with birds, as well as predators. Regardless, resource partitioning is a mechanism by which competition is reduced between organisms.

When competition occurs between two different species, it is known as interspecific competition. If *B. marinus*, the cane toad outcompetes *Rana pipiens*, the North American frog, in obtaining a fly meal, it is an example of interspecific competition. This is the most common form of competition in community ecology. When organisms of the same species compete with each other, it is called intraspecific competition. Mate competitions between two deer or when two hemlock trees struggle for limited light and water represent intraspecific competition.

## Predator–Prey Relationships

Nature can appear cruel to human society. A fast cheetah stalks and kills a cute, little fawn. A California King snake, *Lampropeltis getula*, swallows a mouse whole, seemingly without mercy. We feel sorry for the creature that we like – the one which loses. However, this relationship between species in a community is vital.

The connection between the two organisms is called the predator-prey relationship. Predation is essential for energy flow and the survival of many species in a community. Predation occurs when an organism of one species – the **predator** – stalks and kills an organism of another species – the **prey**. Predators obtain required energy from the parts of its prey. When *L. getula* stalks a mouse, for example, it has adapted, over millions of years of evolution as shown in Chapter 10, strategies and structures to consume small animals. The answer to the problem in our story is to find a predator that is able to withstand the toxin of the cane toad. It kills crocodiles and freshwater turtles when they eat them. There are many species that prey successfully on the cane toad but they

**Resource partitioning**

The condition where two competitors coexist in the same area and use resources in different ways.

**Character displacement**

The phenomenon where organisms evolve characteristics to help them to partition resources.

**Interspecific competition**

The competition between two different species.

**Intraspecific competition**

The competition between organisms of the same species.

**Predator-prey relationship**

The connection between two organisms of unlike species.

**Predation**

When one organism stalks and kills an organism of another species.

© Joel Kempson/Shutterstock.com

**Figure 17.12**    A corn snake is a predator that swallows its prey whole. This corn snake is eating a mouse.

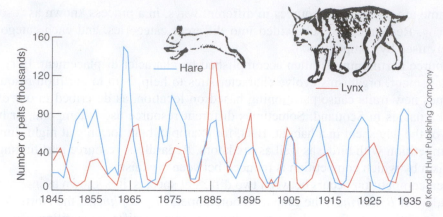

**Figure 17.13**  Predation. The lynx, a predator, stalks and kills its prey, a snowshoe hare; graph of population change of lynx and hare

are not native to Australia. Ecologists are searching for a suitable predator for the cane toad. However, they should be careful not to introduce a new invasive species. Snakes are an excellent predators to rodents (Figure 17.12), but most snake species cannot eat the cane toad.

The most notable predator–prey relationship takes place between the lynx and the hare. The lynx, a predator in Figure 17.13, stalks and kills its prey, a snowshoe hare. Lynx and hare populations fluctuate, dependent in part on one another. When plotting the frequency of individuals of each population over time, Figure 17.13 shows the changes that occur for each species in tandem with one other. As the lynx preys on the hare population, lynx increase because they are exploiting the resource. Then, as they use up the hare (hare population declines) as a food source, they too experience population decreases. With fewer lynx, more rabbits survive. Fewer lynx predators allow their numbers to thrive. Afterward, the lynx again have more food available from the increase in hare population. This fluctuation in frequencies make the predator and the prey dependent on each other.

## Defenses Evolve

However dependent, prey always lose because they are killed and eaten by the predators. The prey's reproductive success is reduced in its interactions with predators. Therefore over time, many prey species evolved a series of defenses to combat predators. Quills on a porcupine or poisonous glands in a cane toad repel predators and save a prey's life. There are several defense mechanisms used by prey. They are divided into two types: those that constitute **physical prey defenses** such as chemicals and structures; and those that are **behavioral prey defenses**, which comprise the actions a prey takes to ward off predators.

At the same time, predators have evolved counter-measures to prey. They have developed (and continue to evolve) structures and strategies that combat changes in prey. It is a coevolutionary arms race between the two. As discussed in Chapter 9, the ongoing coevolution of the Passiflora flower and the Heliconius butterfly represents an arms race in defenses. In some cases, organisms evolved to help each other to defend against mutual enemies. Several species of ants become attracted to substances on conifers. Ants on conifers defend against other insects that eat the tree's needles. The conifers provide a defense against predators for ants. Together, they evolved strategies to help each other defend against predation. Let's explore the prey defenses that developed across other species.

# Physical Prey Defenses

## Mechanical defenses

Mechanical defenses are structures, such as quills on porcupines or shells surrounding a turtle, which serve as passive defenses against predators. They require no work and no confrontation to deter predation. A turtle shell is tough and prevents the soft-bodied turtle from many deadly encounters.

## Camouflage

Armored shells of turtles also usually blend in with the environment. When organisms become less visible in their environments, they are said to use camouflage to avoid being seen. The walking stick, an insect that resembles a branch on a plant, easily blends in with its surroundings. This structure allows a passive defense for walking sticks, as depicted in Figure 17.14.

**Camouflage**

The act by which organisms become less visible in their environments to avoid being seen.

## Warning Coloration

Often a bright-colored organism indicates that it is poisonous. The azure poison dart frog, for example, is orange and spotted to warn predators to beware. Warning coloration is called aposematic coloration and serves to deter predators. However, the poison-color

**Aposematic coloration**

Warning coloration that serves to deter predators.

(a)        (b)

(c)

**Figure 17.14**    Physical prey defenses. a. Turtle in a shell. b. Aposematic coloration of azure poison dart frog. c. Viceroy and Monarch butterfly. d. A walking sick *Diapheromera fermorata*.

(d)

**Figure 17.14**   *(continued)*

link is not necessarily always present. The cane toad, *B. marinus*, in our story is drab, but still poisonous. Plants from the noncolorful genus *Strychnos* produce the toxin strychnine that kills many vertebrates.

**Mimicry**

The resemblance of one organism to another

When a palatable species capitalize on aposematic coloration, it mimics the poisonous organism. This is called mimicry. For example, the Monarch butterfly is poisonous to mammals and bird, containing a toxin that affects heart rate. The Viceroy butterfly resembles the Monarch, both featuring similar orange–black patterns and get protection by appearing similar (Figure 17.14c).

## Behavioral Prey Defenses

### Group Behavior

**Group behavior**

A passive adaptation technique to predation.

The way an organism responds to stimuli or acts comprises their behavioral defenses (Figure 17.15). There are both passive and active forms of behavioral prey defenses. By traveling in groups, prey reduce their individual risks. Usually, when a predator enters the scene, some prey respond, giving warning to the other to flee. In another way, the large numbers of cane toads, for example, satisfy a predator's appetite and allow others to go unharmed. Group behavior is a passive adaptation to predation (Figure 17.15 c and d). It saves many by sacrificing the few. In Chapter 20, we will explore how the genetic predisposition to group thinking influences societal behaviors.

### Alarm Call

**Alarm call**

A warning signal made by an animal or bird about a predator or when startled.

Sometimes in a group, there is an alarm call, which is signaled by one member, that a predator has been spotted. An owl may howl or a bird may chirp in an alarm call. This enables the other members of the group to **hide and flee** or to **fight back**.

The first is a passive defense, and many animals adapt this strategy when encountering humans, for example. Cattle will travel in herds, fish in schools, and birds in a flock. Often, mainland animals, as opposed to island ones, adapt a passive retreat defense from humans. On the island in our story, **hiding and fleeing**, in fact any fear from humans is absent in cane toads. These organisms did not evolve the behavior against humans because on islands their species never encountered humans until recently. The second strategy is an active defense. When organisms **fight back**, they use their defenses to ward off predators. As we discussed in Chapter 10, ostriches are flightless but have strong wings to beat back the predators, when they are unable to run away (Figure 17.15b).

(a)

(b)

(c)

(d)

**Figure 17.15**  Behavioral prey defenses. a. Red carpenter ant attacks a gnat. b. Ostrich will use its wings to protect her eggs. c. Cattle in herds. d. A flock of pigeons.

Sometimes a behavioral defense will be active to a point of suicide. For example, often an enemy insect group attacks a colony of the Malaysian exploding ant, *Camponotus saundersi. C. saundersi* soldiers march up to the enemy. In line and in unison, the *C. saundersi* ants contract poison glands in their abdomens, squirting formic acid onto the predators.

In the process, *C. saundersi* soldiers die, with their abdomens exploding. It is an example of devotion to the colony, as suicide ants lose their lives to protect against predators. We will explore the role of genetics in forming social systems in organisms further in Chapter 20.

## Plants and Herbivory

Have you ever looked at the leaves on a tree in early spring compared with the late summer or fall? Early in the growing season, leaves are fresh and undisturbed, but by the end of the season, they change. They become filled with holes, laden with fungus growths, and ultimately appear ugly. The changes are due to herbivory, or the consumption of plants and plant parts by other organisms. In herbivory, the plant may or may not die as a result. Herbivory is commonly thought of as cattle grazing on grasses; but other organisms, such as fungi and bacteria, feed on plants. Herbivory is a form of predation and it kills many plants.

Some plants evolved defenses against herbivory. Humans are unable to eat most forms of grasses because they contain silica, making hardened blades. These plants are too tough

**Herbivory**

The consumption of plants and plant parts by other organisms.

(a)                                                                    (b)

**Figure 17.16**   a. *Toxicodendron radicans*. b. A rash on a human from poison ivy.

for humans and many animals to consume. Some organisms, such as the shrub oak, evolved its defense by having most of its mass underground in the form of roots. This way, even if herbivores devour the plant, most of its energy is underground and ready to resprout. The best known evolved defense against herbivory, however is the poison from the poison ivy plant, *Toxicodendron radicans*. It contains a toxin that binds to T-helper cells. As you may recall from Chapter 14, T-helper cells begin the specific immune response. When the toxins of *T. radicans* combines with T-helper cells, they initiate an allergic reaction. These manifest as skin rashes commonly seen after contact with poison ivy (Figure 17.16).

## Symbiosis

**Symbiosis**

A relationship formed between two different organisms living in a close, intimate association.

Many relationships within communities form close bonds, making them interdependent. Many organisms live in close, intimate association forming a relationship called symbiosis.

There are three types of symbiosis:

**Commensalism**

The type of symbiosis which occurs when one organism benefits and the other is unharmed by the relationship.

1) Commensalism, which occurs when one organism benefits and the other is unharmed by the relationship (Figure 17.17). When epiphytes or vine-like plants grow on trees, they do not harm the supporting plant but do not help it either. Instead, the epiphyte's motive is to gain height and obtain its limiting resource – sunlight. They are stealing a spot on the tree but not giving anything back. In animals, barnacles are small marine creatures that latch onto skin. When they adhere to a whale's skin, they are hitchhikers and gain a "free ride" on the whale. The whale receives nothing in return for the trip but is not harmed as the weight of the barnacle is negligible.

**Mutualism**

A relationship in which both organisms benefit.

2) Mutualism occurs when both organisms benefit in a relationship. Mutualism is commonly found in nature. Both organisms have a stake in the association, making it a stable strategy for survival. In the Alder tree, a special type of bacteria (which will be discussed in Chapter 19) called nitrogen-fixing bacteria live in the root nodules of the tree. The nodule of the tree provides protection in a "home" for the bacteria. The bacteria provide accessible nitrogen for the Alder tree. Without nitrogen from the bacteria, most of it is unavailable to the tree because it is a gas. Other forms of mutualism are closer to home. The bacteria in our large intestines, described in Chapter 12, provide us with vitamin K and our colons provide a safe and anaerobic home for the enteric bacteria. Mutualism has numerous examples in nature.

**Figure 17.17**    Commensalism: Spanish moss on a tree to capture sunlight. While the tree obtains no benefits, the moss species is able to get greater access to sunlight and therefore food.

3) **Parasitism** is the symbiotic relationship in which one organism benefits and the other is harmed. Parasites kill more people than by predators or by competition. As discussed in Chapter 5, roughly 1 million people die each year from the malaria parasite, *Plasmodium* alone; and yet only a handful are attacked by sharks. The contrast shows that organisms unseen have more impact on human society that we often realize.

Parasites in the animal phylum Nematode (discussed in Chapter 10), for example, the hookworm, *Ancylostoma duodenale*, lives in human intestines and survives on blood (Figure 17.18). However, parasitism, as by a hookworm, is different from predation. Parasitism does not seek to kill its host, merely weakening it as it drains away the host's resources. Predation always seeks to kill its prey. If the parasite kills its host, it too dies, so it pays for the parasite to prevent too much abuse of its host. Many times though, and eventually, parasitism does kill a host organism.

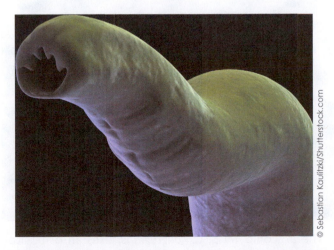

**Figure 17.18**    The hookworm, *Ancylostoma duodenale*, is a parasite in the intestines of humans.

**Figure 17.19**    Brood parasitism: Chestnut-headed oropendolas (*Psarocolius wagleri*) suffer brood parasitism by giant cowbirds (*Molothrus oryzivorus*).

Fungi are the ultimate parasites that evolved to lose their own chlorophyll and leaves. They live off the dead, as described in Chapter 8, and cannot make their own food. Fungi evolved a strategy to completely rely, as a parasite, on other organisms.

Other forms of parasitism occur in nature. Brood parasitism takes place when a bird species lays eggs in another bird's nest (Figure 17.19). The foreign eggs then hatch and are raised by the host mother. The real mother avoids the costs of rearing her young. It is a form of stealing because it saps the energy from the host mother. In cowbirds, females can lay up to 30 eggs a season because they are free of parental care for these newborns. It is a successful parasitic strategy that improves the reproductive success of cowbirds.

Some parasitoids take parasitism to new levels. They lay their eggs within other species. When these eggs hatch, they eat the host organism from the inside out, using it for developmental energy. The braconid wasp, *Cotesia congregatus*, for example, uses an ovipositor (long tube) to lays its eggs within the body of the tomato hornworm (Figure 17.20). When wasp eggs hatch within the hornworm, they gnaw at the

**Brood parasitism**

A form of social parasitism in which a bird species lays eggs in another bird's nest.

**Parasitoid**

An organism living that spends some period of its development on or in a host organism and later kills its host.

**Parasitism**

The symbiotic relationship in which one organism benefits and the other is harmed.

**Figure 17.20**    The braconid wasp, *C. congregatus*, uses an ovipositor to lay its eggs within the body of the tomato hornworm. The eggs of the wasp hatch and eat the tomato hornworm (shown in figure above).

caterpillar hornworm until they reach its exterior. The wasp eggs devour their host completely and emerge as adults from a dead caterpillar carcass.

The many community interactions presented in this section should be understood as complex and sometimes overlapping (Figure 17.21). While mutualism may occur, for example, between species, another interaction is usually taking place simultaneously. A braconid wasp may have within it a parasitoid living in its internal cavities. It is estimated that 25% of insect species are parasitoids, forming many interactions within their communities.

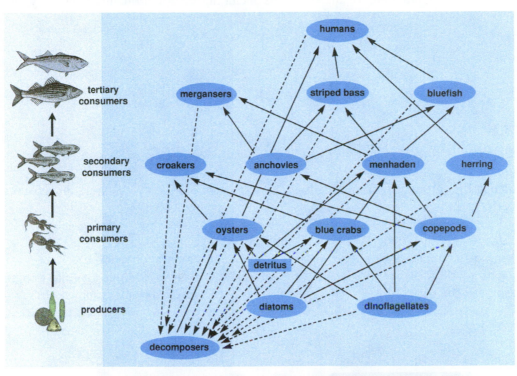

**Figure 17.21**  Multiple relationships in a community: arrows depict relationships between populations within a river community in the waters below these boaters. From *Biological Perspectives*, 3ʳᵈ ed by BSCS.

# Summary

Invasive species take over regions in which they lack natural predators and have abundant resources. There are over 1,500 known invasive species adversely affecting regions around the world. Ecologists study these regions by analyzing population demographics and by looking at how populations interact within a community. Populations grow depending upon the conditions to which they are adapted in their environments. They are limited in growth by scarcity of resources in a given area. We can predict future growth of populations based on their age structure, life history, fertility rates, and survivorship curves. Organisms play differing roles within their community. Many kinds of interactions result from this including competition, predator–prey, and types of symbiosis.

---

## CHECK OUT

**Summary: Key Points**

- Invasive species, without natural predators, grow logistically and exhaust resources for other species in a community.
- Most populations grow at first exponentially and are then limited by a scarcity in resources at an ecosystem's carrying capacity.
- An opportunistic life history capitalizes on high numbers of offspring, and an equilibrial life history emphasizes care of the young.
- Organisms in a community have a niche, which may be fundamental or realized.
- Interactions within a community include the antagonistic, such as competition and predation and parasitism; and the cooperative, such a mutualism and, to a lesser extent, commensalism.

---

## KEY TERMS

| | |
|---|---|
| abiotic factors | competitive exclusion principle |
| age structure diagram | deaths |
| alarm call | demographics |
| aposematic coloration | density-dependent factors |
| biosphere | density-independent factors |
| biotic factors | ecological footprint |
| biotic potential | ecological niche |
| births | ecology |
| brood parasitism | ecosystem |
| camouflage | emigrants |
| carrying capacity, K | equilibrial life history, K-selected |
| character displacement | strategy |
| commensalism | exponential |
| community | exponential model of population growth |
| competition | fertility rate |

fundamental niche
group behavior
habitat
herbivory
immigrants
interspecific competition
intraspecific competition
life history
logistic model of population growth
mimicry
mutualism
opportunistic life history, r-selected strategy

parasitism
parasitoid
population
population density
population ecology
population growth
population size
predation
predator–prey relationship
realized niche
resource partitioning
survivorship curve
symbiosis

# Multiple Choice Questions

**Reflection questions:**

1. The cane toad grows:
   a. inversely
   b. exponentially
   c. diversely
   d. proportionally

2. Populations are:
   a. groups of communities
   b. smaller than groups
   c. larger than ecosystems
   d. the unit of study in ecology

3  When a population of squirrels hits its _____, it shows a _____ growth curve.
   a. exponent; exponential
   b. exponent; logistic
   c. carrying capacity; exponential
   d. carrying capacity; logistic

4. A population of field mice has a chance of death equal at all of their ages, with predation a steady possibility. Their survivorship curve would appear as Type:
   a. I
   b. II
   c. III
   d. IV

5. Before humans hit their carrying capacity, they are likely to experience:
   a. violence
   b. disease
   c. starvation
   d. growth

6. Which represents a logical order, in the development of new niches for organisms in a high population density and resource scarcity?
   a. Competition → resource partitioning → character displacement
   b. Competition → character displacement → resource partitioning
   c. Character displacement → resource partitioning → competition
   d. Resource partitioning → character displacement → competition

7. Which is an example of an abiotic factor in a forest community?
   a. Squirrel droppings
   b. Sunlight levels
   c. Competition between wolves
   d. Competition between wolves and dogs

8. Which takes place when both organisms benefit in a relationship?
   a. Mutualism
   b. Commensalism
   c. Predation
   d. Parasitism

9. Which correctly MATCHES terms in community ecology?
   a. Predator – mutualism
   b. Herbivory – competition
   c. Parasite – commensalism
   d. Predator – prey

10. Herbivory is a form of:
    a. predation
    b. mutualism
    c. commensalism
    d. all of the above

## Short Answers

1. Invasive species are exotic to new areas and growth rapidly. Give two reasons why an invasive species is able to take advantage of a new area.

2. Define the following terms: population size and population density. List one way each of the terms that differ from each other in relation to their: a. importance in predicting competition in a population; b. importance in predicting resource use in an area; and c. relationship with each other.

3. Some ecologists argue that "there is no true form of commensalism." Define commensalism and give an example of it in nature. Do you agree with this statement? Defend your argument.

4. Draw a sketch of an age structure diagram in a growing population. Give an example of a nation with this type of age structure diagram. What does a high fertility rate tell you about the future of this population?

5. Explain the difference between a fundamental niche and a realized niche.

**6.** How does competitive exclusion result in: a. character displacement and b. extinction? Describe the pathway that a community takes to lead to each result.

**7.** Define aposematic coloration. Give an example of this in nature.

**8.** For question #7, how does mimicry act as a physical prey defense? Give an example in nature.

**9.** Describe the predator–prey relationship. Does it pay for the predator to kill its prey? Does it pay for predators to kill off its prey population?

**10.** Define brood parasitism and parasitoids. How are the two types of parasitism similar? How are they different? Give an example of each.

## Biology and Society Corner: Discussion Questions

**1.** In the past 25 years, China has implemented a One Child Policy, which restricts couples by placing sanctions on them for having more than one child. However, it is now allowing two children per couple. What changes have taken place societally in China as a result of this policy? What do you recommend for solutions? What do you predict for future generations, as Chinese society favors sons?

**2.** In the story, the cane toad was introduced into Australia in 1935 to combat beetles. Beetles destroy crops and thus the food supply. The scientists who introduced the cane frog argued that at least their methods were natural, unlike methods of animal control used today. Do you agree or disagree with these scientists? Explain your answer.

**3.** To reduce our ecological footprint, some scientists argue that the United States would see a reduction in its standard of living. Our resource use is the highest in the world, at this time. Would you be willing to lower your standard of living to reduce our ecological footprint? Why or why not?

**4.** Unsustainable human population growth is an environmental threat. However, two scholars argue about the effects of human population growth. Julian Simon, an economist, contended that human innovation and technological advance will increase the carrying capacity. Paul Ehrlich argued against Simon, citing limited resources and predicted logistic growth curves for humans. Research these two opposing sets of viewpoints. On the basis of your research, which side do you take? Why?

5. Jeanne Calment lived for 122 years. Would you like to live for that long? How long would you want to live? What factors play a role in your decision making?

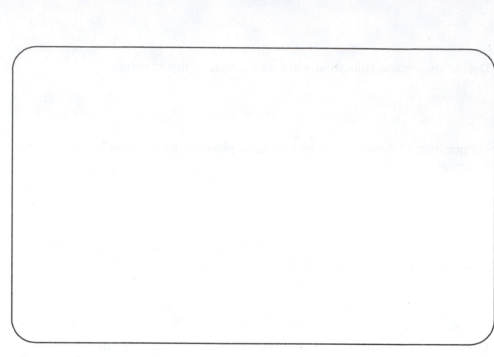

Figure – Concept map of Chapter 17 big ideas

# Ecosystems and Biomes

# 18

The hitchhiker

He meets his destiny in the desert of California

He makes his way out west

Through the grasslands of Iowa

Through the forests of Pennsylvania

© Kendall Hunt Publishing Company

## CHECK IN

**From reading this chapter, students will be able to:**

- explain how humans benefit from the different biomes but negatively impact their own ecosystems.
- define biome, ecosystem, topography, rain shadow desert, tropical rainforests, savannas, deserts, chaparral, temperate grasslands, temperate deciduous forests, taiga, tundra, polar ice caps, estuary, limnology, producer, herbivore, consumer, omnivore, energy pyramid, biomass, ecological succession, climax community, and colonizer.
- define the term biome and explain how local topography affects biomes.
- list and compare each of the nine major biomes of the world.
- compare the three types of aquatic biomes.
- trace the flow of energy as it moves through an ecosystem.
- explain how ecological succession returns an ecosystem to a stable community.

# The Case of the Hitchhiker

I woke up excited and ready to start a new life. I packed my bags and had my last breakfast with my mother in our apartment in the upper east side of New York. I was moving to San Francisco, driving across country this morning. My mother was worried about the trip and the long distance, but I could not wait to see the world.

Last night at my graduation party, my old science professor Mike made a great prediction: "The world is your oyster!" . . . and it really was. I accepted a new job in California last week, hired by a small environmental consulting firm. My first project was to study how new road tarring projects affect chaparral in southern California. As I began driving, I was uncomfortable with the thought that enough tar has been placed on roads in the United States in the past 50 years to cover the state of Ohio completely!

**Biome**

A large community of flora and fauna occupying a major habitat.

Chaparral is a type of biome, or set of ecosystems that occur across large areas of the world. There are several types of biomes. In *chaparral*, there are hot and dry summers, and plants are adapted for drought resistance. We were studying the effects of tarring on scrub oak populations in the California chaparral. I was real lucky.

I packed the car, said my goodbyes to the family, and started the drive over the George Washington Bridge from New York City into New Jersey when an accident just missed me. "Whoa, what a sight! A seven car pileup and it looked like people were going to be really hurt or dead. Dude, I am lucky" I thought.

It was there that I saw him for the first time – the hitchhiker – he was an older man, wearing a gray suit much like in the 1930s. He was plain, with a long stark face and almost, non-descript. I ignored him and drove by but as I looked in the mirror, he smiled at me. I hated him.

I tried to forget about the oddness of the hitchhiker; with his strange look and smile. I loved nature, which is why I majored in ecology in college. I looked forward to going through the *temperate forest* biome of Pennsylvania. I drove through the endless mountain range and saw large maples and pines in the temperate forests. These areas have plentiful water (121 cm or 48 inches/yr) and are able to sustain large plants. I stopped and picked my favorite flower, a daisy in a patch of thyme. However, behind a tree, I caught a glimpse of the hitchhiker – the same man in New Jersey! He smiled again and stuck out his thumb. I thought of confronting the man, but I stepped on the gas pedal instead.

I was shaken up – was he following me? Why? How? I was panicked. I drove for hours and I enjoyed the beautiful deciduous trees – the foliage colors and its majestic look – I assumed that I must be imagining the hitchhiker. I could not call my mother – she would worry; and maybe think that I was crazy. No, I needed to keep driving and see the sights. I did not stop to get a hotel, as planned.

I drove all the way from Ohio to the Iowa prairies, North America's *temperate grasslands* biome. "Here the open lands will clear my mind and I'll forget the apparition," I thought. The thick layers of soil from grasses growing make beautiful crop land. It was open, on I-80 for as far as the eye could see, and only rows and rows of corn fields whisked by me. I loved it. There was a clear and open sky, which you could never get in a city, always surrounded by buildings.

There is not enough precipitation (less than 75 cm or 30 inches/yr) to sustain many large trees in temperate grasslands. The grasslands in the summer are hot, but I know this biome has cold winters. However, again, in between the rows of corn, I saw the hitchhiker popping out, with a smile and a thumb.

As I went to pieces and moved across Colorado, where the farmland gave way to mountains, I thought a change in biome would end this stalker's persistence. I went up and up hoping to get away from the hitchhiker. High in the Rockies, I felt free from the world. At 5,000 feet, the temperature dropped and with a new biome called *taiga*. I knew from my studies that the taiga had moderate amounts of rainfall mostly in the summer. I really enjoyed touching the cone-bearing pines and shrubs. They were adapted for the cold winters and low water levels – but not me, I liked it hot in California.

However, the cold did not keep him out – the hitchhiker looked coldly at me this time, impatient and waiting. He stood by a giant Alder tree. I simply wanted to make it to San Francisco and drove without sleeping, after 36 hours of driving. I could not care – I would not care. I would keep going and see this world. I passed the mountains and slowly the region changed to *desert*, an area where there was almost no precipitation (25 cm or 10 inches/yr). There was little vegetation and little life in the desert. The lower the rainfall a desert has, the fewer plants and animals present.

At this point, I felt a dreadful acceptance. I stopped at a place where there was almost no life – in the Mojave Desert in southern California. I felt the hitchhiker, although I did not see him. I would have it out with that hitchhiker!

I called my mother, and I asked to speak with her. The voice quickly answered at the other end, which I did not recognize, and it said "I am sorry, she is unavailable; there has been a terrible accident – her son died on the George Washington Bridge the day before yesterday."

I opened my car door and walked to the hitchhiker, standing by a cactus. He smiled and reached out his hand whispering, "I wanted to give you just a little more time to enjoy the beauty that this world has to offer."

## CHECK UP SECTION

In the suspense story above, the main character experiences, among other things, many biomes while driving across the United States. Temperature and rainfall are abiotic factors that help determine the kinds of plants and animals inhabiting an area and comprising a biome.

There are many threats of biome loss throughout the globe. Tarring of surfaces creates many fragmented ecosystems within biomes, for example. Choose a particular biome and research the effects of tarring roads on biome biology: types of organisms and abiotic factors. (a) Propose a plan for combating the effects of tarred roads. (b) What are the benefits and drawbacks of the solutions you propose?

– "The Case of the Hitchhiker" is based on a radio play written by Louise Fletcher, first presented on November 17, 1941, broadcast of the *Orson Welles Show* on CBS Radio.

# Major Biomes of the World

## What Are Biomes?

The character in our story saw several of the Earth's biomes as he made his final drive through the United States. As a large nation, the United States encompasses most of the world's biomes when including Hawaii and Alaska. Let's travel through the world to experience the biology of the biomes. Each is influenced by abiotic factors that enable growth of different forms of life. Life evolved to inhabit these regions, over long periods of time. Changes in biome biology have historically threaten the existence of many of these species, so well adapted for certain conditions.

As stated in the story, a **biome** is a set of ecosystems that occur across large areas of the world. Biomes are the largest ecosystems in the world. As you recall from Chapter 17, ecosystems comprise the communities in a region, including how they interact with the environment. Biomes contain innumerable ecosystems that have similarities in types of organisms and abiotic factors.

**Ecosystem**

A system of interacting organisms and their physical environment.

An ecosystem is an arbitrary unit because it is only a human construct. One ecosystem may be a spider web, its relations with prey, and the abiotic factors within its world of a cave. Another ecosystem may include all of the desert area of Death Valley, in California. An ecosystem's boundaries are determined by how we define them. When an ecosystem is defined, it is by humans in order to study it. When defining large and expansive ecosystems, biomes are created.

**Tropical rainforest**

A type of forest characterized by tall, dense trees in an area that experiences high annual rainfall.

Ecologists differ on the number of biomes in the world. We shall accept that there are nine biomes: (1) tropical rainforests, (2) savannas (tropical grasslands), (3) deserts, (4) chaparral, (5) temperate grasslands, (6) temperate deciduous forests, (7) taiga, (8) tundra, and (9) polar ice caps. Our character in the story drove past and "enjoyed" the sights in four of these biomes (3, 5, 6, and 7). In the next section, we will give the salient features of biomes, each depicted in Figure 18.1.

Biomes are found over large areas and have characteristic plants and species. Each of the biomes has plants, animals, and abiotic factors that are similar. A desert looks like a desert because they all have little water and similar type vegetation such as cactuses. As the driver in our story persisted, he noticed the changes in scenery based on the different plants and animals in each biome. He also noticed that abiotic features of each biome – temperature, water, altitude, and sunlight – were similar within a biome. In fact, as the driver went up in altitude as he drove on, the biome changes. Elevation and latitude create similar effects on plant and animal lives in a biome.

These abiotic conditions determine the plant and animal life in an area. The most important abiotic factors to plants are light and water. Organisms adapted for a particular biome are found in that biome. For example, C4 and CAM plants show how organisms adapted to dry conditions are found in deserts. Plant adaptations to hot and arid biomes you may revisit in Chapters 4 and 9.

In another example, in order for angiosperms to grow tall, they must have plentiful water. As our character in the story drove, he saw that trees were found in the wetter, temperate forest biome; and grasses on the drier prairies. Trees require more water to persist in a biome. Recall from Chapter 9, transpiration pull requires ample water to allow plants to carry out photosynthesis. Angiosperms are often large in size and require plentiful water to

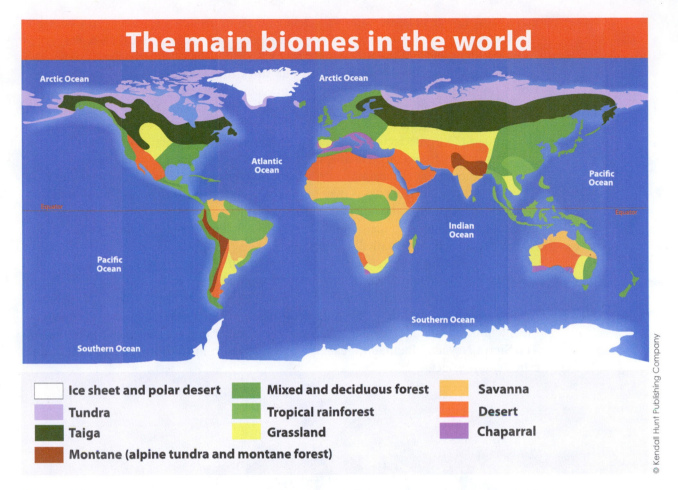

**Figure 18.1**    a. The Major Land Biomes of the World. Biomes occupy large regions of the globe and usually extend onto more than one continent.

move all the way up to the top of the tree. Very few angiosperms were seen by our character in the grasslands. There, the precipitation level could no longer support angiosperm growth. Thus, grasslands have few trees, overall except near rivers, lakes, and wetlands, which provide extra water. These are examples of how the biota of a biome reflects abiotic conditions.

## Topography Affects Land Areas

The Earth's features determine the environmental conditions in each biome: the amount of solar radiation it receives, the circulation of atmospheric winds, and the presence of valleys and land ridges. For example, the direction of the wind determines which species of birds fly in a region. The angle of the Sun's rays hitting a biome also plays a vital role in its temperature. Most of us know that the areas closer to the equator of the Earth are warmer than those closer to the poles. These are examples of the Earth's influences, which have impacts on the weather to determine ecosystems and biomes. These are characteristics of the Earth which will be discussed further in Chapter 19, the biosphere.

Some conditions of the Earth that are regional (not pertaining to the entire Earth) also affect the formation of biomes and ecosystems. Local topography, or the physical features, of the land such as mountains and valleys has dramatic effects on climate in areas. A drive one mile up a steep incline road could change the weather 10°C. As altitude increases, the air pressure in an area decreases because there are fewer air

**Topography**

The physical features of the land such as mountains and valleys.

molecules. Lower pressure causes the temperature to go down (recall from Chapter 2 that temperature is a measure of average kinetic energy of molecules, so fewer molecules moving translates to lower temperature).

As a result, when the character in our story drove up the Rocky Mountains in Colorado, he noted the biome shift from grassland to taiga. Taiga is a biome usually found in wide regions closer to the poles. His drive up in altitude resembled a movement toward the poles of the Earth, getting colder along the way. The changing topography also shifted the biotic factors found in the areas. On the mountains, he saw Alder and other coniferous trees that were adapted for colder climates and less light.

Large mountains also cause deserts that form when ocean winds cross over them. As moist air from the ocean passes over high mountain ranges, it rises and cools at higher altitudes. Cool air loses moisture and causes precipitation in those areas. Afterward, with little moisture remaining, there is reduced rainfall on the other side of the mountain. This region becomes deserts, called rain shadow deserts (Figure 18.2).

**Rain shadow desert**

The dry region of land on one side of a mountain; has very little precipitation.

The deserts form in the shadow of the mountains, as a result of those mountains. They always form on the downwind side of a mountain chain. If the wind comes from the west, as in the United States, rain shadow deserts will form on the east side of a mountain. Our final scene in the story takes place in a rain shadow desert, as shown by the hitchhiker and our character walking away together one final time in Figure 18.2d. The Sierra Nevada Mountains create the Mojave Desert in California as a result of this rain shadow effect.

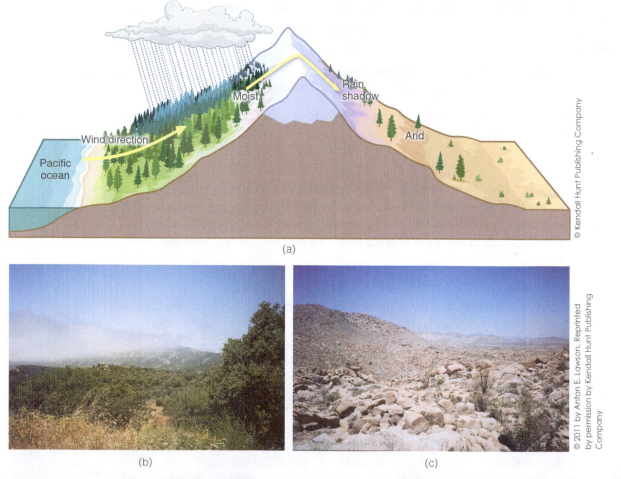

**Figure 18.2**   Rain Shadow Desert: a. How it is created? b and c. Soils on the windward side of a mountain receive much more rainfall than on the leeward side. A. B&C From *Biology: An Inquiry* Approach, 3rd Edition by Anton E. Lawson. d. Photo of the Mojave Desert.

(d)

**Figure 18.2** *(continued)*

Smaller topographical changes can also have large effects on climate and create mini-biomes. Characteristics of cities change the local temperatures. On average, cities are 1°C to 6°C hotter than rural areas. Our character in the story landed a job to study the effects of asphalt. The ever-expanding tarring of our nation's biomes is an ecological threat. The tarring effect increases temperatures, especially in cities. Asphalt absorbs heat as a dark object and raises ground surface temperatures (Figure 18.3). One method of scientists to combat asphalt heat absorption is to use materials that are lighter in color. For example, silver coatings are replacing black tar in roof repair and restoration in buildings in cities. This silver color reflects light and lowers heat absorption on building roofing. In turn, the electric needed to cool top floors of buildings is reduced.

Tall buildings in cities also have the effect of channeling winds into pockets along city sidewalks. This increased surface wind speed changes the dynamic of ecosystems in cities. Birds, for example, have more difficulty flying along certain windy sites, changing their habitat availability. Insect (bird prey) populations are therefore able to thrive,

**Figure 18.3** Asphalt on a Series of City Roofs. Asphalt in the city absorbs heat, changing the microclimate of an urban community. Cities are a few degrees warmer than comparative rural areas.

taking advantage of reduced predation in these city areas. Local changes in topography have large impacts on the ecosystems within these mini-biomes.

# A Drive through the Biomes

## Terrestrial Biomes

Let's take a tour of the biomes, like the character in our story, looking at their biotic and abiotic characteristics that make each unique. Figure 18.1 shows photos and geographic regions of the world's terrestrial biomes from around the globe. Each of the biomes contains life based on the environmental conditions (temperature, light, and rainfall) in which it lives. Rainfall and temperature are the best predictors for where a biome will best develop (Figure 18.4).

1) **Tropical rainforests**. These occur most along the equator of the Earth. It is the biome richest in biodiversity, containing upward of 50 million species. Tropical rainforests house 50% of all living species but comprise only about 2% of the Earth's landmass, as seen in Figure 18.1. They occur in Central America, parts of South America, Africa, and southeast Asia.

**Canopy layer**

Is the upper layer that is made of the leaves of treetops, in forest ecology.

There is lush plant vegetation because rainforests have abundant precipitation (360 cm or 148 inches per year). Without water as a limiting abiotic factor, the next most important is sunlight. It is estimated that only 2% of sunlight actually reaches the tropical rainforest floor. Most is filtered through the thick canopy layer, which is made of the animals and leaves and branches of treetops (Figure 18.5).

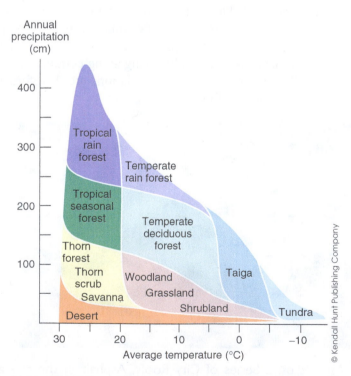

**Figure 18.4** Using Rainfall and Temperature to Identify Biomes. Temperature and rainfall are the two most important factors determining a biome's location.

**Figure 18.5** Many Organisms Live within the Layers of Tropical Rainforests. The canopy is the thickest upper layer of vegetation, containing much of the species diversity. This group of parrots occupy the canopy of a tropical rainforest.

Competition in rainforest plants is fierce and is particularly directed at limited sunlight. Tropical plants have large, broad leaves adapted to capture low levels of light. Height is important for plants, and many vine species are found in rainforests. Some include epiphytes, shown in Chapter 17 to form commensalistic relationships with trees. Epiphytes climb to great heights to capture sunlight as well as water. They are sometimes called "air plants" because they obtain their water from modified leaf structures or from roots suspended in the air. As it rains, epiphyte leaves and roots absorb water.

Logging, mining, and farming have encroached on tropical lands and caused loss of many of its species. Biodiversity loss is a serious threat to the environment and to human society. Some tropical rainforest species are endemic, meaning that they are unique only to those areas. As discussed in Chapter 7, once a species is extinct, it can never be returned to the Earth. Loss of these species also means the loss of the benefits they have to humans. Consider taxol, a cancer therapy, derived from the Pacific Yew tree, *Taxus brevifolia*. As these trees become increasingly endangered, access to taxol also does.

In deforestation, trees are cut and vegetation is burned (called the slash-and-burn technique) with farmland replacing the forest. However, only a thin layer tropical rainforest soil is useful to farming. Most of the minerals and organic material are in the vegetation and not in the soils. Soon, rainforest regions become unproductive and this contributes to desertification, forming of new desert biome. As shown in Figure 18.6, dust blowing from the newly formed deserts of the Sahara creates environmental hazards. The increasing desertification process ruins lands and does not become replaced with farmland long term. Local populations point to the need for farmland and the money for its lumber as reasons to remove tropical rainforests. The farmland quickly turns to wasteland and the local population is left with fewer resources. As such, as shown in Figure 18.7, human populations in Borneo are building homes in rainforests cleared by city developers. In the long run, tropical rainforest destruction will have serious negative consequences.

2) Savannas (tropical grasslands). Some regions along the equator experience less rainfall than the rainforests and are called savannas (Figure 18.8). These are tropical grasslands with moderate precipitation (less than 150 cm or 60 inches), more than deserts but less than rainforests. They occur mostly in Sub-Saharan Africa and in pockets across South America and Australia.

**Desertification**

The process by which fertile lands experience rapid depletion of flora and fauna becoming a desert.

**Savanna**

Regions along the equator that experience less rainfall than the rainforests.

**Figure 18.6**    The Expanding Desert. This ship has been left behind after the contraction of the Aral Sea in Asia. Water was used excessively and the sea shrank markedly over the past century. Inhabitants of this area can barely survive the lack of resources and dry conditions.

**Figure 18.7**    Deforestation of Tropical Rainforests. This photo shows numerous roads being built in Borneo, Malaysia, directly into the rainforest, making way for human settlement for its growing population.

**Figure 18.8**    Savanna of Serengeti, Africa. These wildebeests occupy many areas of this biome.

Because they contain mostly grasses as plant life, savannas support many animals well adapted for herbivory. Zebras, buffalo, and wildebeests prevail along with smaller plant-eating animals, such as grasshoppers, ants, beetles, and termites.

Savannas are prone to fires because lightening quickly turns into infernos along the dry grasses. Most plants of the savanna have adapted to the frequent fires with a high root-to-shoot ratio. In these plants, most of their mass is underground, located in the root system. When fires blaze through savannas, they may destroy the upper part of the plant but it survives.

3) Deserts. These regions have the lowest rainfall of all the biomes (less than 25 cm or 10 inches per year). Plant life is very sparse in deserts; however, the higher the amount of annual rainfall is, the greater will be the population density of plant species in deserts. Some deserts receive almost zero precipitation, such as the Namib Desert of southeast Africa receiving less than 7 cm or 2.5 inches annually.

**Root-to-shoot ratio**

The dry weight of the root divided by the dry weight of the shoot.

**Desert**

A dry, barren area with little or no rainfall.

When we think of deserts, we often imagine them as hot all through the day and night hours. Deserts are hot only during the day and sometimes their temperatures reach over 58°C (136°F). However, they become very cold at night. They have the most extreme temperatures of all the biomes. Temperatures can fluctuate up to 55°C in the course of a day.

Deserts lack water and vegetation, which are able to retain heat at night and prevent surfaces from heating too quickly in the day. A tree will shade the area beneath it but also retain the heat as it is lost. A sandy soil of the desert simply exchanges heat without water's bonds to make and break. Thus, because of lack of water, deserts lose the ability to moderate temperatures. Most of the heat absorbed in the daytime is radiated very quickly away at night. Recall that water's properties, and their unique bonding features, were discussed as a stabilizing life force, in Chapter 2 of this text.

The biota of the desert is adapted for water scarcity. Plants either lose their leaves in the very hot seasons or they have hard, thick cuticles to prevent water loss. Cactuses are a good example of adapting to this biome (Figure 18.9). They have long roots, which reach deep into the ground to obtain water, and thick cuticles and are adapted to keep photosynthesis (and open stomata) short using their CAM method, as described in Chapter 4. Many desert plants are annuals, which thrive only for one season and produce seeds to avoid the dry conditions.

Desert animals are also adapted for water scarcity. For example, the camel drinks large amounts when water is available and can survive for weeks without a drink. Other animals burrow deeply to avoid the hot temperatures of the day.

**Figure 18.9** Desert Biome. The Sonoran Desert of North America. Little rainfall makes many species of cacti, including the treelike saguaro (*Carnegiea gigantea*) in the photo well adapted for dry conditions.

## IS IT WARMER IN COASTAL BIOMES THAN IT IS IN INLAND BIOMES?

Coastal biomes are surrounded by water and therefore have modified temperature changes. Water, held together by many hydrogen bonds, are formed and broken with heat. Hydrogen bonds absorb energy when temperatures warm and release energy when temperatures cool. Water in any biome creates stability in temperature.

When weather warms it is also initially cooler on coastal biomes than in the interior ones. Often the weather shows wide variations in the inland biomes, such as the American Midwest and the plains states, which are both far from large water masses. These biomes experience much wider temperature variations. Highs and lows are unable to be moderated by water's hydrogen bonding potential. Inland biomes usually have colder winters and warmer summers.

**Chaparral**

A type of biome characterized by shrubs and rodents in dry conditions, around the Mediterranean and southern California.

4) Chaparral. This ecosystem type is often overlooked as a cross between desert and grassland, but a focus of our story as a desired destination by our character.. However, it has its own unique biome; it is found only on dry coasts. It occurs along the Mediterranean and the southwestern coast of North America. Chaparral is seen frequently in movies about the Roman era in Europe, Africa, and Asia Minor. It was probably the climate of the Roman Empire, as it follows along the borders of the old regime.

Chaparral is characterized by mild winters and hot and dry summers. It has more rainfall than desert but less than tropical grasslands. Chaparral is dry and therefore has frequent fires, much like tropical grasslands. Its plant life is fire adapted as well and has high root-to-shoot ratios. Plants appear shrubby with small, leathery leaves (Figure 18.10). Large trees do not grow in chaparral and most of the animals are small rodents.

**Temperate grassland**

Are terrestrial biomes whose main vegetation is grass and shrubs.

5) Temperate grasslands. In our story, corn fields predominated in the temperate grasslands of the American Midwest. Before humans cultivated the Midwest, there were no crops and only grasses and small shrubs prevailed. The temperate grasslands have moderate levels of rainfall (less than 75 cm or 30 inches), greater than deserts and less than savannas (Figure 18.11). This limited amount of precipitation inhibits the growth of woody shrubs and trees. Temperature grasslands are also called prairies. They occur in the Midwest of the United States, in much of eastern Europe and in scattered portions of Africa, Asia, and Australia (Figure 18.12).

**Detritus**

Dead organic matter that forms as grasses die.

The biome is used throughout the world as farmland. Grasslands are the breadbaskets of the world, feeding the human population. There is a rich layer of topsoil in grasslands. It contains a generous layer of detritus, or dead organic matter that forms as grasses die. This layer, also called sod, contains needed organic matter and minerals to sustain the growth of crops.

As described earlier, grasslands are usually located inland and have warm summers and cold winters. Rodents such as prairie dogs and other small mammals live as burrowing organisms to avoid predators on the prairie. Large animals adapted for herbivory are

**Figure 18.10**    Chaparral Biome. Chaparral of the Santa Lucia Mountains of California has many drought-resistant shrubs, which occupy most of the plant community.

**Figure 18.11**    Temperate Grassland Biome. Tall prairie in eastern Kansas. The hitchhiker followed our story's character into the grasslands.

**Figure 18.12**    Bison on a Prairie. They graze to a point where herbivory keeps many plants from growing to full size.

native to the region, including the North American Bison. Grazing by bison also limits trees and shrubs, keeping the vegetation almost exclusively grasses.

**Temperate deciduous forest**

A forest type characterized by leaf-shedding trees.

**6)** Temperate deciduous forests. Our story began with a drive through the north-eastern U.S. temperate forests. They are also located in the western North American coast, eastern Asia, eastern Australia, Europe, and New Zealand as well as pockets in South America.

The story depicted abundant plant life in temperate deciduous forests supported by ample rainfall (121 cm or 48 inches). While there are cold winters, there are also warm summers, long enough for a full growing season for most angiosperms. Trees lose their leaves to preserve water during winter months. The trees are called **deciduous**, from the Latin word "*deciduus*" meaning "to fall" because plants lose their leaves. When the autumn arrives, the foliage is beautiful as chlorophyll dissipates in leaves, revealing the carotenoids (yellow) and anthocyanin (red) pigments, which are colorful (Figure 18.13).

However, plant life is not nearly as abundant as compared with tropical rainforests. While temperate forests have about only 90 dominating tree species per hectare (2.5 acres), tropical rainforests have over 450 species. A temperate forest comprises an upper canopy of beech, maple, oak, hemlock, and hickory. Its lower layers of shrubs include berry plants, mountain laurel, herbs, ferns, and mosses. Animal life is also diverse, with large animals such as deer and coyotes and smaller ones including porcupines, rabbits, woodchucks, and beavers.

**Taiga**

A swampy, subartic forest dominated by conifers.

**7)** Taiga. When our driver reached higher altitudes, he encountered taiga, a biome characterized by long, cold winters and lower levels of rainfall. They dominate in higher latitudes of the globe, including most of Canada, Scandinavia, and Russia.

Taiga contains plants that include mostly cone-bearing (coniferous) evergreen trees, including hemlock, fir, and spruce (Figure 18.14). Evergreens are able to withstand the cold temperatures. They do not need to lose leaves during the long winter months. This way, energy is not lost in leaf abscission (drop) as in deciduous plants. Plants occur in stands across taiga. Some trees thrive, such as Alders, seen in our story. They evolved an advantage over other plants: they contain nitrogen-fixing bacteria in their root nodules. This mutualistic relationship allows Alders to exploit taiga's nutrient and mineral-poor soils because their bacteria provide it.

© Ruth Peterkin/Shutterstock.com

**Figure 18.13**    Autumn in New York. A temperate deciduous forest has many colors as the chlorophyll dissipates from the leaves of trees leaving other pigments to show their colors.

**Figure 18.14** The Taiga Hills in Canada. Coniferous forests occupy large, colder regions of the northern hemisphere.

Taiga has long periods of sunlight in the summers and very short periods (less than 6 hours) in the winters. Thus, during summer months, crops grow rapidly with excess sunlight for photosynthesis. Yields are high in these regions and crops are productive for short seasons.

The animal life found in taiga is also adapted for cold temperatures. All have thick fur coats to survive the long winters. Large animals such as the caribou, elk, and moose predominate. Their predators are wolves and bears, which also migrate to taiga for food.

8) Tundra. Along the top of the globe, tundra encompasses the higher latitudes located only in very cold conditions and minimal sunlight. Northern portions of Canada, Russia, and Alaska are regions with tundra. Tundra covers almost 20% of the Earth's landmass (Figure 18.15).

**Tundra**

A tree-less area near the North Pole.

Annual precipitation is very low in tundra regions (less than 25 cm or 10 inches per year). Some small trees and shrubs do grow along the sides of lakes and streams because of access to water. Otherwise, rainfall limits plant growth. While vast, its cold temperatures are also not favorable to plant growth. A layer of permanently frozen subsoil, called permafrost, extends all through the tundra. Permafrost remains frozen at all times of the year, preventing larger trees and shrubs from rooting into the soil.

**Permafrost**

A layer of permanently frozen subsoil.

**Figure 18.15** Tundra Has Short Growing Seasons, Cold Temperatures, and Permafrost Limits Plant and Animal Lives. Tundra occurs across the northernmost regions of the globe.

Thus, tundra appears barren, with only small grasses and lichens able to withstand the weather. However, in summer months, life appears. In the summers, herbs grow rapidly along with small grasses. Migrating birds and large animals such as the caribou and musk ox visit the biome. They find food and isolation away from predators. Perhaps our driver in the story could have escaped his predator, the hitchhiker in this biome as well?

**Polar ice caps**

Dome-shaped ice sheets that slope in all directions from the north and south poles.

**9)** Polar ice caps. The ice caps cover the very reaches of the north and south poles, including all of Antarctica. They are almost one mile in thickness and extend across almost 98% of Antarctica. These areas are comprised of not land but almost all ice. It is estimated that more than half of the Antarctic surface does not have actual land beneath it. Almost all of the arctic is merely ocean below (Figure 18.16).

There is very little precipitation (4 cm or 2 inches per year) on the ice caps and extremely cold temperatures, reaching to $-89°C$ ($-129°F$) at the South Pole. The ice caps are quite barren, but its inhabitants include bacteria, small lichens, and a few mammals, such as whales, polar bears, and penguins. There are about 100 species of mosses, 25 species of liverworts, and only 2 flowering plants that are found in Antarctica. Plant growth is accomplished in only a few weeks in the summer. The Antarctic Peninsula has the most moderate climate, home to most of these organisms (Figure 18.17).

**Figure 18.16**    Polar Ice Caps. They are composed primarily of ice layers with very little life.

**Figure 18.17**    Antarctic Is Comprised of Almost All Ice with Very Few Plants and Animals. Most life in Antarctica exists along the Antarctic (Palmer) Peninsula, shown in the photo above.

There are human, scientific research stations in different regions at the ice caps at both poles. Some estimates of minerals and oil deposits in the polar ice caps have stimulated political jockeying, claiming rights to mine, and drill in the North Pole. Russia recently announced intentions to extract resources from its northern polar ice caps. In Antarctica, the exploitation of resources has been prohibited by the Protocol on Environmental Protection to the Antarctic Treaty of 1998. Several nations have laid claim to the areas of the continent, but the treaty agrees that it will be used for "peace and science" only. Many resources are expected to be found beneath the ice in Antarctica (Figure 18.18).

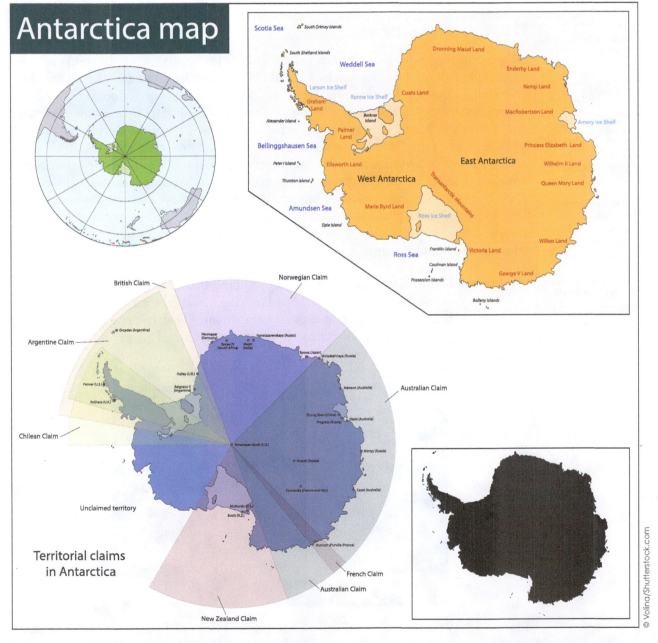

© Volina/Shutterstock.com

**Figure 18.18**   National Claims for Antarctic Resources. Several nations claim ownership of the regions in Antarctica, sometimes with conflicting boundaries.

## Aquatic Biomes

There are many environmental concerns, most a result of scarce resources. Frequently, in history, shortages in resources have stimulated crises in human societies: salt shortages, energy shortages, lack of food, lack of metals, lack of minerals, to name a few. However, the greatest shortage of the future may become the "water crisis." Water's scarcity has limited population growth since the ancients and our society may be headed for a water scarcity crisis. There are three types of water or aquatic biomes: freshwater, estuaries, and marine biomes.

Water is the most important nutrient in every living organism. As discussed in other chapters, organisms cannot survive long without a constant intake of water. It may appear to many readers that water is not scarce nor is very abundant. However, you will see that it is the scarcest of our resources. While roughly three quarters of the Earth's surface is covered in water, not much is accessible.

## Freshwater Biomes

**Limnology**

The study of freshwater systems.

**Productivity**

The production rate of new biomass by a person or community.

**Oligotrophic**

Lakes with low levels of productivity and abundance of dissolved oxygen.

Only 2% of that water is freshwater, readily useable to organisms. All the rest is saltwater, which is toxic when ingested by humans. Most of the world's freshwater is found in only certain regions. For example, the Great Lakes of North America contain over 21% of the Earth's surface freshwater. The largest amount of freshwater in one region is found in Lake Baikal, Russia (Figure 18.19). It contains about 20% of the world's freshwater. Many regions of the Earth have little or no available water for their societies. Regions of Africa, Australia, and Asia have very little access to freshwater and experience frequent droughts. The tragedy of droughts occurs in many parts of the world and limits food supplies. World hunger problems are linked to water scarcity needs because crops need water.

Limnology is the study of freshwater, and the importance of clean water to human society cannot be overstated. It divides freshwater systems into two types: ponds and lakes; or rivers and streams.

1) **Ponds and lakes.** These are still or standing bodies of water. Ponds are shallower than lakes but both may be classified based on their level of growth, or productivity of aquatic populations within them. Algae, bacteria, and other plant life are measured to determine a lake's productivity. Oligotrophic lakes, such

© Mikhail Markovskiy/Shutterstock.com

**Figure 18.19** Lake Baikal in Russia Contains the Most Freshwater of Any Single Lake on the Earth, with Approximately 20% of the World's Freshwater.

as Lake Baikal, have very little growth of organisms in its waters. They are often deep and clear but not very productive. Eutrophic lakes contain many growths of organisms and therefore have a high rate of productivity. They are usually shallow and murky waters. A comparison of the two lake types is given in Figure 18.20.

Lakes and ponds have three layers, each of which has different characteristics (Figure 18.21). The upper layer is called the epilimnion, in which sunlight first hits. Algae known as phytoplankton grow here, often shading out those organisms beneath their layer in eutrophic lakes. The middle portion of lakes is called the metalimnion. Floating organisms feed on phytoplankton in this layer, called zooplankton. Often algal blooms occur in this layer, which can disrupt aquatic ecosystems. The lowest layer is called the hypolimnion, in which there is little productivity because light does not reach so far deep. Temperatures are coldest in this layer during the warm seasons and warmest during the cold season. Aquatic life is able to move to and from the hypolimnion to find the right water temperature.

**Eutrophic**

Shallow and productive lakes.

**Epilimnion**

The top-most layer of lakes and ponds.

**Metalimnion**

The middle portion of lakes.

**Hypolimnion**

The lower-most layer of lakes and ponds.

2) **Rivers and Streams**. These are moving bodies of water (Figure 18.22). They contain more dissolved oxygen than ponds and lakes because their waters are moving, exchanging gases with the air. It is an open system and it obtains many of its nutrients from drainage from the surrounding land. Rainfall washes in organic materials (as well as pollutants), which increase the productivity of rivers and streams.

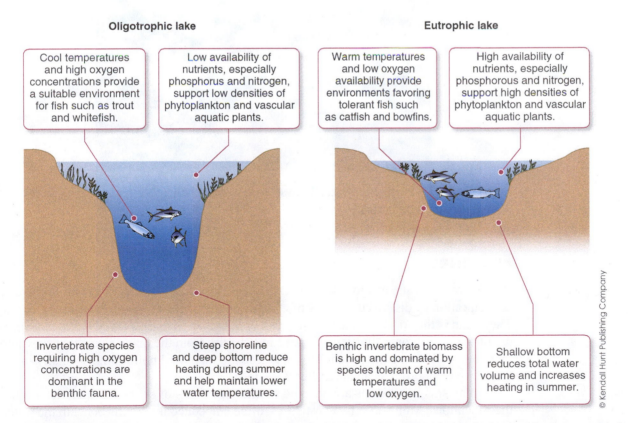

**Oligotrophic lake**

Cool temperatures and high oxygen concentrations provide a suitable environment for fish such as trout and whitefish.

Low availability of nutrients, especially phosphorus and nitrogen, support low densities of phytoplankton and vascular aquatic plants.

Invertebrate species requiring high oxygen concentrations are dominant in the benthic fauna.

Steep shoreline and deep bottom reduce heating during summer and help maintain lower water temperatures.

**Eutrophic lake**

Warm temperatures and low oxygen availability provide environments favoring tolerant fish such as catfish and bowfins.

High availability of nutrients, especially phosphorous and nitrogen, support high densities of phytoplankton and vascular aquatic plants.

Benthic invertebrate biomass is high and dominated by species tolerant of warm temperatures and low oxygen.

Shallow bottom reduces total water volume and increases heating in summer.

© Kendall Hunt Publishing Company

**Figure 18.20** Oligotrophic and Eutrophic Lakes. Oligotrophic lakes have little productivity and few organisms compared with eutrophic lakes.

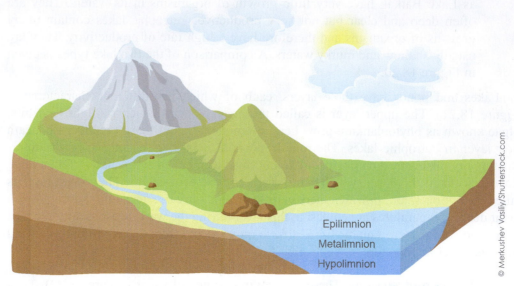

**Figure 18.21**   Layers of Lakes: Epilimnion, Metalimnion, and Hypolimnion. Each layer has different characteristics based on light penetration and temperature and dissolved gases.

**Figure 18.22**   In Rivers, Water Flows More Slowly than in Narrower Streams. Water flows quickly through this Swiss river.

## Estuaries

When rivers and streams join saltwater, they are classified as **estuaries**. The salinity, or concentration of dissolved solutes, of estuaries varies with its distance from the ocean. The ocean is the source of the salinity, as it is saltwater. Estuaries are also the most productive aquatic biome. Nutrients deposit in estuaries from the many rivers that flow into them before entering the ocean. This also leads to increased pollution, especially in the form of phosphorous as an agricultural waste.

Organisms in estuaries are adapted to different levels of salinity, which determines their locations. Mollusks inhabit estuaries, often embedded in its soils beneath the water. Oysters are commonly found in estuaries, along with other populations that feed on oysters – worms, crabs, and snails, for example.

Some organisms wander the estuaries, including several crustacean species such as lobsters, crabs, and fish. A great deal of fishing occurs in estuaries because these organisms are delicacies. Overfishing is a serious concern, disrupting ecosystems and driving some species to extinction in areas.

## Marine Biomes

All bodies of water that are saltwater are considered marine biomes. As stated earlier, about 98% of the Earth's water is contained within marine biomes. The study of its biology is called marine biology. The oceans have an average depth of about 4 kilometers (2.5 miles). The most limiting abiotic factors in marine biomes are light and food. Light penetrates into the water most efficiently to about 100 meters (325 feet). This means that beneath this layer, limited plant life can exist. Therefore, oxygen is present mostly in these upper layers, reaching 200 meters (650 feet) in the ocean.

There are three layers of marine biomes (Figure 18.23). The intertidal zone occurs between the high and low tides along the coasts. It is a harsh environment, with continual change and crashing waves. Nonetheless, the seashore has plenty of living creatures. The next neritic zone extends from the shore to about 100 kilometers (30 miles), reaching a depth of about 200 meters (650 feet). This zone lies above the jutting of land called the continental shelf. Coral reefs are found in this region, the most beautiful and richest ecosystem on the Earth. Once past the continental shelf, the open-sea zone forms the rest of the ocean. Here it reaches its great depths. The top 200 meters of the open-sea zone is called the photic zone. It contains most of the photosynthetic life – namely, phytoplankton. There is a great diversity of life in the oceans. Beyond this layer, the mesopelagic zone has little light. Many strange organisms that exhibit bioluminescence have

**Marine biology**

The study of saltwater organisms.

**Intertidal zone**

The area that is above water level at low tide and below water level at high tide.

**Neritic zone**

The zone of ocean where sunlight reaches the ocean floor.

**Open sea zone**

The main body of ocean or sea.

**Photic zone**

The part of ocean where sunlight penetrates sufficiently and influences the growth of living organisms.

**Mesopelagic zone**

The ocean layer that receives little sunlight.

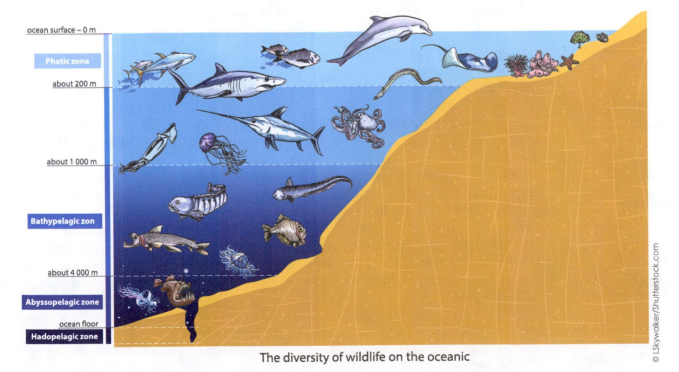

ocean surface – 0 m
Photic zone
about 200 m
about 1 000 m
Bathypelagic zon
about 4 000 m
Abyssopelagic zone
ocean floor
Hadopelagic zone

© LSkywalker/Shutterstock.com

The diversity of wildlife on the oceanic

**Figure 18.23** There Is Great Diversity in the Zones of the Marine Biome. Most life in the open sea occurs within the photic zone, where light drives photosynthesis and plant growth. The slope of the land is exaggerated in the figure to save space.

**Abyssal zone**

A deep layer near the bottom of the ocean.

evolved in this zone. This region is very dark, and these organisms use bioluminescence to communicate. Below 1500–2000 meters is the abyssal zone, meaning "bottomless." In this zone, there is tremendous pressure from water above and no light. Here, surprisingly some organisms thrive. Benthic organisms, those feeding at the bottom of the ocean, live off of material that dies and falls to the ocean floor. Others are chemosynthetic bacteria, which obtain energy from sulfur and hydrogen from hot springs and deep sea vents.

# Ecosystems

## Ecosystems Make Up Biomes

The biomes presented in this chapter are comprised of numerous smaller environmental areas. These areas are known as **ecosystems**, comprised of communities interacting with their environment. The mosses growing along the Antarctic Peninsula or the Alder trees of taiga containing roots with bacteria are each a smaller subset of much larger biomes – each is an ecosystem.

As you sit in the backyard of your garden and read this text or open the refrigerator and view a moldy piece of cheese, you should think of the word "ecosystem." Ecosystems are all around us and shape our interactions. As we pass through them in the next section, much in the same way our character drives through the biomes, they become more appreciated.

As stated earlier, an ecosystem is a community of organisms in an area interacting with their abiotic environment. Many ecosystems comprise a biome. However, what happens in an ecosystem? How is energy transferred? What relationships do communities have with their non-living world? Do these relationships follow certain patterns?

## Energy Flow through Ecosystems

Energy is trapped, transferred, and lost as it moves through organisms in the ecosystem (Figure 18.24). Roughly 1% of sunlight reaching the Earth is transformed by photosynthesis into the food's chemical energy. This process starts the flow of energy through the environment. Organisms that carry out photosynthesis are called producers. Plants and phytoplankton, or marine algae, are the principal producers in natural ecosystems. They possess chloroplasts to carry out photosynthesis and convert sunlight into food, as described in Chapter 4. Over 100 billion metric tons of carbon is made by producers annually.

**Producer**

Organisms that carry out photosynthesis.

**Herbivore**

Organisms that eat plants.

**Carnivore (secondary consumer)**

Organisms that eat herbivores.

**Tertiary consumer**

Carnivores that eat carnivores.

**Omnivore**

Organisms that eat both plants and meat.

Producers nourish those organisms that eat them, known as herbivores. Herbivores are also known as primary consumers. Herbivory was discussed in Chapter 17, as a form of predation. Herbivores kill producers and eat them to obtain some of their energy. Cattle, sheep, insects, and humans are examples of herbivores. Some herbivores need the help of other organisms to digest the plant parts. These plant parts evolved against herbivory, such as cellulose plant cell walls. Herbivores contain microorganisms in them live within their guts and digest plant parts. This way, more plant parts provide energy for herbivores.

Organisms that eat herbivores are called secondary consumers or carnivores. Carnivores are meat eaters – they eat meat by killing herbivores. Spiders, wolves, cats, frogs, and toads are herbivores. They eat their prey to obtain some of the chemical bond energy in those organisms.

Carnivores that eat carnivores are called tertiary consumers. This is generally as high as the pathway of energy flow goes. Humans are unique in the food chain. They are both herbivores and carnivores – we are omnivores, meaning that we eat both meat and

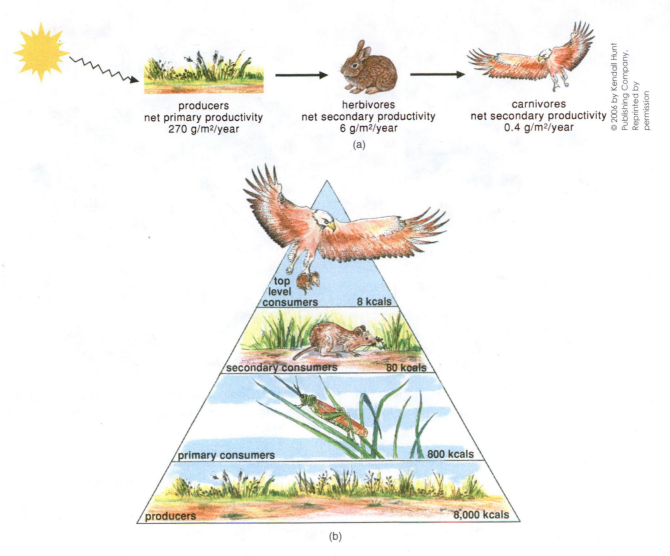

**Figure 18.24** Flow of Energy in an Ecosystem. Energy is trapped, transferred, and lost as it moves through living organisms in the ecosystem. Note the loss of energy in the associated pyramid of energy. At the bottom, 8000 kcal is reduced to only 8 kcal by the top of the energy pyramid. From *Biological Perspectives*, 3rd ed by BSCS.

plants. Our diets derive about 1/3 of its energy from animals and animal products and 2/3 from producers.

Each level of eating is called a trophic level. A trophic level is a group of organisms with a type of feeding style. For example, grasses are autotrophs because they are all photosynthesizers. Tertiary consumers are the "top" trophic level of all the feeders because they all eat other carnivores.

Movement through trophic levels represent the energy flow from one type of feeder to another. A pathway of energy flow is known as a food chain. The pathway through which energy flows in every system is the same: from producer to herbivore to secondary consumer to tertiary consumer.

Sometimes food chains overlap within a community, forming a series of pathways called a food web. Food webs better represent a community's transfer of energy because some organisms, such as humans, are omnivores and may occupy more than one trophic level. As such, complex feeding structures develop, as shown in Figure 18.25. The figure shows the many organisms involved in energy transfer within a forest ecosystem.

**Trophic level**

A group of organisms with the same feeding style.

**Food chain**

The pathway through which energy flows in every system: from producer to herbivore to secondary consumer to tertiary consumer.

**Food web**

A series of pathways formed when food chains overlap within a community.

*© Kendall Hunt Publishing Company*

**Figure 18.25** Food Web of an Eastern Deciduous Forest. The many complex relationships are simplified in comparison with nature's intricate mechanisms.

---

**Decomposer**

Organisms that break down once living organic matter into energy.

**Scavenger**

Animals that feed on dead or decaying matter.

Ultimately, energy continues to flow from living systems after they die. Decomposers, such as bacteria or fungi and scavengers, including vultures and worms break down once living organic matter into energy. Decomposers return materials back into the environment, recycling organic material, as will be discussed in another chapter. The start of the transfer of energy is, of course, photosynthesis and the end is bacteria of decay.

## Energy Pyramids: Not Cutting Out the Middle Man

There are many more grasses in temperate grasslands than there are bison. There are usually more palms in a tropical forest than there are cheetahs. Both bison and cheetahs are higher on the food chain than palms and grasses. High status on the food chain has drawbacks. Producers, such as grasses, get their energy directly from sunlight. As a Bison grazes on grasses, it uses energy to do this. Energy is lost along the way as heat. There is a net loss of about 90% of energy due to metabolism and is lost in feces. Metabolism loses energy as heat; and feces are undigested matter, which is also unused by organisms. Together, very little of the energy trapped from sunlight moves along an ecosystem, with most dissipated.

**Energy pyramid**

A pictorial representation showing the net loss of energy as it travels up a food chain.

**Biomass**

Is the total organic matter in an ecosystem.

We say that there is a net transfer of only 10% of energy per jump in trophic level. Think of each trophic level jump as having a "middle man," who takes a 90% cut in your profits every time it deals with you. It is a terribly inefficient system but it works the same in every ecosystem. Each time a middle man is involved, energy is taken out of the system. Thus, more energy is concentrated at the bottom of a food chain compared to the top in any ecosystem. The energy pyramid, shown in Figure 18.26, is a pictorial representation showing the net loss of energy as it travels up a food chain.

The losses across trophic levels may also be measured in biomass. Biomass is the total organic matter in an ecosystem. It is the sum total of the weight of all the plants and

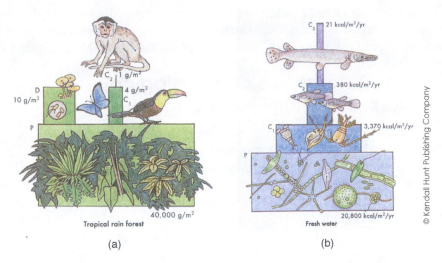

**Figure 18.26**   Energy Pyramid: The Middle Man Cuts Energy Flow.

animals in an ecosystem. Suppose that a bison grazes on those grasses. It adds biomass, but not as much as you might think. If a bison eats 100 pounds of grass, it only gains about 10 pounds of weight. Those consuming bison only obtain 10% of the biomass of their prey. With each succeeding level in the food chain, only 10% of biomass is conserved. This is known as the 10-percent rule for calculating net transfer of energy in ecosystems. The pyramid of biomass in Figure 18.27 shows the net loss of biomass along food chains.

## Vegetarians Cut Out the Middle Man

Ecologically, it is more efficient to consume organisms closer to the bottom of the food chain. Vegetarians, who consume primarily producers, do not experience the higher losses in energy seen at higher trophic levels. A more sustainable strategy for eating is to tap lower levels of the food chain. This way, biomass (or food) is more easily conserved, saving food resources.

Most of the developing nations rely on this strategy. However, the obesity epidemic experienced in the United States and other developed countries is linked with diets higher in

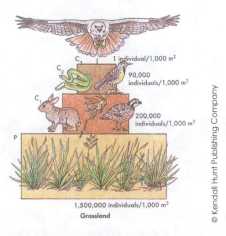

**Figure 18.27**   Pyramid of Biomass. Higher tropic levels in a tropical forest in Silver Springs, FL show far less biomass than lower levels.

the food chain. Eating a large biomass of carrots, for example, does not translate into the same weight gain as eating the same biomass of steak. Eating lower on the food chain may conserve energy resources in the ecosystem and is healthier for us (Figure 18.28).

## Ecosystem Disturbance and Ecological Succession: Communities Change over Time

**Ecological succession**

The process by which nature reclaims an ecosystem after it has been disturbed.

When ecosystems are disturbed, they do not simply die off, never to return. Ecosystems undergo changes when there are natural disasters: volcanoes, floods, earthquakes, fire, and lightening. Humans also create disturbance in building projects and wars. However, nature slowly but surely reclaims the areas, and we see a return to a state of normalcy.

Ecological succession is the process by which nature reclaims an ecosystem after it has been disturbed (Figure 18.29). It occurs slowly over many years of time. Ecological succession follows certain, predictable stages of change. The first stage, after a

**Figure 18.28** Eating Lower on the Food Chain (such as Fruits and Vegetables of Producers) Conserves Energy and Taps into the More Renewable Resources of an Ecosystem.

**Figure 18.29** The Lichens and Mosses in This Beautiful Ecosystem (Primary Ecological Succession) from a Once-glaciated Area Will Eventually Become a Thriving Forest (and Climax Community).

disturbance, is colonization of an ecosystem. The area may be bare rock or barren soil with no life. Only organisms such as lichens and bacteria are able to survive. These organisms are able to grow in the harsh conditions found after a major disturbance.

Colonizers are the first stage in primary succession. Primary succession is the series of stages that start succession, in which there is no life and no soil in an ecosystem. Its organisms prepare the soil for future organisms. Lichens for example secrete acids and break down rock into soils and minerals. Continual breaking down of materials forms thicker and richer soils. After lichens and bacteria produce soils rich enough to support higher plants, seeds from those plants germinate. Mosses arrive early on, and then herbs and small shrubs. Trees arrive and eventually those which grow taller and tolerate shade better win out over the others.

Colonizers and small shrubs are eventually replaced. While they were the only organisms able to exploit the ecosystem when it was nutrient poor, it is no longer the best competitor. Larger plants usually win out over smaller plants and longer-lived species tend to persist. The winning organisms endure in a stable and self-sustaining community, known as a climax community. Theoretically, the climax community is the highest level of organization for an ecosystem. Its populations "won" the battle and the payoff is that they remain forever in the climax community.

However, disturbances are a guarantee in nature. Secondary succession occurs when an established ecosystem is disturbed but some life and soil remains behind. While primary succession takes thousands of years, secondary succession usually proceeds quickly, within decades or a century. During secondary succession, a disturbance changes the ecosystem. The changes to an abandoned farm field are an example of secondary succession. When it is no longer farmed, disturbance ends and the natural biota return. First, smaller plants and animals colonize the field and eventually shrubs and bushes take over. Seeds from larger nearby trees land and germinate. Seeds rise to a new population of trees, which outcompete the smaller plants for sunlight and water. Those former plants die away and the trees then crowd and compete with each other. It is an intense fight for survival, except that it is not obvious because it occurs over many decades. White pine and cedar trees are not very shade tolerant, and get overgrown by hemlock, beech, and maple if the old field is in a temperate deciduous forest. Those plants are able to tolerate the shade well and can grow without disadvantage from the bottom of the forest floor.

Disturbance occurs continually in biomes and ecosystems, but each time the communities return. Climax communities are only climax when they are undisturbed. Most of the time in nature, this is not the case. Ecosystems follow the predicted stages of ecological succession, succeeding pioneer species shown in Figure 18.29.

Some changes are difficult to undo. The massive amount of encroachment into natural areas has led to the formations of pockets of isolated ecosystems. These pockets are called fragmented meta-populations. These populations are cut-off from the rest of their community due to human development. The character is our story begins his trip excited to combat tarring of roads, which are a main cause of fragmented ecosystems. For example, many malls have constructed circular ramps that connect with local roads and highways, as shown in Figure 18.30. They place decorative, small ponds in the center of the circular road. Of course, as you recall from Chapter 10, amphibians breed in ponds. The circular road acts as a death trap for those entering the pond to mate and those born and leaving the pond (Figure 18.30). Amphibian deaths have risen rapidly as an effect of this trend in modern landscaping.

**Colonization**

A process by which a species spreads to new areas.

**Primary succession**

The series of stages that begin ecological succession, including lichens and mosses in a disturbed area.

**Climax community**

The highest level of organization for an ecosystem.

**Secondary succession**

The process that occurs when an established ecosystem replace organisms and soils of primary succession.

**Fragmented meta-population**

Pockets of isolated ecosystems due to massive amount of encroachment into natural areas.

**Figure 18.30**    Vanity Ponds in Malls and Surrounding Housing Developments Have High Incidences of Amphibian Deaths due to Their Placement within Roads.

## Summary

Biomes are ecological regions, which occur over wide areas of the Earth. Humans both derive life's resources from the biomes of the Earth and enjoy their beauty. Sometimes, human activities, such as overuse of resources, negatively impact biomes. Within biomes, local topography effects smaller changes in abiotic factors to produce regional climates. The nine major biomes of the world are most influenced by light, altitude, and water availability. The aquatic biomes differ in their salinity and locations across the globe. Within all of these biomes are ecosystems, which transfer energy through its organisms. When an ecosystem is disturbed, it undergoes a process of reclamation, returning it back to normal.

### CHECK OUT

**Summary: Key Points**

- Biomes play a vital role in providing a habitat for organisms but the human exploitation of their resources sometimes threaten their functioning.
- Biomes are discrete sets of ecosystems, which occur across large regions of the Earth.
- Topography in local areas has effects on climate in biomes.
- The nine major biomes of the world are influenced by three important abiotic factors – light, altitude, and water availability.
- The three types of aquatic biomes are freshwater, estuary, and marine systems.
- Energy moves from producers to herbivores and finally to consumers in an ecosystem.
- Ecological succession, over time, changes soils and organisms within an ecosystem after disturbances.

## KEY TERMS

| | |
|---|---|
| abyssal zone | mesopelaegic zone |
| biomass | metalimnion |
| biome | neritic zone |
| canopy layer | oligotrophic |
| carnivore (secondary consumer) | omnivore |
| chaparral | open sea zone |
| climax community | permafrost |
| colonization | photic zone |
| decomposer | polar ice caps |
| desert | primary succession |
| desertification | producer |
| detritus | productivity |
| ecological succession | savanna |
| ecosystem | scavenger |
| energy pyramid | secondary succession |
| epilimnion | taiga |
| estuary | tertiary consumer |
| eutrophic | topography |
| food chain | trophic level |
| food web | tundra |
| fragmented meta-population | rain shadow desert |
| herbivore | root-to-shoot ratio |
| hypolimnion | temperate deciduous forest |
| intertidal zone | temperate grassland |
| limnology | tropical rainforest |
| marine biology | |

# Multiple Choice Questions

**1.** How do fragmented meta-populations form?

   **a.** through high birth rates

   **b.** through human activity

   **c.** with community interdependence

   **d.** with population interdependence

**2.** Rain shadow deserts always form on the _____ side of a mountain chain.

   **a.** north

   **b.** west

   **c.** upwind

   **d.** downwind

**3.** Which is NOT a characteristic of deserts?

   **a.** hot temperatures

   **b.** stable temperatures

   **c.** cold temperatures

   **d.** all of these are desert characteristics

**4.** Freshwater biomes represent _____ % of the Earth's available water.

   **a.** 2

   **b.** 10

   **c.** 50

   **d.** 99

**5.** A layer of permafrost is found in:

   **a.** taiga

   **b.** tundra

   **c.** deserts

   **d.** estuaries

**6.** Which term includes all of the others?

   **a.** ecological succession

   **b.** climax community

   **c.** colonization

   **d.** primary succession

**7.** Which represents a logical order, from start to end, of the flow of energy in an ecosystem?

   **a.** producer → herbivore → consumer → bacteria

   **b.** consumer → producer → consumer → bacteria

   **c.** herbivore → producer → bacteria → consumer

   **d.** herbivore → consumer → bacteria → producer

8. A scientist discovers a tree has fallen in a stand of pines. He notices new organisms now growing in the open space. Which variables determine the direction of this process?
   a. soil
   b. populations
   c. minerals
   d. all of the above

9. Which of the following biomes is most likely to have a fire?
   a. chaparral
   b. temperate deciduous forests
   c. estuaries
   d. tropical rainforests

10. Which biome contains the greatest biodiversity?
    a. chaparral
    b. temperate deciduous forests
    c. estuaries
    d. tropical rainforests

## Short Answer

1. Describe how humans create fragmented populations within ecosystems.

2. List two types of saltwater biomes. Which is likely to provide more protection for its organisms?

3. Define the following terms: producer and herbivore. List one way each of the terms differ from each other in relation to their a. function; b. role in transfer of energy in an ecosystem; and c. relationship with each other in the environment.

4. Explain how a rise in altitude is the same as a rise in latitude on the globe, ecologically.

5. Draw an energy pyramid for a forest ecosystem. Be sure to label each trophic level and give a plausible example for each level.

6. Describe how 2400 pounds of biomass, when eaten by an elephant, becomes only 240 pounds in her body.

7. Draw a sketch of a food web using arrows to show the flow of energy through the ecosystem. Use the following organisms in your sketch: hawk, dove, cow, corn, grass, bacteria (actinomycota), and human.

8. Explain the process of primary succession from bare rock to beech tress. Use arrows to trace the flow of the animal and plant organism changes.

9. Explain why a high root-to-shoot ratio is important in chaparral.

10. Compare and contrast the characteristics of grasslands and temperate deciduous forest. Be sure to include one way the biomes have similarities and one way they are different.

## Biology and Society Corner: Discussion Questions

1. Deforestation is a serious environmental hazard. People in economically disadvantaged nations claim this is not the case. They feel that exploiting rainforest resources is their only way out of poverty. Research the issue. Form a plan for local farmers, within a developing nation, to implement other methods to improve their situation.

2. Describe how estuaries are easily exploited for its resources? How can safeguards be introduced to prevent this overuse? Research the issue in Chesapeake Bay to make a complete answer.

3. You invite your friend to your home for a meal. Your friend is a vegetarian but offends your father, after declining to eat the steak dinner served. Who is right and who is wrong? What should you do to help? List reasons why a person may be a vegetarian, in answering this question. List reasons why your father may be offended.

4. The United States built Las Vegas right in the middle of a desert in Nevada. Defend the move to do this. Research those techniques that are able to change desert into arable land. Criticize the location of Las Vegas, ecologically.

5. The Protocol on Environmental Protection to the Antarctic Treaty of 1998 showed cooperation among nations. However, several nations still lay claim to the regions on the continent and have not given those claims up. Suppose that a nation sends troop to occupy the Antarctic Peninsula. Write a plan to diffuse the situation.

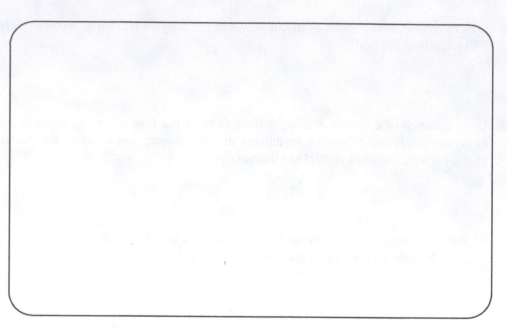

Figure – Concept Map of Chapter 18 Big Ideas

# Biosphere: Life Links to the Earth

**19**

Comet in the sky

Charred trees in the forest

Meteorite from Tunguska

Villager interviewed

A man (Jerry?) looking up at sky through telescope

## CHECK IN

**From reading this chapter, you will be able to:**

- describe extraterrestrial threats to human society.
- define the biosphere and describe abiotic conditions affecting its climate.
- define and describe terms list.
- trace the flow of key chemicals through the ecosystem including water, carbon, nitrogen, and phosphorous.
- describe negative human impacts on the biosphere.

# The Case of the Big Blast

The morning of June 30, 1908, Tunguska, Russia: "It was a visitation by the god, Ogdy!" exclaimed the Siberian villager as he told the story of the great blast. "High in the sky a heavenly body split it into two." Just say the word "Tunguska" and the people in the area shiver. "The gods hated Siberia and struck them down with their might" continued the man. He was one of the few eyewitnesses to the event. Local people were scared to speak of it for years. The villager continued, "The sky was on fire and stones were falling everywhere from the sky."

The incident was a mega-blast in the sky over Tunguska. The witness spoke of the event only years afterward, while interviewed by Western scientists. They lived in fear of the gods, who many locals believed sent the blast to punish them. The recount of this event is described by locals as "the day the sky blew apart." It is now ancient folklore but the ruins are still in the mountains of Tunguska in Siberia. Superstitions and fears of the gods returning with a vengeance led local Siberians to keep quiet about the incident.

The morning of June 30, 2008, Boise, Idaho, the United States: Jerry had gone to the 100-year commemoration of the Tunguska explosion at his local asteroid club. Jerry was an amateur astronomer who studied and followed all of the asteroid sightings. However, the Tunguska event was his favorite.

He read about the asteroid's impact: The blast happened in the mountains and destroyed over 800 square miles of forest and over 80 million trees. No one was reported killed because the area was so remote but its impact could be felt throughout the Asian continent. All the way in London, seismic waves registered on the Richter scale – earthquakes from the blast.

It is hypothesized that a large asteroid hit the Earth's atmosphere at the site of Tunguska, Russia in the Siberian forests. Scientists believe that the asteroid was about 120 feet across and exploded in the sky above Tunguska. When the space rock exploded, it caused ash and debris to emanate from the site. The asteroid traveled at a speed of 33,500 miles per hour. It is estimated that it weighed 220 million pounds. It must have heated up to 44,500°F. While it flew through the air, it exploded into many fragments, which is what the villagers saw causing the sky to be on fire.

Jerry's breath was taken away as he read about the event. As an amateur astronomer, he was impressed to learn that it had the strength of 185 Hiroshima bombs. Jerry was waiting for the next time Haley's comet would visit the Earth, once every 76 years. Its next sighting would be 2062 . . . Jerry hoped that it would not hit the Earth.

**CHECK UP SECTION**

In the story, an asteroid created a major ecological disturbance. Threats from extraterrestrial sources are real but the probability of an actual impact is small.

What kinds of global disturbances are caused by such an event in the biosphere? Research the asteroid impact event associated with the hypothesis for why the dinosaurs became extinct 65 million years ago.

# The Earth, the Sun, and Atmosphere

## The Earth's Boundaries for Life

The disturbance in the story shows what could happen when a space body invades the Earth's thin layer of sky. This layer of global ecosystem, which contains all life on the planet, is called the biosphere. The Earth's biosphere boundaries are not easily measured. This zone of life on the Earth extends from the polar caps to equatorial zones (Figure 19.1). Boundaries above the Earth's surface reach over a mile high in the atmosphere. Here, birds can reach their highest flight over the tallest mountains. Below the surface soils and in the ocean, single-celled microbes reside miles deep in waters of the

**Biosphere**

The layer of global ecosystem, which contains all life on the planet.

(a)

(b)

(c)

**Figure 19.1** Life on the Earth Exists in All Regions from Polar to Tropical. The biosphere contains life on land, underground, in waters, and in the sky.

**Atmosphere**

The gaseous layer surrounding a planet.

**Hydrosphere**

All of the water on Earth's surface.

**Lithosphere**

Earth's outer part.

Mariana Trench. The biosphere integrates the Earth's living things and their interactions with water, rock, and air reservoirs – the hydrosphere, lithosphere, and atmosphere. Human activities that disrupt these processes and the life they support will be considered here in this chapter.

## Atmosphere: A Layer of Protection

The **atmosphere** is the gaseous layer surrounding the Earth held in place by gravity. Oxygen (21%) and nitrogen (78%) together make up 99% of gases in the atmosphere, as described in Chapter 2. Organisms that respire aerobically depend on atmospheric oxygen. The remaining 1% of atmospheric gases is made up of a variety of trace gases including argon, carbon dioxide, neon, and helium. Photosynthetic organisms as described in Chapter 4 require these trace components of carbon dioxide to drive the biological food webs. Water vapor is present in the atmosphere in varying amounts across the globe at any given time. Trace amounts of air pollutants including methane, ozone, chlorofluorocarbons (CFCs), dust particles and pollen, and microorganisms are present too. As a whole unit, the atmosphere that envelopes the planet serves to protect and moderate extreme forces acting on the Earth in just the right manner to allow for life to exist.

One essential service provided to the biosphere by the Earth's atmosphere is the protection from damaging high-energy solar radiation and deadly amounts of cosmic rays from space. Visible light and some infrared radiation do penetrate the atmosphere. The Earth's surface and lower atmosphere are warmed by this low-energy radiation from the Sun. Energy that reaches the Earth is only a very tiny fraction, less than one billionth, of the Sun's total energy, as discussed in Chapter 18. Nonetheless, each day, a tremendous amount of energy from the Sun arrives at the Earth's surface. Of the energy that reaches the Earth, about 30% is rapidly reflected back into space, mainly by surfaces with high albedo or reflectivity such as clouds, snow, and ice (Figure 19.2). The remaining solar energy that is absorbed into the atmosphere is responsible for weather and climate patterns, drives water and chemical element cycles, and powers life on the planet beginning with photosynthesis.

**Albedo**

A proportion of solar energy that is reflected from the Earth back to space.

**Figure 19.2**     A Photo of the Earth from Space. Note the bright white areas of polar ice and cloud systems. Surfaces with high albedo, such as ice and clouds strongly reflect the Sun's energy.

# Solar Radiation: Heat from the Sun

We know that the Earth is generally coldest at its Polar Regions and warmest near the equator. This latitudinal variation in temperature is due to the planet's spherical shape (Figure 19.3). When sunlight strikes the Earth near the equator, it is nearly perpendicular and traveling the shortest possible distance directly through the atmosphere. These rays move into a concentrated surface area, delivering the most intense heating power. Contrast this with sunlight reaching the Earth's Polar Regions. Here solar energy reaches the Earth at an oblique angle, travels a further distance through the radiation-absorbing atmosphere to the curved surface. These rays are distributed over a larger surface area. Energy reaching polar areas is thus more diffuse resulting in lower surface temperatures.

Small alterations in the makeup of the atmosphere can have significant impact on the Earth's climate by affecting incoming solar radiation. For example, large volcanic eruptions can emit large amounts of gas and ash into the upper atmosphere. Historically, some of these events have lowered global temperatures by a degree or two. Volcanic gases and fine ash can block or reflect portions of solar energy from reaching the Earth's surface up to a year or more. The 1815 eruption of Mount Tambora in Indonesia contributed to the 1816 "Year Without a Summer" where extreme cooling conditions were noted in the Northern Hemisphere. Reports of late frosts destroying crops and summer snows in New England and Europe were common. As in the chapter's opening story, destruction from large-scale atmospheric events such as the asteroid impact in Russia creates larger effects such as climate and ecosystem changes in the biosphere.

# Seasonal Changes in Temperature

The Earth tilts on its axis of spin at 23.5° from a line perpendicular to its plane of orbit. This tilt remains the same as the Earth orbits the Sun. It is responsible for the seasonal changes we experience here on the planet. The Northern Hemisphere tilts toward the Sun between March 21 and September 22 receiving more concentrated sunlight and longer days (Figure 19.4). For the other half of the year (September 22 to March 21), the Northern Hemisphere is tilted away from the Sun. It receives less concentrated sunlight and shorter days. Northern and Southern Hemisphere orientations are opposite each other in relation to the Sun for the same given time periods. Hence, when the Northern

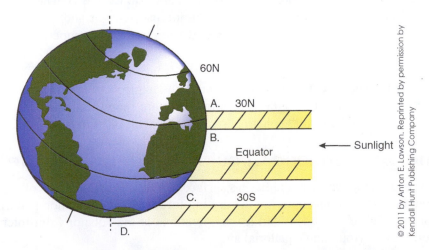

**Figure 19.3**    Solar radiation varies with latitude due to Earth's spherical shape. The sun's radiation travels the shortest most direct route to the equator providing more intense heating. From *Biology: An Inquiry Approach*, 3rd ed by Anton E. Lawson.

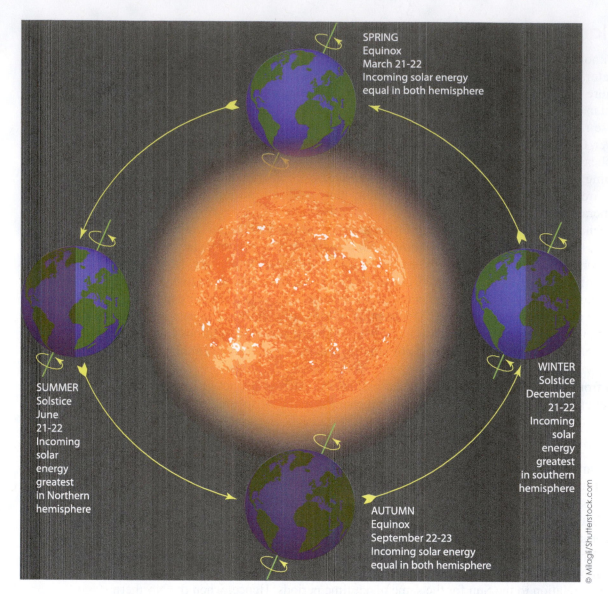

**Figure 19.4**   Seasonal Variation of Solar Radiation. The tilt of the Earth on its axis remains the same as it orbits the Sun. In June, the Northern Hemisphere receives more intense sunlight and in December, the Southern Hemisphere receives more. These changes account for seasonal variation in temperature and hours of daylight.

Hemisphere is experiencing summer, the Southern Hemisphere is experiencing winter. Seasonal variations in day length and temperature are more pronounced at the Polar Regions because of the Earth's spherical shape and the tilt of its axis.

## Global Atmospheric Circulation Affects Climate

**Intertropical convergence zone**

The low pressure area at the equator resulting from the expansion and rising up of hot air mass.

Differences in temperature at the Earth's surface due to varying amounts of solar radiation reaching different locations are the driving force behind circulation patterns in the atmosphere (Figure 19.5). Hot surface temperatures at the equator from intense solar radiation heats surrounding equatorial air.

The hot air mass expands and rises up leaving a low-pressure area at the equator referred to as the intertropical convergence zone (ITCZ) or doldrums. Aloft, the warm air travels away from the equator. The moving air cools as it travels and is unable to hold the same high level of water vapor. Moisture exits the air mass as rain, quenching areas of tropical rain forest.

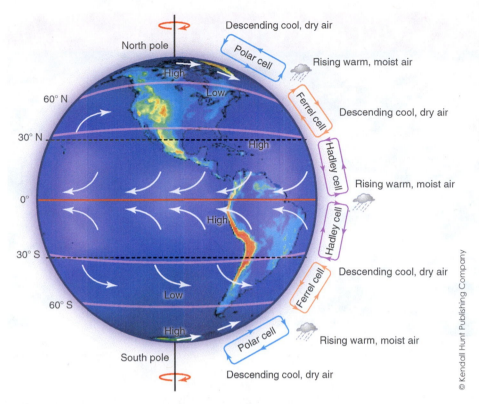

**Figure 19.5** Global Atmospheric Circulation. Varying solar radiation by latitude warms equatiorial air masses. Upward flow of these air masses brings about global patterns of air movement idealized in three circulation cells per hemisphere. Air near Earth's surface is deflected from north-south circulation by the planets rotation creating east or west blowing patterns at different latitudes. Surface winds directions are shown with white arrows.

Eventually, the air sinks down toward the surface at around 30° latitude north and south. The cool dry area descending around 30° latitudes absorbs moisture from the land below. Desert biomes are often found at these 30° latitudes as a result of these circulation patterns. The bulk of descending air flows back along the surface toward the low-pressure zone at the equator.

Air patterns show a similar upward movement at higher latitudes around 60° nearer to the poles. Cold dry polar air sinks at the polar extremes and flows toward lower latitudes generally beneath the warm air aloft flowing in a polar direction. These continually mobile air circulation patterns transfer heat from the equator toward Polar Regions. Air currents return polar air back in an equatorial direction, cooling the surface below as it travels. Air circulation moderates temperatures across the surface of the planet. The nature of global air flow patterns is such that some mixing of the entire atmosphere does take place. Anything held in the atmosphere – dust, ash, pollen, pollutants, and aerosols – can be spread globally as a result.

## Winds: Movement Under Pressure

Differences in atmospheric pressure and the Earth's rotation are contributors to the intricate horizontal movements of the Earth's atmosphere known as winds. Gases that make up the atmosphere put pressure on the Earth because they have weight. This pressure can vary due to the changes in temperature, altitude, and humidity. Generally, stronger winds result from larger differences in pressure. Winds move from the areas of high pressure to the areas of low pressure.

**Wind**

The intricate horizontal movements of Earth's atmosphere caused by differences in atmospheric pressure and Earth's rotation.

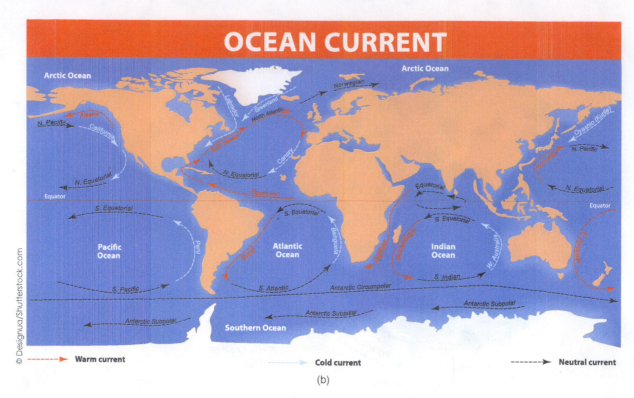

**Figure 19.6** Surface Ocean Currents. Surface ocean currents are primarily wind driven, moving clockwise in the Northern Hemisphere and counter-clockwise in the Southern Hemisphere.

**Coriolis Effect**

The deflection of a moving object with respect to the Earth's rotation.

**Trade winds**

Winds that blow above and below the equator.

**Westerlies**

Winds that blow from the west.

**Polar easterlies**

Dry, cold winds that are deflected to the west like the trade winds.

**Heat capacity**

The amount of heat energy required to raise the temperature of an amount of substance.

The Coriolis effect describes deflection based on rotation. The Earth rotates from west to east. Wind patterns are deflected as a result of this spin. The Northern Hemisphere winds swerve slightly to the right of expected flow along pressure gradients and the Southern Hemisphere wind to the left. These deflections point the winds of atmospheric circulation blowing from 30° latitude toward the equator in a western direction (Figure 19.5).

Since winds are named by the direction from which they blow, winds above and below the equator are known as the northeasterly trade winds in the Northern Hemisphere and southeasterly trade winds in the Southern Hemisphere. Similarly, winds moving poleward from 30° are deflected by Coriolis forces to blow toward the east. Naming them from their point of origin, they are termed westerlies. The cool dry air returning from the poles is deflected to the west like the trade winds and is termed the polar easterlies.

As described in Chapter 18, local geography combined with winds can affect rainfall and create desert areas. Moisture-rich air masses from wet areas often release water as rain while they rise over mountain ranges. The region past the mountains receives little moisture. These rain shadow areas often form deserts downwind of coastal mountain ranges.

# Hydrosphere: Global Transport and Climate Control

## The Earth's Waters

Water's moderating effect on weather and climate stems from its chemical property of high heat capacity. Heat capacity describes the amount of energy it takes to raise an amount of substance a given change in temperature. This means that water has a huge ability to absorb and hold on to heat, about 10,000 times greater than the atmosphere. Because of this, temperature fluctuations in water bodies are much less than those of the air around us. Vast amounts of heat are absorbed from the Sun and held by large water bodies. Temperatures in the areas near large water bodies are moderated by this heat-sink effect of water. Winter air temperatures will be raised as the water slowly releases stored heat to the surrounding air masses. Tropical areas will similarly be cooled by sea breezes from nearby water bodies.

The vast majority, 97%, of the Earth's water is found in oceans, recall from Chapter 18. Ocean water contains an average of 3.5% dissolved salts that add to its density. Freshwater has low concentrations of dissolved salts. Some rivers estuaries such as the Hudson and the Mississippi form a salt wedge at the intersection of the two. Less dense freshwater is sectioned atop the saline ocean water in a wedge shape. This wedge formation fades in estuaries where strong tidal forces mix the waters. The remaining 3% of water on the planet is freshwater. A breakdown of freshwater reservoirs finds 69% is locked away in frozen polar icecaps and glaciers and another 30% is held underground (Figure 19.7). Only the remaining 1% or so is easily accessible in rivers, and lakes, as you may recall from Chapter 18.

**Freshwater**

Water with low concentrations of dissolved salts.

**Salt wedge**

A wedge-shaped intrusion of sea water into a fresh-water estuary.

## Ocean Circulation

Major currents in the oceans affect terrestrial temperatures along coastal regions. Surface ocean currents are driven by winds and temperature. Masses of water flowing at the ocean surface are affected by trade winds and temperate westerlies. Water flows generally in the direction of prevailing winds forming major surface currents. Rotation of the Earth, gravity from the moon, location of landmasses, and topography or shape of the ocean basins also impact the movements of ocean currents. Generally, oceanic circulation moves clockwise in the Northern Hemisphere and counterclockwise in the Southern Hemisphere (Figure 19.6). Notable spills of ship cargos, such as Nike sneakers and Yellow duck bathtub toys in the 1990s, have been used to study the flow of surface ocean currents in the North Pacific. Beachcombers on shore picked up and logged location and dates for such items found thousands of miles away from the spills from Oregon to Vancouver. Data entered into computers provided a model for seasonal flows in the North Pacific.

In addition to ocean surface currents, the circulation of global deep ocean water also has an influence on the Earth's climate. This vast underwater current is driven by density differences in addition to temperature. Dense salty waters cooled by arctic air in

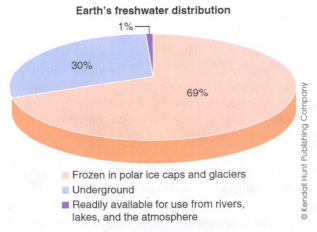

**Figure 19.7**  Freshwater Distribution on the Earth.

**Figure 19.8**    Global Ocean Circulation. Warm water from the South Atlantic Ocean moves to the North Atlantic as the Gulf Stream flow. As these waters are cooled by arctic air, the dense salty waters of the North Atlantic sink down into a global submarine river. Heat, nutrients, and pollutants are moved oceanwide by these deep ocean currents.

the North Atlantic sink deep into the ocean (Figure 19.8). This water flows south in deep under-ocean currents like a submarine river to the Indian and Pacific Oceans where it then rises to the surface. Surface currents of warmer water are wind driven to replace the sinking waters. This moves warmer water from the South Atlantic to the North Atlantic. These warm surface currents moving northward as the Gulf Stream flow keeps Europe and the eastern US waters warmer than expected. Without this warming flow, the United States would experience a cooler climate more like that of Canada. Global distribution throughout the oceans takes place as part of this massive conveyer belt-like flow. Heat and nutrients, along with pollutants, are moved in this way.

## Ocean–Atmospheric Interactions: El Nino

**El Nino**

A band of unusually warm ocean water that develops off the western coast of South America.

El Nino refers to a band of unusually warm ocean water that develops off the western coast of South America (Figure 19.9). The name El Nino refers to the "boy child" recalling Jesus the Christ child. It is in December around the Christmas season that these periodic warming effects in the southern Pacific were first noted. The pattern creates a disruption in ocean currents and causes dramatic changes in weather resulting in floods and droughts in far-reaching regions. The event is characterized by a 0.5°C (0.9°F) fluctuation in ocean temperature over the tropical central Pacific Ocean. The duration of the warming anomaly ranges approximately 1–2 years. El Nino events occur at intervals between two and seven years.

The initial forces that drive El Nino events are not fully understood by scientists but a series of notable changes are associated with event cycles. During the event, trade winds blowing west in the south Pacific weaken and may even reverse direction. Upwelling of normally cold nutrient-rich waters along the western coast of South America (Figure 19.6) is blocked by warm Pacific surface waters being pushed eastward to

## THE EL NIÑO PHENOMENON

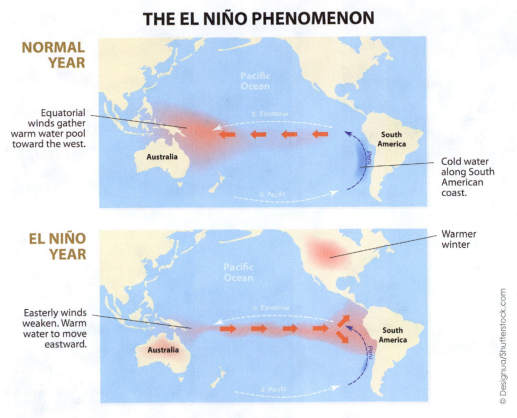

**NORMAL YEAR**

Equatorial winds gather warm water pool toward the west.

Pacific Ocean

South America

Cold water along South American coast.

**EL NIÑO YEAR**

Easterly winds weaken. Warm water to move eastward.

Pacific Ocean

Warmer winter

South America

© Designua/Shutterstock.com

**Figure 19.9** A Strong Warming Band across the Pacific Ocean Reaches the Western Coat of South America during El Nino Events.

the coast by strong surface currents. These warmer waters are usually held back from the South American coast by the western blowing trade winds. The lack of nutrient-rich waters limits plankton growth and can dramatically reduce local fish populations. The air along South America's western coast is usually cool and dry but the warm waters of El Nino add moisture to air masses. Storms and floods result in the usually arid areas. Areas at the western reaches of the Pacific, Indonesia, and India experience extreme drought. Such changes in weather and fish populations have dramatic effects on the lives of people in countries that border the Pacific. Many rely heavily on fishing and agriculture to live (Figure 19.10).

In the case of La Nina "girl child" events, nearly reverse events occur (Figure 19.9). Increased upwelling of deep ocean waters occur along the western South American coast. Nutrient loading and fish populations boom during La Nina events. Increasing droughts occur along west coastal South America and stormy wet weather toward Indonesia. Climate effects from the El Nino/La Nina events extend globally and social impacts are felt by changes in incidences of epidemic diseases. The cycle is linked to increased risks in mosquito-borne diseases such as malaria, dengue fever, forms of encephalitis, and fungal diseases in the areas that prone to flooding from these events.

# Biogeochemical Cycles

Biogeochemistry encompasses the chemical, physical, geological, and biological processes and reactions that govern the workings of the natural environment. Vital chemicals cycle through the Earth and affect the availability of environmental resources.

**La Nina**

Cooling of the ocean surface off the western South American coast.

**Biogeochemistry**

A scientific discipline that encompasses chemical, physical, geological, and biological processes and reactions that govern the workings of the natural environment.

(a)

(b)

(c)

(d)

**Figure 19.10** Devastating Impacts of El Nino Events Include: a. stormy weather and flooding in Peru b. drought in Indonesia, c. increase in mosquito-borne disease in flooded areas, and d. decline in fish population off the coast of Peru.

Environmental resources determine where organism will live. If an area has limited nitrogen in the soil, for example, many plants will not grow there.

Biogeochemical cycles move life's necessities into and out of storage reservoirs for availability and usage. Let's take a closer look at how water, carbon, nitrogen, and phosphorous move between land water and air in the biosphere. In each biogeochemical cycle, materials are circulated and stored within the ecosystem.

Each chemical – $H_2O$, C, N, and P – has varying time it spends stored in any given reservoir. Residence time tells us how long something is retained in a given storage reservoir in the biogeochemical system. Matter stored in rock in the Earth's crust may have long residence periods of millions of years, for example. Ocean waters can have residence times for carbon on the order of hundreds of years. Shorter residence times occur in food webs and the atmosphere.

**Residence time**

The average length of time that tells how long something is retained in a given storage reservoir in the biogeochemical system.

# Water Cycle

Water is necessary for life and its functions. In fact, recall from Chapter 2 that humans are made up of over 65% water. We are very much a part of the water and other biogeochemical cycles. All life continually requires inputs of water to keep us alive. Osmoregulation in Chapter 16 demonstrates the difficulty in maintaining life on land, given our links to water needs.

The Earth's water is in constant motion, allowing water to become available for living systems. The water cycle moves water through various terrestrial, aquatic, and atmospheric regions as seen in Figure 19.11. The Sun's energy is the main driver of the water or hydrologic cycle.

When heat from the Sun reaches wet surfaces (oceans, rivers, and lakes), water is converted from liquid to gas through evaporation. On very humid days, the atmosphere holds more water in gas form – water vapor. You can feel the "stickiness" from this water vapor in the air. Water vapor moves through the atmosphere where it condenses to small water droplets on particles, ice, or dust. Water droplets form clouds and eventually water falls back to the Earth as precipitation in the form of rain, snow, sleet, or hail. Rain may fall directly onto water bodies or land.

On land, gravity leads water to flow downhill via overland flow or surface runoff. Eventually, water reaches rivers or lakes and continues on toward oceans. A fraction of this runoff may infiltrate or soak into soils and move into groundwater flow. Groundwater can be stored below ground in aquifers (the saturated areas of bedrock) or continue moving toward lakes or oceans. Cold regions may accumulate snow into ice caps or glaciers where water may reside for thousands of years. Temperate regions release snowmelt to surface runoff as they warm. Runoff finds its way into streams or groundwater for flow toward oceans and aquifers. Groundwater can also stay near the surface and seep into nearby rivers and lakes, or emerge upward from a land opening such as a freshwater spring.

Biogeochemical cycles work together to distribute the chemicals necessary for life functions on the Earth. The water cycle is the essential driver of other biogeochemical cycles including carbon, nitrogen, and phosphorous. Water moves these elements out of

**Water cycle**

The process by which water continuously moves on, above, and below Earth's surface.

**Surface runoff**

The flow of water over the land surface.

**Aquifer**

Saturated areas of bedrock.

**Water Cycle**

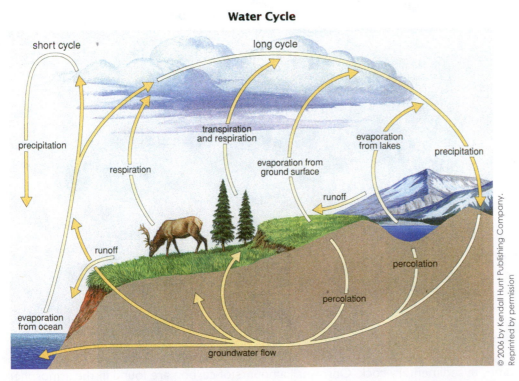

**Figure 19.11** The Water Cycle. Energy from the Sun drives water's movement across air, land, and ocean reservoirs. Most of the Earth's water is held in the oceans. From *Biological Perspectives*, 3rd ed by BSCS.

long-term storage in the Earth through weathering or breaking them down. Rivers then serve as a major form of transport, moving available elements to locations where living organisms can utilize them.

## Carbon Cycle

Carbon serves as a backbone for biological molecules including carbohydrates, proteins, and lipids. Terrestrial and marine food webs will cycle carbon in the atmosphere and oceans. Photosynthesis and respiration, described in Chapter 4, move these chemicals with relatively short residence periods.

Carbon fixation during photosynthesis moves carbon dioxide out of the atmosphere and into plants for availability in the food web (Figure19.12). Respiration returns carbon to the atmosphere as $CO_2$. Atmospheric carbon exists largely as carbon dioxide ($CO_2$). Carbon in the atmosphere is only a tiny fraction of the Earth's total carbon storage. Impact on the biological world due to photosynthesis from this small quantity, however, is massive. The structure of living things is made mostly of this carbon.

Ocean waters are the second largest carbon reservoir on the Earth. Ocean waters hold carbon largely in the form of dissolved bicarbonate ions ($HCO_3^-$). Diffusion takes place between the ocean surface and atmosphere. Carbon in the ocean is available for biological uptake by marine organisms. Some organisms incorporate carbon into shells or coral reefs in the form of calcium carbonate ($CaCO_3$). As these shelled organisms die and sink to the bottom of the ocean, they are buried with sediments. Ocean sediments

**Carbon Cycle**

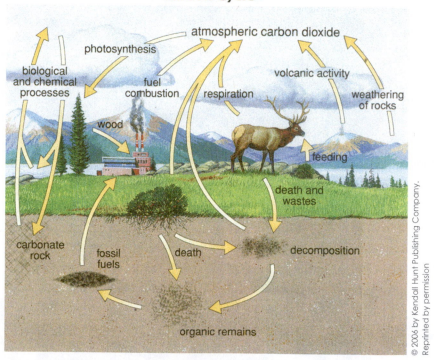

**Figure 19.12**   The Carbon Cycle. The vast majority of carbon is stored for long times in sedimentary rock. Only small amounts of carbon are found in the atmosphere as carbon dioxide, but changes in atmospheric carbon can have wide-ranging effects. From *Biological Perspectives*, 3rd ed by BSCS.

are compacted with massive pressure from weight of the ocean water above. Over time, they become the sedimentary rock we know as limestone.

Carbon-rich sediments from decomposition of organic waste compressed into layers of the Earth form fossil fuel deposits of coal, oil, and natural gas. These resources are mined by humans for use as electricity from coal-burning power plants and fuel for automobiles. Combustion of fossil fuels or burning of wood releases and returns a portion of stored carbon back to the atmosphere as $CO_2$.

Sedimentary rock of the Earth's crust is the planet's largest carbon reservoir. Carbon stored here can be locked away from use for millions of years. It is later released through either volcanic emissions or geological uplifting events. Uplifting events push the Earth's crustal plates toward one another leading to upward movement of rocks or mountain building.

## Greenhouse Effect and Global Climate Change

The greenhouse effect is the process of trapping heat energy in the atmosphere (Figure 19.13). Without any greenhouse effect, the Earth would be diminished to a frozen rock, too cold for living organisms. Greenhouse gases are gases in the atmosphere that slow the release of heat from the planet to space by absorbing and re-emitting long-wave radiation heat back to the surface. Greenhouse gases in the Earth's atmosphere include carbon dioxide, methane, nitrous oxide, ozone, water vapor, and CFCs. Water vapor and carbon dioxide ($CO_2$) are the largest contributors to the Earth's greenhouse effect.

**Greenhouse effect**

The process of trapping heat energy in the atmosphere.

**Greenhouse gases**

The gases in the atmosphere that slow the release of heat from the planet to space by absorbing and re-emitting long wave radiation heat back to the surface.

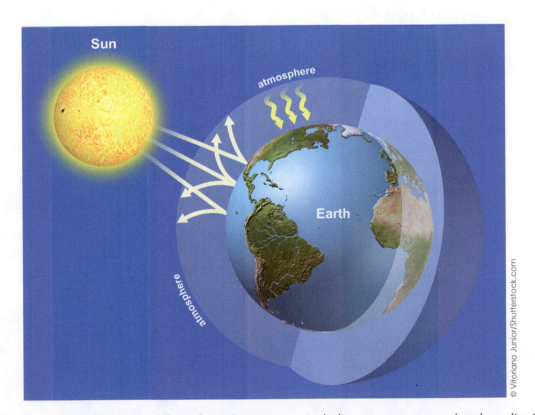

**Figure 19.13** The Greenhouse Effect. Greenhouse gases including water vapor and carbon dioxide absorb and trap some of the Sun's heat and reflect it back to the Earth's surface with a warming effect. Greenhouse gases in the Earth's atmosphere trap warmth from the Sun necessary for our survival. Changes in global greenhouse gas concentrations have effects on climate.

Atmospheric concentrations of both water vapor and carbon dioxide have natural variations based on temperature and vegetation. Over geologic time, there have been ice ages and warming periods on the Earth. Long-term climate variations are part of the planet's natural cycles. Solar cycles, changes in ocean currents, volcanic eruptions, and even asteroid collisions such as the Tunguska event in our story have altered climate in geologic history.

Carbon dioxide is released to the atmosphere when fossil fuels and wood are burned. Human activities producing $CO_2$ emissions from burning fossil fuels have soared in the past 150 years since the Industrial Revolution (Figure 19.14). Levels of $CO_2$ in the atmosphere have risen in tandem with average global temperatures over this time period.

Clearly, human activities such as fossil fuel combustion are adding to the atmospheric levels of $CO_2$. Another contributor to this effect is deforestation. Deforestation across the globe results in tree loss. Fewer trees mean less plant biomass is available to uptake atmospheric $CO_2$. The many possible changes to the Earth's climates due to rises in greenhouse gases in the atmosphere are referred to by scientists as global climate change.

Scientists see these possible changes as the cause for concern if current warming trends continue rising. Many populations cannot adapt evolutionarily when climate shifts occur rapidly or at extremes. Climate change outside of the Earth's normal patterns might not allow ecosystems time to adjust. Possibilities include melting polar ice caps and glaciers. Rapid sea level rise could theoretically alter huge coastal areas. These massive ecosystem changes could result in species loss if organisms are unable to adapt.

**Global climate change**

The many possible changes to Earth's climates due to rises in greenhouse gases in the atmosphere.

## Nitrogen Cycle

Biological organisms rely on the element nitrogen to form necessary nitrogenous bases that make up DNA, proteins, and enzymes critical to life functions. Our atmosphere is abundant in nitrogen. In fact, about 78% of the air we breathe is nitrogen. Nitrogen

© getfile/Shutterstock.com

**Figure 19.14**  Greenhouse Gases From Fossil Fuel Burning. Carbon dioxide emissions to the atmosphere have increased since the Industrial Revolution.

found in the atmosphere is in molecular form ($N_2$). Atmospheric nitrogen is bonded so tightly that plants and animals are unable to metabolize it directly. The nitrogen cycle incorporates soil microbes that transform nitrogen from the atmosphere into a form usable by plants and animals.

Nitrogen fixation is the process of converting molecular nitrogen ($N_2$) into ammonia ($NH_3$) and ammonium ions ($NH_4^+$). These forms are more readily available to plants for biological uptake. Some nitrogen fixation takes place in the atmosphere from volcanic action and lightning. Most nitrogen is biologically fixed by bacteria and *archae* in soils (Figure 19.15). These microbes are abundant and widely available to do the work of nitrogen fixation. Some microbes form symbiotic relationships with specific plant species including peas, beans, clover, and alders as discussed in Chapter 17. These plants have special protective nodules on their roots. Root nodules provide a secure holding place for nutrients and safe housing for the microbes in exchange for the nitrogen they fix. Nitrogen-enriched plants later return nutrients to the soils as they decompose.

Organic decomposition occurs as bacteria and fungi break down nitrogen from living organisms, which have died. This process of ammonification converts organic nitrogen into $NH_3$ and dissolved $NH_4^+$ forms ready for uptake. Some ammonia ($NH_3$) escapes back to the atmosphere as a gas. During nitrification, bacteria in soils produce nitrites ($NO_2^-$), which are then converted by other bacteria into plant-usable nitrates ($NO_3^-$). Some soil bacteria can also denitrify or return nitrates back to molecular nitrogen ($N_2$) gas. Soil gas can readily escape back to the atmosphere. These nitrogen forms found in soils are water soluble. They can be lost or leached away into groundwater and river systems. Nitrogen runoff that reaches coastal waters is available for uptake in marine food webs.

**Nitrogen fixation**

The process of converting molecular nitrogen ($N_2$) into ammonia ($NH_3$) and ammonium ions ($NH_4^+$).

**Ammonification**

The process in which organic nitrogen is converted into ammonia and dissolved $NH_4^+$ forms ready for uptake.

**Nitrification**

The process by which bacteria in soils produce nitrites.

**Nitrogen Cycle**

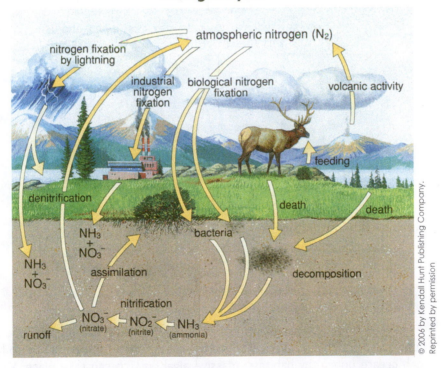

**Figure 19.15** Nitrogen Cycle. Most of Earth's nitrogen is found in the atmosphere. Nitrogen fixing microbes in the soil move nitrogen from its atmospheric form to one available for uptake by plants. After plant uptake it can enter terrestrial food webs. The many different soil microbes can convert nitrogen between many forms. From *Biological Perspectives*, 3rd ed by BSCS.

## Eutrophication

Movement of nitrogen in the global system can take place mechanically through processes of weathering and runoff. Widespread use of lawn and agricultural fertilizers in recent decades, runoff of agricultural waste, sewage discharges, and burning fossil fuels are all human activities that have altered the natural nitrogen cycle. Runoff and point source discharges of such wastes to rivers have increased the amount of nitrogen and another limiting nutrient, phosphorous, in waters.

Extra nutrient loading left unchecked can lead to massive algal growth, known as eutrophication (Figure 19.16). As large colonies of algae die, bacterial decomposition of their remains occurs. Available oxygen levels in the waters may become severely depleted. Low oxygen levels can result in the death of fish and loss of marine life as observed in the Mississippi River Delta's so-called Dead Zone.

**Eutrophication**

The massive algal growth resulting from unchecked extra loading of nutrients.

## Phosphorous Cycle

Phosphorous, like nitrogen, is a limiting nutrient for plant growth. Phosphorous is biologically required for energy transfers as a functional component of ATP. Phosphorous is also found in nucleic acids, as the sugar–phosphate backbone of DNA, and in phospholipids, as a component of cell membranes. Thus, phosphorous is vital for the growth of producers. Fertilizers for large agricultural operations contain added phosphorous to promote vigorous plant growth needed for high-yield crops.

Since phosphorous compounds do not commonly exist in a gaseous phase in the Earth's atmosphere, there is no atmospheric reservoir. The Earth's crust is the primary reservoir for phosphorous found in sedimentary rocks in the form of phosphates ($PO_4^{3-}$).

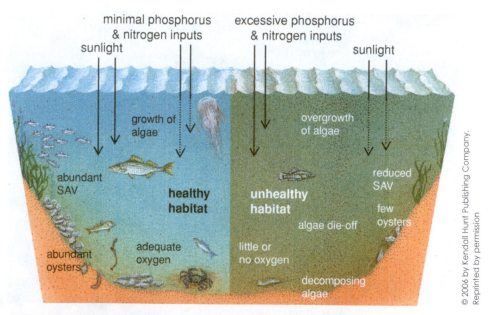

**Figure 19.16** Eutrophication. Nitrogen and phosphorous are needed for healthy functioning of aquatic systems. Excess inputs of these nutrients due to human activities can create unhealthy habitats. Massive algal growth can take place. When large numbers of algae die, oxygen is limited and can result in significant loss of aquatic life. From *Biological Perspectives*, 3rd ed by BSCS.

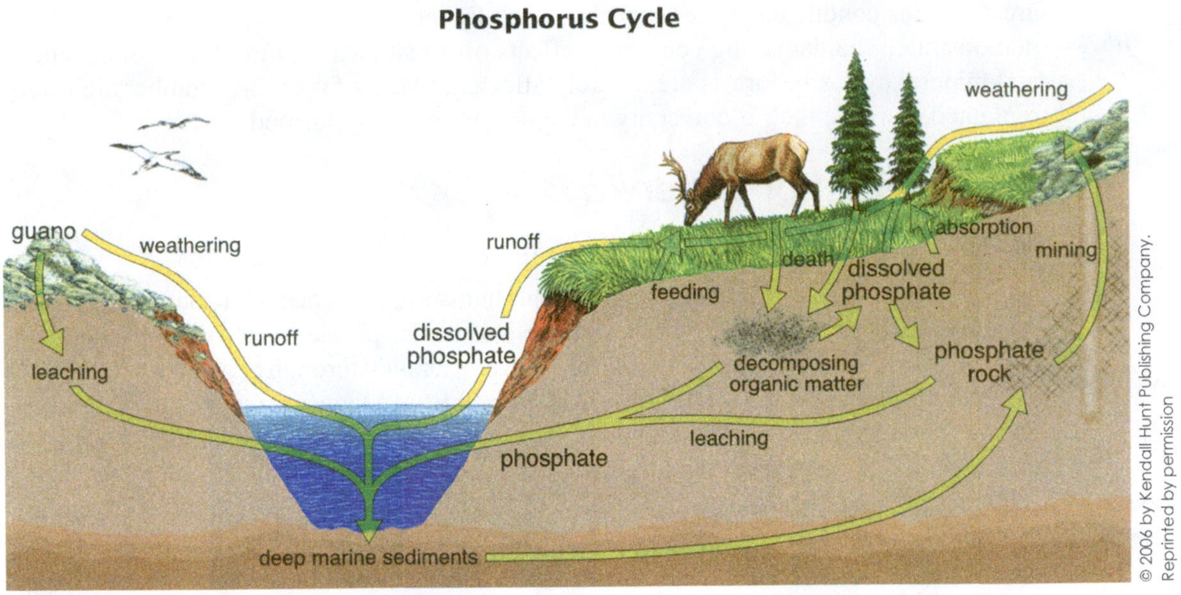

**Phosphorus Cycle**

**Figure 19.17** Phosphorous Cycle. Rocks and marine sediments are Earth's largest reservoir of phosphorous. Most phosphorous moves through the cycle bonded to oxygen as inorganic phosphate ($PO_4^{3-}$) until it enters a food web. From *Biological Perspectives*, 3rd ed by BSCS.

Small amounts of phosphates are gradually released into the environment over time. This occurs as rocks are weathered into small particles and moved by erosion into soils and streams for biological availability (Figure 19.17). Plants absorb the nutrient through their roots in terrestrial food webs. Excretion, death, and decomposition return phosphorous from the food web back to the land. Runoff carries some terrestrial phosphorous to oceans. Phosphates in oceans may enter the marine food web, but eventually deposit as sediments on the marine floor. Millions of years will pass before these marine sediments from sedimentary rock can be moved upward to the Earth's surface for recycling again.

# Human Influences on the Biosphere

## Deforestation

Removal of forested areas or deforestation can result in drastic changes to the individual ecosystems. In our story, massive deforestation of 800 miles was caused by the explosion of the Tunguska asteroid in 1908. However, more significantly in present times, humans have been removing forest lands in many biomes across the globe, in ways described in Chapter 18.

As human populations grow, lands are cleared for agricultural use and cities are built. Loss of tropical rain forest is believed to contribute significant amounts of carbon dioxide to the atmosphere, contributing to the greenhouse effect described earlier. Deforestation affects the biosphere – climate, water, soil, and biodiversity.

When forest areas are cut, the plant transpiration process stops adding moisture to the air. Exposed soils tend to dry out. Best practices for forestry harvesting can limit devastating effects. Avoiding clear cutting can preserve forest pathways utilized by local wildlife populations rather than cutting them off, in isolation. Erosion problems and landslides of forest soils can be limited if some plants and trees are left in a cut

**Deforestation**

Removal of forested areas.

area. Studies conducted at the Hubbard Brook Experimental Forest in New Hampshire demonstrated the damaging ecological effects of forest clear cutting. Soil composition and mineral cycles in forests are severely affected. Forests farmed for lumber are often replanted, but lose their biodiversity in the process of being farmed.

# Engineering of Waterways

## *Mississippi and Atchafalaya Rivers*

Engineering of natural waterways is an obvious human interference with natural water ecosystems. Changes that result from these alterations can be problematic. Still, there is much to be gained in the way of flood control and power generation through engineering solutions.

**Watershed**

An area that collects flow, runoff, and precipitation from a region into a specific body of water.

The Mississippi watershed drains surface water from a large area in the interior United States. A watershed is an area that collects flow, runoff, and precipitation from a region into a specific body of water. For example, the Mississippi River in the United States flows south to the Gulf of Mexico primarily due to the flow from high to low elevation – there are mountains bordering the river's watershed. The Mississippi River changes its course into the Gulf of Mexico around every thousand years. Each new course adds to the Mississippi Delta complex that makes up the state of Louisiana's unique coastline (Figure 19.18).

Delta lobes shift when the river is captured by a tributary, with a steeper and shorter route to the gulf waters. Abandoned lobes compact and make up the bayous we know today. In the mid-twentieth century, migration of the Mississippi River from its current channel to the Atchafalaya River appeared likely. Flood concerns, navigation interests, and economic structure led the Army Corps of Engineers to build a control structure in 1963 to maintain the balance of water flow. Confinement of the river to human-built canals in the lower Mississippi is changing the ecology of the natural river. Muddy waters of the Mississippi are now shunted out to sea. Soils are no longer deposited in land areas downstream. Wetland or saturated areas are no longer being created. Wetlands require slow, meandering rivers that allow sediments and vegetation to build up over time.

**Wetland**

A land that consists of swamps or marshes.

Wetlands serve as natural buffers from heavy flows and hurricanes. Our attempts at engineered flood control are eliminating the river's natural flood regulation systems. The

© AridOcean/Shutterstock.com

**Figure 19.18**    A. The Mississippi River Delta. The Mississippi River changes its course into the Gulf of Mexico about every thousand years. These changes are responsible for Louisiana's uniquely shaped coastline today.

natural waters of the Mississippi have been greatly altered in terms of salinity balances, erosion, and ecological effects.

## Three Gorges Dam

The recently completed Three Gorges Dam Project in Hubei Province, China is the largest hydropower station the world (Figure 19.19a). Over 10 trillion gallons of water are held behind a massive cement wall. Annual energy output is rated at nearly 100 billion kilowatt-hours, enough to supply 10% of China's electricity needs. The dam is 1.3 miles (2.3 km) wide across the Yangtze River and 610 ft (186 m) high. The Project is designed to prevent flooding in the middle and lower reaches of the Yangtze River.

Floods in the river system have killed hundreds of thousands of people in the past century alone. The vast reservoir of water held behind the dam allows for controlled releases to regulate floodwaters. Water from the reservoir is also available for irrigation of farmlands. Widened shipping lanes and a system of locks allow for improved navigation. This should bring more commerce and ease of transportation to the interior of China. Tens of millions of people living in interior China are expected to experience economic benefits, added jobs and better quality of life.

In the wake of the power, flood control, and economic benefits offered by the dam, some social and ecological problems remain. Large tracts of productive farmlands, scenic gorges, and historical sites were forever flooded for the project. Over 1000 villages were submerged (Figure 19.19b) and 1.3 million people permanently displaced from their homes. Rare and endangered populations of river dolphin, sturgeon, and cranes are further threatened by ecological changes due to construction. Sewage and pollution normally washed out to sea are held up by the dam. Soil runoff combined with water retention creates excess silt that clogs power generation equipment. Furthermore, the project overlies seismic zones. There is concern that the excess weight of water could trigger earthquake events.

(a)    (b)

**Figure 19.19**    a. China's Three Gorges Dam Project Is the Largest Hydropower Station in the World. b. Over a Thousand Villages Were Submerged and More than a Million People Relocated for the Project.

### CAN HUMANS CAUSE EARTHQUAKES?

The Earth's crust is made up of moveable tectonic plates that do shift naturally. A release of energy from movement in the Earth's crust is called an earthquake. Geological research shows that human-induced seismicity or vibrations in the Earth's crust do take place.

**Earthquake**

A release of energy from movement in Earth's crust.

**Seismicity**

Vibrations in Earth's crust.

Small seismic events from surface loading – like water pressure from tons of water behind a dam – are documented. These mini earthquakes might hold potential to stimulate larger ones. A large 2008 earthquake – magnitude of 7.9 – in China's Sichuan Province is suspected to be linked to filling of the nearby Zipingpu Dam.

Mining activities also link to seismic events. The US Geological Survey (USGS) reports a number of small earthquakes – magnitudes up to 2.8 – linked to hydraulic fracturing or fracking activities. Hydraulic fracturing involves injection of high pressure water and chemicals into the ground. The process breaks through shale rock formations to allow for extraction of trapped natural gas.

## Pollution

**Pollution**

The introduction of any contaminant into the environment that causes a harmful change.

The introduction of any contaminant introduced into the environment that causes a harmful change is considered to be pollution. A pollutant can take a natural form such as noise, light, smell, or heat. Pollutants also take form as garbage, sewage, nuclear radiation, and chemicals. Human activities produce vast amounts of waste. Our waste takes up so much space that we sometimes find ourselves unsure where to dispose of it (Figure 19.20). Landfills are overflowing with trash. Mining operations and industries discharge liquid wastes into rivers and oceans. Smokestacks from industry and electric plants release air pollution into

(a)

(b)

(c)

**Figure 19.20**   Pollution Takes Many Forms in Our World. a. Household wastes crowd a landfill. b. Pollution to rivers, lakes, and oceans damage our waters. c. Smokestack emissions from energy and industry add to air pollution.

**Figure 19.21** Renewable Energy Solutions of Wind, Solar, and Hydropower Offer Freedom from Dependence on a Limited Supply of Fossil Fuels but Their Own Challenges As Well.

our atmosphere. Acid rain formed from polluting gases in the atmosphere is slowly changing the chemistry of global soils and waters. In some areas, acid rain has killed vegetation or corroded bridges and statues. Its effects on the environment are far reaching and varied.

Ideally, waste should be limited to the highest degree possible. We can do our part to limit waste production by reducing the amount of materials we consume, reusing items that still have useful life, and recycling unusable items or materials for repeat use. Industries are being restricted by government regulations to limit pollution to regulated levels for certain listed pollutants from discharge points into water ways and the atmosphere. Coal-burning facilities have added "scrubbers" to smoke stacks to filter out and limit some of their pollution. Natural gas collected by hydraulic fracturing methods is marketed as a cleaner burning fuel with less greenhouse emissions.

Plenty of pollution still abounds, but new technologies are helping to limit some of its harmful effects. Renewable energy solutions focus on expanding solar, wind, and hydropower to release us from our dependence on fossil fuels (Figure 19.21). Each of these renewable solutions comes with limitations for where they can be used and unique challenges for large-scale energy production. Ethanol is replacing some of the traditional petroleum-derived fuel being put in our cars. It is important to note that all new solutions to our energy and waste problems come with their own set of downfalls and environmental setbacks that should be carefully weighed out to produce the most sustainable technologies.

## Bioaccumulation/Biomagnification

Some chemical pollutants are particularly toxic. They can lead to negative neurological or reproductive effects in biological populations. Lipid- or fat-soluble substances can

**Bioaccumulate**

Accumulation of substances in an organism.

bioaccumulate. They stay in organisms because they are not metabolized or excreted. Over time, this leads to higher and higher concentrations of the substance held inside the organism's tissues.

As an example of bioaccumulation, follow methylmercury through an aquatic system. Methylmercury is a toxic form of mercury available for biological uptake. Low levels of methylmercury, 0.001 ppb (parts per billion), are taken in from seawater by zooplankton. Methylmercury is stored in the zooplankton rather than excreted. Small fish then eat the zooplankton. These small fish have their own burden of methylmercury from surrounding waters. They add to their own levels of methylmercury over time by eating the zooplankton. As they grow and eat more and more zooplankton, they store more and more methylmercury in their own tissues. Methylmercury is being accumulated and concentrated in tissues of the small fish (Figure 19.22).

Big fish consume the small fish and the process is magnified. The methylmercury levels grow larger with each move to a higher trophic level in the food chain. As the toxin moves up successive trophic levels of a food chain, the increase in concentration is known as biomagnification. A shark eating the big fish may be found to have high levels of methylmercury nearing 500 ppb. Biomagnification demonstrates that even in waters with low pollution levels, toxic levels of chemicals can be retained in fish.

**Biomagnification**

The increasing concentration of a particular substance in organisms at the top of the food chain.

A shark is the top carnivore in the example food chain. However, it is important to note that *all* at the top trophic levels of the food web are at risk from biomagnification of toxic chemicals including eagles, polar bears, and humans. Methylmercury discharged to bay waters from a chemical factory in Minamata, Japan during 1956 lead to severe mercury poisoning of thousands of people, cats, and other wildlife. Locals consumed fish from the polluted waters as a primary staple of their diet. An outbreak of neurological effects in the local populations, some resulting in death, proved later to result from toxic levels of methylmercury.

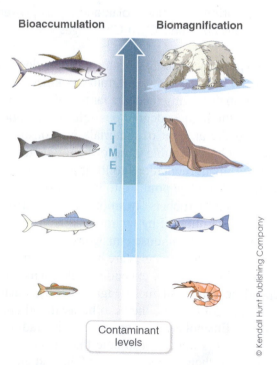

© Kendall Hunt Publishing Company

**Figure 19.22**    Bioaccumulation and Biomagnification of Methylmercury in an Aquatic System. Methylmercury levels in small fish tissue increase over time or bioaccumulate as the fish live and eat in the aquatic system (left). Methylmercury levels increase or biomagnify with each move to a higher trophic level in the food chain (right).

As a result of events such as the Minamata disaster, the US FDA makes fish consumption recommendations for pregnant or nursing mothers as well as young children who are most susceptible to the neurological effects of mercury poisoning. Suggestions restrict consumption of top-level predators such as sharks and swordfish. They advise consumption limited to two meals per week of low-mercury seafood such as salmon, light tuna, catfish, or shrimp. For fish with a higher likelihood of increased mercury levels such as albacore, one meal per week is the recommendation. Check advisories in your own area for recommendations on locally caught fish. Other chemicals subject to biomagnification include heavy metals such as arsenic and lead; the insecticide DDT; PCBs, used widely as coolants and insulating fluids for electrical transformers; and PCDD/Fs or dioxins, formed during low-temperature burning.

## Ozone

Ozone ($O_3$) is a pollutant affecting the Earth in two atmospheric zones. Ozone in the atmosphere near the Earth's surface is a form of air pollution. It is formed by reactions between air pollutants released from fossil fuel combustion and sunlight. Ozone found in the lower atmosphere or troposphere in urban areas is referred to as smog. This can lead to respiratory problems for humans.

**Ozone**

A gas that is a pollutant in Earth's lower atmosphere, but acts as a protector from UV radiation in the upper atmosphere.

Ozone higher up in the Earth's stratosphere or upper atmosphere serves as a protective barrier for the planet. The ozone layer in the stratosphere protects the Earth from damaging UV solar radiation. The whole ozone layer that wraps the Earth has shown some overall thinning over recent decades. Polar Regions, however, show a significant seasonal decrease in ozone levels, particularly over Antarctica (Figure 19.23). This depletion in ozone concentration is sometimes referred to as a hole in the ozone layer.

Concern here is that without this protective layer, damaging effects from enhanced UV radiation can occur. Resulting effects include a rise in skin cancer rates caused by increased UV exposure, as described in Chapter 15, and damage to plants. CFCs (chlorofluorocarbons) are chemicals used in older Styrofoam, aerosol sprays, and coolants. CFCs have been shown to react in a way that lowers ozone amounts in the upper atmosphere. Attempts to reduce the use of CFCs have been made in recent decades with hopes to limit further deterioration of the planet's valuable ozone layer.

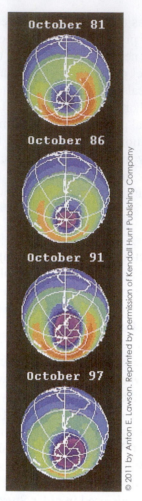

**Figure 19.23**    The Ozone Layer in the Earth's Upper Atmosphere Shields the Planet from Damaging UV Light. Ozone concentrations over Antarctica have decreased in recent decades. Purple and dark blue areas represent the ozone hole. From *Biology: An Inquiry Approach*, 3rd ed by Anton E. Lawson.

## Summary

Many biotic and abiotic systems interact in the Earth's biosphere. The Earth, solar energy, and biogeochemical distribution of planetary resources make life on the Earth possible and sustain ecosystems. Winds, weather, and ocean circulation are driven by the Sun. Biogeochemical cycles continually move the elements necessary for life to the regions where they are available for biological uptake. Many factors, some of them are human induced, can disrupt these environmental cycles resulting in damage to living communities and ecosystems. We can take steps to help by limiting waste and pollution, using best practices, striving for sustainable solutions, and being environmentally conscious of impacts on the biosphere.

# CHECK OUT

**Summary: Key Points**

- The biosphere is made up of all areas on the Earth that contain life.
- Climate is affected by solar radiation, the tilt and orbit of the Earth, atmospheric make up, and air and ocean circulation patterns.
- Energy from the Sun powers the water cycle.
- Water drives biogeochemical cycles and transport for carbon, nitrogen, and phosphorous.
- Carbon dioxide, a greenhouse gas, results from burning fossil fuels.
- Nitrogen and phosphorous are limiting nutrients for biological growth.
- Human influences on the biosphere can have negative effects on the biosphere – pollution, deforestation, engineering waterways, ozone depletion, and climate change.

## KEY TERMS

albedo
ammonification
atmosphere
aquifer
bioaccumulate
biogeochemistry
biomagnification
biosphere
Coriolis effect
deforestation
earthquake
El Nino
eutrophication
global climate change
greenhouse effect
greenhouse gases
freshwater
heat capacity
hydrosphere

La Nina
intertropical convergence zone
lithosphere
nitrification
nitrogen fixation
ozone
polar easterlies
pollution
residence time
salt wedge
seismicity
surface runoff
trade winds
water cycle
watershed
wetland
wind
westerlies

# Multiple Choice Questions

**Reflection questions:**

1. Where is the majority of the Earth's carbon stored?

   a. plants
   b. atmosphere
   c. sedimentary rock
   d. oceans

2. Depletion of the Earth's ozone layer has resulted in human health risks by:

   a. increased effects from solar radiation.
   b. increased effects by asteroid impacts.
   c. decreased effects from solar radiation.
   d. decreased effects by asteroid impacts.

3. Which has the LEAST albedo effect in the Earth's biosphere?

   a. plants
   b. snow
   c. ice
   d. cloud

4. Which region of the biosphere receives the most direct sunlight?

   a. 30° latitude
   b. 60° latitude
   c. Polar Regions
   d. equator

5. In 1816, the year without a summer was caused by:

   a. volcano ash
   b. asteroid dust
   c. ozone depletion
   d. dust bowls

6. Which represents a logical order, in biogeochemical cycling, in the movement of nitrogen through the biosphere?

   a. atmosphere → soil bacteria → animal → plant
   b  atmosphere → soil bacteria → plant → animal
   c. soil bacteria → atmosphere → plant → animal
   d. animal → atmosphere → plant → soil bacteria

7. Which transports chemicals through their cycles on the Earth?

   a. water
   b. air
   c. soil
   d. all of the above

8. Which is NOT a contributor to the greenhouse effect?
    a. nitrogen gas
    b. carbon dioxide gas
    c. water vapor
    d. methane gas

9. Which correctly MATCHES pollutant with its effect?
    a. ozone – thyroid disease
    b. nitrogen gas – greenhouse effect
    c. phosphorous – eutrophication
    d. carbon dioxide – biomagnification

10. Which is an effect on the western South American coast, caused by El Nino?
    a. poor fishing
    b. nutrient upwelling
    c. droughts
    d. plankton growth

## Short Answer

1. List three dangers from extraterrestrial sources that could threaten the life on the Earth. Which threat is most likely?

2. Define the following terms: albedo and greenhouse effect. List one way each of the terms differ from each other in relation to their a. role in the biosphere; b. relationship with each other.

3. What is the function of the ozone layer in protecting life on the Earth?

4. Draw a sketch of the nitrogen cycle, using arrows to show four key organisms transferring nitrogen through the biosphere. Which organism makes nitrogen available to living organisms?

5. List and discuss three negative effects that a dam construction has on its surrounding ecosystems?

6. Pretend that you are a molecule of carbon dioxide, recently placed into a plant. Describe three pathways you might take from the plant to the rest of the biosphere. What process placed you into the plant in the first place? Where will you spend most of your time in the biosphere?

7. Define biomagnification in living organisms. Choose a chemical that biomagnifies and trace it through a food chain.

8. For question #7, how can you change your diet to reduce the effects on biomagnification in your own life? Explain why?

9. Describe the contributing abiotic factors to wind movements in the biosphere. Be sure to use the following terms in your explanation: air pressure, temperature, Coriolis effect, and rotation.

10. Describe three methods that limit the negative ecological effects of deforestation. How are they different? Which would you advise as least costly?

## Biology and Society Corner: Discussion Questions

1. In China, over 1000 villages were flooded, their population relocated due to damming of rivers. This massive resettlement was not well received by the local population. Form a four talking point argument explaining the benefits of damming the rivers to the citizens of the flooded towns. What are ecological drawbacks to your argument?

2. Diets high in fish, including swordfish and albacore, have heart healthy benefits. Fish contain a good proportion of unsaturated fats, omega 3 fatty acids, and vitamins. However, they also contain higher levels of biomagnified mercury in their tissues. Develop a plan to limit the negative effects of toxic mercury with the health benefits of eating fish in your diet.

3. Recently, an asteroid traveled within miles of hitting the Earth. Some scientists suggest that such a collision will occur, as in Tunguska, within the next few years. Should you be concerned? Why or why not? Also research the chances predicted for types of extraterrestrial impacts. Name two ways to prepare for or prevent such an impact, if any.

4. On October 31, 2012, a hurricane flooded of the Greater New York seacoast destroyed thousands of homes and whole neighborhoods. Its costs for rebuilding are in the billions of dollars. Some insurance companies refuse to insure homes along the coast. Do you think that insurance companies should have the right to refuse insurance on people's homes along the coast? Why or Why not?

5. Hydraulic fracturing, also known as "hydrofracking," drills into the Earth to obtain natural gas from shale rock layers. It is controversial because the chemicals used to dissolve rock may leak into layers of the lithosphere and pollute. Proponents of hydrofracking cite increasing energy needs and less greenhouse gases as benefits. Research the pros and cons of hydrofracking. Which side do you agree with more? Why?

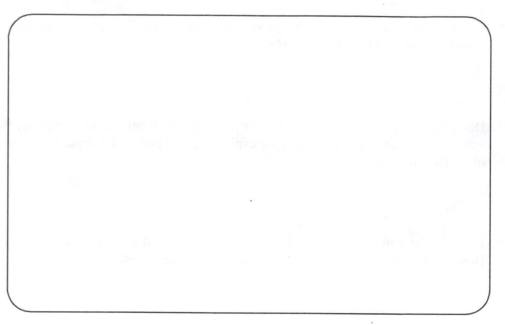

Figure – Concept Map of Chapter 19 Big Ideas

# UNIT 6
# Biology and Society

# The Evolution of Social Behavior: Sociobiology

**20**

## ESSENTIALS

An ant hill is a vibrant, social organization of organisms

Ants move in a line using pheromones to direct their often-complex activities

Do we destroy ourselves?

A party is enjoyable until someone gets hurt

Humans can be violent as well as destructive to other organisms. The ant's perspective – humans can be cruel

# The Case of the Nuclear Ant Hill

I was glued to the news on the TV in my backyard. While I was never into politics, this time there were events going on. North Korea has threatened to launch a nuclear strike against the United States. How could *they* bother *us*? We always heard about nuclear war but no one really took it seriously. It was only a worry from the 1950s; from our grandparents' days, when there were communists and an "enemy" behind the Iron Curtain.

However, I had better things to do than think about boring politics. This afternoon, I am hosting a clam bake in my backyard. I invited my coworkers from the office. While I did not like most of them, I thought it would be good politics. I would have them over and maybe bond a bit. We'll see . . .

I started barbequing in the backyard and noticed a new, large ant hill right in front of my grill. It was blocking the way but I could manage around it for now, but it had to go. The ants must have constructed this monstrosity overnight. It was two feet high and a couple of feet across. I had heard that ants were industrious but they were a problem.

I took a shovel from the garage to remove the ant hill. As I placed the spade into the dirt bordering the colony, I noticed something strange: these ants were lined up, single file, working on some project. They looked serious as they moved a large piece of meat that I must have dropped while barbequing.

I decided to take a closer look at the ants. I brought my magnifying glass up to the line of ants on the ant hill. I examined the ants and noticed, "The meat was way bigger than any of the ants – maybe by 100 times – but seriously, those ants moved the piece with ease." The ants were very orderly, as they swiftly brought the food back into their hole. I was amazed at how they cooperated with each other. *Their* workplace was great – the ants were helping each other. I did not see the usual complaining or disrespect or underhandedness that I suffer through every day in the office.

It was at that point that I decided to let the ant colony live – I would show it mercy, one of my fellow living creatures. This may sound crazy, but I guess at that moment, I yearned to be an ant.

My coworkers came to the yard, boisterous as always. We exchanged the usual pleasantries and I offered them a drink or two. One of my "friends" noticed the ant hill, remarking "What's, this?" Before I could respond, he kicked it across the yard.

A loud beep on the TV interrupted the party. Then, after a few minutes, an announcement was made, "This is an emergency . . ."

## CHECK UP SECTION

The juxtaposition of the ant hill's cooperation with human society is a point of the story. Animal social systems have differing characteristics to compare with human society. Research the ecology of ants. What are the benefits and drawbacks of living in their colony? Do you think the character in our was story right "to yearn to be an ant"?

# Defining Sociobiology

A theme threading through this text explores society's relationship with biology. Stories in each chapter touch on biological matters facing humans: life, death, struggles for survival, and community relations. Medical advances, the human encroachment on ecosystems, and the importance of healthy soil chemicals to human survival all represent the ways biology and society intertwine.

Understanding our society can be complex to study. In this chapter, we tease out the factors underlying human and animal behaviors within societies. There are many questions with complex answers. To start off with, what features of human society make us similar to and different from other organisms? Why do we enjoy other living creatures and yet harm their biosphere, where we all live? Our social structure defines how our species lives and interacts with its environment.

The answer to these questions lies in our behaviors – how humans act toward each other, for example. Any action taken by an organism is defined as its behavior. Behaviors evolved and are subject to natural selection pressure, just like any other trait in an organism. Behaviors evolved in human society to help our survival, and they also adapted to suit members within animal social systems. An animal social system is the social unit of any animal, its organization, and workings. We will explore the biology of animal social systems and compare them with human society.

The evolution of behavior may be studied in the same way we look to adaptations in studying evolution in Chapter 7. Biologists look at how a behavior improves survival in organisms. By doing this, we attempt to answer the questions: How hostile are humans to each other compared with other organisms? Are they more like ants or like warriors? Do other organisms, such as the ants in our story, *really* get along better than humans? The story posits that this is the case, but should a biology text make such an assertion? The basis of the claim is that both human and ant social structure are dependent on the same biological principles.

Humans, as animals, are driven by the same rules underlying all biological systems. This text has explored these characteristics of life all through the chapters. Human behavior and its resulting social order are no exceptions to the tenets. The themes of survival, evolution, and the importance of genetics, for example, are vital in establishing social organization.

**Behaviors**

Any action taken by an organism.

# Animal Behavior

Biological principles help us to understand how organisms behave in a society. They help us predict social behaviors – in animals and in humans – by applying their parameters. This study is termed animal behavior. Animal behavior is the branch of biology that studies the ways in which animals act within their environment. Ethology, before the 1950s, was the precursor to the modern study of animal behavior. It looked at how

**Animal behavior**

The branch of biology that studies the ways in which animals act within their environment.

organisms behave based on their physical processes. Dissection of nervous system and endocrine glands helped ethologists study patterns in behaviors.

The father of ethology Konrad Lorenz was one of the first scientists to argue that behaviors, just like biological structures, evolved to benefit organisms. Thus, behaviors are "tools" that organisms use to survive. Fleeing or fighting, cooperating or acting selfishly – these are all behaviors that have adapted through evolution.

Lorenz focused on geese and their behaviors within social groups. In Figure 20.1, he is shown, famous for the geese he raised. He has a flock following him after their birth. They imitated Lorenz as if he were their own mother. He was the first organism they saw after their birth, so they followed his movements.

The study of the physical and physiological adaptations of behavior advanced to become the study of modern animal behavior. The study of the ways that *groups* of animals act is called sociobiology (or behavioral ecology). Sociobiology is the focus of this chapter. We close the book with sociobiology because it brings the application of biology back to the society. We can better understand biology and our place in the biosphere by understanding how our social structure evolved.

Sociobiology answers the mystery questions to why humans and other animals behave the way they do. In our story, the ant hill is portrayed as a group with good sets of behaviors, defined here as "helping one another." Humans are contrasted as a group that has bad sets of behaviors, defined here as "a selfish or harmful act." Behaviors change in a species over time – or evolve – to benefit an organism's reproductive success (RS), defined as the number of live young it produces. The RS of an organism depends on the choices it makes to survive. Ants cooperate on their hill in our story to help them to survive. Humans also make choices that impact their survival. If a behavior helps an organism's RS, it is likely to persist in a population and become established. These behaviors are called adaptive behaviors because they have evolved to help organisms adapt to their environments and improve their RS.

The story ends with the possibility that a nuclear war is breaking out. Do other organisms kill their own species *en masse*? It is rare to find other organisms kill their own species, except by accident. Our ultimate destruction may be implied at the end of the story – intraspecific competition and killing. This makes humanity bad. But then,

**Sociobiology**

The study of the ways that groups of animals act.

**Reproductive success (RS)**

Is the number of live young ones an organism produces.

**Adaptive behaviors**

The behavior that helps an organism's reproductive success by helping it to persist in a population and become established.

**Figure 20.1**  Konrad Lorenz: Father of Ethology. After imprinting on Lorenz, his geese followed him into the water, imitating him while swimming.

what is a "good" or "bad" behavior? Are human behaviors following the "rules" of RS? We should first define the types of behaviors organisms display.

## Types of Behaviors

Behaviors encompass movements, sounds, mating, and communication – all of the acts an organism performs. Some behaviors are genetically determined, called innate behaviors. Organisms are born with innate behaviors, compelling them to act in certain ways. A bird migrates from taiga to warmer climates during the winter. This innate behavior was established the day the bird was born. It did not need to be told to migrate and did not do it because its parents taught it. When raised in isolation, a bird will still migrate. Other examples of innate behaviors include single-celled organisms. A *Euglena* will travel toward light, a behavior called phototaxis, also found in plants as described in Chapter 9. Innate behaviors, such as butterfly migration are shown in Figure 20.2.

**Innate behaviors**

Behaviors that are genetically determined.

## Learning

Not all behaviors are innate. Are we born knowing how to solve an algebra equation? Do we know how to cook an omelet, without ever having been shown? Some behaviors are developed as we experience the world. These are learned behaviors, which result when experience alters an organism's response. Learning is the process of taking in information and using it to perform tasks and "think" about situations. Learning in animals is a controversial area of study because there is debate on whether animals have self-awareness and a capacity to think so deeply.

**Learned behaviors**

Behavior that develops through experience.

Learning is defined as a change in behavior as a result of a stimulus. You touch a hot iron and it hurts; in the future, you avoid hot items. This example shows a change in behavior based on an unpleasant stimulus (the burning of the iron). This is a simple example of learning.

Learned behaviors are beneficial to organisms because they help them adapt to changes in surroundings. Learning evolved to help organisms survive better. Learning tailors an organism's responses to unique circumstances in its environment. By adapting

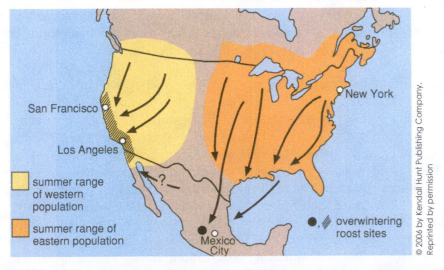

**Figure 20.2** Monarch Butterfly Migration Is an Example of Innate Behaviors. They are not taught to go south for the winter but still form groups and travel each year away from the cold weather. From *Biological Perspectives*, 3rd ed by BSCS.

**Figure 20.3** This Chihuahua Puppy Is Learning Good Behavior.

© Anneka/Shutterstock.com

**Behaviorism**

The theory that human and animal responses are measured to determine an organism's learning.

**Classical conditioning**

An association between a new stimulus and a natural stimulus.

to changing conditions, an organism is more likely to improve its RS. Thus, learning, by altering responses based on experience, is itself an adaptive behavior. Both innate behaviors and learned behaviors influence an organism's actions. Learning occurs as soon as an animal is born (Figure 20.3).

# Behaviorism

Behaviorism looks at the responses organisms make to stimuli. It studies the way in which organisms, particularly people, learn from their experiences. When an organism is presented with a stimulus, it responds in predicted ways. There are five classifications of learned behavior, according to behaviorism: imprinting, habituation, classical conditioning, operant conditioning, and insight.

## Imprinting

During **imprinting**, the learning that occurs is just after an organism's birth. Timing is of the essence, because at first a newborn is most impressionable to imprint. The stimuli must be presented in the very early developmental stages of an organism's life. When a bird is born, for example, it forms an attachment with its mother upon seeing her for the first time. They bond, in part, through feeding, and the baby chicks imitate their mother's behaviors.

Konrad Lorenz, depicted in Figure 20.1, grew gosling eggs in an incubator, which hatched and began imprinting Lorenz's habits and behaviors. The behavior is adaptive because baby goslings know nothing about their world. They are vulnerable and their survival increases as they learn the "dos and don'ts" from their mothers. The photo shows Konrad Lorenz with his baby goslings learning from him, imitating his movements.

There is evidence that psychological and physiological responses are also imprinted early in human life. From Chapter 14, we know that immune system imprinting occurs when T-cells recognize "self" from "non-self" antigens. However, behaviors are also imprinted. Most behaviorists agree that human infants use imprinting to form psychological bonds. Infants begin imitation of those around them, mimicking their behaviors early on. Movements of lips, making noise, and imitating sounds are all example of imprinting in humans. Children learn from imitating older children and adults, as shown in Figure 20.4. Learning at early ages begins the complex pathway to adult behavior in humans.

(a)  (b)

**Figure 20.4** a. Imitation by Children. Imprinting occurs when babies and children learn to imitate others. It is a way of learning about and assimilating to the world. b. Sometimes higher level learning, as in a game of chess, is taught by parents.

## Habituation

The process of "getting used to" stimuli is called habituation. Habituation happens when stimuli become familiar, over a period of time, after which organisms no longer respond to them. In humans, we habituate to many changes in the course of our lives. After moving to a city from the country, the sounds of car alarms and street noise slowly extinguish. Habituation allows us, almost unconsciously, to adapt to changing environments.

Birds that are raised in a town, with people around, habituate to their coexistence. They are not "wild" and afraid of humans as in nature. The process of habituation is gradual and often unnoticed by an organism. Figure 20.5 gives an odd example of habituation in the Serengeti Grasslands of Africa.

**Habituation**

The process of "getting used to" stimuli.

## Classical Conditioning

An association between a new stimulus and a natural stimulus is called **classical conditioning**. It is the classic example of how organisms connect new information and

**Figure 20.5** Habitation in the Serengeti Grasslands of Africa. These elephants are used to the appearance of vans and no longer run and hide from them. This photo shows a unique elephant traffic jam.

© John Erickson/Shutterstock.com

**Figure 20.6**  Pavlov's Experiment Is an Example of Classical Conditioning. His dog salivates at the sound of a bell even though this stimulus does not include food.

experiences with their innate behaviors. For example, if a person gets food poisoning, they are likely to associate that food with the pain and suffering of the new experience. Pain is a natural stimulus and the food is the cause. If someone loves pizza but becomes violently ill after eating a slice, he or she is likely to stay away from pizza for a long time. The connection between food and illness is particularly strong.

In an example of classical conditioning, Ivan Pavlov trained dogs to salivate whenever a bell rang. He would bring a juicy steak into the room and allow them to eat it. Their innate response was to salivate with the presentation of steak (and its odor). Pavlov then coupled the steak with the ring of a bell (the new stimulus). Repeated coupling of the new stimulus (the bell) with the steak caused the dogs to salivate whenever they heard the bell even without a steak present. The example in Figure 20.6 shows how learning is also coupled with our innate behaviors.

## Operant Conditioning

**Operant conditioning**

A complex set of behaviors during which an organism learns to respond to stimuli to produce a desired effect.

A more complex set of behaviors occurs during operant conditioning, in which an organism learns to respond to stimuli to produce a desired effect. It is trial-and-error learning, whereby organisms use their natural associations to obtain a positive stimulus. The "operant" is the stimulus that produces an effect.

In an experiment by B.F. Skinner, a famous behaviorist, rats were placed in a box with a series of levers. This box is now called the *Skinner box*, which contained rats in a box with levels. Rats learned to press levers that enabled them to obtain food. The operant was the lever that gave the desired food. The behavior (pressing the right level) led to a desired positive stimulus (the desired food). The lever and the food were associated together to produce the learned response of pressing the right levers.

Of course, animals also learn by operant conditioning in nature. In Figure 20.7, the pelican chick pecks his mother's beak to beg for food. His mother gives him food each time, positively reinforcing this behavior. Quickly, the chick learns to be more accurate in his begging for food from his parent.

In nature, operant conditioning also avoids negative responses. When a crocodile, for example, eats a cane toad, it becomes violently ill. The crocodile will learn to avoid

**Figure 20.7**   Operant Conditioning. A pelican chick learns to peck his mother more efficiently through operant conditioning.

cane toads in the future and start eating other animals to get the desired food. Figure 20.8 shows the learning process. This is the basis for aposematic coloration in Chapter 17, which associates negative outcomes eating organisms with bright coloration.

## Insight

The most complex and poorly understood form of learning occurs during insight. **Insight** is the recognition and mental solving of a problem before attempting it. The complexity of forethought and planning is obvious when the solution happens. However, what occurs within the mind remains a black box. Reasoning or the process of figuring out novel problems is the highest level of human thought. There are levels of reasoning and the highest forms use analysis and evaluation of information.

No other mammal, except chimpanzees, demonstrates insight besides humans (Figure 20.9b). In experiments observing chimpanzees, they are noted for being able to solve problems. In a classic experiment by Wolfgang Kohler in the 1920s, he placed a box and some poles in a room with bananas suspended in the air. On their own, the chimpanzees figured out how to construct a structure to reach up to the bananas.

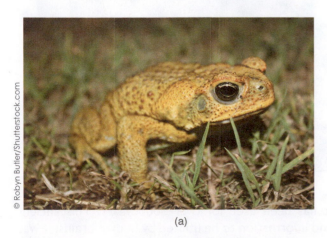
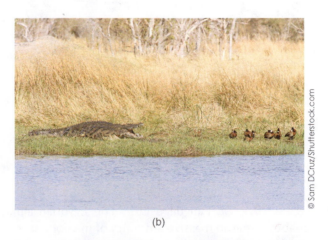

(a)                                        (b)

**Figure 20.8**   a. This Cane Toad Is Not Attacked by a Crocodile. The crocodile learns that the toads are toxic and avoids them as a source for food. b. Instead, crocodiles find birds tasty.

# Cognitivism

**Cognitivism**

The theory that studies the ways individuals use mental abilities to learn.

**Language**

A method of communication.

Another approach to learning explores its mental processing. Cognitivism studies the ways individuals use information to learn. It has become popular in educational research to apply learning in schools. Cognitivism describes organisms as able to process information, store it in their brains, and then retrieve it at a later time. Applying knowledge and using it in novel situations is the focus of cognitive science.

The development of language is an enormous undertaking in humans and a focus of cognitivism. Language requires many of the behaviors described in this section. Humans must remember and apply numerous words to create thought, make associations, and learn numerous words. The development of the alphabet allows this to occur (Figure 20.9). The average human has automaticity with 13,000 words. These construct all of the ideas to help humans communicate with each other and pass down their thoughts. Cognitive psychology is a branch of cognitivism that studies how this happens.

# Sociobiology and Society

Sociobiology explores the social behaviors organisms demonstrate in their groups. Forming societies has advantages and disadvantages. Ants in our story helped each other to gather food. Chimpanzees groom each other to remove debris, dirt, and insects from their fur. In Chapter 17, the benefits of group behavior showed better defense against predators. Alarm calls in prairie dogs, for example, tell others in the group that predators are present. This helps a group to defend against them. The use of language, especially in human societies, further enhances cooperation and benefits of living together. Social organisms seek each other out and desire to form bonds. Group behavior can help members, as shown in Figure 20.10.

Living in groups has its drawbacks for its members, depending upon the type of species and the social construction of the society. For example, in most groups, population density increases as compared with living singularly. This leads to greater competition for mates, food, and territories. Sex is also a resource in an animal society, influencing

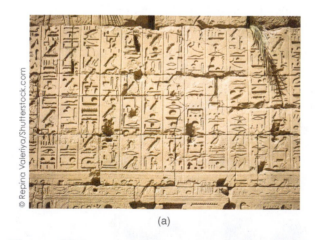

(a)

(b)

**Figure 20.9**   a. Language Use. The Egyptian alphabet has been used for over 3000 years. This ancient alphabet required massive amounts of memory storage and information to be utilized by human brains. It was the start of higher order communication among humans. b. Chimpanzees also show cognitive processes, as shown in the figure in which a chimpanzee (*Pan troglogytes*) uses tools to get fruit from a box.

**Figure 20.10** Social Behavior in Zebras. The stripes on zebras together in a herd confuse their predators, saving some from attack.

social systems. Sexual selection drives many social behaviors. Recall that sexual selection was introduced in Chapter 7.

When groups form, behaviors occur that set up a social structure. A battle for territory and mates sometimes leads to a dominance hierarchy or rank order within a society. High-ranking organisms dominate over lower-ranking organisms. In *Corvus monedula*, a type of jackdaw bird the dominance hierarchy leads to behaviors in which the highest jackdaws support the lowest. Complex systems emerge, which foster discontent between several lower groups by the higher group. This is done to keep the lower levels fighting so that the highest jackdaws may keep their coveted positions. A jackdaw social group is shown in Figure 20.11.

We will explore some of these societies in the next section, evaluating them as they occur in nature. We will look at how they compare with human society. Ultimately, was our character in the story right to yearn to be another organism?

**Dominance hierarchy**

A rank order within a society.

**Figure 20.11** Jackdaws in a Group. Dominance hierarchy can be vicious at times, but maintains order in the society.

# Aggression

There is often aggression, or hostility, between organisms. Humans experience it in groups but it is also intrinsic to animal social systems. However, most sociobiologists agree that intraspecific aggression (between members of the same species) usually chases off and does not kill rivals. Some organisms show intimidation displays with other members of the same species to chase them away, as seen in Figure 20.12. Others, when a battle appears to be lost, avoid further fighting by submissive behavior. Examples of submissive behavior include hiding fangs or concealing claws. With these strategies, animals of the same species almost never fight to the death. It would destroy their own related genes, as species share almost 99% of their DNA among members.

In our story, human conflict reaches a climax with the threat of nuclear war. Human population is threatened by the members of our own species. These situations are unfamiliar in other species. Perhaps because we have evolved complex thoughts to pre-meditate and invent devices, we are different from other species? Perhaps we are different innately – more aggressive? Alternatively, maybe our society corrupts us, leading to such behaviors? Regardless, humans are 99.9% identical genetically, so killing each other is akin to killing oneself. It violates biological principles that dictate self-preservation and genetic survival.

Then, why do murder and warfare occur in human society, and not intentionally, as in most other animal societies? Lorenz argues that society makes humans excessively aggressive. Others argue that the selfish behaviors are inherent within. Richard Dawkins developed a selfish-gene hypothesis that states that we are merely vessels holding our genes. The genes are selfish and want to reproduce to increase its own RS (Figure 20.13). This hypothesis contends that selfish genes drive all human behavior. In this next section, we will explore these two opposing sets of viewpoints.

# Human and Animal Kindness

Let's revisit the character's contention in the story: Animals really are kinder than humans. This is true, surficially. It pays for many animal species to cooperate with each other, forming social groups. Many animals cannot survive long without being in a society. Ants in our story, when taken out of the colony, die very quickly with no purpose and no society.

Cooperation is the cornerstone factor in the success of any animal society. However, what drives organisms to help one another? Recall our biological principle – each member behaves to improve its RS. The same principles apply to the ants in the colony.

**Figure 20.12**    A Cat often Engages in Intimidation Displays to Prevent Further Aggression.

**Figure 20.13**    Are We Selfish in Our Inner Core?

In our story, then why did all of these ants help carry the food, without even one ant taking a quick bite for itself? It is a form of altruism, when one organism helps without receiving an individual benefit. The opposite is selfishness, when one organism harms the other for its own benefit. However, ants are not really altruistic. They work together to improve their RS. The goal of any organism is to increase its RS, as discussed earlier. Ants are no exception. If a behavior increases an organism's RS, then it will be favored in an ecosystem. If not, it will be eliminated.

Then, how can altruism be favored? Altruism gains nothing for an individual giving to another, such as an ant giving food to her queen. Is there true altruism or are all animals, including humans, selfish? Ecologists contend that humans and animals are both very selfish. They cooperate based only on principles to benefit their own genes. It may be that our character in the story is naive to the ways of nature.

## Kin Selection

Let's look at kin selection to better understand the ants in our story. When helping is observed in nature, it is often between kin or family members. Kin selection occurs when individuals help each other because they share genes in common – they are related. For example, your mother, father, brother, sister, and children are related to you so you are more likely to help them, according to kin selection. (Figure 20.14).

As described earlier, our selfishness drives us to help our own genes get to the next generation – to have increased RS. Our immediate kin are 50% identical to us, which means that by helping them we are actually helping ourselves by the amount we are related.

According to kin selection, the amount of help we give depends upon the degree to which we are related to our kin. One's aunt, uncle, nephew, and niece share 25% of the same genes in common. Thus, one might help each of them, but only half as much as our immediate kin. These predictions are not without flaws. Perhaps you dislike your cousin – this complicates matters. You may be inclined not to help them out. Humans are complex but still, the drive of biological principles is present. We are always obligated to help based on our selfish genes. This hypothesis contends that the closer the relatedness is, the stronger will be the helping behavior between kin. Do you think that this is so?

The selfish gene hypothesis is a cold and calculating view of human and animal kindness. It is suspicious and looks for the genetic payoff to helping. Let's take a look at family that has helping as a cornerstone behavior.

**Altruism**

A type of behavior in which one organism helps without receiving an individual benefit.

**Selfishness**

A behavior in which one organism harms the other for its own benefit.

**Kin selection**

Natural selection in which individuals help each other because they share genes in common.

**Figure 20.14** Kim Selection. Does a mother only help her children because she helps herself? Sociobiology asks the question.

Family is an important factor in all human society. We take our girlfriend or boyfriend to "meet the parents" because kin are important. Family is set up, according to the selfish-gene hypothesis, to help each other's genes. Because the members of the unit are more related, it evolved, much like an ant colony, into a helping system. By helping family members, we are helping our own genes that are shared in common. Family is a social construct to act as a vehicle to accomplish this, according to Dawkins.

Even human laws are based on family importance. Inheritance of assets moves down family lines to preserve wealth in tandem with genetic relatedness. Interesting, in a recent study of genetics, after 14 generations removed from our ancestors, we are no more related to them than a stranger. The advice: spend the money and do not save for generations down the road to preserve you wealth genetically!

## What about Helping in Unrelated Organisms?

Some animals help each other who are not related. It is pleasant to imagine that ants are kind and considerate of each other because they are "just nice." Self-sacrificing organisms often have a reason behind their behaviors. To illustrate, vampire bats, *Desmodus rotundus*, appear to help one another in an altruistic way. They are small creatures, each weighing between 15 and 50 grams (0.5–1.7 ounces). Vampire bats are native to South America and the areas of Central America. *D. rotundus* are flying mammals with a bad reputation – they suck blood. (Figure 20.15).

Upon closer evaluation of their society, however, they help each other when one of their members misses a meal. They are communal roosters, meaning that they leave their nests as a group to feed. Usually, they make their journeys in the night to forage for food. When vampire bats bite, they make a quick and painless insertion into their victim. *D. rotundus* feed for up to half an hour (about 20 milliliters of blood) at any one time on its victim. They have a voracious appetite and want to latch on for extended time. To do this, they evolved an anticoagulant in their saliva to keep the victim's blood flowing.

After they return to their nests, they are frequently observed regurgitating blood to each another. By regurgitating blood, they give away some of their own food to benefit

**Figure 20.15** Vampire Bats, *Desmodus rotundus*. While appearing altruistic, they actually only help each other to get something back in return – blood when they need it.

the other bat. It is an example of altruism. The most surprising part is that vampire bats do not discriminate – whether kin or not, they help their fellow bat.

It appears to be altruism but upon an even closer analysis, helping behavior is linked to something more selfish: It is a "You scratch my back, I'll scratch yours" mentality. Because these bats are so small, missing a meal is deadly, with a steady proportion of bats always missing a meal at any one night's feeding. Eventually, every bat will be in the same position and need some regurgitated blood. Thus, by regurgitating blood, such a strategy will pay off in the future. Therefore, the next time a bat needs a blood meal, it will be saved by what is called reciprocal altruism, in which all members of a group cooperate altruistically so that each survive better. This system of reciprocal altruism depends upon members not cheating and mechanisms to detect cheaters; otherwise, the whole set up would fall apart.

**Reciprocal altruism**

A behavior in which an organism helps another and the second organism returns the favor either to the benefactor or his/her progeny.

## Debate on the Nature of Animal Society

OK, so bats are not really all that loving to each other. What about human kindness to each other? Don't strangers help each other, despite being unrelated? What about human acts of kindness in the world? Dogs may show love to their master but is there an ulterior motive? There is debate on the nature of goodness and badness in animal and human society.

Konrad Lorenz (1903–1989) and the current Richard Dawkins, both introduced earlier in this chapter, offer opposing viewpoints. In the selfish gene hypothesis, Dawkins argues that all life is extremely selfish by nature. He claims that our genes puppet our behavior and that it is their selfish desire to reach the next generation that forms human society. He adds the caveat that society should ". . . try to teach generosity and altruism, because we are born selfish." Dawkins sees social structure in all organisms, including ants in our story, driven by this selfishness. He would view the line of ants in the story as a set of selfish vessels doing the will of genes. They help each other to benefit the colony and the colony is only an extension of the selfish genes within (Figure 20.16).

Lorenz argues that behaviors are adaptive and that they benefit the organism. However, his view looks at animals are more caring about each other. He views behaviors such as grief in geese and love of music as beyond mere benefits to the individual – these

**Figure 20.16**    Are Humans Inherently Good or Evil? A debate about our inner depths. Richard Dawkins

behaviors do not benefit an individual. It shows that we are more complex than mere selfishness.

It is not humans but society that leads to animal selfishness, in Lorenz's view. His approach depicts humans and animals as basically good. Aspects of society, such as high density and crowding, he views as corrupting humans. He points to examples of other animal social systems that restrain their aggression except when crowded. Lorenz cites fish in a tank, which become aggressive but in nature, given free space, they are passive. Lorenz blames, in several of his books, the crowded capitalist system as unnatural to animals, as the reason for our excessive aggression (Figure 20.17).

## Group Cooperation vs. Selfish Genes

**Eusociality**

Is the highest level of social organization, in which organisms cooperate and their members have sterile castes.

Lorenz points out kindness in various animal social systems, much as the character in our story notices cooperation on the ant hill. Insect members of the order Hymenoptera (ants, bees, and termites) all exhibit cooperative systems. These societies have traditionally

**Figure 20.17**    People Crowded in a City May Not Be Living in a Natural Situation. A busy street in New York City is shown in this image.

**Figure 20.18**   Eusociality. The order of animals, Hymenoptera (ants, bees, and wasps) exists in highly organized colonies.

been looked upon as a role model of good group behavior. However, how altruistic are the ants in the colony? Finally, let's look at how the ant society is constructed.

Hymenoptera species all have similar characteristics: they exhibit eusocial behavior. Eusociality is defined as the highest level of social organization, in which organisms cooperate and their members have sterile castes (Figure 20.18). These castes do not reproduce and instead give their life for the colony. It is the ultimate in altruism – or is it?

An ant soldier will attack its enemy to save the group in a seemingly selfless act, ripping itself apart, as described in Chapter 17. This act occurs despite the fact that the ant will die within a few hours. As stated, worker ants are sterile and instead serve a queen master. All of this points to the kind of selflessness missing in humans.

However, Dawkins argues that this selflessness is actually selfish. Dawkins posits that the behaviors of these ants are solely based on their desire to pass their own genes onto the next generation.

Dawkins cites the haplodiploidy of the hymenopterans: the queen gives birth by parthenogenesis, which recall from Chapter 16 is a virgin birth (Figure 20.19). She produces all of the males in the colony in this way. The males are thus haploid (contain half of the full set of DNA) and are all identical to one another. After some basic genetic calculations, it is determined that females in the colony are 75% related to each other. Their father contributes all of his genes to his children. This makes the offspring identical on the paternal side. Gene differences only arise from the mother.

Hymenopteran, such as the ants in the story, are therefore more likely to help one another because of their high degree of genetic relatedness. It pays to help each other out – ants are actually helping 75% of themselves, since eusocial insect females are 75% identical to each other. They are like one large family, more related to each other than any other family. In humans, immediate family members (non-twins) are at a maximum 50% related, as shown earlier in the chapter. It would be expected that human helping would be less than in ants. The more related organisms are, the more they tend to help one another. This is a selfish reason why ants are so cooperative – they are really helping themselves by helping each other.

The motivation behind a behavior is important in determining whether it is a truly helping or based on selfishness. If someone walks a stranger across the street, only to steal a wallet, was the behavior a true act of kindness? Natural laws of genetics and ecology may have more influence on human society than expected. Should the character

**Haplodiploidy**

Sex-determining method in which females are developed from fertilized eggs (diploid) and males are developed from unfertilized eggs (haploid).

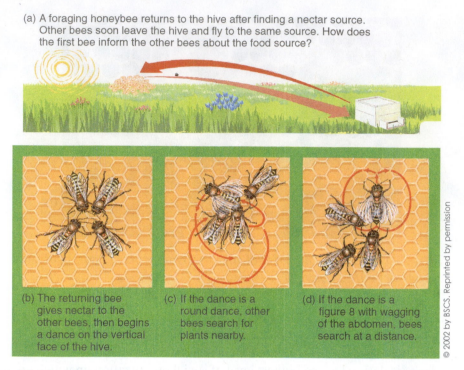

(a) A foraging honeybee returns to the hive after finding a nectar source. Other bees soon leave the hive and fly to the same source. How does the first bee inform the other bees about the food source?

(b) The returning bee gives nectar to the other bees, then begins a dance on the vertical face of the hive.

(c) If the dance is a round dance, other bees search for plants nearby.

(d) If the dance is a figure 8 with wagging of the abdomen, bees search at a distance.

© 2002 by BSCS. Reprinted by permission

**Figure 20.19**   Ants Help Each Other. From *BSCS Biology: An Ecological Approach,* 9th Edition by BSCS.

in our story really envy the citizens on the ant hill colony – the polite and helpful Hymenopterans? Are humans able to overcome their genes through acts of kindness? Will humans eventually destroy themselves? . . .

## Summary

Social structure in humans and in animals mirrors the principles of biology discussed in this text. Animal social systems include innate and learned behaviors. Learned behaviors help animals to adjust to changing environments. Some animals organize into social systems that have both positive and negative consequences for its members. Ultimately, social systems seek to benefit more than harm the RS of its members, taken as a whole unit. Social systems are set up based on a drive to improve RS among its members. Organisms and social systems operate to accomplish this task.

### CHECK OUT

**Summary: Key Points**

- Human society often follows the same principles of survival, competition, and aggression as animal social systems.
- Animals behave either innately or through learning by interacting with others.
- Learned behaviors include imprinting at early ages, habituation or getting used to stimuli, classical conditioning, operant conditioning, and insight.
- The selfish-gene hypothesis claims that all organisms act selfishly to extend their genes to new generations.

## KEY TERMS

adaptive behaviors
aggression
altruism
animal behavior
behaviorism
behaviors
classical conditioning
cognitivism
dominance hierarchy
eusociality
habituation
haplodiploidy

innate behaviors
intimidation displays
kin selection
language
learned behaviors
operant conditioning
reciprocal altruism
reproductive success (RS)
selfish-gene hypothesis
selfishness
sociobiology
submissive behavior

# Multiple Choice Questions

1. Both human and animal social systems seek to improve:

    a. selfishness
    b. altruism
    c. parasitism
    d. reproductive success

2. A pigeon gets accustomed to people in an urban area. It no longer is afraid of them. This is an example of which type of learned behavior?

    a. adaptation
    b. habituation
    c. imprinting
    d. operant conditioning

3. In question #2 above, the same pigeon gets attacked by cats whenever it enters Mr. McGreeley's backyard. This is an example of:

    a. adaptation
    b. habituation
    c. imprinting
    d. operant conditioning

4. Sisters in ant colonies are _____ % related genetically.

    a. 1
    b. 50
    c. 75
    d. 99

5. The statement, "Humans are basically good but society is the problem" is a statement from:

    a. Dawkins
    b. Lorenz
    c. Skinner
    d. Pavlov

6. Which term includes all of the others?

    a. social organization
    b. dominance hierarchy
    c. reciprocal altruism
    d. cooperation

7. A whale, born knowing how to defend itself is an example of _____ behavior.

    a. imprinted
    b. innate
    c. learned
    d. habituated

**8.** Social organization in vampire bats was shown to be:
   **a.** selfish
   **b.** altruistic
   **c.** uncooperative
   **d.** individualistic

**9.** Cognitivism studies:
   **a.** mental processes
   **b.** responses
   **c.** negative stimuli
   **d.** positive stimuli

**10.** A change in behavior as a result of stimuli is:
   **a.** learning
   **b.** non-social
   **c.** innate
   **d.** all of the above

## Short Answer

**1.** Describe one way human society is the same and one way it is different when compared with animal social systems.

**2.** List three ways learning occurs in animal social systems.

**3.** How does association play a role in classical conditioning?

**4.** List and define the two types of animal behaviors. How do they both work to help organisms?

**5.** Explain how reciprocal altruism is demonstrated in vampire bats.

**6.** Explain how altruism differs from selfishness, using the term reproductive success.

7. Give one example of psychological imprinting and one example physiological imprinting.

8. What dangers exist in habituation? Do you think that this plays a role in car accidents? Explain.

9. Explain why insight is so difficult to measure, as compared with other behaviors.

10. Compare and contrast the goals of submissive displays and intimidation displays. Be sure to include one way the behaviors have goals in common and one way the behaviors are different.

## Biology and Society Corner: Discussion Questions

1. How are dominance hierarchies seen in the work place? Compare it to the jackdaw social system described in this chapter. Are human employment systems very different from those found in animal social systems? Explain your answer.

2. The selfish gene hypothesis and Lorenz's views oppose each other. After reading this chapter, form an argument supporting or refuting the following statement: "Humans are selfish to the core." By the end of the chapter, did you agree with the character in the story, that you too wish you were an ant? Why or why not?

3. A controversial statement was made by a scientist in the news recently: "Nations that are more homogeneous have better cooperation." Based on the readings in this chapter, do you agree or disagree with this assertion? Why or why not?

4. The use of behaviorism in schools often looks only at the test results (an expressed behavior) and not at the total person, according to many education groups. In America today, testing in schools evaluates the scores and makes decisions. Educators argue that scores on tests do not look deeper into the whole person. Do you agree or disagree? Why?

5. What if ants in the hill in our story were only related by 10%? Predict how its society would change. Explain your answer.

Figure – Concept Map of Chapter 20 Big Ideas

# GLOSSARY

**Abiotic factors**   The non-living factors in an environment.

**Abscess**   Collection of pus that is walled-off and builds within the tissues of the body.

**Abyssal zone**   A deep layer near the bottom of the ocean.

**Accessory organs**   Association of organs that produce, store and/or release chemicals to carry out the processes of the breakdown of food.

**Acid**   The resulting solution in which water yields more hydrogen into its surroundings.

**Acrosome**   A caplike structure covering the top end of the sperm, containing enzymes that digest layers surrounding the egg.

**Actinomycetes**   A type of bacteria having filament strands and resembles fungi. They are decomposers, recycling dead organic matter.

**Action**   The direction in which muscles move when they contract.

**Action potential**   A change in the electric potential across the plasma membrane that occurs when a cell is stimulated.

**Activation energy**   The minimum amount of energy that must be possessed by the reacting species to undergo a specific reaction.

**Active artificial**   The condition that occurs when a vaccine activates the immune system to produce antibodies in a response to.

**Active natural**   A type of immunization that occurs when the immune system defends itself and enables future defense against pathogens.

**Active site**   Special shapes on enzymes that allow for binding to other chemicals.

**Active transport**   Is the movement of substances against a concentration gradient, from a lower concentration to a higher concentration, which requires cellular energy.

**Adaptation**   Populations of living things change a result of their surroundings and evolve or change as a group.

**Adaptive behavior**   The behavior that helps an organism's reproductive success by helping it to persist in a population and become established.

**Adaptive radiation**   The changes that occur in a group of organisms to fill different ecological niches.

**Adenine**   A purine base that is a base, a component of RNA and DNA.

**Adenosine triphosphate (ATP)**   A special nucleotide that holds readily available energy for cell functions.

**Adipose**   A body tissue used for fat storage.

**Adrenal cortex**   The exterior portion of the adrenal gland that secretes steroid hormones.

**Adrenal glands**   Glands that sit atop both kidneys and control a variety of body functions.

**Adrenal medulla**   The inner part of the adrenal gland.

**Adult**   Fully grown or developed.

**Aerobic**   Occurring in the presence of oxygen or require oxygen to live.

**After-birth**   A brief relaxation period followed by smaller contractions that propel the placenta out of the uterus after child birth.

**Age structure diagram**   A graphical illustration that is used to predict future patterns of growth or declines of various age groups in a population.

**Aggression**   Hostility between organisms.

**Air sacs (alveoli)**   Tiny sacs within the lungs where exchange of oxygen and carbon dioxide gases takes place.

**Alarm call**   A warning signal made by an animals or birds about a predator or when startled.

**Albedo**   A proportion of solar energy that is reflected from the Earth back to space.

**Albinism**   Is a noncontagious disease that is genetically inherited and results in a lack of pigmentation.

**Aldosterone**   A hormone produced by the cortex of adrenal gland.

**Algae**   Are multicellular organisms that are photosynthetic, and they contain a variety of pigments such as chlorophyll a and b (green), carotenoids (yellow-orange), phycobillins (red and blue), and xanthophyll (brown).

**Alimentary canal**   The tube and associated organs of digestion, which includes all of the parts of the digestive system that contribute to the breakdown of food.

**All-or-none response**   All sarcomeres contract in a muscle fiber or none at all.

**Alleles**   An alternative form of the same trait.

**Allergen**   Antigens that cause allergies.

**Allergy**   An inappropriate immune reaction to antigens that are otherwise not harmful.

**Allopatric speciation**   The process of development of new species when there is a physical barrier separating members of a group of organisms.

**Alpha cells of pancreas**   Cluster of cells found in the pancreas, which make glucagon.

**Alternation of generations**   The life cycle of plants in which haploid and diploid phases of their lives exist for survival and reproduction.

**Altitude sickness**   The condition in which the altitude affects a person's ability to breath.

**Altruism**   A type of behavior in which one organism helps without receiving an individual benefit.

**Alzheimer's disease**   Progressive mental deterioration brought on by aging. The disease is associated with protein masses or tangles that form plaques along the nerves in the brains of the victims.

**Amino acid**   The building blocks of proteins.

**Ammonification**   The process in which organic nitrogen is converted into ammonia and dissolved $NH_4^+$ forms ready for uptake.

**Amnion**   The fluid sac that protects a fetus.

**Amniotes**   Terrestrial animals that develop their eggs encapsulated within an amniotic sac.

**Amoebocytes**   A type of sponge cell that transports food through the sponge body.

**Amphibians**   Vertebrates that live a portion of their lives in water and another portion on land.

**Amplexus**   Mating behavior seen in frogs and toads.

**Amygdala**   A section of brain associated with fear, panic, and aggression.

**Anabolism**   A series of reactions that build up complex molecules, using stored energy.

**Anaerobic respiration**   A series of reactions that form alcohol from sugar.

**Anal canal**   Terminal part of the large intestine.

**Anaphase**   A cell division stage in which chromosomes split into two identical groups and move toward the opposite poles of cells.

**Anaphase I**   The stage of cell division in meiosis in which homologous chromosomes separate.

**Anatomical position**   The position that describes a specific way of positioning for a human body.

**Anatomy**   The study of the structure of body parts.

**Anemia**   The condition in which blood lacks normal oxygen carrying capacity.

**Angioplasty**   A surgical procedure to widen obstructed arteries or veins.

**Angiosperm**   Are flowering plants with seeds developed in an ovary.

**Angiotensin**   A hormone that promotes aldosterone secretion in blood and causes blood pressure to rise.

**Animal behavior**   The branch of biology that studies the ways in which animals act within their environment.

**Animalcules**   The dated term for a microscopic animal, we now know of as microorganisms.

**Animals**   Living organisms that feed on organic matter (other living creatures) for survival. Animals are multicellular eukaryotes and motile in nature.

**Anion**   Negatively charged ion.

**Anorexia nervosa**   An eating disorder characterized by a loss of appetite for food and a fear or refusal to maintain normal body weight.

**ANOVA (Analysis of Variance)**   Is a powerful statistical method that compares the means of three or more groups.

**Anterior**   At the front of or situated before.

**Anther**   A reproductive structure that holds pollen grains.

**Anthropoids**   A higher primate, including monkeys.

**Anti-codon**   A sequence of three nucleotides in transfer RNA molecules.

**Antibiotic**   Any chemical that stops the growth of microorganisms.

**Antidiuretic hormone (ADH)**   A hormone released by the pituitary gland that helps in water retention in the body.

**Antigens**   Any molecule or cell part that initiates an immune response.

**Antioxidants**   The substances that eliminate molecules with extra electrons, thus preventing damage to body structures.

**Anvil**   A tiny bone in the middle ear.

**Aortic arches**   The simple hearts of segmented worms.

**Apical meristem**   Meristems that are found at the tips of roots and in shoot buds to begin primary growth.

**Apical surface**   The free side of all epithelial tissues.

**Apoptosis**   Cell death that occurs as a part of an organism's growth.

**Aposematic coloration**   Warning coloration that serves to deter predators.

**Appendicular skeleton**   The portion of skeleton that consists of bones of the limbs, the pelvic girdle, and the pectoral girdle.

**Apple shape**   A body shape that is characterized by excess body fat in the abdominal region.

**Aqueous humor**   The clear fluid present between the cornea and lens of the eyes.

**Aquifer**   Saturated areas of water within bedrock.

**Arachnids**   An arthropod characterized by having eight legs.

**Arachnoid**   The middle layer of meninges.

**Archaea**   Microorganisms that are similar to bacteria in size and structure but different in molecular organization.

**Archaebacteria**   Ancient forms of bacteria, with only a few surviving branches.

**Areolar**   Packaging type tissue.

**Arrector pili muscle**   Small muscles attached to hair follicles in skin.

**Arrhythmia**   A condition in which the heart beats in an abnormal rhythm.

**Arteriole**   A small branch of artery leading to a capillary.

**Arteriosclerosis**   A chronic condition characterized by abnormal thickening of vessel walls.

**Artery**   A vessel that carries oxygenated blood away from the heart to cells, tissues, and organs.

**Arthropods**   Invertebrates with a specialized segmented body and a protective external skeleton, or exoskeleton, and jointed appendages.

**Articulation**   The area of connection between two bones.

**Asexual reproduction**   The process in which a single individual produces new offspring, without genetic material contributed from a partner.

**Asexuality**   The lack of sex drive.

**Asthma**   A respiratory condition characterized by inflammation of the respiratory passageways.

**Atherosclerosis**   Condition in which saturated fats are linked to heart disease and hardening of the arteries occurs.

**Atmosphere**   The gaseous layer surrounding a planet.

**Atom**   The smallest component of any element that retains the unique properties of that element.

**Atomic mass**   The mass of an atom is the combined weights of the subatomic parts that have weight.

**Atomic number**   The number of protons in the nucleus of an atom.

**Atoms**   Are the smallest units of matter that can exist and maintain the properties of the larger sample.

**Atrioventricular (AV) node**   Small mass of neuromuscular fibers located at the base of the interatrial septum.

**Atrium**   An entry chamber of the heart from which blood is passed to the ventricles.

**Auscultation**   Listening to sounds produced within the body.

**Autogenous model**   The model that states that eukaryotes developed directly from a prokaryote by compartmentalization of functions of the prokaryote plasma membrane.

**Autoimmune disease**   A disease in which the immune system overreacts to its own cells.

**Autonomic nervous system**   System of involuntary nerves.

**Autorythmic**   Cardiac muscle cells that beat independent of the nervous system.

**Autosomal dominant**   The patterns of inheritance of single-gene traits in which the dominant allele gets expressed.

**Autosomal recessive**   The patterns of inheritance of single-gene traits in which both recessive alleles are present for a person to get the recessive trait.

**Axial skeleton**   The portion of skeleton that consists of the skull bones, ribs, sternum, and vertebrae.

**Axon**   A long thread-like structure of the nerve cell that transmits information to other cells away from the cell body.

**B-cell**   A type of white blood cell that produces antibodies.

**Bacillus**   Rod-shaped bacteria.

**Backbone**   Set of nervous tissue surrounded by bones for protection.

**Bacteria**   Single-celled microorganisms that are found everywhere.

**Bacteria**   Single-celled organisms that have cell walls but lack an enclosed nucleus and organelles.

**Basal cell carcinoma**   A type of skin cancer, which is relatively common, but very rarely kills its victims.

**Basal layer**   The lowest layer of epithelial layers.

**Basal metabolic rate (BMR)**     The minimal rate of energy used by an organism at complete rest.

**Base**     The resulting liquid in which water absorbs more hydrogen from its surroundings.

**Basement membrane**     A thin extracellular membrane underlying the epithelium of many organs.

**Basophil**     A type of white blood cell associated with allergies.

**Behaviorism**     The theory that human and animal responses are measured to determine an organism's learning.

**Behaviors**     Any action taken by an organism.

**Beta cells of pancreas**     Cluster of cells found in the pancreas, which make insulin.

**Bicarbonate (HCO3-)**     A buffer; within the digestive system it neutralizes the acidic chyme entering the intestines.

**Bilateral symmetry**     The property of being roughly identical upon surface observation when a line is drawn down their middle.

**Bilirubin**     A substance produced by the digestive system during the breakdown of RBCs.

**Binary fission**     The process by which a cell divides directly in half.

**Binomial nomenclature**     Naming convention for living creatures, in which organisms are a given unique scientific name, composed of two parts. The first indicates the genus and the second the species.

**Bioaccumulate**     Accumulation of generally harmful substances in an organism.

**Biodiversity**     Variety of life forms in a particular habitat.

**Biogeochemistry**     A scientific discipline that encompasses chemical, physical, geological, and biological processes and reactions that govern the workings of the natural environment.

**Biogeography**     The way species are distributed.

**Biological literacy**     Is the ability to interpret, negotiate, and make meaning from the many aspects of knowing about life to make decisions and use biology and its technology.

**Biology**     Study of living creatures.

**Biomagnification**     The increasing concentration of a particular substance in organisms at the top of the food chain.

**Biomass**     Is the total organic matter in an ecosystem.

**Biome**     A large community of flora and fauna occupying a major habitat.

**Biophilia**     The affinity human beings share with other living creatures.

**Bioprocessing**     The process of building up (anabolism) and breaking down (catabolism) of macromolecules.

**Biosphere**     All of the different ecosystems of the Earth interacting with their environment make up the biosphere.

**Biotechnology**     The branch of science that uses biological knowledge and procedures to produce goods and services for human use and financial profit.

**Biotic factors**     Factors that comprise the living things in a habitat.

**Biotic potential**   Maximum possible growth achieved by organisms under ideal conditions.

**Bipolar cells**   A neuron that has two processes.

**Birds**   Warm-blooded vertebrates characterised by feathers, wings, beak with no teeth, scaly legs, and typically by being able to fly.

**Births**   New born additions.

**Bivalves**   The property of having two shells hinged together.

**Blastula**   A hollow ball of cells at the early stage of development.

**Blood**   A red fluid that connects different parts of the body by providing nourishment and removing wastes.

**Body landmarks (surface regions)**   Terms for the specific location of the lesions on the human body to give detail to their descriptions.

**Bolus**   A rounded mass of food with the digestive tract.

**Bone**   Is the substance that has a solid form, with calcium salts embedded within fibers of its extracellular matrix.

**Bone marrow**   A compartment within bones that stores stem cells.

**Bony fishes**   Type of fishes that have skeletons composed of bone.

**Bowman's capsule**   A cup-like sac surrounding the glomerulus within the nephron.

**Brain**   A part of central nervous system that functions as the command center of the body.

**Brainstem**   The portion of brain that consists of pons, midbrain, and medulla oblongata.

**Bronchial tubes**   Tubes that let air in and out of the lungs.

**Brood parasitism**   A form of social parasitism in which a bird species lays eggs in another bird's nest.

**Budding**   A form of asexual reproduction in which new organisms develop from a bud as a result of cell division at one specific site.

**Bulbourethral gland**   A small gland located at the base of the male reproductive organ and donates small amounts of fluid to the semen at the end of ejaculation.

**Bulimia**   An eating disorder characterized by abnormal and constant craving for food alongside purging.

**Bulk transport**   Is the movement of large amounts of material across the plasma membrane.

**Bundle of His**   A collection of heart muscle cells that transmit electrical impulses from the AV node to the interventricular septum and ventricles.

**C3 pathway**   The most common form of photosynthesis that uses a 3-carbon molecule in the Calvin cycle.

**C4 pathway**   A method used by plants to pull carbon dioxide into the Calvin Cycle more easily.

**Calcitonin**   A hormone made by the thyroid directs calcium ion uptake by the bones.

**Calcium carbonate**   A naturally occurring chemical compound , making up by coral skeletons.

**Calories**   The amount of energy required to raise 1 gram of water by °1 Celsius.

**Calvin cycle**   A set of chemical reactions absorbing carbon dioxide and making glucose, taking place in chloroplasts during photosynthesis.

**CAM pathway**   A type of photosynthesis working at night and exhibited by plants that inhabit warm and dry areas.

**Cambrian explosion**   A evolutionary event during which rapid diversification of multicellular animal life occurred.

**Camouflage**   The act by which organisms become less visible in their environments to avoid being seen.

**Cancer**   A tumor caused by an uncontrolled division of cells.

**Canines**   The front teeth on the side, long and narrow, evolved to tear and pull foods.

**Canopy layer**   Is the upper layer that is made of the leaves of treetops, in forest ecology.

**Capillary**   A tiny blood vessel that connects arteries and veins.

**Capillary bed**   The whole system of capillaries of the body.

**Capsid**   The protein coat that surrounds structure of a typical virus.

**Carbaminohemoglobin**   One of the forms in which carbon dioxide exists in blood.

**Carbohydrate**   Organic compounds providing "instant energy" for living tissues.

**Carbon fixation**   The conversion process of carbon dioxide to organic compounds by living organisms.

**Carbon monoxide poisoning**   A potentially fatal condition that occurs when carbon monoxide binds to hemoglobin, replacing oxygen.

**Carbonic acid-bicarbonate buffering system**   A set of reactions that regulate the pH of blood.

**Cardiac muscle**   The muscle found only in the heart and beats spontaneously to pump blood throughout the body.

**Cardiac sphincter**   The muscle surrounding the opening between the stomach and esophagus.

**Cardiovascular system**   The system comprising the heart and blood vessels

**Carnivore -(secondary consumer)**   Organisms that eat herbivores.

**Carpals**   Any bones of the wrist.

**Carpel**   An organ found at the center of a flower and bears one or more ovules.

**Carrying capacity (K)**   The maximum number of individuals an environment is able to sustain in the long term.

**Cartilage**   A dense connective tissue that provides cushioning support in vertebrates.

**Cartilaginous fishes**   Are fishes that have a skeleton composed of the flexible but solid connective tissue, cartilage.

**Catabolism**   A series of reactions that break down complex molecules to yield energy.

**Catastrophism**   The theory that explained that new species formed after sudden and violent catastrophes.

**Cation**   Positively charged ion.

**Cell**   The structural and functional unit of an organism.

**Cell body**    The central part of the neuron that contains the machinery of the cell, with organelles and a nucleus that directs nerve functions.

**Cell cycle**    The life span phases a cell goes through.

**Cell respiration**    A series of energy-producing reactions that convert food energy into ATP.

**Cell-mediated immunity**    An immune response that is based on on antigen-specific T lymphocytes.

**Cellular respiration**    The process through which most organisms break down food sources into useable energy.

**Cementum**    A glue-like substance holds the tooth's ligament that connects the dentin to the underlying bones of the face.

**Central canal**    A central tube through which water enters the arms of a starfish.

**Central core**    The foundational part of an organism that helps regulate basic life processes.

**Central Dogma**    A theory that explains how inherited material gives rise to all our unique structures, functions, assets, and liabilities.

**Central nervous system (CNS)**    The part of the nervous system consisting of the brain and the spinal cord.

**Centriole**    Minute cylindrical organelles found in animal cells, which serve in cell division.

**Cephalopods**    A group of mollusks characterised by a large head, eyes, and a ring for sucker-bearing tentacles.

**Cerebellum**    The posterior part of the brain, involved in coordination.

**Cerebrum**    The largest region of the brain.

**Cervical vertebrae**    The top seven vertebrae of the spinal column that form the neck.

**Chaparral**    A type of biome characterized by shrubs and rodents in dry conditions, around the Mediterranean and southern California.

**Character displacement**    The phenomenon in which organisms evolve characteristics to help them to partition resources.

**Characteristics of life**    The features (adaptation, order, response to stimuli, growth development, and use of energy, homeostasis, reproduction, metabolism, diversity) that differentiate between life and nonlife.

**Chemical bond**    Relationship between atoms, involving the exchange of electrons.

**Chemistry**    Study of matter, its properties and its interactions.

**Chemoautotroph**    Bacteria that use inorganic chemicals as energy.

**Chemoreceptor**    A sensory cell that is stimulated by chemicals.

**Chlorophyll a**    A special pigment molecule in a photosystem that does not move its electrons back to the ground state.

**Chloroplast**    A part of plant that contains chlorophyll and conducts photosynthesis.

**Cholecystokinin (CCK)**    A hormone that slows peristalsis.

**Chondrocyte**    A cell within cartilage.

**Chordates**   Are animals with a spinal cord or spinal cord-like structure.

**Chorion**   Fetal membrane that nourishes the fetus and becomes a part of its placenta.

**Chromatin**   Is a complex of macromolecules found in cells and consist of protein, RNA, and DNA.

**Chromoplast**   An organelle that contains any plant pigment other than chlorophyll.

**Chromosome**   A thread-like structure formed when a cell divides and its DNA coils more tightly to histones.

**Chyme**   Acidic ball of food within the digestive tract.

**Cilia**   Are short extensions that help cells move.

**Ciliary body**   A part of the eye located between the choroid and iris.

**Circular genome**   Genetic material in a circular form found in prokaryotes.

**Class**   A group of related orders.

**Classical conditioning**   An association between a new stimulus and a natural stimulus.

**Cleavage**   Division of cells in the early embryo stage.

**Climax community**   The highest level of organization for an ecosystem.

**Clitoris**   A site of external female stimulation.

**Clonal selection**   The process by which T-cells or B-cells divide rapidly to produce large numbers of their own type of immune cells on encountering an antigen.

**Clotting factors**   Are proteins that undergo a series of chemical reactions that halt bleeding.

**Cnidarian**   An aquatic invertebrate that comprises coelenterates.

**Cnidocysts**   Stinging cells present in Cnidarians.

**CNS (central nervous system)**   The part of the nervous system that consists of the brain and spinal cord.

**Coccus**   Round-shaped bacteria.

**Coccyx**   A small triangular-shaped bone located at the base of the spine.

**Cochlea**   A spiral-shaped cavity of inner ear.

**Codon (triplet)**   Normal genetic code in which a sequence of three nucleotides codes for a specific amino acid.

**Coelenterons**   The open body cavity present in Cnidarians and opens to the outside environment, in which digestion occurs.

**Coelom**   An open body cavity that separates the organs of the annelid.

**Cognitivism**   The theory that studies the ways individuals use mental abilities to learn.

**Cohesion (or cohesive forces)**   The force that is formed when water molecules stick together due to hydrogen bonding.

**Collar cells**   A type of sponge cell that has beating flagella move water through the internal cavity of the sponge.

**Collecting duct**   A collecting tube that receives urine from several nephrons.

**Colonization**    A process by which a species spreads to new areas.

**Commensalism**    The type of symbiosis which occurs when one organism benefits and the other is unharmed by the relationship but does not benefit.

**Community**    A group of living organisms living in the same area or having a particular characteristic in common.

**Compact bone**    A portion of bone that is dense and contains very little open space.

**Companion cells**    Are specialized parenchyma cells found in the phloem of flowering plants.

**Compartmentalization**    The formation of cellular spaces, each separate from one another.

**Competition**    The activity that occurs when organisms strive for the same limited resources.

**Competitive exclusion principle**    A principle, which states that organisms will compete with each other in an area until one goes extinct.

**Complementarity**    The specific coupling of bases.

**Compound light microscope**    Microscope that uses two sets of lenses (an ocular and an objective lens).

**Concentration, higher and lower**    The presence of a certain amount of a specific substance in a solution or mixture in a certain concentration. Any solution containing fewer dissolved particles is lower concentration.

**Cones**    A form of photopigment that sends impulses to the brain that give color perception.

**Conjugation**    The process of exchange of genetic material through pili in bacteria.

**Connective tissue**    Tissue that binds and supports different parts of the body.

**Contact inhibition**    Cell's normal ability to come into contact with its neighbors while dividing and this inhibits its growth based on the limited spacing around it.

**Control center**    An operational center for a group of related activities.

**Control group**    A group in a study or experiment not receiving treatment by researchers and used as a benchmark to measure how other tested subjects do.

**Copulation**    The act of placing a male structure into a female's reproductive tract.

**Coral bleaching**    The loss of algae from corals, and resulting coral death.

**Corals**    Marine invertebrates that live in large colonies composed of limestone skeletons.

**Coriolis Effect**    The deflection of a moving object with respect to the Earth 's rotation.

**Cork cambium**    A tissue found in a plant's stem and is responsible for thickening stems and roots.

**Cornea**    Transparent part of the eye.

**Coronary artery**    An artery supplying blood to the heart.

**Coronary artery bypass graft (CABG)**    A type of surgery that improves blood flow in the heart.

**Coronary circuit**    The system in which some vessels branch off and resend blood toward the heart.

**Corpus callosum**    An attachment area that connects the two hemispheres of the cerebrum.

**Corpus luteum**   Hardened mass of an ovarian follicle.

**Correlation**   Relationship between two variables.

**Cotyledon**   The first leaf formed during the initial development of embryos in a seed.

**Covalent bond**   Bonds that result from the equal sharing of electrons between atoms.

**Cranial bones**   The bone that enclose the brain.

**Cristae**   A fold in the inner membrane of the mitochondria.

**Critical thinking**   The analysis and evaluation of an issue to form a judgment

**Critical threshold potential**   The critical level (-55mV) at which the entire neuron fires a nerve impulse across its membrane.

**Crossing over**   The exchange of genes between chromosomes.

**Crustaceans**   An arthropod characterized by having five sets of appendages.

**CT scan**   Computerized axial tomography that produces detailed images of internal organs.

**CTP (connective tissue proper)**   A set of tissue types that act as package materials in the body.

**Cuboidal**   A type of epithelial cell appearing square in shape.

**Cyanobacteria**   Are photosynthetic bacteria that contains bacteriochlorophyll.

**Cytochrome**   Heme proteins that contain heme groups and are responsible for ATP generation through electron transport.

**Cytokinesis**   The division of cell cytoplasm following mitosis or meiosis.

**Cytology**   The study of cell parts.

**Cytoplasm**   A semisolid liquid that holds organelles suspended within it.

**Cytosine**   A type of base found in DNA.

**Cytotoxic T-cells**   Killer T-cells stimulated by T-helper cells to kill specific invaders.

**Data analysis (Qualitative and Quantitative)**   The process of evaluating information that is obtained by investigation. The reporting and use of non-numerical data is qualitative data analysis while reporting and use of numerical data is quantitative data analysis.

**Deaths**   The end of life; those leaving permanently.

**Decomposer**   Organisms that break down once living organic matter into energy.

**Dedifferentiation**   Is the loss of the specialized functions that normal cells perform.

**Deep**   A surface marking that is considered the opposite of superficial and away from a surface.

**Deep vein thrombosis (DVT)**   The condition that occurs when a blood clot forms in one of body's large veins, most commonly in legs.

**Deforestation**   Removal of forested areas.

**Dehydration synthesis**   A process in which hydroxyl and hydrogen atoms are removed from two organic compounds that merges them into one (covalent) bond.

**Demographics** The data collected by ecologists about the statistics of a population of species.

**Dendrite** A thread-like structure of the nerve cell that receives signals from other cells

**Dense irregular** Connective tissue composed of irregularly arranged fibers, such as the dermis of the skin.

**Dense regular** Connective tissue composed mostly of collagen fibers such as ligaments and tendons.

**Density-dependent factors** Factors that limit the population size, whose effects are dependent on the number of individuals of a population.

**Density-independent factors** Factors that limit population size, whose effects do not depend on the number of individuals of a population.

**Dentin** The bony tissue of a tooth.

**Deoxygenated blood** Blood that lacks oxygen.

**Deoxyribonucleic acid (DNA)** A long macromolecule containing the information code that directs cellular activities in living organisms.

**Deoxyribose** The sugar backbone found in DNA.

**Dependent variable** The results of an experiment.

**Depression** Blood or nerve openings in human bones.

**Dermal papillae** A wavy layer of the skin which is also responsible for human fingerprints.

**Dermis** The middle layer of the skin, containing most of its organs and sense receptors.

**Desert** A dry, barren area with little or no rainfall.

**Desertification** The process by which fertile lands experience rapid depletion of flora and fauna becoming a desert.

**Desmosome** One of the three types of connections between cells, is a cell structure specialized for cell-to-cell adhesion.

**Detritus** Dead organic matter that forms as grasses die.

**Deuterostomes** Animals belonging to the group Deuterostomia, in which the body cavity first forms from the back, or anus region.

**Developmental anatomy** The study of changes in structures of an organism since its birth.

**Diabetes Type I** An autoimmune disease resulting from an attack on the pancreatic cells making insulin.

**Diabetes Type II** A medical condition as a result of resistance to insulin by cells.

**Dialysis** A treatment for kidney impairment.

**Diatoms** A single-celled algae that are a major producer of oxygen via photosynthesis.

**Dicot** Angiosperms which produce seeds that contain an embryo with two seed leaves.

**Diffraction** The random scattering of light.

**Diffusion**   The net movement of molecules from higher concentration to a lower concentration.

**Digestion**   The process in which food breaks down mechanically and chemically. Mechanical digestion changes only the size of food particles, making them smaller and easier to digest, while chemical digestion changes the structure of the substances being digested.

**Dioecious**   Flowers that have only a male or a female part, with stamen and carpals on separate plants.

**Diploid (2N)**   The full complement of chromosomes in all body cells (except sex cells).

**Directional selection**   The process that occurs when individuals at one extreme of the range of variation in a population have a higher degree of fitness.

**Directional terms**   Are words that describe a location or position on the human body.

**Disaccharide**   A class of sugars formed when two monosaccharaides combine.

**Disease**   An imbalance in the proper working of a tissue, organ, or organ system.

**Disruptive selection**   The process in which individuals at extremes of the variation spectrum experience higher fitness than at the middle.

**Distal**   Situated away from the point of attachment.

**Diversity**   The adaptation and evolving of organisms showing a great deal of variety.

**DNA**   A long macromolecule containing the information code that directs cellular activities in living organisms; see deoxyribonucleic acid.

**DNA ligase**   A type of enzyme that joins DNA strands together.

**DNA polymerase**   Special enzymes that add new bases onto the exposed DNA strands.

**Domain**   A division of organisms ranking above a kingdom in the systems of classification based on similarities in DNA and not based on structural similarities.

**Dominance hierarchy**   A rank order system developed within an animal society.

**Dominant**   The trait that covers up other forms of the characteristic.

**Dopamine**   A neurotransmitter type that plays an important role in a number of different brain functions.

**Down regulation**   Decrease in the number of effective receptors on cell surfaces.

**Duodenum**   The top portion of the small intestine.

**Dura mater**   The outer layer of meninges.

**Ear drum**   The membrane separating the outer ear from the middle ear.

**Earthquake**   A release of energy from movement in Earth's crust.

**Echinoderms**   A group of marine animals with a spiny skin and an endoskeleton.

**Ecological footprint**   The amount of resources used by a specified group.

**Ecological niche**   The role an organism plays in its environment.

**Ecological succession**   The process by which nature reclaims an ecosystem after it has been disturbed.

**Ecology**   The study of the interactions between organisms and their environments.

**Ecosystem**   The interaction of organisms with their physical environment.

**Ectoderm**   Outer layer of the gastrula, which develops into the skin and nervous system.

**Ectotherm**   Organisms that rely on their environment to set their internal temperature.

**Effector**   A muscle that moves or a gland that sends out chemicals to carry out a response.

**Egg**   The female reproductive cell in plants and animals.

**El Nino**   A band of unusually warm ocean water that develops off the western coast of South America.

**Elastic cartilage**   A type of cartilage that is composed of large amounts of elastin fiber, which is able withstand pulling forces.

**Elasticity**   The ability of a muscle cell to resume its original length after a contraction.

**Electromagnetic energy**   A type of energy released into space by stars (sun).

**Electron**   A negatively charged subatomic particle found in the orbit.

**Electron transport chain (ETC)**   A chemical reaction in which electrons are transferred from a high-energy molecule to lower-energy molecule.

**Electronegativity**   The ability of an atom to attract electrons to itself.

**Element**   Substances that cannot be broken down by ordinary chemical means.

**Elongation**   One of the three phases of transcription in which nucleotides are added to the growing RNA chain.

**Embolus**   A floating thrombus.

**Embryo**   An unborn offspring in the early stages of development; before 12 weeks gestation in humans.

**Embryology**   The study of anatomy before birth; looks at structures of developing embryos and fetuses

**Emigrants**   Organisms leaving an area.

**Emphysema**   A condition in which lungs lose their elasticity and air sacs are hardened, unable to properly exchange gases.

**Enamel**   The strong covering protecting the teeth.

**Endocrine cell**   Any cell that secretes a hormone.

**Endocrine system**   Glands and parts of glands that produce internal chemicals called hormones that cause a response in another organ or tissue.

**Endocytosis**   The process of moving materials into the cell.

**Endoderm**   Inner layer of gastrula forming the digestive and respiratory tracts.

**Endometrium**   A membrane that lines the womb.

**Endoplasmic reticulum (ER)**   Is a system of interconnected membranes that form canals or channels throughout the cytoplasm of a cell.

**Endorphins**   A type of neurotransmitter that improves mood and inhibits pain and depressive feelings.

**Endoskeleton**   A living, internal but hard structure found in animals and humans.

**Endosperm**   The nutritive tissue found inside the seeds of flowering plants.

**Endospore-forming bacteria**   Are Gram-positive, flagellated rods that form endospores to endure harsh, dry conditions.

**Endosymbionts**   Any organism living in the body or cells of another organism.

**Endosymbiotic theory**   The theory that states that some organelles in eukaryotes were descendants of ancient bacteria that were absorbed by larger cells.

**Endothelium**   The squamous epithelial tissue lining the chambers of heart and blood vessels.

**Endotherm**   Organisms that generate heat produced internally by cell respiration to maintain a stable internal body temperature.

**Energy pyramid**   A pictorial representation showing the net loss of energy as it travels up a food chain.

**Entropy**   Randomness or any increase in disorder.

**Enzyme**   Specialized protein that speeds up chemical reactions.

**Epidermal cells**   A type of sponge cell that covers and protects sponges.

**Epidermis**   The outer layer of the skin.

**Epididymis**   An elongated organ that stores sperm and transports them from the testes.

**Epiglottis**   The flap of elastic cartilage covering the trachea.

**Epilimnion**   The top-most layer of lakes and ponds.

**Epinephrine**   A hormone secreted by adrenal medulla.

**Epithelial tissue**   Tissue made of cells that either covers other tissues or cells that produce hormones or other materials for export.

**Equilibrial life history, K-selected strategy**   A type of life history that occurs when parents invest in extended care to their young, live a long time and have few offspring.

**Equilibrium**   The even level of dispersion of substances.

**Esophagus**   Food tube which connects the mouth with the stomach.

**Essential amino acids**   Amino acids that are not synthesized by the body.

**Estuary**   An area where rivers and streams join saltwater.

**Euglena**   A green single-celled, motile freshwater organism.

**Eukarya**   One of the three domains of the biological classification system.

**Eukaryote**   Organisms that contain organelles and a distinct, true nucleus with genetic material contained therein.

**Euler's Buckling equation**   An engineering formula to show where a cylinder is most likely to fail. When applied to a femur, equation shows a thicker area in the region predicted to break.

**Eusociality**   Is the highest level of social organization, in which organisms cooperate and their members have sterile castes.

**Eutherians**   Mammals that develop their embryos internally and nourish them using a placenta.

**Eutrophic**   Shallow and productive lakes.

**Eutrophication**   The massive algal growth resulting from unchecked extra loading of nutrients.

**Evolution**   The change in gene frequencies in a population, over time.

**Excited state**   A state of a physical system that is higher in energy than in its normal state.

**Excretion**   The process of eliminating wastes from organisms.

**Exocytosis**   The process of moving materials outside the cell.

**Exon**   A segment of RNA or DNA that contains information coding for a protein.

**Exoskeleton**   A rigid outer covering of an animal.

**Experiment**   A planned intervention that analyzes the effects of a particular variable.

**Experimental group**   A group in a study or experiment that receives the test variable.

**Expiration**   The process of expelling air to the outside world.

**Exponential model of population growth**   A model that depicts the increase of population growth at a constant rate.

**Exponential period**   A time of rapid and unchecked growth.

**Extant**   Are organisms that exist today.

**External fertilization**   The fertilization process that occurs outside the bodies of animals.

**Extinct**   Are organisms that have died off.

**Extinction**   The state in which a species is lost forever, with no remaining organisms to maintain its population reproductively.

**Extracellular matrix**   Is a collection of proteins and carbohydrates found in every connective tissue type.

**Facial bones**   The bones that attach to the muscles of the face.

**Facilitated diffusion**   A type of passive transport requiring a carrier protein.

**Family**   A group of genera consisting of organisms related to each other.

**Fascicle**   Bundle of muscle fibers.

**Fat-soluble vitamins**   Includes D, A, E, and K, which accumulate in fatty tissues in the body.

**Fermentation**   A special kind of anaerobic respiration yielding low amounts of energy from sugars, when oxygen is not present.

**Fertility rate**   Is the average number of children born to females in a population.

**Fertilization**   Is the process in which male and female sex cells unite.

**Fibers**   Threadlike structure embedded in connective tissue, giving strength and support.

**Fibroblast**   Specialized cells that produce and maintain connective tissue.

**Fibrocartilage**   Cartilage that contains many elastic and collagen fibers in its extracellular matrix.

**Fibrosis**   The scarring or thickening of connective tissue.

**Fibrous protein**   Structural compound that does not dissolve in water but remains solid support in parts of organisms.

**Fibrous root**   A root system made up of numerous branching roots and gives increased exposure to water in soils.

**Filament**   Thread-like structure supports the anther.

**Filtration**   The process by which substances in blood are separated out by pressure through kidneys.

**First law of thermodynamics**   A law that states that energy can be changed from one form to another but cannot be created or destroyed.

**Flagella**   Are long, whip-like extensions on a cell's surface that help in the movement of cells.

**Flame cells**   Specialized excretory cells found in certain invertebrates.

**Flatworms**   Any worm belonging to the phylum Platyhelminthes.

**Fluid mosaic model**   A model that describes the structure of cell membranes.

**Folic acid**   A water-soluble vitamin and a very important nutrient in a fetus's brain and spinal cord development.

**Follicle**   A small ovarian sac that contains a maturing ovum.

**Food chain**   Interactions of organisms with each other through the transfer of nutrients and energy.

**Food Plate (MyPlate guided diet)**   A nutrition guide modeled for healthy eating in the United States.

**Food pyramid**   A pyramid-shaped graphic representation that represents the optimal number of servings to be taken each day.

**Food web**   A network of interdependent and interlocking food chains.

**Formed elements**   The cells and cell fragments formed within the blood and have a definite shape.

**Fossil record**   One of the four sources of evidence for evolution, which shows organisms of the past in rock layers.

**Fragmentation**   The process in which a piece of a parent breaks off and forms a new organism.

**Fragmented meta-population**   Pockets of isolated ecosystems due to massive amounts of encroachment into natural areas.

**Free radicals**   Molecules with extra electrons that cause damage to body structures.

**Freshwater**   Water with low concentrations of dissolved salts.

**Frontal lobe**   The anterior part of the brain that is responsible for much of human personality, intelligence, and skeletal movements.

**Frontal plane**   The plane that divides the front (anterior) and back (posterior) regions of the body.

**Functional group**   A group of atoms that are involved in reactions.

**Fundamental niche**   The area and resources that an organism is theoretically able to utilize.

**Fungi**   Eukaryotic organisms that secrete chemicals to break down other living or once-living materials.

**G₁ phase**   A period in the cell cycle in which a cell grows rapidly in size, forming new organelles and proteins for future daughter cells.

**G₂ phase**   A period in the cell cycle in which growth of the cell's cytoplasm and organelles is completed and final preparations for mitosis takes place.

**G3P**   Also known as glyceraldehyde 3-phosphate, is a chemical substance occurring as a product of the Calvin Cycle.

**Gall bladder**   A small organ on the underside of the liver, which stores bile.

**Gametes**   Reproductive cells which include eggs and sperm.

**Gametophytes**   Haploid organisms that produce the gametes.

**Gap (communicating) junction**   Are channels that run from one cell into another to allow rapid transport helping cells communicate with other cells.

**Gastro-esophageal reflux disease (GERD)**   A chronic condition caused when acids repeatedly escape into the esophagus.

**Gastropods**   Mollusks with an enlarged foot.

**Gastrovascular cavity**   The primary organ of digestion found in Cnidaria and Platyhelminthes.

**Gastrula**   The entire mass of cells of an embryo developing after the blastula stage.

**Gastrulation**   A developmental stage in which three distinct germ layers are formed.

**Gene**   A portion DNA sequence serving as the basic unit of heredity, coding for a polypeptide (protein).

**Gene expression**   The ability of a gene to carry its information to the rest of a cell and perform its directives.

**Gene regulation**   The ability to shut certain genes off and turn some genes on.

**Gene technology**   The technology that modifies plants, bacteria and animals to create products for society.

**Gene therapy**   The process in which genes are inserted into an organism to treat its disease.

**Genetic engineering**   The process in which an organism's genes are manipulated in a way other than is natural.

**Genetic variation**   The sum total of gene differences in a population. An evolutionary reason for sex is to increase genetic variety in a species.

**Genetically -modified organism**   Are organisms in which DNA is genetically altered via genetic engineering techniques.

**Genotype**   The genetic makeup of a cell.

**Genus**   A group of individuals of the same species.

**Geotropism**   The growth of shoots upward and roots downward in response to gravity, results also from plant chemicals.

**Germ theory**   The theory that places a focus on sterile techniques to prevent microbial disease spread, led to important improvements in medicine.

**Germinate**   To begin to grow.

**Gestation**    The process of development in the womb.

**Giant cells**    Enlarged macrophages.

**Glands**    Collections of epithelial cells that secrete a product such as hormones.

**Global climate change**    The many possible changes to Earth's climates due to rises in greenhouse gases in the atmosphere.

**Globular protein**    A type of protein that is water soluble.

**Glomerulus**    A ball of blood vessels at the start of a nephron.

**Glucagon**    A hormone produced by pancreas, which causes the liver to convert stored glycogen into glucose, sent into the blood and thus available for cells to use.

**Glycolysis**    Is a sequence of chemical steps in which glucose is rearranged to form two molecules of pyruvic acid, or pyruvate.

**Gold Foil experiment**    Also called Rutherford's gold foil experiment, is a series of experiments that showed an atom's structure.

**Golgi apparatus**    Is the processing plant of the cell city that refines the materials passing through it.

**Gradient**    The difference between higher and lower concentration areas.

**Gram stain**    A dying technique that identifies bacteria as being one of two categories.

**Gram-negative bacteria**    A group bacteria that lose the crystal violet dye in Gram staining method of bacterial differentiation.

**Gram-positive bacteria**    A group of bacteria that retains the dye in Gram staining method of bacterial differentiation.

**Greenhouse effect**    The process of trapping heat energy in the atmosphere.

**Greenhouse gases**    The gases in the atmosphere that slow the release of heat from the planet to space by absorbing and re-emitting long wave radiation heat back to the surface.

**Gross anatomy**    The study of body parts that can be seen without use of microscopy.

**Ground state**    The lowest state of energy of a particle.

**Ground tissue**    Tissues that are neither vascular nor dermal and support a plant's structure and store and produce food.

**Group behavior**    A passive adaptation technique to minimize the effects of predation.

**Guanine**    A purine base that functions as a fundamental constituent of RNA and DNA.

**Gymnosperm**    Plants with seeds that do not develop in an ovary, usually cone producing.

**Habitat**    The space an organism occupies, including all of the factors with which an organism interacts.

**Habituation**    The process of "getting used to" stimuli.

**Hair follicle**    A structure from which hair grows.

**Hair root**    Part of hair embedded in a hair follicle.

**Halophiles**    Are organisms that grow or live in very salty conditions.

**Hammer**   A bone that is the outermost of the three small bones in the middle ear.

**Haplodiploidy**   Sex-determining method in which females are developed from fertilized eggs (diploid) and males are developed from unfertilized eggs (haploid).

**Haploid (N)**   The half number of chromosomes of the parent.

**Heart**   A specialized muscular organ that propels blood through the body of vertebrates.

**Heart attack (myocardial infarction)**   The condition in which heart muscle is damaged from the sudden blockade of coronary artery by blood clot.

**Heat capacity**   The amount of heat energy required to raise the temperature of an amount of substance.

**Helicase**   The enzyme that untwists the double helix so that replication can occur.

**Hemagglutinin**   A type of protein that enables Myxovirus to bind with its host.

**Hemophilia**   A condition of uncontrolled bleeding in which blood clots occur too slowly.

**Herbivore**   Organisms that eat plants.

**Herbivory**   The consumption of plants and plant parts by other organisms.

**Heredity**   The passing of characteristics from parent to offspring.

**Hermaphrodite**   A person or animal having both male and female reproductive parts.

**Herpes simplex I**   An inflammatory skin disease characterized by the formation sores around the lips.

**Herpes simplex II**   A sexually transmitted virus and disease that is characterized by genital sores.

**Heterotrophs**   Also called consumers, these organisms acquire energy by eating other organisms.

**Heterozygous**   The condition in which alleles a pair are different from each other.

**High blood pressure**   A chronic elevation of pressure above the normal 120/80, for a consistent period of time.

**Hippocampus**   Part of the brain involved in short term memory inputs, acting as a conduit to long term memory storage in the cerebrum.

**Histamines**   Chemicals that bring more blood to a site of infection by vasodilation of the vessels surrounding the area.

**Histology**   The study of tissues.

**Histone**   Group of basic proteins in chromatin, around which DNA coils.

**Homeostasis**   Maintaining a steady set of environmental conditions.

**Homeotherm**   Organisms that maintain a stable internal body temperature.

**Hominids**   A primate belonging to the family Hominidae.

**Homologous structures**   Similar structures found in different species.

**Homologous**   The chromosome partners in a diploid cell.

**Homology**   Common ancestry.

**Homozygous**   The condition in which a pair of alleles is the same.

**Hormones**   Chemical messengers that cause change or direct activity in another area of the body.

**Humerus**   Upper arm bone.

**Humoral immunity**   A form of immunity in which plasma cells and B lymphocytes produce antibodies.

**Hyaline cartilage**   A type of cartilage that is composed of large amounts of collagen fibers, giving it strength.

**Hydra**   Freshwater organisms with a set of tentacles on the outside of their coelenterate opening.

**Hydrochloric acid (HCl)**   An aqueous solution of hydrogen chloride.

**Hydrogen bond**   Are fleeting bonds that form between hydrogen atoms and atoms of different structures. These bonds are based on attraction between positive and negative charges.

**Hydrolysis**   The breakdown of a compound due to its reaction with water.

**Hydrophilic**   Compounds that have the tendency to dissolve in or mix with water.

**Hydrophobic**   Compounds that do not dissolve in water (also called, water fearing).

**Hydrosphere**   All of the water on Earth's surface.

**Hydrostatic skeleton**   A type of skeleton found in earthworms and jellyfish, which use water and muscles for support and movement.

**Hydroxyapatite**   An essential component and major ingredient of normal bone.

**Hyper-disease theory**   The theory that states that a microbe evolved rapidly to kill off other living creatures during the time period.

**Hypersexuality**   The condition in which one has many sex partners.

**Hyperthyroidism**   An overactive thyroid, which results in too much thyroxine, causing nervousness, excess energy, sometimes enlarged eyes and irregular heart rates.

**Hypertonic**   A cell having a higher concentration of solutes as compared with its surrounding environment.

**Hypha**   Each of individual threads that make up the fungal mycelium.

**Hypodermis**   The deepest layer of skin, composed mostly of fat.

**Hypolimnion**   The lower-most layer of lakes and ponds.

**Hypothalamus**   The region below the thalamus.

**Hypothesis**   A possible or proposed explanation based on limited evidence for a natural phenomenon.

**Hypothyroidism**   An underactive thyroid, which results in weight gain, intolerance to cold and higher cholesterol.

**Hypotonic**   A cell having a lower concentration of solute as compared with its surrounding environment.

**Ileum**   The third and final portion of the small intestine.

**Imbibition**   The process in which germination starts with the massive influx of water into the seed.

**Immigrants**   New organisms moving in from other areas.

**Immune system**   A system that includes the set of disease-fighting factors that protect against pathogens.

**Immunization**   The technique that uses weakened or dead pieces of disease-causing agents to strengthen immunity against a disease.

**Immunodeficiency**   The most serious malfunction of the immune system; occurs when it does not work efficiently.

**Impaired kidney function**   The failure of kidneys to function properly.

**Incisors**   The four front teeth evolved and adapted for cutting and tearing.

**Incomplete dominance**   A genetic situation in which one allele does not completely dominate another allele.

**Independent variable**   A variable that is altered by the experimenter.

**Inflammation**   A series of events which identify, recruit and attack invading cells, causing swelling.

**Ingestion**   Consumption of a substance by living organisms.

**Initiation sequence**   A sequence of bases that starts the unwinding of DNA during replication.

**Innate behaviors**   Behaviors that are genetically determined.

**Inner cell mass**   A central set of cells of the blastula, which develop to become a new, whole organism.

**Innervation**   The distribution of nerves to a muscle.

**Inquiry**   Critical thinking used behind science to arrive at the truth.

**Insects**   Small invertebrates with a head, thorax, abdomen, six legs, and one or two pairs of wings.

**Insertion**   Movable end of a muscle; the bone or part that moves when a muscle contracts.

**Insoluble fiber**   Fibers that do not dissolve in water and serve as roughage to cleanse the intestines.

**Inspiration**   The process of taking of air into the lungs.

**Insulin**   A hormone produced in the pancreas that regulates the amount of glucose in blood.

**Integral protein (transmembrane or carrier protein)**   A type of membrane protein permanently embedded within the biological membrane.

**Interdisciplinary**   Involves two or more areas of knowledge.

**Interferon**   Small proteins which bind with receptors on neighboring cells.

**Intermediate fiber**   The smallest fibers of the cytoskeleton, which circulate materials within a cell.

**Intermembrane space**   The area of a mitochondrion found outside of cristae.

**Internal fertilization**   The fertilization process that occurs inside the bodies of animals.

**Interneuron**   A neuron that transmits impulses between other neurons.

**Interphase**   The stage in cell development in between two successive mitotic or meiotic divisions

**Interspecific competition**   The competition between two different species.

**Interstitial cells**   Cells that produce androgens, sex hormones.

**Intertidal zone**   The area that is above water level at low tide and below water level at high tide.

**Intertropical -convergence zone**   The low pressure area at the equator resulting from the expansion and rising up of hot air mass.

**Intervertebral disc**   Pads of fibrocartilage that separate individual vertebrae.

**Intestinal cells**   Cells lining the GI tract.

**Intimidation displays**   Threat display that makes an organism look large and intimidating.

**Intracellular digestion**   The process of breakdown of substances within the cytoplasm of a cell.

**Intracellular parasite**   Living organisms that invade host cells and live within them.

**Intracellular transport**   A process in which microfilaments circulate materials within cells.

**Intraspecific competition**   The competition between organisms of the same species.

**Intron**   A nucleotide sequence removed by RNA splicing.

**Invagination**   The process of being folded back on itself to form a pouch (not given in bold in text).

**Invertebrates**   Animals that lack a backbone.

**Ionic bond**   Bonds that result from complete transfer of electrons from one atom to another.

**Iris**   The colored part found around the pupil of the eye.

**Iron deficiency anemia**   A condition characterized by lack of healthy RBCs in blood.

**Isotonic**   Even concentration of solute and water on either side of the plasma membrane.

**Isotope**   Are atoms of the same element having different atomic masses.

**Jaundice**   A disease characterized by yellow coloration of the skin.

**Jejunum**   The second part of the small intestine.

**Jellyfish**   Free-swimming marine creatures that have a central cavity making them appear as cup-like.

**Keratin**   A protein that is the principal constituent of nails, hair, and skin tissues.

**Keratinocyte**   An epidermal cell that produces keratin granules.

**Kidney failure**   A medical condition in which the filtration rate falls to 50 percent or below.

**Kidney**   Organ that filters blood, removes wastes, and at the same time conserves needed materials including water.

**Kin selection**   Natural selection in which individuals help each other because they share genes in common.

**Kin selection**   The theory that evolution favors helping between family members or kin to augment the transmission of their related genes.

**Kingdom**   The highest grouping under which living organisms are classified.

**Krause's corpuscle**   A bulbous cell that senses cold.

**Krebs cycle**   A series of enzyme-catalyzed reactions forming an important part of aerobic respiration in cells.

**La Nina**   Cooling of the ocean surface off the western South American coast.

**Labor**   The opening of the cervix and uterus contractions leading to the birth of the baby.

**Lactation**   Formation of milk by mammary glands.

**Lacteal**   A lymphatic vessel of the small intestine that absorbs digested fats.

**Lacuna**   An open space containing a chondrocyte in cartilage.

**Langerhans cell**   Special white blood cells that reside within the skin.

**Language**   A method of communication.

**Larvae**   The active immature form of an insect.

**Larynx**   The part of throat containing the vocal cords.

**Larynx**   Voice box.

**Lateral meristem**   A type of meristem that is found along the sides of stems and roots which gives rise to secondary growth.

**Lateral**   Of or relating to the side.

**Law of dominance**   The idea that a dominant trait covers up another.

**Law of independent assortment**   The idea which tells that each pair of alleles is sorted independently when sperm and egg are formed.

**Law of segregation**   The hypothesis that states that there are two separate, discrete alleles that could be inherited separately.

**Leaf abscission**   Loss of leaves.

**Learned behaviors**   Behavior that develops through experience.

**Lens**   A very hard structure of the eye but flattens to focus the rays of light passing through it.

**Lichens**   Green algae or cyanobacteria living in association with fungi.

**Life history**   Series of changes an organism undergoes during its lifetime.

**Ligament**   Band of tissue that anchor bones to bones.

**Light reactions**   A reaction that traps energy from sunlight using special pigments.

**Light-independent reactions**   Chemical reactions that convert carbon dioxide into glucose.

**Limbic system**   A group of brain structures found on both sides of the thalamus, along the inner core of the brain.

**Limbs**   An arm or a leg of a person or animal.

**Limnology**   The study of freshwater systems.

**Lipid**   Neutral fats, phospholipids, and steroids found in food and in living systems.

**Lithosphere**   Earth's outer part.

**Liver**   A large glandular organ found in the abdomen of vertebrates.

**Lobe-finned fishes**   A smaller group of fishes having a developed pelvis, primitive lungs and muscular fins — precursors to life on land.

**Logistic model of population growth**   A model that depicts the decrease of population growth rate with the increasing number of individuals as they reach a certain point.

**Loop of Henle**   A large tube in the nephron that descends and then ascends.

**Lumbar vertebrae**   The five vertebrae that make up the lower back

**Lungs**   A pair of breathing organs.

**Lymph**   The fluid of the body that carries excess liquids and cells not normally transported by the circulatory system.

**Lymphatic system**   The series of vessels, nodes and their organs carrying lymph.

**Lymphocyte**   White blood cells that work to specifically target invaders.

**Lyse**   Breakdown of cell membrane.

**Lysogenic life cycle**   A reproduction cycle during which a virus inserts its genes into a host and waits for a time in the future to destroy the host.

**Lysosome storage disease**   A group of 30 known inherited human diseases associated with the abnormal functioning of lysosomes.

**Lysosome**   A small sac filled with digestive, hydrolytic enzymes enclosed in a membrane.

**Lytic life cycle**   A reproduction cycle which results in a virus's immediate destruction of a host cell.

**Macrobiology**   The study of how organisms interact with each other and within the environment

**Macromolecules**   Molecules containing large number of atoms, which are the building blocks of living things.

**Macronutrients**   Macromolecules that possess energy within their bonds.

**Macrophage**   The largest of the white blood cells.

**Macrophage--presentation**   The display of antigens by a white blood cell.

**Magnification**   Is the amount by which an image size is larger than the object's size.

**Malaria**   An infectious disease spread by mosquitos carrying a parasite that invades red blood cells and reproduces in them.

**Malignant**   The ability of a cancerous cell to spread.

**Mammals**   Are homeotherms that produce milk to feed their young.

**Mandible**   Jawbone.

**Manipulation**   Manual movement of anatomical parts to either help treat symptoms or diagnose the cause of a disease or injury.

**Mantle**   One of the three-point body plans of mollusks that secretes the outer shell.

**Marine biology**   The study of saltwater organisms.

**Marsupials** A type of mammal in which young ones are born immature and continue to develop in a pouch.

**Mast cell** A type of immune cell.

**Matrix** The inside space within the cristae.

**Matter** Anything that has mass and occupies space.

**Mechanical breathing** The process of moving air into and out of the lungs

**Mechanoreceptors** A sense organ responding to physical changes.

**Medial** Situated in the middle.

**Medulla oblongata** The inner part of the brain.

**Medusa stage** Cnidarians in their free swimming stage.

**Meiosis** A special form of cell division in which the newly produced daughter cells contain only half the number of chromosomes of the parent.

**Meiosis I** The process of cell division by which homologous chromosomes separate and new cells are haploid.

**Meiosis II** The stages in which sister chromosomes are separated.

**Meissner's corpuscle** A receptor that senses light touch.

**Melanin** The pigment that gives color to human eyes, hair, and skin

**Melanocyte** Cells that make the skin pigment, melanin.

**Melanoma** The most dangerous form of skin cancer.

**Melatonin** A hormone produced by the pineal gland that makes a person sleepy.

**Membrane** A sheet-like structure that acts as a boundary in an cell.

**Memory cell** Type of lymphocytes that continue to defend against pathogens long after they are gone.

**Menarche** Beginning of menstruation.

**Mendelian characteristic (single-gene trait)** Traits that are determined by instructions on a single gene.

**Meninges** A series of protective membranes that surround the spinal cord nerves.

**Menopause** The time when ovulation and menstruation cease.

**Menstrual cycle** The process of ovulation in women and other female primates.

**Menstruation** Discharge of blood from the uterus.

**Meristem** A formative plant tissue responsible for growth whose cells divide to form plant tissues and organs.

**Mesenchyme** Embryonic connective tissue that gives rise to all the connective tissues.

**Mesoderm** The middle layer of gastrula that develops into the body's organs and muscles.

**Mesoglea**    The gelatinous filling found in between the two cell layers in the bodies of Cnidarians and sponges.

**Mesopelagic zone**    The ocean layer that receives little sunlight.

**Metabolism**    Chemical processes occurring in a living organism that are necessary for life maintenance.

**Metalimnion**    The middle portion of lakes.

**Metamorphosis**    A complete change of physical form.

**Metaphase**    A phase in which homologous chromosomes line up at the middle of the nucleus, the equator, attaching to the spindle fibers.

**Metaphase II**    The stage of meiosis in which chromosomes line up singly and then the two sister chromatids separate and move to opposite poles of the cell.

**Metastasize**    The process in which cancer cells spread to other parts of the body.

**Methanogens**    Organisms that react to oxygen as a poisonous substance.

**Microfilament**    A cytoskeletal fiber used for muscle movement.

**Micronutrients**    Chemical substance required in small quantities, namely vitamins and minerals.

**Microscopic anatomy**    The study of structures too small to be seen with the naked eye

**Microtubule**    Is a larger filament structure that helps whole cells move.

**Microvilli**    Smaller villi; used to increase surface area for absorption within intestines.

**Midbrain**    The short part of the brainstem above the pons.

**Middle ear**    Middle ear bones found inside the eardrum.

**Mimicry**    The resemblance of one organism to another.

**Minerals**    Are inorganic substances that form ions in the body, which help to perform many functions.

**Mitochondria**    Is the organelle that makes energy for a cell.

**Mitral (bicuspid) valve**    A heart valve located between the left atrium and left ventricle.

**Molars**    A grinding tooth found at the back of the mouth and suited for grinding and chewing foods.

**Molecular genetics**    A new field that united biology, chemistry and genetics, to study inheritance at the chemical level.

**Molecules**    Atoms bonded together.

**Mollusk**    Invertebrates, chiefly marine, characterized by a soft unsegmented body and an external hard shell.

**Molt**    To shed the outer covering.

**Monocot**    Angiosperms-produced seeds that contain embryo with one seed leaf.

**Monoecious**    Flowers that have both male and female parts.

**Monohybrid cross**    The mating between two organisms, each having both characteristics for a particular trait.

**Monosaccharide**   Ring-shaped structures that are the building blocks of carbohydrates.

**Monotremes**   Primitive mammals that lay eggs.

**Morphology**   A particular structure or shape.

**Morula**   A solid ball of cells formed by cleavage of a fertilized ovum.

**Motor neuron**   A nerve cell that brings nerve impulses from the brain and spinal cord to a muscle.

**Mouth**   An opening in the lower part of the face.

**MRI**   A technique that uses radio waves and magnetic field to generate detailed images of tissues and organs.

**Multiple alleles**   A series of three or more alternative forms of a gene, out of which only two can exist in a normal, diploid individual.

**Murmur**   An abnormal sound made by blood during the heartbeat cycle.

**Muscle fibers**   The functional muscle cell.

**Muscle tissue**   A type of tissue that is composed of cells that are able to contract.

**Muscular foot**   One of the three-point body plans of mollusks, used for movement.

**Muscular pump**   A collection of skeletal muscles that aid the heart in blood circulation.

**Mutualism**   A relationship in which both organisms benefit.

**Mycelium**   The mass of filaments that form the vegetative part of a fungus.

**Mycoplasma**   The smallest known bacterium.

**Myelin sheath**   Pads of insulation that prevent the action potential from weakening.

**Myofibrils**   A rod-like protein structure in a muscle cell.

**Myxovirus**   Any group of RNA-containing viruses.

**NADH**   Nicotinamide adenine dinucleotide is a naturally occurring biological compound, which is converted to energy (not given in bold in text).

**NADPH**   Nicotinamide adenine dinucleotide phosphate is used as reducing agent in reactions.

**Natural selection**   The process whereby organisms better adapted to their environment survive and produce more off spring.

**Necrosis**   Death of tissue.

**Negative feedback**   A key mechanism that regulates the physiological functions in living organisms.

**Nematocysts**   Barbed threads found in tentacles of Cnidarians.

**Neolithic diet**   The prehistoric nutrition system.

**Nephridia**   Excretory organs found in many invertebrates.

**Nephron**   The functional unit of the kidney.

**Neritic zone**   The zone of ocean where sunlight reaches the ocean floor.

**Nerve cord**   A dorsal tubular cord of nervous tissue present in chordates.

**Nerve impulse**   Neuron message signals that are actually a flow of charged ions.

**Nerve network**   Set of nerve cells that help Cnidarians respond to stimuli.

**Nervous system**   Network of nerve cells that transmits messages from one part of the body to another.

**Nervous tissue**   An excitable tissue specialized to send, store, and receive ionic impulses

**Neuraminidase**   A protein found in Myxovirus that digests through mucous membranes.

**Neuroglia**   Helper nerve cells present in the nerve tissue.

**Neuron**   A nerve cell.

**Neurotransmitter**   Special chemicals that carry a nerve impulse to new cells.

**Neurulation**   The process by which the ectoderm folds to become the brain and spinal cord.

**Neutral fat**   A fat that is composed of three large fatty acids joined together by a short-chained glycerol molecule.

**Neutron**   Particles with zero charge found in the nucleus.

**Neutrophils**   A type of WBC that are the most abundant in mammals and are first to arrive at an invasion.

**Nitrification**   The process by which bacteria in soils produce nitrites.

**Nitrogen fixation**   The process of converting molecular nitrogen ($N_2$) into ammonia ($NH_3$) and ammonium ions ($NH_4^+$).

**Nitrogenous base**   A nitrogen containing molecule having the same chemical properties as a base.

**Nociceptors**   A sense organ responding to pain.

**Non-amniotes**   Terrestrial animals that develop their eggs in the absence of an amniotic sac.

**Nonessential amino acids**   Amino acids made by the human body and thus are not required in a diet to survive.

**Notochord**   A flexible rod of nerve tissue that develops in all chordates.

**Nuclear envelop**   The double membrane that protects the nucleus.

**Nucleic acid**   The genetic material of a cell.

**Nucleoli**   A small, dense round structure found in the nucleus of a cell.

**Nucleoside triphosphate**   A molecule that contains a nucleoside bound to three phosphates.

**Nucleotide**   The basic functional unit of a DNA molecule.

**Nucleus**   The central and the most important part of a cell and contains the genetic material.

**Null hypothesis**   The hypothesis that asserts that there is no effect or change due to a potential treatment.

**Nutrients**   The substances that the body uses to obtain energy and to maintain the body's activities, such as growth, repair, and reproduction

**Obesity**    The state of being overweight with a BMI greater than 30.

**Obligatory parasite**    Organisms that are unable to live outside of a host cell.

**Oblique plane**    A plane running at an angle to the organ or organisms.

**Observation**    The act of obtaining information from a primary source.

**Occipital lobe**    The posterior lobe of the brain.

**Octet rule**    A chemical rule that reflects how atoms react to attain eight electrons in their valence shell.

**Olfaction**    The sense of smell.

**Oligochaetes**    Aquatic and terrestrial worms.

**Oligotrophic**    Lakes with low levels of productivity and abundance of dissolved oxygen.

**Omnivore**    Organisms that eat both plants and meat.

**Oncogene**    A normal gene that under certain circumstances can cause cancer.

**Oncovirus**    Any virus that carries a gene associated with cancer.

**Oogenesis**    A process which takes place in cells of the ovaries.

**Open sea zone**    The main body of ocean or sea.

**Operant conditioning**    A complex set of behaviors during which an organism learns to respond to stimuli to produce a desired effect.

**Opportunistic life history, r-selected strategy**    Type of life history when parents have many young and invest very little in each.

**Opsonization**    The process by which antibodies often coat the invading pathogen to enable macrophages to attach more easily.

**Order**    Used in classification as a group of families.

**Organ System**    A group of organs working together performing a united function.

**Organelle (subcellular structure)**    Structures that function within cells in a discrete manner

**Organelles**    Structures that carry out specific functions within cells.

**Organism**    Living creature formed as a whole by organ systems.

**Organs**    Specialized body parts that carry out specific functions for organisms.

**Origin**    The location (bone) at which muscles attach.

**Osmoregulation**    The process by which organisms control their fluid intake along with dissolved solute balances.

**Osmosis**    The process of diffusion of water through a semipermeable membrane that allows the transfer of only some substances.

**Osteoarthritis**    A disease in which bones and their joints deteriorate, usually because the cushioning cartilage in between these bones wears out.

**Osteoblast**    Special bone building cells.

**Osteoclast**    Bone destroying type cell.

**Osteoporosis**    Thinning and weakening of bones.

**Outer ear**    Pinna and eardrum.

**Oval window**    An oval-shaped opening that is the start of the inner ear.

**Ovarian cycle**    The monthly cycle by which eggs are developed and released from the ovary between puberty and menopause.

**Ovary**    A female reproductive organ containing ovules in which eggs develop.

**Oviduct (fallopian tube)**    Tube through which an ovum passes from an ovary.

**Ovulation**    The process of producing and discharging eggs from ovary.

**Ovule**    The female gametophyte.

**Ovum**    A mature egg.

**Oxygen revolution**    The biologically induced appearance of dioxygen in Earth's atmosphere 2.5 billion years ago.

**Oxyhemoglobin**    A bright red complex of oxygen and hemoglobin present in oxygenated blood.

**Oxytocin**    A hormone released by the pituitary gland that enhances the stimulation of muscle contraction in the uterus.

**Ozone**    A gas that is a pollutant in Earth's lower atmosphere, but acts as a protector from UV radiation in the upper atmosphere.

**Pacinian corpuscle**    A receptor that senses deep pressure.

**Palpation**    The act of feeling with one's hand.

**Pancreas**    A diffuse gland located near the stomach.

**Pancreatic juice**    A secretion of the pancreas that contains enzymes that digest all of the macromolecules.

**Papillomavirus**    A group of viruses that cause papillomas or warts.

**Paracrine regulators**    Chemicals that bind to receptors on neighboring cells to elicit a response.

**Parasitism**    The symbiotic relationship in which one organism benefits and the other is harmed.

**Parasitoid**    An organism living that spends some period of its development on or in a host organism and later kills its host.

**Parasympathetic nervous system**    Opposing set of nerves that are stimulated when the body calms down, under relaxing conditions.

**Parathyroid hormone (PTH)**    Hormone produced by the parathyroid glands help in regulating the amount of phosphorous and calcium in the body.

**Parenchyma cells**    The typical plant cells that carry out most of the metabolism in plants.

**Parietal lobe**    One of the four major lobes of the brain that contains an area concerned with higher levels of thought (speech, sensation, and sensory integration).

**Parthenogenesis**   A virgin birth.

**Parturition**   The contractions and dilation of the cervix during child birth; the act of giving birth.

**Passive artificial**   The condition that occurs when a medicine gives immunity to a patient without stimulating their immune system.

**Passive natural**   The condition that occurs when immunity is obtained naturally but without the work of the immune system.

**Passive transport**   The movement of substances across cell membranes without the need of energy expenditure by the cell.

**Pathogen**   Any disease causing organism.

**Pear shape**   A body shape characterized by extra weight around the hips.

**Pedigree**   Are diagrams of genetic relationships among family members through different generations; they are used to trace gene flow through a family.

**Penicillin**   An antibiotic obtained from the molds of the Penicillium genus.

**Penis**   The male reproductive organ.

**Pepsin**   An enzyme produced in the stomach which breaks down protein.

**Pepsinogen**   The inactive precursor to pepsin.

**Peptidoglycan**   Are a type of protein found in bacterial cell walls.

**Peripheral nervous system (PNS)**   The portion of the CNS that is outside the brain and the spinal cord.

**Peripheral protein**   Is a protein that adheres only temporarily to the biological membrane with which it is associated.

**Peristalsis**   The involuntary muscular contractions of the digestive tract by which contents are forced onward.

**Permafrost**   A layer of permanently frozen subsoil.

**pH scale**   A numeric scale that specifies the acidity or alkalinity of an aqueous solution.

**Phagocytosis**   The movement of solid particles into a cell.

**Pharyngeal slits**   Openings in the pharynx that develop into gills in some chordates.

**Pharynx**   A tube that starts behind the nose and mouth connecting to the esophagus.

**Phenotype**   The observable traits of an organism.

**Pheromone**   A chemical that travels between different organisms to interconnect them.

**Phloem**   A series of tubes that carry sugars and dissolved organic materials down a plant.

**Phospholipid**   A lipid composed of both a charged phosphate group and fatty acid chains.

**Photic zone**   The part of ocean where sunlight penetrates sufficiently and influences the growth of living organisms.

**Photo pigments**   Special pigments found in the retinal rods and cones.

**Photolysis**   The process in which water is split to yield free electrons and hydrogen ions to replace electrons lost from a photosystem.

**Photon**   Discrete unit of light energy that when hits a pigment in chlorophyll transfers its energy to electrons in the pigment

**Photosynthesis**   The process by which green plants (plus some algae and bacteria) use sunlight to synthesize nutrients from water and carbon dioxide.

**Photosystems**   A light capturing bundle of pigments which absorbs light for photosynthesis.

**Phototrophic anaerobic bacteria**   A group of bacteria that do not release oxygen in their photosynthetic-like processes because the photolysis of water does not occur.

**Phototropism**   A tropism in which the growth of a plant is toward sunlight.

**Phylum**   Number of similar classes grouped together.

**Physiology**   The study of the function of body parts.

**Phytoplankton**   All the aquatic organisms that absorb carbon dioxide and release oxygen into the atmosphere.

**Pia mater**   The inner layer of meninges.

**Pigment**   A naturally occurring special chemical that absorbs and reflect light.

**Pili**   Surface hairs that allow bacteria to bind with each other.

**Pineal gland**   A small gland located deep within the brain.

**Pinna**   Projecting part of the external ear.

**Pinocytosis**   The mechanism by which cells ingest extracellular fluid and its contents.

**Pituitary gland**   The master gland of the endocrine system that sends messages to stimulate all of the other glands.

**Placenta**   An organ inside the mother's body that provides food and removes the waste of a developing organism.

**Plants**   Living organisms that are able to obtain food by converting sunlight's energy to chemical energy through the process of photosynthesis.

**Plasma**   The straw-colored liquid that makes up 55 percent of the blood.

**Plasma (cell) membrane**   A biological membrane that separates the cell's interior from the outside environment.

**Plasma cell**   A type of lymphocyte produces antibodies.

**Plastid**   Organelle, found only in plants and algae, which store special substances.

**Platelets**   Are chips of cells that form clots within vessels to prevent blood loss.

**Pleiotropy**   The condition in which one gene affects more than one trait.

**PNS (peripheral nervous system)**   The portion of the nervous system situated outside the brain and spinal cord.

**Polar covalent bond**   The unequal sharing of electrons between atoms.

**Polar easterlies**   Dry, cold winds that are deflected to the west like the trade winds.

**Polar ice caps**   Dome-shaped ice sheets that slope in all directions from the north and south poles.

**Pollen grains**   The male gametophyte.

**Pollination**   Movement of pollen from one plant to another.

**Pollution**   The introduction of any contaminant into the environment that causes a harmful change.

**Polyatomic ion**   A special kind of ion, composed of more than one atom, forming a charge.

**Polychaetes**   A marine annelid worm.

**Polygenic traits**   Are traits with patterns of inheritance determined by more than one gene and influenced by the environment.

**Polyp stage**   The stage in which Cnidarians are sessile.

**Polypeptide**   A long string of amino acids formed as molecules of protein adds amino acids.

**Polysaccharide**   The combination of three or more monosaccharides.

**Pons**   Part of the brainstem that helps relay impulses from cortex to cerebellum.

**Population**   A group of organisms of the same species living in a given area.

**Population density**   A measurement of population that reveals the number of organisms per area of land in an ecosystem.

**Population ecology**   The study of a population of organisms and how it interacts with its environment.

**Population genetics**   The study of patterns of gene flow from one group to another and within groups.

**Population growth**   The increase in the number of individuals inhabiting a place.

**Population**   Organisms of the same species inhabiting a specific area.

**Population size**   A measurement of population that gives the number of organisms in a population.

**Porifera**   A phylum of aquatic invertebrates that is comprise of sponges.

**Porphyria**   An inherited disease which is characterized by abnormal metabolism of the blood hemoglobin.

**Positive feedback**   A key regulatory mechanism that enhances the original stimulus.

**Post-anal tail**   An extension of the spinal cord that extends beyond an animal's normal digestive tract at some point in development.

**Posterior**   Backside.

**Powerstroke**   Movement of muscle filaments using ATP during the contraction of muscle.

**Pre-molars**   Teeth situated between canine and molar teeth and suited for grinding and chewing foods.

**Prebiont**   A sphere of organic material that led to first living cells.

**Predation**   When one organism stalks and kills an organism of another species.

**Predator-prey relationship**   The connection between two organisms of unlike species.

**Primary electron acceptor**   An electron acceptor in a substance that can be reduced by gaining an electron from some other particle

**Primary succession**   The series of stages that begin ecological succession, including lichens and mosses as first organisms in a disturbed area.

**Prions**   A small, infectious particle believed to be the smallest disease-causing agent.

**Producer**   Organisms that carry out photosynthesis.

**Productivity**   The production rate of new biomass by a person or community.

**Projection**   Sites for muscle attachment or joint connections.

**Prokaryote**   Organisms that lack a distinct nucleus and organelles.

**Proliferation**   Rapid growth of cells by producing new parts.

**Prophase**   A stage that is characterized by chromatin being packaged into chromosomes in a cell's nucleus.

**Prophase I**   Also called the first stage of meiosis I, in which homologous chromosomes in proximity to each other exchange genetic material through a process called crossing over.

**Prophase II**   The first stage of meiosis II.

**Prostate gland**   A gland located just below the urinary bladder that secretes a milky and basic fluid to buffer the effects of the acidic female environment sperm will first encounter.

**Protein**   The most common macromolecule in living systems.

**Prothallus**   The gametophyte generation of ferns.

**Protime**   A blood test that measures the rate at which blood clots.

**Protista**   A diverse group composed of both single-celled and multi-celled organisms.

**Proton**   A subatomic particle found in the nucleus, which is positively charged.

**Protostomes**   Organisms that form their mouth first.

**Protozoan**   A group of single-celled protists that resemble animals.

**Proximal**   Situated close to a point of attachment.

**Pulmonary artery**   The artery that carries blood from the right ventricle to the lungs.

**Pulmonary circuit**   The connection between the heart and lungs.

**Pulmonary embolism**   The condition in which an embolus lodges in the lungs.

**Pulmonary vein**   A vein that carries oxygenated blood from lungs to the heart's left atrium.

**Pulp cavity**   A cavity within the dentin containing blood vessels and nerves.

**Pupa**   The stage between the larval and adult stage, in a cocoon.

**Pupil**   The opening in the center of iris.

**Purkinje fibers**   Specialized heart muscle fibers that carry electrical impulses controlling the contraction of ventricles.

**Pus**   A yellow fluid emerging from a site of infection.

**Pyloric sphincter**   Muscle fibers around the stomach opening between it and the duodenum.

**Pyrogene**   A chemical that causes fever.

**Radial symmetry**   Symmetry that describes any organism that is structured so that when a line is drawn down the middle of it, at any orientation, both sides are identical.

**Radiant energy**   A type of energy travelling by waves or particles.

**Rain shadow desert**   The dry region of land on one side of a mountain; has very little precipitation.

**Ray-finned fishes**   Fishes characterized by skeletal rays emanating from their central backbones.

**Reabsorption**   The process by which some materials that were filtered out by the kidneys are returned back into the blood.

**Realized niche**   The area and resources that an organism is actually able to use.

**Receptor**   A special protein that monitors or receives information from the environment.

**Receptor-mediated endocytosis**   A form of endocytosis that requires a specific binding of a receptor protein to cell membrane.

**Receptor-hormone match**   Match making between the receptors of target cells and their respective hormones.

**Recessive**   The trait that is covered up by a dominant trait.

**Reciprocal altruism**   A behavior in which an organism helps another and the second organism returns the favor either to the benefactor or his/her progeny.

**Recognition protein**   A protein type functioning as binding site for hormones.

**Recombinant DNA technology**   The process by which DNA is extracted from nuclei of organisms and treated with restriction enzymes.

**Rectum**   The final part of the large intestine.

**Red blood cells**   Blood cells that contain hemoglobin and carry oxygen to and from the tissues.

**Referred pain**   Pain that is not at the site of its cause.

**Regeneration**   Replacement of damaged tissues with an original tissue type.

**Regulation**   Control over functions of the body.

**Regurgitation**   The most common valve problem; a backflow of blood from a valve.

**Relaxation**   A state of freedom from skeletal muscle tension and anxiety.

**Renal capsule**   A fibrous layer surrounding the kidney, affording it some protection.

**Renal cortex**   The outer portion of the kidney.

**Renal medulla**   The inner portion of the kidney.

**Renal pelvis**   A hollow funnel that removes liquid from the kidney and into the ureters.

**Replication fork**   Molecules of DNA with both its sides exposed for adding bases.

**Reproduction**   The process of making new offspring.

**Reproductive isolation**   The inability to mate.

**Reproductive success (RS)**   Is the number of live young an organism produces.

**Reproductive system**   The system that includes all of the pelvic structures and it products related to forming new organisms.

**Reptiles**   Cold-blooded vertebrates that crawl or creep.

**Residence time**   The average length of time that tells how long something is retained in a given storage reservoir in the biogeochemical system.

**Resilience**   The ability of the lungs to inflate and deflate continuously to function properly.

**Resolution**   Is the ability to see two close objects as separate.

**Resource partitioning**   The condition where two competitors coexist in the same area and use resources in different ways.

**Respiration**   The process of taking up of oxygen gas from the environment and the release of the waste gas, carbon dioxide.

**Respiratory acidosis**   A condition that occurs when the lungs and heart do not sufficiently transport needed gases within the body, leading to the development of acidic blood.

**Respiratory burst:**   The rapid release of hydrogen peroxide and superoxide radical from neutrophils.

**Respiratory pump**   The movement of blood when muscles contract in the chest and abdominal cavity during normal breathing.

**Respiratory system**   The system by which oxygen is taken into the body from the environment and carbon dioxide is eliminated.

**Response to stimuli**   Ability to react to the various changes of the environment.

**Resting potential**   The potential of a cell that does not exhibit the activity resulting from a stimulus; -70mV.

**Reticular**   A connective tissue that traps foreign invaders such as bacteria; found in lymph nodes and the spleen.

**Retrovirus**   A virus containing RNA and the enzyme reverse transcriptase.

**Reverse transcriptase**   An enzyme that generates complementary DNA from an RNA template.

**Rhabdovirus**   A bullet- or rod-shaped RNA virus found in plants and animals.

**Rhinovirus**   The most common viral infectious agent that causes the common cold in humans.

**Rhodopsin**   One type of photo-pigment found in rods that responds to light by changing shape and generating a nerve impulse and sending it to the brain.

**Ribonucleic acid (RNA)**   A nucleic acid present in living cells, used in ribosomes and in the processes of transcription and translation of proteins.

**Ribose**   The sugar backbone found in RNA.

**Ribosome**   Small, spherical organelle that is the site protein synthesis.

**Rigor mortis**   Stiffening of the body that happens a few hours after death.

**RNA processing**   The process in which cap and tail is added to mRNA before it leaves the nucleus (for protection).

**Rods**   One form of photopigment that sends impulses to the brain that give black-and-white perception.

**Root cap**   A section of tissue at the tip of a plant root.

**Root system**   The parts of plants below the surface.

**Root-to-shoot ratio**   The dry weight of the root divided by the dry weight of the shoot.

**Round worms**   A nematode worm infesting the intestine of mammals.

**rRNA**   RNA component of ribosome.

**RUBISCO**   An enzyme present in chloroplast of plants that helps absorption of carbon dioxide.

**RuBP**   The first chemical in the Calvin Cycle, which combines with carbon dioxide.

**Ruffini's corpuscle**   A receptor that senses heat.

**Rugae**   Series of folds produced by folding the wall of an organ.

**S phase**   A period in the cell cycle in which DNA is replicated.

**Sacrum**   A large, wedge-shaped bone located between the two hip bones of the pelvis.

**Sagittal plane**   A vertical plane that divides the left and right side of an organism.

**Salivary amylase**   An enzyme present in the saliva that chemically breaks down starch into smaller polysaccharides.

**Salivary glands**   The gland that secretes saliva.

**Salt wedge**   A wedge-shaped intrusion of sea water into a fresh-water estuary.

**Sarcolemma**   A nerve's cell membrane.

**Sarcomere**   A series of contractile units that make up the myofibrils.

**Saturated fat**   Neutral fats that are literally saturated with as many hydrogen atoms as is possible in the carbon skeleton.

**Savanna**   Regions along the equator that are warm but experience less rainfall than the rainforests.

**Scales**   Dermal or epidermal structures that form the external covering of reptiles, fishes, and certain mammals.

**Scanning -electron -microscope (SEM)**   An electron microscope that looks at the surfaces of objects in detail by focusing a beam of electrons on the surface of the object.

**Scavenger**   Animals that feed on dead or decaying matter.

**Science literacy**   The comprehension of scientific concepts, processes, values, and ethics, and their relation to technology and society.

**Scientific method**   A procedure that has characterized natural science for centuries.

**Sclerenchyma**   Stringy and elongated cells with thick cell walls.

**Scrotum**   A sac beneath the penis, holding the testes.

**Sea anemone**   Water-dwelling animals that are brightly colored and fix themselves onto rocks.

**Second law of thermodynamics**   A law that states that all reactions within a closed system lose potential energy and tend toward entropy.

**Secondary active transport**   The movement of substances using stored energy.

**Secondary growth**   Growth in vascular plants emanating from two lateral meristems resulting in wider branches and stems.

**Secondary succession**   The process that occurs when established ecosystems replace organisms and soils of primary succession.

**Secretin**   A digestive hormone secreted by the duodenum.

**Secretion**   The removal of unwanted or unneeded substances from the blood.

**Seed**   An embryonic plant with its own internal and protected supply of water and nutrients, which led to another division of plants: seedless and seeded.

**Segmented worms**   Worms characterized by cylindrical bodies segmented both externally and internally.

**Segments**   The repeating chambers or units found in annelids.

**Seismicity**   Vibrations in Earth's crust.

**Selectively permeable**   A condition in which the membrane allows some materials to pass through cells but not others.

**Selfish-gene hypothesis**   The hypothesis that states that organisms are merely vessels holding their genes.

**Selfishness**   A behavior in which one organism harms the other for its own benefit.

**Semen**   Male reproductive fluid.

**Semi-conservative model**   A mode by which DNA replicates as half-new and half-old DNA.

**Semicircular canals**   Part of the inner ear filled with a fluid substance.

**Semilunar valves**   A valve of the heart that prevents backflow into vessels.

**Seminal vesicle**   A gland situated behind the bladder and above the prostate gland in males.

**Seminiferous tubules**   Highly coiled structures within the testes.

**Senescence**   The process of aging.

**Sensation**   Information received by the neurons.

**Sensory neuron**   Neurons that bring information from the external environment, toward the brain and spinal cord.

**Septum**   A partition that separates two chambers of tissue in an organism.

**Serotonin**   A type of neurotransmitter that improves mood and inhibits pain and depressive feelings.

**Sessile**   Immobile.

**Set point**   The normal value at which a variable physiological state stabilizes.

**Sex chromosome**   The final smallest pair of the 23 pairs of chromosomes in humans.

**Sex-linked**   One of the three possible patterns of inheritance of single-gene traits in which the X chromosome determines the characteristic.

**Sexual reproduction**   The process in which two individuals contribute genetic material to their offspring.

**Sexual selection**   Is the natural selection not based on a struggle for survival but instead based on a struggle for the opposite sex.

**Shoot system**   The system that consists of stem, leaves, lateral buds, flowering stems, and flowering bud.

**Sickle cell anemia**   A disease that leads to abnormally shaped red blood cells, poor oxygen carrying capacity, and a host of complications such as blood clots and organ damage.

**Sieve-tube members**   Cells that transport sap through phloem vessels, are alive at maturity, unlike xylem cells.

**Significance level**   The percentage chance that the results of a study are wrong.

**Simple**   An epithelial tissue that is only one cell layer in thickness.

**Sinoatrial (SA) node**   The center that controls the heart beats.

**Skeletal muscle**   Long muscles that are found attached to bones of the skeleton and move the bones.

**Skin cancer**   A condition characterized by the abnormal growth of cells of the skin.

**Sliding filament theory**   The theory that explains muscle contraction.

**Slime molds**   Organisms that live freely as single cells but form multicellular reproductive structures upon reaching a certain size.

**Small intestine**   The portion of digestive tract that lies between the colon and stomach.

**Smooth muscle**   Muscle tissue that provides support and propels movement of food through the organs in which it is found.

**Sociobiology**   The study of the ways that groups of animals act.

**Sodium—potassium pump**   An integral protein that uses ATP energy to move sodium ions out of the cell and brings potassium ions into the cell.

**Soluble fiber**   Fibers that dissolve in water.

**Solute**   The component in a solution that is dissolved in the solvent.

**Solvent**   Substance that does the dissolving.

**Somatic cells**   The body cells other than gametes (or sex cells).

**Somatic nervous system**   Part of the PNS that controls the voluntary movements in the body.

**Special connective tissue**   A unique connective tissue that has either a rigid or a liquid extracellular matrix.

**Specialized cells**   Cells that carry out a particular function.

**Speciation**   The process by which natural selection drives one species to split into two or more species.

**Species**   A group of individuals similar enough to be able to reproduce with one another to produce live, fertile young.

**Species-specific**   Limited to or found in one species.

**Specific-immunity**   The third line of defense in the human immune system; involves targeting of antigens by immune system.

**Sperm activation**   The process by which sperm are additionally mobilized by calcium and even more able to fertilize the egg.

**Spermatogenesis**   The process by which sperm are produced by spermatogonia cells in the testes.

**Spinal cord**   A long cord of nerve tissues that connect the brain to the other parts of the body.

**Spirillum**   Spiral-shaped bacteria.

**Spongy bone**   Tissue found inside the bones that resemble a sponge; it contains many open spaces.

**Spontaneous generation**   The idea that states that life appeared from nowhere.

**Sporophyte**   The diploid organism in plants, producing spores.

**Squamous cell carcinoma**   A type of skin cancer characterized by a flaky, reddened area.

**Squamous**   Flat-shaped epithelial cells.

**Stabilizing selection**   Occurs when individuals at mean or average range of variation in a population have higher fitness.

**Stamen**   Male reproductive structure in flowering plants.

**Staph**   The prefix given to bacteria that are found in clusters.

**Statistics**   The study of the collection, organization, analysis and interpretation of data.

**Stem cells (Pleuripotential)**   Are specialized cells that are able to develop into many types of cells, given particular conditions.

**Steroid**   A type of fat that stabilizes the structure of cell membranes.

**Stigma**   A sticky flat surface on which pollen grains land in flowering plants.

**Stimulus**   Something that causes an organ or cell to react.

**Stirrup**   The innermost bone of the middle ear.

**Stomach**   An internal organ sac that holds and digests food before entering the small intestines.

**Stomata**   A minute pore found in the epidermis of a plant's leaf or stem through which gas and water pass.

**Stratified**   An epithelial tissue that is two or more layers thick.

**Stratum basale**   The deepest layer of the epidermis, mitotic.

**Stratum corneum**   The outermost layer of the epidermis; it is thickest and is composed of dead cells.

**Stratum granulosum**   A thin layer of granular cells in the epidermis located between stratum lucidum and stratum spinosum.

**Stratum lucidum**   A clear layer of dead skin cells in the epidermis located between stratum corneum and stratum granulosum.

**Stratum spinosum**   A layer in the epidermis located between stratum granulosum and stratum basale.

**Strep**   The prefix given to bacteria that are found in chains.

**Striations**   Alternating patterns of proteins in the skeletal muscle.

**Stroke**   The sudden diminution of brain cells due to lack of oxygen caused by obstruction or rupture of a blood vessel of brain.

**Style**   The part of carpel that extends to ovules in which eggs develop in flowering plants.

**Submissive behavior**   Willing to submit to avoid further fighting, when a battle appears to be lost.

**Substrate**   A compound on which an enzyme acts.

**Superficial**   An anatomical term referring to a surface region or area of the body.

**Surface markings**   The distinctive features found on human bones.

**Surface runoff**   The flow of water over the land surface.

**Surfactant**   A special chemical that helps to keep the alveoli open by reducing the surface tension of fluid within the lungs.

**Survivorship curve**   A graph that gives number of survivors in a population over time.

**Sweat glands**   A tubular gland that secretes sweat.

**Swim bladder**   Organ that is present in many bony fishes and helps them maintain buoyancy.

**Symbiosis**   A relationship formed between two different organisms living in a close, intimate association.

**Sympathetic nervous system**   A part of the nervous system that increases heart rate, stimulates muscles, and raises blood pressure.

**Sympatric speciation**   The emergence of new species while living within the same geographical areas.

**Synapse**   The gap or region separating neurons from other cells or each other.

**Systemic circuit**   The vessel connection between the heart and body cells.

**T-cell**   A type of lymphocyte that matures in the thymus.

**T-helper cell**   A specific type of T-cell that attaches to the macrophage to start specific-immunity.

**T-suppressor cell**   Type of immune cells "demilitarize" an immune response when an immune response ends.

**Taiga**   A swampy, subartic forest dominated by conifers.

**Tape worms**   Parasitic flatworms that live in the intestines of people and animals.

**Taproot**   Large vertical root that burrows downward, anchoring the plant.

**Target cell**   Any cell having a specific receptor for an antibody, hormone, or antigen.

**Taxonomy**   The science of classifying the vast biodiversity.

**Telomerase**   An enzyme that rebuilds the DNA ends of cancer cells.

**Telomere**   A compound structure found at the end of a chromosome.

**Telophase**   A phase during which time there is a reversal of the events occurring during prophase.

**Telophase I**   The stage meiosis resulting in the forming of a set of new cells.

**Telophase II**   The last stage in the second meiotic division of meiosis resulting in a new set of haploid cells.

**Temperate deciduous forest**   A forest type characterized by leaf-shedding trees.

**Temperate grassland**   Are terrestrial biomes whose main vegetation is grass and shrubs.

**Temporal lobe**   One of the four major lobes of the brain that contains an area concerned with hearing and visual sensing as well as language comprehension.

**Tendon**   Strong fibrous tissue that anchors muscles to bones.

**Termination**   The phase in which RNA polymerase will reach a sequence of DNA that tells it to stop.

**Tertiary consumer**   Carnivores that eat carnivores.

**Testcross**   A known homozygous recessive organism is mated with a dominant organism.

**Testes**   Organs that produce sperm.

**Thalamus**   The rounded area underneath the corpus callosum.

**Thalassemia**   The condition in which a faulty or absent hemoglobin chain makes the molecule fragile and less able to carry oxygen.

**The bends**   The condition in which nitrogen gas accumulates in a diver's blood.

**Thermacidophiles**   Organisms that thrive in strongly acidic environments at high temperatures.

**Thermodynamics**   The science of energy transformations that explains the flow of energy through environment and in cells.

**Thermoreceptors**   A sense organ responding to temperature.

**Thigmotropism**   Any plant growth response to touch.

**Thoracic vertebrae**   The 12 vertebrae along the thorax along the center of the vertebral column.

**Thrombin**   An important enzyme in blood that facilitates clotting of blood by converting fibrinogen to fibrin.

**Thrombosis**   Clots forming in the wrong places (an unbroken vessel).

**Thymine**   A pyrimidine base that is found in DNA but not RNA.

**Thyroid stimulating hormone (TSH)**   Hormone produced by the pituitary gland and stimulates the thyroid gland.

**Thyroxin**   A hormone that increases metabolism throughout the human body.

**Tight junction**   A specialized cell junction that fuses areas together to prevent leaking and acts as a sealant.

**Tissues**   Groups of cells having similar structure and performing similar functions.

**Topography**   The physical features of the land such as mountains and valleys.

**Trabeculae**   Small spindles that make up the spongy bone.

**Trachea**   A tube-like portion of the respiratory tract that connects to the lungs.

**Tracheids**   Elongated cells found in the xylem of vascular plants that conduct the transport of water and mineral salts.

**Tracheophyte**   Vascular plant; plants having a well developed vessel system.

**Trade winds**   Winds that blow above and below the equator.

**Transcription**   The first step of gene expression in which information in a DNA strand is copied into mRNA by RNA polymerase.

**Transduction**   The process that occurs when a virus invades a prokaryote, inserting its genes into the host.

**Transformation**   The process in which a newly inserted DNA from the environment changes or transforms a bacterial cell into a new genotype.

**Transitional epithelium**   A type of tissue that consists of multiple layers of epithelial cells, which looks at times cuboidal and at other times squamous in shape; responds to pressure as in the bladder.

**Translation**   The synthesis of protein from the information contained in a molecule of mRNA.

**Transmission electron -microscope (TEM)**   A type of electron microscope that magnifies structures within a cell. .

**Transpiration**   Loss of water from leaves by evaporation through stomata.

**Transpirational pull**   The process in which water and minerals are transported upwards through xylem from roots because of an upward force or pull.

**Tricuspid valve**   A heart valve between the right atrium and ventricle and keeps blood from flowing back into the atrium.

**Trimester**   A normal pregnancy divided roughly into three parts.

**tRNA**   Small RNA molecules that carry amino acids to ribosomes for protein synthesis.

**Trophic level**   A group of organisms with the same feeding style.

**Tropical rainforest**   A type of forest characterized by tall, dense trees in an area that experiences high annual rainfall.

**Tropomyosin**   A protein rope that plays an important role in muscle contraction.

**Troponin**   A protein found in all muscle.

**Tube feet**   Small suction cup-shaped feet that are used for holding prey.

**Tundra**   A tree-less area near the North Pole.

**Turgor pressure**   The pressure exerted against the walls of a plant cell when water enters the water vacuoles of plant cell.

**Twitch**   The time period comprising a contraction and relaxation.

**Ultrasound**   A technique that emits high frequency sound waves and creates images based on the echos received back from the body part.

**Universal donor**   A person of blood type O who may donate blood to any other blood group because the blood group contains no antigens on its red blood cells.

**Universal recipient**   A person of blood type AB who may receive blood from any other blood group because the blood group contains all antigens on its red blood cells.

**Uracil**   A pyrimidine base that is one of the fundamental components of RNA.

**Ureter**   A tube connected to each kidney responsible for transporting urine from the renal pelvis to the urinary bladder.

**Urethra**   The duct by which urine is removed from the body.

**Urinalysis**   An analysis that tests urine for normal and abnormal substances.

**Urinary bladder**   An organ that holds the urine.

**Urinary system**   The system that is responsible for eliminating wastes.

**Urogenital system**   The system comprising the reproductive organs and the urinary system.

**Uterus**   A thick, muscular organ in which a fetus develops.

**Vacuole**   Single membrane structures that hold materials in a cell.

**Vagina**   A muscular tube leading from the outside to the cervix of the uterus in female mammals.

**Valence electrons**   Electrons present in the outermost shell of an atom; these electrons are responsible for the chemical reactivity of atoms.

**Variable region**   Regions that vary from antibody type to antibody type.

**Variation**   Differences that are inherited from generation to generation.

**Varicose veins**   The condition in which valves are incompetent within the legs leading to the formation of blood pools.

**Vascular cambium**   One of the lateral meristems that produces xylem and phloem.

**Vascular tissue**   Tissues that transports water, minerals and food throughout a plant.

**Vasoconstrict**   Narrowing of blood vessels.

**Vasodilate**   Widening of blood vessels.

**Vein**   A blood vessel that carries deoxygenated blood back to the heart after it has picked up wastes and carbon dioxide from body cells.

**Vena cava**   A large vein that carries deoxygenated blood into the heart.

**Ventricle**   A chamber of heart that receives blood from the atrium.

**Venules**   Small veins connecting capillaries with larger systemic veins.

**Vertebrates**   Animals having a backbone.

**Vessel element**   A cell type found in xylem.

**Vestigial organs**   Structures that once had a purpose but no longer appear to be functional.

**Villi**   Small folds or projections lining the walls of the small intestine.

**Visceral mass**   One of three-point body plans of mollusks that contain the internal organs.

**Vitamin D**   A fat-soluble vitamin that promotes that is essential for the absorption of calcium.

**Vitreous humor**   Posterior chamber of the eyes.

**Vocal cords**   Two elastic cords stretching across the upper end in the larynx.

**Vomer**   A small bone found inside the nose.

**Water**   An essential nutrient, comprising between 60% and 80% of the volume of cells, and is the medium in which all cell reactions take place.

**Water cycle**   The process by which water continuously moves on, above, and below Earth's surface.

**Water-soluble vitamins**   Includes B-vitamins and vitamins C and K, can be taken in large doses and do not become toxic because they are eliminated through the urine.

**Water-vascular system**   A set of internal channels that circulate water through echinoderm bodies, enabling gas exchange and waste removal.

**Watershed**   An area that collects flow, runoff, and precipitation from a region into a specific body of water.

**Westerlies**   Winds that blow from the west.

**Wetland**   A land that consists of swamps or marshes.

**White blood cells**   Blood cells that help body fight infections.

**Wind**   The intricate horizontal movements of Earth's atmosphere caused by differences in atmospheric pressure and Earth's rotation.

**Wolff's law**   The phenomenon which states that bones grow and remodel according to the forces placed upon them.

**X chromosome**   A sex chromosome that is found twice in females and singly in males (not given in bold in text).

**X-rays**   A form of EM radiation that visualizes dense structures within the body.

**Xylem**   A series of tubes conducting water and dissolved minerals up a plant.

**Y chromosome**   A sex chromosome that is found only in males (not given in bold in text).

**Yellow marrow**   Hollow cavities within bones filled with fat.

**Zona pellucida**   A layer of proteins that surround and protect the ovum.

**Zone of cell division**   One of the zones of development in which mitosis occurs in a slow but protected manner.

**Zone of differentiation**   One of the zones of development in which cells become one of the three types of plant tissues.

**Zone of elongation**   One of the zones of development in which cells elongate.

**Zoology**   The branch of biology that is dedicated to the study of animals and their characteristics.

**Zygomatic bone**   Bone that forms an important part of the cheeks.

**Zygote**   A diploid cell that is produced when the haploid sperm nucleus fuses with the egg's haploid nucleus.

# INDEX